FOOD ENZYMOLOGY

Volume 1

FOOD ENZYMOLOGY

Volume 1

Edited by

P. F. Fox

University College Cork
Cork, Ireland

ELSEVIER APPLIED SCIENCE
LONDON and NEW YORK

ELSEVIER SCIENCE PUBLISHERS LTD
Crown House, Linton Road, Barking, Essex IG11 8JU, England

Sole Distributor in the USA and Canada
ELSEVIER SCIENCE PUBLISHING CO., INC.
655 Avenue of the Americas, New York, NY 10010, USA

WITH 53 TABLES AND 72 ILLUSTRATIONS

British Library Cataloguing in Publication Data
Food enzymology.
1. Food. Enzymes
I. Fox P. F.
664.06

ISBN 1-85166-615-X V.1
ISBN 1-85166-616-8 V.2

Library of Congress Cataloging-in-Publication Data
Food enzymology.
p. cm.
Includes bibliographical references and indexes.
ISBN 1-85166-615-X (v. 1). — ISBN 1-85166-616-8 (v. 2)
1. Enzymes. 2. Food—Composition. I. Fox, P. F.
TX553.E6F66 1991
664—dc20

Typeset and printed by The Universities Press (Belfast) Ltd.

Preface

Throughout its history, the development of enzymology has been very closely associated with food science, or, perhaps, vice versa. In many respects, food science might be regarded as applied enzymology. With a few minor exceptions, our food is derived from animal or plant tissues or excretions and, consequently, its constituents are synthesized via enzyme-catalyzed reactions. These reactions do not cease after the animal has been slaughtered, or the plant harvested, but continue more or less actively to cause changes in foods which may be desirable or detrimental to quality, depending on the circumstances. Food constituents are susceptible to alteration by enzymes produced by contaminating microorganisms, which may lead to food spoilage. On the other hand, enzymes produced by certain microorganisms may cause desirable changes in foods and such organisms are widely exploited in the production of a very wide range of fermented foods. Many food processing operations are concerned with controlling or preventing the activity of indigenous or microbial enzymes, e.g. heat treatment, dehydration and chilling/freezing. Food constituents are amenable to modification by exogenous (added) enzymes which may be used to induce specific changes, perhaps resulting in entirely new food products; the food industry is probably the principal industrial user of enzymes.

Not surprisingly, various aspects of the importance of enzymes in foods and food processing have been the subjects of an extensive scientific and technical literature, including several text books, for example, Whitaker (1972, 1974), Reed (1975), Schwimmer (1981), Birch *et al.* (1981), Kruger *et al.* (1987) and Stauffer (1989) and chapters in many others. However, most of these books focus on the enzymes rather than on the foods containing them. This book is an attempt to review food enzymology more from the viewpoint of the food scientist or technologist rather than that of the enzymologist. The book, consisting of thirty chapters, in two volumes, attempts to assess the importance of indigenous, microbial and exogenous enzymes in a wide range of foods based on milk, meat, cereals, fruits and vegetables including beverages and food ingredients.

A chapter is devoted to the importance of enzymes in food analysis, both as indicators of quality and as analytical reagents. Two major new aspects of enzymology, immobilized enzymes and genetic engineering, are also considered.

Certain gaps in the coverage will be apparent due to the failure of some chapters to materialise—most noticeably those on enzymes in baking, enzymes in the starch and sugar industries and the enzymatic modification of lipids. It is hoped to include these in a later volume.

P. F. Fox
University College, Cork

Contents

VOLUME 1

VOLUME 2

List of Contributors

VOLUME 1

J. B. ADAMS
Campden Food & Drink Research Association, Chipping Campden, Gloucestershire GL55 6LD, UK

J. M. ALONSO
Instituto de Agroquímica y Technología de Alimentos (CSIC), Jaime Roig 11, 46010 Valencia, Spain

A. T. ANDREWS
AFRC Institute of Food Research, Reading Laboratory, Shinfield, Reading RG2 9AT, UK

R. B. ARIE
Agricultural Research Organization, The Volcani Center, Institute for Technology and Storage of Agricultural Products, PO Box 6, Bet Dagan 50250, Israel

G. BENGTSSON-OLIVECRONA
Department of Medical Biochemistry and Biophysics, University of Umea, S-90187 Umea, Sweden

L. BJÖRCK
Department of Animal Nutrition and Management, Swedish University of Agricultural Sciences, Box 7024 Ultuna, S-75007 Uppsala, Sweden

G. CAMPBELL-PLATT
Department of Food Science and Technology, University of Reading, PO Box 226, Whiteknights, Reading RG6 2AP, UK

J. CHAMARRO
Instituto de Agroquímica y Technología de Alimentos (CSIC), Jaime Roig 11, 46010 Valencia, Spain

P. E. COOK
Department of Food Science and Technology, University of Reading, PO Box 226, Whiteknights, Reading RG6 2AP, UK

J. F. D. DEAN
Plant Hormone Laboratory and Plant Molecular Biology Laboratory, Beltsville Agricultural Research Center (W), ARS, US Department of Agriculture, Beltsville, Maryland 20705, USA

M. J. Desmazeaud
INRA, Centre de Recherches de Jouy-en-Josas, Station de Recherches Laitières, Domaine de Vilvert-78350, Jouy-en-Josas, France

D. Dubourdieu
Université de Bordeaux, Institut d'Oenologie, Bordeaux, France

N. Y. Farkye
Food Chemistry Department, University College, Cork, Republic of Ireland

P. F. Fox
Department of Food Chemistry, University College, Cork, Republic of Ireland

J. L. Garcia-Martinez
Instituto de Agroquímica y Technología de Alimentos (CSIC), Jaime Roig 11, 46010 Valencia, Spain

J.-C. Gripon
INRA, Centre de Recherches de Jouy-en-Josas, Station de Recherches Laitières, Domaine de Vilvert-78350, Jouy-en-Josas, France

M. B. Grufferty
Department of Food Chemistry, University College, Cork, Republic of Ireland

E. Harel
Department of Botany, The Hebrew University of Jerusalem, Jerusalem 91904, Israel

J. S. Hawker
Department of Botany, The University of Adelaide, GPO Box 498, Adelaide, South Australia 5001, Australia

M. Knee
Department of Horticulture, The Ohio State University, 2001 Fyffe Court, Columbus, Ohio 43210, USA

G. Lamberet
INRA, Centre de Recherches de Jouy-en-Josas, Station de Recherches Laitières, Domaine de Vilvert-78350, Jouy-en-Josas, France

A. K. Mattoo
Plant Hormone Laboratory and Plant Molecular Biology Laboratory, Beltsville Agricultural Research Center (W), ARS, US Department of Agriculture, Beltsville, Maryland 20705, USA

A. M. Mayer
Department of Botany, The Hebrew University of Jerusalem, Jerusalem 91904, Israel

M. Mazelis
Department of Food Science and Technology, University of California, Davis, California 95616, USA

V. Monnet
INRA, Centre de Recherches de Jouy-en-Josas, Station de Recherches Laitières, Domaine de Vilvert-78350, Jouy-en-Josas, France

N. M. O'Brien
Department of Nutrition, University College, Cork, Republic of Ireland

T. P. O'Connor
Department of Food Chemistry, University College, Cork, Republic of Ireland

T. Olivecrona
Department of Medical Biochemistry and Biophysics, University of Umea, S-90187 Umea, Sweden

R. E. Paull
Department of Plant Molecular Physiology, University of Hawaii at Manoa, 3190 Maile Way, Honolulu, Hawaii 96822, USA

W. Pilnik
Department of Food Science, Agricultural University, PO Box 8129, 6700 EV Wageningen, The Netherlands

N. C. Price
Department of Molecular and Biological Sciences, University of Stirling, Stirling FK9 4LA, UK

D. S. Robinson
Procter Department of Food Science, University of Leeds, Leeds LS2 9JT, UK

T. Sørhaug
Department of Dairy and Food Industries, Agricultural University of Norway, 1432 Ås-NLH, Norway

L. Stepaniak
Department of Dairy and Food Industries, Agricultural University of Norway, 1432 Ås-NLH, Norway

L. Stevens
Department of Molecular and Biological Sciences, University of Stirling, Stirling FK9 4LA, UK

J. C. Villettaz
Ecole d'Ingénieurs du Valais, Food Technology Department, Rte du Rawyl 47, 1950 Sion, Switzerland

A. G. J. Voragen
Department of Food Science, Agricultural University, PO Box 8129, 6700 EV Wageningen, The Netherlands

J. R. Whitaker
Department of Food Science and Technology, University of California, Davis, California 95616, USA

VOLUME 2

H. V. Amorim
Fermentec S/C Ltda, Rua 13 de Maio 768, 13400 Piracicaba SP, Brazil

S. Arai
Department of Agricultural Chemistry, The University of Tokyo, Bunkyo-Ku, Tokyo 113, Japan

M. E. Bailey
Department of Food Science and Nutrition, University of Missouri, 21 Agriculture Building, Columbia, Missouri 65211, USA

P. S. Dimick
Department of Food Science, The Pennsylvania State University, University Park, Pennsylvania 16802, USA

L. W. Doner
Eastern Regional Research Center, US Department of Agriculture, 600 East Mermaid Lane, Philadelphia, Pennsylvania, USA

S. J. B. Duff
Division of Biological Sciences, National Research Council of Canada, Ottawa, Canada K1A 0R6

R. E. Feeney
Department of Food Science and Technology, 1480 Chemistry Annex, University of California, Davis, CA 95616, USA

R. FitzGerald
Microbiology Department and National Food Biotechnology Centre, University College, Cork, Republic of Ireland

M. Fujimaki
Department of Agricultural Chemistry, The University of Tokyo, Bunkyo-Ku, Tokyo 113, Japan

J. E. Kruger
Canadian Grain Commission, Grain Research Laboratory, 1404-303 Main Street, Winnipeg, Manitoba, Canada R3C 3G8

A. S. Lopez
CEPLAC/CEPEC, CxP 07, 45600 Itabuna BA, Brazil

A. W. MacGregor
Canadian Grain Commission, Grain Research Laboratory, 1404-303 Main Street, Winnipeg, Manitoba, Canada R3C 3G8

B. A. Marchylo
Canadian Grain Commission, Grain Research Laboratory, 1404–303 Main Street, Winnipeg, Manitoba, Canada R3C 3G8

M. Melo
Fermentec S/C Ltda, Rua 13 de Maio 768, 13400 Piracicaba SP, Brazil

P. Monsan
BioEurope, 4 Impasse Didier Daurat, Z. I. de Montaudran, 31400 Toulouse, France

F. A. Murdock Jr.
Department of Food Science and Nutrition, University of Missouri, 21 Agriculture Building, Columbia, Missouri 65211, USA. Present address: Department of Animal and Food Science, River Falls, Wisconsin, USA

W. D. Murray
Division of Biological Sciences, National Research Council of Canada, Ottawa, Canada K1A 0R6

M. Nuti
Dipartimento di Biotecnologie Agrarie, Università degli Studi di Padova, Padova, Italy

F. O'Gara
Microbiology Department and National Food Biotechnology Centre, University College, Cork, Republic of Ireland

T. Osuga
Department of Food Science and Technology, 1480 Chemistry Annex, University of California, Davis, California 95616, USA

F. Paul
BioEurope, 4 Impasse Didier Daurat, Z. I. de Montaudran, 31400 Toulouse, France

F. G. Priest
International Centre for Brewing and Distilling, Heriot-Watt University, Riccarton, Edinburgh EH14 4AS, UK

J. C. Slaughter
International Centre for Brewing and Distilling, Heriot-Watt University, Riccarton, Edinburgh EH14 4AS, UK

H. E. Swaisgood
Department of Food Science, University Biotechnology Program, North Carolina State University, Raleigh, North Carolina 27695–7624, USA

M. R. Ullah
Tocklai Experimental Station, Jorhat, Assam, India. Present address: BSB College, Jorhat-15, Assam, India

J. R. Whitaker
Department of Food Science and Technology, University of California, Davis, California 95616, USA

1

Enzymes: Structure and Function

N. C. Price & L. Stevens
Department of Biological and Molecular Sciences, University of Stirling, Stirling, FK9 4LA, UK

GENERAL INTRODUCTION

The use of enzymes in the food industry occurred long before there was any awareness of enzymes as such. The oldest uses of enzymes in food are those associated with the use of yeast in brewing and baking, and date back more than 3500 years, whereas the recognition of catalysts in living organisms dates from just over 100 years. It is interesting that the name, enzyme, itself and the first named enzyme, diastase, both have a close relationship to foods. The term 'enzyme' was first used by Kühne in 1878 from the Greek *ενζομοζ* meaning 'in yeast'. At the time the term was used to distinguish what were described as 'organised ferments' (meaning whole microorganisms) and 'unorganised ferments' (meaning extracts or secretions from whole organisms). However, before this date there was an awareness that material from living organisms displayed catalytic activity. As early as 1833, Payen and Persoz used the term 'diastase' for the extract of malt which catalysed the degradation of starch to sugar. Again, the derivation is Greek, *διαστασιζ,* meaning 'to make a breach'. In this case, it referred to the ability to make a breach in the outer envelope of starch granules.

Many of the early studies on enzymes made use of the extensive knowledge of yeast physiology acquired through the food and drinks industry. In 1897, Büchner demonstrated that filtrates of yeast extracts could catalyse fermentation. This was a highly significant discovery since it meant that enzymes could be studed *in vitro,* and this facilitated their subsequent purification and analysis. Many of the early kinetic analyses by Michaelis and Menten and their predecessors used data obtained from hydrolytic enzymes associated with digestion. It was in an effort to define precisely the nutritional requirements of the human diet that vitamins were discovered. This led, in particular, to the discovery of the multiplicity of B vitamins, and to the eventual understanding of their roles as enzyme cofactors. Thus, there has been a type of symbiotic relationship between the food industry and enzymology; the food and its importance has provided the focal point at which to start the enzymology, whereas at the present time the increased understanding and use of enzymes is significantly benefiting food technology. For further details of the history of enzymology see Fruton (1972), Dixon *et al.* (1979) and Richmond (1986).

The most striking properties of enzymes are their catalytic activity and their specificity. Both of these attributes were apparent long before the nature of enzymes was realised. A third property which is equally important with certain enzymes is their ability to be regulated. This includes allosteric mechanisms, and reversible and irreversible covalent modification. Allosteric mechanisms and reversible covalent modification are most important with intracellular enzymes, whereas irreversible covalent modification, such as zymogen activation, may be important with extracellular enzymes. Until recently it was possible to say categorically that all enzymes so far isolated have been found to be proteins. This statement now has to be qualified. The precursor of ribosomal RNA undergoes self-splicing. In other words, the RNA can catalyse the cleavage of the internucleotide bond. The RNA catalyst is referred to as a ribozyme (Cech & Bass, 1986). It still remains the case, however, that almost all catalytic activity in cells is due to proteinaceous enzymes and it seems unlikely that the picture will change dramatically in the future.

The high catalytic power of enzymes can be seen by the comparisons shown in Table 1, in which the rates of reactions catalysed by enzymes and by inorganic catalysts are compared. Enzymes may cause rate enhancements of as much as 10^{14}-fold. It is difficult to obtain extensive comparative data, since the reactions in the absence of enzymes are often so slow as to be difficult to measure. The chemical basis of the catalytic power of enzymes is not fully understood, although a number of factors that contribute to it have been clarified, and are discussed in the section entitled 'Origins of rate enhancement', p. 33.

One of the earliest models to explain the catalytic activity and the specificity

TABLE 1
Examples of the Catalytic Power of Enzymes

Substrate	*Catalyst*	*Temperature (°C)*	*Rate Constant, k (mol/litre s)*
Amide (hydrolysis)			
benzamide	H^+	52	$2{\cdot}4 \times 10^{-6}$
benzamide	OH^-	53	$8{\cdot}5 \times 10^{-6}$
benzoyl-L-tyrosinamide	α-chymotrypsin	25	14·9
Urea	H^+	62	$7{\cdot}4 \times 10^{-7}$
(hydrolysis)	urease	21	$5{\cdot}0 \times 10^{6}$
$2H_2O_2 \rightleftharpoons 2H_2O + O_2$	Fe^{2+}	22	56
	catalase	22	$3{\cdot}5 \times 10^{7}$

of enzymes was the 'lock and key' model proposed by Emil Fischer in 1894. In this model, the enzyme was believed to have a shape complementary to that of the substrate(s). Only substrates having the correct geometry, or very close to it, could fit the active site, like the correct key fitting into a lock. This theory held sway until about the 1950s. By this time much had been discovered about the nature of proteins. It was realised that they were not rigid templates but had a significant degree of flexibility. This led Koshland to propose his induced fit theory. In this, he envisaged that the active site of the enzyme changes as it formed several weak interactions with the substrate. It was only those substrates that could bind to the enzyme and induce the appropriate conformational change that would be good substrates. Spectroscopic and other evidence for such conformational changes has been obtained. Koshland was also able to show that strong binding to the active site was not on its own sufficient to induce high catalytic activity.

Enzymes show different degrees of specificity. The lowest type of specificity is *bond specificity*. In this case, the enzyme is only specific for the type of bond with which it interacts. This group includes a number of degradative enzymes such as peptidases, phosphatases and esterases, which are capable of catalysing the hydrolysis of a range of substrates which contain the bond in question. It is rare for enzymes concerned with biosynthesis to exhibit such low specificity. The second type of specificity is known as *group specificity* where the range of substrates is much more restricted, as for example to the phosphorylation of hexoses in the case of hexokinase.

$$\text{hexose} + \text{ATP} \rightleftharpoons \text{hexose-6-phosphate} + \text{ADP}$$

Another example would be acyl-CoA dehydrogenase which acts on a range of acyl-CoAs.

$$\text{acyl-CoA} + \text{NADP}^+ \rightleftharpoons \text{2,3-dehydroacyl-CoA} + \text{NADPH} + \text{H}^+$$

The third type is the most specific, known as *absolute specificity* where each enzyme will catalyse a reaction with only a single substrate (or a single pair of substrates in a bimolecular reaction). Examples are fumarate hydratase and enolase. The former will only act on fumarate in one direction and L-malate in the other, and the latter will only act on 2-phospho-D-glycerate in one direction and phosphoenolypyruvate in the other.

$$\text{fumarate} + \text{H}_2\text{O} \rightleftharpoons \text{L-malate}$$

$$\text{2-phospho-D-glycerate} \rightleftharpoons \text{phosphenolpyruvate} + \text{H}_2\text{O}$$

ENZYME CLASSIFICATION AND NOMENCLATURE

As enzymes were isolated and partially characterised they were often given trivial names which reflected their source (e.g. papain) or other properties (e.g. old yellow enzyme). The need for a systematic classification and naming procedure for each enzyme became more pressing as the number of characterised enzymes grew. The present classification scheme is based on the reactions catalysed, rather than on the structure of the enzyme or some other property. Each enzyme is given a systematic name and number. The systematic name is generally rather cumbersome, so that most enzymes are known by their recommended name, which is sometimes a trivial name, and sometimes a truncated version of the systematic name. However, to avoid any ambiguity it is important that every enzyme should be defined by its 'Enzyme Number' (EC no.) in all original publications so that its identity is clear. We describe here briefly the classification system with some examples, but full details can be found in IUB (1984).

Each enzyme-catalysed reaction is classified into one of six groups.

(1) Oxidation–reduction reactions, catalysed by *oxidoreductases.*
(2) Group transfer reactions, catalysed by *transferases.*
(3) Hydrolytic reactions catalysed by *hydrolases.*
(4) Elimination reactions in which a double bond is formed, catalysed by *lyases.*
(5) Isomerisation reactions, catalysed by *isomerases.*
(6) Reactions in which two molecules are joined at the expense of an energy source (usually ATP), catalysed by *ligases* (also known as *synthetases*).

Within each major group there are subgroups, sub-subgroups and then a number of enzymes within each sub-subgroup. This is indicated in the EC no., for example, α-amylase has the EC no. 3.2.1.1, indicating that it belongs to the group *hydrolases,* the sub-group 2 which includes all glycosidases, and sub-subgroup 1 which includes only *O*-glycosyl compounds as substrates. In general, the subgroup defines the type of bond attacked, and sub-subgroup further defines bond specificity.

Many of the enzymes discussed in the subsequent chapters of this volume belong to group 3, the hydrolases. It therefore seems appropriate to illustrate the classification system further by examples from that group. Group 3 contains all the hydrolases. There are 11 sub-divisions within the group which define the atoms forming the bond that is split, e.g. C—O, C—N, C—C, P—N, S—N and C—P. The largest of these groups is the first group, which is sub-divided into enzymes that hydrolyse esters (EC 3.1), enzymes that hydrolyse glycosides (EC 3.2) and enzymes that hydrolyse ethers (EC 3.3).

The type of ester hydrolysed by subgroup 3.1 enzymes is further divided into, e.g. carboxylic esters, phosphomonoesters, phosphodiesters. The nucleases are separated from the phosphodiesterases, although they are, in reality, a particular group of phosphodiesterases. In recent years there has been an enormous increase in the number of restriction endodeoxyribonucleases characterised, largely driven by the need for a range of such enzymes in genetic engineering. The restriction endodeoxyribonucleases are currently fitted into a single subgroup EC 3.1.21 and all are included under three EC nos 3.1.21.3, 3.1.21.4 and 3.1.21.5. There are now well over 500

TABLE 2
Examples of Restriction Enzymes

Enzyme	*Restriction Site*	*Bacterial Strain*
EcoRI	G↓AATTC	*Escherichia coli* RY13
BamHI	G↓GATCC	*Bacillus amyloliquefaciens* H
BsuRI	GG↓CC	*Bacillus subtilis* R
HhaI	GCG↓C	*Haemophilus haemolyticus*
HindIII	A↓AGCTT	*Haemophilus influenzae* Rd
SmaI	CCC↓GGG	*Serratia marcescens* Sb
DpnI	GA↓TC	*Diplococcus pneumoniae*

enzymes (Old & Primrose, 1989) within these three EC nos. The reason for this is that they all cleave endolytically a phosphodiester link joining adjacent deoxyribonucleotides in DNA. Their specificity lies in recognising different bases on either side of the bond cleaved (Table 2). The restriction endonucleases are named after the bacterium from which they were isolated, with the first three letters giving the name of the bacterium, e.g. Eco from *Escherichia coli,* and the fourth letter indicating the strain. If more than one restriction endonuclease has been isolated from one strain this is indicated by a Roman number. The vast majority of these enzymes recognise a sequence of four or six nucleotides in DNA which has two-fold rotational symmetry.

The enzymes that catalyse the hydrolysis of oligo- and polysaccharides belong to group EC 3.2. Here some of the recommended names are more informative than others, e.g. α- and β-amylases are clearly enzymes that hydrolyse amylose, but the name itself gives no indication that α-amylase catalyses the endohydrolysis of 1,4 α-D glycosidic linkages. The prefix, endo- indicates that bond cleavage occurs at random throughout the length of the polymer, in this case whereever there is a 1,4 α-D link. Exohydrolysis, on the other hand, indicates that hydrolysis occurs starting from the ends of the polymer and continues until a branch point or alternative link is reached. β-Amylase catalyses the exohydrolysis of 1,4 α-D glucoside links to remove successive maltose units from the non-reducing end. Other names which give no indication of the bond cleaved within the substrate are: inulinase, cellulase, lysozyme, isopullulanase, isoamylase, agarase and κ-carrageenase. Many of the more recently discovered carbohydrate-hydrolysing enzymes have more informative names, e.g. glucan 1,3-α-glucosidase, which catalyses the hydrolysis of terminal 1,3-α-D glucosidic links in 1,3-α-D-glucan; mannan endo-1,6-β-mannosidase, which catalyses the random hydrolyis of 1,6-β-D mannosidic linkages in unbranched mannans; and exo-poly-α-galacturonosidase, which catalyses the hydrolysis of pectic acid from the non-reducing end, releasing digalacturonate.

The subgroup EC 3.4, which includes all enzymes that cleave peptide bonds, has proved difficult to classify further. The group is subdivided into exopeptidases and proteinases. The proteinases are not divided on the basis of their specificity, because many of these proteinases have wide specificity which in many cases is difficult to define precisely. Instead, they are divided on a different basis, namely, according to the nature of their active sites. The main groups are serine proteinases, EC 3.4.21, cysteine proteinases EC 3.4.22, aspartic proteinases, EC 3.4.23, and metalloproteinases, EC 3.4.24 (see the section entitled 'Analogies between the mechanisms of different proteinases', p. 38).

ENZYME PURIFICATION

Introduction

In designing a suitable procedure for the isolation and purification of an enzyme, a number of questions have to be considered. These include

(1) Is the enzyme of interest required in a pure state, i.e. free from contaminating enzymes, proteins and other macromolecules?
(2) Do any contaminating enzymes (or any associated activities of the enzyme of interest) pose problems as far as the use of the enzyme is concerned?
(3) Can enzymes of the required purity be produced economically on the scale required? Attention may also be paid to the stability of the enzyme under the conditions in which it is to be used. It may be possible to stabilise the enzyme by adsorption or immobilisation on an inert matrix and this may allow the enzyme to be re-used.

Enzyme purification for industrial use has rather different goals from the purification of enzymes for research purposes in the laboratory. In the latter case, small-scale preparations are generally performed, yielding milligram quantities of enzymes of high (>95%) purity, which can then be characterised in terms of structure, kinetics, mechanism, etc. For industrial purposes, however, questions (1) and, especially, (3), above are of great importance; it is of little benefit to produce highly purified enzyme by a costly, elaborate, process when all that is required is a relatively crude preparation, perhaps with only certain contaminants removed. Methods which are of considerable value in the laboratory, such as column chromatography or the use of affinity matrices, are often expensive and impractical to scale up.

The choice of source is important. Enzymes which are secreted by cells (i.e. extracellular enzymes) are much more convenient to purify than those which occur inside cells. Extracellular fluids are likely to be much simpler in terms of composition and therefore easier to fractionate than cell extracts. In addition, the need to disrupt cells is dispensed with; this is an important consideration in the case of plant, fungal or bacterial cells which contain strong cell walls. Extracellular fluids can also be harvested continuously, if required. It should be pointed out, however, that the concentration of any given enzyme in the extracellular fluid can be very low and it may be necessary to perform a preliminary concentration step. Concentration could be achieved by precipitation with ammonium sulphate, by the formation of a specific complex, e.g. the precipitation of α-amylase by glycogen (Schramm & Loyter, 1966), or by adsorption, e.g. on an ion-exchanger such as (diethylaminoethyl) (DEAE)–Sephadex.

Recently, recombinant DNA techniques have been developed in order to increase the production of a particular protein by cloning the gene corresponding to that protein and expressing it in a suitable vector (Old & Primrose, 1989). Although the manipulations involved can be difficult, the enormous degrees of amplification of protein production which can be achieved make it very likely that such methods will be used more frequently in the future.

Procedures involved in enzyme purification

The general approach to enzyme purification can be summarised as follows:

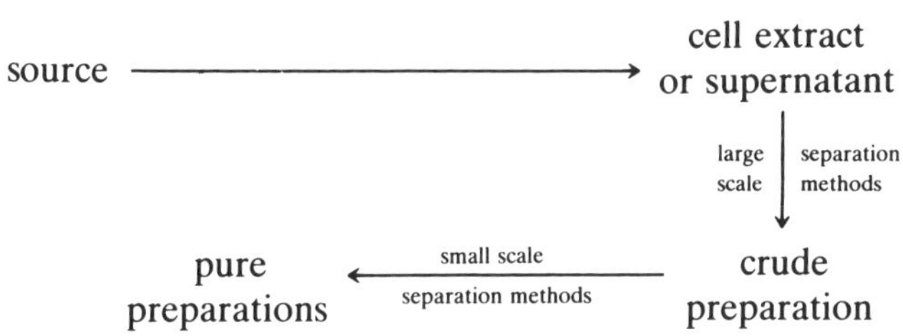

The extent to which the purification is carried out will depend on the answers to the questions mentioned in the previous section. Some of the methods which are available to separate mixtures of enzymes are listed in Table 3; they are grouped according to the property of the enzyme on the basis of which separation is achieved. Further details about the practical and theoretical aspects of these methods can be found in the books by Price and Stevens (1989) and Scopes (1987).

Assessment of the success of a purification procedure

The ultimate aim of a purification procedure is to produce an enzyme of the highest possible purity with the maximum possible yield of activity relative to the initial extract or supernatant. Analytical techniques such as sodium dodecylsulphate–polyacrylamide gel electrophoresis (SDS–PAGE) (Laemmli, 1970) can be used to assess the degree of contamination by other proteins. However, problems can arise when the enzyme of interest contains polypeptide chains of different relative molecular mass (M_r), since in such cases multiple bands are seen on electrophoresis of pure enzyme. In such cases, electrophoresis of the enzyme under non-denaturing conditions (i.e. in the absence of SDS) can be used to assess homogeneity. The yield of enzyme can be determined using a suitable assay for enzyme activity (see the section

TABLE 3
Separation Methods Used in Enzyme Purification

Property of Enzyme	*Method*	*Scale of Method*	
		Large	*Small*
Size or mass			
	Centrifugation	√	√
	Gel filtration		√
	Dialysis/ultrafiltration		√
Polarity			
	Ion-exchange chromatography	√	√
	Chromatofocusing		√
	Electrophoresis		√
	Isoelectric focusing		√
	Hydrophobic chromatography		√
Solubility			
	Change in pH	√	
	Change in ionic strength (addition of salt)	√	√
	Decrease in dielectric constant (addition of organic solvent)	√	
Specific structural features	Affinity chromatography		√
	Affinity elution		√
	Dye-ligand chromatography	√	√
	Immunoadsorption		√
	Covalent chromatography		√

[a] 'Large scale' means that the method can be conveniently used on the 100 mg scale or greater.

entitled 'Enzyme assays', p. 20.) It is also possible to assay for other enzyme activities to measure the degree of contamination by these enzymes.

In a laboratory preparation procedure, the purification is continued until homogeneous enzyme of the highest specific activity (catalytic activity per unit weight of protein) is obtained. Since there is no way of calculating the specific activity of an enzyme from first principles, the goal is normally said to have been attained when further purification steps lead to no further significant increase in specific activity. For some enzymes, it has been possible to develop 'active site titration' methods in which the amount of active enzyme in a preparation can be determined directly (Kézdy & Kaiser, 1970); the use of such a method in the case of papain is described in the section entitled 'The purification of papain', p. 10. 'Active site titration' is a very useful check on the purity of a preparation, since many of the other analytical techniques (e.g. SDS–PAGE) do not distinguish between active and inactive enzyme, provided that the polypeptide chain remains intact.

Examples of enzyme purification

Two examples of enzyme purification are described which illustrate some of the ways in which the various separation techniques are applied. The enzymes are D-xylose isomerase and papain (which represent intracellular and extracellular enzymes, respectively).

Purification of D-xylose isomerase
D-Xylose isomerase (EC 5.3.1.5) catalyses the reaction

$$\text{D-xylose} \leftrightarrows \text{D-xylulose}$$

The enzyme is also known as glucose isomerase since it also acts on D-glucose (converting it to D-fructose), although D-xylose is the preferred substrate. The enzyme is used in the food industry to produce glucose–fructose syrups which have found extensive applications as sweetening agents in a variety of processed foods.

Xylose isomerase activity can be determined in a coupled assay system (see the section entitled 'Coupled assays', p. 23.) involving D-xylulokinase, pyruvate kinase and lactate dehydrogenase as coupling enzymes. The oxidation of NADH is conveniently monitored spectrophotometrically at 340 nm.

Procedures for the isolation of xylose isomerase from a variety of sources such as *Lactobacillus brevis* are described by Yamanaka (1966). *L. brevis* is grown on a medium containing 1% peptone, 0·2% yeast extract and a variety of salts, including manganese sulphate ($MnSO_4$). The enzyme requires manganese ions (Mn^{2+}) for activity and inclusion of 0·05% $MnSO_4$ leads to a 10-fold increase in production of the enzyme.

After harvesting by centrifugation, the cells are ground with alumina to disrupt the cell wall, the cell contents extracted with 0·02 M tris-HCl buffer at pH 7·0, and centrifuged to yield a clear supernatant (the crude extract).

The crude extract is then treated with manganese chloride ($MnCl_2$) (final concentration, 50 mM) which precipitates nucleic acids and some unwanted proteins. After centrifugation, the supernatant is fractionated by addition of ammonium sulphate ($(NH_4)_2SO_4$). The fraction precipitating between 45 and 95% saturation contains the xylose isomerase. This precipitate is redissolved in 0·01 M tris-HCl buffer, pH 7·0.

After addition of Mn^{2+} to a final concentration of 5 mM, the solution is treated at 55°C for 10 min and then cooled. The heat treatment denatures and coagulates other proteins; those are removed by centrifugation, and the supernatant again fractionated with $(NH_4)_2SO_4$.

In a final step, the solution is subjected to ion-exchange chromatography on DEAE–Sephadex, and the protein eluted by increasing concentrations of potassium chloride (KCl). Xylose isomerase elutes at a salt concentration of 0·15–0·20 M. The active fractions are pooled and the enzyme concentrated by precipitation with $(NH_4)_2SO_4$.

TABLE 4
Purification of D-Xylose Isomerase from *L. brevis*

Step	*Total Protein (mg)*	*Enzyme Activity (units)*[a]	*Specific Activity (units/mg)*	*Yield*[b] *(%)*
Crude extract	7 500	8 500	1·13	100
$MnCl_2$ treatment	4 900	9 950	2·04	(117)
$(NH_4)_2SO_4$ fractionation	1 920	6 850	3·56	81
Heat treatment	990	3 960	4·00	47
Ion-exchange on DEAE Sephadex	109	1 040	9·48	12

[a] 1 unit is defined as the amount of enzyme required to produce 1 μmol D-xylulose per min.
[b] The yield is calculated as the recovery of activity from the crude extract. The small degree of activation on treatment with $MnCl_2$ is presumably associated with the requirement of the enzyme for Mn^{2+}.

The progress of the purification is recorded as shown in Table 4. The final yield is 109 mg enzyme from 110 g wet weight of harvested cells (46 litres of culture medium), and the overall purification 8·4 fold.

The purification of papain

Papain (EC 3.4.22.2) is a proteinase which preferentially cleaves a peptide chain at either —Arg↓, —Lys↓ or —Phe—X↓. It can also catalyse the hydrolysis of ester bonds and a convenient substrate is benzoyl-L-arginine ethyl ester; hydrolysis can be monitored in a pH-stat by measuring the uptake of alkali. An alternative small substrate is benzoyl-L-arginine-*p*-nitroanilide; hydrolysis of the amide bond liberates *p*-nitroaniline which can be monitored by its absorbance at 405 nm (see the section entitled 'The use of chromogenic substrates', p. 22). Papain has been extensively used as a meat tenderiser (see the section entitled 'The use of papain in meat tenderising', p. 34) because of its ability to degrade connective tissue proteins.

The purification of papain would appear to be much simpler than that of xylose isomerase, since papain is an extracellular enzyme secreted by the papaya tree. However, the preparation of fully active papain has proved difficult because of the tendency of the essential thiol group of Cys-25 (see the section entitled 'The mechanism of papain', p. 37) to undergo oxidative damage.

The usual starting material is dried papaya latex; this can be dispersed by grinding with sand and celite and extracted with a buffer containing cysteine (which helps to protect the integrity of the essential cysteine side chain).

This crude extract can be fractionated by addition of $(NH_4)_2SO_4$; the material precipitated at 40% saturation is known as the papain-rich fraction. However, using the active site titrant, 2,2′-dipyridyldisulphide, the fraction was found to contain a substantial amount of inactive enzyme (Brocklehurst *et al.*, 1973). Fully active papain can be prepared by covalent chromatography in which this 'papain-rich' fraction is applied to a column containing Sepharose–(glutathione–2–pyridyldisulphide). Papain becomes covalently attached to the Sepharose conjugate as a result of a thiol–disulphide interchange reaction and the formation of a mixed disulphide between the thiol group of Cys-25 and the bound glutathione. Unbound material is removed by washing with buffer. In the second step, papain is displaced from the conjugate by the addition of buffer containing 50 mM L-cysteine which causes a second thiol–disulphide interchange reaction. The active fractions are concentrated by precipitation with ammonium sulphate and the enzyme is separated from materials of low M_r (e.g. L-cysteine) by gel filtration on Sephadex G-25. Active papain prepared by this means has $1{\cdot}05 \pm 0{\cdot}05$ mol rapidly reacting thiols per mol protein as judged by active site titration, whereas some commercially available preparations of enzyme had thiol contents of only about half this value.

The active site titrant, 2,2′-dipyridyldisulphide, can also be used to assess the extent of contamination of papain preparations by the related enzyme chymopapain (Baines & Brocklehurst, 1978). If the content of rapidly reacting thiol groups is the same at pH 4 and at pH 8, then the peparation is free of chymopapain. If, however, the value obtained at pH 8 is greater, contamination by chymopapain is indicated.

ENZYME STRUCTURE

Introduction

This section will review the principal features of enzyme structure. A number of excellent, detailed, texts can be consulted for further information (Schulz & Schirmer, 1979; Creighton, 1983; Fersht, 1985).

Four distinct levels of structure are required to characterise an enzyme. *Primary structure* refers to the linear sequence of the 20 different amino acids in the polypeptide chain. This sequence is specified by the sequence of bases in the appropriate mRNA and hence by the sequence of bases in the appropriate segment of DNA (noting that post-transcriptional modifications, such as excision of introns, may occur). After translation of the mRNA, the polypeptide chain may be subject to a variety of post-translational modifications such as limited proteolysis, glycosylation, disulphide bond formation, etc. The lengths of polypeptide chains in enzymes range from about 90 to about 2300 amino acids (i.e. the M_r values range from about 10 000 to 250 000).

The remaining levels of structure refer to different aspects of the three-dimensional structure (or conformation) of an enzyme. *Secondary structure* refers to local interactions between amino acids in close proximity in the sequence to generate regular structural features (helices, sheets, turns, etc). *Tertiary structure* refers to the overall folding of the polypeptide chain so as to cause interactions between amino acids remote from one another in the sequence. *Quaternary structure* refers to the arrangement of polypeptide chains of an enzyme which contains more than one such chain (e.g. lactate dehydrogenase consists of four such chains, each of M_r 35 000).

Determination of structure

The primary structure of an enzyme can be determined by the 'classical' strategy in which the polypeptide chain is fragmented using proteases such as trypsin or chemical cleavage reactions (e.g. CNBr cleaves on the *C*-terminal side of methionine). After separation, the fragments are sequenced individually. The overall sequence can be deduced by preparing a new set of overlapping fragments which allow the individual sequences to be ordered. A number of developments such as the introduction of automated sequencing machines capable of determining the sequence of up to 40 or so amino acids from the *N*-terminus (Edman & Begg, 1967; Hunkapiller *et al.*, 1983), and the application of mass spectrometry to determine the sequences of fragments (Morris, 1979; Williams *et al.*, 1982), have helped to speed up this classical approach, but despite these, sequencing is often a very time-consuming operation.

Since the late 1970s, developments in the classical, 'direct' approach to protein sequencing have been overtaken by those in the methods of sequencing bases in DNA (Old & Primrose, 1985). If the gene coding for the enzyme of interest can be isolated or the cDNA (complementary DNA) obtained from the corresponding mRNA by the action of reverse transcriptase, the amino acid sequence of the enzyme can be 'read off' from the base sequence of the DNA using the genetic code. This 'indirect' method of protein sequencing is usually at least an order of magnitude more rapid than the 'direct' method, but has some disadvantages, namely that it can be difficult to isolate the required segment of DNA and post-transcriptional modifications to the polypeptide chain are not detected.

The secondary, tertiary and (where appropriate) quaternary structures of an enzyme can be determined by X-ray crystallography (Blundell & Johnson, 1976). The diffraction pattern of X-rays by crystals of the enzyme can be analysed to yield a three-dimensional map of the electron density of the molecule. Interpretation of this electron density map in terms of atomic structure requires the covalent structure (amino acid sequence) of the enzyme. Determination of the three-dimensional structure of an enzyme by X-ray crystallography remains a difficult task requiring several major obstacles to be overcome successfully, despite recent developments such as the use of more intense synchrotron X-ray sources and more rapid computation of electron density maps from diffraction data.

The three-dimensional structure of an enzyme obtained by X-ray crystallography must be related to the structure of the enzyme in solution. Although enzymes in solution show more flexibility because of the lack of constraints imposed by the packing of molecules in the crystals, it is clear from a number of observations that the crystal structure is likely to be a very good guide to the time-average structure in solution (Creighton, 1983; Fersht, 1985). Recent developments in high resolution nuclear magnetic resonance have made the determination of the solution structures of small enzymes (mol.wt $< 15\,000$) a realistic objective. The results obtained with lysozyme show that the structure in solution is indeed similar to that observed in the crystal (Redfield & Dobson, 1988).

The forces involved in maintaining the folding structures of enzymes

The forces involved in maintaining the three-dimensional structures of globular protein such as enzymes are weak non-covalent interactions. As well as forces between different parts of the enzyme, account must also be taken of the forces between the enzyme and the aqueous solvent.

Experiments, in which the refolding of denatured enzymes has been studied, have shown that disulphide bonds are formed by oxidation of pairs of cysteine side chains when the three-dimensional structure of the enzyme has already been formed so that these groups are brought into juxtaposition. Thus, disulphide bonds, if they occur, tend to confer stability on the folded structure of an enzyme (see section entitled 'Enzyme stability', p. 40) rather than direct the acquisition of this structure. The bond energy of a disulphide bond (approximately 200 kJ/mol) is much greater than those of the non-covalent interactions described below.

The non-covalent interactions in enzymes are usually considered as being of four types—hydrogen bonds, electrostatic forces, van der Waals forces, and hydrophobic forces. These interactions are discussed in detail by Creighton (1983) and Fersht (1985). (To some extent, the division between the various forces is rather arbitrary since ultimately all attractive forces arise from favourable interactions between nuclei and electrons.)

Hydrogen bonds

These bonds are formed between an electronegative atom and a hydrogen atom attached to another electronegative atom. For biological systems, O and N are the relevant electronegative atoms. Hydrogen bonds are relatively weak (20 kJ/mol *in vacuo*) and their strength is highly dependent on geometrical factors such as the linearity of the nuclei and the internuclear distance between the two electronegative atoms concerned. (Indeed, this distance is used to detect the presence of such bonds in the X-ray structure.) Hydrogen bonds are thus important in specifying the correct folded structure of an enzyme and the correct interaction between the enzyme and other molecules such as substrate. It is probable, however, that they contribute relatively little to the stability of the folded state of an enzyme, since in the unfolded state alternative hydrogen bonds could be made to the solvent.

Electrostatic forces

Attractive electrostatic forces (also known as salt bridges) arise from the interaction between groups which bear opposite charges, e.g. the $—CO_2^-$ and $—NH_3^+$ groups of aspartic acid (or glutamic acid) and lysine, respectively. The magnitude of such forces depends on the distance between the charged groups and on the dielectric constant (D) of the intervening medium. For water, $D = 80$, so such forces are much weaker in an aqueous medium than in, say, benzene for which $D = 2 \cdot 3$. Recently, some progress has been made in determining the effective value of D in proteins. The range of values of D obtained ($30 < D < 60$) suggest that in general, electrostatic forces make only a relatively small contribution to the stability of the folded structures of enzymes (Gilson & Honig, 1987; Sternberg *et al.*, 1987).

van der Waals forces
These weak forces (≤10 kJ/mol) occur when molecules or groups of atoms come into close contact with one another (<0·4 nm). Attractive forces result from favourable interactions between dipoles which may be permanent or transient and which can be induced by the presence of a neighbouring dipole. The contribution of van der Waals forces to the stability of a folded structure is difficult to estimate but is probably small.

Hydrophobic forces
Water is almost a uniquely bad solvent for hydrocarbon solutes such as methane or hexane. The reason for this behaviour is to be found in the structure of water, which consists of transient clusters of hydrogen-bonded molecules. If a hydrocarbon is introduced into the aqueous medium, a reorganisation of water molecules occurs around this hydrocarbon; this makes the system more ordered. The consequent decrease in entropy would be unfavourable on thermodynamic grounds, and hence the hydrocarbon is forced out of contact with water into the organic phase. The hydrocarbon is termed 'hydrophobic', although it would be more accurate to describe water as 'hydrocarbon-hating', rather than the hydrocarbon as 'water-hating' (Tanford, 1978).

The hydrophobicities of the various amino acid side chains can be assessed by measuring the tendencies of the amino acids to partition between water and an organic solvent such as octanol. Using these measures, the side chains of valine, leucine, isoleucine, phenylalanine and tryptophan are very hydrophobic and should thus be buried in the interior of an enzyme away from the aqueous solvent. Calculations show that this burying of the non-polar side chains makes by far the major contribution to the stability of the folded compared with the unfolded structure (Creighton, 1978). It should be emphasised, however, that the stabilisation is due to the burying of these side chains away from water, rather than to any interaction between the non-polar side chains themselves.

General features of enzyme structures

An analysis of the structures of over 100 enzymes and other proteins which have been determined by X-ray crystallogaphy has revealed a number of general features (Creighton, 1983). Some of these are listed below.

(1) The interiors of globular proteins, such as enzymes, are closely packed structures, with approximately 75% of the interior volume occupied by atoms, as measured by their van der Waals radii. There are usually only a small number of cavities, which can be occupied by water molecules.

(2) Structural elements such as the planar amide bond, α-helix, β-sheet and β-turn (or bend) generally have similar geometries to those observed in model polypeptide compounds, although some distortions are noted. Thus, the α-helix can be distorted towards a 3_{10} helix and β-sheets are generally twisted rather than planar with a right-handed twist of 0–30° between strands. In general, a very high proportion of the amino acids in enzymes occur in recognisable secondary structural features. The average values for a number of proteins are: α-helix, 38%; β-sheets, 20%; β-turn, 32% (Creighton, 1983); but it should be emphasised that there is considerable variation in these proportions between different enzymes (see (4) below).

(3) Non-polar amino acid side chains are generally buried in the interior of enzymes away from the aqueous solvent (see the section entitled 'Hydrophobic forces', p. 14) whereas ionised side chains are on the exterior and able to interact with the solvent. Exceptions to this general rule are usually associated with specific functions, e.g. the buried side chain of Asp-194 in chymotrypsin is involved in an interaction with the α-amino group of Ile-16 which is crucial for activation of the zymogen. Polar groups in the interior of enzymes (e.g. the —NH— or $\rangle$C=O groups of the polypeptide chain) are generally paired in hydrogen bonds either with each other or with water.

(4) Most larger enzymes (of $M_r > 20\,000$) exist as a number of distinct folded structural units or domains. The domains often consist of regular arrangements of secondary structural elements such as α-helix or β-sheet known as 'supersecondary structures'. It is possible to classify proteins into four main types on the basis of the pattern of secondary structural elements (Levitt & Chothia, 1976). All α proteins (such as myoglobin) have only α-helices, usually in contact with each other. All β proteins (such as chymotrypsin) have mainly β-sheet structures in which the β-strands pack against each other or form a continuous cylindrical sheet or barrel. In ($\alpha + \beta$) proteins (such as thermolysin), the α-helix and β-sheet structural segments occur in separate parts of the molecule. (α/β) Proteins (such as kinases and dehydrogenases) have mixed or alternating segments of α-helix and β-sheet, with a helix usually occurring in each of the segments of polypeptide chain connecting the β-strands.

Richardson (1981) has presented a more detailed classification of protein structures based on patterns within domains rather than in the whole molecule. An example of this is shown in the case of papain (see the section entitled 'An example of enzyme structure', p. 16).

Enzymes of M_r greater than about 50 000 usually consist of a number of individual polypeptide chains or subunits. The forces involved in associations between subunits are of the weak non-covalent type, similar to those involved in maintaining the folded tertiary structures of enzymes (see the section entitled 'The forces involved in maintaining the folded structures of enzymes', p. 13). The presence of multiple subunits can offer greater possibilities for control of enzyme activity (see the section entitled 'Kinetics of one-substrate reactions', p. 25).

Representation of enzyme structures

The complex, irregular structures adopted by enzymes and other globular proteins pose considerable problems of representation. Three-dimensional scale models built from the coordinates of the electron density map are the most complete representation; however such models are cumbersome and expensive to assemble. Recent developments in computer technology have made possible the display of such complex structures; a suitable 'molecular graphics facility' allows attention to be focused on the environments of particular side chains, the geometry of the active site, and inter-subunit contact regions, for example.

For two-dimensional representation, it is preferable to simplify the structure, e.g. by stripping off the side chains so as to reveal the folding of the polypeptide chain. Figure 1 shows the folding of the chain in papain by linking the positions of the α-carbon atoms of each residue. In this type of representation it is not always easy to detect regular elements of secondary structure and this has encouraged a 'structural cartoon' approach in which such elements are emphasised. α-Helices are shown as cylinders or ribbons and strands of β-sheets are shown as thick arrows. Figure 2 shows the structure of adenylate kinase in such a 'cartoon' representation; the alternating helix–sheet pattern in this α–β protein can be seen clearly. The helix–sheet motif (or a variation on it) is found in proteins which bind nucleotides such as kinases and dehydrogenases. It seems probable that different enzymes of these types may have arisen by the fusion of a gene coding for this 'nucleotide-binding domain' with another gene coding for the 'substrate binding domain' which is of variable structure according to the reaction catalysed (Creighton, 1983).

An example of enzyme structure

Papain has been chosen as an example of enzyme structure because of its widespread use in the food industry, mainly as a meat tenderiser. (In the section entitled 'the mechanism of papain', p. 37, the mechanisms of action of the enzyme will be discussed in relation to its structure.)

The α-carbon chain diagram (Fig. 1) shows that the enzyme (212 amino acids with three disulphide bonds and one cysteine side chain) consists of two domains (Drenth *et al.*, 1971). The left-hand domain contains residues 11–111 and 209–212 and the right-hand domain residues 1–10 and 112–208. Thus, the connection between the two domains is in the middle of the polypeptide chain and at both ends. The terminal segments (1–10 and 209–212) make a number of contacts (hydrogen bond, electrostatic, and hydrophobic) with the rest of the respective domain.

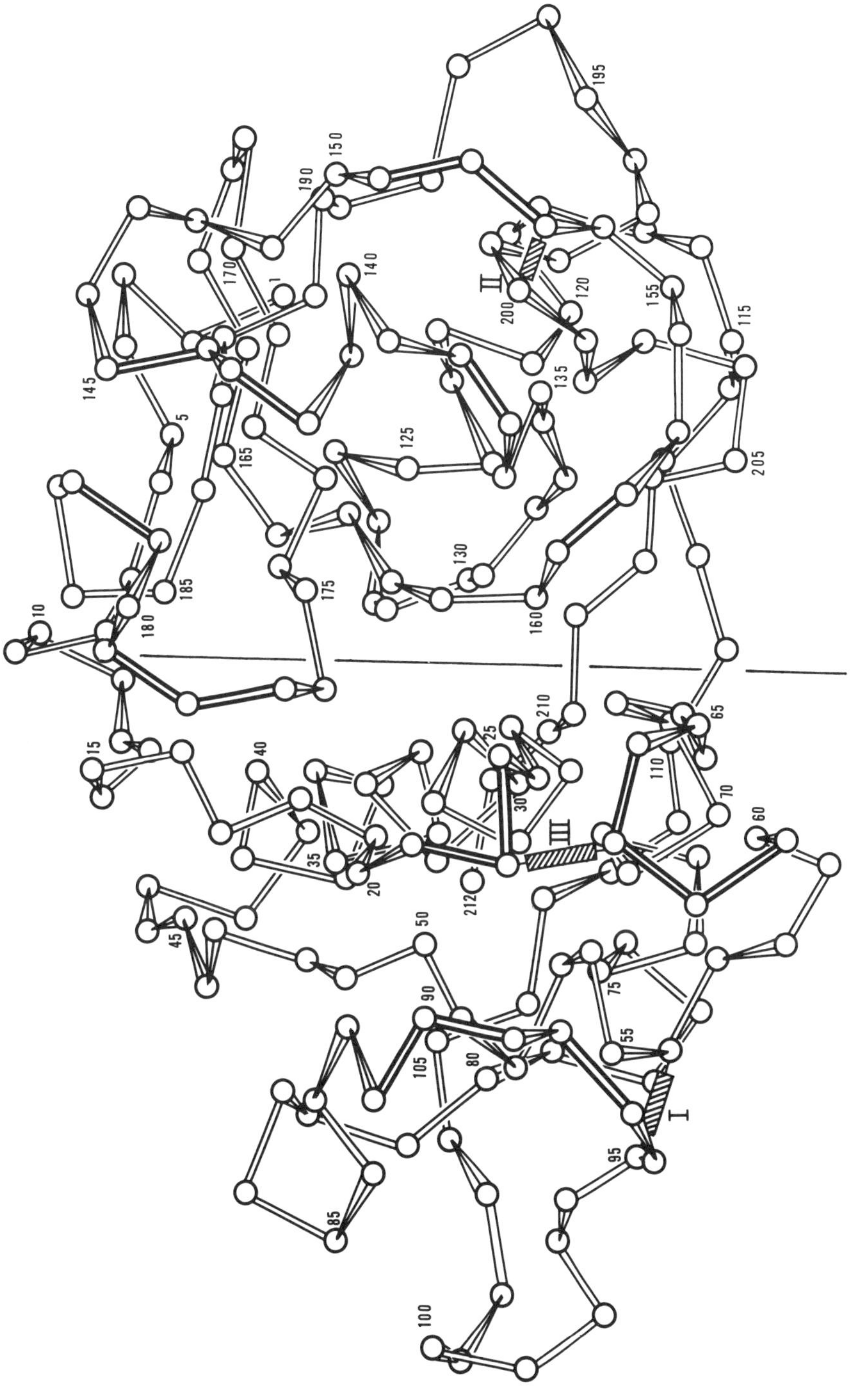

FIG. 1. The folding of the polypeptide chain in papain. The heavy line separates the two domains of the molecule. Disulphide bonds are shown as hatched rectangles. (From Drenth *et al.* (1971), reproduced by permission of Academic Press Inc., Orlando, FL, USA.)

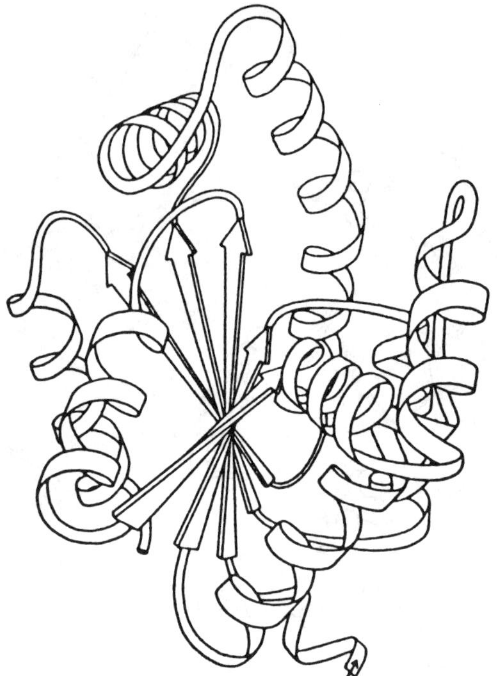

FIG. 2. Structural cartoon of adenylate kinase. In this representation, α-helices are shown as ribbons and strands of β-sheets as thick arrows. (From Richardson (1981) reproduced by permission of J. S. Richardson.)

There are also a large number of interactions (hydrogen bond, electrostatic, and hydrophobic) between the two domains. The structural stability of each domain is due, at least in part, to an extensive hydrophobic core. The folding of the polypeptide chain is quite different in the two domains as can be appreciated from the structural cartoon diagrams shown in Fig. 3. The left-hand domain contains mainly α-helices and is classified as a Greek key helix bundle (Richardson, 1981). In contrast, the right-hand domain consists mainly of a β-barrel structure with two flanking α-helices.

Between the domains is situated the active site region as a 2 nm long groove or cleft. The important catalytic side chains Cys-25 and His-159 are on opposite sides of this cleft.

Intracellular and extracellular enzymes

Most of the first enzyme structures to be elucidated by X-ray crystallography were those of small hydrolytic extracellular enzymes (lysozyme, ribonuclease, chymotrypsin, etc); the majority of the enzymes used in the food industry are of this type. More recently, the structures of a number of intracellular enzymes

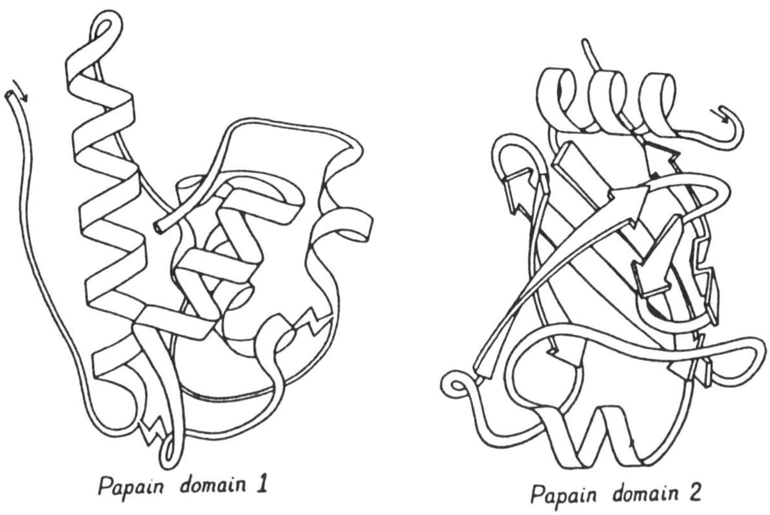

FIG. 3. Structural cartoon of the two domains of papain. (From Richardson (1981) reproduced by permission of J. S. Richardson.)

have been determined. Although the basic principles involved are the same, there are two general differences between the two types.

(1) In general, disulphide bonds are confined to extracellular (secreted) enzymes whereas free thiols are found in intracellular enzymes. Formation of the disulphide bonds is a post- (or co-) translational event occurring in the lumen of the endoplasmic reticulum in eukaryotes or in the periplasm of prokaryotes.
(2) In general, extracellular enzymes consist of a single polypeptide chain whereas many intracellular enzymes are oligomeric, i.e. consist of multiple subunits.

These general differences have the following consequences:

(1) The formation of disulphide bonds leads to enhanced stability against unfolding of the polypeptide chain. This could be important for extracellular enzymes which might encounter a wide range of conditions in contrast to the relatively stable conditions within the cell.
(2) The occurrence of multi-subunit structures in intracellular enzymes can give rise to greater possibilities for control of catalytic activity via conformational changes relayed between the subunits. This is particularly important in intracellular enzymes because it facilitates the integrated control of the various metabolic pathways in the cell.

ENZYME ASSAYS

Introduction

The presence of an enzyme in an extract is established by detection of its catalytic activity. Thus, for any studies on an enzyme a suitable method of assay is a prerequisite. The advantages and disadvantages of particular methods of assay will depend on the type of investigation. For example, if an enzyme is being purified using a number of stages, the prime requirement is a method of detection which can easily be applied to a large number of samples, and a high level of accuracy may be less important. On the other hand, if the kinetic parameters of an enzyme are being determined to support a particular proposed mechanism, then high accuracy and reproducibility are extremely important.

When an enzyme is incubated with its susbstrate(s), the concentration of the substrate(s) will fall and that of the product(s) rise. Enzyme assays, therefore, either measure the decrease in substrate concentration or the increase in product concentration as a function of time. When measuring the activity of an enzyme in an extract, it is usual to perform preliminary experiments to determine the optimum conditions for assay. This entails determination of the optimum substrate concentration, pH, temperature, ionic strength, and requirement for cofactors or activators (The optimum temperature often depends on the length of time over which the assay is made, since at high temperatures, although the reaction rate is increased, so is the rate of inactivation of the enzyme.) For most studies, it is the initial rate that is measured, since the rate of any back-reaction or any product inhibition can then be disregarded. The rate of the reaction will be related to the substrate concentration and this is usually set sufficiently high so that the decrease in substrate concentration during the period of assay will be low and the concentration of substrate may be considered as constant. In enzyme assays it is necessary to perform controls. An enzyme blank (i.e. enzyme omitted) is normally carried out to check that the non-enzymic reaction is either negligible or can be corrected for. When using crude enzyme extracts it is sometimes useful to incubate the enzyme extract without the added substrate to measure activity in the presence of any endogenous substrate.

Methods of assay

The simplest type of assay is one in which the substrate and the products differ in some property, usually the absorption spectrum, which can be used to measure either the disappearance of the substrate or the appearance of the product. For example, any dehydrogenase which uses NAD^+ or $NADP^+$ as cosubstrate, e.g. lactate dehydrogenase,

$$\text{lactate} + NAD^+ \rightleftharpoons \text{pyruvate} + NADH + H^+$$

can be monitored using the difference in absorption spectra at 340 nm (NAD(P)H absorbs at 340 nm, $NAD(P)^+$ does not). Often, however, there may be no readily measurable difference in the absorbance of the substrate and product. It is then sometimes possible to measure the substrate or product by its conversion to some coloured product. For example, a phosphatase that catalyses the hydrolysis of glucose-6-phosphate may be detected by reacting the inorganic phosphate formed with ammonium molybdate and then reducing the phosphomolybdate formed to a blue-coloured complex:

$$\text{glucose-6-phosphate} + H_2O \rightleftharpoons \text{glucose} + P_i$$

Enzyme assays may be continuous or discontinuous. With a continuous assay the disappearance of the substrate or appearance of the product is monitored continuously, as, for example, in the case of lactate dehydrogenase where the production of NADH is measured at 340 nm. In a discontinuous assay, the enzyme and substrate are incubated for a fixed time, the reaction is then stopped and the product or substrate concentration measured. The example of glucose-6-phosphatase would require a discontinuous assay, since the method of colour development inhibits further enzyme activity. Continuous assays, where feasible, are generally preferred since they give more information. It is possible to confirm that the initial rate is linear for the period of measurement. With a discontinuous assay, additional experiments are necessary to check that the reaction rate is linear for the incubation period. Discontinuous assays can, on occasions, be less time consuming, if for example a large number of column fractions are being monitored for activity to find the peak fractions.

A change in the absorbance in the visible or near UV spectrum is perhaps the most common method of assaying enzymes. There are, however, several other methods. If there is a difference in the fluorescence spectrum of the substrate and product (e.g. NAD^+ and NADH, in which only the latter fluoresces), this may be used. In general, methods based on fluorescence are more sensitive that those based on differences in the absorption spectrum. If an enzyme-catalysed reaction is accompanied by the release or uptake of a proton, then this may form the basis of an assay, e.g. the reaction catalysed by urease which entails a net uptake of protons.

$$CO(NH_2)_2 + 3H_2O \rightleftharpoons CO_2 + 2NH_4{}^+ + 2OH^-$$

The enzyme may be assayed by continuous titration to maintain a constant pH. A similar titration method may also be used to assay papain (see the section entitled 'The purification of papain', p. 10).

A highly sensitive method of assay is to use a radioactively labelled substrate. This generally will entail the detection of the radioactive product. For example, the activity of ornithine decarboxylase in most tissues is far too low to be able to detect as CO_2 by gravimetric trapping. In a crude extract it would also not be possible to distinguish CO_2 which had arisen from ornithine or from other metabolites.

$$\text{L-ornithine} \rightleftharpoons \text{putrescine} + CO_2$$

By using carboxyl-labelled ornithine it is possible to devise a very sensitive assay which entails trapping and measuring $^{14}CO_2$. Sometimes the use of radioactive substrates entails the separation of the substrate and product by solvent extraction, e.g. phosphatidylcholine–cholesterol acyltransferase.

$$^{3}\text{H-cholesterol} + \text{phosphatidylcholine} \rightleftharpoons {}^{3}\text{H-cholesterolester} + \text{1-acylglycerophosphocholine}$$

For further details of methods of assay see Dixon *et al.* (1979) and Tipton (1978).

Assay of hydrolytic enzymes

Many of the natural substrates for enzymes belonging to group 3, the hydrolases, are high M_r polymers, e.g. enzymes that catalyse the hydrolysis of polysaccharides, proteins and polynucleotides. These can generally be assayed by using differences in the properties of the polymers and the monomers or oligomers. For example, proteolytic enzymes can be assayed by incubating with a protein substrate and then stopping the reaction using a protein precipitant, such as trichloroacetic acid (TCA) or perchloric acid (PCA). These agents will precipitate all peptides greater than about 10–15 amino acids in length. The lower M_r peptides and the amino acids can then be measured in the supernatant. This may be done by using ninhydrin, which reacts with all amino groups, or by measuring $A_{280\,\text{nm}}$ which detects the aromatic amino acids. A similar procedure can be adopted with enzymes that catalyse the hydrolysis of polynucleotides. Again, the reaction may be stopped using TCA or PCA and the mononucleotides and oligonucleotides detected by their $A_{260\,\text{nm}}$.

The hydrolysis of polysaccharides can be measured either by the increase in reducing sugars or by the decrease in the viscosity of the polymer as it is degraded. Sometimes, the polymeric substrates are modified to increase the sensitivity of detection. For example, when assaying proteases the substrates hide powder azure, azocoll or azocasein are sometimes used. These are native proteins, collagen or casein, that have been chemically modified to include a chromophoric group. After degradation, the peptides containing the chromophore are soluble in TCA and are easily quantified. Cellulose is a difficult substrate to use owing to its insolubility, and a modified form, carboxymethylcellulose, which is more readily dispersed, is often used as an alternative.

The use of chromogenic substrates

The methods just described for group 3 enzymes are all useful for detection of the enzymes concerned, but they have definite limitations in kinetic analysis. In each there are many bonds in the substrate which may be cleaved, and in the case of heteropolymers the regions adjacent to the bonds may differ. In some of these assays, a product is detected only when it has been degraded below a certain size. During the course of incubation the enzyme will, in a sense, have a choice of attacking the initial remaining substrate, or of attacking a range of intermediate-sized substrates. These factors make the interpretation of kinetic data more difficult. It is therefore often desirable to use a low M_r substrate which contains a susceptible bond similar to that in the polymeric substrate. Such low M_r substrates are generally designed to be chromogenic. Much of the kinetic analysis on ribonuclease was carried out using 2′,3′-cyclic CMP. Ribonuclease cleaves the 2′ester bond and this results in an increase in $A_{284\,nm}$. For measurement of proteinases and glycosidases the chromophore used most commonly is either the nitrophenyl or the nitroanilide derivative. For example, trypsin, which catalyses the hydrolysis of peptide bonds on the carboxyl side of arginine or lysine residues, can be readily assayed using benzoylarginine *p*-nitroanilide. In each case, the nitrophenol or nitroaniline released can be detected by its $A_{405\,nm}$. A similar range of fluorogenic substrates is also available, the most common being derivatives of 4-methylumbelliferone.

The use of chromogenic substrates with proteinases has made it possible to define more precisely their specificity. For example, many proteinases may have binding sites which recognise not just the peptide bond cleaved but also the amino acid residues adjacent or close to the bond cleaved. The specificity can be recognised by assaying the proteinase on a range of dipeptide and tripeptide nitroanilides. By measurement of k_{cat}/K_m (see the section entitled 'Significance of the kinetic parameters for an enzyme', p. 28), the enzyme's preference for a particular substrate can be quantified.

Coupled Assays

With a number of enzyme-catalysed reactions there may not be an obvious difference in the absorption spectra of the reactants and products. These reactions may often be coupled to a second, or second and third reaction in which there is a measurable spectral change. The coupling enzyme most frequently used is a dehydrogenase for which NAD^+ or $NADP^+$ is co-substrate. Two examples of coupled enzyme assays are those for pyruvate kinase and fructose bisphosphatase:

(1) Pyruvate kinase
phosphoenol pyruvate + ADP ⇌ pyruvate + ATP
coupled with lactate dehydrogenase:

$$\text{pyruvate} + \text{NADH} + \text{H}^+ \rightleftharpoons \text{lactate} + \text{NAD}^+$$

(2) Fructose bisphosphatase
fructose bisphosphate + H_2O ⇌ fructose-6-phosphate + P_i
coupled with phosphohexoseisomerase:

$$\text{fructose-6-phosphate} \rightleftharpoons \text{glucose-6-phosphate}$$

and glucose-6-phosphate dehydrogenase:

$$\text{glucose-6-phosphate} + \text{NADP}^+ \rightleftharpoons \text{6-phosphogluconate} + \text{NADPH} + \text{H}^+$$

For coupled assays, it is necessary to have an excess of the coupling enzymes and substrates so that the rate-limiting step is always that of the enzyme being assayed. In practice, this means that the V_{max} for the coupling reaction must be greater than that of the primary reaction. When starting such a coupled assay there will be a lag while the concentration of the product of the primary reaction builds up to a steady state level (Fig. 4). The lag should be kept as short as possible, and this is done by having a large excess of the coupling enzyme. For a detailed theoretical treatment of the kinetics of coupled reactions, see Cornish-Bowden (1979), Suelter (1985), Brooks & Suelter (1989). A point to note whenever using coupled assays is that changing any of the parameters which affect the primary enzyme, e.g. substrate, cofactor, activator or inhibitor concentrations, temperature or pH may also affect the coupling enzymes in such a way that their V_{max} is not in excess of that of the primary reaction and thus the rate measurement would be invalidated. In practice, this can be checked by doubling the concentrations of the coupling enzymes to see if the rate is increased.

Units of activity

The units of enzyme activity are usually expressed as either μmol substrate consumed, or product formed, per min ('units' or 'international units') or mol substrate consumed, or product formed, per s ('katal' in the SI system). The specific activity of an enzyme preparation is usually expressed as units/mg or

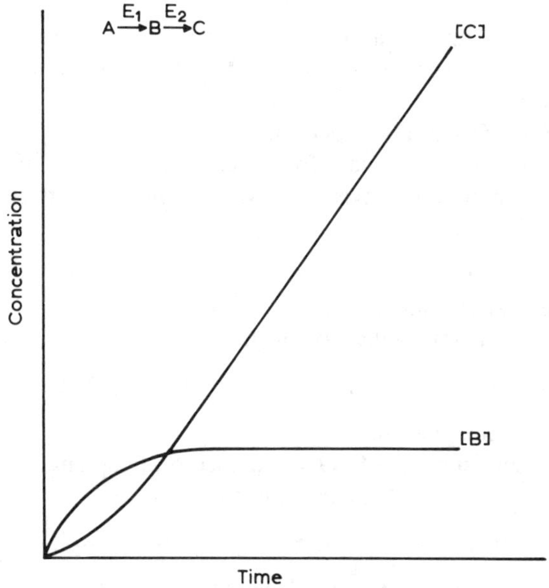

FIG. 4. The changes in concentration of the (B) intermediate product and (C) the final product with time in an enzyme coupled reaction.

katal/kg. For a purified enzyme for which M_r is known, the catalytic activity can be expressed as the turnover number (see the section entitled 'The kinetics of one-substrate reactions', p. 25).

ENZYME KINETICS

Introduction

In this section some of the more important aspects of the interpretation of enzyme kinetic data will be discussed. A number of texts can be consulted for further details (Laidler & Bunting, 1973; Cornish-Bowden, 1979; Dixon *et al.*, 1979; Engel, 1981).

The following two sections deal with 'steady state kinetics' where the enzyme is present at a much lower molar concentration (usually 0·001–1%) than that of the substrate(s) acted on. Experiments carried out under such conditions are relatively easy to perform (since data can be collected over a period of several minutes) and to interpret (since in the equations it is possible to set the concentration of free substrate equal to the total concentration of substrate). Steady state measurements are of value in characterising a particular enzyme activity and in elucidating the mechanism of an enzyme-catalysed reaction (see the section entitled 'Enzyme mechanism', p. 33). However, it is important to realise that *in vivo* and in some industrial applications, enzymes can be present at concentrations of the same order of magnitude as their substrates. The study of enzyme-catalysed reactions under these conditions (pre-steady state kinetics) is dealt with on p. 31 and the modifications needed to the equations involved are also described on p. 31. Finally, a brief reference will be made to the kinetics of immobilised enzyme systems which are sometimes used in industrial processes (p. 32).

Kinetics of one-substrate reactions

Although the study of actions involving only one substrate helps to introduce most of the important concepts, it must be remembered that only the reactions catalysed by hydrolases (if water is in large excess), isomerases and lyases can be considered to fall into this category. The simplest scheme

$$E + S \underset{k_{-1}}{\overset{k_1}{\rightleftharpoons}} ES \xrightarrow{k_{cat}} E + P$$

involves enzyme (E) and substrate (S) combining reversibly to form a complex (ES) which then breaks down irreversibly to yield product (P) and regenerate enzyme.

Using either the equilibrium assumption (in which the breakdown of ES to yield $E+P$ is assumed not to disturb the $E+S \rightleftharpoons ES$ equilibrium) or the steady state assumption (in which the concentration of ES is assumed to remain constant), an equation of the form shown in eqn (1) is obtained. This is often known as the Michaelis–Menten equation.

$$v = \frac{V_{max}[S]}{K_m + [S]} \tag{1}$$

in which v is the observed velocity at substrate concentration $[S]$, V_{max} is the maximum velocity (at infinite $[S]$), and K_m is a constant (Michaelis constant) which corresponds to the $[S]$ at $v = V_{max}/2$.

When the simple scheme is modified to include an additional intermediate, ES', the equation obtained is of the same form as eqn (1), but the significance of K_m in terms of the rate constants of the elementary steps of the reaction is changed.

If the molar concentration ($[E]_0$) of enzyme active sites is known, V_{max} can be written as $k_{cat}[E]_0$, where k_{cat} is the catalytic rate constant or turnover number of the enzyme.

Equation (1) shows that v has a hyperbolic dependence on $[S]$ (Fig. 5). It is difficult to estimate V_{max} (and hence K_m) visually from such a plot, although a number of computer programs are now available to derive the 'best fit' values of these parameters from the v, $[S]$ data (Cleland, 1979). For graphical determination of V_{max} and K_m, various rearrangements of eqn (1) are made; these are depicted in Fig. 6. Each plot has its advantages and disadvantages. Thus, the Lineweaver–Burk plot (Fig. 6(a)), which is probably the most widely used, has the advantage that the variables v and $[S]$ are plotted separately on the two axes. However, in this plot the distribution of errors is highly non-uniform, and this has led many workers to prefer the use of the Eadie–Hofstee and Hanes plots (Figs 6(b) and (c)) in which the distribution of errors is more uniform. The 'direct linear plot' (Fig. 6(d)) is of a different type and can be used for direct analysis of v, $[S]$ data as they are being obtained.

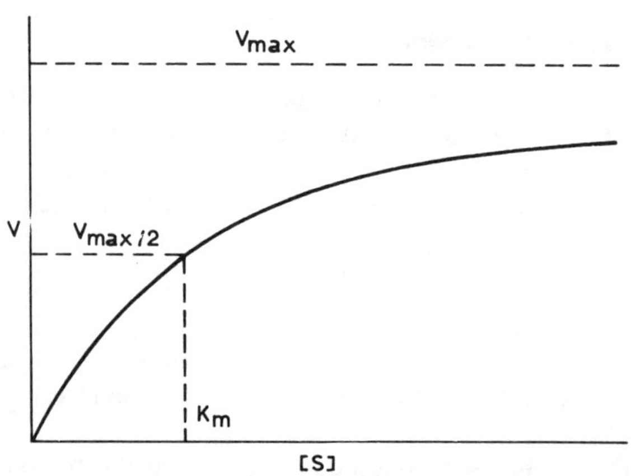

FIG. 5. Dependence of velocity (v) on substrate concentration ($[S]$) according to eqn (1).

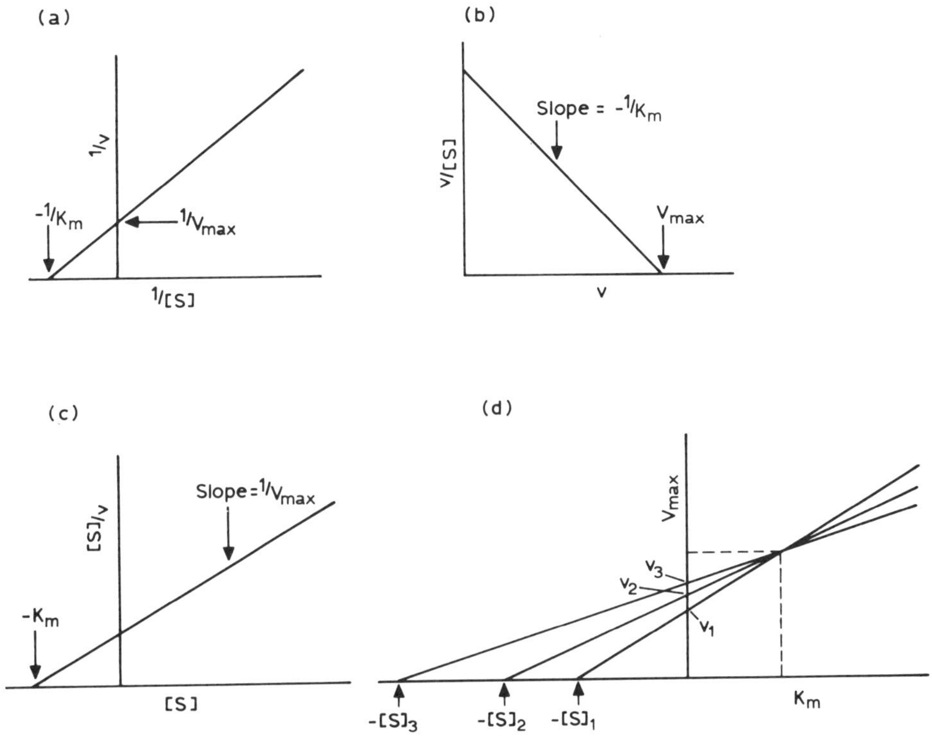

FIG. 6. Graphical methods for determination of the parameters V_{max} and K_m in eqn (1). (a) Lineweaver–Burk plot, (b) Eadie–Hoftee plot, (c) Hanes plot, (d) direct linear plot. In (d) experimental values of v and $[S]$ are plotted directly on the V_{max} and K_m axes as shown. The point of intersection of the straight lines drawn through the individual sets of values gives the values of V_{max} and K_m. In practice the lines do not usually intersect at a single point and the best fit values of V_{max} and K_m are taken to be the median of the values from the various points of intersection (see Cornish-Bowden, 1979).

The 'best fit' values of K_m and V_{max} are the medians of the values obtained from the points of intersection of the various lines (Eisenthal & Cornish-Bowden, 1974).

The kinetic-behaviour of a number of intracellular enzymes has been found not to be hyperbolic (eqn (1) and Fig. 5), but sigmoidal (Fig. 7). Over a particular range of substrate concentrations, the velocity of this latter type shows a more sensitive response to changes in substrate concentrations than the former. A number of mathematical models to account for sigmoidal behaviour have been proposed; most of these depend on the transmission of conformational changes between the subunits in the multi-subunit (oligomeric) enzyme. The activity of many of these enzymes can be also regulated by the binding of other ligands. The lack of structural similarity between substrate and regulator makes it likely that the interaction between the two sites is

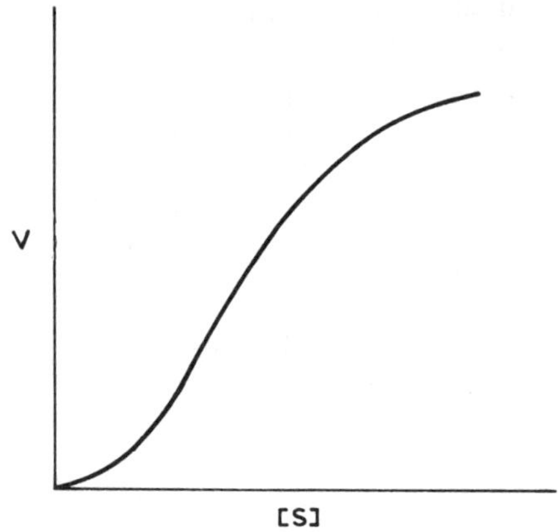

FIG. 7. A sigmoidal plot of velocity (v) against substrate concentration ($[S]$). Such a plot is observed, for instance, in the case of aspartate carbamoyltransferase from *E. coli* when the velocity is plotted against the concentration of L-aspartate.

indirect (i.e. mediated via conformational changes in the enzyme); the term 'allosteric' has been applied to this type of indirect interaction. For a fuller discussion of sigmoidal kinetics and allosteric effects and their importance in metabolic regulation, the reviews by Newsholme & Start (1973) and Tipton (1979) should be consulted.

Significance of the kinetic parameters for an enzyme

The parameters K_m and V_{max} for an enzyme-catalysed reaction can indicate the role of that enzyme in metabolism. Knowledge of the prevailing concentration of substrate and the maximum catalytic capacity allows the contribution of the enzyme to the overall turnover of substrate to be estimated. Thus, the liver isoenzyme IV of hexokinase (glucokinase) can be shown to be responsible for the phosphorylation of the extra blood glucose following a meal (Newsholme & Start, 1973).

The parameters V_{max} (or more precisely k_{cat}) and K_m can be used to give a quantitative measure of the specificity of an enzyme for a given substrate and of the efficiency of an enzyme. The higher the ratio k_{cat}/K_m, the greater the preference (or specificity) of the enzyme towards that substrate. The upper limit of k_{cat}/K_m corresponds to the diffusion-controlled encounter rate between enzyme and substrate (around 10^9 mol/litre s). Some enzymes (such as triosephosphate isomerase) approach this limit and have been described as being 'almost perfectly evolved'. For a fuller discussion see Fersht (1985).

Inhibition of enzyme-catalysed reactions

Inhibitors of enzyme-catalysed reactions can be classified as reversible or irreversible. In the former case the effect can be reversed by lowering the concentration of inhibitor, e.g. by dilution or dialysis of the mixture of enzyme and inhibitor. The comments below are confined to this category. Irreversible inhibitors can be thought of as simply reducing the effective concentration of enzyme.

Inhibitors are classified according to their effects on the parameters V_{max} and K_m. *Competitive inhibitors* increase the apparent K_m but leave V_{max} unchanged; thus, the effects of the inhibitor can be competed out by increasing the concentration of substrate. Competitive inhibitors often, but by no means always, possess structural analogy with the appropriate substrate and hence presumably bind at the active site of the enzyme. *Pure non-competitive inhibitors* decrease V_{max} but leave K_m unaffected. *Uncompetitive inhibitors* cause decreases in both K_m and V_{max}. Pure non-competitive and uncompetitive inhibition are only rarely observed in one-substrate kinetics, but are more common in multi-substrate systems. Inhibitors which do not conform to the limiting cases mentioned are said to cause *mixed inhibition,* and are classified as predominantly competitive or predominantly uncompetitive according to their relative effects on the kinetic parameters (IUB, 1981).

The kinetics of two-substrate reactions

Enzymes which belong to the classes oxidoreductases, transferases and ligases by definition catalyse reactions involving more than one substrate. The algebraic treatment of these mechanisms uses the same concepts as for one-substrate reactions, although the equations derived are more complex.

Generally, for two-substrate reactions, the mechanisms can be divided into ternary complex or enzyme substitution ('ping-pong') types as depicted below (A and B represent the substrates, P and Q the products).

Ternary complex

ordered $E + A \rightarrow EA$

$EA + B \rightarrow EAB$

but $E + B \not\rightarrow EB$

random $E + A \rightarrow EA$ or $E + B \rightarrow EB$

$EA + B \rightarrow EAB$ $\quad EB + A \rightarrow EAB$

Enzyme substitution

$$E + A \rightarrow E' + P$$

$$E' + B \rightarrow E + Q$$

The equation which applies to the ternary complex mechanism is

$$v = \frac{V_{max}[A][B]}{K_A{}'K_B + K_B[A] + K_A[B] + [A][B]} \tag{2}$$

and for the enzyme substitution type

$$v = \frac{V_{max}[A][B]}{K_B[A] + K_A[B] + [A][B]} \tag{3}$$

(eqn (3) is thus a special case of eqn (2) when $K_A' = 0$)

In eqns (2) and (3), K_A and K_B represent the individual Michaelis constants for A and B, respectively, in the presence of saturating concentrations of the other substrate. K_A' in eqn (2) does not have such a simple meaning, though in terms of the mechanisms of the reactions, K_A', K_A and K_B represent combinations of rate constants of individual steps in the reaction.

The values of the parameters V_{max}, K_A', K_A and K_B can be derived by computer fitting procedures (Cleland, 1979) or by graphical methods involving primary and secondary plots. For the ternary complex mechanisms, the lines in

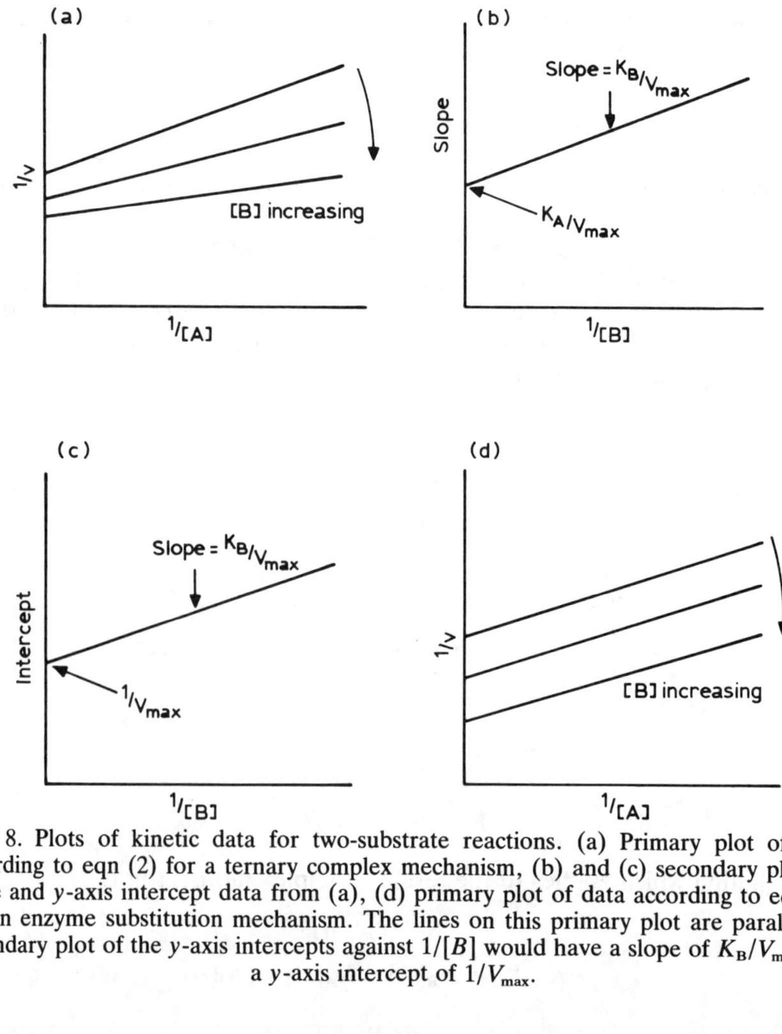

FIG. 8. Plots of kinetic data for two-substrate reactions. (a) Primary plot of data according to eqn (2) for a ternary complex mechanism, (b) and (c) secondary plots of slope and y-axis intercept data from (a), (d) primary plot of data according to eqn (3) for an enzyme substitution mechanism. The lines on this primary plot are parallel. A secondary plot of the y-axis intercepts against $1/[B]$ would have a slope of K_B/V_{max} and a y-axis intercept of $1/V_{max}$.

the primary plots (Fig. 8(a)) are convergent. The secondary plots of the slopes and *y*-axis intercepts of the primary plots can then be used to derive the parameters (Figs 8(b) and (c)).

The primary plot for the enzyme substitution mechanism yields a set of parallel lines (Fig. 8(d)). It is, thus, an easy task to distinguish this mechanism from the ternary complex ones. However, in order to distinguish between the ordered and random ternary complex mechanisms further experiments are necessary, involving product inhibition patterns, substrate binding measurements, and studies of isotope exchange at equilibrium (Engel, 1981).

From these types of observations it has been possible to show that, for instance, the reactions catalysed by creatine kinase and glycogen phosphorylase proceed via random ternary complex mechanisms and those catalysed by lactate dehydrogenase and malate dehydrogenase via ordered ternary complex mechanisms. Transaminases, yeast phosphoglycerate mutase and nucleoside diphosphate kinase catalyse reactions which proceed via the enzyme substitution mechanisms.

Effects of substrate and enzyme depletion

When enzyme and substrate are present at comparable concentrations, as is often the situation *in vivo,* the equations of the type eqns (1)–(3) have to be modified to take account of the fact that the concentration of free substrate can no longer be set equal to the total concentration of substrate. A full discussion of the modifications required has been given by Griffiths (1979). For instance, the value of K_m is equal to the total substrate concentration at which $v = V_{max}/2$ minus half of the total concentration of enzyme active sites. (This is true at all values of $[E]$, but the correction is negligible under the usual steady stage conditions, where $[S] \gg [E]$).

Griffiths (1979) also discusses the effects of mutual depletion (i.e. depletion of a significant fraction of both E and S on formation of the ES complex) on the kinetic behaviour of enzymes both *in vitro* and *in vivo*. The derivation of equations governing the flux through metabolic pathways can be somewhat simplified by assuming that a steady state exists, in which the concentrations of the intermediates in the pathway remain constant. (This condition is found to be obeyed *in vivo* in most cases, to a good approximation).

Pre-steady-state kinetics

In pre-steady-state kinetics, attention is focused on those steps in an enzyme-catalysed reaction prior to establishment of the steady state (where the concentration of the ES complex remains constant). Since events which occur on the time scale of seconds or less are involved, specialised rapid mixing and observation techniques are employed. The stopped flow technique (Halford, 1974; Fersht, 1985) is applicable to processes which occur on a time scale of a few milliseconds or longer, and can be used to identify enzyme-containing complexes on the reaction pathway and determine rate constants for their formation and decay (see the section entitled 'Kinetic studies', p. 35); this information is not generally available from steady-state kinetics. In pre-steady-state experiments it is necessary for the enzyme to be present at concentrations sufficiently high to be detected by the appropriate property (absorbance, fluorescence etc); in general, enzyme and substrate(s) are present at comparable concentrations, of the order of 10 μM.

The kinetics of immobilised enzyme systems

The study of enzymes bound to solid supports is of interest because such systems have considerable application in industrial processes. Immobilisation allows the enzyme to be recovered and re-used and may also serve to increase the stability of the enzyme under the conditions employed. In addition, immobilised enzyme can serve as models for understanding the behaviour of membrane-bound enzymes *in vivo*.

The effects of immobilisation on the kinetic behaviour of enzymes can be complex and have been discussed by Laidler and Bunting (1973) and Goldstein (1976). Four types of effects could operate:

(1) Conformational and steric effects, in which the enzyme is in a conformationally altered state when attached to the support.
(2) Partitioning effects, in which the substrate concentration in the matrix is different from that in the bulk solution, as a result of electrostatic or hydrophobic interactions between the matrix and the substrate.
(3) Micro-environmental effects on the intrinsic catalytic properties of the enzyme, in which the micro-environment provided by the support affects the interaction between enzyme and substrate.
(4) Diffusional effects, in which the rate of diffusion of substrate or product from the catalytic site becomes limiting. This could occur in the case of very rapid enzyme-catalysed reactions or when the substrate is bulky.

The effects of these various factors have been explored in a number of cases (Goldstein, 1976). Thus, the action of chymotrypsin on small substrates (e.g. *N*-acetyl-L-tyrosine amide) has been compared for the free enzyme and for the enzyme attached to charged supports. Although the apparent K_m does not change significantly on immobilisation of the enzyme, the value of k_{cat} is increased by a negatively charged matrix and decreased by a positively charged matrix. The results have been interpreted in terms of the effects on a kinetically significant step in the catalytic process in which two positively charged side chains (believed to be those of His-57 and Arg-145) come into close proximity. A negatively charged matrix would screen these interactions, leading to increased catalytic efficiency, whereas a positively charged matrix would have the opposite effects.

ENZYME MECHANISM

Introduction

In this section we shall discuss the mechanism of enzyme-catalysed reactions. The term mechanism can be defined as the description of the sequence, rates of interconversion, and structures of enzyme-containing complexes as substrate(s) is (are) converted to product(s). It is fair to state that, at present, elucidation of the mechanism has not been achieved in any particular case to the extent that the degree of rate enhancement can be understood in quantitative terms. However, using a variety of approaches, substantial progress has been made towards this goal in several cases. The texts by Jencks (1969), Walsh (1979), Creighton (1983) and Fersht (1985) should be consulted for further details.

Origins of rate enhancement

Enzymes are remarkably powerful catalysts, i.e. cause high degrees of rate enhancement. In a number of unimolecular reactions the enhancement factor ranges from 10^5 to as high as 10^{14} (Creighton, 1983); indeed many reactions cannot be observed to any significant extent over prolonged time periods in the absence of enzyme. The catalytic power and the specificity shown by enzymes have their origin in the precise interactions which occur between enzyme and substrate. From the study of the mechanisms of organic reactions a number of factors have been identified which contribute to the rate enhancement; they can cause a decrease in the free energy of activation of the reaction, by affecting the enthalpy or entropy component or both.

These factors are also important in enzyme-catalysed reactions, though not necessarily all in any single enzyme. They can be summarised as

(1) Proximity and orientation effects, i.e. the bringing together of reactants in the correct relative orientation for reaction.
(2) Acid–base catalysis, where proton donation and/or abstraction facilitates bond cleavage.
(3) Covalent catalysis, in which an intermediate is rapidly formed and broken down.
(4) Strain or distortion, in which part of the energy released on binding of substrate is involved in distorting its geometry so as to convert the substrate to a more reactive form.
(5) Changes in environment, where the micro-environment provided by the side chains at the active site may facilitate the reaction, e.g. by stabilising an intermediate.

In some cases, a metal ion may also be involved; this would provide a powerful electrophilic centre which could serve to polarise a bond and render it more susceptible to attack by nucleophiles. For a further discussion of the importance of these effects, the texts by Creighton (1983) and Fersht (1985) should be consulted.

Methods for studying mechanisms

It is obvious from the definition given earlier that there are structural and kinetic aspects of the description of a mechanism. Complementary information must therefore be obtained from a variety of approaches, as indicated below.

Structural studies

By far the most detailed information on enzyme structure comes from X-ray crystallography (see 'Determination of structure', p. 12). In order to locate the active site it would be desirable to study the enzyme–substrate complex directly. However, this is usually not possible since the collection of diffraction data takes a considerable time (up to several hours) during which conversion to products and diffusion from the active site will have occurred. This difficulty is usually overcome by studying a complex of the enzyme with a competitive inhibitor (assumed to bind at the active site) or a very poor substrate. In the case of enzymes that catalyse reactions involving two substrates, such as lactate dehydrogenase, it is possible to study the structure of binary complexes (e.g. E^{NADH}) or ternary complexes involving an inhibitor (e.g. $E^{\mathrm{NADH}}_{\mathrm{oxamate}}$, where oxamate is a competitive inhibitor with respect to pyruvate).

Two approaches have been developed recently to circumvent the problems associated with the unstable nature of the enzyme–substrate complex. Firstly, by working at low temperatures (down to −50 to −60°C) the reaction may be slowed sufficiently to allow such complexes to be observed on the appropriate time scale. This 'cryoenzymology' approach usually involves mixed solvent systems such as methanol–water or dimethylsulphoxide–water to prevent freezing and it is necessary to check that the principal features of the reaction mechanism are the same as in an aqueous system. Using this approach, it has been possible to stabilise and study a tetrahedral intermediate in the papain-catalysed hydrolysis of a *p*-nitroanilide substrate (Angelides & Fink, 1979). The second approach involves the application of synchrotron sources to the study of enzyme catalysis. The much higher intensity of radiation compared with conventional X-ray sources, means that the time required for data collection is dramatically reduced; in principle, it is possible to study structural changes or catalytic processes which occur on the millisecond time-scale. The synchrotron approach has been used to study the structure of intermediates in the glycogen phosphorylase-catalysed reaction of a relatively poor substrate (Hajdu *et al.*, 1988).

Identification of side chains at the active site

The nature of the side chains present at the active site of an enzyme can be inferred by specific chemical modification or by the more recent technique of site-directed mutagenesis. In both types of experiment the effect of the perturbation on the catalytic activity is determined; if activity is lost or decreased it can be concluded (at least tentatively) that the side chain which is modified is involved in the activity. The usual problem in chemical modification experiments is in obtaining selectivity; interpretation of such experiments is clearly easiest if only one particular type of side chain has been modified and then only a small number (ideally one) of that type. Although a great deal of ingenuity has been employed to design a number of 'specific' reagents (e.g. 5,5′-dithiobis-(2-nitrobenzoate) for cysteine side chains), it is not always possible to predict how any given reagent will modify an enzyme, and careful analysis of the product must be undertaken to establish the degree of specificity (Means & Feeney, 1971; Cohen, 1970). In some cases, side chains at the active sites of enzymes have enhanced reactivity which is essential for their role in catalysis (e.g. Ser-195 in chymotrypsin) and this can give rise to fortuitous specific modification. An alternative approach to obtaining specific modification is to design 'affinity' or 'active-site directed' reagents which incorporate some structural features of the substrate to direct the reagent to the active site. A suitably placed reactive group in the reagent then brings about the modification of the active site side chains(s) (de la Mare *et al.*, 1972).

In order to check that modification has occurred at the active site it is necessary to show that at least two criteria are fulfilled; namely (i) that inactivation is stoichiometric with the degree of modification, and (ii) that substrate(s) or competitive inhibitors protect against inactivation. It must be emphasised that even if these criteria are fulfilled it is still possible that the side chain concerned is only close to the active site and not directly involved in the catalytic mechanism.

Site-directed mutagenesis involves the use of a deliberately mismatched oligonucleotide as a primer for DNA synthesis. To apply the technique, it is necessary to isolate the gene coding for the enzyme of interest and to have a suitable system to allow the gene to be expressed. Details of the technique are given by Fersht (1985) and Rossi & Zoller (1987). Site-directed mutagenesis allows an amino acid at any position in a polypeptide chain to be replaced by any other amino acid, and thus an almost infinite number of possible mutant enzymes could be produced. Clearly, most information will be derived from experiments in which the side chains to be mutated are those identified as important from other studies (e.g. X-ray crystallography). The ability to change single amino acid side chains in a rational and systematic fashion offers superb opportunities for the 'dissection' of an active site. The work of Fersht and his colleagues (Fersht, 1987) on the substrate binding and catalytic properties of tyrosyl tRNA ligase is an excellent illustration of the power of the technique.

Kinetic studies

In 'Enzyme kinetics' (p. 25) it was mentioned that kinetics (steady-state and pre-steady-state) could be used to determine the sequence of complexes on the reaction pathway and, in favourable circumstances, determine the rate constants for the interconversion of the complexes. Variation of enzyme activity with pH can also be used to identify the types of side chains involved in the catalytic mechanism, e.g. the two histidines (His-12 and His-119) in the active site of ribonuclease (Findlay *et al.*, 1962). However, the large shifts of the pK_a of amino acid side chains due to the micro-environmental effects of the enzyme can complicate this type of assignment considerably (Creighton, 1983).

The nature of the active site of an enzyme can be probed by systematically varying the structure of the substrate. Using this approach, it was shown that the active site of papain contains seven 'subsites' extending over some 2·5 nm (Berger & Schechter, 1970)

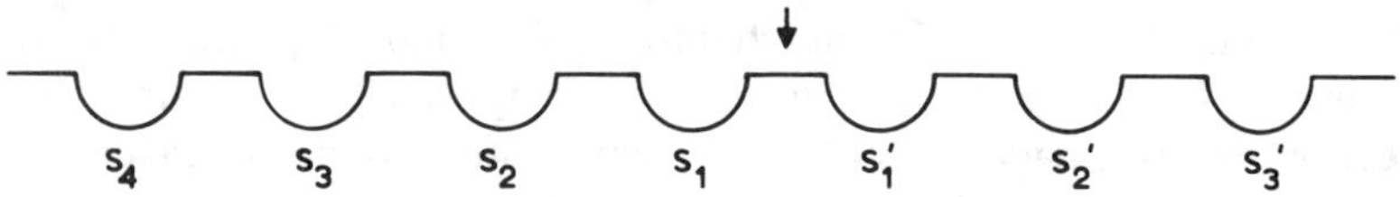

The site of hydrolysis lies between the amino acids occupying sites S_1 and S_1'. Subsite S_2 specifically interacts with the side chain of L-phenylalanine, S_1 shows a preference for lysine over alanine and S_1' is stereospecific for L-amino acids with a preference for the hydrophobic side chains of leucine and tryptophan.

The types of products formed on hydrolysis of oligosaccharides have been used to ascertain the pattern of subsites in α-amylases. Thus, the action of α-amylase from *Bacillus amyloliquefaciens* on oligosaccharides of D-glucose could be interpreted in terms of an active site capable of accommodating up to nine D-glucosyl units with the catalytic groups positioned between the third and fourth subsites from the reducing end of the substrates (Robyt & French, 1963). By contrast, α-amylase from pig pancreas seemed to possess only five subsites for binding D-glucosyl units with the catalytic groups between the second and third subsites from the reducing end. In technological terms, the differences in the actions of the α-amylases are important because of the uses of the various products. Trisaccharides are normally the largest units capable of being fermented by yeasts, so it is desirable to identify α-amylases which will give efficient breakdown of starch to di- and trisaccharides. In this respect, α-amylase from *B. licheniformis* is superior (Norman, 1981) to that from *B. amyloliquefaciens* and also has the advantage of greater heat stability and activity over a wider range of pH.

The use of pre-steady-state kinetics in the investigation of reaction mechanisms is illustrated by studies on alcohol dehydrogenase from horse liver. Stopped-flow experiments showed that when enzyme was mixed with NAD^+ and ethanol, there was a rapid burst of NADH production, which was complete within about 30 ms. The concentration of NADH formed in this burst phase was equal to the concentration of active sites and from its fluorescence properties it was shown that this rapidly-formed NADH was enzyme-bound rather than free. A minimal mechanism as follows, could be proposed:

$$\underbrace{E \xrightarrow{\text{NAD}^+,\ \text{ethanol}} E^{\text{NAD}^+}_{\text{ethanol}} \longrightarrow E^{\text{NADH}}_{\text{acetaldehyde}}}_{\text{fast}} \underbrace{\xrightarrow{\text{acetaldehyde, NADH}} E}_{\text{slow}}$$

From further studies the rate constants for some of the individual steps in the reaction could be determined (Shore & Gutfreund, 1970).

In the above case the nature of the intermediate ($E^{\text{NADH}}_{\text{acetaldehyde}}$) was inferred from its spectroscopic properties. If the rate of breakdown is sufficiently slow it may be possible to isolate and characterise the intermediate. Thus, in the papain-catalysed hydrolysis of *N-trans*-cinnamoylimidazole, a stable thioester intermediate could be isolated and the site of labelling of the enzyme was shown to be the thiol side chain of Cys-25 (Glazer & Smith, 1971). This side chain was also modified by a number of haloacids and by chloromethylketone substrate analogues; in each case, labelling led to loss of enzyme activity.

Using the various approaches outlined in this section, it has proved possible to prepare plausible mechanisms for a number of enzymes (Fersht, 1985). As an example, the mechanism of papain is discussed in the following section.

The mechanism of papain

Papain (EC 3.4.22.2) is one of a number of thiol proteinases, i.e. protein-hydrolysing enzymes dependent on the integrity of a thiol group. Other examples from plant sources are ficin, bromelain and actinidin which are all similar to papain. The thiol proteinases from bacterial and mammalian sources are, however, considerably different from the plant enzymes.

Papain will also catalyse the hydrolysis of ester and amide bonds and for detailed kinetic work, smaller ester or amide substrates are more convenient than proteins.

The X-ray structure of papain, with its two domains, has been described (p. 16). The substrate binding site has been located by studying enzyme–inhibitor complexes, and is located in the cleft between these two domains. Subsite S_2 (which shows specificity for the hydrophobic side chain of L-phenylalanine) is formed by a deep depression lined with hydrophobic side chains (e.g. Pro-68 and Trp-69 from one domain; Val-133, Val-157 and Phe-207 from the other). Some of the other subsites indicated by the specificity studies (see the previous section) have been identified in the X-ray structure (Drenth *et al.*, 1976). The side chain of Cys-25 implicated in acylenzyme intermediate formation is also in this cleft.

The enzyme mechanism involves an acylenzyme intermediate

$$\underset{}{E+S} \rightleftharpoons \underset{\substack{\text{Michaelis}\\ \text{complex}}}{ES} \rightleftharpoons \underset{\substack{\text{acyl}\\ \text{enzyme}}}{ES'+P_1} \begin{cases} \xrightarrow{H_2O} E + P_2 \\ \xrightarrow[\text{ROH}]{} E + P_3 \end{cases}$$

Formation of the acylenzyme is associated with release of the first product, P_1. The acylenzyme can be broken down by reaction with incoming nucleophiles such as water or added alcohols. Both the formation and breakdown of the acylenzyme involve tetrahedral intermediates (see the next section).

A plot of enzyme activity (or better k_{cat}/K_m) against pH shows a bell-shaped curve with an optimum at pH around 6, with pK_a values of 4·3 and 8·2 for the two ionising groups concerned. The pK_a of 8·2 has been attributed to the side chain of Cys-25; the lower pK_a was originally thought to be due to a carboxyl side chain but is now considered to be due to the imidazole side chain of His-159. The low pK_a of this side chain (about two units lower than in the free amino acid) is presumably a consequence of its being partially buried in a hydrophobic environment (Glazer & Smith, 1971). The close proximity of the side chains of Cys-25 and His-159 has been shown by the fact that they can be cross-linked by the bifunctional reagent 1,3-dibromoacetone.

From the pH–rate studies, the active form of the enzyme appears to be

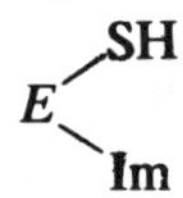

(Im = imidazole)

however, this would also exist as the ion pair

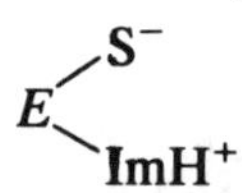

and current evidence favours the latter as the dominant tautomeric form (Fersht, 1985). The nature of the ion pair can be probed by various electrophilic reagents such as derivatives of 2-pyridyldisulphide (Brocklehurst, 1982).

Analogies between the mechanisms of different proteinases

The mechanism of papain, as a thiol proteinase, has a number of analogies with the other classes of proteinases (serine-, aspartic- and metallo-proteinases). In each case, the same fundamental chemical problem is addressed, namely cleavage of the stable amide bond. The requirements for catalysis are three fold.

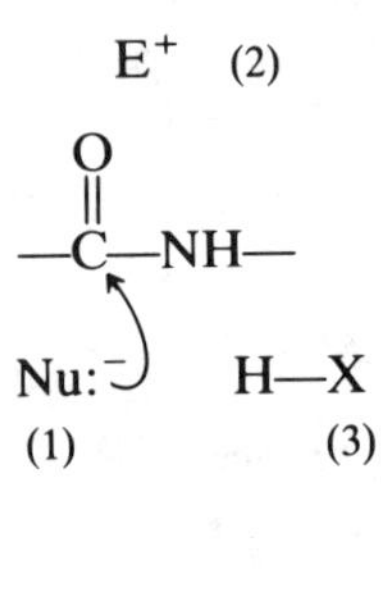

(1) a nucleophilic group, Nu^-, to attack the carbonyl group to lead to the formation of a tetrahedral intermediate

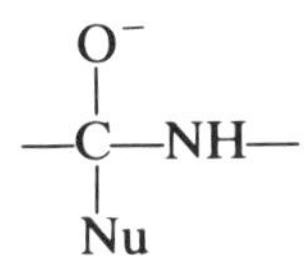

(2) an electrophilic species, E^+, in the neighbourhood of the amide carbonyl group, in order to increase the polarisation of the group and stabilise the incipient negative charge on the oxygen atom in the tetrahedral intermediate.

(3) a proton-donating species, *X*—H, to make the —NH moiety a better leaving group. (Note that Nu^- and E^+ need not necessarily carry full charges).

In the case of papain, the roles of Nu^- and *X*—H are played by the side chains of Cys-25 and His-159, respectively. Polarisation of the carbonyl group is achieved by hydrogen bonding to the polypeptide chain —NH— group of Cys-25 and the side chain of Gln-19 (Drenth *et al.*, 1976).

A similar chemical pattern can be discerned in the other classes of proteinases. Thus, in the serine proteinases, such as chymotrypsin, the role of Nu^- is played by a serine side chain (Ser-195 in chymotrypsin) and in metalloproteinases, such as carboxypeptidase, a metal ion (usually Zn^{2+}) serves as the electrophile. Thus, the different classes of enzymes have adopted different solutions to the same chemical problem. The differences in the specificities both between and within the classes arise from differences in the detailed structures of the active sites; many of these differences can now be understood from the results of X-ray crystallography. Chymotrypsin, trypsin and elastase are serine proteinases of similar overall structure but with pronounced differences in specificity. The structures of the substrate binding pockets explain these differences; trypsin has a negatively charged side chain (Asp-189) at the bottom of the pocket which allows it to recognise the positively charged side chains (lysine and arginine) of potential substrates. In chymotrypsin, residue 189 is serine which has a neutral side chain, consistent with the fact that chymotrypsin acts on substrates with bulky hydrophilic side chains (e.g. phenylalanine). In elastase, access to the binding pocket is hindered by the bulky side chains of Val-216 and Thr-226; this explains why elastase acts on substrates containing small hydrophobic side chains (e.g. alanine) (Hartley & Shotton, 1971).

The use of papain in meat tenderising

Papain is widely used to tenderise cuts of meat by digesting connective tissue and other proteins. The ready inactivation of papain by modification of the side chain of Cys-25 and the subsequent reactivation by treatment with thiol-containing compounds is exploited in the application of the enzyme (Dransfield & Etherington, 1981).

A concentrated solution of papain is injected into the jugular vein of the animal 10–30 min prior to slaughter to allow the blood circulation to achieve an even distribution of the enzyme in the muscle tissue. The enzyme is injected in an inactivated form in which the side chain of Cys-25 is oxidised to form a disulphide derivative. Papain remains in this oxidised form in the living animal; if slaughter did not occur, the enzyme would be lost by excretion via the kidneys. After slaughter, however, the continuing glycolysis tends to deplete the muscles of oxygen so that free thiol compounds and other reducing agents accumulate, allowing papain to become activated. The activation process is slow in chilled meat but occurs rapidly on warming.

It is important to note that the injection of active papain into the circulation of a living animal would cause activation of the complement system by the 'alternative pathway'; this would lead to severe shock, haemorrhage and death. The ready inactivation and reactivation of thiol proteinases such as papain is clearly essential for its successful application; it would be difficult, if not impossible, to design strategies for the use of the other classes of proteinases in this process.

ENZYME STABILITY

Introduction

A number of factors contribute to the stability of an enzyme (Creighton, 1983; Baldwin & Eisenberg, 1987). One of these is whether the enzyme is a pure preparation or is one of a number of components present in the extract. With a pure preparation it is the intrinsic stability of the enzyme which is of importance, whereas in an extract all the other components have to be considered. For an enzyme to maintain activity it has to maintain its native structure. Activity may be lost by a change in the three-dimensional structure (this is usually termed denaturation) or by the cleavage of peptide bonds (degradation). Cleavage of peptide bonds is generally brought about by attack by proteinases, although under extremes of pH and temperature the enzyme may be completely degraded non-enzymically. Generally, when considering enzyme stability, denaturation is the main focus, and this section is restricted to this aspect. We are normally concerned with enzymes in solution. It is of course possible to freeze-dry an enzyme preparation thereby removing all the water. The enzyme will then show no detectable activity but on rehydration it may resume full activity. This aspect will be considered in the section entitled 'Water activity', p. 45.

The stability of the native conformation of an aqueous solution is due to a small difference between large but opposing factors, and so the conformation may be considered as only marginally stable. Enzymes are stabilised by hydrogen bonds, hydrophobic interactions, internal salt bridges and in some cases disulphide bonds. A change in the magnitude of any of these could cause denaturation. Denaturation can be brought about by numerous factors, including heat, extremes of pH, organic solvents, inorganic or organic solutes, detergents and radiation. The mechanism of denaturation and the effects of these agents are not fully understood.

Denaturation is usually a cooperative process; complete denaturation occurs through a narrow range of concentrations of the denaturing agent, e.g. staphylococcal nuclease is fully denatured over a range of 0·3 pH units. In the case of small single-domain proteins the evidence from calorimetry and from the temperature dependence of the equilibrium constant (K) suggests a two-state process in which the level of any intermediates is negligible.

$$\text{native} \rightleftharpoons \text{denatured} \qquad K = [\text{denatured}]/[\text{native}]$$

For multi-domain proteins, such as papain, evidence suggests that the individual domains unfold independently. Denaturation involves a large increase in entropy, but a decrease in enthalpy resulting from the cleavage of bonds, including hydrogen bonds. The process of denaturation is generally reversible and is often studied by the addition of guanidinium chloride and its subsequent removal by dialysis.

Some of the agents that cause denaturation are considered in the next section.

Denaturing agents

Heat

Enzymes are denatured by an increase in temperature, and denaturation is completed over a narrow range of temperatures. The temperature at which denaturation occurs varies between different enzymes, depending on their structure. The effect of protein structure on denaturation temperature has been studied by comparing enzymes from mesophilic organisms with those from thermophiles. Many enzymes from mesophilic organisms are inactivated at temperatures above 40°C, whereas thermophilic organisms which grow at temperatures greater than 55°C, and sometimes as high as 90°C, have enzymes which generally show greater heat stability. Ferredoxin and glyceraldehyde 3-phosphate dehydrogenase are two of the enzymes used in this comparison. Ferredoxin is a small protein involved in redox reactions. Ferredoxin from a mesophile, *Clostridium pasteurianum,* and from a thermophile, *C. thermosaccharolyticum*, differ markedly in their heat stability. The former is completely inactivated at 70°C for 2 h whereas the latter retains 90% activity under the same conditions. When the structures of the two are compared, the salt bridges on the surface of the proteins are the only source of additional stability. The tetrameric glyceraldehyde 3-phosphate dehydrogenases from lobster, from the thermophile, *Bacillus stearothermophilus,* and from the extreme thermophile, *Thermus aquaticus,* have been compared. The enzyme from *B. stearothermophilus* shows full activity up to between 60 and 70°C whereas that from *T. aquaticus* is stable up to 95°C. There are higher proportions of internal hydrophilic interactions and additional salt bridges in the enzyme from the extreme thermophile compared with the other two. There are also additional hydrophobic interactions at the interface between the subunits which also increase the stability of the tetramer from the extreme thermophile. When the structure of lysozyme from a temperature-sensitive mutant of T4 phage and that from the corresponding wild type are compared, the main difference found is a reduction in the non-polar or hydrophobic interactions for the mutant enzyme (Alber & Matthews, 1987). All these studies emphasise the importance of hydrophobic interactions and salt bridges in stabilising enzyme against heat denaturation.

Changes in pH

Enzymes are denatured by extremes of pH. This is generally as a result of protonation and deprotonation of certain amino acid residues, which may cause the disruption of salt bridges and hydrogen bonds. In addition, placing a charge on an internal residue, e.g. tyrosine, may increase its polarity and disrupt hydrophobic interactions.

Organic solvents

The effect of organic solvents such as ethanol, acetone, chloroform or phenol on enzymes is generally to promote or cause denaturation. These solvents are believed to interact with non-polar groups on the enzyme, which are generally present in the interior. The enzyme becomes unfolded and denatured. Although most monohydric alcohols at concentrations greater than 10% tend to denature enzymes, a number of polyhydric alcohols, e.g. glycerol and ethylene glycol, tend to have a stabilising influence. Enzymes can often be stored without loss of activity below 0°C after the addition of glycerol.

Solutes

Two organic solutes which have been used extensively in protein denaturation studies are urea and guanidinium chloride. At one time it was thought that their denaturing properties were due to their ability to hydrogen bond with the proteins. However, this now seems less likely since they do not hydrogen bond to proteins any more strongly than does water. Both urea and guanidinium chloride increase the solubility of non-polar groups, such as certain amino acid side chains, in water (Creighton, 1983). This can decrease the strength of hydrophobic interactions by as much as one third.

Certain anions have a marked effect on denaturation, which is related to their ability to disrupt the structure of water. Their ability to do this is in decreasing order (Hofmeister series), i.e. $SCN^- > ClO_4^- > I^- > Cl^- >$ citrate > acetate > SO_4^{2-}. It is interesting to note that guanidinium thiocyanate is a more powerful denaturant than guanidinium chloride and that guanidinium sulphate shows negligible denaturing properties. These differences are due to the combined effects of the guanidinium cation and the other anions. Although most proteins are denatured by 8 M urea or 6 M guanidinium chloride, some proteins which are stabilised by disulphide bridges or divalent metal ions require additional agents for denaturation. Pronase, a proteinase from *Streptomyces griseus,* is stable in 6 M guanidinium chloride unless EDTA is added, and the proteinase, thermolysin, is active in 8 M urea. Both enzymes are stabilised by divalent metal ions. Carboxypeptidase is not completely unfolded in 10 M urea unless 2-mercaptoethanol is also added to reduce disulphide bonds.

Detergents

Detergents show strong denaturing properties and they are effective at much lower concentrations than urea or guanidinium salts. SDS is able to induce complete denaturation at concentrations of less than 10 mM, whereas urea or guanidinium salts are generally effective only at concentrations at least an order of magnitude higher. The action of detergents is due to strong binding of the detergent to the protein in its denatured state compared to the native state, thus shifting the equilibrium strongly in favour of denaturation.

Importance of disulphide bonds in enzyme stability

The presence of disulphide bonds influences the rate and extent of denaturation. In general, proteins that are stabilised by disulphide bonds are more difficult to denature fully than those lacking them (Ghélis & Yon, 1982; Jaenicke, 1987). The energy of the disulphide bond relative to two thiol groups depends on the redox potential of the environment. The free energy of formation of the disulphide bond varies with its location within the protein, with the equilibrium constant for the formation of disulphide bonds in the interior of a protein being significantly higher than for disulphide bonds on the surface.

Comparisons have been made of the rates of renaturation of proteins with and without disulphide bonds. Proteins lacking disulphide bonds occur most frequently in the all α, α/β and $\alpha+\beta$ classes (Levitt & Chothia, 1976) whereas those with disulphide bonds occur most frequently in all β, and $\alpha+\beta$ classes. The small proteins, lacking disulphide bonds, which have been most studied, e.g. staphyloccal nuclease, T4 lysozyme, cytochrome c and carbonic anhydrase, can all be competely denatured and renatured. Refolding is generally rapid, e.g. staphylococcal nuclease refolds with a $t_{1/2}$ of 0·1–0·2 s. The unfolding and refolding of proteins with disulphide bonds can be measured either when the disulphide bonds remain intact, or after their reduction and reoxidation. The time for refolding is much longer in the latter case. When elastase is denatured and renatured it regains complete activity within seconds if the disulphide bonds are not reduced, whereas if the bonds are reduced there is only 50% recovery of activity within 80 min (Ghélis &

Yon, 1982). Bovine pancreatic trypsin inhibitor, which is a small protein of 58 residues with three disulphide bonds, requires one day for complete recovery of activity by air oxidation to enable the disulphide bonds to reform. Although the time for renaturation is considerably longer if disulphide bonds have to reform, the observation that the correct (native) disulphide bonds do reform, as in, for example, ribonuclease and bovine pancreatic trypsin inhibitor, is evidence that the primary structure of the protein governs the folding, so that native disulphide bonds eventually form. The pathways of refolding have been studied in both the above proteins and it has been shown that non-native disulphide bonds form during refolding, but eventually reshuffle to form the native bonds, the process being catalysed by protein disulphide isomerase.

Protein engineering and enzyme stability

The advent of site-directed mutagenesis (see the section entitled 'Identification of side chains at the active site', p. 34) has made it possible to replace amino acids at specified positions in the sequence and determine their effect on the properties of the protein. A potential application of this technique is that it may be used to improve the properties of an enzyme so that it acts as a better catalyst, or has enhanced stability (Hartley *et al.*, 1987; Wells *et al.*, 1987). An enzyme which has been much studied is subtilisin. By suitable genetic manipulation, it is possible for *B. subtilis* to secrete more than 1 g/litre subtilisin into the growth medium. The enzyme is used extensively in detergents and in food processing. Two types of modification have been attempted to improve its stability. Although the enzyme is relatively stable, it is susceptible to oxidation which causes loss of activity. The sensitivity to oxidation is due to Met-222 which becomes oxidised to methionine sulphoxide. Mutants were prepared in which this amino acid was replaced in turn by the other 19 amino acids; the only one which resulted in an increase in activity was cysteine, but this did not give the necessary resistance towards oxidation. Replacing the methionine by either alanine or serine, although reducing its activity to 53 and 35%, respectively, resulted in the necessary resistance to oxidation. Another replacement is to introduce a disulphide bond; subtilisin lacks disulphides, and it was thought that the introduction of disulphide bonds might give the enzyme additional stability. Two different replacements have been carried out by introducing two pairs of cysteine residues (Ser-24→Cys, Ser-87→Cys and Thr-22→Cys, Ser-87→Cys) but neither resulted in increased stability. This is in contrast to the observations with T4 lysozyme. In this case, a cysteine residue was introduced into position 3 using recombinant DNA techniques. A disulphide bond was then formed by mild oxidation between Cys-3 and the already existing Cys-97. The engineered enzyme had the same specific activity as that from the wild type T4 but showed greater thermal stability; the enzyme from the wild type had $t_{1/2} = 11$ min at 67°C, whereas in the engineered mutant enzyme, $t_{1/2} = 28$ min (Perry & Wetzel, 1984).

Immobilised enzymes

There are several reports that enzyme immobilisation increases stability but this is not universally the case. Out of 50 cases studied, 60% led to stabilisation, 16% caused destabilisation and 24% had no effect (Klibanova, 1979). Immobilisation can be achieved in several ways (Price & Stevens, 1989). The increased stability may be due to a combination of factors. Entrapping an enzyme in a matrix may make it inaccessible to proteolytic attack. Entrapment, covalent attachment to the support, and in some cases crosslinking, will generally make the enzyme less easy to unfold, and so, provided that the enzyme does not lose activity by being immobilised, it may show greater stability.

ENVIRONMENTAL FACTORS

Introduction

When measuring enzyme activity in the laboratory, the assay is usually performed using a dilute solution of the enzyme in the presence of an excess of the substrate(s) in a buffered solution close to the pH optimum for the enzyme concerned. Enzymes in their natural environment often encounter very different conditions. It is important, therefore, to have some appreciation of the effects of these environmental factors on enzyme activity. Enzyme-catalysed reactions in food cause changes, some of which cause deterioration of the food, e.g. the rancidity of fats or the autolysis of vegetables prior to freezing, whereas others are desirable, e.g. fermentation of grape juice or the autolysis of meat by hanging to bring about tenderisation. More recently, enzyme preparations have been added to foods in the course of their production or processing. Proteinases are used to improve extraction and clarification of fruit juices. Chymosin (also known as rennin) or fungal proteinases are used to clot milk in the process of cheesemaking. Papain is added to meat as a tenderising agent. To prevent cereal grains from deterioration during storage, their moisture content is lowered so as to minimise the growth of xerotolerant fungi; the low water content prevents appreciable enzyme activity. In most of these examples, the enzymes act under conditions which differ significantly from those used in laboratory assays. The various environmental factors that affect enzyme activity are considered in the next sections.

Water activity

Most enzymes require an aqueous medium in which to exhibit catalytic activity. Water is necessary for the maintenence of the enzyme structure, as the medium in which the substrates diffuse to the active site, and the products diffuse from the active site, and in some cases, as one of the reactants in the enzyme-catalysed reaction. Solutions of proteins in water can be considered as having three categories of water molecules (Cooke & Kuntz, 1974). Type I water molecules are those which have the same properties as in the pure solvent, and do not have any appreciable interaction with the proteins in solution. Type II water molecules are loosely bound to the protein, and their movement is restricted by their proximity to the protein. Type III water molecules are tightly bound to the protein forming an integral part of the protein structure (In food science, the reverse numbering is often used to designate the types of water, i.e. type I water is tightly bound, whereas type III has the properties of the pure solvent water.) The proportion of each of these varies with the degree of hydration of the tissue or preparation concerned. For example, the average water content of meat at slaughter is about 75%; this figure includes all three types of water, although the proportion of type I water is not uniform throughout the tissue. The mitochondrial matrix may have a protein concentration as high as 500 mg/cm^3 (Srere, 1980) and will thus have relatively little type I water. A cereal grain during storage may have between 5 and 20% water (Roberts & Roberts, 1972), and therefore have no type I water, and the amount of type II varies with the moisture content.

The effects of reducing the level of available water have been studied with a number of enzymes, either by controlled hydration of dry enzymes, or by the addition of organic solutes to enzymes solutions. The hydration of lysozyme has been studied by IR spectroscopy, heat capacity calorimetry and diamagnetic susceptibility (Careri *et al.*, 1980). When hydration has reached 0·07 g water/g protein, the water is tightly bound to the proteins by hydrogen bonding. Enzyme activity is detectable at 0·25 g water/g protein, although monolayer coverage of the surface is not complete until about 0·38 g water/g protein. The thermodynamic water activity, a_w, can be decreased by the addition of various solutes, e.g. glycerol, sucrose, fructose or sorbitol, and this causes a similar reduction in lysozyme activity (Larreta-Garde *et al.*, 1988). In a separate study in which the regain of enzyme activity on hydration was studied with glucose 6-phosphate dehydrogenase, phosphoglucose isomerase, hexokinase and fumarate hydratase, activity was detectable at between 0·15–0·2 g water/g protein (Stevens & Stevens, 1977). The activities of alcohol dehydrogenase and polyphenol oxidase after the addition of organic solutes suggest that the observed suppression in enzyme activities is caused by the solutes binding to the water surrounding the enzyme, rather than binding to the enzyme itself (Zaks & Klibanov, 1988). All these studies suggest that no enzyme activity is detectable when only type III water is present, but it becomes measurable once a significant amount of type II water is also present. Studies have also been made to detect enzyme activities in dried foods which are held at particular vapour pressures to effect a controlled hydration. When the cysts of the brine shrimp *Artemia* are exposed to increasing vapour pressure it is found that below 0·6 g water/g metabolic activity becomes restricted, and below 0·3 g water/g it is not detectable (Clegg, 1978). The studies on the dependence of water for activity using foods and cysts are broadly in agreement with those on isolated enzymes.

Effects of solutes on enzyme activity

Both xerophilic and halophilic organisms face problems of low water activity in the external environment; this might be expected to cause the movement of water out of cells. However, both types of organisms have a compensating mechanism. In the case of halophilic bacteria, the intracellular environment maintains a high ionic strength by retaining a high concentration of electrolytes, in particular potassium chloride (KCl). In halophilic algae and in xerophilic fungi (particularly osmophilic yeasts) the solute concentrations within the cell are maintained by a high level of low M_r organic solutes. These are generally polyhydric alcohols such as glycerol or erythritol, but may also be amino acids or other carbohydrate derivatives. In extreme halophilic bacteria, the intracellular salt concentration may be between 3·0 and 5·3 M, mainly KCl (Brown, 1976). Such high salt concentrations might be expected to inactivate certain enzymes. The optimum salt concentrations for enzymes from extreme halophilic bacteria are about 4·0 M for membrane bound enzymes, but 0·5–1·0 M for many cytoplasmic enzymes. The intracellular salt concentration might thus be expected to partially inhibit enzyme activity. This seems to be the case, although provided the dominant salt is KCl, it causes only weak inhibition at high concentrations. Many enzymes from halophilic bacteria lose activity if dialysed against dilute buffers and thus require a certain salt concentration to maintain stability. In a number of cases, the proteins from halophiles have been compared with those from other bacteria. Those from halophiles generally show lower hydrophobicity, by having a higher proportion of acidic amino acid residues. The hydrophobic residues present are generally the weaker hydrophobic amino acids, e.g. alanine and threonine. Since hydrophobic interactions are strengthened by higher salt concentrations (Creighton, 1983) it is therefore probable that the weaker hydrophilic interactions involving, e.g. alanine and threonine are strengthened by the presence of high salt in the intracellular environment. A particularly striking example of the differences in protein structure is that of ferredoxin from *Halobacterium halobium* which is 22 residues longer than from other species of bacteria; nine of these extra residues are acidic amino acids.

The xerophilic and halophilic algae use the alternative mechanism of generally increasing the concentration of intracellular organic solutes. The enzymes from these organisms, e.g. the halophilic alga *Dunaliella,* unlike those from halophilic bacteria, are sensitive to high salt concentrations. The organic solutes which they synthesise are often referred to as 'compatible solutes'. A general property of these solutes is believed to be that they have a much greater affinity for water than for the intracellular proteins and hence have little effect on the stability of the latter. For further details see Brown (1976) and Jaenicke (1981).

Effects of extremes of pH

Most organisms that inhabit aqueous environments encounter pH values near neutrality. However, a minority are exposed to extremes of pH, highly acidic ones being more common than alkaline pHs. This is particularly the case where sulphides are present in the medium, since these may become oxidised by sulphur bacteria to sulphuric acid. The extremes that have been discovered are *Cyanidium cladarium* growing at pH ≈ 0 and species of *Rhizobium* surviving at pH = 12. However, it is rare for an organism to have a requirement for an extreme pH for growth. Although organisms can grow at extremes of pH, it has been generally found that their intracellular environment is near neutrality. It seems unlikely, therefore, that any intracellular enzyme has to function at an extreme of pH. This does not apply to extracellular enzymes. A number of extracellular proteinases capable of acting at high pH are known, particularly those produced by alkali-requiring bacilli. These organisms will grow at pH values over 10 and the extracellular proteinases they produce are active and stable at pH 12 (Aunstrup, 1979). The molecular basis of their stability is not clear, although they have pI values ≈11 (reflecting a high preponderance of basic amino acids) which are higher than those of proteinases such as subtilisin (pI = 9·7).

The gastric juice is an environment which has a pH below 1, and in which a number of digestive enzymes are active. Pepsin has a pH optimum of 2, and is dependent on the correct ionisation states of two catalytically important carboxyl groups, having pK_a values of 1 and 4. The enzyme is denatured rapidly above pH 6, due to rupture of hydrogen bonds involving carboxyl groups. Below pH 6 and at its pH optimum it is stabilised largely by hydrophobic interactions (Fruton, 1971).

Effect of environmental temperature

Organisms can be classified into psychrophiles, mesophiles and thermophiles according to the range of temperatures in which they grow. The ranges are generally considered to be (i) below 25°C, (ii) 25–40°C, and (iii) above 40°C, respectively. Microorganisms as a group show a wider range of growth temperatures than plants and animals. There has been considerable interest in the enzymes from thermophiles in recent years for two reasons: their study may throw light on the factors which govern the thermostability of proteins, and they may be useful in industrial processes in which they can operate at higher temperatures without inactivation. Most of the data on thermophiles suggest that they are able to function at higher temperatures as a result of the greater stability of their macromolecules, rather than because of a higher turnover rate of macromolecules having the same stability as mesophiles. A number of enzymes from thermophiles have been isolated and shown to have high thermostability; a comprehensive list is given in Jaenicke (1981).

In the section entitled 'Heat', p. 41, the differences in the structures of enzymes from thermophiles and mesophiles were discussed. Not all enzymes studied in thermophiles have been found to have high heat stability, but it is assumed that those that do not may have a high rate of turnover *in vivo*. Where comparisons have been made between the catalytic efficiencies at 37°C of enzymes from thermophiles and mesophiles, the former are generally less efficient. However, the catalytic efficiencies of enzymes from thermophiles measured at ≈55°C are usually comparable with those from thermophiles at 37°C. Both kinetic and thermodynamic parameters have been studied extensively for lactate dehydrogenase from poikilothermic and homoiothermic animals and from mesophilic and thermophilic bacteria (Jaenicke, 1981). The idea that enzymes from thermophilic organisms must have a higher degree of rigidity than their mesophilic counterparts seems unlikely since there are many examples of allosteric enzymes in thermophiles and this implies a high degree of conformational flexibility.

Several of the enzymes used in food processes show an unusually high degree of thermostability, for example the preparations known as 'alcalase' and 'neutrase'. The former is an extracellular proteinase preparation from *Bacillus licheniformis* and the latter from *Bacillus subtilis*; both show optimal activity between 50 and 60°C (Petersen, 1981). Many of the carbohydrate hydrolysing enzymes used in food processing have impressive heat stabilities: the α-amylases from *Bacillus amyloliquefaciens* and *Bacillus lichenformis* have optima at 70°C and 95°C, respectively (Norman, 1981). The temperature–activity profiles of a number of thermophilic enzymes are given in Fig. 9.

Whereas thermophilic organisms have evolved so that a wide range of their

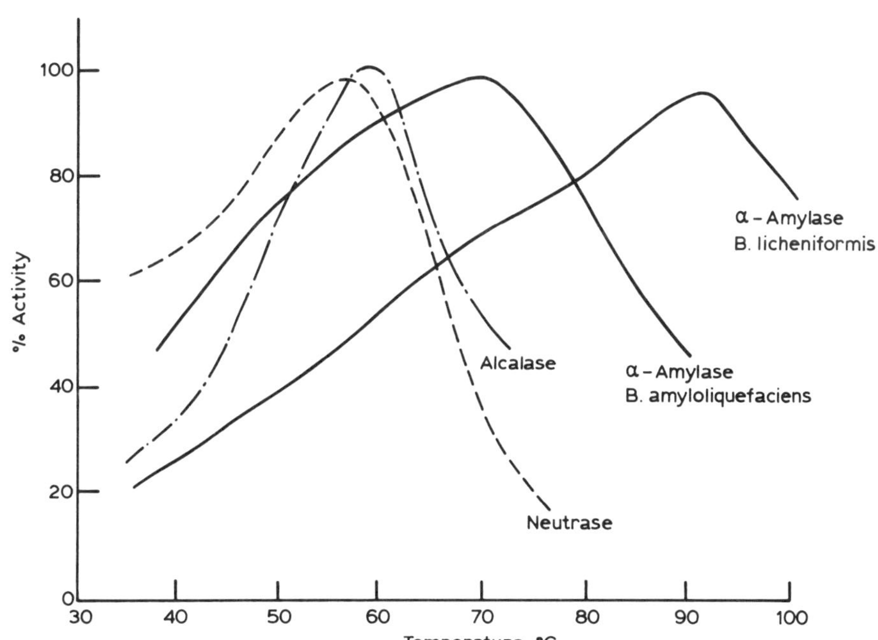

FIG. 9. The influence of temperature on enzyme activity (Norman, 1981; Petersen, 1981).

macromolecules differ from those in mesophiles, the same does not seem to hold for psychrophiles. Psychrophiles appear unable to grow at moderate temperatures because of the temperature sensitivity of a single macromolecule. Evidence in favour of this is that psychrophiles can be transformed into mesophiles by mutation or transduction. The temperature sensitive enzyme varies with the particular psychrophile, but the most common group is the aminoacyl tRNA ligases (Malcolm, 1969). Pyruvate carboxylase and triose-phosphate isomerase are the temperature sensitive enzymes in *Candida gelida* (Grant *et al.*, 1968) and *Clostridium* strain 69 (Shing *et al.*, 1975), respectively.

In the course of food processing and storage, enzymes will encounter a wide range of conditions, e.g. pH values as low as 2 in fruit juices and preservation in acetic acid, temperatures below 0°C during storage, and up to 100°C during liquefaction of starch, high concentrations of sodium chloride in salted fish and meats, and low water activities in flour, whole grains and dried fruits. Evidence from the study of enzymes in model systems in a wide range of environments helps to understand enzymic changes which may occur during food processing and storage. This provides a rationale for controlling enzyme activity in foods and food processing where it is sometimes required to suppress unwanted activity and in other processes to enhance activity.

REFERENCES

Alber, T. & Matthews, B. W. (1987). In *Protein Engineering,* ed. D. L. Oxender & C. F. Fox. Alan R. Liss, New York, USA, pp. 289–97.

Angelides, K. J. & Fink, A. L. (1979). *Biochemistry,* **18,** 2363–9.

Aunstrup, K. (1979). In *Enzyme Technology,* ed. L. B. Wingard, E. Katchalski-Katzir & L. Goldstein. Academic Press, New York, USA, pp. 27–69.

Baines, B. S. & Brocklehurst, K. (1978). *Biochem. J.,* **173,** 345–7.

Baldwin, R. L. & Eisenberg, D. (1987). In *Protein Engineering,* ed. D. L. Oxender & C. F. Fox. Alan R. Liss, New York, USA, pp. 127–48.

Berger, A. & Schechter, I. (1970). *Phil. Trans. R. Soc. London,* **B257,** 249–64.

Blundell, T. L. & Johnson, L. N. (1976). *Protein Crystallography.* Academic Press, New York, USA.

Brocklehurst, K. (1982). *Methods Enzymol.,* **87,** 427–69.

Brocklehurst, K., Carlsson, J., Kierstan, M. P. J. & Crook, E. M. (1973). *Biochem. J.,* **133,** 573–84.

Brooks, S. P. J. & Suelter, C. H. (1989). *Anal. Biochem.,* **176,** 1–14.

Brown, A. D. (1976). *Bacteriol Rev.,* **40,** 803–46.

Careri, G., Grattan, E., Yang, P. H. & Rupley, J. A. (1980). *Nature,* (London), **284,** 572–3.

Cech, T. R. & Bass, B. L. (1986). *Ann. Rev. Biochem.,* **55,** 599–629.

Clegg, J. S. (1978). In *Dry Biological Systems,* ed. J. H. Crowe & J. S. Clegg. Academic Press, New York, USA, pp. 117–53.

Cleland, W. W. (1979). *Methods Enzymol.,* **63,** 103–38.

Cohen, L. A. (1970). In *The Enzymes* (3rd edn) ed. P. D. Boyer, Academic Press, New York. **1,** 147–211.

Cooke, R. & Kuntz, I. D. (1974). *Ann. Rev. Biophys. Bioeng.,* **3,** 95–126.

Cornish-Bowden, A. (1979). *Fundamentals of Enzyme Kinetics.* Butterworths, London, U.K.

Creighton, T. E. (1978). *Prog. Biophys. Molec. Biol.*, **33,** 231–97.
Creighton, T. E. (1983). *Proteins: Structures and Molecular Properties.* Freeman, New York, USA.
de la Mare, S., Coulson, A. F. W., Knowles, J. R., Priddle, J. D. & Offord, R. E. (1972). *Biochem. J.*, **129,** 321–31.
Dixon, M., Webb, E. C., Thorne, C. J. R. & Tipton, K. F. (1979). *Enzymes* (3rd edn). Longman, London, UK.
Dransfield, E. & Etherington, D. (1981). In *Enzymes and Food Processing,* ed. G. G. Birch, N. Blakebrough & K. J. Parker. Applied Science Publishers, London, UK, pp. 177–94.
Drenth, J., Jansonius, J. N., Koekoek, R. & Wolthers, B. G. (1971). *Adv. Prot. Chem.* **25,** 79–115.
Drenth, J., Kalk, K. H. & Swen, H. M. (1976). *Biochem.*, **15,** 3731–8.
Edman, P. & Begg, G. (1967). *Eur. J. Biochem.*, **1,** 80–91.
Eisenthal, R. & Cornish-Bowden, A. (1974). *Biochem. J.* **139,** 715–20.
Engel, P. C. (1981). *Enzymes Kinetics* (2nd edn). Chapman and Hall, London, UK.
Fersht, A. R. (1985). *Enzyme Structure and Mechanism* (2nd edn). Freeman, New York, USA.
Fersht, A. R. (1987). *Biochem.*, **26,** 8031–7.
Findlay, D., Mathias, A. P. & Rabin, B. R. (1962). *Biochem. J.*, **85,** 139–44.
Fruton, J. S. (1971). In *The Enzymes* (3rd edn). ed. P. D. Boyer, Academic Press, New York. **3,** 119–164.
Fruton, J. S. (1972). *Molecules and Life.* Wiley Interscience, New York, USA.
Ghélis, C. & Yon, J. (1982). *Protein Folding.* Academic Press, New York, USA.
Gilson, M. K. & Honig, B. H. (1987). *Nature* (London), **330,** 84–6.
Glazer, A. N. & Smith, E. L. (1971). In *The Enzymes* (3rd edn). ed. P. D. Boyer, Academic Press, New York. **3,** 501–46.
Goldstein, L. (1976). *Methods Enzymol.*, **44,** 397–443.
Grant, D. W., Sinclair, N. A. & Nash, C. H. (1968). *Can. J. Microbiol.*, **14,** 1105–10.
Griffiths, J. R. (1979). *Biochem. Soc. Trans.*, **7,** 429–39.
Hajdu, J., Acharya, K. R., Stuart, D. I., Barford, D. & Johnson, L. N. (1988). *Trends Biochem. Sci.*, **13,** 104–9.
Halford, S. E. (1974). In *Companion to Biochemistry* (*Vol. 1*), ed. A. T. Bull, J. R. Lagnado, J. O. Thomas & K. F. Tipton. Longman, London, UK, pp. 197–226.
Hartley, B. S. & Shotton, D. M. (1971). In *The Enzymes* (3rd edn). ed. P. D. Boyer, Academic Press, New York. **3,** 323–73.
Hartley, B. S., Anderton, T. & Shaw, P.-C. (1987). In *Chemical Aspects of Food Enzymes,* ed. A. T. Andrews. Special Publication No. 63, Royal Society of Chemistry, London, UK, pp. 120–36.
Hunkapiller, M. W., Hewick, R. M., Dreyer, W. J. & Hood, L. E. (1983). *Methods Enzymol.*, **91,** 399–413.
IUB (1981). *Enzyme Nomencalture.* Recommendations of the Nomenclature Committee of the International Union of Biochemistry. Reprinted in *Eur. J. Biochem.*, **128,** 281–91.
IUB (1984). *Enzyme Nomenclature.* Recommendations of the Nomenclature Committee of the International Union of Biochemistry, Academic Press, Orlando, FL, USA.
Jaenicke, R. (1981). *Ann. Rev. Biophys. Bioeng.*, **10,** 1–67.
Jaenicke, R. (1987). *Prog. Biophys. Mol. Biol.*, **49,** 117–237.
Jencks, W. P. (1969). *Catalysis in Chemistry and Enzymology.* McGraw-Hill, New York, USA.
Kézdy, F. J. & Kaiser, E. T. (1970). *Methods Enzymol.*, **19,** 3–20.
Klibanova, A. (1979). *Anal. Biochem.*, **93,** 1–25.
Laemmli, U. K. (1970). *Nature* (London), **227,** 680–5.

Laidler, K. J. & Bunting, P. S. (1973). *The Chemical Kinetics of Enzyme Action* (2nd edn). Clarendon Press, Oxford, UK.
Larreta-Garde, V., Xu, Z. F., Lamy, L., Mathlouthi, M. & Thomas, D. (1988). *Biochem. Biophys. Res. Commun.*, **155,** 816–22.
Levitt, M. & Chothia, C. (1976). *Nature* (London), **261,** 552–8.
Malcolm, N. L. (1969). *Nature* (London), **221,** 1031–3.
Means, G. E. & Feeney, R. E. (1971). *Chemical Modification of Proteins.* Holden Day, San Francisco, CA, USA.
Morris, H. R. (1979). *Phil. Trans. R. Soc. London,* **A293,** 39–51.
Newsholme, E. A. & Start, C. (1973). *Regulation in Metabolism.* Wiley, London, UK.
Norman, B. E. (1981). In *Enzymes and Food Processing,* ed. G. G. Birch, N. Blakebrough & K. J. Parker. Applied Science Publishers, London, UK, pp. 15–50.
Old, R. W. & Primrose, S. B. (1989). *Principles of Gene Manipulation* (4th edn). Blackwell, Oxford, UK.
Perry, L. J. & Wetzel, R. (1984). *Science,* (New York), **226,** 555–7.
Petersen, B. R. (1981). In *Enzymes and Food Processing,* ed. G. G. Birch, N. Blakebrough & K. J. Parker. Applied Science Publishers, London, UK, pp. 149–76.
Price, N. C. & Stevens, L. (1989). *Fundamentals of Enzymology* (2nd edn). Oxford University Press, Oxford, UK.
Redfield, C. & Dobson, C. M. (1988). *Biochem.*, **27,** 122–36.
Richardson, J. S. (1981). *Adv. Prot. Chem.*, **34,** 167–339.
Richmond, C. (1986). *Trends Biochem. Sci.*, **11,** 528–30.
Roberts, E. H. & Roberts, D. L. (1972). In *Viability of Seeds,* ed. E. H. Roberts. Chapman and Hall, London, UK, pp. 424–37.
Robyt, J. F. & French, D. (1963). *Arch. Biochem. Biophys.*, **100,** 451–67.
Rossi, J. & Zoller, M. (1987). In *Protein Engng,* ed. D. L. Oxender & C. F. Fox. Alan R. Liss, New York, USA, pp. 51–63.
Schramm, M. & Loyter, A. (1966). *Methods Enzymol.*, **8,** 533–7.
Schulz, G. E. & Schirmer, R. H. (1979). *Principles of Protein Structure.* Springer, New York, USA.
Scopes, R. K. (1987). *Protein Purification: Principles and Practice* (2nd edn). Springer, New York, USA.
Shing, Y. W., Akagi, J. M. & Himes, R. H. (1975). *J. Bacteriol.*, **122,** 177–84.
Shore, J. D. & Gutfreund, H. (1970). *Biochem.*, **9,** 4655–9.
Srere, P. A. (1980). *Trends Biochem. Sci.*, **5,** 120–1.
Sternberg, M. J. E., Hayes, F. R. F., Russell, A. J., Thomas, P. G. & Fersht, A. R. (1987). *Nature* (London), **330,** 86–8.
Stevens, E. & Stevens, L. (1977). *Biochem. J.*, **179,** 161–7.
Suelter, C. H. (1985). *A Practical Guide to Enzymology.* Wiley Interscience, New York, USA.
Tanford, C. (1978). *Science* (New York), **200,** 1012–18.
Tipton, K. F. (1978). *Techniques in Protein and Enzyme Biochemistry.* Elsevier/North Holland Biomedical Press, New York, USA, **Vol. B112**, pp. 1–56.
Tipton, K. F. (1979). In *Companion to Biochemistry* (Vol 2). ed. A. T. Bull, J. R. Lagnado, J. O. Thomas & K. F. Tipton. Longman, London, UK, pp. 327–82.
Walsh, C. (1979). *Enzymatic Reaction Mechanisms.* Freeman, New York, USA.
Wells, J. A., Powers, D. B., Bott, R. R., Katz, B. A., Ultsch, M. M., Kossiakoff, A. A., Power, S. D., Adams, R. M., Heyneker, H. H., Cunningham, B. C., Miller, J. V., Graycar, T. P. & Estell, D. A. (1987). In *Protein Engineering,* ed. D. L. Oxender & C. F. Fox. Alan R. Liss, New York, USA, pp. 279–87.
Williams, D. H., Bradley, C. V., Santikarn, S. & Bojesen, G. (1982). *Biochem. J.*, **201,** 105–17.
Yamanaka, K. (1966). *Methods Enzymol.*, **9,** 588–93.
Zaks, A. & Klibanov, A. M. (1988). *J. Biol. Chem.*, **263,** 8017–21.

2

Indigenous Enzymes in Milk

Editor's comment
This chapter is divided into six more-or-less independent self-contained sections, covering the principal indigenous enzymes in milk.

I. General Introduction

A. T. Andrews
AFRC Institute of Food Research, Reading Laboratory, Shinfield, Reading, RG2 9AT, UK

II. Lipase

T. Olivecrona & G. Bengtsson-Olivecrona
Department of Medical Biochemistry and Biophysics, University of Umea, S-90187 Umea, Sweden

III. Proteinase

P. F. Fox
Food Chemistry Department, University College, Cork, Republic of Ireland

IV. Phosphatases

A. T. Andrews
AFRC Institute of Food Research, Reading Laboratory, Shinfield, Reading, RG2 9AT, UK

V. Lactoperoxidase

L. Björck
Department of Animal Nutrition and Management, Swedish University of Agricultural Sciences, Box 7024 Ultuna, S-75007 Uppsala, Sweden

VI. Other enzymes

N. Y. FARKYE
Food Chemistry Department, University College, Cork, Republic of Ireland

I. GENERAL INTRODUCTION

Probably all the constituents of blood, both soluble and cellular, are also found in milk, albeit at lower concentrations. Of course, there are in addition other components which are unique to milk, including all the major proteins (i.e. the caseins and the principal whey proteins, α-lactalbumin and β-lactoglobulin). Bovine milk contains approximately 32 g protein/litre, of which the caseins (α_{s1}, α_{s2}, β and κ), β-lactoglobulin and α-lactalbumin represent 24–25, ~4 and ~2 g/litre, respectively. This leaves only 1–2 g/litre for all other proteins; by far the most important of these, quantitatively, are serum albumin and the immunoglobulins. The proteose–peptone fraction, often regarded as part of the whey protein (non-casein) fraction, is in fact composed mainly of peptides produced from caseins through the action of proteinases, particularly plasmin, the principal indigenous proteinase in good-quality milk. Because of its origin, the proteose–peptone content of milk is highly variable, but it may be as high as 1–3 g/litre. All the other proteins in milk represent less than 1% of the total protein, and most of them are present at less than 1 mg/litre.

Over 60 different enzymes have been reported to be indigenous to milk. However, since only some have been isolated and characterized, doubts must remain about the truly indigenous nature of many of the enzymes which have been described only in terms of identifiable activities and limited studies of properties. Some enzymes in milk may well originate from contaminating microorganisms.

Lactoperoxidase was probably the first enzyme reported in bovine milk (Arnold, 1881). During the first part of this century the number of indigenous enzymes reported gradually increased so that in a review, Shahani (1966) stated that at least 19 enzymes had been purified, isolated or, at least, definitely identified. By 1971 this number had increased to 55–60 (Got, 1971), but there was little further increase during the next decade (Blanc, 1982). The reasons for these fluctuations in the fortunes of enzyme hunters are obvious. In earlier times it was often difficult to detect an enzymic activity, and the chemistry of enzyme-catalysed reactions was only slowly and painstakingly being unravelled. Enzymology advanced rapidly after World War II and many low-molecular-weight artificial chromogenic or radioactive substrates were developed. These facilitated measurement of enzyme kinetics and greatly improved the ease and sensitivity of enzyme detection and assay. At this time, the catalogue of enzymes identified in all types of biological material increased enormously.

After this exploratory phase came a period of consolidation, which continues to date, when merely detecting an enzyme and superficially assessing some of its properties is no longer considered to be an adequate objective *per se*. By using the most sensitive modern methods of detection, the number of indigenous enzymes in milk could probably be expanded considerably, but it is likely that all the enzymes present in milk in reasonably large amounts have already been reported and a further extensive search for traces of hitherto undetected enzymes would be rather pointless, even if successful. This probably explains why few recent papers have been concerned with the discovery of new enzymes in milk. Instead, many recent studies on milk enzymes are concerned with their origin, their biological or physiological significance as indicators of performance or state of health, or with their relevance in food processing, nutrition and product quality or stability. Considerations such as these preclude most of the enzymes reported in milk as being of real practical importance, although there will be occasions when, for specific purposes, it is necessary to study them or take cognizance of their activities.

Occurrence and distribution of indigenous milk enzymes

The great majority of work on milk enzymes has been performed with bovine milk, but all milks contain enzymes. With the exception of bile-activated lipase in the milks of humans and other primates, it is probably true that there are no well-substantiated reports of enzyme activities in other milks that have not also been found in bovine milk, and conversely, when an activity previously reported in bovine milk is sought in the milks of other species, it is nearly always found. Although there may be marked quantitative differences, the complement of enzymes, and their properties, in the milks of all mammals are very similar. This is not surprising when it is remembered that the enzymes in all milks originate via almost identical pathways, i.e. either from mammary tissue cells, blood plasma or leucocytes. Since mammals are quite closely related phylogenetically, their emergence in the evolutionary tree being a relatively recent event, interspecies differences in proteins and enzymes and their properties are essentially confined to those which can be related to relatively conservative amino-acid substitutions, deletions or insertions.

Enzymes are not distributed homogeneously throughout milk. While many are found principally in the serum (whey) phase of skim milk (e.g. catalase, lactoperoxidase, ribonuclease, γ-glutamyltransferase, N-acetyl-β-D-glucosaminidase, α-mannosidase, etc.), others are associated with the casein micelles (e.g. proteinases and lipase). The cream phase is the principal source of xanthine oxidase. One would expect to find most enzymes of cellular origin in a cellular fraction (centrifugal pellet), but frequently this is not the case, and the distribution of enzymes in milk is often variable and difficult to reconcile with their origin. For example, plasmin, the main proteinase, is of blood-serum origin, but rather than being found in the soluble (whey) fraction it is associated with the casein micelles. Lysosomal enzymes (e.g. acid phosphatase and catalase) are found largely in the milk serum rather than in any particulate fraction, although up to 25% of the catalase activity, for example, has been reported to be in the cream phase (Kitchen *et al.*, 1970). This is a good demonstration of the fact that, while most enzymes are found predominantly in one of the phases of milk, many are not confined to a single phase. This may reflect a genuine redistribution of the enzyme, but in some cases it undoubtedly reflects differences in the behaviour of isoenzymes, or even of totally distinct enzymes from different sources but with the same activity. For example, proteinases can originate from both blood serum and leucocytes, although the properties of enzymes from the two sources are very different, and catalases are present in both cytoplasm and lysosomal particles. Redistribution of some enzymes occurs as a result of milk processing, especially due to redistribution of the natural fat globule membrane.

Factors that influence enzyme activities in milk

To add further confusion, neither the enzyme complement nor distribution are constant but are influenced by several factors. These include not only the type of animal from which the milk originates but also its breed, age, lactation number, diet and nutritional status, stage of lactation (especially early or late) and health, particularly with regard to mammary gland infections (mastitis). While most milks, qualitatively, have a generally similar complement of enzymes, quantitative differences can be considerable. For example, according to Chandan *et al.* (1968) the level of lysozyme in human milk (400 μg/ml) is over one thousand-fold higher than in bovine, ovine or caprine milks (0·1–0·25 μg/ml), whereas bovine milk contains more ribonuclease and lipase than the others. However, considerable caution must be exercised when considering this sort of comparative data because often only a few animals are considered, and it is important that other factors which may influence milk enzyme composition are comparable. Since non-enzyme proteins show considerable interspecies variations (e.g. human milk does not contain β-lactoglobulin or α_{s1}-casein, which are major proteins in bovine, buffalo, caprine or ovine milks), it seems reasonable to assume that even when all influencing factors are taken into account, considerable interspecies differences in enzyme levels will remain. As would be expected, interbreed differences within a single species are very much smaller, but may still be significant. For example, Holbrook and Hicks (1978) reported higher levels of superoxide dismutase in the milk of Jersey cows (1·27 U/ml) than in Holstein milk (0·92 U/ml), and xanthine oxidase levels were also slightly higher (0·078 and 0·065 U/ml, respectively).

Generally, dietary factors have minor effects on milk enzyme profiles and levels, although there are reports that dietary supplementation with protein or fat can increase the levels of lipase, esterase, xanthine oxidase, acid and alkaline phosphatases in human milk (Karmarkar & Ramakrishnan, 1959;

Belavady, 1960, 1962) but this may only be relevant when women are severely undernourished to begin with.

Lactational effects on milk enzyme levels are far greater than breed or nutritional effects. Colostrum just before and shortly after parturition is very different from mature mid-term milk in almost all its major constituents, and it is not surprising that these secretions also differ with respect to enzymes, both in total amounts and in relative proportions. According to Wuthrich *et al.* (1963), the activities of a number of enzymes, including α-amylase, catalase, lactoperoxidase and acid phosphatase, are higher in colostrum than in mature milk; only lipase levels are lower. While the levels of various enzymes in colostrum are elevated compared with milk, the extent and time-course of this elevation are very variable. For example, at parturition, β-glucuronidase levels are more than tenfold higher than normal but decrease to normal values after 3 days (Kiermeier & Gull, 1968), while lactoperoxidase is at a low level initially but increases to a maximum after 3–5 days before gradually decreasing again (Kiermeier & Kuhlmann, 1972). Both acid and alkaline phosphatases are greatly elevated in colostrum, but decline to normal values at very different rates: 10 days for alkaline phosphatase, but 40 days for acid phosphatase (Kiermeier & Meinl, 1961). This may reflect the different origins of the enzymes (alkaline phosphatase is believed to originate in the plasmalemma of mammary gland cells, whereas acid phosphatase is a lysosomal enzyme), or the variations in isoenzyme behaviour, with some isoenzymes returning to normal faster than others. Once the enzyme composition has reached that typical of mature milk, there is generally little change until very late in lactation when the somatic cell content of milk gradually increases so that those enzymes associated with somatic cells also increase slightly (e.g. acid phosphatase, acid proteinase, *N*-acetyl-β-D-glucosaminidase, catalase). The permeability of mammary gland cell walls for soluble components may also increase slightly in late lactation, and consequently enzymes of blood-serum origin (e.g. plasmin) also increase. In some respects, therefore, the enzyme profile of very-late-lactation milk tends to approach that of colostrum. All these lactational trends become more marked with increasing lactation number and, hence, age. Generally, as lactation number increases, so does cell-wall permeability, and the concentrations of both soluble and cellular constituents of blood origin in milk increase, including the levels of enzymes from those sources. The somatic-cell count of mid-lactation milk from a healthy cow in its first lactation is approximately $2–4 \times 10^4$ cells/ml, but by the seventh lactation, $2–4 \times 10^5$ cells/ml are more typical.

Mastitis

Apart from very marked differences between colostrum or late-lactation milk and mid-lactation milk, the most striking changes are caused by disease, most commonly and most dramatically by mastitis. The occurrence of mastitis is accompanied by very pronounced changes in the concentrations of nearly all milk components (for reviews, see Haenlein *et al.*, 1973; Newstead, 1973; Newbold, 1974; Randolph *et al.*, 1974; Kitchen, 1981). Probably the most significant feature of mastitis is an increase in the permeability of mammary gland cell walls so that the concentration of virtually all blood constituents increases in the milk. The whey protein fraction of milk increases markedly due to the influx of blood proteins, especially serum albumin and the immunoglobulins, while the concentrations of α-lactalbumin and β-lactoglobulin change little (Anderson & Andrews, 1977). High levels of plasminogen also enter the milk and this is rapidly activated to plasmin (De Rham & Andrews, 1982; Schaar & Funke, 1986). Other proteinases, including probably thrombin and proteinases of cellular origin, also increase greatly, and consequently proteolysis can be very rapid and extensive in severely mastitic milk (Barry & Donnelly, 1981; De Rham & Andrews, 1982; Andrews, 1983).

The influence of mastitis on milk composition varies with the severity and stage of the disease. Compositional changes are much greater during severe and clinical infections than in relatively mild or subclinical cases. In addition to the passage of soluble blood constituents into milk, the population of somatic cells also increases. Milk from healthy cows in mid-lactation typically contains 2×10^4 to 2×10^5 cells/ml, while typical bulk supplies contain $5–6 \times 10^5$ cells/ml due to contributions from animals with subclinical mastitis. While the diagnosis of clinical mastitis does not always precisely parallel somatic-cell count, erythrocytes are not usually observed in milk with total cell counts less than about 2×10^7 cells/ml, when mammary tissue damage may be severe. The time-course of a typical infection may involve a latent period of subclinical infection, of varying length (perhaps several weeks), followed by a rapid transition to an acute infection which provokes diagnosis and treatment, leading to a gradual recovery phase. The population of somatic cells in milk differs during infection, with polymorphonuclear leucocytes predominating in early and acute phases, and other cell types, including those of mammary-tissue origin (e.g. macrophages and eosinophiles), being more numerous during the recovery phase. Not surprisingly, therefore, changes in milk proteins, including the complement of enzymes present, also depends on the stage of infection. Therefore, it is not possible to be precise about the influence of mastitis on milk enzymes beyond saying that, qualitatively, enzymes of blood origin, both soluble and cellular, will be elevated, the extent depending upon both the severity and stage of infection. Few, if any, enzymes are likely to be diminished in activity. Surprisingly, it is often difficult to detect significant levels of bacterial enzymes, although this may be due largely to the difficulty of distinguishing between an isoenzyme of bacterial origin and enhanced levels of similar isoenzymes of mammalian origin.

Benefits of indigenous milk-enzyme activities

One positive use of enzymes in relation to mastitis is in diagnosis, and a number of attempts have been made to use catalase, arylesterase and various glycosidase activities for this purpose; the most successful so far is *N*-acetyl-β-D-glucosaminidase activity (Kitchen *et al.*, 1978, 1980). A few years ago there was considerable research into the use of various enzymes to detect a range of udder disorders, but less importance has been placed on this recently. There was also some interest in possible beneficial roles of milk enzymes in the development of desirable gut flora and in enhancing immune defence mechanisms in the newborn (both humans and other animals), but again interest in this area appears to have waned. Studies in cell biology and in biosynthetic pathways within the mammary gland, including the synthesis, transport and secretion of enzymes, have gained in popularity. However, by far the most frequent rationale for investigating milk enzymes includes an appraisal of their technological significance. This often concerns problems of flavour, texture or stability in dairy products, but the use of enzymes as bacteriocidal agents or as indices of thermal treatments (Andrews *et al.*, 1987) are also of technological interest. Table 1 summarizes the most important enzymes from these viewpoints.

The enzymes in bovine milk have been studied intensively, and the literature has been reviewed at regular intervals (Whitney, 1958; Shahani, 1966; Kitchen *et al.*, 1970; Got, 1971; Groves, 1971; Jodar, 1972; Dwivedi, 1973; Shahani *et al.*, 1973; Teadlecova & Hanus, 1973; Johnson, 1974; Fox & Morrissey, 1981; Kitchen, 1985). Some of the enzymes in human milk have also been

TABLE 1
The Most Important Enzyme Activities in Bovine Milk

Enzyme	*EC Number*	*Action*
Proteinase (plasmin, thrombin, cathespin D)	3.4.21.7	Stability and textural defects in milk and dairy products Bitter peptide formation
Lipase	3.1.1.3	Off-flavour production and rancidity
Alkaline phosphatase	3.1.3.1	Index of pasteurisation
Acid phosphatase	3.1.3.2	
Lactose synthetase	2.4.1.22	Lactose biosynthesis
γ-Glutamyltranspeptidase	2.3.1.1	Amino-acid uptake in protein synthesis
Superoxide dimutase	1.15.1.1	Antioxidant Free radical inhibitor
Xanthine oxidase	1.2.3.2	Pro-oxidation
Catalase	1.11.1.6	Pro-oxidation Mastitis diagnosis
Lactoperoxidase	1.11.1.7	Antibacterial agent Pro-oxidation
N-Acetylglucosaminidase	3.2.1.30	Mastitis diagnosis
Lysozyme	3.2.1.17	Antibacterial agent
Ribonuclease	3.1.27.5	
Leucine aminopeptidase	3.4.11.1	
Rhodanase	2.8.1.1	
Sulphydryl oxidase	1.8.3	Reduction of cooked flavours

thoroughly investigated: reviews include Shahani *et al.* (1980), Hamosh *et al.* (1985) and Hernell and Lonnerdal (1989).

In the following sections, the molecular properties and significance of the principal enzymes in bovine milk are reviewed with some interspecies comparisons.

REFERENCES

Anderson, M. & Andrews, A. T. (1977). *J. Dairy Res.*, **44**, 223–35.

Andrews, A. T. (1983). *J. Dairy Res.*, **50**, 57–66.

Andrews, A. T., Anderson, M. & Goodenough, P. W. (1987). *J. Dairy Res.*, **54**, 237–46.

Arnold, C. (1881). *Arch. Pharm.*, **219**, 41–62.

Barry, J. G. & Donnelly, W. J. (1981). *J. Dairy Res.*, **48**, 437–46.

Belavady, B. (1960). *Indian J. Med. Res.*, **48**, 654–60.

Belavady, B. (1962). *Indian J. Med. Res.*, **50**, 100–3.

Blanc, B. (1982). *Le Lait*, **62**, 350–95.

Chandan, R. C., Parry Jr, R. M. & Shahani, K. M. (1968). *J. Dairy Sci.*, **51**, 606–7.

De Rham, O. & Andrews, A. T. (1982). *J. Dairy Res.*, **49**, 587–96.

Dwivedi, B. K. (1973). *CRC Crit. Rev. Food Technol.*, **3**, 457–78.

Fox, P. F. & Morrissey, P. A. (1981). In *Enzymes and Food Processing*, ed. G. G. Birch, N. Blakeborough & K. J. Parker. Applied Science Publishers, London, UK, pp. 213–38.

Got, R. (1971). *Ann. Nutr. Alim.*, **25**, A291–311.

Groves, M. L. (1971). In *Milk Proteins: Chemistry and Molecular Biology* (Vol. II), ed. H. A. McKenzie. Academic Press, New York, USA, pp. 367–418.

Haenlein, G. F. W., Schultz, L. H. & Zikakis, J. P. (1973). *J. Dairy Sci.*, **56**, 1017–24.

Hamosh, M., Freed, L. M., Jones, J. B., Berkow, S. E., Bitman, J., Mehta, N. R., Happ, B. & Hamosh, P. (1985). In *Human Lactation: Milk Components and Methodologies*, ed. R. G. Jensen & M. C. Neville. Plenum Press, New York, USA, pp. 251–66.

Hernell, O. & Lonnerdal, B. (1989). In *Protein and Non-Protein Nitrogen in Human Milk*, ed. S. A. Atkinson & B. Lonnerdal. CRC Press, Boca Raton, FL, pp. 67–75.

Holbrook, J. & Hicks, C. L. (1978). *J. Dairy Sci.*, **61**, 1072–7.

Jodar, L. G. (1972). *Rev. Esp. Lech.*, **83**, 3–18.

Johnson, A. H. (1974). In *Fundamentals of Dairy Chemistry* (2nd edn), ed. B. H. Webb, A. H. Johnson & J. A. Alford. AVI Publishing Co. Inc., Westport, CT, pp. 1–57.

Karmarkar, M. G. & Ramakrishnan, C. V. (1959). *J. Nutr.*, **69**, 274–6.

Kiermeier, F. & Gull, J. (1968). *Z. Lebensm. Unters. Forsch.*, **138**, 205–8.

Kiermeier, F. & Kuhlmann, H. (1972). *Munchner Medizinische Woehenschrift*, **114**(49), 2144–6. Cited from *Dairy Sci. Abstr.*, **35** (1973) 5338.

Kiermeier, F. & Meinl, E. (1961). *Z. Lebensm. Unters. Forsch.*, **114**, 189–202.

Kitchen, B. J. (1981). *J. Dairy Res.*, **48**, 167–88.

Kitchen, B. J. (1985). In *Developments in Dairy Chemistry—3*, ed. P. F. Fox. Elsevier Applied Science Publishers, London, pp. 239–79.

Kitchen, B. J., Taylor, G. C. & White, I. C. (1970). *J. Dairy Res.*, **37**, 279–88.

Kitchen, B. J., Middleton, G. & Salmon, M. C. (1978). *J. Dairy Res.*, **45**, 15–20.

Kitchen, B. J., Middleton, G., Durward, I. G., Andrews, R. J. & Salmon, M. C. (1980). *J. Dairy Res.*, **63**, 978–83.

Newbold, F. S. H. (1974). In *Lactation* (Vol. 2), ed. B. L. Larson & V. R. Smith. Academic Press, New York, pp. 268–88.
Newstead, D. F. (1973). *NZ J Dairy Sci. Technol.*, **8,** 52–5.
Randolph, H. E., Erwin, R. E. & Richter, R. L. (1974). *J. Dairy Sci.*, **57,** 15–18.
Schaar, J. & Funke, H. (1986). *J. Dairy Res.*, **53,** 515–28.
Shahani, K. M. (1966). *J. Dairy Sci.*, **49,** 907–20.
Shahani, K. M., Harper, W. J., Jensen, R. G., Parry Jr, R. M. & Zittle, C. A. (1973). *J. Dairy Sci.*, **56,** 531–43.
Shahani, K. M., Kwan, A. J. & Friend, B. A. (1980). *Amer. J. Clin. Nutr.*, **33,** 1861–8.
Teadlecova, M. & Hanus, J. (1973). *Die Nahrung*, **17,** 565–77.
Whitney, R. McL. (1958). *J. Dairy Sci.*, **41,** 1303–23.
Wuthrich, S., Richterich, R. & Hostettler, H. (1963). *Z. Lebensm. Unters. Forsch.*, **124,** 336–48.

II. LIPASE

Introduction

It has long been known that bovine milk contains substantial amounts of a lipase, which usually expresses virtually no activity in the milk, but which, when activated, can make the milk rancid and unsuitable for consumption (Deeth & Fitz-Gerald, 1975; Downey & Murphy, 1975; Olivecrona & Bengtsson, 1984; Chillard *et al.*, 1986). The same enzyme is present at the endothelial surfaces of blood vessels where it hydrolyses triglycerides in blood lipoproteins (Olivecrona & Bengtsson, 1984; Olivecrona & Bengtsson-Olivecrona, 1987). It is, therefore, called lipoprotein lipase (LPL), and has attracted much interest with respect both to regulation of its activity in various tissues, and to its kinetic properties. Milk contains as much (or more) LPL on a weight basis as other tissues do (Olivecrona & Bengtsson, 1984; Olivecrona & Bengtsson-Olivecrona, 1987). It is possible to obtain 1 mg of pure LPL from 1 liter of bovine milk. To obtain the same amount of enzyme one would have to process several kilograms of hearts or of adipose tissue. LPL from milk has, therefore, become a much used model enzyme. Several reviews (Jackson, 1983; Olivecrona & Bengtsson, 1984; Olivecrona & Bengtsson-Olivecrona, 1987; Garfinkel & Schotz, 1988), including a dedicated book (Borensztajn, 1987), have recently been published on aspects of this enzyme. The perspective here will be to present the basic properties of the enzyme as they are known from lipoprotein research, and to use this information to discuss the enzyme in milk.

Function of LPL in lipoprotein in metabolism

LPL is attached to the vascular endothelium through binding to the glycan chains of heparan-sulfate proteoglycans (Fig. 1). This locates the enzyme some distance out from the membrane, where it can readily interact with plasma lipoproteins. The lipoprotein binds temporarily to the heparan-sulfate–LPL complex, but it is only the lipolytic products that move into tissue cells. The lipoprotein particle is not consumed but returns to the circulating blood, depleted in triglycerides. Local changes in the LPL activity of tissues is the mechanism by which the body directs the flow of lipoprotein triglycerides (Cryer, 1987). The mammary gland is a prime example of this. Its LPL activity is low before and during pregnancy, increases shortly before parturition and remains high throughout lactation (Scow & Chernick, 1987). In contrast, adipose tissue LPL activity decreases at parturition and remains low throughout lactation. By these changes in LPL activity, blood triglycerides are diverted to the mammary gland for synthesis of milk lipids.

Does LPL have a function in milk?

Until recently it was believed that LPL leaks out from the mammary gland, where it is produced in abundance (Scow & Chernick, 1987). This hypothesis is

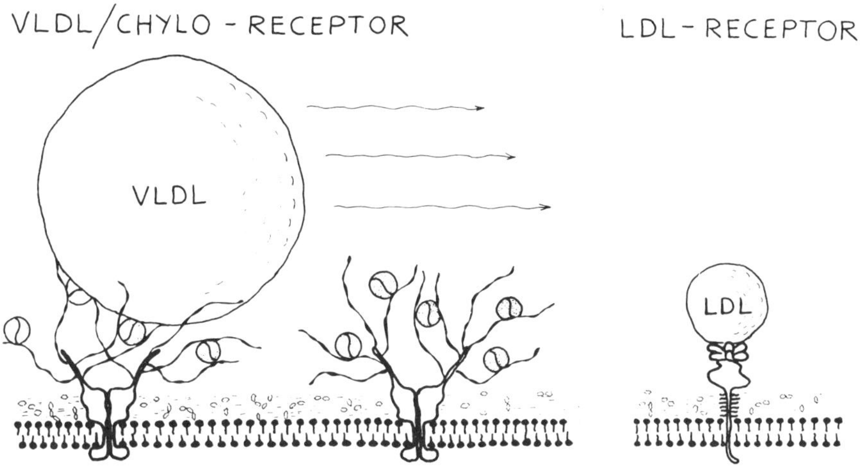

FIG. 1. Schematic view of LPL at the vascular endothelium. (a) Heparan sulfate proteoglycans are intercalated in the plasma membrane of an endothelial cell. Each core protein of the proteoglycan carries several polysaccharide chains. The attached LPL molecules (small spheres) are in a position where they can freely interact with very-low-density lipoprotein (VLDL) and chylomicrons. Several important aspects are not illustrated in this simplified drawing: (i) other enzymes are bound to endothelial heparan sulfate (antithrombin, diamine oxidase, a superoxide dismutase, etc.); (ii) apolipoproteins B and E on the lipoprotein also bind to the heparan-sulfate chains; (iii) chylomicrons are much larger than VLDL and probably interact simultaneously with several heparan-sulfate proteoglycans. (b) The low-density lipoprotein (LDL) receptor also has its ligand-binding domain outside the glycocalyx. This is accomplished by a domain with *O*-linked glycan chains, which spans the glycocalyx (Yamamoto *et al.*, 1984). (From Olivecrona & Bengtsson-Olivecrona (1987), reproduced by permission of Evener Press, Chicago, USA.)

now in question. Mehta *et al.* (1982) studied the appearance of LPL in human milk. Milk was collected from one breast by a pump, and the milk which simultaneously dripped from the other breast was also collected. Several intracellular enzymes were present in higher amounts in the pumped milk. LPL activity was, however, higher in the drip milk. This argues against the idea that LPL appears in milk because of tissue damage.

In guinea pigs, LPL accounts for about 0·1% of the protein in milk (Wallinder *et al.*, 1982), and about 0·1% of the proteins synthesized in the mammary gland (Semb & Olivecrona, 1986). This suggests that LPL is transferred into milk as efficiently as are other milk proteins. Immunohistochemical studies showed more LPL immunoreaction in milk within secretory vesicles, alveoli and ducts, than at the vascular endothelium (Camps *et al.*, 1990). This is in accord with the early studies of McBride and Korn (1963) and Robinson (1963), who found that most of the LPL activity in mammary glands is associated with the milk. It is hard to reconcile these data with the view that LPL has no useful purpose in milk. Some possible functional roles are listed below.

- A recent study shows that heparin-like molecules at the surface of the intestinal mucosa bind pancreatic lipase and cholesterol esterase/non-specific lipase (Bosner *et al.*, 1988). The implication is that most of the intestinal lipolysis takes place at the brush border, and that this facilitates absorption of the lipolysis products. If so, the process would be similar to the lipolysis of lipoprotein triglycerides at the surface of blood capillaries. Since LPL binds strongly to heparin-like molecules (Olivecrona & Bengtsson-Olivecrona, 1989), it is tempting to suggest that the LPL from milk might attach to the intestinal wall and contribute to lipid digestion/absorption.
- It has been demonstrated that LPL can catalyze transfer of lipids (e.g. cholesteryl esters, tocopherol) into cells (Stein *et al.*, 1983; Traber *et al.*, 1985). This transfer activity requires that the lipase binds to the cell surface, but is not dependent on the hydrolytic activity of the lipase. Perhaps milk LPL has such a role in the intestine.
- Milk fat globules are not efficiently attacked by pancreatic lipase. To initiate lipolysis it is necessary that some other lipase hydrolyzes a small fraction of the triglycerides. This function is usually attributed to gastric lipase (Bernbäck *et al.*, 1989). The possibility that LPL can also initiate intestinal lipolysis has not been explored.
- Hernell and collaborators (Hernell and Bläckberg, 1986, 1988, Hernell, *et al.*, 1986) have shown that fatty acids generated in human milk by LPL have a powerful antiparasitic function.

In what state is LPL present in milk?

In bovine (Downey & Murphy, 1975; Sundheim & Bengtsson-Olivecrona, 1986) and guinea-pig milk (Wallinder *et al.*, 1982), more than 80% of the LPL is associated with casein micelles. This binding has been studied in some detail (Downey & Murphy, 1975). The casein micelles are large aggregates of phosphoproteins held together by Ca^{2+} bridges and by hydrophobic interactions. The aggregates are disrupted into smaller 'soluble' protein complexes when Ca^{2+} is removed by dialysis or by addition of a Ca^{2+} chelator. This brings LPL into solution. The lipase can also be released from the casein micelles by increasing the salt concentration or by addition of heparin (Sundheim & Bengtsson-Olivecrona, 1985). This indicates that the binding is primarily electrostatic. In human milk, a significant fraction of the LPL is associated with the fat globules (Hernell & Olivecrona, 1974). This may relate to the much lower amount of casein in human than in bovine milk.

Bovine milk is usually drawn at intervals of 8–16 h. At body temperature, the lipase loses activity in milk with a half-life of about 4 h. Some of this may be attributed to proteolysis, but most of the inactivation occurs by other mechanisms. This is evident from the experiment in Fig. 2. LPL activity decreased by about 70% in 6 h at 37°C, but there was no loss of enzyme protein as measured by immunoassay. Thus, milk contains a mixture of active

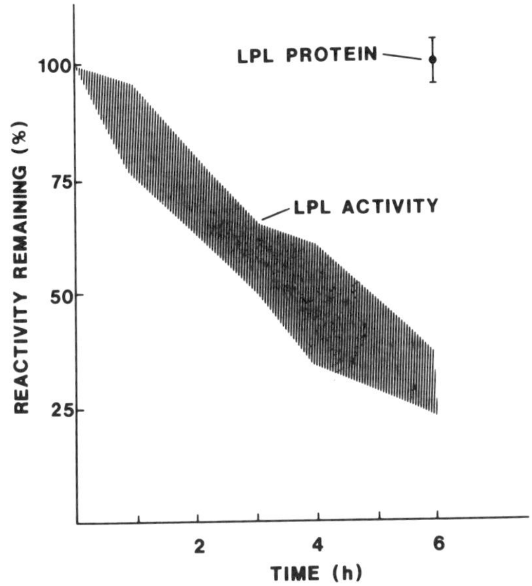

FIG. 2. Decrease of LPL activity in bovine milk during storage at 37°C. Milk samples were obtained from 10 cows and were kept at 37°C. Samples were taken for assay of LPL activity and for radioimmunoassay of LPL protein at the indicated times. The first samples (time 0) were taken within 2 h of milking. Values are expressed as percent of initial. The hatched area is the range within which values for LPL activity fell. (From Olivecrona & Bengtsson-Olivecrona (1987), reproduced by permission of Evener Press, Chicago, USA.)

and inactive LPL. The inactive lipase has lower affinity for heparin–Sepharose and is therefore removed during purification.

Purification of LPL from milk

In most procedures for the purification of milk LPL, the first step is batch adsorption to heparin–agarose. The gel is washed with 0·7–0·9 M NaCl, and the enzyme is then eluted by 1·5–2 M NaCl. Posner *et al.* (1983) introduced an alternative first step in which a lipid emulsion (Intralipid) is added to the skim milk and an Intralipid–LPL complex is collected by centrifugation. Delipidation of the complex yields a powder which can be stored for long periods of time at −20°C. Further purification is usually by adsorption to heparin–agarose, followed by gradient elution with NaCl.

The ability to bind to heparin is not unique for LPL but is shared by several plasma proteins, small amounts of which are always present in milk. Particularly troublesome is antithrombin (Olivecrona & Bengtsson-Olivecrona, 1987). The content of antithrombin is usually low in bovine milk; however, occasional batches do contain significant amounts. On gradient elution, antithrombin elutes slightly before LPL and can give rise to either a double

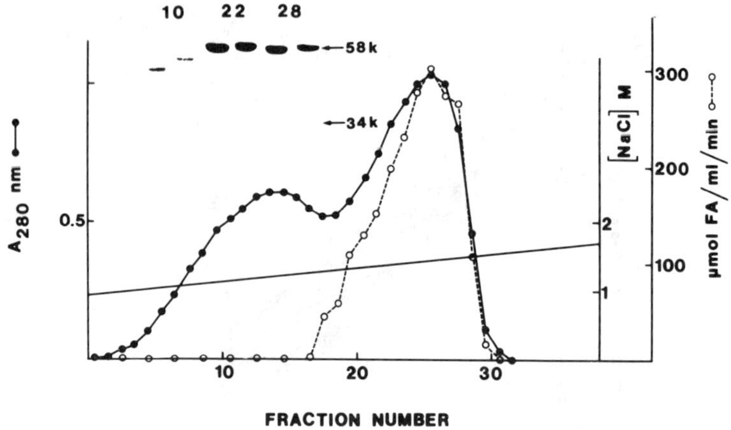

FIG. 3. Partial separation of LPL from antithrombin by chromatography on heparin–agarose. This was an unusual batch of bovine milk which contained antithrombin. The protein peak preceding and partly overlapping the peak of LPL activity was collected and further analyzed. Its *N*-terminal amino-acid sequence (20 residues determined in collaboration with Prof. Hans Jörnvall, Stockholm, Sweden) was identical to that of antithrombin. (●) A_{280}, (○) LPL activity. The inset shows an SDS/PAGE of fractions from the column. In each pair, the sample to the right was reduced before electrophoresis. (From Olivecrona & Bengtsson-Olivecrona (1987), reproduced by permission of Evener Press, Chicago, USA.)

peak of protein or a skewed protein peak (Fig. 3). The fractions most likely to be free of antithrombin are from the trailing portion of the peak of lipase activity. Alternatively, one can rechromatograph the preparation on heparin which has been *N*-desulfated and then acetylated. This results in a much reduced affinity for antithrombin. With guinea-pig milk this is the only way to obtain pure LPL (Wallinder *et al.*, 1982).

Mild digestion with trypsin or plasmin cleaves LPL into two major fragments (Bengtsson-Olivecrona *et al.*, 1986) (Fig. 4). Chymotryptic digestion reveals the presence of another exposed segment. Similar fragments are present in most preparations of LPL from bovine milk. Several years ago, it was found that antisera raised against these components react with LPL, demonstrating that they are proteolytic degradation products (Bengtsson & Olivecrona, 1981). More recently, Socorro & Jackson (1985) have demonstrated that monoclonal antibodies recognize both intact LPL and several proteolytic fragments which are present in milk or are produced during purification; the use of phenylmethylsulfonylfluoride (PMSF) was advocated to prevent proteolysis during purification.

Structural aspects

cDNAs have been cloned for human (Wion *et al.*, 1987), mouse (Kirchgessner *et al.*, 1987), guinea pig (Enerbäck *et al.*, 1987) and bovine LPL (Senda *et al.*,

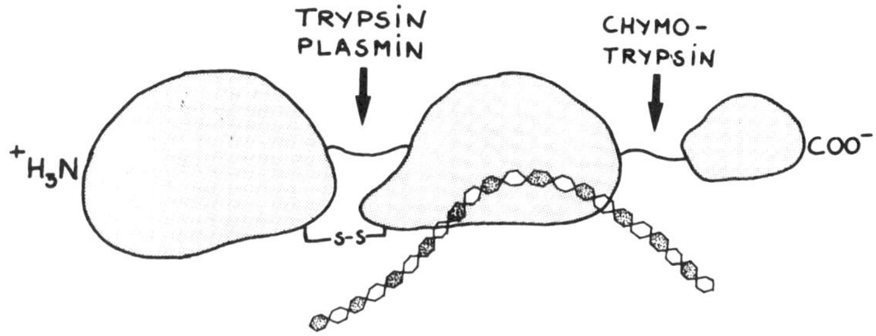

FIG. 4. Exposed segments of LPL as deduced from studies with limited proteolysis. This schematic figure shows the two exposed segments in the bovine-LPL subunit which are attacked by certain proteases. The fragments which are generated by trypsin and plasmin are held together by disulfide bonds. A heparan-sulfate chain is illustrated as a string of saccharide units to show that the binding site is on the putative middle domain. (From Olivecrona & Bengtsson-Olivecrona (1987), reproduced by permission of Evener Press, Chicago, USA.)

1987). The deduced amino-acid sequences show a highly conserved structure which is also related to the hepatic heparin-releasable lipase and to pancreatic lipase (Persson *et al.*, 1989) (Fig. 5). Thus, these three lipases are members of a gene family and have evolved from a common ancestral lipase. Bovine LPL has 450 amino-acid residues (Senda *et al.*, 1987). There are probably five disulfide bridges (Persson *et al.*, 1989). Guinea-pig LPL has three *N*-linked oligosaccharide chains (Semb & Olivecrona, 1989), whereas murine LPL has only two (Vannier & Ailhaud, 1989). The overall subunit weight is about 55 000. Studies with limited proteolysis indicate the protein has two, perhaps three, independently folded domains (Bengtsson-Olivecrona *et al.*, 1986). The only functional site that has thus far been localized is the heparin-binding region, between residues 229 and about 370. In this segment, 16 of 60 residues are either lysine or arginine. The rest of the molecule has a more normal distribution of amino-acid residues. but the agglomeration of basic residues in the heparin-binding region gives the molecule a high isoelectric point, above 9, and a positive net charge at physiological pH.

Studies with ultracentrifugation and gel filtration have shown that active LPL (from bovine milk) is an elongated, noncovalent dimer of two identical subunits (Iverius & Östlund-Lindqvist, 1976; Osborne *et al.*, 1985), and studies by target analysis (radiation inactivation) have shown that the enzyme remains dimeric when it is in complex with heparin, when it is bound to lipid with or without apolipoprotein CII (apo-CII) present, and when it is in complex with detergents (Olivecrona *et al.*, 1985). Each subunit has a heparin-binding site, and a heparin chain of sufficient length can simultaneously satisfy both sites on the dimer (Clarke *et al.*, 1983). This results in cooperative binding with enhanced affinity. It is not known if any of the other functions of LPL depend on its dimeric structure. There is an apo-CII binding site on each subunit, but in this case two apo-CII molecules are needed to satisfy the sites, and there is

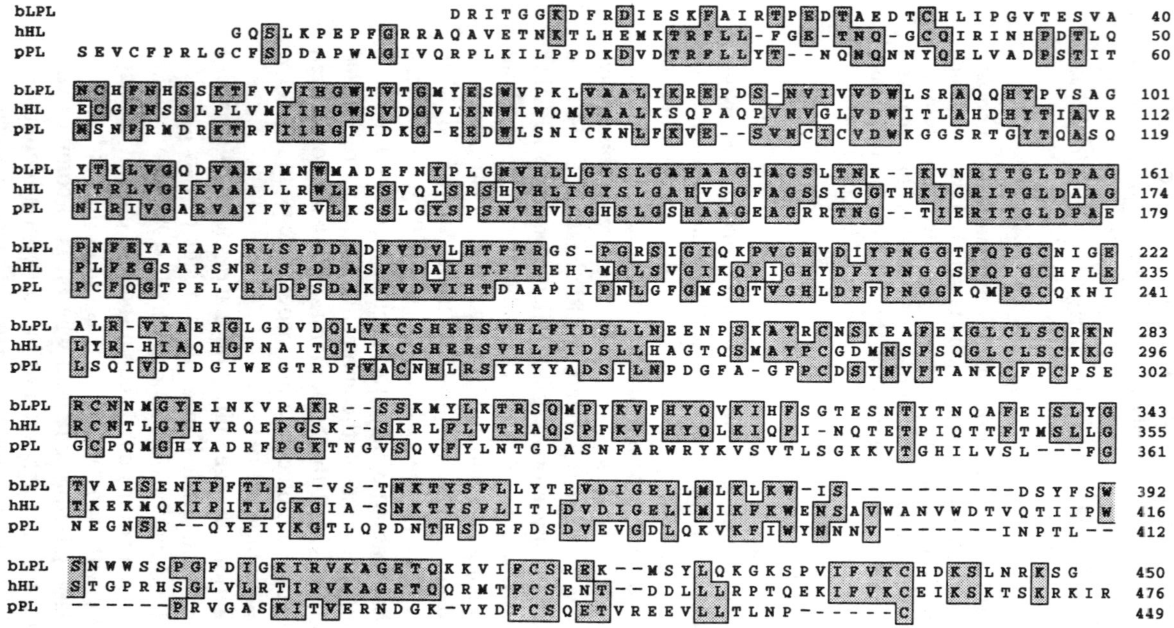

FIG. 5. Sequence homologies in the lipase family. This figure shows an alignment of the sequences of bovine lipoprotein lipase (bLPL: Senda *et al.*, 1987; Yang *et al.*, 1989), human hepatic lipase (hHL: Stahnke *et al.*, 1987, Ameis *et al.*, 1990) and porcine pancreatic lipase (pPL: De Caro *et al.*, 1981). The recent resolution of the three dimensional structure of human pancreatic lipase (Winkler *et al.*, 1990) shows that Ser-152 is in the active site. Centered around this is a 10-residue segment (146–155) which is homologous in a large number of lipases (Persson *et al.*, 1989). There are cysteines in bLPL and hHL which correspond to disulfide bridges between residues 237–261, 285–296, 299–304 and 433–449 in pPL. Cysteines forming a disulfide bridge 29–42 in bLPL (Yang *et al.*, 1989) are also present in hHL, but not in pPL, which, on the other hand, has two unique disulfides (4–10, 90–101). The heparin-binding site in bLPL has been located to the segment 262–316 (Persson *et al.*, 1989). Of these 54 residues, 15 are Arg or Lys which could interact with negative charges on heparin/heparan-sulfate. In the corresponding segment of hHL, which binds to heparin but not so strongly, there are 10 basic residues, whereas in pPL there are only 6 basic residues here. The evolutionary relationship between the lipases is not definitely established, but it appears likely that a first gene duplication gave rise to a digestive lipase (pancreatic lipase) and a lipoprotein-metabolizing lipase, which has later specialized into LPL and hepatic lipase by a second gene duplication (Ameis *et al.*, 1990; Olivecrona & Bengtsson-Olivecrona, 1990).

no cooperation between them (Clarke & Holbrook, 1985). The dimeric structure *per se* is probably not required for catalytic activity, since pancreatic lipase is known to be active in the monomeric form (Verger, 1984). The dimeric structure is, however, required to maintain conformational stability in LPL (Osborne *et al.*, 1985). Dissociation of the dimeric enzyme leads to refolding of the monomer to a catalytically inactive form.

The instability of LPL has made studies on its properties and its interactions difficult. LPL is unstable also under physiological conditions; it has been suggested that this represents a built-in mechanism to self-destruct (Osborne *et al.*, 1985). This would be an important property for an enzyme whose activity should decrease rapidly on the appropriate hormonal signal, even though the enzyme has been secreted from the hormone-sensitive cells.

Active site

LPL is a rather non-specific esterase/lipase (Olivecrona & Bengtsson-Olivecrona, 1987). Of physiologically relevant substrates, it readily attacks the 1,3-positions in tri-, di-, and monoglycerides and the 1-position in glycerophospholipids. It does not hydrolyze cholesteryl esters or sphingolipids at significant rates. With triglycerides as substrate, the fatty acid chain length seems to be of only minor importance for the reaction rate. Thus, the rate of tributyrin hydrolysis is about 0·6, and that of trioctanoin hydrolysis is about 1·3 of that for triolein.

LPL catalyses not only cleavage of ester bonds, but also their formation; it is a transacylase (Bengtsson & Olivecrona, 1980*b*). In this reaction, fatty acids are a substrate for the enzyme, and are incorporated into di- and triglycerides. Mono- and diglycerides are efficient acyl acceptors, but glycerol is not. Lipolysis in milk leads to preferential release of short- and medium-chain fatty acids (Chilliard *et al.*, 1986). Why this is so is not clear. In general, the shorter fatty acids are at the sn-3 position and there are long-chain fatty acids at the sn-1 and -2 positions in milk triglycerides. One possibility is that the triglycerides orient at the lipid–water interface such that the shorter fatty acids are favorably positioned for hydrolysis. Another possibility is that hydrolysis is random with respect to chain length, but that some of the long-chain fatty acids are reincorporated into glycerides by the LPL-catalyzed 'reverse reaction', whereas the shorter fatty acids leave the droplets.

There are several reports that LPL shows a bell-shaped profile for activity versus pH with an optimum between 8 and 9. However, there are also reports that the activity continues to rise with pH to at least 10·5 (Rapp & Olivecrona, 1978; Bengtsson & Olivecrona 1982*b*). The implication is that the decrease in activity at higher pH values is due to some factor other than the active-site reaction. This has been explored using milk fat globules as the substrate (Bengtsson & Olivecrona, 1983). Binding of LPL to the globules was found to be a pH-dependent process with an optimum around pH 8, and the activity curve was a composite of the pH effect on the binding step and the pH effect on the active-site reaction. This is the explanation for the bell-shaped curves often observed.

Lipid-binding site

LPL has been shown to bind to a variety of lipid structures, such as triglyceride droplets, liposomes and lipoproteins (Olivecrona & Bengtsson, 1984; Olivecrona & Bengtsson-Olivecrona, 1987). This binding is mediated by a lipid-binding site which is separate from the active site. The binding is rapid and reversible, and results in stabilization of the enzyme.

Hamilton and Small (1981) have demonstrated that, up to about 2·8 mol%, triolein can be solubilized in bilayers of egg-yolk phosphatidylcholine. A similar amount of triglyceride is probably present in the surface layer on lipoproteins and on milk fat globules. It is presumably these surface-located molecules that are attacked by the lipase. As they are hydrolyzed they are replenished by new triglycerides coming from the core of the particle. Some of the phospholipids are also hydrolyzed by the enzyme, both in natural lipoproteins (Scow & Egelrud, 1976; Fitzharris *et al.*, 1981) and synthetic emulsions (Bengtsson & Olivecrona, 1980*a*), but the rate, only a few percent, is much less than for triglyceride hydrolysis.

Activator protein

For optimal activity, LPL requires apo-CII (for discussion, see Olivecrona & Bengtsson-Olivecrona, 1987). *In vivo,* apo-CII is supplied as an integral part of the substrate lipoproteins, chylomicra and very-low-density lipoproteins (VLDL), and its interaction with LPL presumably takes place with, or after, binding of the lipoprotein particle to the enzyme. There is no evidence that activation by soluble apo-CII would precede binding of the lipoprotein particle to LPL at the vascular endothelium, or that apo-CII would remain with the enzyme when the lipoprotein particle detaches after a round of lipolysis. As the core of the lipoprotein is depleted of triglyceride, excess surface material is shed from the particle and taken up into the high-density lipoprotein (HDL) fraction (Havel *et al.*, 1973). These lipoproteins then serve as a reservoir of apo-CII from which the activator can move to newly secreted chylomicra/VLDL.

In vitro, LPL can be made to hydrolyze model substrates efficiently without apo-CII (Bengtsson & Olivecrona, 1980*b*). Thus, the activator is not required for the catalytic reaction as such. The mechanism by which apo-CII enhances the reaction is under active investigation in several laboratories (discussed in Jackson 1983; Olivecrona & Bengtsson 1984; Olivecrona & Bengtsson-Olivecrona, 1987). Current information indicates that apo-CII ensures that LPL gets to a proper place at the interface so that the active site has ready access to substrate molecules. This is accomplished by the formation of an apo-CII–LPL complex at the lipid–water interface. In some systems, the enzyme itself can bind to the interface but can not adequately reach the substrate molecules. The role of apo-CII here is to bring the enzyme into the right orientation/conformation. How stringent the requirement is for precise orientation depends on the physical properties of the interface and of the substrate molecules. In other systems, the enzyme itself can not bind to the interface, and the primary role of apo-CII is to bring the enzyme to the interface. Combinations of these situations could explain why activation varies so much with the system studied.

Biologically packaged lipid droplets which lack apo-CII are not attacked efficiently by LPL. One illustration is the massive hypertriglyceridemia in patients with genetic defects in apo-CII (Breckenridge *et al.*, 1982). Another illustration is that milk fat globules are normally not attacked by LPL. If, however, apo-CII is added, rapid lipolysis ensues (Bengtsson & Olivecrona, 1982*a*). The concept that emerges is that apo-CII is a safety mechanism built into the LPL system to direct its action to certain lipoproteins and prevent it from acting on other lipid droplets.

Human apo-CII is a single-chain polypeptide of 79 residues (Jackson *et al.*, 1977; Hospattankar *et al.*, 1984). The structure of the canine analogue, as deduced from its cloned cDNA, is highly homologous (Datta *et al.*, 1987). The functional roles of different parts of the molecule have been explored by the use of fragments, obtained by selective cleavage of human apo-CII or by solid-phase synthesis. These studies indicate that the molecule has two functional parts (Smith & Pownall, 1984). Residues 1–52, corresponding to the *N*-terminal two thirds of the molecule, appear to be involved with binding to lipoproteins through interaction with surface phospholipids. This region in the molecule has the potential to form two or three amphipathic helixes (Smith & Pownall, 1984; Datta *et al.*, 1987). The *C*-terminal third of apo-CII is suggested to be involved with binding to, and activation of, LPL.

Fojo *et al.* (1986) have shown that human apo-CII is secreted as a carbohydrate-containing pro-protein which undergoes deglycosylation and proteolytic cleavage in plasma. There are four isoforms of apo-CII in human plasma. Two of these are glycosylated and contain sialic acid. The major isoform is the 79 amino acid, unglycosylated protein. It can undergo proteolytic cleavage of a six-amino-acid peptide from its *N*-terminal. This cleavage is analogous to that involved in maturation of pro-apo-AI. The relative effectiveness of the various isoforms as activators of LPL is unknown.

The bovine analogue to apo-CII has been isolated (Astrup & Bengtsson, 1982) and part of its sequence has been published (Astrup, 1987). The sequence has recently been completed (Bengtsson-Olivecrona & Sletten, 1990) and shows that the bovine activator is homologous to the human and canine proteins, with some interesting differences. The major isoform isolated was the cleaved variant and no evidence for glycosylated variants was obtained.

It has been repeatedly suggested that a cause of spontaneous lipolysis in bovine milk could be that the activator protein is transferred to the milk. It is certainly true that addition to milk of even small amounts of purified activator protein (Bengtsson & Olivecrona, 1982*a*), of isolated lipoproteins (Clegg,

1980) or of whole serum (Driessen & Stadhouders, 1982) leads to rapid lipolysis. Hamosh (1982) has pointed out that this may be the rationale behind the ancient Jewish rule forbidding the mixing of meat and milk. The presence in some milk samples of material which can enhance LPL activity has been reported (Cartier & Chilliard, 1986), but no direct demonstration of substantial amounts of apo-CII, or fragments thereof, has been made. This remains an open question.

Effects of other proteins

The activity of LPL is inhibited by a variety of proteins. The mechanism of this inhibition has not been explored in detail. It is probably non-specific, since inhibition is seen with many different proteins (Cartier & Chilliard, 1986). In no case has a specific protein–protein interaction between the lipase and an inhibitory protein been demonstrated. The common denominator seems to be that the proteins bind to the surface of the substrate droplets.

A major factor in the resistance of bovine milk fat globules to lipolysis is the presence of proteins which inhibit LPL action (Deeth & Fitz-Gerald, 1975; Sundheim & Bengtsson-Olivecrona, 1987*b*). The globule is enveloped by a surface layer of polar lipids and proteins; these proteins are probably an important impediment to the binding of LPL to the droplets. If isolated droplets are physically disturbed, or if some of their proteins are extracted, binding of LPL to the fat globules is substantially increased, with resultant lipolysis (Bengtsson & Olivecrona, 1982*b*). There are also proteins in the skim-milk fraction which impede LPL action, both when tested in milk systems (Sundheim & Bengtsson-Olivecrona, 1987*c*) or with artificial lipid emulsions (Cartier & Chilliard, 1986). These proteins, which are present in the so-called proteose–peptone fraction, have been partially separated (Cartier & Chilliard, 1986; Sundheim & Bengtsson-Olivecrona, 1987*c*).

Inhibition by inorganic salts

One of the first characteristics noted for LPL was that its activity is inhibited by high salt concentration. This property has been used to differentiate LPL from other tissue lipases as well as from the hepatic (heparin-releasable) lipase in post-heparin plasma. However, there is considerable variation in the extent of inhibition obtained between different laboratories and also in the same laboratory using different assay conditions, different enzyme sources or enzyme from the same source but at different degrees of purity. It now appears that the effect of salt is exerted on the physical state of the enzyme molecule, rather than on the catalytic reaction as such (Bengtsson & Olivecrona, 1983).

Fatty-acid binding—product inhibition

LPL is subject to strong product inhibition (Bengtsson & Olivecrona, 1980*b*; Olivecrona & Bengtsson-Olivecrona, 1987; Posner & DeSanctis, 1987). A mechanism shared with many other lipases is that fatty acids can act as substrates, and di- and monoglycerides can act as acyl acceptors. If the fatty acids are not removed or complexed, the lipase will spend much of its time in the non-productive pursuit of hydrolyzing and then reforming glycerides.

A mechanism more specific for LPL is that the enzyme binds fatty acids (Fielding, 1968; Bengtsson & Olivecrona, 1980*b*). This has at least two effects relevant to product inhibition.

(1) Fatty acids interfere with the binding of the enzyme to the lipid droplets, and the enzyme which remains with the droplets can be sequestered in complexes with fatty acids in the surface film on the particle or in bilayer extensions of the surface film.
(2) Fatty acids abolish the effects of the activator protein.

Product inhibition is probably an important mechanism for control of LPL action *in vivo,* so that products are not formed more rapidly than they can be taken up by the tissue. In contrast, inhibition by fatty acids does not seem to be a limiting factor for lipolysis in bovine milk. In the first place the amounts released are relatively small and the short-chain fatty acids which predominate are water-soluble. Furthermore, Ca^{2+} are available to complex the fatty acids, and the proteins in milk may also bind some fatty acids. Addition of albumin does not promote lipolysis in bovine milk (Bengtsson-Olivecrona, G. unpublished).

Interactions with heparin and related molecules

Injection of heparin releases LPL into the circulating blood by formation of soluble LPL–heparin complexes. This has led to many studies on the interaction between heparin and the enzyme, and the effects thereof (reviewed in Olivecrona & Bengtsson-Olivecrona, 1989). It is now clear that the enzyme has a high affinity for heparin, but that heparin is not necessary for its activity. The physiological function of the heparin-binding site appears to be to attach the enzyme to heparan–sulfate proteoglycans at the vascular endothelium.

The effects of heparin on the activity of milk LPL are complex, and depend on the nature of the system studied. The activity against simple substrates is not markedly changed, indicating that binding of heparin has little or no effect on the active site itself. Nonetheless, addition of heparin to milk usually results in marked stimulation of lipolysis (Sundheim & Bengtsson-Olivecrona, 1985). The mechanism appears to be that heparin dissociates the enzyme from the casein micelles with the formation of soluble enzyme–heparin complexes, which, in turn, more readily bind to, and hydrolyze, the fat globules.

Heparin-like polysaccharides are present in milk, but it is not known to what extent they influence lipolysis (Olivecrona & Lindahl, 1969, Shimizu *et al.*, 1981). It is of interest to note that the content of such polysaccharides is higher in human than in bovine milk (Shimizu *et al.*, 1981), and that there is also less binding of LPL to casein and more lipolysis in human milk (Hernell & Bläckberg, 1986).

Lipolysis in milk

Freshly drawn bovine milk contains only small amounts of free fatty acids. If the milking and storage procedures are appropriate, the milk can be kept for several days with little further development of free fatty acids. If this were not so, the modern dairy industry would not exist. Hydrolysis of as little as 1–2% of the milk triglycerides to fatty acids gives a rancid or 'lipolyzed' flavor to the milk. This makes the milk unpalatable for consumption. Concentration of the reaction products in the cream may, in turn, impair the flavor quality of products manufactured from milk.

It is now generally agreed that lipolysis is largely due to the LPL indigenous to the milk. It is only after prolonged storage or after poor hygienic handling that bacterial lipases become important. Virtually all bovine milks, with the possible exception of colostrum (Driessen, 1976), contain substantial amounts of lipolytic activity, but there seems to be only a weak correlation between the (potential) lipase activity of the milk and its susceptibility to lipolysis (Sundheim, 1988). If the conditions in milk were optimal for lipase activity, hydrolysis would proceed rapidly. A typical lipase activity for bovine milk at 37°C, pH 7, with activator added, is 100 nmol fatty acid released per min per

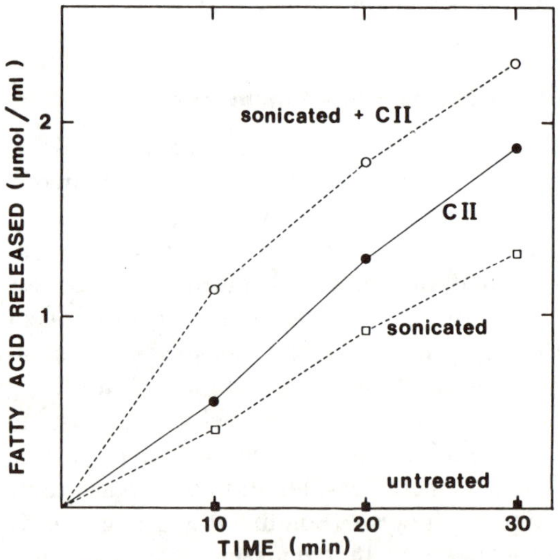

FIG. 6. Induction of lipolysis in fresh bovine milk by addition of apolipoprotein CII (apo-CII) and/or by mechanical treatment. Milk was collected in a thermos at 30°C. The experiment was then started within 1 h. Four incubations were run: A, untreated milk, B, milk which was sonicated before incubation, C, milk to which apo-CII was added immediately before incubation, and D, sonicated milk to which apo-CII was added. Note that there was no lipolysis in the untreated milk, but addition of apo-CII caused immediate, rapid lipolysis. Lipolysis could also be induced by disruption of the organized structure of the milk fat globules (sonication), and combination of sonication and apo-CII resulted in the most rapid lipolysis.

ml milk. At this rate, the milk would become unpalatable in a few minutes. Thus, the question is not why some samples of milk become lipolyzed, but rather why most milk samples do not.

Lipolysis, as it occurs in dairy practice, is often differentiated into spontaneous and induced (Deeth & Fitz-Gerald, 1975). Spontaneous lipolysis is initiated in some milk samples simply by prompt cooling of the fresh milk. In contrast, lipolysis can be induced in all milk samples by rough mechanical treatment (Fig. 6). If the lipase was not destroyed by prior pasteurization, the conventional homogenization of the milk would result in rapid lipolysis and make the milk unusable. Induced lipolysis was not a big problem when milk was collected by hand milking and transported to the dairy the same day: it has become more of a problem with the modern pipeline milking machines and holding of the milk for several days at the farms before transportation to a central dairy.

At the biochemical level, induced lipolysis is presumably equivalent to disruption of the organized structure of the fat-globule membrane; this allows lipase to bind to the fat globules and hydrolyze their lipids (Deeth & Fitz-Gerald, 1975; Bengtsson & Olivecrona, 1982*a*). Spontaneous lipolysis is less well understood, but again lipolysis correlates to binding of lipase to the fat globules (Sundheim & Bengtsson-Olivecrona, 1986, 1987*a*). It should be emphasized again that there is an abundance of lipase in milk; the question is why most milks are so resistant to lipolysis. It has been demonstrated that this resistance of the milk fat is partly due to factors related to the structure of the milk fat globule itself and partly to factors in the skim-milk fraction. Methods to assess these factors have been worked out in two laboratories (Cartier & Chilliard, 1986; Sundheim & Bengtsson-Olivecrona, 1987*b, d*).

ACKNOWLEDGEMENT

The original work on lipolysis in bovine milk reported in this review was supported by the Swedish Board for Technical Development (Grant No. 03186P).

REFERENCES

Ameis, D., Stahnke, G., Kobayashi, J., McLean, J., Lee, G., Büscher, M., Schotz, M. C. & Will, H. (1990). *J. Biol. Chem.*, **265,** 6552–5.
Astrup, H. N. (1987). *Norw. J. Agric. Sci.*, **1,** 45–8.
Astrup, H. N. & Bengtsson, G. (1982). *Comp. Biochem. Physiol.*, **72B,** 487–91.
Bengtsson, G. & Olivecrona, T. (1980*a*). *Eur. J. Biochem.*, **106,** 549–55.
Bengtsson, G. & Olivecrona, T. (1980*b*). *Eur. J. Biochem.*, **106,** 557–62.
Bengtsson, G. & Olivecrona, T. (1981). *Eur. J. Biochem.*, **113,** 547–54.

Bengtsson, G. & Olivecrona, T. (1982*a*). *FEBS Lett.*, **147,** 183–7.
Bengtsson, G. & Olivecrona, T. (1982*b*). *Biochim. Biophys. Acta,* **712,** 196–9.
Bengtsson, G. & Olivecrona, T. (1983). *Biochim. Biophys. Acta,* **751,** 254–9.
Bengtsson-Olivecrona, G. & Sletten, K. (1990). *Eur. J. Biochem.*, **192,** 515–21.
Bengtsson-Olivecrona, G., Olivecrona, T. & Jörnvall, H. (1986). *Eur. J. Biochem.*, **161,** 281–8.
Bernbäck, S., Bläckberg, L. & Hernell, O. (1989). *Biochim. Biophys. Acta,* **1001,** 286–93.
Borensztajn, J. (1987). *Lipoprotein Lipase.* Evener Press, Chicago, USA.
Bosner, M. S., Gulick, T., Riley, D. J. S., Spilburg, C. A. & Lange III, L. G. (1988). *Proc. Nat. Acad. Sci. USA,* **85,** 7438–42.
Breckenridge, W. C., Alaupovic, P., Cox, D. W. & Little, A. A. (1982). *Atherosclerosis,* **44,** 223.
Camps, L., Reina, M., Llobera, M., Vilaró, S. & Olivecrona, T. (1990). *Amer. J. Physiol.*, **258,** C673–C681.
Cartier, P. & Chilliard, Y. (1986). *J. Dairy Sci.*, **69,** 952–955.
Chilliard, Y., Chazal, M.-P. & Cartier, P. (1986). *Revue des ENIL,* **112,** 12–19.
Clarke, A. R. & Holbrook, J. J. (1985). *Biochim. Biophys. Acta,* **827,** 358–68.
Clarke, A. R., Luscombe, M. & Holbrook, J. J. (1983). *Biochim. Biophys. Acta,* **747,** 130–7.
Clegg, R. A. (1980). *J. Dairy Res.*, **47,** 61–70.
Cryer, A. (1987). In *Lipoprotein Lipase,* ed. J. Borensztajn. Evener Press, Chicago, USA, pp. 277–327.
Datta, S., Li, W.-H., Ghosh, I., Luo, C.-C. & Chan, L. (1987). *J. Biol. Chem.*, **262,** 10588–93.
de Caro, J., Boudouard, M., Bonicel, J., Guidoni, A., Desnuelle, P. & Rovery, M. (1981). *Biochim. Biophys. Acta,* **671,** 129–138.
Deeth, H. C. & Fitz-Gerald, C. H. (1975). *Int. Dairy. Fed. Doc.*, **86,** 24.
Downey, W. K. & Murphy, R. F. (1975). *Int. Dairy Fed. Doc.*, **86,** 19–23.
Driessen, F. M. (1976). *Neth. Milk Dairy J.*, **30,** 186–96.
Driessen, F. M. & Stadhouders, J. (1982). *Neth. Milk Dairy J.*, **28,** 130–8.
Enerbäck, S., Semb, H., Bengtsson-Olivecrona, G., Carlsson, P., Hermansson, M.-L., Olivecrona, T. & Bjursell, G. (1987). *Gene,* **58,** 1–12.
Fielding, C. J. (1968). *Biochim. Biophys. Acta,* **159,** 94–102.
Fitzharris, T. J., Quinn, D. M., Goh, E. H., Johnson, J. D., Kashyap, M. L., Srivastava, L. S., Jackson, R. L. & Harmony, J. A. K. (1981). *J. Lipid Res.*, **22,** 921–33.
Fojo, S. S., Taam, L., Fairwell, T., Ronan, R., Bishop, C., Meng, M. S., Hoeg, J. M., Sprecher, D. L. & Brewer Jr, H. B. (1986). *J. Biol. Chem.*, **261,** 9591–4.
Garfinkel, A. S. & Schotz, M. C. (1988). In *Lipoproteins,* ed. A. M. Gotto Jr. Elsevier, Amsterdam, The Netherlands, pp. 335–57.
Hamilton, J. A. & Small, D. M. (1981). *Proc. Nat. Acad. Sci. USA,* **78,** 6878–82.
Hamosh, M. (1982). *Adv. Pediatr.*, **29,** 33–67.
Havel, R. J., Kane, J. P. & Kashyap, M. L. (1973). *J. Clin. Invest.*, **52,** 32–8.
Hernell, O. & Bläckberg, L. (1986). In *Composition and Physiological Properties of Human Milk,* ed. J. Schaub. Elsevier, Amsterdam, The Netherlands, pp. 165–75.
Hernell, O. & Bläckberg, L. (1988). In *Biology of Human Milk,* ed. P. Goyens & L. Å. Hansson. Vol. 15, Nestlé Nutrition Workshop series, Raven Press, New York, USA, pp. 159–70.
Hernell, O. & Olivecrona, T. (1974). *J. Lipid Res.*, **15,** 367–74.
Hospattankar, A. V., Fairwell, T., Ronan, R. & Brewer Jr, H. B. (1984). *J. Biol. Chem.*, **259,** 318–22.
Iverius, P.-H. & Östlund-Lindqvist, A.-M. (1976). *J. Biol. Chem.*, **251,** 7791–5.

Jackson, R. J. (1983). In *The Enzymes* (Vol. 16). Academic Press, New York, USA, pp. 141–81.
Jackson, R. L., Baker, H. N., Gilliam, E. B. & Gotto, A. M. (1977). *Proc. Nat. Acad. Sci. USA,* **74,** 1942–5.
Kirchgessner, T. G., Svenson, K. L., Lusis, A. J. & Schotz, M. C. (1987). *J. Biol. Chem.,* **262,** 8463–6.
McBride, O. W. & Korn, E. D. (1963). *J. Lipid Res.,* **4,** 17–20.
Mehta, N. R., Kones, J. B. & Hamosh, M. (1982). *J. Pediatr. Gastroenterol.,* **1,** 317–26.
Olivecrona, T. & Bengtsson, G. (1984). In *Lipases,* ed. B. Borgström & H. Brockman. Elsevier, Amsterdam, The Netherlands pp. 206–61.
Olivecrona, T. & Bengtsson-Olivecrona, G. (1987). In *Lipoprotein Lipase,* ed. J. Borensztajn. Evener Press, Chicago, USA, pp. 15–58.
Olivecrona, T. & Bengtsson-Olivecrona, G. (1989). In *Heparin,* ed. D. Lane & U. Lindahl. Edward Arnold, London, UK, pp. 335–61.
Olivecrona, T. & Bengtsson-Olivecrona, G. (1990). *Curr. Opin. Lipidol.,* **1,** 116–21.
Olivecrona, T. & Lindahl, U. (1969). *Acta Chem. Scand.,* **23,** 3587–9.
Olivecrona, T., Bengtsson-Olivecrona, G., Osborne Jr, J. C. & Kempner, E. S. (1985). *J. Biol. Chem.,* **260,** 6888–91.
Osborne, Jr, J. C., Bengtsson-Olivecrona, G., Lee, N. & Olivecrona, T. (1985). *Biochemistry,* **24,** 5606–11.
Persson, B., Bengtsson-Olivecrona, G., Enerbäck, S., Olivecrona, T. & Jörnvall, H. (1989). *Eur. J. Biochem.,* **179,** 39–45.
Posner, I. & DeSanctis, J. (1987). *Biochemistry,* **26,** 3711–17.
Posner, I., Wang, C.-S. & McConathy, W. J. (1983). *Arch. Biochem. Biophys.,* **266,** 306–16.
Rapp, D. & Olivecrona, T. (1978). *Eur. J. Biochem.,* **91,** 379–85.
Robinson, D. S. (1963). *J. Lipid Res.,* **4,** 21–3.
Scow, R. O. & Chernick, S. S. (1987). In *Lipoprotein Lipase,* ed. J. Borensztajn. Evener Press, Chicago, USA, pp. 149–56.
Scow, R. O. & Egelrud, T. (1976). *Biochim. Biophys. Acta,* **431,** 538–49.
Semb, H. & Olivecrona, T. (1986). *Biochim. Biophys. Acta,* **878,** 330–7.
Semb, H. & Olivecrona, T. (1989). *J. Biol. Chem.,* **264,** 4195–200.
Senda, M., Oka, K. & Brown, W. V., Qasba, P. K. & Furuichi, Y. (1987). *Proc. Nat. Acad. Sci. USA,* **84,** 4369–73.
Shimizu, M., Uryu, N. & Yamauchi, K. (1981). *Agric. Biol. Chem.,* **45,** 741.
Smith, L. C. & Pownall, H. J. (1984). In *Lipases,* ed. B. Borgström & H. L. Brockman. Elsevier, Amsterdam, The Netherlands, pp. 263–305.
Socorro, L. & Jackson, R. L. (1985). *J. Biol. Chem.,* **260,** 6324–8.
Stahnke, G., Sprengel, R., Augustin, J. & Will, H. (1987). *Differentiation,* **35,** 45–52.
Stein, O., Friedman, G., Chajek-Shaul, T., Halperin, G., Olivecrona, T. & Stein, Y. (1983). *Biochim. Biophys. Acta,* **750,** 306–16.
Sundheim, G. (1988). *J. Dairy Sci.,* **71,** 620–6.
Sundheim, G. & Bengtsson-Olivecrona, G. (1985). *J. Dairy Sci.,* **68,** 589–93.
Sundheim, G. & Bengtsson-Olivecrona, G. (1986). *J. Dairy Sci.,* **69,** 1776–83.
Sundheim, G. & Bengtsson-Olivecrona, G. (1987*a*). *J. Dairy Sci.,* **70,** 499.
Sundheim, G. & Bengtsson-Olivecrona, G. (1987*b*). *J. Dairy Sci.,* **70,** 1815.
Sundheim, G. & Bengtsson-Olivecrona, G. (1987*c*). *J. Dairy Sci.,* **70,** 2040.
Sundheim, G. & Bengtsson-Olivecrona, G. (1987*d*). *J. Dairy Sci.,* **70,** 2467.
Traber, M., Olivecrona, T. & Kayden, H. J. (1985). *J. Clin. Invest.,* **75,** 1729–34.
Vannier, C. & Ailhaud, G. (1989). *J. Biol. Chem.,* **264,** 13206–16.
Verger, R. (1984). In *Lipases,* ed. B. Borgström & H. Brockman. Elsevier, Amsterdam, The Netherlands, pp. 83–150.

Wallinder, L., Bengtsson, G. & Olivecrona, T. (1982). *Biochim. Biophys. Acta,* **711,** 107–13.
Winkler, F. K., D'Arcy, A. & Hunziker, W. (1990). *Nature,* **343,** 771–4.
Wion, K. L., Kirchgessner, A. J., Lusis, A. J., Schotz, M. C. & Lawn, R. M. (1987). *Science,* **235,** 1638–41.
Yamamoto, T., Davis, C. G., Brown, M. S., Schneider, W. J., Casey, M. L., Goldstein, J. L. & Russel, D. W. (1984). *Cell,* **39,** 27–38.
Yang, C. Y., Gu, Z. W., Yang, H. X., Rohde, M. F., Gotto Jr, A. M. & Pownall, H. J. (1989). *J. Biol. Chem.,* **264,** 16822–7.

III. PROTEINASE

A number of indigenous proteinases occur in bovine milk (Reimerdes, 1983). The most thoroughly studied is milk alkaline proteinase (MAP), the properties and significance of which have been reviewed (Humbert & Alais, 1979; Fox, 1981; Visser, 1981; Grufferty & Fox, 1988*a*). Proteolytic activity has also been reported in human (see Hernell & Lonnerdal, 1989), buffalo (Mathur & Sharma, 1979), goat (Varshney & Mathur, 1979) and equine (Visser, 1981) milks.

In fact, MAP is plasmin (EC 3.4.21.7) from blood (Kaminogawa *et al.*, 1972). Most (85–90%) of the plasmin in bovine milk exists as its precursor, plasminogen, which can be activated by urokinase (Kaminogawa *et al.*, 1972; Eigel *et al.*, 1979; De Rham & Andrews, 1982*a*; Korycka-Dahl *et al.*, 1983; Richardson, 1983*a*).

MAP is associated with casein in milk and is concentrated in rennet casein (Richardson & Elston, 1984) or casein micelles prepared by centrifugation (Reimerdes & Klostermeyer, 1974). Plasminogen, also, is associated mainly with the casein (De Rham & Andrews, 1982*a*; Korycka-Dahl *et al.*, 1983; Richardson, 1983*a*) and both can be dissociated from the micelles by making milk to 0·1 M with lysine or 6-aminohexanoic acid, suggesting that they associate with casein via lysine residues. MAP may also be dissociated from the micelles by ethanol (Warner & Polis, 1945), 1 M NaCl (Grufferty & Fox, 1988*d*) or by acidification to pH 4·0 (McMeekin *et al.*, 1959). According to Richardson and Elston (1984), MAP dissociates from the micelles at pH values $<$ pH 5·7; however, Grufferty and Fox (1988*d*) found no dissociation of MAP at pH values $\geqslant$4·8 on holding for up to 4 h; it dissociated completely at pH values $<$pH 4·6.

The milk fat globule membrane (MFGM) possesses proteolytic activity comparable to bovine casein micelles on an equivalent protein basis (Hofmann *et al.*, 1979); various criteria indicate that the principal proteinase in MFGM is plasmin.

Isolation of MAP

MAP has been purified to varying degrees (up to 17 000-fold) by acid extraction and $(NH_4)_2SO_4$ precipitation (Warner & Polis, 1945; Zittle, 1965; Yamauchi *et al.*, 1969), and chromatography on one or more of the following: DEAE cellulose, CM cellulose, DEAE Sepharose 6B, CM Sephadex, lysine–Sepharose 4B, Sephadex G-100 or G-200 (Kaminogawa *et al.*, 1971; Dulley, 1972; Reimerdes & Klostermeyer, 1974; Halpaap *et al.*, 1977; Snoeren & van Riel, 1979).

MAP is a serine proteinase with trypsin-like activity; it is inhibited by DFP, Cu, Zn, Hg, soya bean trypsin inhibitor and tosyllysine chloromethylketone, partly by Mg and Ca, but not by EDTA, α,α'-dipyridyl, ρ-CMB or iodoacetate (Kaminogawa *et al.*, 1969, 1971, 1972; Chen & Ledford, 1971; Andrews, 1983*a*).

Kaminogawa *et al.* (1971, 1972) reported that the mol. wt of MAP, determined by gel filtration, is 100 000; however, based on the amino-acid sequence of plasmin, its mol. wt is 83 200 or 81 000 (Castellino & Powell, 1981). MAP is maximally active at ~pH 7·5 and 37 °C (Warner & Polis, 1945; Chen & Ledford, 1971; Kaminogawa *et al.*, 1972; Dulley, 1972). It is stable over the pH range 1·5–9·0 (Chen & Ledford, 1971) or 4–9 (Kaminogawa *et al.*, 1972). It heat stability is discussed later.

Activation of plasminogen in blood and milk

The fibrinolytic system in blood consists of five main components: plasminogen, plasmin, plasminogen activators, plasmin inhibitors and inhibitors of plasminogen activators. The components of this system and the mechanism of plasminogen activation were reviewed by Grufferty and Fox (1988*a*).

Plasminogen activators and inhibitors in bovine milk

De Rham and Andrews (1982*a*) reported that the level of plasmin in freshly drawn milk is negligible, but increases owing to plasminogen activation during storage. Activation of plasminogen in milk and casein solutions was confirmed by Richardson (1983*a*). These results suggest that the entire fibrinolytic system is present in milk.

Plasminogen activators in bovine milk are associated with the casein micelles, while plasmin inhibitors occur only in the serum (Reimerdes *et al.*, 1976; Korycka-Dahl *et al.*, 1983). Richardson (1983*a*) estimated that the plasminogen-activator activity in milk is equivalent to 0·5 Ploug units of urokinase/ml. The activator retained full activity following heating for 10 min at 70°C, but lost ~50% of its activity in 10 min at 80°C.

Honkanen-Buzalski and Sandholm (1981), using gel filtration, showed that half the trypsin inhibitor activity of colostrum was due to a fraction of mol. wt = 13 000, the remainder being due to species with a mol. wt ~65 000. Gel-fitration profiles of blood plasma and mastitic milk on Sephacryl S-300 were identical, suggesting that the trypsin inhibitors in mastitic milk are derived from blood. The major fraction had a mol. wt of ~67 000, with a minor one of mol. wt ~700 000, which probably correspond to α_2-antiplasmin (the principal plasmin inhibitor in blood) and α_2-macroglobulin, respectively.

The plasmin activity in milk is increased 30–40% by pasteurization, possibly due to inactivation of a plasmin inhibitor (Noomen, 1975). However, this seems unlikely, since the heat stabilities of plasmin, plasminogen and trypsin inhibitor in milk are similar (Richardson, 1983*a*). Plasminogen activation in milk is increased by pasteurization (De Rham & Andrews, 1982*a*; Richardson, 1983*a*), suggesting that the increased plasmin activity in pasteurized milk is due to inactivation of an inhibitor of plasminogen activator, rather than of a plasmin inhibitor (Richardson, 1983*a*).

Proteolysis of milk proteins by MAP

β-Casein is hydrolyzed more quickly than α_{s1}-casein in milk, Na-caseinate or isolated form by indigenous MAP or added plasmin (Chen & Ledford, 1971; Kaminogawa *et al.*, 1972; Noomen, 1975; Eigel, 1977*a,b*; De Rham & Andrews, 1982*a*). It is hydrolyzed by MAP or plasmin to γ-caseins (Kaminogawa & Yamauchi, 1972*a*, 1974; Eigel, 1977*a*; Snoeren & van Riel, 1979) and certain proteose peptones (PP) (Andrews, 1978*a,b*; Eigel & Keenan, 1979; Andrews & Alichanidis, 1983). The γ_1-, γ_2- and γ_3-caseins correspond to residues 29–209, 106–209 and 108–209 of β-casein, respectively (Groves *et al.*, 1972, 1973). PP5 consists of two peptides corresponding to β-casein f1–105 and HO7, while PP8-fast represents β-casein f1–28 (Andrews, 1978*a,b*). Eigel and Keenan (1979) suggested that PP8-slow is β-casein f29–105, but band 12 of the PP fraction corresponds to this sequence (Andrews & Alichanidis, 1983).

The PP fraction of milk contains at least 38 components, 25 of which could be attributed to proteolysis of the caseins by MAP (Andrews & Alichandis, 1983); 52% of the total PP was derived from β-casein, 29% from α_{s1}-casein, 9% from α_{s2}-casein and 4% from κ-casein; the remaining 6% were attributed to the action of other indigenous proteinases. However, Le Bars and Gripon (1989) reported that most of the pH 4·6-soluble N produced by plasmin from whole casein is derived from α_{s2}-casein. Schaar (1985) showed a statistically significant relationship between plasmin activity and the PP content in the milks of 81 cows; on average, only 23% of this fraction arose from post-secretory plasmin activity during storage at 5°C for 72 h, the remainder being produced within the secretory cells. Donnelly and Barry (1983) made a similar observation in regard to γ-casein.

Plasmin produces several peptides from α_{s1}-casein (Eigel, 1977*b*; Andrews & Alichanidis, 1983), some of which appear to be λ-casein (Aimutis & Eigel, 1982). α_{s2}-Casein is rapidly hydrolyzed by plasmin and MAP to several peptides (Snoeren & van Riel, 1979; Andrews & Alichanidis, 1983); eight of the bonds hydrolyzed have been identified (Le Bars & Gripon, 1989). Chen and Ledford (1971) reported that κ-casein is resistant to proteolysis by MAP, but Andrews and Alichanidis (1983) found it to be hydrolyzed quite rapidly by plasmin.

β-Lactoglobulin (β-lg) and α-lactalbumin (α-la) are not hydrolyzed by MAP (Chen & Ledford, 1971; Yamauchi & Kaminogawa, 1972); in fact, β-lg inhibits proteolysis of casein by MAP. Snoeren *et al.* (1980) and Grufferty and Fox (1986) reported that inhibition is due primarily to denatured β-lg, probably via thiol-disulphide interchange between the activated sulphydryl group of denatured β-lg and the disulphide bonds of MAP.

Factors affecting plasmin activity in milk

Stage of lactation

The concentrations of plasmin and plasminogen in milk increase in late lactation (Richardson, 1983*b*; Schaar, 1985). Korycka-Dahl *et al.* (1983)

concur with respect to plasminogen, but found that the concentration of plasmin did not change significantly. The elevated level of plasmin/plasminogen in late lactation milks is reflected in reduced levels of β-casein with concomitant increases in γ-casein, which may represent 30% of β-casein (Davies & Law, 1977; Barry & Donnelly, 1980; Donnelly & Barry, 1983).

Breed and age of cow

The γ-casein content of Friesian milks (Barry & Donnelly, 1980) is more than twice that of Ayrshire milk (Davies & Law, 1977), indicative of higher MAP activity in the former. Milk from Jersey cows has lower plasmin activity than that from Friesians (Richardson, 1983*b*; Schaar, 1985) or Swedish Red and White cows (Schaar, 1985).

The increase in γ-casein, with the concomitant decrease in β-casein, in late lactation is especially marked in milks from older cows (Davies & Law, 1977), and plasmin activity in milk increases with lactation number (Schaar, 1985).

Mastitis

The levels of plasmin and plasminogen in milk increase during mastitic infection (Barry & Donnelly, 1981; De Rham & Andrews, 1982*b*; Andrews, 1983*b*; Schaar & Funke, 1986), and there is a partial transfer of both from the casein micelles to whey (Schaar & Funke, 1986). Barry and Donnelly (1981) suggest that most of the proteolytic activity in mastitic milk is due to plasmin, with proteinases from leucocytes representing only a small proportion. However, De Rham and Andrews (1982*b*) reported that only one-third of the proteolytic activity in mastitic milk was due to plasmin, another third was associated with the casein micelles but was not inhibited by SBTI (soybean trypsin inhibitor), while the final third was in the serum and was inhibited by SBTI. Andrews (1983*b*) showed the presence of several proteinases with broad pH and temperature optima and no clear-cut inhibitory pattern; freezing and thawing, which disrupt somatic cells, increased proteolysis, suggesting that some of the proteinases in mastitic milk are from somatic cells.

Leucocytes contain plasminogen activators which they can secrete in inflamed cells (Christman *et al.*, 1977), and which may be partly responsible for the increased plasmin activity in mastitic and late-lactation milks.

Low-temperature storage

Reimerdes and Herlitz (1979) reported that the formation of γ-casein was greater in milk stored at 4°C than at 10 or 26°C, and attributed this to dissociation of both β-casein and MAP from the micelles. However, Donnelly and Barry (1983), using ^{14}C-labelled β-casein, showed that the conversion of β-casein to γ-casein in milk was 20 times faster at 37°C than at 4°C; and Grufferty and Fox (1988*d*) found that MAP activity in casein micelles prepared from milks stored at 4 or 20°C for 16 h was the same.

Heat stability of MAP

Isolated MAP at pH 7·0 was completely inactivated in 10 min at 80°C (Kaminogawa *et al.*, 1972). The heat stability of isolated MAP (Dulley, 1972), and of MAP in casein systems (Grufferty & Fox, 1988*c*), decreased with increasing pH in the range 3·5–9·2.

While pasteurization of milk at 72°C for 15 s reduces MAP activity by 10% (Korycka-Dahl *et al.*, 1983) or 17% (Richardson, 1983*a*), it leads to increased activity during subsequent storage (Noomen, 1975; De Rham & Andrews, 1982*a*; Andrews, 1983*a*; Andrews & Alichanidis, 1983; Korycka-Dahl *et al.*, 1983; Richardson, 1983*a*), probably owing to the inactivation of inhibitors of plasminogen activator (Richardson, 1983*a*).

A heat treatment of 142°C for 18 s was necessary to completely prevent proteolysis in UHT milk during storage, while heating at 120°C for 15 min was necessary for in-bottle sterilized milk (Driessen & van der Waals, 1978). The heat stability of porcine plasmin in phosphate buffer (pH 7·0) was increased when casein was added at 25 mg/ml, but in skim milk, with a similar casein concentration, it had the same heat stability as in buffer (Alichanidis *et al.*, 1986). The heat stability of MAP was lower in milk than in a micellar casein dispersion (Grufferty & Fox, 1988*c*; Rollema & Poll, 1986). When β-lg was added to porcine plasmin in phosphate buffer (Alichanidis *et al.*, 1986), or to a micellar casein dispersion (Grufferty & Fox, 1988*c*; Rollema & Poll, 1986), the heat stability of the enzyme was reduced, suggesting that the active sulphydryl group of denatured β-lg may interact with and inactivate MAP via thiol-disulphide interchange.

The heat stability of MAP was higher in a micellar than in a non-micellar casein system, and the heat stability of plasmin in buffer was increased by lysine, suggesting that the binding of MAP to casein via lysine residues increases its heat stability (Grufferty & Fox, 1988*c*).

Significance of MAP activity in dairy products

Several physicochemical changes occur in milk, cheese and casein solutions as a result of proteolysis by MAP.

Viscosity

A decrease in the viscosity of UHT-sterilized custard during storing coincided with casein hydrolysis by MAP (Driessen *et al.*, 1981). Although most of the MAP is dissociated from casein on acidification to pH 4·6, commercial acid caseins do contain plasmin which reduces the viscosity of caseinates prepared from them (Warner & Polis, 1945; Richardson, 1982; Grufferty & Fox, 1988*b*). Na-caseinates prepared in late lactation had low viscosity (Southward & Dolby, 1968), presumably due to elevated MAP activity. Most of the MAP in milk is present in rennet casein (Richardson & Elston, 1984), and may cause proteolysis and textural changes in cheese analogues during storage.

Non-casein nitrogen
MAP forms pH 4·6-soluble N in Na-caseinate (Warner & Polis, 1945) or micellar casein dispersion (Grufferty & Fox, 1988*b*) during storage. The N content of rennet whey increased following storage of milk, presumably due to MAP; the increase was even larger when the milk was pasteurized before storage (Noomen, 1975). The PP fraction, which is soluble at pH 4·6, is probably mainly responsible for these increases.

Ethanol stability
Even extensive proteolysis of a micellar casein dispersion by MAP did not affect its ethanol stability (Grufferty & Fox, 1988*b*). Therefore, proteolysis by MAP in late-lactation milk does not appear to be responsible for its low ethanol stability, which is probably due mainly to a high concentration of calcium (Donnelly & Horne, 1986).

Heat stability
Although the heat stability of milk shows considerable seasonal variation (see Fox, 1982; Van Boekel *et al.*, 1989), it appears that these variations are due mainly to changes in the concentrations of urea and soluble calcium; MAP activity is probably of little consequence.

Cheesemaking characteristics
The cheesemaking properties of milk (rennet coagulation time (RCT), curd strength and curd syneresis) deteriorate with advancing lactation, and the resulting cheese has a high moisture content (White & Davies, 1958; O'Keeffe *et al.*, 1982; O'Keeffe, 1984). It has been suggested that these changes may be due to proteolysis by MAP, but Pearse *et al.* (1986) and Grufferty and Fox (1988*b*) showed that the RCT of micellar casein systems increased only after extensive proteolysis by MAP or plasmin.

Late-lactation milks with poor curd-forming properties had higher contents of γ-casein and other casein fragments, and lower levels of α_{s1}-, β- and κ-caseins, than milks with good curd-forming properties (Okigbo *et al.*, 1985). However, the rennet coagulability of the former could be improved by acidification or addition of $CaCl_2$, suggesting that high MAP activity is not the only factor responsible for the poor curd-forming properties of late-lactation milk.

Donnelly *et al.* (1984) reported that the moisture content of Cheddar cheeses was directly related to the extent of pre-hydrolysis of cheesemilk by trypsin (which has similar specificity to plasmin), and concluded that proteolysis of caseins by MAP contributes significantly to the high moisture levels in cheese made from late-lactation milk. However, according to Pearse *et al.* (1986), normal syneresis occurred in a model milk in which ~50% of the casein had been hydrolyzed by plasmin.

Cheese ripening

Swiss and Cheddar cheeses contain 6–13 and 3–4·6 μg plasmin/g, respectively, and there are higher levels of γ-casein in 1-month-old Swiss cheese than in 3-month-old Cheddar, indicating that MAP is very active in Swiss cheese during ripening (Richardson & Pearce, 1981). β-Casein is hydrolyzed to γ-caseins more quickly in Gouda than in Cheddar, possibly because its higher pH and moisture content favour MAP activity (Creamer, 1975). High cooking temperatures appear to increase the plasmin activity in cheese (Farkye & Fox, 1990). However, an extensive study (see Visser, 1977) indicated that MAP contributes less to the formation of soluble N in Gouda than rennet or starter proteinases, and did not contribute to bitter-flavour development. Stadhouders (1960) considered that MAP plays no role in the ripening of Edam, but Reiter *et al.* (1969) hold the opposite view. Proteolysis by MAP in Romano-type cheese was indicated by gel electrophoresis (Guinee & Fox, 1984).

In a simulated Meshanger-type cheese, MAP activity was dependent on pH, NaCl concentration and ripening temperature, and was normally slight in this cheese (Noomen, 1978).

Age gelation of UHT milks

This subject has been reviewed by Harwalkar (1982). While Nakai *et al.* (1964) and Cheng and Gelda (1974) failed to demonstrate a proteinase in UHT milks or creams, Snoeren *et al.* (1980) isolated MAP from UHT milk by affinity chromatography. Several studies (e.g. Bjorck, 1973; Corradini, 1975; Snoeren *et al.*, 1979) have shown that proteolysis does occur in UHT milk during storage, and that β-casein is hydrolyzed more quickly than α_{s1}-casein with the formation of γ-caseins, suggesting MAP activity.

When inhibitors of MAP were added to UHT milk, no proteolysis or gelation occurred after 9 months at 20°C, although extensive proteolysis occurred in the control which gelled after approx. 60 days (De Koning *et al.*, 1985). However, in concentrated UHT milk in which no plasmin or plasminogen was detected, gelation occurred after 41 days at 30°C. It was concluded that gelation of unconcentrated UHT-sterilized milks is due to proteolysis, but that concentrated UHT milks gelled via a non-enzymic, physicochemical mechanism. Manji *et al.* (1986) found that milk sterilized by a direct UHT process gelled during storage, whereas indirectly heated UHT milk did not; although no correlation was observed between the shelf-life of the directly sterilized milk and the extent of proteolysis, this milk contained higher levels of plasmin and plasminogen than indirectly sterilized milk. Plasminogen was activated during storage and was probably mainly responsible for proteolysis. A later study (Manji & Kakuda, 1988) showed that UHT-treated concentrated milk gelled via a physicochemical mechanism, supporting the hypothesis of De Koning *et al.* (1985).

Other indigenous proteinases in milk

Slight differences in the breakdown pattern of caseins in milk, with and without added plasmin, during storage at 37°C, the failure of SBTI to totally inhibit proteolysis in milk (Andrews, 1983*a*) and only 25 of the 38 peptides in the PP fraction being attributable to the action of MAP on individual caseins (Andrews & Alichanidis, 1983), all suggest that other proteinases besides MAP occur in milk.

Acid milk proteinase

Kaminogawa and Yamauchi (1972*b*) isolated an acid proteinase from milk with a yield of 10% and a 2100-fold increase in specific activity. The enzyme had a pH optimum of 4·0 and a mol. wt ~36 000; it was not inhibited by EDTA, α,α'-dipyridyl, iodoacetic acid or DFP, and only ~10% by ρ-CMB. Its properties were similar to those of cathepsin D. α_{s1}-Casein was more susceptible than β-casein, and its specificity on individual caseins was generally similar to that of chymosin (Kaminogawa *et al.*, 1980). The acid proteinase is associated with the casein in milk and may be active during cheese ripening.

Thrombin

Reimerdes (1983) described the isolation of two serine proteinases from milk by affinity chromatography on lysine–Sepharose: one enzyme was similar to plasmin, while the properties of the other suggested that it might be thrombin.

Aminopeptidases

Using 4-nitroanilides of free and substituted amino acids, Reimerdes (1983) reported that there are aminopeptidases in milk serum specific for the 4-nitroanilides of alanine, leucine and lysine. Milk and blood sera showed similar peptidase profiles, suggesting that the aminopeptidases in milk originate from blood.

Proteinases from leucocytes in milk

At the levels found in milk, MAP was 2–8 times more proteolytic than leucocytes isolated from blood and added to caseinate at a level of 10^6/ml (Grieve & Kitchen, 1985). Pasteurization (65°C for 30 min) of milk completely inactivated the leucocyte enzyme(s). Homogenization caused cell disruption and should release intracellular enzymes; however, no degradation of casein was observed during storage of this milk at 5°C for 12 days. Gel electrophoresis showed that the leucocytes from blood hydrolyzed the individual caseins in the order α_{s1}- > β- > κ-casein, in contrast to MAP which hydrolyzed β- > α_{s1}- > κ-casein. Verdi *et al.* (1987) reported that the proteinase activity associated with elevated somatic-cell counts increased the breakdown of both α_{s1}- and β-caseins during storage.

REFERENCES

Aimutis, W. R. & Eigel, W. N. (1982). *J. Dairy Sci.*, **65**, 175–81.

Alichanidis, E., Wrathall, J. H. M. & Andrews, A. T. (1986). *J. Dairy Res.*, **53**, 259–69.

Andrews, A. T. (1978*a*). *Eur. J. Biochem.*, **90**, 59–65.

Andrews, A. T. (1978*b*). *Eur. J. Biochem.*, **90**, 67–71.

Andrews, A. T. (1983*a*). *J. Dairy Res.*, **50**, 45–55.

Andrews, A. T. (1983*b*). *J. Dairy Res.*, **50**, 57–66.

Andrews, A. T. & Alichanidis, E. (1983). *J. Dairy Res.*, **50**, 275–90.

Barry, J. G. & Donnelly, W. J. (1980). *J. Dairy Res.*, **47**, 71–82.

Barry, J. G. & Donnelly, W. J. (1981). *J. Dairy Res.*, **48**, 437–46.

Bjorck, L. (1973). *Milchwissenschaft*, **28**, 291–3.

Castellino, F. J. & Powell, J. R. (1981). *Meth. Enzymol.*, **80**, 365–78.

Chen, J. H. & Ledford, R. A. (1971). *J. Dairy Sci.*, **54**, 763 (Abstract).

Cheng, W. S. & Gelda, C. S. (1974). *J. Dairy Sci.*, **57**, 1502–4.

Christman, J. K., Silverstein, S. C. & Acs, G. (1977). In *Proteinases in Mammalian Cells and Tissues*, ed. A. J. Barrett. North Holland Publishing Company, New York, USA, pp. 91–149.

Corradini, C. (1975). *Milchwissenschaft*, **30**, 413–16.

Creamer, L. K. (1975). *J. Dairy Sci.*, **58**, 287–92.

Davies, D. T. & Law, A. J. R. (1977). *J. Dairy Res.*, **44**, 447–54.

De Koning, P. J., Kaper, J., Rollema, H. S. & Driessen, F. M. (1985). *Neth. Milk Dairy J.*, **39**, 71–87.

De Rham, O. & Andrews, A. T. (1982*a*). *J. Dairy Res.*, **49**, 577–85.

De Rham, O. & Andrews, A. T. (1982*b*). *J. Dairy Res.*, **49**, 587–96.

Donnelly, W. J. & Barry, J. G. (1983). *J. Dairy Res.*, **50**, 433–41.

Donnelly, W. J. & Horne, D. S. (1986). *J. Dairy Res.*, **53**, 23–33.

Donnelly, W. J., Barry, J. G. & Buchheim, W. (1984). *Ir. J. Food Sci. Technol.*, **8**, 121–30.

Driessen, F. M. & van der Waals, C. B. (1978). *Neth. Milk Dairy J.*, **32**, 245–54.

Driessen, F. M., van Hooydonk, A., Streuper, A. & Kingma, F. (1981). *Neth. Milk Dairy J.*, **35**, 121–31.

Dulley, J. R. (1972). *J. Dairy Res.*, **39**, 1–9.

Eigel, W. N. (1977*a*). *Int. J. Biochem.*, **8**, 187–92.

Eigel, W. N. (1977*b*). *J. Dairy Sci.*, **60**, 1399–403.

Eigel, W. N. & Keenan, T. W. (1979). *Int. J. Biochem.*, **10**, 529–35.

Eigel, W. N., Hofmann, C. J., Chibber, B. A. K., Tomich, J. M., Keenan, T. W. & Mertz, E. T. (1979). *Proc. Nat. Acad. Sci. USA*, **76**, 2244–8.

Farkye, N. Y. & Fox, P. F. (1990). *J. Dairy Res.*, **57**, 413–18.

Fox, P. F. (1981). *Neth. Milk Dairy J.*, **35**, 233–53.

Fox, P. F. (1982). In *Developments in Dairy Chemistry—1: Proteins*, ed. P. F. Fox. Applied Science Publishers, London, UK, pp. 189–228.

Grieve, P. A. & Kitchen, B. J. (1985). *J. Dairy Res.*, **52**, 101–12.

Groves, M. L., Gordon, W. G., Kalan, E. B. & Jones, S. B. (1972). *J. Dairy Sci.*, **55**, 1041–9.

Groves, M. L., Gordon, W. G., Kalan, E. B. & Jones, S. B. (1973). *J. Dairy Sci.*, **56**, 558–68.

Grufferty, M. B. & Fox, P. F. (1986). *J. Dairy Res.*, **53**, 601–13.

Grufferty, M. B. & Fox, P. F. (1988*a*). *J. Dairy Res.*, **55**, 609–30.

Grufferty, M. B. & Fox, P. F. (1988*b*). *NZJ Dairy Sci. Technol.*, **23**, 95–108.

Grufferty, M. B. & Fox, P. F. (1988*c*). *NZJ Dairy Sci. Technol.*, **23**, 143–52.

Grufferty, M. B. & Fox, P. F. (1988*d*). *NZJ Dairy Sci. Technol.*, **23**, 153–63.

Guinee, T. P. & Fox, P. F. (1984). *Ir. J. Food Sci. Technol.*, **8,** 105–14.
Halpaap, I., Reimerdes, E. H. & Klostermeyer, H. (1977). *Milchwissenschaft,* **32,** 341–6.
Harwalkar, V. R. (1982). In *Developments in Dairy Chemistry—1: Proteins.* ed. P. F. Fox. Applied Science Publishers, London, UK, pp. 229–69.
Hernell, O. & Lonnerdal, B. (1989). In *Protein and Non-Protein Nitrogen in Human Milk,* ed. S. A. Atkinson & B. Lonnerdal. CRC Press, Boca Raton, FL, USA, pp. 67–75.
Hofmann, C. J., Keenan, T. W. & Eigel, W. N. (1979). *Intern. J. Biochem.*, **10,** 909–17.
Honkanen-Buzalski, T. & Sandholm, M. (1981). *J. Dairy Res.*, **48,** 213–23.
Humbert, G. & Alais, C. (1979). *J. Dairy Res.*, **46,** 559–71.
Kaminogawa, S. & Yamauchi, K. (1972*a*). *Agric. Biol. Chem.*, **36,** 255–60.
Kaminogawa, S. & Yamauchi, K. (1972*b*). *Agric. Biol. Chem.*, **36,** 2351–6.
Kaminogawa, S. & Yamauchi, K. (1974). *Agric. Biol. Chem.*, **38,** 2343–8.
Kaminogawa, S., Yamauchi, K. & Tsugo, T. (1969). *Jap. J. Zootech. Sci.*, **40,** 559–65.
Kaminogawa, S., Sato, F. & Yamauchi, K. (1971). *Agric. Biol. Chem.*, **35,** 1465–7.
Kaminogawa, S., Mizobuchi, H. & Yamauchi, K. (1972). *Agric. Biol. Chem.*, **36,** 2163–7.
Kaminogawa, S., Yamauchi, K., Miyazawa, S. & Koga, Y. (1980). *J. Dairy Sci.*, **63,** 701–4.
Korycka-Dahl, M., Ribadeau Dumas, B., Chene, N. & Martal, J. (1983). *J. Dairy Sci.*, **66,** 704–11.
Le Bars, D. & Gripon, J.-C. (1989). *J. Dairy Res.*, **56,** 551 (Abstract).
Manji, B. & Kakuda, Y. (1988). *J. Dairy Sci.*, **71,** 1455–63.
Manji, B., Kakuda, Y. & Arnott, D. R. (1986). *Indian J. Dairy Sci.*, **69,** 2994–3001.
Mathur, M. P. & Sharma, R. K. (1979). *Indian J. Dairy Sci.*, **32,** 177–81.
McMeekin, T. L., Hipp, N. J. & Groves, M. L. (1959). *Arch. Biochem. Biophys.*, **83,** 35–43.
Nakai, S., Wilson, H. K. & Herreid, E. O. (1964). *J. Dairy Sci.*, **47,** 754–7.
Noomen, A. (1975). *Neth. Milk Dairy J.*, **29,** 153–61.
Noomen, A. (1978). *Neth. Milk Dairy J.*, **32,** 26–48.
O'Keeffe, A. M. (1984). *Ir. J. Food Sci. Technol.*, **8,** 27–37.
O'Keeffe, A. M., Phelan, J. A., Keogh, K. & Kelly, P. (1982). *Ir. J. Food Sci. Technol.*, **6,** 39–47.
Okigbo, L. M., Richardson, G. H., Brown, R. J. & Ernstrom, C. A. (1985). *J. Dairy Sci.*, **68,** 1887–92.
Pearse, M. L., Linklater, P. M., Hall, R. J. & MacKinlay, A. G. (1986). *J. Dairy Res.*, **53,** 477–80.
Reimerdes, E. H. (1983). *J. Dairy Sci.*, **66,** 1591–1600.
Reimerdes, E. H. & Herlitz, E. (1979). *J. Dairy Res.*, **46,** 219–21.
Reimerdes, E. H. & Klostermeyer, H. (1974). *Milchwissenschaft,* **29,** 511–23.
Reimerdes, E. H., Klostermeyer, H. & Sayk, E. (1976). *Milchwissenschaft,* **31,** 329–34.
Reiter, B., Sorokin, Y., Pickering, A. & Hall, A. J. (1969). *J. Dairy Res.*, **36,** 65–76.
Richardson, B. C. (1982). *NZJ Dairy Sci. Technol.*, **17,** 277–82.
Richardson, B. C. (1983*a*). *NZJ Dairy Sci. Technol.*, **18,** 233–45.
Richardson, B. C. (1983*b*). *NZJ Dairy Sci. Technol.*, **18,** 247–52.
Richardson, B. C. & Elston, P. D. (1984). *NZJ Dairy Sci. Technol.* **19,** 63–7.
Richardson, B. C. & Pearce, K. N. (1981). *NZJ Dairy Sci. Technol.*, **16,** 209–20.
Rollema, H. S. & Poll, J. K. (1986). *Milchwissenschaft,* **41,** 536–40.
Schaar, J. (1985). *J. Dairy Res.*, **52,** 369–78.
Schaar, J. & Funke, H. (1986). *J. Dairy Res.*, **53,** 515–28.
Snoeren, T. H. M. & van Riel, J. A. M. (1979). *Milchwissenschaft,* **34,** 528–31.

Snoeren, T. H. M., van der Spek, C. A., Dekker, R. & Both, P. (1979). *Neth. Milk Dairy J.*, **33,** 31–9.
Snoeren, T. H. M., van Riel, J. A. M. & Both, P. (1980). *Zuivelzicht,* **72,** 42–3.
Southward, C. R. & Dolby, R. M. (1968). *J. Dairy Res.*, **35,** 25–30.
Stadhouders, J. (1960). *Neth. Milk Dairy J.*, **14,** 83–110.
Van Boekel, M. A. J. S., Nieuwenhuijse, J. A. & Walstra, P. (1989). *Neth. Milk Dairy J.*, **43,** 97–127.
Varshney, G. C. & Mathur, M. P. (1979). *Indian J. Biochem. Biophys.*, **16,** 375–8.
Verdi, R. J., Barbano, D. M., Dellavalle, M. R. & Senyk, G. F. (1987). *J. Dairy Sci.*, **70,** 230–42.
Visser, F. M. W. (1977). *Neth. Milk Dairy J.*, **31,** 265–76.
Visser, S. (1981). *Neth. Milk Dairy J.*, **35,** 65–88.
Warner, R. C. & Polis, E. (1945). *J. Amer. Chem. Soc.*, **67,** 529–32.
White, J. C. D. & Davies, D. T. (1958). *J. Dairy Res.*, **25,** 267–80.
Yamauchi, K. & Kaminogawa, S. (1972). *Agric. Biol. Chem.*, **36,** 249–54.
Yamauchi, K., Kaminogawa, S. & Tsugo, T. (1969). *Jap. J. Zootech. Sci.*, **40,** 551–8.
Zittle, C. A. (1965). *J. Dairy Sci.*, **48,** 771 (Abstract).

IV. PHOSPHATASES

Alkaline phosphatase (EC 3.1.3.1)

Alkaline phosphatase is without doubt the most important indigenous milk enzyme from a technological and dairy-food-products viewpoint, due to the almost universal adoption of the phosphatase test as an index of the efficiency of HTST pasteurization (Kay & Graham, 1935; Aschaffenburg & Mullen, 1949). This is because alkaline phosphatase is always present in raw milk at easily measured levels, and its heat-stability profile closely follows that required for adequate pasteurization. Since milk is a proven vector for a number of pathogenic bacteria, including *Salmonella* (Anon., 1984), *Campylobacter* (Anon., 1985) and *Listeria* (Hayes *et al.*, 1986), the test is of very great significance to public health as a means of policing the thoroughness of heat treatments or the addition of raw milk to heated or unheated products. The test has proved to be very reliable and useful, although there have been cases where false positive readings, suggesting inadequate heating, have been observed and attributed to a partial reactivation of the enzyme (e.g. Lyster & Aschaffenburg, 1962; Peereboom, 1969). This usually occurs when milk or cream is exposed briefly to high temperatures (about 80–180°C for a few seconds) followed by storage at between 5 and 40°C (the optimum being 30°C), but not after treatments at lower temperatures, such as those typical of pasteurization. Reactivation occurs more frequently in cream than in milk and is promoted by Mg^{2+} and Zn^{2+}, but it never exceeds 1% of the initial activity. However, if the milk is preheated and Mg^{2+}, glycerophosphate or low levels of β-lactoglobulin then added to promote reactivation, up to 30% of the enzyme activity could be restored (Lyster & Aschaffenburg, 1962). When reactivation does occur, the enzyme can become detectable after about 6 h and activity increases progressively, so early post-heating measurement to assess the efficiency of pasteurization is recommended (Linden, 1979).

Alkaline phosphatase occurs in the milks of all mammals, although the levels present may vary considerably, and a variant of the phosphatase test in which sample size is altered to allow for this variation has been applied successfully to caprine milk (Williams, 1986). There are also many variations of the test for application to bovine milk, usually developed on grounds of speed and convenience in particular products or situations. The Scharer Rapid Phosphatase Test is particularly widely used in the US and, when combined with a pasteurization-type heating step, has been used to distinguish alkaline phosphatases of mammalian and bacterial origins, the former being heat-inactivated and the latter not (e.g. Pratt-Lowe *et al.*, 1987). In ultrafiltered milks there is some indication that alkaline-phosphatase activity is influenced by lactose content, being lower at high protein and low lactose levels (Mistry, 1989), so the reliability of the phosphatase test may be questionable under such conditions. β-Lactoglobulin has also been reported to have an inhibitory effect on alkaline-phosphatase activity (Lorient & Linden, 1976), but the effect is not large and probably of no real technological importance. There are also small differences in the inhibitory powers of the genetic variants of β-lactoglobulin (Jasinska *et al.*, 1985).

The level of alkaline phosphatase in milk is rather variable, one early study (Haab & Smith, 1956) showing a 40-fold variation between individual animals and a 2-fold variation in the activity in bulk milks collected during different seasons. In spite of this, there appears to be little correlation between enzyme level and either the breed or the diet of the cow, and some of this variation may be due to lactational effects, since levels which are very high in colostrum fall to a minimum 1–2 weeks after parturition and then rise again, reaching a more or less constant level at about 25 weeks post-partum (Jenness & Patton, 1959; Kiermeier & Meinl, 1961). There is an inverse correlation with milk yield but little consistent relationship to fat content, although this has been suggested frequently. Another source of variation may be the considerable elevation in alkaline-phosphatase activity in the milk of cows with mastitis (Bogin & Ziv, 1973; Singh & Ganguli, 1975).

In milk, alkaline phosphatase is associated with microsomal particles and is particularly enriched in the cream phase, since the microsomes are adsorbed onto the surface of fat globules. About 30–50% is found in the cream phase and the rest in the skim phase. The usual route for extraction is to churn the milk or cream and recover the enzyme from the buttermilk, generally employing a butanol-extraction stage, as the enzyme is stable to organic solvents (Morton, 1953; Linden *et al.*, 1974) and this treatment dissociates the protein–fat complexes. This is usually followed by gel filtration (Sephadex G-200) and ion-exchange chromatography (DEAE–cellulose) for further purification (e.g. Linden *et al.*, 1974; Linden & Alais, 1976; Kuzuya *et al.*, 1980), but affinity chromatography on concanavalin-A–Sepharose-4B has also been used (Kuzuya *et al.*, 1982) to give a final purification to apparent homogeneity. This enzyme was purified almost 7200-fold over that present in the original milk to give a preparation with an activity of 43·1 units/mg of protein. In their work, Linden *et al.* (1974) did not use affinity chromatography, but achieved a similar purification factor (7440) and claimed an enzyme preparation homogeneous by all their criteria with a specific activity of 96 units/mg. Since the assay conditions and definition of units were very similar in both cases, there is a slight discrepancy here that remains to be resolved.

Peereboom (1968) and some other workers have described a number of alkaline-phosphatase isoenzymes, but the majority of workers have found no real evidence for this, and it now seems probable that they merely represent the same enzyme either in different states of aggregation or enzyme bound to membrane fragments or other proteins or lipids. However, there is still conflicting evidence as to whether the enzyme is a glycoprotein or not, since some of the more thorough analytical work on the enzyme (Morton, 1953; Linden & Alais, 1976; Linden *et al.*, 1974) failed to mention the presence of sialic acid or any other carbohydrate, while Sharma and Ganguli (1969) claim that it is a sialoglycoprotein and Peereboom (1968) showed that the electrophoretic mobility of the larger 'isoenzyme' was drastically altered by neuraminidase treatment. This, of course, could be due to the liberation of sialic acid by the neuraminidase from the accompanying bound material rather than from the enzyme itself. The efficacy of affinity chromatography on concanavalin-A–Sepharose as a purification step (Kuzuya *et al.*, 1982) argues strongly in favour of a glycoprotein structure for the enzyme.

As its name implies, alkaline phosphatase is a phosphomonoesterase with optimum activity in the alkaline pH range (9–10·5). As with many phosphomonoesterases, it has a wide range of substrate specificity and will hydrolyse most phosphate ester bonds, releasing inorganic phosphate, including AMP, glycerophosphate, phosphates of glucose, serine and threonine and also phosphoproteins, but it will not cleave ADP, ATP, pyrophosphates or polyphosphates. Phosphoserine is hydrolyzed most rapidly at pH 8·5–9·0, but pH 6·5–7·0 is best for cleaving the phosphate groups off casein (Zittle & Bingham, 1959; Linden & Alais, 1976; Lorient & Linden, 1976). The catalytic mechanism proceeds via a transient phosphoryl-enzyme intermediate involving a reactive serine hydroxyl group, and when *p*-nitrophenylphosphate is used as substrate at pH 10·0 and 25°C, the catalytic-centre activity was found to be 2700 s^{-1} and $k_m = 4{\cdot}7 \times 10^{-4}$ M (Barman & Gutfreund, 1966), the latter being in good agreement with values given by Linden and Alais (1976) and Kuzuya *et al.* (1980). Barman and Gutfreund (1966) also found that inorganic orthophosphate was a strong competitive inhibitor of *p*-nitrophenylphosphate hydrolysis ($K_1 = 0{\cdot}3$ mM), which was a factor of 10 greater than the value (3·0 mM) reported by Kuzuya *et al.* (1980) at a similar pH (9·6) but higher temperature (37°C).

Many of the molecular properties of bovine milk alkaline phosphatase are summarized in Table 2. The enzyme is clearly a dimer containing four or five Zn atoms per molecule. It is both active and stable in solution in a purified state at alkaline pH values. While Zn ions are essential for activity, Mg^{2+} are also strong activators, so it is usual to incorporate 1 mM Mg^{2+} in all buffers for assay. The other metal ions shown as activators in Table 2 are much less effective. Not surprisingly for a metallo-enzyme, metal chelators are powerful inhibitors. The amino-acid composition (Linden *et al.*, 1974) shows high contents of serine, glutamic acid, alanine and lysine, with a low content of aromatic and hydrophobic amino acids.

Milk alkaline phosphatase, as pointed out by Kitchen (1985), has many similarities to the mammary-gland enzyme described by O'Keeffe and Kinsella (1979). Both enzymes are dimers, consisting of monomers having molecular weights of about 80 000, and both have similar K_m values (0·7 mM) for *p*-nitrophenylphosphate. The pH optimum of the mammary enzyme has been reported to vary according to substrate concentration, being close to 10 with 2 mM *p*-nitrophenylphosphate and 9 at 0·1 mM substrate. This effect has not been reported for the milk enzyme, but nevertheless, both enzymes do have optima in the same pH range. The mammary enzyme is a glycoprotein with almost 30% carbohydrate and binds strongly to concanavalin-A-bound to

TABLE 2
Bovine Milk Alkaline Phosphatase

Property	*Value*
Molecular weight (native)	160 000–190 000
Molecular weight (subunit)	85 000
Sedimentation constant	6·0 S
pH optimum	9·8–9·9
Metal-ion content	4·9 ± 0·6 g atoms/mol, Zn^{2+}
Specific activity	43–96 units/mg
pH stability range (5°C, 24 h)	7·0–10·0
(33°C, 24 h)	8·0–9·0
Activators	Mg^{2+}, Mn^{2+}, Co^{2+}, Cd^{2+}, Ni^{2+}
Inhibitors	Ca^{2+}, Be^{2+}, Cu^{2+} Chelators, oxidizing agents, orthophosphate (K_i = 3·0 mM) L-phenylalanine (K_i = 3·7 mM); L-leucine (K_i = 0·2 mM)
k_m (*p*-nitrophenylphosphate, pH 9, 37°C)	0·5–0·7 mM
V_{max} (*p*-nitrophenylphosphate, pH 9, 37°C)	210 μmol/min mg^{-1}
Thermal stability (*D*-value 60°C)	27·2 min
(*D*-value 63°C)	8·3 min

agarose and, although the milk enzyme has not been clearly shown to be a glycoprotein, affinity chromatography on the same immobilized ligand has been used in its purification by Kuzuya *et al.* (1982). It seems most probable, therefore, that the mammary gland itself is the source of the milk enzyme, and that these two enzymes are the same.

Acid phosphatase (EC 3.1.3.2)

This enzyme is also a phosphomonoesterase and, like the alkaline phosphatase, it is distinguished operationally by the pH range over which it is active. Although the levels of acid phosphatase in milk are lower than those of the alkaline enzyme, it may also be of technological importance as far as dairy products are concerned for three reasons. Firstly, it has much greater thermal stability, so that in pasteurized products it is the only active phosphatase; secondly, it is most active at pH values (about 5) typical of many food products; and thirdly, it is very active against phosphoprotein substrates such as the caseins of milk. As substrates for acid phosphatase, the caseins should act as competitive inhibitors of the hydrolysis of *p*-nitrophenylphosphate by the enzyme, and this was found to be so (Andrews, 1974), the strength of their inhibitory activity being directly proportional to their phosphate content. α_{s1}-Casein (8 phosphate groups per molecule) is a more powerful competitive inhibitor than β-casein (5 phosphates/molecule) and κ-casein (1 phosphate/molecule) is least effective. The caseins in milk occur as multimolecular aggregates consisting of mixtures of all four caseins (α_{s1}-, α_{s2}-, β- and κ-caseins) and these aggregates associate further to form micellar particles that may contain up to 5×10^4 or more casein molecules. While both hydrophobic and electrostatic bonds contribute to micelle formation and stability, another very important interaction involved is interchain bridging via phosphate groups on the casein polypeptide chains and calcium ions. The importance of this is readily seen by the dramatic change in the transparency of milk when a chelator such as ethylenediaminetetraacetic acid (EDTA) is added. This chelates the calcium ions and disintegrates the micelles with an accompanying sharp drop in light scattering due to the micellar particles. The resulting caseins still remain in an aggregated form (hydrophobic and electrostatic bonds remain) but as much smaller particles which are much less effective at scattering light. Cleavage of the phosphate groups off the serine residues of the caseins by phosphatase action has very similar consequences by likewise disrupting calcium-phosphate bridging. Thus, phosphatase activity has been shown to reduce the formation of stable casein complexes (Pepper & Thompson, 1963; Bingham *et al.*, 1972, 1976; Knoop & Peters, 1975), and although micellar-type aggregates may be formed, these are both fewer in number and larger when dephosphorylated as compared to the unmodified casein studied, suggesting weaker intermolecular bonding and a 'looser' structure.

Since many of the technologically important properties of milk depend upon the presence of casein micelles and the maintenance of their integrity, it is not surprising that the possible significance of acid phosphatase has been discussed in relation to the gelling of UHT milk (Andrews & Pallavicini, 1973), coagulation of milk by chymosin (Zittle, 1964), curd structure and strength (Ohmiya *et al.*, 1983; Pearce *et al.*, 1986), and in cheese maturation and flavour development (Schormuller *et al.*, 1960; Dulley & Kitchen 1973; Andrews & Alichanidis, 1975*a*). Proteolysis of the caseins during cheese maturation is extensive, and a number of phosphopeptides are among the products formed. Schormuller *et al.* (1960) suggested that these might be resistant to proteolysis and contribute to bitter-flavour defects so that the synergistic activity of proteinases and phosphatases is required for correct flavour development. Although starter organisms also contribute phosphatases, the extent of this depends on starter strain and amounts may be small, so the predominant phosphatase activity in typical hard cheeses can be attributed to milk acid phosphatase. In some soft cheeses this may not be so (Larsen & Parada, 1988). The pH of most cheeses is close to the optimum for milk acid phosphatase which is very stable in a cheese environment, so that dephosphorylation of caseins and phosphopeptides certainly occurs, although possibly only slowly (Dulley & Kitchen, 1973), but the importance of the enzyme to cheese flavour profiles has not been established. Most bitter peptides so far identified in food protein hydrolysates are strongly hydrophobic, so it seems improbable that any large proteinase-resistant phosphopeptides, which are likely to be very hydrophilic, would contribute to bitter flavour.

When first reported in milk (Mullen, 1950), the acid phosphatase was identified as the most heat-stable indigenous milk enzyme, and subsequent work has suggested that plasmin is probably the only more stable milk enzyme identified to date (Alichanidis *et al.*, 1986). Figure 7 shows a comparison of the heat stability of bovine milk acid phosphatase with the stabilities of a number of other indigenous milk enzymes that may have potential value as indicators of various heat treatments given to milk and dairy products (Andrews *et al.*, 1987). The acid phosphatase survives normal pasteurization conditions and heating at pH 6·7 for nearly 5 s at 100°C or 20 s at 90°C is required for 90% inactivation (Andrews, 1974). As much as 15–20% of the initial activity may survive even UHT processing conditions (Andrews & Pallavicini, 1973). The heat stability is rather pH dependent, however, and the enzyme is even more stable than this at acid pH values (e.g. 4·9–5·2) typical of many dairy products. Thermal denaturation followed first-order kinetics, and addition of casein substrates provided no substrate protection (Andrews, 1974). In spite of its considerable thermal stability, Mullen (1950) reported that the enzyme was quite rapidly inactivated by visible (but not UV) light radiation (30 min

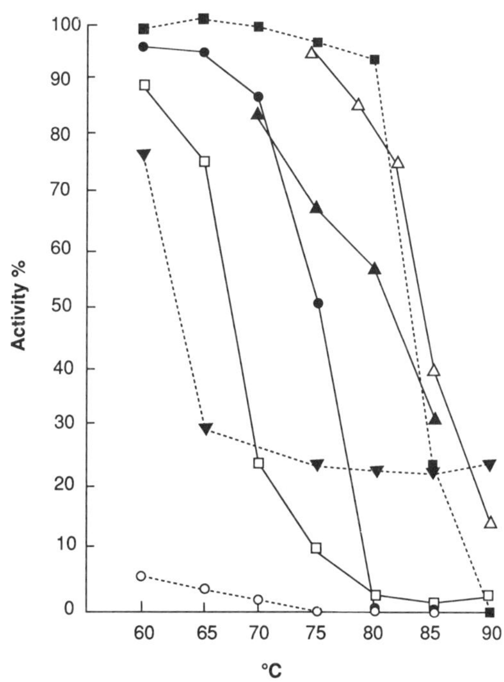

FIG. 7. Heat stability of bovine milk enzymes. Whole milk samples (50–100 μl) sealed in thin-walled glass capillary tubes were heated for 15 s at the indicated temperatures before cooling and assay of residual enzyme activity. Lipoprotein lipase, ○; α-L-fucosidase, ▼; *N*-acetyl-β-glucosaminidase, □; γ-glutamyl transpeptidase, ●; xanthine oxidase, ▲; α-D-mannosidase, ■; acid phosphatase, △. (Reproduced with permission from Journal of Dairy Research).

exposure, 15 cm from an 80 W quartz mercury-vapour lamp, caused up to 80% inactivation).

The level of acid-phosphatase activity in normal bovine milk usually lies in the range 0·0026–0·0037 IU/ml using *p*-nitrophenylphosphate as substrate at pH 4·0 (Kitchen *et al.*, 1970), although values would have been approximately twice this if they had been measured at the pH optimum (4·9). Other workers have found values slightly higher (Bingham & Zittle, 1963) or lower (Andrews & Pallavicini, 1973) than this, but whatever the level it is clear that it is almost two orders of magnitude lower than for alkaline phosphatase (Kitchen *et al.*, 1970). This difference in activity would of course be very much less for phosphoprotein substrates against which the acid enzyme is particularly active. The distribution of acid phosphatase in milk is somewhat similar to that of alkaline phosphatase in that, while most of the enzyme is present in the skim-milk phase, the specific activity is greater (in IU/ml) in the cream phase (Bingham *et al.*, 1961; Kitchen *et al.*, 1970). With acid phosphatase there is clearly a relatively rapid equilibrium since all the enzyme is quite quickly bound to acidic ion-exchange resins when these are added with gentle stirring and then allowed to settle (Bingham & Zittle, 1963; Andrews & Pallavicini, 1973). This, in fact, forms the extremely efficient first step of virtually all purification protocols, and this single step can easily give a 300-fold increase in specific activity (IU/mg protein). Subsequent gel filtration gave a further 30-fold increase in specific activity and, when followed by further chromatography on ion-exchange, cellulose phosphate and affinity (immobilized α_{s1}-casein) columns, a 10^{-6}-fold purification (to 30·2 IU/mg), giving apparently homogeneous enzyme, was obtained (Andrews, 1976*a*). Thus, in spite of the low levels present initially, acid phosphatase is one of the most completely purified and characterized of the milk enzymes.

Bovine milk acid phosphatase is a phosphomonoesterase with a broad specificity and hydrolyses aromatic phosphomonoesters such as *p*-nitrophenylphosphate and phenolphthalein phosphate, pyrophosphates including ADP and ATP, polyphosphates and the phosphoserine residues of phosphoproteins. However, it does not appear to hydrolyse aliphatic phosphomonoesters such as AMP, *β*-glycerophosphate (a substrate of many other phosphatases) or free serine phosphate (Bingham & Zittle, 1963). Thus, its substrate specificity is very different from that of milk alkaline phosphatase. When *p*-nitrophenylphosphate is used as substrate, available evidence is consistent with the rapid formation of an enzyme–substrate complex which is hydrolyzed to release initially *p*-nitrophenol and then the free phosphate moiety (Andrews & Pallavicini, 1973). A histidine residue is close to or part of the active site. This was deduced from the influence of pH on K_m, and from the inhibition of the enzyme by iodoacetate, which reacts with both histidine and thiol groups, but not by other more specific thiol-modifying reagents.

The molecular properties of the enzyme are shown in Table 3, from which it can be seen that it is obviously a single polypeptide chain molecule with a molecular weight of about 42 000, determined by a variety of methods under

TABLE 3
Molecular Properties of Bovine Milk Acid Phosphatase

Molecular weight (native)	42 500
Molecular weight (subunit)	41 500
Sedimentation constant (native), $S_{20,w}$	2·8
Partial specific volume ($\bar{v}$), calculated	0·728
Extinction coefficient (ε) at 280 nm	9·4
Isoelectric point (pI)	7·9
pH optimum	4·9
pH stability range	3·0–9·0
Activators	Ascorbic acid, β-mercaptoethanol, dithiothreitol L-cysteine, thioglycollic acid
Inhibitors	Ag^{+}, Cu^{2+}, Hg^{2+}, Sn^{2+}, Zn^{2+}, Fe^{3+}, Al^{3+}, F^{-}, oxidising agents
K_m (*p*-nitrophenylphosphate, pH 4·8, 37°C)	0·81 mM
(*p*-nitrophenylphosphate, pH 5·2, 37°C)	1·41 mM
K_m (total casein, pH 5·2, 37°C)	2·84 mM

both denaturing and non-denaturing conditions, including gel filtration, sedimentation equilibrium and SDS–PAGE. In general, the enzyme is activated slightly by reducing agents, unaffected by metal chelators, tartrate (an inhibitor of some phosphatases) or thiol blocking reagents and inhibited by oxidizing agents and a range of heavy metal ions. It is also particularly strongly inhibited by F^{-} ($\kappa_i = 0{\cdot}3$ mM), which acts as a non-competitive inhibitor of *p*-nitrophenylphosphate hydrolysis, and by Al^{3+} ($\kappa_i = 0{\cdot}12$ mM), and less strongly by orthophosphate ($\kappa_i = 2{\cdot}2$ mM), sodium pyrophosphate ($\kappa_i = 1{\cdot}1$ mM) and sodium hexametaphosphate ($\kappa_i = 1{\cdot}9$ mM), all of which are competitive inhibitors.

Compositional analysis of the purified enzyme (Andrews, 1976*a*) showed that the enzyme is a glycoprotein containing 2 residues of galactose, 2 of mannose and 4 of *N*-acetylglucosamine per molecule. Amino-acid analysis revealed the absence of methionine and relatively high, but not exceptionally so, levels of lysine and arginine, as would be expected for a protein with a slightly basic isoelectric point (pI). The contents of proline and histidine were also rather low.

While studies on normal bovine milk indicate that a substantial proportion of the acid phosphatase, perhaps almost one third (Bingham *et al.*, 1961; Kitchen *et al.*, 1970), is associated with the cream phase and probably bound to the milk fat globule membrane, no evidence has ever suggested that this enzyme is in any way different to that present in the skim phase. Indeed, many of the properties of skim and cream enzymes are similar (Bingham *et al.*, 1961), and the observations that virtually all the acid-phosphatase activity can be extracted from milk with an ion-exchange resin, and then behaves as a homogeneous enzyme component throughout subsequent purification steps, argues in favour of a weak and readily-reversible interaction between part of the skim milk phase enzyme and the milk fat globule membrane. This contention is supported by the fact that enzyme-activity staining of both extracts and of whole milk itself following PAGE reveals only a single band of acid phosphatase (e.g. Andrews & Pallavicini, 1973).

However, the situation is rather different in the milk of animals with mastitis. Total levels of acid phosphatase activity are elevated to an extent depending upon the stage and severity of the infection, but frequently 5–10-fold, and it appears that this increase is due not to changes in the level of the enzyme present in normal milk but to the appearance of isoenzymes of leucocyte origin (Anderson *et al.*, 1975; Andrews & Alichanidis, 1975*b*). Most of these isoenzymes are bound to particles and cellular components or fragments of different sizes and only a small proportion is extractable with ion-exchange resin treatments. The latter appears to be identical to a blood-serum isoenzyme which may also contribute to levels in milk. The properties of all these isoenzymes differ from the normal milk enzyme, and they are probably of little technological significance because they are almost completely inactivated by typical pasteurization conditions (Andrews, 1976*b*).

REFERENCES

Alichanidis, E., Wrathall, J. H. M. & Andrews, A. T. (1986). *J. Dairy Res.*, **53**, 259–69.

Anderson, M., Brooker, B. E., Andrews, A. T. & Alichanidis, E. (1975). *J. Dairy Res.*, **42**, 401–17.

Andrews, A. T. (1974). *J. Dairy Res.*, **41**, 229–37.

Andrews, A. T. (1976*a*). *Biochim. Biophys. Acta*, **434**, 345–53.

Andrews, A. T. (1976*b*). *J. Dairy Res.*, **43**, 127–31.

Andrews, A. T. & Alichanidis, E. (1975*a*). *J. Dairy Res.*, **42**, 327–39.

Andrews, A. T. & Alichanidis, E. (1975*b*). *J. Dairy Res.*, **42**, 391–400.

Andrews, A. T. & Pallavicini, C. (1973). *Biochim. Biophys. Acta*, **321**, 197–209.

Andrews, A. T., Anderson, M. & Goodenough, P. W. (1987). *J. Dairy Res.*, **54**, 237–46.

Anon. (1984). *Center for Disease Control, Morbid Mortal. Weekly Rep.*, **33**, 505–7.

Anon. (1985). *Center for Disease Control, Morbid Mortal. Weekly Rep.*, **35**, 311–14.

Aschaffenburg, R. & Mullen, J. E. C. (1949). *J. Dairy Res.*, **16**, 58–67.

Barman, T. E. & Gutfreund, H. (1966). *Biochem. J.*, **101**, 460–6.

Bingham, E. A. & Zittle, C. A. (1963). *Arch. Biochem. Biophys.*, **101**, 471–7.

Bingham, E. A., Jasewicz, L. & Zittle, C. A. (1961). *J. Dairy Sci.*, **44**, 1247–56.

Bingham, E. A., Farrell, H. M. & Carroll, R. J. (1972). *Biochem.*, **11**, 2450–4.

Bingham, E. W., Farrell, H. M. & Dahl, K. J. (1976). *Biochim. Biophys. Acta*, **429**, 448–60.

Bogin, E. & Ziv, G. (1973). *Cornell Vet.*, **63**, 666–8.

Dulley, J. R. & Kitchen, B. J. (1973). *Aust. J. Dairy Technol.*, **29**, 114–6.

Haab, W. & Smith, L. M. (1956). *J. Dairy Sci.*, **39**, 1644–50.

Hayes, P. S., Feeley, J. C., Graves, L. M., Ajello, G. W. & Fleming, D. W. (1986). *Appl. Environ. Microbiol.*, **51**, 438–44.

Jasinska, B., Kleczkowski, K. & Michalak, W. (1985). *J. Dairy Sci.*, **68**, 2172–5.

Jenness, R. & Patton, S. (1959). *Principles of Dairy Chemistry*, Wiley & Sons, New York, USA, pp. 189–203.

Kay, H. D. & Graham, W. R. (1935). *J. Dairy Res.*, **6**, 191–203.
Kiermeier, F. & Meinl, E. (1961). *Z. Lebensm. Unters. Forsch.*, **114**, 189–202.
Kitchen, B. J. (1985). In *Developments in Dairy Chemistry* (Vol. 3), ed. P. F. Fox. Elsevier Applied Science, London, UK, pp. 239–79.
Kitchen, B. J., Taylor, G. C. & White, I. C. (1970). *J. Dairy Res.*, **37**, 279–88.
Knoop, A. M. & Peters, R. H. (1975). *Milchwissenschaft*, **30**, 680–5.
Kuzuya, Y., Kanamaru, Y. & Tanahashi, T. (1980). *Jap. J. Zootech. Sci.*, **51**, 501–10.
Kuzuya, Y., Kanamaru, Y. & Tanahashi, T. (1982). *Jap. J. Zootech. Sci.*, **53**, 45–9.
Larsen, R. F. & Parada, J. L. (1988). *Sci. des Aliments*, **8**, 285–94.
Linden, G. (1979). *Milchwissenschaft*, **34**, 329–32.
Linden, G. & Alais, C. (1976). *Biochim. Biophys. Acta*, **429**, 205–13.
Linden, G., Mazeron, P., Michalowski, J. B. & Alais, C. (1974). *Biochim. Biophys. Acta*, **358**, 82–90.
Lorient, D. & Linden, G. (1976). *J. Dairy Res.*, **43**, 19–26.
Lyster, R. L. J. & Aschaffenburg, R. (1962). *J. Dairy Res.*, **29**, 21–35.
Mistry, V. V. (1989). *J. Dairy Sci.*, **72**, 1112–7.
Morton, R. K. (1953). *Biochem. J.*, **55**, 795–800.
Mullen, J. E. C. (1950). *J. Dairy Res.*, **17**, 288.
Ohmiya, K., Sugano, S., Yun, S-E. & Shimizu, S. (1983). *Agric. Biol. Chem.*, **47**, 535–42.
O'Keeffe, R. B. & Kinsella, J. E. (1979). *Int. J. Biochem.*, **10**, 125–31.
Pearce, M. J., Linklater, P. M., Hall, R. J. & Mackinlay, A. G. (1986). *J. Dairy Res.*, **53**, 477–80.
Peereboom, J. W. C. (1968). *Neth. Milk Dairy J.*, **22**, 137–52.
Peereboom, J. W. C. (1969). *Milchwissenschaft*, **24**, 266–9.
Pepper, L. & Thompson, M. P. (1963). *J. Dairy Sci.*, **46**, 764–7.
Pratt-Lowe, E. L., Richardson, T. & Geiger, R. M. (1987). *J. Dairy Sci.*, **70**, 1804–6.
Schormuller, J., Lehmann, K. & Belitz, H. D. (1960). *Z. Lebensm. Unters. Forsch.*, **111**, 180–90.
Sharma, R. S. & Ganguli, N. C. (1969). *Indian J. Dairy Sci.*, **22**, 272–6.
Singh, L. N. & Ganguli, N. C. (1975). *Indian J. Dairy Sci.*, **28**, 67–8.
Williams, D. J. (1986). *Aust. J. Dairy Technol.* **41**, 28–30.
Zittle, C. A. (1964). *J. Dairy Sci.*, **47**, 202–5.
Zittle, C. A. & Bingham, E. A. (1959). *J. Dairy Sci.*, **42**, 1772–80.

V. LACTOPEROXIDASE

Introduction

Peroxidases (EC 1.11.1.7) are a group of enzymes that are widely distributed in nature. They occur both in plants and animals. The best characterized peroxidase is derived from horseradish. The basic chemical reaction catalysed by peroxidases can be written as:

$$AH_2 + H_2O_2 \rightarrow A + 2H_2O$$

where AH_2 and A are reduced and oxidized forms of a suitable electron donor, respectively. A wide variety of organic and inorganic compounds can serve as electron donors, but substrate specificity varies between different types of peroxidase.

Various mammalian tissues contain different peroxidases, i.e. glutathione peroxidase, thyroid peroxidase, eosinophil peroxidase, myeloperoxidase and lactoperoxidase. Peroxidase activity in bovine milk was observed as long ago as the beginning of this century by Hansen (1924), who also was able to demonstrate a correlation between the antibacterial activity of bovine milk and peroxidase activity. The following year, Thurlow (1925) was able to separate milk peroxidase from oxidase activity by ammonium-sulphate fractionation. However, it was not until 20 years later that lactoperoxidase was obtained in pure, crystalline form (Theorell & Åkesson, 1943) and more detailed chemical characterizations could be made. About the same time, Agner (1941) isolated a peroxidase from empyema from patients suffering from pulmonary tuberculosis. This enzyme was initially called verdoperoxidase, but was later renamed myeloperoxidase when it was discovered that lactoperoxidase was also greenish.

Distribution of lactoperoxidase

Today, lactoperoxidase has been demonstrated to be present in milks from many species, e.g. bovine, porcine, caprine, murine and human. Bovine milk is rich in lactoperoxidase, i.e. $\approx 30\ \mu g/ml$ (Polis & Schmukler, 1953). More recent investigations are difficult to compare, since different assay methods have been used. The highest lactoperoxidase activity found is in guinea-pig milk, which contains 10–15 times more than bovine milk (Stephens *et al.*, 1979).

The type of peroxidase present in human milk has been disputed. The peroxidase activity of human milk was first analysed by Gothefors and Marklund (1975), who found the level of lactoperoxidase to be about 5% of that normally present in bovine milk. More recently, Moldveanu *et al.* (1982) concluded that human milk peroxidase is derived from milk leucocytes. This has in turn been refuted by the findings of Langbakk and Flatmark (1984), who isolated a peroxidase from human colostrum with chromatographic and immunoreactive properties similar to those of bovine lactoperoxidase. These results have been confirmed by Langbakk and Flatmark (1989), who reported that human colostrum contains lactoperoxidase at a level corresponding to 0·004% of the total protein content in crude colostrum.

Lactoperoxidase has also been shown to be present in other secretions. Morrison and Allen (1963), by use of immunochemical methods, demonstrated the presence of lactoperoxidase in the salivary gland. Studies on other human secretions have subsequently demonstrated the presence of lactoperoxidase in several mucus glands/secretions, i.e. in cervical mucus (Shindler *et al.*, 1976), nasal glands (Wanatabe, 1978) and tears (Haeringen *et al.*, 1979). Lactoperoxidase is also present in the salivary glands of pigs (Morrison & Steele, 1968), monkeys (Mäkinen *et al.*, 1978), rats (Mäkinen *et al.*, 1980), guinea pigs and hamsters (Kumlien, 1972).

The nomenclature of these peroxidases is somewhat confused today. Although the peroxidases present in saliva and uterine mucus are very similar to bovine lactoperoxidase, it has been suggested that, until further chemical characterizations have been made, they should be referred to as salivary and uterine peroxidase, respectively (Pruitt & Tenovuo, 1985).

Chemical structure of lactoperoxidase

Lactoperoxidase is a glycoprotein with a molecular weight of approximately 78 000 Da and containing one haem group (Theorell & Pedersen, 1944). The iron content has been found to be 0·0680–0·0709% and the carbohydrate content 9·9–10·2% (Carlström, 1969*a*). The amino acid composition is given in Table 4.

The identity and type of binding of the haem moiety in lactoperoxidase has been discussed. The inability of acidic acetone to extract the haem was taken as evidence for a covalent haem–protein bond, which was identified tentatively as an ester or amide bond (Hultqvist & Morrison, 1963; Morell & Clezy, 1963). More recent investigations have demonstrated that no covalent haem–protein bond exists and that the prosthetic group is protohaem IX (Sievers, 1979).

Dumonet and Rousset (1983) reported having isolated a peak by ion-exchange chromatography which they suggested was a non-haem lactoperoxidase present in bovine milk. This molecule was said to have the same apparent molecular weight but, unlike lactoperoxidase, it had no absorbance at 412 nm and no enzymic activity; it showed partial immunochemical identity with lactoperoxidase. Ekstrand *et al.*, 1984 (unpublished) were not able to confirm these results, but found similarities between this compound and lactoferrin. Whether an apo-form of lactoperoxidase is present remains to be proven by further investigations.

Several different fractions of lactoperoxidase have been demonstrated in bovine milk. Polis and Schmukler (1953) separated milk lactoperoxidase into fractions A and B, fraction A being the more acidic. A total of ten different fractions of lactoperoxidase, designated A 1, A 2/I, A 2/II, A 3, B 1, B 2/I, B

TABLE 4
Amino-Acid and Carbohydrate Composition of Lactoperoxidase B 1, Expressed to the Nearest Integral Number per 78 000 g of Protein (Carlström, 1969*a*)

Aspartic acid	71
Glutamic acid	61
Amide nitrogen	62
Threonine	32
Serine	33
Proline	42
Glycine	40
Alanine	37
Valine	28
Isoleucine	27
Leucine	68
Methionine	11
Cysteine	0
Cystine	8
Tyrosine	15
Phenylalanine	30
Tryptophan	15
Lysine	34
Histidine	14
Arginine	37
Mannose	26
N-Acetylglucosamine	14
N-Ac-galactosamine	4
Sialic acid	0

2/II, B 3, B 4 and B 5, were subsequently identified (Carlström, 1969*b*). Paul and Ohlsson (1985) have, by chemical analysis of the different fractions, proposed the following relationships:

$$\begin{array}{ccccccccccc} B\,1 & \rightarrow & B\,2/I & \rightarrow & B\,2/II & \rightarrow & B\,3 & \rightarrow & B\,4 & \rightarrow & B\,5 \\ \downarrow & & \downarrow & & \downarrow & & \downarrow & & & & \\ A\,1 & \rightarrow & A\,2/I & \rightarrow & A\,2/II & \rightarrow & A\,3 & & & & \end{array}$$

According to this pattern, B 1 represents the intact, or least degraded, form of lactoperoxidase. Its amino acid composition gives a molecular weight of 78 431 Da and an iron content of 0·071%. The horizontal changes within both groups occur under conditions known to favour deamination of aspargine or glutamine residues, while each of the vertical shifts (B → A) is due to the loss of four glucosamine, six mannose and one galactosamine residues.

The primary structure of lactoperoxidase is unknown. A recent study has given some information on the secondary structure, i.e. lactoperoxidase contains 65% β-structure, 23% α-structure and 12% uncoordinated structures

(Sievers, 1979). Some information is also available regarding the tertiary structure. Electrophoretic studies on lactoperoxidase after pretreatment with 1% SDS and 1% mercaptoethanol showed a single component, indicating that lactoperoxidase consists of a single peptide chain. The presence of eight disulphide bonds contribute to the rigidity of the molecule (Sievers, 1980).

Isolation and analysis of lactoperoxidase

Lactoperoxidase can easily be isolated from milk, i.e. skim milk or whey, by use of cation exchangers such as DEAE–Sephadex or CG-50-NH_4. Paul *et al.* (1980) described a method by which the enzyme was absorbed first to an ion exchanger, type CG-50-NH_4 (20 g/litre), and then further purified on phenyl–Sepharose. In recent years, large-scale commercial processes for isolation by ion-exchange chromatography have been developed (Prieels & Peiffer, 1986). Processes based on this technique are in commercial operation today.

Lactoperoxidase activity is traditionally measured by a spectrophotometric technique using a chromogen as electron donor. Originally, various types of benzidine derivatives were used, e.g. dianisidine, *o*-toluidine, but because of the toxicity of these substances, they are seldom used today. Instead, the reagent 2,2′-azino-di-(3-ethylbenzthiazoline-6-sulfonic acid) (ABTS) from Boeringer Mannheim (Germany) has been introduced into peroxidase studies as an ideal reagent. It is relatively non-toxic, is available in purified form and has a high absorbance in the oxidized form. An analytical method using ABTS for determining lactoperoxidase activity was published by Shindler *et al.* (1976). Lactoperoxidase can also be determined by an FPLC technique (Ekstrand & Björck, 1986).

Biological significance of lactoperoxidase

Lactoperoxidase has attracted much interest, as it has been shown to be involved in the antibacterial activity of various secretions. This phenomenon has been studied especially in human saliva and bovine milk. The antibacterial activity has been reviewed in detail by Pruitt and Tenovuo (1985). In the following, therefore, only a brief summary of the antibacterial mechanism and its biological significance will be given.

At the beginning of this century, it was observed that lactoperoxidase activity could be correlated with the antibacterial properties of raw milk (Hansen, 1924). However, it took until the mid 1960s before this phenomenon could be elucidated in detail. Portman and Auclair (1959) demonstrated that lactoperoxidase was involved in the inhibition of certain types of starter cultures in milk. Concurrently, Jago and Morrison (1962) demonstrated that H_2O_2 was necessary for the antibacterial effect. In 1964, Reiter *et al.* (1964) showed that the hydrogen donor involved was thiocyanate ions. The exact nature of the antibacterial agent was obscure until Hoogendoorn *et al.* (1977) and Aune and Thomas (1977) independently showed that hypothiocyanite

($OSCN^-$) was the antibacterial agent formed by the lactoperoxidase-catalysed oxidation of thiocyante. More recent investigations indicate that hypothiocyanite may not be the only antibacterial agent formed. Björck and Claesson (1980) were able to demonstrate that non-enzymatically formed hypothiocyanite did not give the same antibacterial effect as an optimal mixture of lactoperoxidase, thiocyanate and H_2O_2. They concluded that, in addition to hypothiocyanite, other short-lived oxidation products are formed, as originally proposed by Hogg and Jago (1970). Further evidence of this hypothesis has been presented by Pruitt *et al.* (1982). By polarographic studies of the oxidation products, they concluded that the higher oxyacids of thiocyanate are also involved in the antibacterial effect. In conclusion, although hypothiocyanite is unquestionably formed in the lactoperoxidase-catalysed oxidation of thiocyanate and has antibacterial activity, there is plenty of indirect evidence today that some higher oxyacids, i.e. HO_2SCN and/or HO_3SCN, are involved in the antibacterial mechanism.

The biological significance of the lactoperoxidase/thiocyanate/H_2O_2 system is only partially known today. Bovine milk contains an array of antibacterial agents, which are known to give the neonate vital protection against various infectious agents before its own defence systems are developed (Reiter, 1981, 1985). The lactoperoxidase system is one of these agents. Whether this system may be involved in the protection of the mammary gland has also been discussed (Reiter, 1980).

Today, it is well established that it can inhibit certain starter cultures, i.e. those strains which accumulate H_2O_2 through their oxygen metabolism. Most of these strains have today been identified and are no longer in use. It is less well known that some strains of *Lactococcus* are in fact stimulated by the presence of lactoperoxidase and thiocyanate. It has been suggested that this effect is due to the removal of H_2O_2 and alleviation of the direct inhibitory effect of this component (Stadhouders, J., 1986, pers. comm.).

Many bacteria have a longer lag time and multiply slower in raw milk than in pasteurized milk (Langeveld *et al.*, 1973; Stadhouders, 1982). Whether this effect can be attributed to the antibacterial systems in milk is still an open question. Many milk-spoiling Gram-negative organisms, however, are sensitive to the lactoperoxidase system (Björck *et al.*, 1975; Reiter & Härnulv, 1984). Beumer *et al.* (1985) demonstrated that *Campylobacter jenuni* is killed in raw milk and that this effect can be attributed to the lactoperoxidase system.

The more detailed knowledge of the antibacterial mechanisms of the lactoperoxidase system has led to different practical applications. One such is the activation of the system in raw milk, by supplementation with thiocyanate and H_2O_2, in order to prevent bacterial deterioration of non-refrigerated milk during collection and storage in developing countries (Björck *et al.* 1979; IDF, 1988).

Another application of the lactoperoxidase system is in calf-milk replacers as a substitute for low-dose antibiotics. Improved performance of calves by the activated lactoperoxidase system has been demonstrated by Reiter *et al.* (1981)

and Waterhouse and Mullan (1980*a,b*). Consequently, calf-milk replacers containing an activated lactoperoxidase system are marketed in several countries in which the use of feed antibiotics is restricted, e.g. Denmark, Sweden, Finland and Norway.

The availability of lactoperoxidase, through large-scale isolation from skim milk or whey, has opened up several applications for the antibacterial effects of the lactoperoxidase system. Of great interest are several dental products presently being developed. In the future it is likely that several other applications will be developed.

REFERENCES

Agner, K. (1941). *Acta Physiol. Scand.* (2 Suppl.), **8,** 8–62.
Aune, T. M. & Thomas, E. L. (1977). *Eur. J. Biochem.*, **80,** 209–14.
Beumer, R. R., Noomen, A., Marys, J. A. & Kampelbackher, E. H. (1985). *Neth. Milk Dairy J.*, **39,** 107–14.
Björck, L. & Claesson, O. (1980). *J. Dairy. Sci.*, **63,** 919–22.
Björck, L., Rosén, C-G., Marshall, V. & Reiter, B. (1975). *Appl. Microbiol.* **30,** 199–204.
Björck, L., Claesson, O. & Schulthess, W. (1979). *Milchwissenschaft,* **34,** 726–9.
Carlström, A. (1969*a*). *Acta Chem. Scand.*, **23,** 185–202.
Carlström, A. (1969*b*). *Acta Chem. Scand.*, **23,** 171–84.
Dumonet, C. & Rousset, B. (1983). *J. Biol. Chem.*, **258,** 14166–72.
Ekstrand, B. & Björck, L. (1986). *J. Chromatogr.*, **382,** 429–33.
Gothefors, L. & Marklund, S. (1975). *Infect. Immun.*, **111,** 1210–5.
Haeringen, N. J. van, Ensink, F. T. E. & Glasius, E. (1979). *Exper. Eye Res.*, **28,** 343–7.
Hansen, F. W. (1924). *Brit. J. Exper. Pathol.*, **5,** 271–80.
Hogg, D. McC. & Jago, G. R. (1970). *Biochem. J.*, **117,** 779–90.
Hoogendoorn, H., Prissens, J. P., Scholtes, W. & Stoddard, L. A. (1977). *Caries Res.*, **11,** 77–84.
Hultqvist, D. E. & Morrison, M. (1963). *J. Biol. Chem.*, **238,** 2843–6.
IDF (1988). Bulletin no. 234, International Dairy Federation, Brussels, Belgium.
Jago, G. R. & Morrison, M. (1962). *Proc. Soc. Exper. Biol. Med.*, **111,** 585–90.
Kumlien, A. (1972). *Acta Endocrinol.*, **70,** 239–46.
Langbakk, B. & Flatmark, T. (1984). *FEBS Lett.*, **174,** 300–3.
Langbakk, B. & Flatmark, T. (1989). *Biochem. J.*, **259,** 627–31.
Langeveld, L. P. M., Cuperus, F. & Stadhouders, J. (1973). *Neth. Milk Dairy J.*, **27,** 54–65.
Mäkinen, K. K., Bowen, W. H., Dalgard, D. & Fitzgerald, G. (1978). *J. Nutr.*, **108,** 779–89.
Mäkinen, K. K., Kölling, D. & Söderberg, E. (1980). *Int. J. Vit. Nutr. Res.*, **50,** 79–83.
Moldveano, Z., Tenovuo, J., Mestecky, J. & Pruitt, K. M. (1982). *Biochim. Biophys. Acta,* **718,** 103–8.
Morell, D. B. & Clezy, P. S. (1963). *Biochim. Biophys. Acta,* **71,** 157–64.
Morrison, M. & Allen, P. Z. (1963). *Biochem. Biophys. Res. Comm.*, **13,** 490–4.
Morrison, M. & Steele, W. F. (1968). In *Biology of the Mouth,* ed. P. Person. American Association for Advancement of Science, Washington, DC, USA, pp. 89–110.

Paul, K. G. & Ohlsson, P. I. (1985). In *The Lactoperoxidase System: Chemistry and Biological Significance,* ed. K. M. Pruitt & J. Tenovuo. Marcel Dekker Inc., New York, USA, pp. 15–29.
Paul, K. G., Ohlsson, P. I. & Henriksson, A. (1980). *FEBS Lett.,* **100,** 200–4.
Polis, D. & Schmukler, H. W. (1953). *J. Biol. Chem.,* **201,** 475–500.
Portman, A. & Auclair, J. E. (1959). *Le Lait,* **39,** 147–9.
Prieels, J. P. & Peiffer, R. (1986). UK Patent Application GB 2 171 102 A1.
Pruitt, K. M. & Tenovuo, J. (1985). In *The Lactoperoxidase System: Chemistry and Biological Significance,* ed. K. M. Pruitt & J. Tenovuo, Marcel Dekker Inc., New York, USA, pp. 101–22.
Pruitt, K. M., Tenovuo, J., Andrews, P. & McKane, T. (1982). *Biochemistry,* **21,** 562–7.
Reiter, B. (1980). In *Resistance Factors and Genetic Aspects of Mastitis Control,* ed. L. Basselik-Chabielska. Jablonna, Poland, pp. 351–92.
Reiter, B. (1981). In *Immunological Aspects in the Foetus and the Newborn,* ed. H. P. Lambert & C. B. S. Woods. Academic Press, London, UK, pp. 155–95.
Reiter, B. (1985). International Dairy Bulletin No. 191. 1DF, Brussels, Belgium.
Reiter, B. & Härnulv, G. (1984). *J. Food Protect.,* **47,** 724–32.
Reiter, B., Pickering, A., Oram, J. D. & Pope, G. S. (1963). *J. Gen. Microbiol.,* **33,** XII.
Reiter, B., Pickering, A. & Oram, J. D. (1964). An inhibitory system—lactoperoxidase/thiocyanate/hydrogen peroxide—in raw milk. In N. Molin (ed.) *Microbial Inhibitors in Food,* Almqvist & Wiksell.
Reiter, B., Fulford, R. J., Marshall, V., Yarrow, N., Ducker, M. J. & Knutsson, M. (1981). *Anim. Prod.,* **32,** 297–306.
Shindler, J. S., Childs, R. E. & Bardsley, W. G. (1976). *Eur. J. Biochem.,* **65,** 325–31.
Sievers, G. (1979). *Biochim. Biophys. Acta,* **624,** 249–59.
Sievers, G. (1980). *FEBS Lett.,* **127,** 253–6.
Stadhouders, J. (1982). *Kieler Milchw. Forshungsber.,* **34,** 19–28.
Stephens, S., Harkness, R. A. & Cockle, S. M. (1979). *Brit. J. Exper. Pathol.,* **60,** 252–8.
Theorell, H. & Åkesson, Å. (1943). *Arkiv. Kemi Geol. Mineral.,* **17B,** 1–6.
Theorell, H. & Pedersen, K. O. (1944). In *The Svedberg.* Almqvist & Wiksell, Stockholm, Sweden, pp. 523–9.
Thurlow, S. (1925). *Biochem. J.,* **19,** 175–98.
Wanatabe, K. (1978). *Acta Histochem. Cytochem.,* **11,** 151–9.
Waterhouse, A. & Mullan, W. M. A. (1980*a*). *Anim. Prod.,* **30,** 458.
Waterhouse, A. & Mullan, W. M. A. (1980*b*). *Ir. J. Food Sci. Technol.,* **4,** 69.

VI. OTHER ENZYMES

Catalase (EC 1.11.1.6)

Catalase is an oxido-reductase which decomposes hydrogen peroxide according to the following reaction: $2H_2O_2 \rightarrow 2H_2O + O_2$. The enzyme is present in animals, plants, microorganisms (Whitaker, 1972) and in normal milk (Groves, 1971). Its level in bovine milk varies with the disease status of the animal, and is used to indicate mastitis (Kitchen *et al.*, 1970). However, opinions differ on the relationship between catalase activity and raw-milk quality. Two studies have reported positive correlation coefficients of 0·70 (Read *et al.*, 1969) and 0·72 (Kitchen, 1976) between catalase activity and the direct somatic-cell count of raw milk. However, Phillips and Griffiths (1987) found no correlation between catalase activity in raw milk and total bacterial or somatic-cell counts, but did find a significant ($p < 0{\cdot}001$) negative correlation ($r = -0{\cdot}74$) between bacterial-cell count and catalase activity in pasteurized milk.

In bovine milk, catalase is distributed between the cream (~24%) and skim-milk (~73%) fractions (Kitchen *et al.*, 1979). Similarly, ~22% of the catalase activity in caprine milk was found in the cream (Ito & Akuzawa, 1982). Catalase in the cream phase is associated with the milk fat globule membrane (MFGM), with which ~56% of it is firmly bound to insoluble lipoprotein (Ito, 1971). The enzyme has been purified and crystallized from bovine milk (Ito & Akuzawa, 1983*a*). The method for purification involves cream separation and churning to obtain catalase-rich buttermilk, followed by *n*-butanol extraction, $(NH_4)_2SO_4$ fractionation, and chromatography on DEAE–Sephacel and Sephacryl S-300 (Ito & Akuzawa, 1983*a*). From 1200 kg of milk, 120 mg of crystalline catalase, representing 23 000-fold purification, were obtained. The molecular weight of the crystalline enzyme was estimated to be ~225 000 Da by Sephadex G-200 gel filtration. Following treatment with 1% SDS for 48 h, the purified enzyme was resolved by gel filtration or SDS–PAGE into five components with estimated molecular weights of 55 000, 40 000, 34 000, 24 000 and 11 000 Da. Only the 55 000-Da component had catalase activity (Ito & Akuzawa, 1983*b*). Ito and Akuzawa (1983*c*) reported the isolation of two catalase isoenzymes with similar immunochemical properties (Ito *et al.*, 1984) from bovine milk. The isoenzymes had pH optima of 7 and 8, and the former was the more active within the pH range 5–8 (Ito *et al.*, 1984).

Catalase is stable at temperatures below its optimum of 30°C (Ito & Akuzawa, 1983*a*). Heating solutions of crystalline catalase in buffer for 30 min at 60 or 70°C resulted in 75 and 100% inactivation, respectively (Ito & Akuzawa, 1983*a*). The heat stability of catalase in milk appears to differ from that of the crystalline enzyme in buffer. Shidlovskaya (1982) reported 63, 74 and 84% reduction in catalase activity when milk was heated at 77–83°C for 20 s, 88–90°C for 3 min or 91–95°C without holding, respectively. Griffiths (1986) reported 90% reduction of catalase activity when milk was heated at

71°C for 15 s. Shidlovskaya (1982) and Griffiths (1986) also found that the catalase activity in heated milk increased when stored at 4°C.

Catalase is inhibited by Cu^{2+}, Fe^{2+}, Hg^{2+}, Sn^{2+}, CN^- and N_3^- (Ito, 1974). Because the enzyme contains haem iron, its involvement in lipid oxidation has been suggested (see Fox & Morrissey, 1981).

Superoxide dismutase (EC 1.15.1.1)

Superoxide dismutase (SOD), properly called superoxide: superoxide oxidoreductase, catalyses the dismutation of superoxide anion ($O_2^{\overline{\cdot}}$) according to the equation

$$O_2^{\overline{\cdot}} + O_2^{\overline{\cdot}} + 2H^+ \rightarrow 2H_2O_2 + O_2$$

The source of $O_2^{\overline{\cdot}}$ for SOD in milk is xanthine oxidase which produces substantial amounts of $O_2^{\overline{\cdot}}$ under conditions of high pH, high O_2 concentrations and low xanthine concentrations (Fridovich, 1970).

SOD is present in almost all organisms. Three distinct types of SOD are known: (1) Cu/Zn-containing SOD, which is found in eukaryotes; (2) Mn-containing SOD, found in bacteria; and (3) Fe-containing SOD, found in bacteria and a few plants (Fridovich, 1986). The structure, chemistry and biological significance of the SODs have been reviewed (Fridovich, 1974, 1986). The Cu/Zn-containing SOD is a dimer of molecular weight ~32 000 Da (McCord & Fridovich, 1969). The two subunits, which are chemically identical (Evans *et al.*, 1974), are associated by non-covalent interactions (Abernethy *et al.*, 1974) and stabilized by intrachain disulphide bonds (Abernethy *et al.*, 1974). The enzyme contains one Cu^{2+} and one Zn^{2+} per subunit; Cu^{2+} is responsible for electron transfer, while Zn^{2+} stabilizes the enzyme (Fridovich, 1974).

The occurrence of SOD in bovine milk was reported in the mid-1970s (Hicks *et al.*, 1975; Hill, 1975; Asada, 1976). Bovine-milk SOD has similar properties to the Cu/Zn-containing enzyme isolated from bovine erythrocytes, suggesting that the enzyme enters milk from blood (Hill, 1975). According to Hicks (1980), Mn-containing SOD also occurs in milk. SOD is also present in human milk at levels lower than those in bovine milk (Keen *et al.*, 1980). There is lack of agreement on the level of SOD in bovine milk, e.g. Hill (1975) reported ~0·15 mg/litre, Asada (1976) ~0·32 mg/litre, while Korycka-Dahl *et al.* (1979) found ~2·5 mg/litre; the large variation in the reported levels can be attributed to differences in assay techniques.

The levels of SOD in milk are significantly ($p < 0{\cdot}01$) affected by breed of cow (~38% higher activity in Jersey than in Holstein milk), but not by the stage of lactation, age of cow or mastitis (Holbrook & Hicks, 1978). SOD is found mainly in the milk serum (Holbrook & Hicks, 1978) and the enzyme is isolated from milk by a procedure similar to that used to isolate SOD from bovine blood (McCord & Fridovich, 1969), except that acid whey is used as starting material (Hill, 1975).

The Cu/Zn-containing SODs are stable proteins (Fridovich, 1974). Bovine milk SOD survives normal high-temperature short-time (HTST) (71·7°C for 15 s) pasteurization, and is relatively stable even at 75°C (~75% inactivation after 20 min) (Hicks *et al.*, 1979).

The significance of SOD in dairy technology lies in its ability to scavenge the superoxide anion ($O_2^{\cdot -}$) produced by xanthine oxidase in milk stored under conditions where SOD is saturated with O_2 (Hicks, 1980). Holbrook and Hicks (1978) reported low rates of oxidative rancidity in raw milk that was not exposed to light and contained high levels of SOD. The significance of SOD in milk has been reviewed by Hicks (1980).

Xanthine oxidase (EC 1.2.3.2)

Xanthine oxidase (XO) is an oxido-reductase which catalyses the oxidation of hypoxanthine to xanthine and xanthine to uric acid. It also catalyses the oxidation of a wide range of purines, aldehydes and pteridines. The oxidation reactions occur with concomitant reduction of O_2 to H_2O_2. However, at high pH, high O_2 concentration and low xanthine concentration, the very reactive superoxide anion ($O_2^{\cdot -}$) is formed (Fridovich, 1970).

XO also exists as xanthine dehydrogenase (EC 1.1.1.204). The oxidase form can be converted to the NAD^+-dependent dehydrogenase by treatment with thiol-modifying reagents, partial proteolysis or storage of the enzyme at −20°C (Waud & Rajagopalan, 1976). Xanthine dehydrogenase can be converted to XO by treatment with sulphydryl oxidase (Clare *et al.*, 1981).

The enzyme is a dimeric metallo-flavoprotein with a molecular weight of ~300 000 Da (Hart & Bray, 1967). The active dimer contains molybdenom, flavin adenine dinucleotide (FAD), and iron–sulphur (Fe_2–S_2) in the ratio of 1:1:4:4 (Whitaker, 1972). Acidification or heating at 100°C at pH 7 readily releases S from Fe_2–S_2 as H_2S (Whitaker, 1972). The enzyme also contains a mole of phosphorus covalently bound per active-centre molybdenom (Davis *et al.*, 1984). A catalytically inactive desulpho-XO, in which the molybdenom centre contains an oxo-ligand instead of the normal sulphide ligand, occurs in bovine milk (Gutteridge *et al.*, 1978; Bordas *et al.*, 1980). Treatment of the normal enzyme with cyanide (Massey & Edmondson, 1970) or H_2O_2 (Ventom *et al.*, 1988), converts it to the desulpho form. More recently, the occurrence of a demolybdo form of XO (Spanning *et al.*, 1987; Ventom *et al.*, 1988) containing a pterin-like moiety at the molybdenom site has been reported (Ventom *et al.*, 1988).

Bovine milk is a very rich source of XO, containing ~35 mg/litre (Walstra & Jenness, 1984); its occurrence is under genetic control (Zikakis & Treece, 1971). The enzyme is also present in caprine milk (Modi *et al.*, 1959), and, at a low level, in human milk (Zikakis *et al.*, 1976).

XO is distributed between the cream and skim-milk phases of milk (Kitchen *et al.*, 1970). In cream, it is concentrated in the MFGM where it is estimated to account for ~10–20% of the MFGM protein (Mangino & Brunner, 1977;

Keenan *et al.*, 1988). XO is attached to the inner (cytoplasmic) surface of the MFGM by butyrophilin (Valivullah & Keenan, 1989), a protein which strongly binds phospholipids (Freundenstein *et al.*, 1979).

Several methods for isolating XO from bovine milk have been published (Gilbert & Bergel, 1964; Edmondson *et al.*, 1972; Nathan & Hade, 1975, 1978; Mangino & Brunner, 1977; Nishino *et al.*, 1981; Sullivan *et al.*, 1982; Cheng *et al.*, 1988) and a procedure for preparing immobilized milk XO has been described (Tramper, 1987). Graham *et al.* (1989) purified XO from human milk by hydroxyapatite chromatography and raised antibodies to it. Some of the traditional methods for isolation include treatment of MFGM with a proteinase to release the enzyme. However, there is now ample evidence (Waud *et al.*, 1975; Nathan & Hade, 1978; Silver & Zikakis, 1979; Cheng *et al.*, 1988) to suggest that treatment with proteinase during purification results in partial proteolysis of the enzyme. When purified milk XO is treated with chymotrypsin, papain or subtilisin it remains irreversibly in the oxidase form (Battelli *et al.*, 1973), raising the possibility that indigenous milk proteinase may modify XO during its purification from milk (Mangino & Brunner, 1977). However, current evidence (Cheng *et al.*, 1988) suggests that plasmin, the principal indigenous milk proteinase, has limited proteolytic activity on XO. There are at least five genetic variants of XO with isoelectric pH values between 6·9 and 7·6 (Sullivan *et al.*, 1982). The enzyme has been isolated in the form of xanthine-dehydrogenase from milk treated with dithioerythritol (Nakamura & Yamazaki, 1982). The optimum pH for milk-XO activity is in the range 8·0–9·0 (Briley & Eisenthal, 1975; Mangino & Brunner, 1977).

Oster (1971) postulated that XO activity in homogenized milk causes depletion of plasmalogen in cell membranes, leading to atherosclerosis in individuals who drink such milks. This was a cause for concern in the dairy industry, prompting extensive investigations of XO activity in milk and dairy products (Cerbulis & Farrell, 1977; Zikakis & Wooters, 1980). The findings show that commercial processing (i.e. pasteurization and homogenization) causes ~70% (Cerbulis & Farrell, 1977) or ~82% (Zikakis & Wooters, 1980) inactivation of milk XO. More recently, Griffiths (1986) reported that XO survives in milk heated at 75°C for 15 s, but loses ~70% of its activity when milk is heated at 80°C for 15 s. Ruegg *et al.* (1977) reported that during heating, XO undergoes conformational changes at 61·4 and 77·6°C, suggesting that renaturation occurs after heating, since activity survives heating at these temperatures. They also indicated that the conformational change, together with the protective effect of other milk proteins, may account for the thermostability of XO. The investigations on XO activity in dairy products, together with epidemiological studies (American Heart Association, 1981; Clifford *et al.*, 1983; Deeth, 1983), suggest that the claim by Oster (1971) is without foundation. The nutritional implication of XO have been reviewed by McBean and Speckman (1988).

XO has been implicated in the oxidative deterioration of milk and dairy products via the production of superoxide anion ($O_2^{\cdot-}$) during oxidation of its substrates (Aurand *et al.*, 1977). However, Bruder *et al.* (1982) found no evidence to support a role for native bovine milk XO in lipid oxidation. A recent treatise on the role of XO in the oxidative deterioration of milk has been published (Weihrauch, 1988). It has also been suggested that in the presence of purines, XO activity generates H_2O_2 for the lactoperoxidase system in milk (Björck & Claesson, 1979), therefore making it a bactericidal or bacteriostatic agent in milk.

Sulphydryl oxidase

Sulphydryl oxidase (SOX), a Fe/Cu-containing glycoprotein (Janalino & Swaisgood, 1975), was first reported in bovine milk by Kiermeier and Petz (1967*a,b,c*), and has since been detected in the milks of human, goat, pig, rabbit and rat (Clare *et al.*, 1984; Isaacs *et al.*, 1984). Fe is essential for SOX activity (Janalino & Swaisgood, 1975). The carbohydrate moiety aids in solubilizing the enzyme (Swaisgood & Horton, 1980).

The enzyme oxidizes the sulphydryl groups of cysteine, reduced glutathione and milk proteins to their corresponding disulphides using molecular oxygen as electron acceptor (Janalino & Swaisgood, 1975). The stoichiometry of the reaction catalysed has been established (Janalino & Swaisgood, 1975) as

$$2RSH + O_2 \rightleftharpoons RSSR + H_2O_2$$

SOX is a distinctly different enzyme from thiol oxidase (EC 1.8.3.2), both in molecular and biochemical properties (Janalino & Swaisgood, 1975).

The concentration of SOX in bovine milk is ~5–10 mg/litre (Swaisgood & Abraham, 1980). Most, if not all, of the SOX activity in bovine milk is in the skim-milk phase where the enzyme is bound to skim-milk membranes (Sliwkowski *et al.*, 1982), which are derived from the MFGM. Similarly, ~95% of the SOX in human milk is bound to skim-milk membranes (Isaacs, 1985). The activities of SOX in bovine colostrum and milk are similar (Isaacs, 1985).

Methods for isolating SOX at different degrees of purity from skim milk have been described (Janalino & Swaisgood, 1975, 1979; Swaisgood, 1980; Janalino *et al.*, 1981). All the methods use sweet (rennet) whey as starting material because the enzyme is unstable at pH values <6 (Janalino & Swaisgood, 1975; Swaisgood *et al.*, 1987). The enzyme has been immobilized on controlled-pore glass beads (Janalino & Swaisgood, 1978).

In milk, SOX exhibits a high degree of aggregation, making it difficult to determine the molecular weight of the native enzyme (Janalino & Swaisgood, 1975). However, its subunit molecular weight, estimated by SDS gel electrophoresis (Janalino & Swaisgood, 1975) and by light scattering in 5 M guanidinium chloride (Janalino *et al.*, 1980), is ~89 000 Da. The pH and temperature optima for activity are 7·0 and 37°C, respectively (Janalino & Swaisgood, 1975). Its K_m for reduced gluthathione is ~0·3 mM (Swaisgood & Abraham, 1980). For human-milk SOX, an optimum pH of 7·0–7·5 has been reported (Isaacs *et al.*, 1984). SOX from both human and bovine milks are relatively heat stable (Isaacs *et al.*, 1984). Pasteurization (temperature and time not specified, but presumably 72°C for 15 s) of milk reduces its SOX activity by ~60% (Swaisgood, 1980).

Storage of human milk at 4 or −20°C had no effect on its SOX activity (Isaacs, 1985). The human-milk enzyme is resistant to proteases in the gastrointestinal tract but loses ~50% of its original activity when stored for 1 h at pH 2·5 (Isaacs, 1985). Isaacs *et al.* (1984) found SOX activity in the contents of the stomach and small intestine of suckling rats, and suggested that the enzyme may aid protein absorption in the intestine by changing the mucoprotein coating of the small intestine.

Technologically, purified bovine-milk SOX may be added to UHT milk to reduce cooked flavour. Patents (see Swaisgood *et al.*, 1987) for this application have been issued. Physiologically, SOX may have a role of catalysing non-random formation of disulphide bonds during protein biosynthesis.

γ-Glutamyl transferase (EC 2.3.2.2)

γ-Glutamyl transferase, also called γ-glutamyl transpeptidase (GGT), is a glycoprotein which catalyses the transfer of γ-glutamyl residues from γ-glutamyl-containing peptides to other acceptor amino acids or peptides:

$$\gamma\text{-glutamyl peptide} + \text{amino acid} \rightarrow \text{peptide} + \gamma\text{-glutamyl amino acid}$$

Except for proline, all the L-amino acids and their peptides are good acceptors (Meister, 1973) but D-amino acids and their peptides are not (Tate & Meister, 1974).

GGT is present in many mammalian tissues (Meister & Tate, 1976) and in plants and microorganisms (Yasumoto *et al.*, 1978), as well as in secretory products like milk (Majumder & Ganguli, 1972). GGT activity in milk varies during lactation (Barcelos & Weight, 1986). The highest level of the enzyme was found in pre-colostrum, while the lowest level was in peak-lactation milk (Barcelos & Weight, 1986). The activity of GGT in bovine colostrum is ~2–3 times higher than that in milk (Majumder & Ganguli, 1972; Sobiech *et al.*, 1974). Barcelos (1983) found no effect of sub-clinical mastitis on GGT activity in milk, but observed a 75% reduction in its activity in milk from cows with clinical mastitis.

The enzyme is distributed between the cream (23%) and skim-milk (74%) phases (Baumrucker, 1979). About 48% of the activity in milk is associated with milk membranes (41% with the skim-milk membrane and 7% with the MFGM) (Baumrucker, 1979).

Methods for purifying GGT from milk (Sobiech *et al.*, 1974; Baumrucker, 1980) or colostrum (Yasumoto *et al.*, 1978) have been described. The molecular weight of the enzyme purified from bovine colostrum, as determined by gel filtration, is ~80 000 Da (Yasumoto *et al.*, 1978). Active GGT is composed of two subunits with molecular weights (as estimated by SDS–

electrophoresis) of ~57 000 and ~25 500 (Yasumoto *et al.*, 1978; Baumrucker, 1980). The enzyme has an isoelectric pH of 3·85 (Yasumoto *et al.*, 1978). Its optimum pH for activity is 8·0–9·0 (Sobiech *et al.*, 1974; Yasumoto *et al.*, 1978; Baumrucker, 1980), depending on the acceptor amino acid or peptide (Yasumoto *et al.*, 1978).

Purified bovine-milk GGT was stable at ~20°C, but it lost 25% of its activity when heated in buffer at 56°C for 30 min (Sobiech *et al.*, 1974). In milk, the heat stability of GGT is variable, depending on time and temperature of heating. Majumder and Ganguli (1972) reported that heating milk at 62°C for 30 min or at 72°C for 15 s reduced its GGT activity by 18 and 78%, respectively. More recently, Patel and Wilbey (1989) found that most of the enzymic activity survived thermization (65°C for 15 s) of milk, but heating milk at 75°C for 15 s reduced its GGT activity by more than 90%. Therefore, assaying for GGT activity can be a useful tool for detecting milks that had been heated to 70–80°C (Andrews *et al.*, 1987).

Physiologically, GGT is believed to play a role in amino-acid transport into the mammary gland (Baumrucker, 1980; Pocius *et al.*, 1980; Vina *et al.*, 1985). Ingested GGT is readily absorbed across the intestinal wall (Braun *et al.*, 1984), resulting in elevated levels of serum GGT in newborn calves and kids fed colostrum, compared to those fed milk replacers free of GGT activity on their first day of life (Braun *et al.*, 1982; 1984). Therefore, serum GGT levels may be a useful tool for monitoring colostrum intake by kids and calves (Braun *et al.*, 1984).

Ribonuclease (EC 3.1.27.5)

The occurrence of ribonuclease (RNase) in bovine milk was demonstrated by Bingham and Zittle (1962), who also showed that the milk enzyme closely resembles bovine pancreatic RNase. RNase catalyses endonucleolytic cleavage of 3′-phosphomononucleotides and 3′-phosphooligonucleotides ending in polycytidylic or polyuridylic acids.

Bovine milk is a rich source of RNase, containing ~25 mg/litre (Shahani *et al.*, 1973). The enzyme is present in human, caprine and porcine milks (Chandan *et al.*, 1968). One study (Chandan *et al.*, 1968) reported three times more RNase activity in bovine milk than in human milk, while another study (Meyer *et al.*, 1987*b*) found 100 times more RNase activity in the former. Bovine and human colostra have 3–5 times more RNase activity than their corresponding milks (Meyer *et al.*, 1987*b*).

The enzyme enters milk from blood (Chandan *et al.*, 1968), and is found mostly in the milk serum. Skim milk and cream contain 94 and 4%, respectively, of milk-RNase activity, while acid whey has ~88% (Meyer *et al.*, 1987*a*). The enzyme is usually isolated from acid whey by precipitation with 80% $(NH_4)_2SO_4$, followed by solubilization and chromatography on Amberlite-IRC-50 (Bingham & Zittle, 1964) or Sephadex G-50 (Dalaly *et al.*, 1970).

Bovine-milk RNase consists of three isoenzymes: A, B and II-1, in the ratio 72:27:1, respectively (Meyer *et al.*, 1987*a*). The isoenzymes can be resolved by chromatography on phosphocellulose (Meyer *et al.*, 1987*a*). Human-milk RNase consists of six isoenzymes (Meyer *et al.*, 1987*b*), and also resembles bovine pancreatic RNase (Dalaly *et al.*, 1970). The major difference between RNase A and B is in their carbohydrate contents: while RNase A contains no carbohydrate, RNase B is a glycoprotein containing 3·0% glucosamine, 1·2% galactosamine and 5·17% mannose (Bingham & Kalan, 1967). RNase II-1 differs from A and B by its inability to hydrolyse polycytidylate (Meyer *et al.*, 1987*a*).

Milk RNase is a basic protein (Dalaly *et al.*, 1970) with a molecular weight of ~13 600 Da (Bingham & Zittle, 1964). The pH and temperature optima for activity are 7·5 (Bingham & Zittle, 1962) and 37°C (Shahani *et al.*, 1973), respectively. The enzyme is heat stable at low pH values (2–4·5) but is heat labile at high pH (Bingham & Zittle, 1962). At pH 3·5, the enzyme survived heating at 90°C for 20 min but was totally inactivated by the same heat treatment at pH 7·0 (Bingham & Zittle, 1962). Griffiths (1986) reported that heating milk at 80°C for 15 s reduced its RNase activity by ~40%. Meyer *et al.* (1987*a*) found that heating milk at 62·5°C for 30 min, 72°C for 2 min or 121°C for 10 s reduced its RNase activity by 6, 43 and 99·9%, respectively. Milk-based infant formulae have less than 0·3% of the RNase activity of bovine milk (Meyer *et al.*, 1987*a*). This is not surprising because infant formulae are usually processed at high temperatures.

RNase and lysozyme coexist naturally in biological systems (Dalaly *et al.*, 1970); RNase may have a bacteriocidal role in milk (Meyer *et al.*, 1987*b*), possibly by complementing lysozyme activity. RNase activity in milk doubles during mastitis and may be a useful index thereof (Meyer *et al.*, 1987*b*). In human milk, RNase inhibits the activity of RNA-directed DNA-polymerase (McCormick *et al.*, 1974).

α-Amylases (EC 3.2.1.1)

The presence of α- and β-amylases in bovine milk was reported by Richardson and Hankinson (1936). Human milk contains ~5 times more amylase activity than bovine milk (Wuthrich *et al.*, 1964). Guy and Jenness (1958) partially purified α-amylase from bovine milk. Their preparation had starch dextrinizing, liquefying and saccharifying properties, suggesting that both α- and β-amylases were present (Richardson & Hankinson, 1936). α-Amylase has been purified to apparent homogeneity from human colostrum by precipitation with 30–50% $(NH_4)_2SO_4$ and chromatography on Sephadex G-100 (Goussault *et al.*, 1966). The pH optimum for activity of α-amylase purified from human colostrum is 6·4 (Goussault *et al.*, 1967) and that for the enzyme partially purified from bovine milk is 7·4 (Guy & Jenness, 1958). Optimum temperature for activity is 44°C (Guy & Jenness, 1958). Reports on the heat stability of bovine-milk α-amylase differ. Guy and Jenness (1958) observed a 50% decrease in bovine-milk α-amylase activity when skim milk was heated for 30 min at 50°C. Wuthrich *et al.* (1964) reported that pasteurization (74°C for

15 s) and sterilization (115°C for 15 min) of milk reduces its α-amylase activity by 92 and 95%, respectively. More recently, Griffiths (1986) reported that heating milk at 75°C for 15 s reduces its dextrinizing activity by ~45% and its saccharifying activity by ~30%.

Human milk amylases have attracted considerably more attention than bovine milk amylases because of their possible significance in neonatal nutrition. The α-amylases of human milk and saliva are quite similar (Got *et al.*, 1968; Stejskal *et al.*, 1981; Hegardt *et al.*, 1984). Stejskal *et al.* (1981) reported that the biosynthesis of human milk and salivary amylases is controlled on the same gene locus but are subject to different hormone regulators. Using PAGE, Got *et al.* (1968) and Stejskal *et al.* (1981) showed the presence of four isoamylases in human colostrum and milk, respectively, although the enzyme does not enter milk from blood. α-Amylase activity in human colostrum is ~6 times that in human milk (Lindberg & Skude, 1982). The α-amylase activity in human milk is 10–25 times that in blood serum (Stejskal *et al.*, 1981), suggesting that the enzyme does not enter milk from the blood . The stability of milk amylases at low pH, which may be of importance in neonatal nutrition has been studied by Lindberg and Skude (1982), who reported that human milk α-amylase retained full activity when incubated at pH 5·3 for 30 min, but was completely inactivated in 30 min at pH 3. However, Jones *et al.* (1982) found that the enzyme retained full activity after incubation for 2 h at pH 3, but was totally inactivated in 1 h at pH ≤ 2·0. Heitlinger *et al.* (1983) found that purified amylase from human milk lost ~80% of its activity when incubated at pH 3·5 for 2 h, while a crude enzyme lost only ~67% of its activity in 6 h at the same pH. Addition of bovine serum albumin or human-milk proteins to purified α-amylase stabilized it at low pH (Heitlinger *et al.*, 1983), suggesting that the divergence of opinion among authors on the pH stability of α-amylase can be attributed to the purity of the enzyme preparations used.

The relatively high stability of human-milk amylases at low pH values suggests that the enzyme can partially survive in the gastrointestinal tract of infants after ingesting human milk (Jones *et al.*, 1982). Lindberg and Skude (1982) found considerable milk-amylase activity in the duodenal juice of two infants after feeding on human milk. Ca^{2+} increase the resistance of α-amylases to proteolysis (Stein & Fischer, 1958), and human-milk amylases are resistant to proteolysis by pepsin (Heitlinger *et al.*, 1983). It has been suggested that breast-fed premature infants, who require high-calorie diets, should be given starch-containing supplements (Jones *et al.*, 1982), since the α-amylase in breastmilk will contribute to starch digestion with the slow release of glucose in infants with insufficient pancreatic amylase (Lindberg & Skude, 1982). The importance of human-milk α-amylase in neonatal nutrition has been reviewed by Hamosh (1986, 1988).

Lysozyme (EC 3.2.1.17)

Lysozyme (muramidase) is a peptidoglycan *N*-acetylmuramyl hydrolase which cleaves β-(1-4) linkages between *N*-acetylmuramic acid and *N*-acetyl-D-glucosamine residues in peptidoglycan, a constituent of bacterial cell walls. Its structure, chemistry and uses have been reviewed (Proctor & Cunningham, 1988).

The enzyme was first isolated from egg white, which is its richest source. It is found in the milks of many species (Shahani, 1966; Shahani *et al.*, 1973). The highest level, ~79 mg/100 ml, is in equine milk (Jauregui-Adell, 1975); human milk contains ~39 mg/100 ml (Chandan *et al.*, 1968). For bovine milk, values from 0 to 284 μg/100 ml have been reported (Chandan *et al.*, 1965; McKenzie *et al.*, 1985; Skripnichenko, 1987). The levels of lysozyme in human pre-colostrum and colostrum are higher than those in normal milk (Lewis-Jones & Reynolds, 1983). Similarly, its levels in bovine colostra are higher than those in milk (Panfil-Kuncewicz & Kisza, 1976). The concentration of lysozyme in milk increases with advancing lactation (Kumari & Mathur, 1981) and during mastitis (Sokolova *et al.*, 1983; Skripnichenko, 1987), and is reported to be higher in afternoon than in morning milkings (McKenzie *et al.*, 1985).

Several methods for isolating lysozyme from milk have been reported (Jolles *et al.*, 1962; Chandan *et al.*, 1965; Jolles & Jolles, 1967; Brignon & Ribadeau-Dumas, 1982; McKenzie *et al.*, 1985). All the methods have yielded a homogeneous protein. However, White *et al.* (1988) reported that bovine milk lysozyme (BML) is a mixture of two components which differ by 7 amino acids, within the first 39 residues. Milk lysozyme bears close similarities in molecular weight, amino-acid composition and 3-dimensional structure to α-lactalbumin (Brew & Campbell, 1967; Hill *et al.*, 1968; Browne *et al.*, 1969). Phillips *et al.* (1987) observed similarities in the exon–intron organization of the genes for lysozyme and α-lactalbumin, confirming that lysozyme and α-lactalbumin have a common ancestry.

Both human milk lysozymes (HML) and bovine milk (BML) lysozymes are basic proteins but they differ antigenically and serologically (Eitenmiller *et al.*, 1974). Their amino-acid compositions also differ slightly (Eitenmiller *et al.*, 1971). HML has a molecular weight of ~15 000 Da, as determined by sedimentation velocity (Jolles & Jolles, 1967; Parry *et al.*, 1969), and an optimum pH of 6·35 (Parry *et al.*, 1969). BML has a molecular weight of 18 000 Da (Eitenmiller *et al.*, 1971), a pH optimum of 7·9 (Chandan *et al.*, 1965) or 7·5 (McKenzie & White, 1986), and an isoelectric pH of 9·5 (Chandan *et al.*, 1965). The activities of both BML and HML are destroyed by reduction of their intrachain disulphide bonds with 2-mercaptoethanol (Friend *et al.*, 1972). Reoxidation of the reduced enzymes regenerates 84 and 328% of the original activities of HML and BML, respectively (Friend *et al.*, 1972).

HML and BML exhibit pH-dependent temperature sensitivity. They are most stable at acid pH values (3–4) but labile at pH values >7 (Jauregui-Adell, 1975; Eitenmiller *et al.*, 1976). Heating human milk at 62·5°C for 30 min had no effect on its lysozyme activity, but heating at 75°C for 15 min reduced it to ~7% of the original value (Friend *et al.*, 1983). Bovine milk retains ~75% of its lysozyme activity when heated at 80°C for 15 s (Griffiths, 1986). Freeze-drying reduced the lysozyme activity in human milk, but storage of the freeze-dried milk for 6 months at 25°C had no effect on its residual lysozyme activity (Friend *et al.*, 1983). The lysozyme activity in bovine milk decreased during storage at 2°C, but the enzyme was relatively stable in diluted milk as for activity determination, suggesting that the decrease in BML activity observed during cold storage of milk was due to the presence of destabilizing factors in milk and not a consequence of low enzyme stability (McKenzie & White, 1986).

Physiologically, lysozyme functions as a bacteriocidal agent in human milk, but this role in bovine milk is questionable because of the low level present (Kitchen, 1985). Owing to the high levels of lysozyme in human pre-colostrum, it has been suggested that this should be used to feed pre-term and sick newborn infants (Lewis-Jones & Reynolds, 1983). Carlsson *et al.* (1979) reported that antibodies, lactoferrin and lysozyme in human milk act complementarily to prevent gastroenteritis and allergies in infants. The very high lysozyme activity in human milk compared to that in bovine milk has prompted the idea of supplementing bovine milk-based infant formulae with egg-white lysozyme. This and other applications of lysozyme in dairy technology are discussed in Chapter 5.

N-Acetyl-β-D-glucosaminidase (EC 3.2.1.30)

N-Acetyl-*β*-D-glucosaminidase (NAGase) is involved in the catabolism of hexosamine-containing sugars. Specifically, it hydrolyses the terminal non-reducing 2-acetamido-2-deoxy-*β*-D-glucose residues from glycoproteins (Dixon *et al.*, 1979). The enzyme is widely distributed in mammalian tissues and is present in bovine (Mellors, 1968; Kitchen & Middleton, 1976; Kitchen *et al.*, 1978; Mattila *et al.*, 1986; Fox *et al.*, 1988; Miller *et al.*, 1988), human (Oberkotter *et al.*, 1982), caprine (Timms & Schultz, 1985; Maisi & Riipinen, 1988) and porcine (Hurley & Grieve, 1988) milks.

Most of the NAGase activity in bovine milk is in the skim-milk fraction, 94% (Mellors, 1968), 91% (Kitchen *et al.*, 1978) or 63% (Fox *et al.*, 1988). Bovine-milk NAGase exhibits electrophoretic and chromatographic (gel filtration) patterns similar to mammary-gland tissue NAGase (Kitchen, 1985), suggesting that most of the milk enzyme originates in the mammary secretory cells; somatic cells and blood serum contribute 5–15% of the NAGase in milk (Kitchen *et al.*, 1978; Fox *et al.*, 1988).

Bovine milk has two NAGase iso-enzymes, each consisting of two subunits (Kitchen *et al.*, 1979; Kitchen, 1985). The isoenzymes differ in molecular weight and heat stability. The lower-molecular-weight (~120 000 Da) iso-enzyme is heat labile, while the (MW ~240 000 Da) is heat stable (Kitchen *et al.*, 1979; Kitchen, 1985). Similarly, human milk contains two NAGase iso-enzymes, which differ in their heat stabilities (Oberkotter *et al.*, 1982). The optimum pH for bovine-milk NAGase is 4·2 and its K_m for *β*-*N*-acetylglucosaminide is 1 mM (Mellors, 1968); its optimum temperature is ~50°C (Kitchen *et al.*, 1978). The enzyme is completely inactivated when milk is heated at 70°C for 15 s (Griffiths, 1986), and residual NAGase activity can be used to detect milk that had been heated to 65–75°C (Andrews *et al.*, 1987).

NAGase activity in milk is recommended as an index of mastitis (Kitchen & Middleton, 1976; Kitchen *et al.*, 1978; 1980). Positive correlation coefficients of 0·86 (Kitchen *et al.*, 1978), 0·72 (Obara & Komatsu, 1984) or 0·62 (Miller & Paape, 1988) have been obtained between milk-NAGase activity and somatic cell count. NAGase activity in milk varies during lactation and is 10 times higher in dry cow secretions than in milk (Miller & Paape, 1988). Kitchen *et al.* (1978) described a routine assay for milk-NAGase activity as an index of udder infection. However, because of variations in NAGase activity in bovine milk during lactation, Miller and Paape (1988) caution against its use as the sole indicator of udder infections.

β-Glucuronidase (EC 3.2.1.31)

β-Glucuronidase is a lysosomal acid hydrolase which catalyses the following reaction:

$$\beta\text{-D-glucuronoside} + H_2O \rightarrow \text{alcohol} + \text{D-glucoronic acid}$$

The enzyme is present in multiple forms in rat (Owen *et al.*, 1975) and human (Gasper *et al.*, 1975) liver. The active enzyme, with a molecular weight of ~280 000 Da, consists of four identical subunits, each with a molecular weight of ~75 000 Da (Stahl & Touster, 1971). The isoelectric pH of β-glucuronidase is 6·7 (Owen *et al.*, 1975).

Kiermeier and Gull (1968) demonstrated the presence of β-glucoronidase activity in human and bovine milks. To our knowledge, β-glucuronidase has not been purified from milk and there is no information on the molecular properties of the milk enzyme. Activity of the enzyme in milk depends on the stage of lactation and the disease status of the animal. β-Glucuronidase activity is highest in colostrum and decreases between the second and tenth days of lactation (Kiermeier & Gull, 1968). These authors also found that β-glucuronidase activity in milk is proportional to the leucocyte count up to 2×10^6 cells/ml. Perdigon *et al.* (1986) also observed high β-glucuronidase activity in milk samples with somatic-cell counts between 5×10^5 and 2×10^6 cells/ml, and suggested (Perdigon *et al.*, 1988) combining β-glucuronidase activity and somatic-cell count in milk for the diagnosis of mastitis.

HTST pasteurization (71·5°C × 15 s) of milk completely inactivates β-glucuronidase activity, but some activity survives UHT treatment (Kiermeier & Doruk, 1972*a*). According to Kiermeier and Doruk (1972*a*), the heat stability of milk β-glucoronidase is identical to that of alkaline phosphatase at 71·5°C but β-glucuronidase is less stable than alkaline phosphatase at lower temperatures and more stable at high temperatures. In milk, β-glucuronidase activity is inhibited by Cu^{2+}, the inhibitory effectiveness of which is increased by ascorbic acid (Kiermeier & Doruk, 1972*b*).

α-Mannosidase (EC 3.2.1.24)

α-Mannosidase catalyses the hydrolysis of terminal non-reducing α-D-mannose residues from α-D-mannosides. Its activity in milk was first demonstrated by Mellors and Harwalkar (1968), who also found that the enzyme activity was distributed between cream (~3·5%), skim milk (~86·3%) and separator slime (~10%). Attempts to purify the enzyme from milk by chromatography failed; most of the activity in the skim-milk fraction is associated with β-casein (Mellors & Harwalkar, 1968). Dilution of milk with water or 0·1 M NaCl reduces its α-mannosidase activity by 50 and 35%, respectively. For the hydrolysis of *p*-nitrophenyl α-mannoside, the enzyme has an optimum pH of 2·7–2·9 and a K_m of 1·2 mM (Mellors & Harwalker, 1968). It is inhibited by Cu, EDTA and dibasic amino acids (Mellors & Harwalkar, 1968). α-Mannosidase is heat stable: ~90% of α-mannosidase activity survives HTST pasteurization of milk, and heating milk for 15 s at 80, 85 and 90°C reduced its α-mannosidase activity by 7, 76 and 100%, respectively (Andrews *et al.*, 1987).

5′-Nucleotidase (EC 3.1.3.5)

This enzyme is a glycoprotein which catalyses the hydrolysis of 5′-nucleotides with the release of orthophosphate (Pi):

$$5'\text{-ribonucleotide} + H_2O \rightarrow \text{ribonucleoside} + \text{Pi}$$

5′-Nucleotidase is a constituent of liver–plasma membranes (Emmelot & Bos, 1966) and is present in milk on the outer surface of MFGM (Patton & Trams, 1971; Snow *et al.*, 1980). The enzyme has been purified from acid whey (Caulini *et al.*, 1972) and MFGM (Huang & Keenan, 1972*a,b*).

The 5′-nucleotidase purified from acid whey had an estimated molecular weight of ~134 000 Da (Caulini *et al.*, 1972). Huang and Keenan (1972*a*) isolated two 5′-nucleotidase isoenzymes (designated as Fractions V and VI) which differed in their phospholipid content. The enzyme hydrolyses AMP, GMP (Caulini *et al.*, 1972; Huang & Keenan, 1972*a*), IMP (Caulini *et al.* 1972) and CMP (Huang & Keenan, 1972*a*). It does not require a metal ion for activity (Caulini *et al.*, 1972; Huang & Keenan, 1972*a*). Its pH optimum is 7·0–7·5 (Caulini *et al.*, 1972; Huang & Keenan, 1972*a*), but in the presence of Mg^{2+}, its optimum pH is 10·0 (Huang & Keenan, 1972*a*). Fractions V and VI had an optimum temperature of 69°C but activation energies of 43·68 kJ/mol and 40·32 kJ/mol, respectively (Huang & Keenan, 1972*a*). ATP (Caulini *et al.*, 1972; Huang & Keenan, 1972*a*) and $HgCl_2$ (Huang & Keenan, 1972*a*) are inhibitors of 5′-nucleotidase.

The enzyme fully survived in milk heated at 60°C for 15 s, but heating milk for 15 s at 72 or 80°C reduced its activity by 40 and 97%, respectively (Andrews *et al.*, 1987).

The major significance of 5′-nucleotidase in milk is its use as a 'marker' enzyme for cellular plasma-membrane material, in studying biogenesis of MFGM material and the secretory mechanisms of fat globule (Kitchen, 1985).

TABLE 5
Partial List of Minor Enzymes in Milk

EC No.	*Enzyme*	*Reaction Catalysed*	*Source*	*Distribution in Milk*	*Reference*
1.1.1.14	L-Iditol dehydrogenase	L-Iditol + $NAD^+ \rightleftharpoons$ L-sorbose + NADH	—	SM	Neuman *et al.* (1965)
1.1.1.27	Lactate dehydrogenase	L-Lactate + $NAD^+ \rightleftharpoons$ pyruvate + NADH	Mammary gland	SM	Deodhar *et al.* (1964); Karlsson & Kjellberg (1966); Hellung-Larsen (1968); Grigor & Hartmann (1985)
1.1.1.37	Malate dehydrogenase	Malate + $NAD^+ \rightleftharpoons$ oxaloacetate + NADH	Mammary gland	SM	Grigor & Hartmann (1985); Andrews *et al.* (1987)
1.1.1.40	Malic enzyme	Malate + $NADP^+ \rightleftharpoons$ pyruvate + CO_2 + NADPH	Mammary gland	SM	Deodhar *et al.* (1964); Karlsson & Kjellberg (1966); Grigor & Hartmann (1985)
1.1.1.42	Isocitrate dehydrogenase	Isocitrate + $NADP^+ \rightleftharpoons$ 2-oxogluterate + CO_2 + NADPH	Mammary gland	SM	Grigor & Hartmann (1985)
1.1.1.44	Phosphoglucuronate dehydrogenase (decarboxylating)	6-Phospho-D-gluconate + $NADP^+$ $\rightleftharpoons$ D-ribose-5 phosphate + CO_2 + NADPH	Mammary gland	SM	Grigor & Hartmann (1985)
1.1.1.49	Glucose-6-phosphate dehydrogenase	D-Glucose-D-gluconate + $NADP^+ \rightleftharpoons$ D-glucono-1,5-lactone 6 phosphate + NADPH	Mammary gland	SM	Sklavunu-Zurukzoglu *et al.* (1965); Grigor & Hartmann (1985)
1.11.1.9	Glutathione peroxidase	2 Gluthatione + $H_2O_2 \rightleftharpoons$ oxidized gluthatione + $2H_2O$	—	SM	Hojo (1982); Williams (1983)
1.4.3.6	Amine oxidase (Cu-containing)	$RCH_2NH_2 + H_2O + O_2 \rightleftharpoons RCHO + NH_3 + H_2O_2$	—	SM	Patterson *et al.* (1969)
—	Polyamine oxidase	spermine $\rightleftharpoons$ spermidine $\rightleftharpoons$ putrescine	—	SM	Morgan & Toothill (1985)
—	Fucosyltransferase	catalyses the transfer of fucose from GDP L-fucose to specific oligosaccharides and glycoproteins	—	SM	Jarkovsky *et al.* (1970); Prieels *et al.* (1977)
1.6.99.3	NADH dehydrogenase (Cytocrome reductase)	NADH + acceptor $\rightleftharpoons NAD^+$ + reduced acceptor	—	FGM	Martel-Pradal & Got (1972); Jarasch *et al.* (1977)
1.8.1.4	Dihydrolipoamide dehydrogenase (Diaphorase)	Dihydrolipoamide + $NAD^+ \rightleftharpoons$ lipoamide + NADH	—	SM/FGM	Kitchen (1974)

2.4.1.22	Lactose synthetase A protein, UDP-galactose: D-glucose, 1-galactosyltransferase B protein: α-lactalbumin	UDP-galactose + D-glucose ⇌ UDP + lactose	—	SM	Babad & Hassid (1964); Brodbeck & Ebner (1966)
2.4.1.38	Glycoprotein 4-β-galactosyltransferase	UDP galactose + *N*-acetyl-D-glucosaminyl-glycopeptide ⇌ UDP + 4, β-D-galactosyl-*N*-acetyl-D-glucosaminyl glycopeptide	—	FGM	Blaken *et al.* (1982)
2.4.1.90	*N*-Acetyllactosamine synthase	UDP galactose + *N*-acetyl-D-glucosamine ⇌ UDP + *N*-acetyllactosamine	Golgi apparatus		Powell *et al.* (1977)
2.4.99.6	CMP-*N*-acetylneuraminate-*N*-acetyllactosaminide α-2, 3-sialytransferase	CMP-*N*-acetylneuraminate + β-D-galactosyl 1,4-*N*-acetyl-D-glucosaminyl glycoprotein ⇌ CMP + α-*N*-acetylneuraminyl 1-2, 3-β-D-galactasyl-1,4-*N*-acetyl-D-glucosaminyl-glycoprotein	—	SM	Hesford & Berger (1983)
2.5.1.3	Thiamin-phosphate pyrophosphorylase	2-methyl-4-amino-5-hydroxymethyl/pyrimidine diphosphate + 4-methyl-5-(2-phosphonooxyethyl)-thiazole ⇌ pyrophosphate + thiamine monophosphate	—	FGM	Martel-Pradal & Got (1972)
2.6.1.1	Aspartate aminotransferase	L-aspartate + 2-oxogluterate ⇌ oxaloacetate + L-glutamate	Blood	SM	Heyndrickx (1962); Neuman *et al.* (1965)
2.6.1.2	Alanine aminotransferase	L-alanine + 2-oxogluterate ⇌ pyruvate + L-glutamate	Blood	SM	Heyndrickx (1962); Neuman *et al.* (1965)
2.7.7.49	RNA-directed DNA polymerase	*n* Deoxynucleoside triphosphate ⇌ *n* pyrophosphate + DNA_n	—	SM	Schlom *et al.* (1972); Gerwin *et al.* (1973); Roy-Burman *et al.* (1973)
2.8.1.1	Thiosulphate sulphurtransferase	Thiosulphate + cyanide ⇌ sulphite + thiocyanate	—	SM	Garcia Alfonso & Costella Bertran (1953)
3.1.1.8	Cholinesterase	An acylcholine + H_2O ⇌ choline + a carboxylic acid anion	Blood	FGM	Augustinsson (1960); Montoure & Forster (1960)

TABLE 5—*contd.*

EC No.	*Enzyme*	*Reaction Catalysed*	*Source*	*Distribution in Milk*	*Reference*
3.1.3.9	Glucose-6-phosphatase	D-Glucose 6-Phosphate + $H_2O \rightleftharpoons$ D-glucose + Pi	—	FGM	Dowben *et al.* (1967); Martel-Pradal & Got (1972)
3.1.6.1	Arylsulphatase	A phenol sulphate + $H_2O \rightleftharpoons$ a phenol + $\rightleftharpoons$ sulphate	—	—	Proczekaj & Wenclewski (1963); Kitchen (1976)
3.2.1.21	β-Glucosidase	Hydrolysis of terminal non-reducing β-D-glucose residues	Lysosomes	FGM	Anderson (1977)
3.2.1.23	β-Galactosidase	Hydrolysis of terminal non-reducing β-D-galactose residues in β-D-galactosides	Lysosomes	FGM	Anderson & Cheeseman (1975); Griffiths (1986)
3.2.1.51	α-L-Fucosidase	An α-L-fucoside + $H_2O \rightleftharpoons$ an alcohol + L-fucose	Lysosomes	—	Kitchen (1976); Andrews *et al.* (1987)
3.4.11.1	Cytosol aminopeptidase (Leucine aminopeptidase)	Aminoacyl-peptide + $H_2O \rightleftharpoons$ amino acid + peptide	—	SM	Serban *et al.* (1969); Eichel (1981)
3.4.11.3	Cystyl-aminopeptidase (Oxytocinase)	Cystyl-peptidase + $H_2O \rightleftharpoons$ amino acid + peptide	—	SM	Brouet-Yager *et al.* (1974)
3.4.21.4	Trypsin	Hydrolyses peptide bond, preferentially Arg-, Lys-	—	SM	Mizakorimov & Tashmatova (1971); Borulf *et al.* (1987)
3.6.1.1	Inorganic pyrophosphatase	Pyrophosphate + $H_2O \rightleftharpoons$ 2 orthophosphate	—	SM/FGM	Heyndrickx (1963); Kitchen (1974)
3.6.1.3	Adenosinetriphosphatase (Mg^{2+} activated)	ATP + $H_2O \rightleftharpoons$ 2 orthophosphate	—	SM/FGM	Heyndrickx (1963); Patton & Trams (1971); Huang & Keenan (1972*b*)
3.6.1.9	Nucleotide pyrophosphatase	A dinucleotide + $H_2O \rightleftharpoons$ 2 mononucleotides	—	SM/FGM	Patton & Trams (1971); Kitchen (1974)
4.2.1.1	Carbonate dehydratase	$H_2CO_3 \rightleftharpoons CO_2 + H_2O$	—	SM	Garcia Alfonson & Costella Bertran (1953)
5.3.1.9	Glucose 6-phosphate isomerase	D-Glucose-6-phosphate $\rightleftharpoons$ fructose 6-phosphate	—	SM	Heyndrickx (1962)
6.4.1.2	Acetyl-CoA carboxylase	ATP + acetyl-CoA + $HCO_3^- \rightleftharpoons$ ATP orthophosphate + malonyl-CoA	—	FGM	Shriver *et al.* (1989)

SM = skim milk; FGM = fat globule membrane

Aldolase (EC 4.1.2.13)

Aldolase (fructose-bisphosphate aldolase) is a lysosomal enzyme which catalyses the hydrolysis of D-fructose 1,6-bisphosphate to glycerol phosphate and D-glyceraldehyde 3-phosphate. It is present in a variety of animal tissues (Sibley & Lehninger, 1949) and was first demonstrated in milk by Polis and Shmukler (1950), who also found that the aldolase activity in milk was similar to that in blood serum (Sibley & Lehninger, 1949), suggesting that the enzyme enters milk from blood (leucocytes). While Polis and Shmukler (1950) found most of the milk-aldolase activity in the cream layer where it is associated with the MFGM (Erwin & Randolph, 1975), Kitchen *et al.* (1970) found ~66% of the total milk-aldolase activity in the skim-milk phase and only 19% in the cream phase. Aldolase activity is higher in mastitic than in normal milk (Kitchen *et al.*, 1970).

The enzyme is pH-sensitive and is inactivated during isoelectric precipitation of casein (Polis & Shmukler, 1950). The enzyme is unstable in milk at 37°C (Polis & Shmukler, 1950). Approximately 80% of the total aldolase activity in milk remains soluble in 2·8 M (25% saturation) $(NH_4)_2SO_4$ which precipitates the caseins and most of the whey proteins (Polis & Shmukler, 1950). Kitchen *et al.* (1970) found ~44% of the total milk-aldolase activity in the supernatant obtained after making milk to 70% saturation with $(NH_4)_2SO_4$.

Other enzymes

The activities of several 'minor' enzymes have been demonstrated in milk. A partial list of these enzymes is given in Table 5. Though present, these enzymes have not been studied sufficiently to elucidate their molecular properties and significance in milk. Lactose synthetase (2.4.1.22) is included in Table 5. The individual components (A and B proteins) of the lactose synthetase are well characterized, but the system probably has no activity in milk and is therefore not discussed here.

REFERENCES

Abernethy, J. L., Steinman, H. M. & Hill, R. L. (1974). *J. Biol. Chem.*, **249**, 7339–47.

American Heart Association (1981). Advisory statement concerning the claims that consumption of homogenized milk increases the risk of heart disease. Nutrition Program Committee, American Heart Association, Dallas, Texas, USA.

Anderson, M. (1977). *J. Dairy Sci.*, **60**, 1217–22.

Anderson, M. & Cheeseman, G. C. (1975). *IDF Bull. Doc.*, **80**, 11–18.

Andrews, A. T., Anderson, M. & Goodenough, P. W. (1987). *J. Dairy Res.*, **54**, 237–46.

Asada, K. (1976). *Agric. Biol. Chem.*, **40**, 1659–60.

Augustinsson, K.-B. (1960). *Experimentia*, **16**, 411–12.

Aurand, L. W., Boone, N. H. & Giddings, G. G. (1977). *J. Dairy Sci.*, **60**, 363–9.

Babad, H. & Hassid, W. Z. (1964). *J. Biol. Chem.*, **239**, PC946.

Barcelos, J. de A. (1983). Inaugural Dissertation, Tierarzliche Hochschule, Hannover, Germany; *Dairy Sci. Abstr.* (1985) **47,** 113.
Barcelos, J. de A. & Weight, U. (1986). *J. Veter. Med.*, **33,** 308–19; *Dairy Sci. Abstr.* (1986), **48,** 760.
Battelli, M. G., Lorenzoni, E. & Stirpe, F. (1973). *Biochem. J.*, **131,** 191–8.
Baumrucker, C. R. (1979). *J. Dairy Sci.*, **62,** 253–8.
Baumrucker, C. R. (1980). *J. Dairy Sci.*, **63,** 49–54.
Bingham, E. W. & Kalan, E. B. (1967). *Arch. Biochem. Biophys.*, **121,** 317–24.
Bingham, E. W. & Zittle, C. A. (1962). *Biochem. Biophys. Res. Comm.*, **7,** 408–13.
Bingham, E. W. & Zittle, C. A. (1964). *Arch. Biochem. Biophys.*, **106,** 235–9.
Björck, L. & Claesson, O. (1979). *J. Dairy Sci.*, **62,** 1211–15.
Blaken, W. M., Hooghwinkel, G. J. M. & Eijnden, D. H. van den (1982). *Eur. J. Biochem.*, **127,** 547–52.
Bordas, J., Bray, R. C., Garner, C. D., Gutteridge, S. & Hasnain, S. S. (1980). *Biochem. J.*, **191,** 499–508.
Borulf, S., Lindberg, T. & Mansson, M. (1987). *Acta Pediatr. Scand.*, **76,** 11–15.
Braun, J. P., Tainturier, D., Laugier, C., Benard, P., Thouvenot, J. P. & Rico, A. G. (1982). *J. Dairy Sci.*, **65,** 2178–81.
Braun, J. P., Tainturier, D., Bezille, P., Raviart, I. & Rico, A. G. (1984). *Enzyme,* **31,** 193–6.
Brew, K. & Campbell, P. M. (1967). *Biochem. J.*, **102,** 258–64.
Brignon, G. & Ribadeau-Dumas, B. (1982). *Biochimie,* **64,** 231–5.
Briley, M. S. & Eisenthal, R. (1975). *Biochem. J.*, **147,** 417–23.
Brodbeck, U. & Ebner, K. E. (1966). *J. Biol. Chem.*, **241,** 762–4.
Brouet-Yager, M., Kleiner, H., Bogaert, E. & Guff, G. L. A. van (1974). *Clin. Chim. Acta,* **54,** 387–9.
Browne, W. J., North, A. C. T., Phillips, D. C., Brew, K., Vanaman, T. C. & Hill, R. L. (1969). *J. Mol. Biol.*, **42,** 65–86.
Bruder, G., Heid, H., Jorasch, E.-D., Keenan, T. W. & Mather, I. H. (1982). *Biochim. Biophys. Acta,* **701,** 357–69.
Carlsson, B., Cruz, J. R., Garcia, B., Hanson, L. A. & Urrutia, J. J. (1979). In *Nutrition and Metabolism of the Foetus and Infants,* ed. H. K. A. Visser. Martinus Jijhoff, The Hague, The Netherlands, p. 263. (Cited by Proctor, V. A. & Cunningham, F. E. (1988). *CRC Crit. Rev. Food Sci. Nutr.*, **36,** 359.)
Caulini, G., Binotti, I., Allessandrini, M., Ipata, P. L. & Magni, G. (1972). *Bullettino della Societa Italiana di Biologia Sperimentale,* **48,** 890–3.
Cerbulis, J. & Farrell, H. M. (1977). *J. Dairy Sci.*, **60,** 170–6.
Chandan, R. C., Parry Jr, R. M. & Shahani, K. M. (1965). *Biochim. Biophys. Acta,* **110,** 389–98.
Chandan, R. C., Parry Jr, R. M. & Shahani, K. M. (1968). *J. Dairy Sci.*, **51,** 606–7.
Cheng, S. G., Koch, U. & Brunner, J. R. (1988). *J. Dairy Sci.*, **71,** 901–16.
Clare, D. A., Blakistone, B. A., Swaisgood, H. E. & Horton, H. R. (1981). *Arch. Biochem. Biophys.*, **211,** 44–7.
Clare, D. A., Horton, H. R., Stabel, T. J., Swaisgood, H. E. & Lecce, J. G. (1984). *Arch. Biochem. Biophys.*, **230,** 138–45.
Clifford, A. J., Ho, C. Y. & Swenerton, H. (1983). *Amer. J. Clin. Nutr.*, **38,** 327–32.
Dalaly, B. K., Eitenmiller, R. R., Vakil, J. R. & Shahani, K. M. (1970). *Anal. Biochem.*, **37,** 208–11.
Davis, M. D., Edmondson, D. E. & Muller, F. (1984). *Eur. J. Biochem.*, **145,** 237–43.
Deeth, H. C. (1983). *J. Dairy Sci.*, **66,** 1419–35.
Deodhar, A. D., Rajalakshimi, R. & Ramakrishnan, C. V. (1964). *Acta Peadiatr. Stockh.*, **53,** 101–4; *Dairy Sci. Abstr.* (1968), **26,** 602.
Dixon, M., Webb, E. C. (assisted by Thorne, C. J. R. & Tipton, K. F.) (1979). *Enzymes,* 3rd ed. Academic Press, New York, p. 862.

Dowben, R. M., Brunner, J. R. & Philpott, D. E. (1967). *Biochim. Biophys. Acta,* **135,** 1–10.
Edmondson, D., Massey, V., Palmer, G., Beacham, L. M. & Elion, G. B. (1972). *J. Biol. Chem.,* **247,** 1597–604.
Eichel, V. (1981). *Archiv fur Experimentalle Veterinarmedzin,* **35,** 235–44. *Dairy Sci. Abstr.* (1982), **44,** 478.
Eitenmiller, R. R., Friend, B. A. & Shahani, K. M. (1971). *J. Dairy Sci.,* **54,** 762 (Abstr.).
Eitenmiller, R. R., Friend, B. A., Shahani, K. M. & Ball, E. M. (1974). *J. Food Sci.,* **39,** 930–3.
Eitenmiller, R. R., Friend, B. A. & Shahani, K. M. (1976). *J. Dairy Sci.,* **59,** 834–9.
Emmelot, P. & Bos, C. J. (1966). *Biochim. Biophys. Acta,* **120,** 369–82.
Erwin, R. E. & Randolph, H. E. (1975). *J. Dairy Sci.,* **58,** 9–11.
Evans, H. J., Steinman, H. M. & Hill, R. L. (1974). *J. Biol. Chem.,* **249,** 7315–25.
Fox, P. F. & Morrissey, P. M. (1981). In *Enzymes and Food Processing,* ed. G. G. Birch, N. Blakenbrough & K. J. Parker. Applied Science Publishers, London, UK, p. 213–38.
Fox, L. K., Hancock, D. D., McDonald, J. S. & Gaskins, C. T. (1988). *J. Dairy Sci.,* **71,** 2915–22.
Freundenstein, C., Keenan, T. W., Eigel, W. N., Sasaki, M., Stadler, J. & Franke, W. (1979). *Exper. Cell Res.,* **118,** 277–94.
Fridovich, I. (1970). *J. Biol. Chem.,* **254,** 4054–7.
Fridovich, I. (1974). *Adv. Enzymol. and Related Areas of Mol. Biol.,* **41,** 35–97.
Fridovich, I. (1986). *Adv. Enzymol. and Related Areas of Mol. Biol.,* **58,** 61–97.
Friend, B. A., Eitenmiller, R. R. & Shahani, K. M. (1972). *Arch. Biochem. Biophys.,* **149,** 435–40.
Friend, B. A., Shahani, K. M., Long, C. A. & Agel, E. N. (1983). *J. Food Protect.,* **46,** 330–4.
Garcia Alfonso, C. & Castella Bertran, E. (1953). *Le Lait,* **33,** 386–93.
Gasper, J. H., Roozen, K. J., Brot, F. E. & Sly, W. S. (1975). *Arch. Biochem. Biophys.,* **116,** 524–36.
Gerwin, B. I., Ebert, P. S., Chopra, H. C., Smith, S. G., Kvedar, J. P., Albert, S. & Brennan, M. J. (1973). *Science,* **180,** 198–201.
Gilbert, D. A. & Bergel, F. (1964). *Biochem. J.,* **90,** 350–3.
Got, R., Bertagnolo, G., Pradal, M. B. & Frot-Coutaz, J. (1968). *Clin. Chim. Acta,* **22,** 545–50.
Goussault, Y., Got, R. & Marnay, A. (1966). *Protides Biol. Fluids,* **14,** 621–3. H. Peeters (ed.), Elsevier Publishing Co. London, pp. 621–3.
Graham, K., Fleming, J. E., Young, R. & Bensch, K. (1989). *Int. J. Biochem.,* **21,** 715–22.
Griffiths, M. W. (1986). *J. Food Protect.,* **49,** 696–705.
Grigor, M. R. & Hartman, P. E. (1985). *J. Dairy Res.,* **52,** 501–6.
Groves, M. L. (1971). In *Milk Proteins, Chemistry and Molecular Biology* (Vol. 2), ed. H. A. McKenzie. Academic Press, New York, USA, p. 408.
Gutteridge, S., Tanner, S. J. & Bray, R. C. (1978). *Biochem. J.* **175,** 887–97.
Guy, E. J. & Jenness, R. (1958). *J. Dairy Sci.,* **41,** 13–27.
Hamosh, M. (1986). In *Human Milk in Infant Nutrition and Health,* ed. R. R. Howell, F. H. Morriss Jr & L. K. Perkings. Charles C. Thomas, Springfield, IL, USA, pp. 66–97.
Hamosh, M. (1988). *Nestle Nutrition Workshop Series,* **15,** 41–61; *Dairy Sci. Abstr.* (1989), **51,** 415.
Hart, L. I. & Bray, R. C. (1967). *Biochim. Biophys. Acta* **146,** 611–13.
Hegardt, P., Lindberg, T., Borjesson, J. & Skude, G. (1984). *J. Pediatr. Gastroenterol. Nutr.,* **3,** 563–6; *Dairy Sci. Abstr.* (1985), **47,** 116.

Heitlinger, L. A., Lee, P. C., Dillion, W. P. & Lebenthal, E. (1983). *Pediatr. Res.*, **17**, 15–18.
Hellung-Larsen, P. (1968). *Comp. Biochem. Physiol.*, **27**, 703–8.
Hesford, F. J. & Berger, C. G. (1983). *Experientia*, **39**, 648.
Heyndrickx, G. V. (1962). *Ann. Paediatr.*, **198**, 356–62; *Dairy Sci. Abstr.* (1963), **25**, 82.
Heyndrickx, G. V. (1963). *Pediatrics*, **31**, 1019–23.
Hicks, C. L. (1980). *J. Dairy Sci.*, **63**, 1199–204.
Hicks, C. L., Korycka-Dahl, M. & Richardson, T. (1975). *J. Dairy Sci.*, **58**, 796 (Abstr.).
Hicks, C. L., Bucy, J. & Stofer, W. (1979). *J. Dairy Sci.*, **62**, 529–32.
Hill, R. D. (1975). *Aust. J. Dairy Technol.*, **30**, 26–8.
Hill, R. L., Brew, K., Vanaman, T. C., Trayer, I. P. & Mattock, P. (1968). *Structure, Function and Evolution in Proteins*, **21**, 139–54.
Hojo, Y. (1982). *Biol. Trace Element Res.*, **4**, 233–9; *Dairy Sci. Abstr.* (1984), **46**, 629.
Holbrook, J. & Hicks, C. L. (1978). *J. Dairy Sci.*, **61**, 1072–7.
Huang, C. M. & Keenan, T. W. (1972*a*). *Biochim. Biophys. Acta*, **274**, 246–57.
Huang, C. M. & Keenan, T. W. (1972*b*). *Comp. Biochem. Physiol.*, **43B**, 277–82.
Hurley, W. L. & Grieve, R. C. J. (1988). *Veter. Res. Comm.*, **12**, 149–53; *Dairy Sci. Abstr.* (1988), **50**, 755.
Isaacs, C. E. (1985). In *Human Lactation: Milk Components and Methodologies*. ed. R. G. Jensen & M. C. Neville. Plenum Press, New York, USA, pp. 277–82.
Isaacs, C. E., Pascal, T., Wright, C. E. & Gaull, G. E. (1984). *Pediatr. Res.*, **18**, 532–5.
Ito, O. (1971). *Bull. Nippon Veter. Zootech. Coll.*, **20**, 56–61.
Ito, O. (1974). *Bull. Nippon Veter. Zootech. Coll.*, **23**, 45–8.
Ito, O. & Akuzawa, R. (1982). *Bull. Nippon Veter. Zootech. Coll.*, **31**, 198–202.
Ito, O. & Akuzawa, R. (1983*a*). *J. Dairy Sci.*, **66**, 967–73.
Ito, O. & Akuzawa, R. (1983*b*). *Jap. J. Zootech. Sci.*, **54**, 39–45.
Ito, O. & Akuzawa, R. (1983*c*). *J. Dairy Sci.*, **66**, 2468–73.
Ito, O., Akuzawa, R. & Kamata, S. I. (1984). *Jap. J. Zootech. Sci.*, **55**, 722–7.
Janalino, V. G. & Swaisgood, H. E. (1975). *J. Biol. Chem.*, **250**, 2532–8.
Janalino, V. G. & Swaisgood, H. E. (1978). *J. Dairy Sci.*, **61**, 393–9.
Janalino, V. G. & Swaisgood, H. E. (1979). *J. Dairy Sci.*, **62** (Suppl. 1) **64** (Abstr.).
Janalino, V. G., Barnes, C. S. & Swaisgood, H. E. (1980). *J. Dairy Sci.*, **63**, 1969–74.
Janalino, V. G., Clare, D. A. & Swaisgood, H. E. (1981). *Biochim. Biophys. Acta*, **658**, 460–9.
Jarasch, E.-D., Bruder, G., Keenan, T. W. & Fanke, W. W. (1977). *J. Cell Biol.*, **73**, 223–41.
Jarkovsky, Z., Marcus, D. M. & Gollman, A. P. (1970). *Biochem.*, **9**, 1123–8.
Jauregui-Adell, J. (1975). *J. Dairy Sci.*, **58**, 835–8.
Jolles, J. & Jolles, P. (1967). *Biochem.*, **6**, 411–17.
Jolles, P., Zowall, H., Jauregui-Adell, J. & Jolles, J. (1962). *J. Chromatogr.*, **8**, 363–8.
Jones, J. B., Metha, N. R. & Hamosh, M. (1982). *J. Pediatr. Gastroenterol. Nutr.*, **1**, 43–8; *Dairy Sci. Abstr.* (1984), **46**, 455.
Karlsson, B. & Kjellberg, B. (1966). *Naturwissenschaften*, **53**, 90.
Keen, C. L., Lonnerdal, B., Stein, T. N. & Hurley, L. S. (1980). *Biol. Trace Element Res.*, **2**, 221–7; *Dairy Sci. Abstr.* (1982), **44**, 473.
Keenan, T. W., Mather, I. H. & Dylewski, D. P. (1988). In *Fundamentals of Dairy Chemistry*, ed. N. P. Wong, R. Jenness, M. Keeney & E. H. Marth. Van Nostrand Reinhold Co., New York, USA, pp. 511–82.
Kiermeier, F. & Doruk, M. (1972*a*). *Z. Lebensm. Unters-Forsch.*, **150**, 220–4.
Kiermeier, F. & Doruk, M. (1972*b*). *Z. Lebensm. Unters-Forsch.*, **150**, 274–6.
Kiermeier, F. & Gull, J. (1968). *Z. Lebensm. Unters-Forsch.*, **138**, 205–8. *Dairy Sci. Abstr.* (1969) **31**, 96.

Kiermeier, F. & Petz, E. (1967*a*). *Z. Lebensm. Unters-Forsch.*, **132,** 342–52.
Kiermeier, F. & Petz, E. (1967*b*). *Z. Lebensm. Unters-Forsch.*, **134,** 97–102.
Kiermeier, F. & Petz, E. (1967*c*). *Z. Lebensm. Unters-Forsch.*, **134,** 149–56.
Kitchen, B. J. (1974). *Biochim. Biophys. Acta,* **356,** 257–69.
Kitchen, B. J. (1976). *J. Dairy Res.*, **43,** 251–8.
Kitchen, B. J. (1985). In *Developments in Dairy Chemistry* (Vol. 3), ed. P. F. Fox. Elsevier Applied Science Publishers Ltd, London, UK, pp. 239–79.
Kitchen, B. J. & Middelton, G. (1976). *J. Dairy Res.*, **43,** 491–4.
Kitchen, B. J., Taylor, G. C. & White, I. C. (1970). *J. Dairy Res.*, **37,** 279–88.
Kitchen, B. J., Middelton, G. & Salmon, M. C. (1978). *J. Dairy Res.*, **45,** 15–20.
Kitchen, B. J., Masters, C. J. & Winzor, D. J. (1979). *Proc. Aust. Biochem. Soc.*, **12,** 26; *Dairy Sci. Abstr.* (1980) **42,** 241.
Kitchen, B. J., Middelton, G., Durward, I., Andrews, R. D. & Salmon, M. C. (1980). *J. Dairy Sci.*, **63,** 978–83.
Korycka-Dahl, M., Richardson, T. & Hicks, C. L. (1979). *J. Food Protect.*, **42,** 867–71.
Kumari, V. & Mathur, M. P. (1981). *Indian J. Dairy Sci.*, **34,** 385–90.
Lewis-Jones, D. I. & Reynolds, G. J. (1983). *Acta Paediatr. Scand.*, **72,** 13–17.
Lindberg, T. & Skude, G. (1982). *Pediatrics,* **70,** 235–8.
Maisi, P. & Riipinen, I. (1988). *J. Dairy Res.*, **55,** 309–14.
Majumder, G. C. & Ganguli, N. C. (1972). *Milchwissenschaft,* **27,** 296–9.
Mangino, M. E. & Brunner, J. R. (1977). *J. Dairy Sci.*, **60,** 841–50.
Martel-Pradal, M. B. & Got, R. (1972). *FEBS Lett.*, **21,** 220–2.
Massey, V. & Edmondson, D. (1970). *J. Biol. Chem.*, **245,** 6595–8.
Mattila, T., Pyorala, S. & Sandholm, M. (1986). *Vet. Res. Comm.*, **10,** 113–24; *Dairy Sci. Abstr.* (1986), **48,** 402.
McBean, L. B. & Speckman, W. E. (1988). In *Fundamentals of Dairy Chemistry,* ed. N. P. Wong, R. Jenness, M. Keeney & E. H. Marth. Van Nostrand Reinhold Co., New York, USA, pp. 350–60.
McCord, J. M. & Fridovich, I. (1969). *J. Biol. Chem.*, **244,** 6049–63.
McCormick, J. J., Larson, L. J. & Rich, M. A. (1974). *Nature,* **251,** 737–40.
McKenzie, H. A. & White Jr, F. H. (1986). *Anal. Biochem.*, **157,** 367–74.
McKenzie, H. A., Pearce, R. J. & White Jr, F. H. (1985). *Proc. Aust. Biochem. Soc.*, **17,** 132 (Abstr.).
Meister, A. (1973). *Science,* **180,** 33–9.
Meister, A. & Tate, S. S. (1976). *Ann. Rev. Biochem.*, **45,** 559–604.
Mellors, A. (1968). *Can. J. Biochem.*, **46,** 452–5.
Mellors, A. & Harwalkar, V. R. (1968). *Can. J. Biochem.*, **46,** 1351–6.
Meyer, D. H., Kunin, A. S., Maddalena, J. & Meyer, W. L. (1987*a*). *J. Dairy Sci.*, **70,** 1797–803.
Meyer, W. L., Meyer, D. H., Kunin, A. S., Capeless, E. L., Simmons, K. R. & Pankey, J. W. (1987*b*). *J. Dairy Sci.*, **70** (Suppl. 1), 266 (Abstr.).
Miller, R. H. & Paape, M. J. (1988). *J. Dairy Sci.*, **71,** 2508–12.
Miller, R. H., Paape, M. J., Peters, R. R. & Young, M. D. (1988). *J. Dairy Sci.*, **71** (Suppl. 1), 163 (Abstr.).
Mizakorimov, U. M. & Tashmatova, K. Ya. (1971). *Med. Zh. Uzbeck.*, **5,** 45–6; *Dairy Sci. Abstr.* (1972), **34,** 571.
Moidi, U. V., Owen, E. C. & Proudfoot, T. L. (1959). *Proc. Nutr. Soc.*, **18,** 1.
Montoure, J. E. & Forster, T. L. (1960). *J. Dairy Sci.*, **43,** 878–9.
Morgan, D. M. L. & Toothill, V. J. (1985). *Biochem. Soc. Trans.*, **13,** 352–3.
Nakamura, M. & Yamazaki, I. (1982). *J. Biochem.*, **92,** 1279–86.
Nathan, G. R. & Hade, E. P. K. (1975). *Biochem. Biophys. Res. Comm.*, **66,** 108–14.
Nathan, G. R. & Hade, E. P. K. (1978). *Biochim. Biophys. Acta,* **526,** 328–44.
Neuman, V., Kudelka, E. & Sidelarova, K. (1965). *Veter. Med. Praha,* **10,** 349–54; *Dairy Sci. Abstr.* (1965), **27,** 536.
Nishino, T., Nishino, T. & Tsushima, K. (1981). *FEBS Lett.*, **131,** 369–72.

Obara, Y. & Komatsu, M. (1984). *J. Dairy Sci.*, **64,** 1043–6.
Oberkotter, L. V., Koldovsky, O. & Tenore, A. (1982). *Int. J. Biochem.*, **14,** 151–4.
Oster, K. A. (1971). *Amer. J. Clin. Res.*, **2,** 30–5.
Owen, J. W., Gammon, K. L. & Stahl, F. D. (1975). *Arch. Biochem. Biophys.*, **166,** 258–72.
Panfil-Kuncewicz, H. & Kisza, J. (1976). *Zeszyty Naukowe Academii Rolniczo-Technioznej w Olsztynie. Technol. Zwynosci.*, **151,** 105–12; *Dairy Sci. Abstr.* (1977), **39,** 119.
Parry Jr, R. M., Chandan, R. C. & Shahani, K. M. (1969). *Arch. Biochem. Biophys.*, **103,** 59–65.
Patel, S. S. & Wilbey, R. A. (1989). *J. Soc. Dairy Technol.*, **42,** 79–80.
Patterson, D. S. P., Berrett, S. & Cullen, G. A. (1969). *Veter. Record,* **85,** 708–13.
Patton, S. & Trams, E. G. (1971). *FEBS Letters,* **14,** 230–2.
Perdigon, G., Medici, M., Cecilia, M., Nader de Macias, M. E., Haedo, R., Oliver, G. & Pesce de Ruiz Holgado, A. A. (1986). *J. Dairy Sci.*, **69,** 27–31.
Perdigon, G., Medici, M., Oliver, G. & Pesce de Ruiz Holgado, A. A. (1988). *Milchwissenschaft,* **43,** 25–6.
Phillips, J. D. & Griffiths, M. W. (1987). *J. Appl. Bact.*, **62,** 223–6.
Phillips, D. C., Acharya, K. R., Handoll, H. H. G. & Stuart, O. I. (1987). *Biochem. Soc. Trans.*, **15,** 737–74.
Pocius, P. A., Baumrucker, C. R., McNamara, J. P. & Bauman, D. E. (1980). *Biochem. J.*, **188,** 565–8.
Polis, B. D. & Shmukler, H. W. (1950). *J. Dairy Sci.*, **33,** 619–22.
Powell, J. T., Jarlfors, U. & Brew, K. (1977). *J. Cell Biol.*, **72,** 617–27.
Prieels, J.-P., Bayer, T. & Hill, R. L. (1977). *Biochem. Soc. Trans.*, **5,** 838–9.
Proctor, V. A. & Cunningham, F. E. (1988). *CRC Crit. Rev. Food Sci. Nutr.*, **26,** 359–95.
Proczekaj, J. & Wenclewski, A. (1963). *Ginek. polska.*, **34,** 577–82; *Dairy Sci. Abstr.* (1965) **27,** 255.
Read Jr, R. B., Reyes, A. L., Bradshaw, J. G. & Peeler, J. T. (1969). *J. Dairy Sci.*, **52,** 1359–69.
Richardson, G. A. & Hankinson, C. L. (1936). *J. Dairy Sci.*, **19,** 761–72.
Roy-Burman, P., Rongey, R. W., Henderson, B. E. & Gardner, M. B. (1973). *Nature New Biol.*, **244,** 146.
Ruegg, M., Moor, U. & Blanc, B. (1977). *J. Dairy Res.*, **44,** 509–20.
Schlom, J., Spiegelman, S. & Moore, D. H. (1972). *Science,* **172,** 542–4.
Serban, A. M. D., Biener, J. & Diaconescu, C. (1969). *Studii Cerc. Endocr.*, **20,** 147–52; *Dairy Sci. Abstr.* (1971), **33,** 563.
Shahani, K. M. (1966). *J. Dairy Sci.*, **49,** 907–20.
Shahani, K. M., Harper, W. J., Jensen, R. G., Parry Jr, R. M. & Zittle, C. A. (1973). *J. Dairy Sci.*, **56,** 531–43.
Shidlovskaya, V. P. (1982). *Proc. 21st Int. Dairy Congr. Moscow,* **1**(2), 520–1.
Shriver, B. J., Allred, J. B. & Roman-Lopez, C. R. (1989). *Biochem. J.*, **257,** 925–7.
Sibley, J. A. & Lehninger, A. L. (1949). *J. Biol. Chem.*, **177,** 859–72.
Silver, M. R. & Zikakis, J. P. (1979). *J. Dairy Sci.*, **62** (Suppl. 1), 245 (Abstr.).
Sklavunu-Zurukzoglu, S., Mameletzis, C. & Katriu, D. (1965). *Helv. Paediatr. Acta,* **20,** 193–6. *Dairy Sci. Abstr.* (1965) **51,** 40.
Skripnichenko, G. G. (1987). *Doklady Vsesoyuznoi Akademi Sel'skokhozyaistvennykh Nauk,* **4,** 29–32; *Dairy Sci. Abstr.* (1989), **51,** 40.
Sliwkowski, M. B., Swaisgood, H. E. & Horton, H. R. (1982). *J. Dairy Sci.*, **65,** 1681–7.
Snow, L. D., Dross, R. C. & Carraway, K. L. (1980). *Biochim. Biophys. Acta,* **661,** 333–41.
Sobiech, K. A., Ziomek, E. & Szewczuk, A. (1974). *Arch. Immun. Ther. Exper.*, **22,** 645–56; *Dairy Sci. Abstr.* (1976), **38,** 410.

Sokolova, A. P., Sokolova, I. V. & Lyubimor, I. A. (1983). *Soviet Agric. Sci.*, **1,** 34–5; *Dairy Sci. Abstr.* (1984), **46,** 272.
Spanning, R. J. M. van, Wansell-Bettenhaussen, C. W., Oltmann, L. F. & Stouthamer, A. H. (1987). *Eur. J. Biochem.*, **169,** 349–52.
Stahl, P. D. & Touster, O. (1971). *J. Biol. Chem.*, **246,** 5398–5406.
Stein, E. A. & Fischer, E. H. (1958). *J. Biol. Chem.*, **232,** 867–79.
Stejskal, J., Kamaryt, J. & Muchova, M. (1981). *Refrigeration Sci. Technol.*, **2,** 85–9. *Dairy Sci. Abstr.* (1983) **45,** 284.
Sullivan, C. H., Mather, I. H., Greenwalt, D. E. & Madara, P. J. (1982). *Mol. Cell Biochem.*, **44,** 13–22.
Swaisgood, H. E. (1980). *Enzyme Microb. Technol.*, **2,** 265–72.
Swaisgood, H. E. & Abraham, P. (1980). *J. Dairy Sci.*, **63,** 1205–10.
Swaisgood, H. E. & Horton, H. R. (1980). *Ciba Foundation Symposium,* **72,** 205–22.
Swaisgood, H. E., Janolino, V. G. & Skudder, P. J. (1987). In *Methods in Enzymology—136: Immobilized Enzymes and Cells* (Part C), ed. K. Mosbach. Academic Press, Inc., London, UK, pp. 423–31.
Tate, S. S. & Meister, A. (1974). *J. Biol. Chem.*, **249,** 7593–602.
Timms, L. L. & Schultz, L. H. (1985). *J. Dairy Sci.*, **68,** 3363–6.
Tramper, J. (1987). In *Methods in Enzymology—136: Immobilized Enzymes and Cells* (Part C), ed. K. Mosbach. Academic Press Inc., London, UK, pp. 254–62.
Valivullah, M. & Keenan, T. W. (1989). *Int. J. Biochem.*, **21,** 103–7.
Ventom, A. M., Deistung, J. & Bray, R. C. (1988). *Biochem. J.*, **255,** 949–56.
Vina, J. R., Puertes, I. R., Montoro, J. B., Saez, G. T. & Vina, J. (1985). *Biology of the Neonate,* **48,** 250–6.
Walstra, P. & Jenness, R. (1984). *Dairy Chemistry and Physics.* John Wiley & Sons, New York, USA, p. 128.
Waud, W. R. & Rajagopalan, K. V. (1976). *Arch. Biochem. Biophys.*, **172,** 365–79.
Waud, W. R., Brady, F. D., Wiley, R. D. & Rajagopalan, K. V. (1975). *Arch. Biochem. Biophys.*, **169,** 695–701.
Weihrauch, J. L. (1988). In *Fundamentals of Dairy Chemistry,* ed. N. P. Wong, R. Jenness, M. Keeney & E. H. Marth. Van Nostrand Reinhold Co., New York, USA, pp. 244–5.
Whitaker, J. R. (1972). *Principles of Enzymology for the Food Sciences.* Marcel Dekker Inc., New York, USA, pp. 584, 600.
White Jr, F. H., McKenzie, H. A. & Shaw, D. C. (1988). *Biochem. Int.*, **16,** 521–8.
Williams, M. M. F. (1983). *Proc. Univ. Otago Med. School,* **61,** 20–1; *Dairy Sci. Abstr.* (1983) **45,** 938.
Wuthrich, S., Richterrich, R. & Hosteller, H. (1964). *Z. Lebensmitt Untersuch,* **124,** 345–8; *Dairy Sci. Abstr.* (1964), **26,** 601.
Yasumoto, K., Iwami, K., Fushiki, T. & Mitsuda, H. (1978). *J. Biochem.*, **84,** 1227–36.
Zikakis, J. P. & Treece, J. M. (1971). *J. Dairy Sci.*, **54,** 648–54.
Zikakis, J. P. & Wooters, S. C. (1980). *J. Dairy Sci.*, **63,** 893–904.
Zikakis, J. P., Dougherty, T. M. & Biasoto, N. O. (1976). *J. Food Sci.*, **41,** 1408–12.

3

Microbial Enzymes in Cheese Ripening

J.-C. Gripon, V. Monnet, G. Lamberet & M. J. Desmazeaud
INRA, Centre de Recherches de Jouy-en-Josas, Station de Recherches Laitières, Domaine de Vilvert-78350, Jouy-en-Josas, France

INTRODUCTION

Fresh curd obtained by milk clotting and drainage is composed of caseins, fat, carbohydrates and minerals; cheese ripening corresponds to the enzymatic modification of these substrates by which it will progressively acquire the texture and flavour of mature cheese. Enzymes synthesised by curd microorganisms play a major role in these biochemical modifications. There is a very wide variety of cheeses with different appearances and organoleptic characteristics. This great diversity is due to the various techniques used in curd preparation and hence to the establishment of different physico-chemical conditions (pH, water content, salt content, ripening temperature), but also to the utilisation of various microflora. The main role of lactic acid bacteria, which constitute the basal flora of all cheeses, is to transform lactose into lactic acid and hence to acidify the curd. They also act as ripening agents via their enzymes. Mesophilic streptococci are used as starters in most cheeses except in highly cooked cheese in which thermophilic lactobacilli are used together with *Streptococcus thermophilus*. The normal flora of highly cooked cheeses also includes propionic acid bacteria which produce carbon dioxide and cause eye formation in the cheese. Many cheeses are characterised by the activity of their surface microflora. That of the bacterial surface cheeses includes mainly coryneform bacteria (*Brevibacterium linens*), micrococci and yeasts. Surface-mould-ripened soft cheeses are ripened with *Penicillium camemberti*. In contrast, *P. roqueforti* develops within blue-vein cheeses providing them with their typical appearance and flavour. In this chapter, we shall examine the enzymatic changes in the different curd components by describing the origin and characteristics of microbial enzymes involved and their role during ripening.

LACTOSE AND CITRATE METABOLISM

The transformation of carbohydrates into lactate by lactic acid bacteria represents, most likely, the most important fermentation process among those used in the agricultural and food industries. Because of this economic importance, lactose metabolism has been studied extensively, in particular during the last 15 years. The metabolic pathways used vary according to bacterial genus and species. Most of them, belonging to the genera *Lactococcus, Pediococcus* and *Lactobacillus,* are homofermentative and convert lactose into lactate. Others, belonging to the genera *Leuconostoc* and *Lactobacillus,* are heterofermentative and produce equimolar quantities of

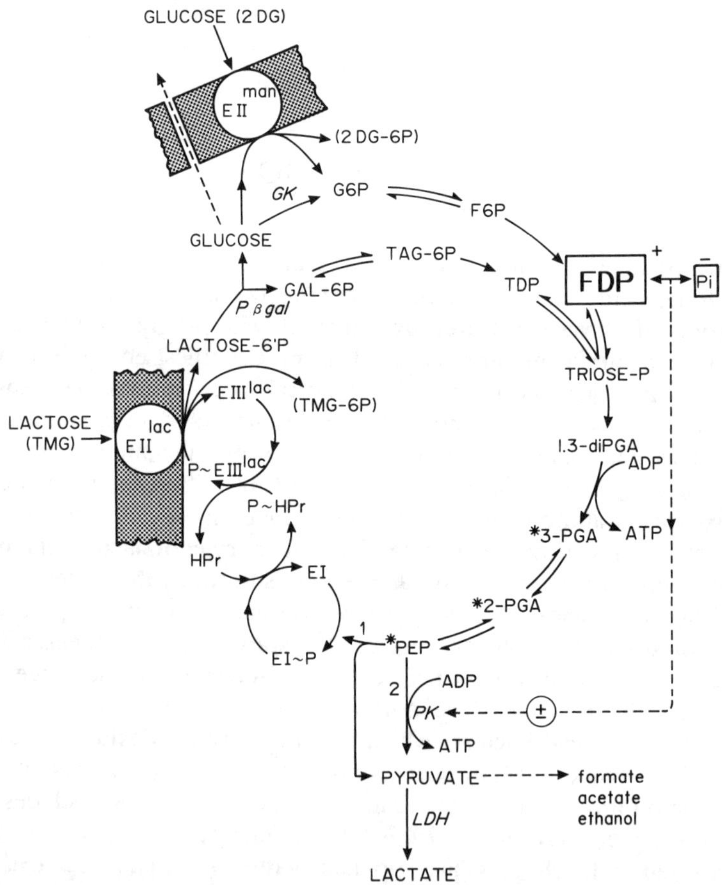

FIG. 1. Diagrammatic representation of the lactose –PTS–glycolysis cycle in *L. lactis.* Numbers 1 and 2 show distribution of phosphoenol pyruvate (PEP) to phosphotransferase system (PTS) or to pyruvate kinase. The symbols (+) and (−), indicate positive or negative effectors of pyruvate kinase; (*) constituent of the PEP-potential in starved cells (From Thompson (1987)).

lactate, carbon dioxide (CO_2) and acetate or ethanol. The mesophilic lactic *streptococci* used in soft cheeses, pressed cheeses or blue-vein cheeses phosphorylate lactose at the expense of phosphoenol-pyruvate (PEP) by a phosphotransferase system (PTS) and generate L-lactic acid. The *Leuconostoc* bacteria, which are operative in the manufacture of some of the same cheeses, ferment lactose by the heterofermentative pathway and produce D-lactic acid. The homofermentative thermophilic lactic acid bacteria used to acidify pressed cooked cheeses produce either L-lactic acid (e.g. thermophilic streptococci) or DL-lactic acid (e.g. *Lactobacillus helveticus*) or D-lactic acid (e.g. *L. lactis*) (Sharpe, 1979). Thermophilic lactic acid bacteria generally transport lactose by a permease system without phosphorylating it.

It may be assumed that metabolism of residual sugars in the curd by the surviving lactic acid bacteria is altered by the particular physico-chemical conditions, i.e. a low pH, and a high salt content which modifies the water activity. Compounds other than lactic acid (heterofermentation) could be produced from pyruvate.

Some species, such as *Lactococcus lactis* subsp. *diacetylactis* and *Leuconostoc lactis,* may utilise citrate and produce various volatile and even aromatic compounds.

Production of lactic acid from lactose, glucose and galactose in cheese

Homofermentative mesophilic streptococci

The fermentation of lactose by group N streptococci involves three steps: (i) transport through the plasma membrane and phosphorylation by a PTS, (ii) lactose catabolism via energy-producing pathways, and (iii) release of lactic acid from the cell.

Transport and phosphorylation. In lactic group N streptococci, lactose is transported across the cell membrane by a PTS which phosphorylates it at the expense of PEP (Fig. 1). The enzymes involved interact according to the following reactions (McKay, 1983):

$$\text{PEP} + \text{Enzyme I} \xrightleftharpoons{Mg^{2+}} \text{P-Enzyme I} + \text{pyruvate}$$

$$\text{P-Enzyme I} + \text{HPr} \rightleftharpoons \text{P-HPr} + \text{Enzyme I}$$

$$\text{P-HPr} + 1/3\ \text{Factor III}^{lac} \rightleftharpoons 1/3\ \text{P-Factor III}^{lac} + \text{HPr}$$

$$1/3\ \text{P-Factor III}^{lac} + \text{lactose} \xrightarrow[Mg^{2+}]{\text{Enzyme II}^{lac}} \text{lactose-6-P} + 1/3\ \text{Factor III}^{lac}$$

where P stands for phosphate. Enzyme I and protein HPr are cytoplasmic compounds required for the transport of many sugars. Factor III^{lac} is a cytoplasmic molecule specific for galactosides. Enzyme II^{lac} is a galactoside-specific membrane-bound compound. Enzyme I and protein HPr are constitutive proteins whereas enzyme II^{lac} and factor III^{lac} are induced during growth in the presence of lactose or galactose (Cogan, 1980). As PEP is both a product and a reactant of sugar fermentation, this generator of high phosphorylated energy plays a key role in the different steps of sugar transport and metabolism. It is located at a fundamental junction which leads either to lactose transport or to ATP synthesis by pyruvate kinase (= ATP: pyruvate-2-0-phosphotransferase, EC 2.7.1.40) (Thompson, 1987). Thus, the cell controls this compound in order to balance sugar transport and metabolism rates. This is achieved via a control of pyruvate kinase activity.

Thus, lactose appears in the cell in the form of glucosyl-β-(1,4)-galactoside-6-phosphate (or lactose-phosphate). Hence, it is ready for hydrolysis by a β-D-phosphogalactosidase.

The free residual glucose in the curd may be transported by another PTS. In this case also, pyruvate kinase occupies a pivotal role in the regulation of the functions of this PEP–PTS system and of the subsequent glycolysis (Thompson, 1978, 1979, 1980, 1987). However, according to Tinson *et al.* (1982), glucose can also be transported by a permease system, the energy for which is supplied by ATP.

It has been well established that in mesophilic streptococci, especially *S. lactis* ML3, free residual galactose is transported by two separate systems (Thompson, 1980), i.e. a PEP–PTS system (Thompson & Thomas, 1977) and a permease system which obtains its energy from the hydrolysis of ATP. The affinity of galactose for the permease system is 10-fold higher than for the PTS system.

Carbohydrate catabolism. The enzymes involved in the catabolism of carbohydrates are indicated in Fig. 2. Thus, in mesophilic streptococci, lactose phosphate is hydrolysed to glucose and galactose-6-phosphate by a phospho-β-galactosidase. The resulting glucose is then broken down via the glycolytic Embden–Meyerhof–Parnas (EMP) pathway. Galactose-6-phosphate is catabolised via the D-tagatose-6-phosphate pathway. Free galactose may be degraded via the Leloir route.

Most enzymes involved in the metabolism of carbohydrates have been purified, in particular phospho-β-galactosidase. In *S. cremoris* HP this enzyme shows some resistance to heat treatments and a maximal activity from pH 5 to 8. Sulphydryl residues interfere with the enzymic activity. Pyruvate kinase of *S. lactis* ML3 was described by Collins and Thomas (1974). It shows maximal activity in exponentially growing cells.

Homofermentative mesophilic streptococci using the EMP pathway in the last step of glycolysis convert pyruvate into lactate and thus regenerate nicotinamide adenine dinucleotide (NAD) from the previously formed reduced nicotinamide adenine dinucleotide (NADH). It is precisely this key step which makes the cycle functional. Thus, all lactic acid bacteria possess a lactate dehydrogenase (LDH) (Garvie, 1980); the LDH–NAD dependent ones are specific for the L(+) isomer of lactic acid and they are strongly activated by fructose-1,6 diphosphate (FDP).

Enzymes of the D-tagatose-6-phosphate pathway were evidenced by Thomas *et al.* (1980) in *S. cremoris* and *S. lactis*. Enzymes involved in the transforma-

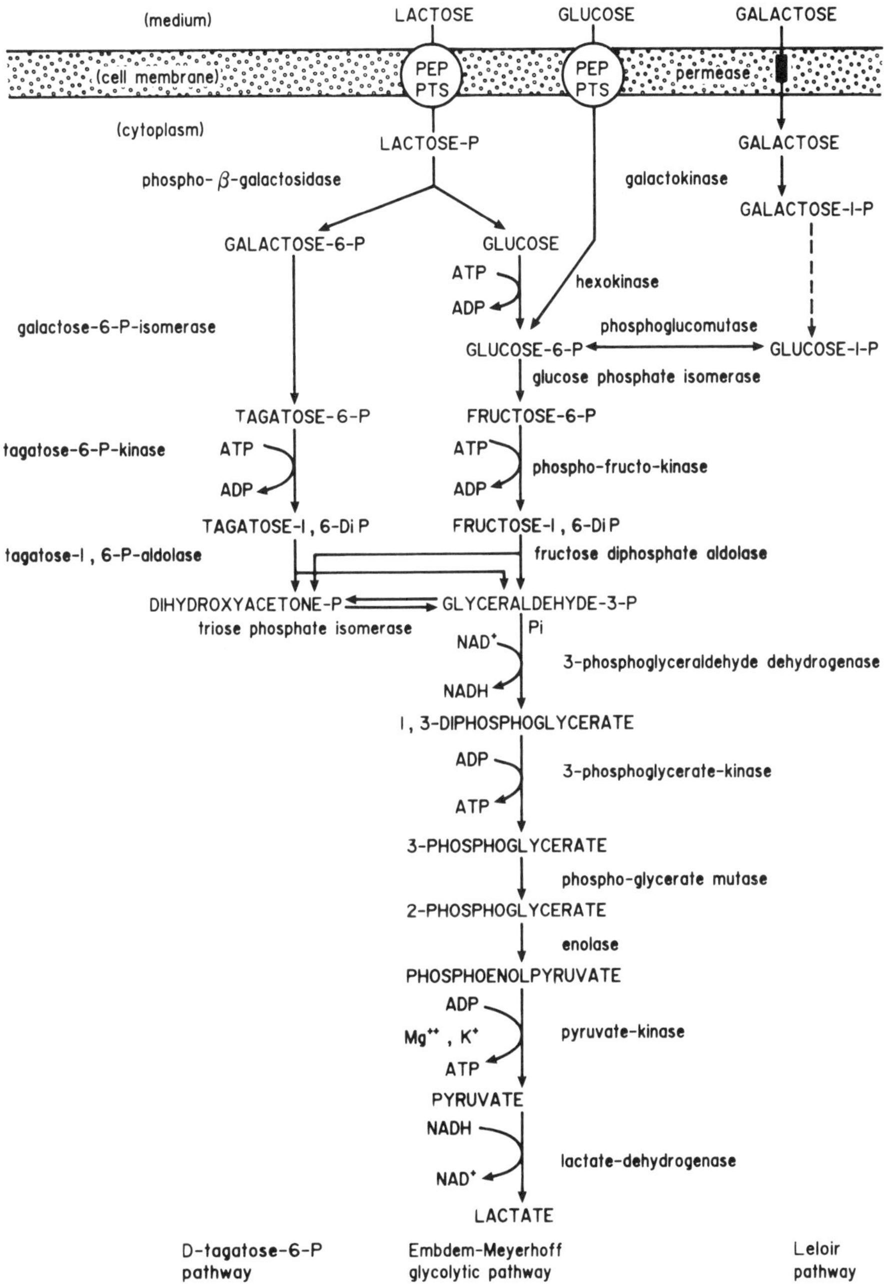

FIG. 2. Transport and pathways of lactose, glucose or galactose metabolism in homofermentative *Streptococci* (adapted from Thompson & Thomas, 1977; Cogan, 1980). P = phosphate; PEP-PTS = phosphoenol pyruvate-dependent–phosphotransferase system.

tion of galactose-6-phosphate into triose phosphates (Fig. 2) are; galactose-6-phosphate isomerase, D-tagatose-6-phosphate kinase and D-tagatose-1,6 diphosphate-aldolase.

Free galactose is utilised by the enzymes of the Leloir pathway, described by Thomas *et al.* (1980) (Fig. 2).

The regulation of carbohydrate metabolism depends on repressive and retroinhibitory mechanisms. Cogan (1980) distinguished between a 'coarse regulation' affecting the synthesis of enzymes and a 'fine regulation' controlling enzymatic activity. As indicated previously, control of pyruvate kinase activity may regulate the intracellular PEP pool and hence couple lactose transport (via the PTS–PEP dependent system) to lactose hydrolysis. LDH, activated by pH and fructose diphosphate or tagatose diphosphate (Thomas, 1976), is also subject to an allosteric regulation. Because of the regulation of this enzyme, pyruvate concentrations remain low during exponential growth and therefore prevent the onset of the other pyruvate-derived metabolic steps. In this case, only lactate is formed. Glycolysis may go on since the reduced coenzyme (NADH) is reoxidated to NAD^+.

Mesophilic streptococci spontaneously produce mutants which cannot utilise lactose, while their growth is normal on glucose. It has been demonstrated that they do not contain either of the two proteins of the PEP–lactose–PTS or phospho-β-galactosidase, all of which are proteins encoded by a DNA plasmid (McKay, 1983; De Vos, 1987).

Efflux of lactic acid by the cells. Konings and Otto (1983), Otto *et al.* (1983) and Kashket (1987) examined the consequences of medium variations on the contribution of released lactate as an energy source to *S. cremoris* metabolism, in particular by study of the proton motive force and the lactate gradient during cell growth. The ratio of lactate gradient to proton motive force decreases during bacterial growth (Fig. 3). This lowered ratio accounts for a decrease in H^+–lactate stoichiometry and hence a reduced contribution of released lactate to metabolism energy. These authors also showed that the internal pH of the growing organism is maintained constant at 6·6 when the pH value of the medium falls, thus implying the formation of a pH gradient. But at the end of the exponential growth, the pH can no longer be kept constant, probably due to lack of ATP.

Heterofermentative Leuconostoc *bacteria*

Leuconostoc bacteria are heterofermentative lactic acid bacteria, present in natural starters or added in the case of blue-vein cheesemaking. They utilise the tagatose-6-phosphate and glycolysis pathways, but also the pentose phosphate route (Desmazeaud, 1983) (Fig. 4) Thus, in these organisms, lactose fermentation induces the formation of equimolar amounts of lactate, ethanol and CO_2. Each mole of fermented lactose gives about two moles of lactate, two moles of ethanol and two moles of CO_2. Traces of formate and acetate can also be produced (Cogan, 1980, 1987; Cogan & Daly, 1987).

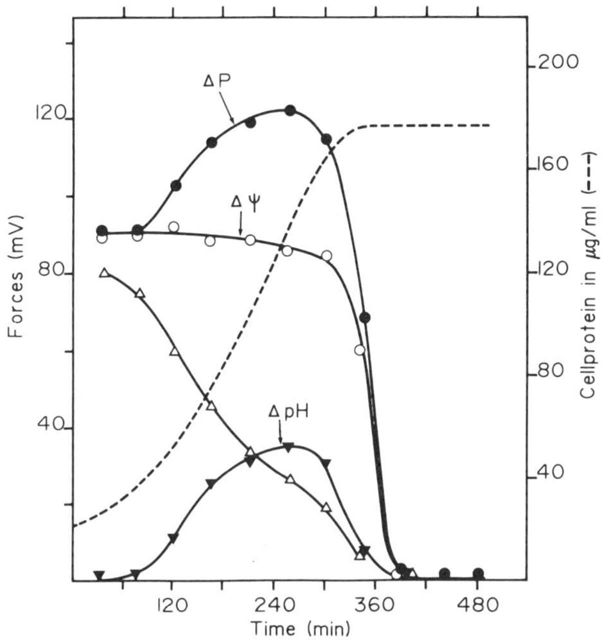

FIG. 3. The proton motive force and the lactate gradient in *S. cremoris* during growth in batch cultures on complex medium with lactose (2 g/litre) as sole energy source. All gradients expressed in mV. (●) Proton motive force, (○) membrane potential, (▲) pH gradient, (△) lactate gradient, (---) cell protein concentration (From Konings & Otto (1983)).

Homofermentative thermophilic lactic acid bacteria used in cooked cheese manufacture

S. thermophilus transports lactose in the free form via a permease system, as evidenced by the presence of β-galactosidase. Indeed, sodium fluoride which inhibits the PTS–PEP system has no effect on lactose uptake. In some strains, lactose permease exhibits a high maximum velocity ($V_{max} = 311$ accumulated μmol/min g dry cell weight) which may account for the high growth rates of this organism in milk (Hutkins & Morris, 1987).

When lactose has entered the cell and has been cleaved to glucose and galactose (by β-galactosidase), only glucose is utilised rapidly. Thus, excessive galactose is released into the milk (Hutkins & Morris, 1987) where it may inhibit the utilisation of residual lactose. The glycolytic enzymes (e.g. hexokinase, aldolase, glyceraldehyde-3-phosphate dehydrogenase, pyruvate kinase and LDH) have been evidenced in *S. thermophilus* (Hemme *et al.*, 1980, 1981). However, some strains are capable of using galactose, especially when the lactose content of the medium (e.g. the curd) decreases.

Only a few data are available on the lactobacilli. Hickey *et al.* (1986) showed that they possess a β-galactosidase rather than a phospho-β-

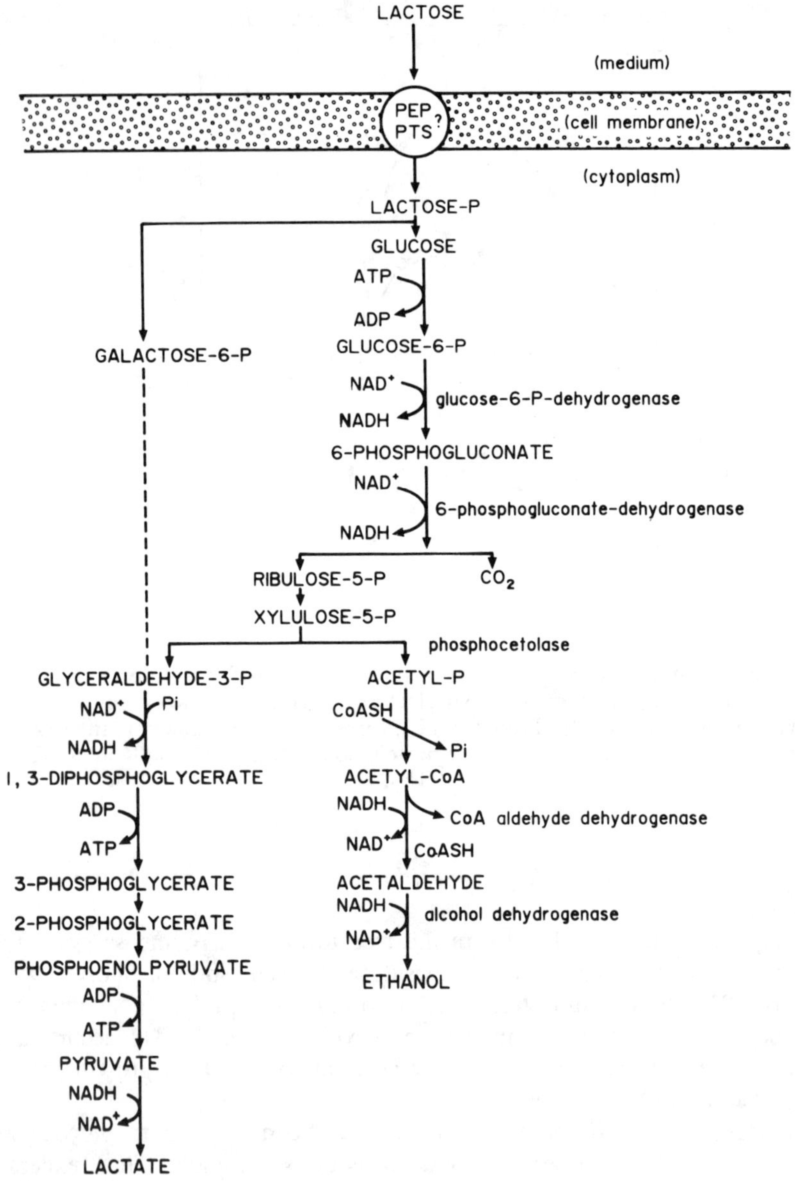

FIG. 4. Lactose catabolism in *Leuconostoc* bacteria (heterofermentative pathway) (adapted from Cogan, 1980).

galactosidase, the activity of which is very low. However, *L. bulgaricus* and *L. helveticus* seem to have a PEP–glucose PTS. Galactose seems to be utilised better than in the case of thermophilic streptococci although galactose$^+$ strains do exist (Hutkins & Morris, 1987). *L. lactis* generates the D(−) isomer of lactic acid while *L. helveticus* produces DL-lactic acid (Accolas *et al.*, 1980). As in the case of mesophilic streptococci, some *L. helveticus* strains may produce slow milk clotting variants characterised by the loss of a plasmid (Morelli *et al.*, 1986).

Behaviour of lactic acid bacteria in the curd

Cheese manufacture with mesophilic streptococci

When mesophilic streptococci are carbohydrate-deficient, 3-phosphoglycerate, 2-phosphoglycerate (2PG) and PEP concentrations increase substantially while the other glycolysis intermediates decrease. The ability of the cells to retain an intracellular 'pool' of PEP and 2PG provides them with a survival potential which enables them to produce the required amount of ATP (Thompson & Thomas, 1977).

By analogy with results obtained on model media, resting *S. lactis* cells in the presence of residual galactose would ferment this sugar into lactate, formate, acetate and ethanol. The intracellular levels of FDP, an LDH activator, and of phosphate trioses, inhibitors of pyruvate lyase, seem to be the major factors involved in the production of compounds other than lactate (Thomas *et al.*, 1980).

Cheese manufacture with thermophilic lactic acid bacteria

Milk clotting by rennet represents the first step in the manufacture of pressed highly cooked cheese. The next step involves cutting the coagulum and stirring it mechanically whereafter the curd grains are heated to 53–56°C. This curd treatment in the vat leads to a rather rapid and extensive syneresis of the grains. Since the lactose content of the curd is low, all of this sugar is fermented into lactic acid. Acidification of the curd during pressing favours its syneresis and if the draining proceeds normally, the cheese will contain 38–39% water and exhibit a pH of 5·20 on removal from the press the day after manufacture (Accolas *et al.*, 1980).

It has been shown that *S. thermophilus* rapidly metabolises all of the residual lactose in the curd grains into galactose and L-lactic acid during the first 10 h after the start of pressing. The lactobacilli which utilise galactose develop later and produce a DL-lactic acid mixture (Turner *et al.*, 1983).

Accolas *et al.*, (1978, 1980) and Auclair and Accolas (1983) described the mechanisms of draining due to this lactic acidification. The multiplication of streptococci during curd treatment in the vat is low. In addition, the lactobacilli show a lesser resistance to heating in the vat which may explain the large difference observed during pressing between the development of *S. thermophilus* and *L. helveticus*. After the start of pressing of the curd, *S. thermophilus* develops rapidly, and after 3 h of pressing, reaches a maximum population in the core of the cheese. The lactobacilli reach their maximum number much later (after 21 h of pressing). These population variations are due to the temperature range between the cheese core and periphery. When the curd is put under the press, its temperature is close to 50°C. As the pressing room temperature is much lower, the curd mass cools, but as it is a poor heat conductor, the temperature of the periphery falls rapidly while it decreases only slowly in the cheese core. This leads to the establishment of a temperature gradient. Hence, lactic fermentation starts earlier at the periphery where the temperature range is favourable to the development and metabolism of lactic acid bacteria. Thus, the lactose content of the periphery decreases more rapidly than that of the core resulting in the diffusion of lactose from the centre to the surface. Conversely, since the temperature of the core is too high during the first hours of pressing the fermentation is delayed. When the temperature of the core becomes favourable again, no more residual lactose is available for the development of starter bacteria.

In model cheese produced without lactobacilli or with strains which do not use galactose, this sugar accumulated in 24-h old cheeses. Only L-lactic acid was produced and the pH remained high (≈5·40). Although galactose was thereafter rapidly metabolised in a warm cellar, the cheese body and eye formation remained atypical (Turner *et al.*, 1983). This confirms the major role played by thermophilic lactobacilli which ferment galactose in pressed cooked cheese and their ability to regulate the pH of the 24-h curd by metabolising this sugar.

Lactate isomerisation and metabolism

Production of the two lactic acid isomers by the same microorganism might be due to the presence of two specific LDHs or of one specific LDH and that a racemase transforms one isomer into another. These racemases do not exist in the mesophilic streptococci of starters which produce only L-lactate. But Cheddar cheese manufactured with these bacteria contains large amounts of L- and D-lactate. Thus, D-lactate production in cheese results from the activity of two LDHs specific to bacteria not included in the starters, but which play an important role during cheese ripening, i.e. *L. casei, L. plantarum* or *Pediococcus pentosaceus* (Cogan & Daly, 1987).

Fermentation of lactate in young cheese by propionic acid bacteria represents a fundamental step in the ripening of pressed cooked cheeses. Propionic fermentation involves a complex cycle via pyruvate. Thus, fermentation of lactate is initiated by its oxidation to pyruvate with the concomitant production of a reduced flavoprotein. A fraction of pyruvate is then oxidatively decarboxylated with the formation of CO_2, acetate, NADH and ATP. The rest is simultaneously reduced to propionate. The conversion of pyruvate to propionate is mediated by malate dehydrogenase and methyl malonyl CoA transcarboxylase (Fig. 5) (Law, 1981; Bergère & Accolas, 1986). The overall reaction leads to propionic and acetic acids as well as to carbon dioxide and ATP.

$$3 \text{ lactate} \rightarrow \text{acetate} + CO_2 + 2 \text{ propionate} + 2 \text{ ATP}$$

This fermentation is thus due to Gram-positive anaerobic short rod-shaped propionic bacteria. Their natural habitat is the rumen and intestine of ruminants. Because of improved hygiene on farms, it is now often necessary to

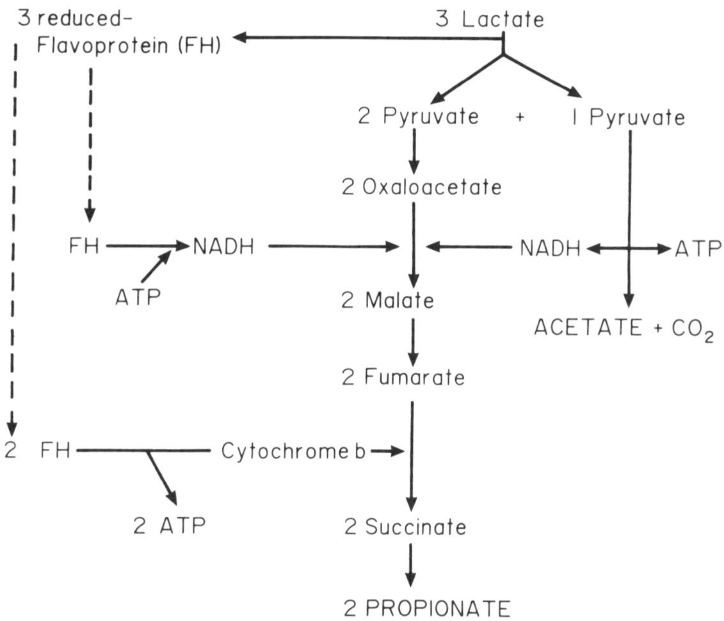

FIG. 5. Pathways leading to propionate, acetate and CO_2 in *Propionibacterium shermani.*

preinoculate the milk with *Propionibacterium frendenreichii* subsp. *shermanii.* The CO_2 produced contributes to the characteristic eye formation in some cheeses. The amount of CO_2 produced and the level of eye formation depend on the length of cheese storage in warm cellars (20–24°C) or on ripening room temperatures. Thus, the length of storage and the temperature used leads to a reduced eye formation in Comté Gruyère or to eyeless cheese in the case of French Beaufort. However, the eye formation process is rather complex. The curd could be considered as a gel with a continuous phase containing the dissolved CO_2 produced by the propionic acid bacteria. The CO_2 concentration increases until saturation and when a critical pressure is reached, gas bubbles are formed around initiating nuclei. The additional production of gas leads to an enlargement of existing eyes rather than to the formation of new ones.

A fraction of lactate may also be metabolised by undesirable bacteria, especially *Clostridium tyrobutyricum,* to butyric and acetic acids, CO_2 and hydrogen. This fermentation causes deterioration of the flavour and leads to an excessively high gas production which may result in defective eye formation and even cheese disruption.

An excessive amount of CO_2 may also be produced by heterofermentative lactobacilli and even by amino acid decarboxylation (Steffen *et al.*, 1987).

In Camembert-type cheese with the presence of *P. camemberti* on the surface, a rapid fall in the concentration of lactate is observed between 5 and 10 days of ripening. This coincides with the growth of *Penicillium.* Degradation of lactic acid leads to neutralisation of the curd, the surface of traditional Camembert reaching a pH of 7·0 during late ripening whereas the core pH is 6·0. In blue-vein cheese, which contains *P. roqueforti,* the pH increase is less spectacular. However, the pH reaches a value of 6·5 in Danish Blue and 6·2 in Bleu d'Auvergne (Gripon, 1987).

This neutralisation of the curd plays a major role in the ripening process. It facilitates the growth of the other microorganisms (micrococci, coryneform bacteria such as *B. linens*). In particular, it creates a medium with a pH close to the optima of most of the enzymes of the bacteria present which accounts for the advanced ripening of this type of cheese. In addition, neutralisation strongly affects the rheological characteristics of the curd and leads to mineral migrations.

Citrate metabolism

Citrate is transported inside the cells by a citrate permease. In *S. lactis* subsp. *diacetylactis* it is functional at pH values lower than 6, its optimum being 5 (Harvey & Collins, 1962). Citrate is cleaved to acetate and oxaloacetate inside the cells by the citrate–lyase complex. This enzyme is inducible by citrate in *Leuconostoc lactis* (Cogan, 1987), but it is constitutive in *S. lactis* subsp. *diacetylactis* (Mellerick & Cogan, 1981). Oxaloacetate is then converted to pyruvate and CO_2 by an oxaloacetate decarboxylase. The second step of decarboxylation results in the formation of acetaldehyde thiamine pyrophosphate (Fig. 6). Two different pathways may lead to diacetyl or acetoin. Magnesium and manganese ions might be involved in different reactions, but this has not yet been accurately stated (Cogan & Daly, 1987). There may be production of 2,3-butylene glycol from acetoin by acetoin-reductase (or 2,3-butylene-glycol dehydrogenase). Citrate represses the synthesis of this enzyme as well as that of diacetyl-reductase which may account for the accumulation of diacetyl or acetoin in *S. lactis* subsp. *diacetylactis* cultures. *Leuconostoc* bacteria metabolise citrate at acid pH values with an optimum of about 5·4. In contrast, *S. lactis* subsp. *diacetylactis* produces acetoin and diacetyl from citrate from the onset of growth.

Kempler and McKay (1981) showed that in *S. lactis* subsp. *diacetylactis,* citrate metabolism is related to a plasmid which codes for the synthesis of citrate permease. Mutants which do not contain this plasmid can not transport citrate into the cell, but they retain the other enzymes (e.g. citrate lyase).

The rather low citrate concentration in milk (8 mM) reduces its metabolism in cheese. However, it plays a role not only in the production of aromatic volatile compounds (e.g. diacetyl), but also in the formation of CO_2. The latter is involved in eye formation in some curds, especially Dutch cheese (Walstra *et al.,* 1987) or blue cheese. In blue-vein cheese, these openings favour the growth of *P. roqueforti* and hence lead to an optimum production of the characteristic aroma. Conversely, this would lead to textural defects in the case of Cheddar cheese.

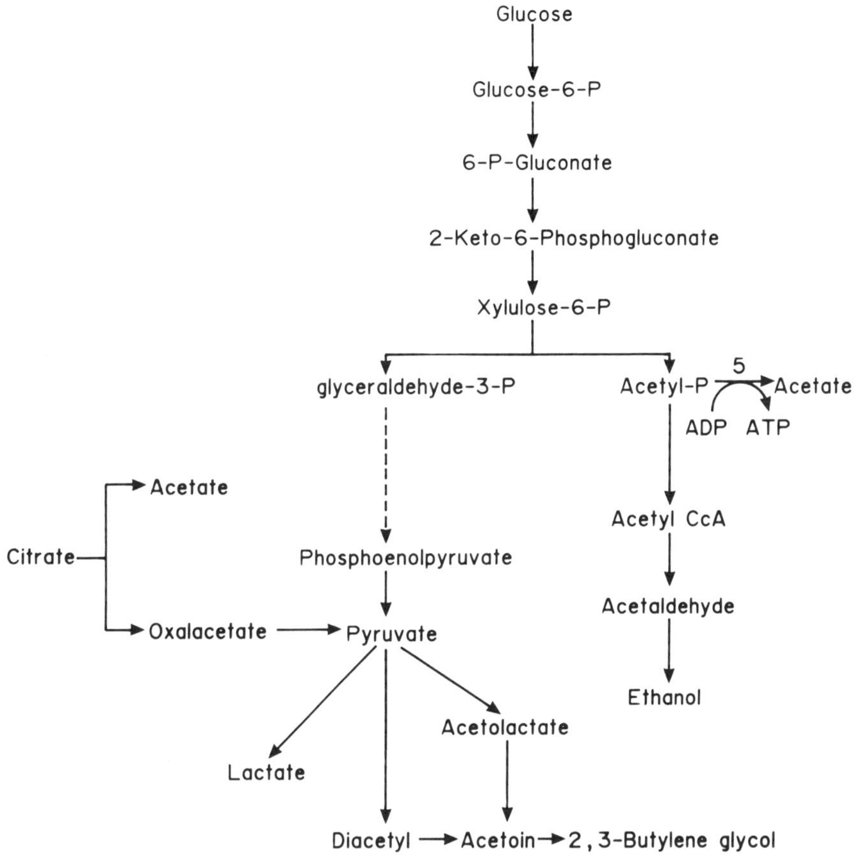

FIG. 6. Outline of the main features of glucose and citrate metabolism in *Leuconostoc* spp. Citrate metabolite are shown in bold and glucose in normal type (From Cogan (1987)).

PROTEOLYSIS AND DEGRADATION OF AMINO ACIDS

Proteolysis plays a major role during cheese ripening. The microbial endopeptidases, together with rennet and endogenous milk protease (plasmin), hydrolyse caseins and modify the curd texture. According to Creamer and Olson (1982), modifications of Cheddar curds are chiefly attributable to α_{s1}-casein degradation. Hydrolysis of β-casein seems to play a less important role. Thus, Alkhalaf *et al.* (1988) observed no marked textural changes in Saint-Paulin cheeses after addition of a microbial protease which hydrolyse mainly β-casein in the curds.

Microbial peptidase activity leads to the development of small peptides and free amino acids. These small degradation products are used as a nitrogen source by the secondary flora and also play an important role in aroma development. Indeed, free amino acids are the precursors of numerous aromatic volatile compounds. Besides, the water soluble, non-volatile fraction, which contains the peptides and free amino acids, has a marked flavour (McGugan *et al.*, 1979). Thus, these products contribute to flavour formation either directly or as carriers of slowly-released volatile compounds (Kowalewska *et al.*, 1985).

Proteolytic enzymes of dairy microorganisms

Lactic acid bacteria

The nutritional requirements of lactic acid bacteria are complex. They need not only a fermentable sugar, but also various amino acids that they are unable to synthesise from a more simple nitrogen source. The amino nitrogen of milk is available in the form of proteins (casein and whey proteins) and compounds of low molecular weight (free amino acids and small peptides). The latter are present at low concentrations and only enable a reduced growth of lactic acid bacteria (Mills & Thomas, 1981). Although weakly proteolytic as compared to other bacterial genera, lactic acid bacteria possess a complex proteolytic system (Desmazeaud, 1983, Law & Kolstad, 1983; Thomas & Pritchard, 1987) composed of enzymes of various type and location. Casein breakdown by this proteolytic system leads to optimal cell growth and hence to good acidification of milk and subsequent development of cheese aroma during ripening.

Cell wall enzymes. Data concerning the nitrogen requirements (Reiter & Oram, 1962; Matthews & Payne, 1975; Law *et al.*, 1976; Cogan, 1980; Payne, 1980; Morishita *et al.*, 1981; Marshall & Law, 1983) and transport systems for peptides and amino acids of the lactic acid bacteria (Leach & Snell, 1960; Law, 1977, 1978; Rice *et al.*, 1978; Akpemado & Bracquart, 1983; Benetaya *et al.*, 1986; Konings *et al.*, 1989) show that these cells are obliged to synthesise proteases and peptidases located on the outer surface of the plasma membrane. Through their joint action, these enzymes release peptides and free amino acids which are used by the cells or which participate in texture changes and development of cheese aroma (Fig. 7).

Production of free endopeptidase in the culture medium has never been demonstrated with certainty (Law & Kolstad, 1983). The only extracellular proteases described to date in lactic acid bacteria are enzymes loosely bound to the cell wall and released either during incubation of the cells in the absence of calcium, at pH 7 and 25°C, or during the formation of spheroplasts. The existence of such a cell wall activity was first demonstrated by Thomas *et al.* (1974) in *Lactococcus lactis* subsp. *lactis* and thereafter isolated from numerous strains of mesophilic streptococci and lactobacilli (Exterkate, 1975, 1976; Vescovo & Bottazzi, 1979; Neviani *et al.*, 1984; Cliffe & Law, 1985; Ezzat *et al.*, 1985, 1987; El Soda *et al.*, 1986*b*; Geis *et al.*, 1986; Monnet *et al.*, 1987*a,b*). The thermophilic streptococci do not seem to possess such an activity. Biochemical (Exterkate, 1975, 1976; Cliffe & Law, 1985; Visser *et al.*,

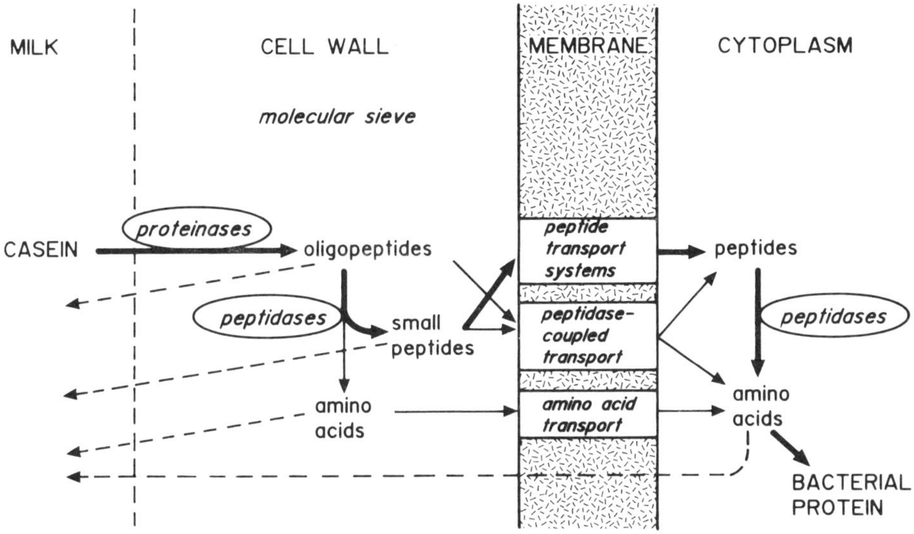

FIG. 7. Location of proteolytic enzymes and utilisation of casein for growth of lactic *Streptococci* in milk.

1986) and immunological investigations (Hugenholtz *et al.*, 1984) suggested the existence of several cell wall proteases. But it is only by the recent advances in the genetics of lactic acid bacteria that the situation has been clarified. A plasmid fragment coding for the overall cell wall protease activity was isolated and sequenced for three strains of mesophilic streptococci (De Vos, 1987; Kok, 1988; Kok & Venema, 1988; Kiwaki *et al.*, 1989). A single gene codes for a single protease which most likely exists in different, more or less hydrolysed forms (Exterkate & de Veer, 1987; Hugenholtz *et al.*, 1987; Kok *et al.*, 1988; Laan *et al.*, 1988). A membrane-bound protein (prt M) is involved in the maturation of the precursor proteinase (Vos *et al.*, 1989).

Purified cell wall proteases from different *Streptococcus* strains show all the same characteristics, i.e. a high molecular weight (80 000–145 000), presence of a serine essential for enzymic activity in the active site, activation in the presence of 1 mM calcium, optimal activity at an acid pH (5·5–6 on casein) and preferential hydrolysis of β-casein. However, although encoded by very similar genes (98% homology for the three sequenced genes), cell wall proteases exhibit different hydrolytic specificities on casein. Some strains only slightly hydrolyse α_{s_1} and α_{s_2}-caseins while these caseins are good substrates for other strains. Furthermore, at least two different types of hydrolytic specificity on β-casein have been observed (Visser *et al.*, 1986). Peptides soluble at pH 4·6 released on hydrolysis of β-casein by the proteases of three mesophilic *Streptococcus* strains (AC1, HP, NCDO 763) have been identified (Monnet *et al.*, 1986, 1989; Visser *et al.*, 1988) and seem to be very similar. These peptides are hydrophobic and potentially bitter. The cleavage sites do not seem to be very specific although these proteases seem to have a special affinity for the hydrophobic areas of the substrate (Monnet *et al.*, 1989).

Production of cell wall proteases is subject to regulation, at least in some strains, and depends on the composition of the culture medium. It is higher in a media poor in peptides and amino acids (Exterkate, 1979; Kok & Venema, 1988).

Several peptidases (aminopeptidases, dipeptidases, tripeptidases) have been isolated from cell wall extracts of streptococci (Law, 1979; Geis *et al.*, 1985), but the majority of these activities are intracellular. They are generally metalloenzymes which are reactivated by magnesium ions and have a mean molecular weight ranging between 25 000 and 40 000. Thus, Exterkate (1984) showed the existence of alanyl, leucyl and lysyl aminopeptidases at the cell wall–membrane interface in a strain of *L. lactis* subsp. *cremoris*. Eggimann and Bachmann (1980) demonstrated a cell wall aminopeptidase in *Lactobacillus lactis*.

Membrane enzymes. Only a few examples have been reported, probably because of methodological problems. A protease bound to the membrane of *L. lactis* subsp. *lactis* was isolated by Cowman *et al.* (1967). Exterkate (1976, 1984) described two proteases capable of hydrolysing synthetic substrates, but not caseins: one was optimally active at 37°C (P37) and the other at 50°C (P50). A pyrrolidone-carboxyl peptidase is located together with protease P50 on the inner surface of the membrane in association with lipids. A loosely membrane-bound glutamate aminopeptidase has also been described (Exterkate, 1979, 1984).

Intracellular enzymes. Intracellular proteolytic enzymes play a major role in protein turnover, activation of zymogens and maturation of de novo synthesised proteins. However, these enzymes are probably involved later in the proteolysis of cheese curd after cell lysis.

Information available about intracellular proteases of lactic acid bacteria is scarce and contradictory, which suggests that there are several of them. The majority of described intracellular proteases are metalloenzymes, active between pH 6 and 7, at 35 and 45°C, and which do not, or only weakly, hydrolyse caseins. They more rapidly break down shorter substrates such as insulin or glucagon or even peptides originating from casein hydrolysis on which they have a thermolysin-type specificity (Desmazeaud, 1974; Desmazeaud & Zevaco, 1976; Yan *et al.*, 1987; Muset *et al.*, 1989). An endopeptidase activity has been detected in the ribosomal fraction of *Lb. casei* as well as a serine protease in that of *L. lactis* subsp. *lactis* (Ichishima *et al.*, 1986). Other proteases differ from the previous ones either by having a particularly low temperature optimum (6°C) in a strain of *L. lactis* subsp. *lactis* (Akuzawa *et al.*, 1984) or by activity on caseins (Exterkate, 1976), or by a high molecular weight in the case of *L. lactis* subsp. *cremoris* (Ohmiya & Sato, 1975) or by the nature of the enzyme, e.g. a thiol enzyme in a strain of *L. lactis* subsp. *lactis* (Cowman & Speck, 1967).

Numerous intracellular peptidases (amino-, di-, and tripeptidases) have been described in the streptococci (Rabier & Desmazeaud, 1973; Mou *et al.*, 1975; Desmazeaud & Jugé, 1976; Cliffe & Law, 1979; Law, 1979; Kolstad & Law, 1985; Kaminogawa *et al.*, 1984) as well as in the lactobacilli (El Soda *et al.*, 1978, 1982*b*; Hickey *et al.*, 1983). They are generally metalloenzymes. It should be noted that lactic acid bacteria possess prolyl-specific peptidases which are essential for the hydrolysis of peptides from proline-rich caseins. Enzymes such as iminopeptidase (Mou *et al.*, 1975) proline iminopeptidase, aminopeptidase P (Hickey *et al.*, 1983), prolyldipeptidase (Desmazeaud & Jugé, 1976), prolidase (Kaminogawa *et al.*, 1984; Booth *et al.*, 1990*a*), X-prolyl dipeptidyl peptidase (Casey & Meyer, 1985; Meyer & Jordi, 1987; Keifer-Partsch *et al.*, 1989; Zevaco *et al.*, 1990; Booth *et al.*, 1990*b*) have been demonstrated in *Streptococci* although their cellular localisation is not always indicated. Recently the gene coding for the X-prolyl dipeptidyl aminopeptidase has been cloned and sequenced for two strains of *Lactococci* (Nardi *et al.*, 1991; Mayo *et al.*, 1991). On the other hand, lactic acid bacteria have no or a very low carboxypeptidase activity as only a single example of this kind of enzyme has been reported in *Lb. casei* (El Soda *et al.*, 1978).

Proteolytic enzymes of penicillia

Blue-vein cheeses are characterised by the development of a mould, *P. roqueforti,* inside the curd. The spores are added to the milk or the curd during manufacture. The cheeses are then spiked to admit air and stimulate the growth of *Penicillium* moulds. Mould-ripened soft cheeses such as Camembert or Brie are made with a surface flora of the white mould. *P. camemberti* (formerly called *P. caseicolum* or *P. candidum*). Because of their great biochemical activity, *P. roqueforti* and *P. camemberti* have a large effect on ripening. These two moulds possess extracellular proteolytic systems which show marked analogies; both synthesise an aspartyle proteinase, a metalloproteinase and several exopeptidases (Cerning *et al.*, 1987).

Aspartyl proteinases are active at acid pH values (pH 3·5 on haemoglobin) and their maximal stability ranges from pH 3·5 to 5·5 (*P. camemberti*) or 6·0 (*P. roqueforti*). Their molecular weights (about 35 000), as well as their amino acid compositions are very similar (Zevaco *et al.*, 1973; Modler *et al.*, 1974; Lenoir *et al.*, 1979). They are inactivated rapidly by the classical aspartyl proteinase reagents (diazoacetyl DL norleucine methylester and 1,2-epoxy-3-*p*-nitrophenoxy propane).

The action of *P. camemberti* and *P. roqueforti* aspartyl proteinases on caseins are very similar. They hydrolyse β-casein and rapidly release five peptides, designated β_{ap1} to β_{ap5}, representing fragments 98–209, 100–209, 30–209, 1–29 and 1–97/99, respectively. They correspond to the preferential cleavage of the Lys-29–Ile-30, Lys-97–Val-98, Lys-99–Glu-100 bonds (Trieu-Cuot & Gripon, 1981; Trieu-Cuot *et al.*, 1982*a*).

The metalloproteinases of *P. camemberti* and *P. roqueforti* also have very similar characteristics. Their molecular weight (close to 20 000) and their amino acid compositions are very homologous. Their pH optima on whole casein are 5·5–6·0. The two enzymes still exhibit high activity at pH 8·5–9·0 and their pH stability range is rather wide (Gripon & Hermier, 1974; Lenoir & Auberger, 1977). They are inhibited by EDTA and contain a zinc atom required for catalytic activity. They have the same broad specificity on the B-chain of insulin (Gripon *et al.*, 1980). A clear rule concerning this specificity is lacking, but it differs definitely from that of thermolysin-type metalloproteinases which cleave peptide bonds possessing a hydrophobic amino acid linked by its amino group. As indicated by bidimensional electrophoresis, these two metalloproteinases have a similar action on caseins; three bonds in β-casein cleaved by the *P. camemberti* enzyme have been identified as Lys-28–Lys-29, Pro-90–Glu-91 and Glu-100–Ala-101 (Trieu-Cuot *et al.*, 1982*b*).

The intracellular proteinases of the two moulds are not as well known and their activity is more restricted than that of the extracellular enzymes. The pH optima of crude extracts of *P. camemberti* and *P. roqueforti* are close to 6·0. Paquet and Gripon (1980) demonstrated acid proteinases of a molecular weight ranging from 33 000 to 44 000 in several strains of *P. roqueforti*. Two fractions with pH optima of 6 and 6·5 have been separated from *P. camemberti* extracts (Auberger *et al.*, unpublished results).

P. roqueforti synthesises a carboxypeptidase that is inhibited by DFP. This serine carboxypeptidase has an acid pH optimum (close to 5·5 on the Z–Glu–Tyr substrate) and a molecular weight close to 110 000. It has a broad specificity and is able to release acid, basic or hydrophobic amino acids from glucagon (Gripon, 1977*a*). However, it only cleaves the Pro-28–Lys-29 bond of the insulin B-chain very slowly compared to other serine carboxypeptidases of fungal origin. A carboxypeptidase bound to the mycelium, but with similar characteristics, has been demontrated (Ichishima *et al.*, 1978). Its specificity is very broad and it is able to release *C*-terminal proline from angiotensin. The characteristics of the extracellular serine carboxypeptidase produced by *P. camemberti* are analogous to those of the *P. roqueforti* enzyme (Auberger *et al.*, 1985). It liberates hydrophobic amino acids particularly well and is able to reduce the bitterness of casein hydrolysates or soya bean proteins (Ahiko *et al.*, 1981). The culture medium and the mycelium of *P. roqueforti* exhibit carboxypeptidase activity with a neutral pH (Gripon & Debest, 1976). This activity has been detected only in *P. camemberti* mycelia; its pH optimum on Z–Glu–Tyr is close to 6·5 (Auberger *et al.*, 1985).

The extracellular aminopeptidase of *P. roqueforti* is a metalloenzyme that is inhibited by EDTA. Its pH optimum is approximately 8·0 and its molecular weight is close to 35 000. The apolar amino acids (leucine, alanine, methionine) are released rapidly; the presence of a glycine in the penultimate position considerably reduces the hydrolysis rate (Gripon, 1977*b*). The presence of two intracellular aminopeptidases has been observed (Paquet & Gripon, 1980). The extracellular aminopeptidase activity of *P. camemberti* is optimal at pH 8·0–8·5 and might be due to two enzymes (Auberger *et al.*, 1982).

Proteolytic enzymes of the secondary flora
The secondary cheese flora is important as it is considered as the means by which different varieties acquire their distinctive characteristics (Law, 1984). The bacteria of the secondary flora are often used in cheesemaking because of other metabolic properties than their proteolytic activity (e.g. lactate fermentation, CO_2 production, lipolytic activity). This is the reason why only few data are available about their proteolytic enzymes, mainly referring to four groups of microorgansims.

Propionic acid bacteria. The studies of Langsrud *et al.* (1977, 1978) demonstrated that these bacteria are capable of producing free proline more easily from casein hydrolysates than from the caseins themselves. This release of proline seems to be due to the activity of intracellular proteolytic enzymes as it occurs only after 7 days of incubation, probably due to cell lysis.

Coryneform bacteria. B. linens possesses strong and varied proteolytic activities. Extracellular and intracellular proteases have been demonstrated using a gelatin substrate (Foissy, 1974) and at least six different peptidase activities have been determined on cell homogenates (Torgersen & Sørhaug, 1978; Sørhaug, 1981). An intracellular aminopeptidase, active on the Leu–pNA substrate, has been described (Foissy, 1978). Although the lysis process of these bacteria remains unknown, the intracellular proteolytic enzymes most likely participate in cheese ripening (Torgersen & Sørhaug, 1978).

Micrococci. Most *Micrococcus* strains preferentially hydrolyse β-casein, but they are able to hydrolyse whole casein after a sufficiently long time (Bhowmik & Marth, 1988). As a matter of fact, it seems that the extracellular proteases produced by a minority of strains preferentially hydrolyse α_s caseins whereas the intracellular extracts degrade β-casein more rapidly in the micellar and not in the purified form (Nath & Ledford, 1972). An extracellular protease, the production of which is induced by peptides (Desmazeaud & Hermier, 1968*a*) and which is responsible for the overall extracellular protease activity of *Micrococcus caseolyticus,* has been purified. It is a neutral metalloprotease with a molecular weight of 38 kDa (Desmazeaud & Hermier, 1968*b*). This protease has a great affinity for casein and preferentially hydrolyses peptide bonds in which hydrophobic amino acids are involved via their amine group (Desmazeaud & Hermier, 1971). Intracellular leucyl-, alanyl-, methionyl aminopeptidase, iminopeptidase and dipeptidase activities have also been demonstrated in several strains of micrococci (Bhowmik & Marth, 1988).

Yeasts. Yeasts (principally of the genera *Kluveromyces, Debaryomyces* and *Saccharomyces*) develop on the surface of numerous soft cheeses. Their proteolytic activity is mainly intracellular. Schmidt (1982) observed caseinolytic activity, with a pH optimum of about 6, in 165 strains isolated from Camembert. The intracellular exopeptidases of *Kluyveromyces lactis* (one carboxypeptidase and three aminopeptidases) have been partly characterised (Desmazeaud & Devoyod, 1974).

Geotrichum candidum is a yeast-form fungus that grows on the surface of Camembert made from raw milk and which synthesises extra- and intracellular proteases with pH optima close to 6·0 (Guéguen & Lenoir, 1976).

Cheese proteolysis

Pressed cheeses

During ripening, the action of the proteolytic enzymes of lactic acid bacteria are superimposed on those of rennet and plasmin. As shown in chapter 5, rennet releases peptides of low and high molecular weight (Gripon *et al.*, 1975; Visser, 1977*a*), but does not produce free amino acids. The role of plasmin is rather limited: in Gouda cheese it was observed that it is responsible for the formation of about 10% of the soluble nitrogen (Visser, 1977*a*). Studies made on model curds with a controlled flora (Desmazeaud *et al.*, 1976; O'Keeffe *et al.*, 1976; Visser, 1977*a*), show that lactic acid bacteria have low endopeptidase activity (much lower than that of rennet) and augment only slightly the amount of soluble nitrogen. In contrast, they markedly modify the composition of soluble nitrogen by considerably increasing the concentrations of amino acids and small peptides. Peptides produced by rennet are thus broken down by the lactic acid bacteria into products of low molecular weight, some of which are aroma precursors or carriers. When intracellular enzyme extracts from streptococci (Law & Wigmore, 1983) or lactobacilli (El Soda *et al.*, 1981, 1982) were added to Cheddar cheeses, the production of amino acids and peptides was more rapid and the cheese aroma more intense. Thus, lactic acid bacteria seem to play a key role in aroma development in Cheddar, Gouda and probably other pressed cheeses (Reiter *et al.*, 1967; Kleter, 1976).

The bitterness defect sometimes found in cheese is due to the presence of peptides particularly rich in hydrophobic amino acids. Rennet, at the level normally used in cheesemaking, is capable of producing such peptides (Visser, 1977*b*), but bacteria also play a role in this process. As previously shown, cell wall proteinases are able to produce bitter peptides from caseins. In Cheddar manufacture, the strains which multiply well at the usual curd cooking temperatures tend to give bitter cheeses because the number of mesophilic *Streptococcus* cells, and hence the amount of cell wall proteinases, is high (Lowrie & Lawrence, 1972). Utilisation of a given proportion of protease-free variants (prt^-) reduces the bitter flavour, even if the bacterial population remains high (Mills & Thomas, 1980). According to Visser *et al.* (1986), the differences between bitter and non-bitter strains in Gouda manufacture are due to casein specificity differences between cell wall proteinases, the bitter strains hydrolysing β-casein better than the other caseins. The difference between bitter and non-bitter strains may also be due to a higher peptidase activity in the latter, enabling them to better hydrolyse bitter peptides (Visser, 1977*b*).

Cooked cheese (Swiss-type cheeses)

Little information is available on proteolysis in high-cooked cheeses. Heating of the curd causes rennet denaturation (Garnot & Mollé, 1988); however, weak residual chymosin activity can be detected (Collin *et al.*, 1988). Plasmin seems to play a more important role than in other cheeses. Its action is stimulated by the rather high curd pH (5·2–5·6) and by the length of ripening; the amount of γ-caseins increases regularly during ripening (Collin *et al.*, 1987). The large amounts of free amino acids and small peptides result from the action of lactic acid bacteria and there seems to be a correlation between flavour intensity and the proportion of nitrogen soluble in phosphotungstic acid (i.e. small peptides and free amino acids) (Berdagué & Grappin, 1988). However, in Swiss Emmental excessive proteolysis seems to be correlated with the secondary fermentation (late eye formation defect) (Lavanchy *et al.*, 1979). Cheeses with this defect have a high iminopeptidase activity which has been associated with propionic acid bacteria (Blanc *et al.*, 1979). Several authors have studied the role of propionic acid bacteria in Gruyère cheese. They are not only responsible for the eye formation, but also participate strongly in aroma development by the release of proline. Langsrud (1977, 1978) attributed the typical fruity aroma of Emmental to this amino acid. Biede and Hammond (1979) emphasised that the non-volatile, water-soluble fraction of Swiss cheeses has a marked flavour. This fraction contains small peptides and free amino acids which seem to be responsible for the brothy-nutty flavour. The sweetness seems to depend on the interactions of calcium and magnesium ions with the peptides and free amino acids. It has been suggested (Kowalenska *et al.*, 1985) that the non-volatile, water-soluble fraction is the origin of volatile flavours, either as a precursor or as a carrier of the slowly released aroma.

Mould-ripened cheeses

Proteolysis in mould-ripened or blue-vein cheeses is particularly extensive. At the end of ripening, the soluble nitrogen content (pH 4·6) of traditional Camembert reaches 35% of total nitrogen (Lenoir, 1962) and that of blue-vein cheese about 50% (Devoyod *et al.*, 1968). In addition, the proportions of free amino acids are large in the latter (up to 10% of total nitrogen) and in ripened traditional Camembert, the sum of free amino acids plus ammonia represents more than 10% of total nitrogen.

Consumption of lactic acid by the mould and production of ammonia by the surface flora lead to a rise in cheese pH (it may reach 7·0 on the Camembert surface and about 6·0 in the core of Bleu d'Auvergne after 1 month of ripening (Lenoir, 1984; Trieu-Cuot & Gripon, 1983). This increase has very marked repercussions on proteolysis. The activity of rennet, the pH optimum of which in the curd is close to 5·0 (Noomen, 1978), probably decreases during late ripening. Conversely, this pH increase favours the activity of plasmin (pH optimum close to 8·0) and its action in mould-ripened cheeses is higher than in mould-free cheeses (Trieu-Cout *et al.*, 1982).

The pH increase also induces a higher proteolytic activity by lactic acid starter bacteria and the secondary flora, whose pH optima are generally closer to the neutral than to the acid zone. However, the role of moulds in casein breakdown is particularly significant. Studies on model curds with a controlled flora composed of only *P. roqueforti* (or *P. camemberti*) demonstrated the intensity of proteolysis caused by these moulds (Desmazeaud *et al.*, 1976) which produce large amounts of peptides of high and low molecular weight and free amino acids. The action of *Penicillium* has also been demonstrated in Camembert where the proteolytic activity in the cheese is high in the external zone but remains low in the internal regions. The activities of aspartyl protease and metalloprotease have been determined and it appears that the two enzymes are produced at the surface, mainly between days 6 and 9 of ripening, i.e. when *Penicillium* appears. They exhibit maximal activity towards day 15 which decreases slowly thereafter, indicating that the two proteases are rather stable under the physico-chemical conditions of the external zone of Camembert (Fig. 8) (Lenoir, 1984).

The action of *Penicillium* proteases can not be detected in the core of the cheese, indicating that migration of these enzymes inside the curd is limited. However, the migration is sufficient to clearly create higher proteolysis in the external part of Camembert cheese (Lenoir, 1963). In the case of blue-vein cheeses, the sodium chloride gradient due to surface salting reduces the development of mycelium in the sub-rind area; however, the largest proportion of the cheese is subjected to the action of the proteolytic enzymes of the mould which accounts for the extensive proteolysis in these cheeses.

The extracellular peptidases of *Penicillium* are active at acid or almost

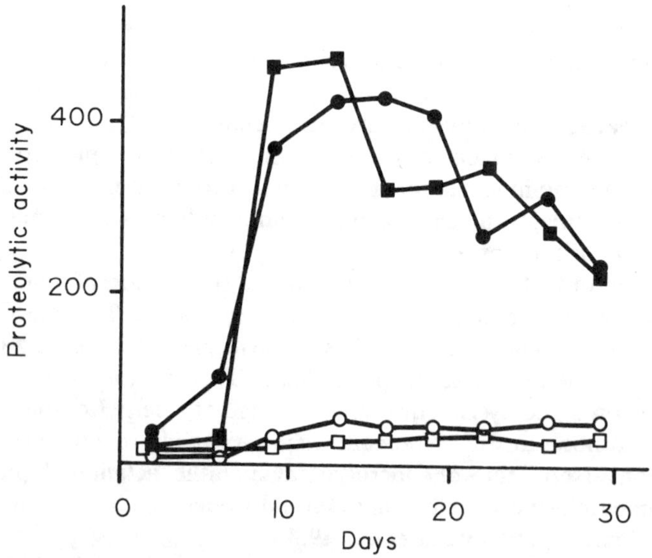

FIG. 8. Changes in the activities of the metalloproteinase and the aspartate proteinase of *P. camemberti* during Camembert cheese ripening. Asparate proteinase (●) surface, (○) interior; metalloproteinase (■) surface, (□) interior (From Lenoir (1984)).

neutral pH values. Some of them (acid carboxypeptidase) have a broad specificity. Synthesis of these enzymes leads to the production of large amounts of free amino acids, which are aroma precursors.

Very large modifications to the texture have been observed at the surface of Camembert. The initially hard and crumbly cheese becomes flexible and mellow, even 'slippery' if the water content of the curd is too high. These changes were formerly attributed to the proteolytic action of *Penicillium*. As a matter of fact they result mainly from the pH rise due to lactic acid consumption by the mycelium. Incubation of young Camembert without *Penicillium* growth in an ammonia atmosphere simulates the pH gradient observed in ripened cheeses and causes by istelf a softening of the cheese (Vassal *et al.*, 1986). According to Noomen (1983), the physico-chemical conditions of the curd alone are not responsible for the textural changes; cleavage of the α_{s1} casein by rennet ($\alpha_{s2} \rightarrow \alpha_{s1}$ I) is also required.

Amino acid degradation

Degradation of amino acids leads to the production of numerous aroma compounds and plays a major role in the development of cheese flavour. A general scheme for the microbial catabolism of amino acids during ripening has been suggested by Hemme *et al.* (1982). Amino acids are broken down by deaminases, carboxylases, transaminases and enzymes capable of modifying the side chains. Among the products obtained, α-ketoacids and amines are transformed into aldehydes, alcohols or acids. The secondary flora generally play an important role in the degradation of amino acids and contribute to providing each cheese with its typical character.

Sulphur compounds are major components of surface moulded or rind-washed cheeses and give them a garlic or cabbage-like taint. *B. linens* strains isolated from dairy products possess a demethiolase which releases methanethiol from methionine (Law & Sharpe, 1978; Ferchichi *et al.*, 1985). Methanethiol is the precursor of other sulphur compounds which may be formed by non-enzymic reactions (Adda *et al.*, 1982). *P. camemberti* (Tsugo & Matsuoka, 1962) and *G. candidum* (Mourgues *et al.*, 1983) also seem to be capable of producing sulphur products in soft cheeses. Methanethiol and other sulphur compounds contribute to the typical aroma of Cheddar cheese (Manning & Robinson, 1973; Manning *et al.*, 1976), but the mechanism of their formation has not yet been fully elucidated (Manning, 1979).

Arginine is broken down in blue-vein cheese mainly by *P. roqueforti* with the production of ornithine and citrulline (Cecchi & Resmini, 1972). Production of γ-aminobutyric acid by decarboxylation of glutamic acid has been observed in surface moulded cheese (Ismail & Hansen, 1972; Do Ngoc *et al.*, 1971). In surface moulded cheese, phenylalanine is degraded to phenylethanol which contributes to the typical aroma of traditional Camembert (Roger *et al.*, 1988) and which is most likely produced by surface yeasts. Strains of *B. linens*

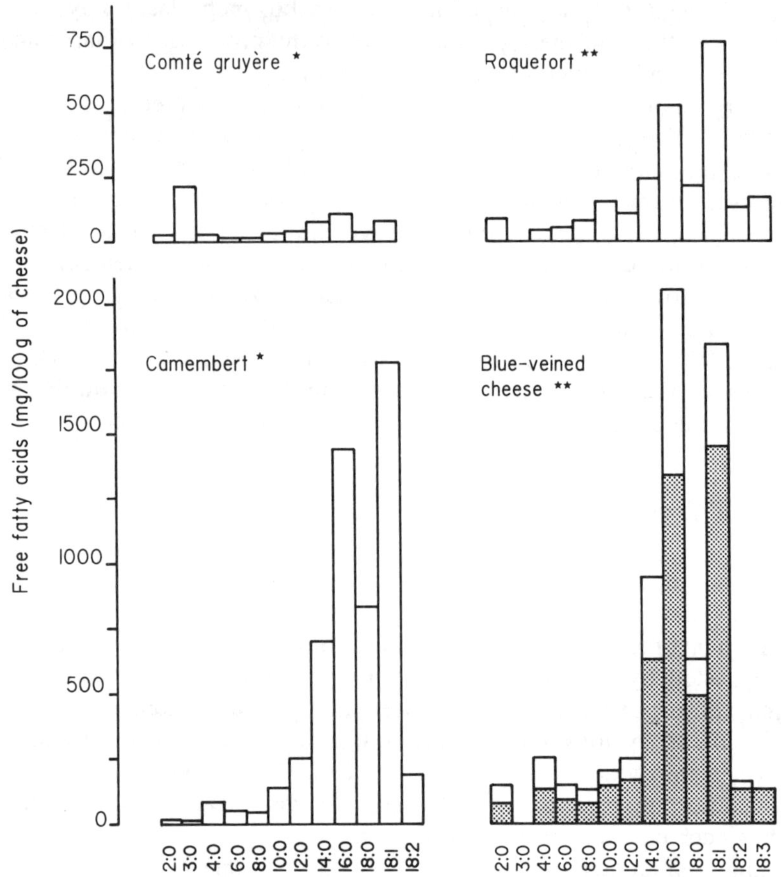

FIG. 9. Comparison of the main free fatty acid profiles of some cheese varieties (adapted from ([a]) Kuzdzal & Kuzdzal-Savoie (1966) and ([b]) Anderson & Day (1965*a*)).

also degrade phenylalanine, but by cleavage of the benzene nucleus by a 3–4, dihydrophenylacetate-2-3-deoxygenase (Lee *et al.*, 1985).

Ammonia is an important component of the aroma of traditional Camembert and results from the deamination of amino acids by the different elements of the flora. The role of *G. candidum* (Greenberg & Ledford, 1979) and *B. linens* (Hemme *et al.*, 1982) has been emphasised.

Decarboxylation reactions lead to the formation of biogenic amines such as tyramine, histamine, phenylethylamine, tryptamine, putrescine and cadaverine (Joosten & Stadhouders, 1987). The amounts of tyramine usually exceed those of histamine and tryptamine (Colonna & Adda, 1976) and the concentrations vary much according to samples (Sen, 1969). In Gouda and Maasdaam cheeses, the concentrations of amines are not very high under good hygienic and pasteurisation conditions (Joosten & Stadhouders, 1987). Thus, it appears that the lactic starters used in the manufacture of these cheeses (mesophilic streptococci, thermophilic lactobacilli and propionic acid bacteria) are not involved in the production of these compounds. The same conclusion was drawn by Antilla *et al.* (1984) for Emmental and Edam cheeses. A variety of microorganisms isolated from dairy products are able to break down one or several amino acids in a culture medium (Edwards & Sandine, 1981). Joosten and Northolt (1987) studied the nature of non-starter bacteria capable of inducing the formation of amines in Gouda cheese. The presence of enterococci does not lead to the formation of amines in Gouda if their population is lower than 10^7 cfu/g. When their number increases, the presence of tyramine has been observed. Contamination of milk by coliform bacteria may lead to the formation of cadaverine and putrescine from lysine and ornithine, respectively. Likewise, addition of mixtures of lactobacilli to Gouda cheese milk causes the formation of tyramine, histamine and putrescine and suggests that these organisms are important agents in the production of biogenic amines in cheese. Growth of Gram-negative bacteria during milk storage in the cold does not cause the formation of amines if the milk is pasteurised prior to cheese manufacture (Joosten & Northolt, 1987).

LIPOLYSIS AND FATTY ACID TRANSFORMATION

The hydrolysis of triglycerides, which constitute more than 98% of cheese fat, is the major biochemical fat transformation during ripening. It leads to the production of free fatty acids, di-and monoglycerides and possibly glycerol.

The lipases (acylglycerol hydrolases, EC 3.1.1.3) which catalyse this hydrolysis can be distinguished from the classical esterases by their activation on water–oil interfaces; their action on esters in aqueous solution is usually very low (Verger, 1985). The reaction rate may also vary widely according to the following factors (i) the substrate, tri-, di- or monoglyceride, (ii) the position of the ester bond, with hydrolysis of primary esters only or non-specific hydrolysis of primary and secondary esters, (iii) preference for some fatty acids according to the degree of unsaturation or the acyl chain length, and (iv) a stereospecificity favouring the sn-1 or sn-3 positions of the sn-glycerides (Jensen *et al.*, 1983).

The di- and monoglycerides formed may also be hydrolysed by other enzymes, i.e. carboxylesterases (carboxylic esterhydrolases, EC 3.1.1.1) which act on the esters in solution, or by di- or monoglyceride lipases. The former are easily demonstrated in microorganisms, but not their actual hydrolytic action in cheeses; the latter do not seem to have been accurately characterised as such in microorgansims. The action of phospholipases does not seem to be important in cheeses.

Free fatty acids contribute to aroma formation in cheese. They have specific flavours (rancid, sharp, 'goaty', soapy, coconut-like), particularly marked for acids between C-4:0 and C-12:0, whose intensity depends on the concentration but also on the distribution between aqueous and fat phases, the pH of the medium, the presence of certain cations (e.g. Na^+, Ca^{2+}) and protein degradation products.

Lipases of dairy microorganisms

Penicillium *lipases*

The two species, *P. roqueforti* and *P. camemberti,* belong to the most lipolytic microorganisms for cheese ripening. It was established a long time ago (e.g. Niki *et al.*, 1966) that *P. roqueforti* has two types of lipolytic activities differing by their pH optimum and the nature of the free fatty acids released from butter oil. The catalytic properties of extracellular alkaline lipase were described by Eitenmiller *et al.* (1970) and more recently another enzyme with a neutral or slightly acid pH optimum has been isolated and its activity characterised (Lobyreva & Marchenkova, 1980, 1981; Lamberet & Menassa, 1983*a*). *P. camemberti* produces a single lipase which has been purified by Lamberet and Lenoir (1976*a,b*). The catalytic properties of the extracellular and mycelium-bound enzymes are similar.

The preferred substrate of the alkaline lipases of the two moulds is tributyrin, their relative action on triolein being about 20% and on butter oil 60% for *P. roqueforti* lipase (Eitenmiller *et al.*, 1970; Menassa & Lamberet, 1982). *P. camemberti* lipase is more active on polyunsaturated oils than on butter oil (Al-Hir & Chandan, 1982). These enzymes have the same pH and temperature optima, pH 9·0–9·5 and 30–35°C, but that of *P. camemberti* is characterised by higher relative values than that of *P. roqueforti* at pH 6·0 as well as at the usual ripening temperatures (Lamberet & Lenoir, 1976*b*; Menassa & Lamberet, 1982). *P. roqueforti* lipase is activated by Mn^{2+} and Mg^{2+} (Eitenmiller *et al.*, 1970) but not that of *P. camemberti* which requires the presence of Ca^{2+} to reach its activity maximum (Lamberet & Lenoir, 1976*b*).

An acid lipase of *P. roqueforti* has been partly purified as a complex with a molecular weight exceeding 500 000 Da; its activity is maximal on tricaproin; it has 20 and 3% of maximal activity on tributyrin and butter oil, respectively. It is stable between pH 3·7 and 6·0 at 30°C and optimal hydrolysis conditions are pH 6·0 and 35–40°C; there is a second activity maximum at pH 3·0. Lowering of the assay temperature markedly increases the relative activity at alkaline pH but reduces it at acid pH. High calcium concentrations (0·1 M) reduce the activity (Lamberet & Menassa, 1983*a*).

Yeast lipases

G. candidum, often present on a variety of cheeses before salting, produces an extracellular lipase that is rather selective towards *cis*-9 unsaturated fatty acids, similar to oleic acid; it shows no positional selectivity (R. G. Jensen, 1974). The extra- and intracellular fractions show the same characteristics (Tahoun, 1987). It is stable at 30°C within a broad pH range (4·5–9·5) and its pH optimum is 5·5–6·5; it retains more than 50% of its maximal activity at pH 8·5 (Tsujisaka *et al.*, 1973).

The action of *G. candidum* on milk fat is characterised by a three-fold increase in the relative amounts of free oleic–free linoleic acids compared to the initial triglycerides; two-to five-fold lower values have been recorded for short or long chain saturated fatty acids (Jensen *et al.*, 1965).

The other yeast species on cheeses exhibit great diversity concerning their lipolytic potentiality, the extracellular nature of the lipase (for instance *Candida mycoderma* and *Debaryomyces kloecheri*, (Hosono & Tokita, 1970) or its intracellular nature (for instance *Saccharomyces cerevisiae,* which is a weakly lipolytic microorganism, Schousboe, 1976). Yeast lipase activity has been comprehensively studied in only a few species isolated from cheeses; however, it appears that the properties of these enzymes are highly varied which has been confirmed by the characterisation of lipases of strains or species of other origins.

The pH optimum can vary from 4·5 for *C. mycoderma* and *D. kloecheri* (Hosono & Tokita, 1970) to almost neutral or alkaline values for *C. paralipolytica* (Ota *et al.*, 1973). Some enzymes like those of *C. rugosa* (Jensen *et al.*, 1983), *D. kloecheri* (Hosono & Tokita, 1970), are not selective towards the position of fatty acids in the triglyceride. Preferences for short-chain fatty acids by *Torulopsis* spp. (Motai *et al.*, 1966) or for C-12 to C-16 by *C. mycoderma* (Hosono & Tokita, 1970) have also been reported. In addition to strains isolated from cheeses, some species, like *Rhodotorula pilimanae,* can produce two enzymes with different pH optima (Muderhwa *et al.*, 1986) while others, such as *C. curvata,* may show a preference towards fatty acids close to that exhibited by *G. candidum* (Montet *et al.*, 1985).

Lipases of the bacterial flora

The lipolytic activity of the lactic and propionic acid cheese flora is very low, but the mesophilic streptococci and the *Leuconostoc* bacteria can be considered as being more active than the lactobacilli and *S. thermophilus* (Oterholm *et al.*, 1968; Searles *et al.*, 1970). In the case of starter cell suspensions, it seems that their actoin on di-and monoglycerides is stronger than on triglycerides which in cheeses would enhance the previous partial lipolysis (Stadhouders & Veringa, 1973). In these bacteria the lipases are intracellular, or cell bound; Umemoto and Sato (1978) showed that the enzyme isolated from *S. lactis* hydrolyses the neutral cell lipids more rapidly than olive oil; it might participate in cell lysis. De Moraes and Chandan (1982) reported optimal activity at pH 9·0 and 45°C for *S. thermophilus* lipase which is three times more active on tributyrine than on butter oil.

In the lactobacilli, the activity is maximal at pH 6·5–7·5 and at 40–50°C (Searles *et al.*, 1970; Chander *et al.*, 1973). Eight species of lactobacilli studied by El Soda *et al.* (1986*d*) hydrolyse the triglycerides from tributyrin to tripalmitin. *L. bulgaricus* and *L. plantarum* are the most active species on butter oil, *L. casei* being the least active one. Like lactobacilli in meat (Papon

& Talon, 1988), there might be a parallel between growth and lipase production.

Data concerning non-lactic acid bacteria are more scarce. The enterococci strains are a little more active than the lactic acid bacteria (Dovat *et al.*, 1970). Only a few comprehensive studies have been devoted to the surface bacterial flora, micrococci or coryneform bacteria; these bacteria have substantial intra- or extracellular activities which probably exhibit highly variable properties (e.g. Ordonez & Ortiz-Apodaca, 1977). The lipase of *Micrococcus freudenreichii* which hydrolyses soluble and insoluble substrates, has been purified (Lawrence *et al.*, 1967). In *B. linens,* the lipase bound to cell wall polysaccharides has also a very broad specificity (Sorhaug & Ordal, 1974).

Cheese lipolysis

The level and type of lipolysis in cheeses are affected by the nature and growth of the flora and by the synthesis and properties of its lipases. Other factors are also involved such as the variation in cheese pH, the contents of water and salt (Godinho & Fox, 1981), the physical state and dispersion of fat and the location of lipolytic microorgansims.

At normal ripening temperatures, the partial crystallisation of triglycerides, which slows down the action of lipases, probably accounts for the lowering of the selectivity of action of yeast lipases observed by Carini *et al.* (1975) between 30 and 15°C.

Fat may remain dispersed (Mozzarella) or tend to agglomerate (Cheddar) during curd preparation; the membrane of the fat globules, which protects the triglycerides, can also be altered due to pH changes and/or mechanical action. The consequences on lipolysis have not yet been clearly established; however, homogenisation of milk, which disperses fat and reduces protection by the membrane, enhances microbial lipolysis in cheeses (e.g. Nielsen & Edelsten, 1974).

The lipolytic action of microorganisms mainly occurs close to the agents since the migration of enzymes, even the extracellular ones, is probably not very marked; yet, their molecular weight is lower (about 24 000 Da for *P. camemberti* lipase, Lamberet & Lenoir, 1976*a*) than that of animal pregastric lipases, migration of which is very limited (Lee *et al.* 1980). Thus, sub-rind lipolysis is high (up to 30% of fatty acids) in a traditional Camembert (Kuzdzal-Savoie, 1968) or in the cavities colonised by *P. roqueforti* in blue-vein cheeses. The amounts of lipases produced by these moulds have been estimated in the aforementioned cheeses (Lamberet & Lopez, 1982; Lamberet & Menassa, 1983*b*). Some partly water-soluble volatile fatty acids may diffuse slowly (Lee *et al.*, 1980).

The highest lipolysis rates have been observed in traditional mould-ripened cheeses (Fig. 9): 5–10% for Camembert (Kuzdzal & Kuzdzal-Savoie, 1966), Roquefort (Anderson & Day, 1965*a*) and up to 20% for some other blue-vein cheeses (Anderson & Day, 1965*b*; F. Jensen 1974). In these cheeses, lipolysis is responsible for more than 95% of the total free fatty acids; from C–4:0, the profiles of free fatty acids differ rather little from those of the triglycerides. However, in Camembert the proportion of oleic acid is a little higher which might be accounted for by the action of *G. candidum* (Kuzdzal-Savoie, 1968) together with that of *Penicillium*. In blue-vein cheeses, the utilisation of different *P. roqueforti* strains leads to different lipolysis rates and also to variable proportions of volatile fatty acids (Stepaniak *et al.*, 1974) which might have an influence on the synthesis of methylketones.

In cheeses containing mainly a flora of lactic acid bacteria, lipolysis remains low, i.e. from 0·5 to 1 or 1·5% according to cheese type or age (Cheddar: Bills & Day, 1964; Gouda: Iyer *et al.*, 1967). This is also the case for high-cooked pressed cheeses (Langler & Day, 1966) where lipolysis may reach 2% in Comté Gruyère (Kuzdzal-Savoie & Kuzdzal, 1971). The proportions of free fatty acids are almost the same as those of triglycerides (e.g. De Man, 1966), but lipolysis only accounts for little of the total amount of free fatty acids (20–50% in mols); in Cheddar and high-cooked pressed cheeses, the most characteristic fraction of volatile fatty acids is derived from the degradation of carbohydrates and lactic acid.

Microbial lipolysis does not play a major role either in Italian cheeses where the typical flavour is due to short chain fatty acids, especially butyric acid, resulting from the action of animal lipases traditionally added during cheese manufacture (e.g. Woo & Lindsay, 1984). In mould-free soft cheeses, lipolysis is variable and does not have any marked influence on flavour (Kuzdzal-Savoie & Kuzdzal, 1971) compared to that resulting from the degradation of caseins and amino acids.

More recently, lower free fatty acids concentrations have been reported for various commercial cheeses, (Vanbelle *et al.*, 1978; Woo *et al.*, 1984) the ripening level of which was probably different, mainly for soft cheeses.

Transformations of fatty acids

Synthesis of esters and the formation of methyl ketones are the most important transformations of free fatty acids in cheese ripening.

Synthesis of esters

This synthesis mainly concerns short- or medium-chain fatty acids and the alcohols involved may be aliphatic (ethanol), aromatic (phenylethanol) or thiol groups (methanethiol). A variety of enzymes such as carboxylesterases or arylesterases (aryl-ester hydrolases, EC 3.1.1.2) with variable substrate specificities are present in yeasts (e.g. Schmidt & Lamberet, 1981), lactic acid bacteria (e.g. Harper *et al.*, 1980; El Soda *et al.*, 1986*a,c*), microococci and coryneform bacteria (Brandl & Stahl, 1973; Sørhaug & Ordal, 1974) or D-group streptococci (Dovat *et al.*, 1970). These enzymes, revealed by their hydrolytic action, can synthesise esters in a slightly aqueous-alcoholic medium without cofactors (yeasts: Schermers *et al.*, 1976; lactic and acid bacteria:

Hosono *et al.*, 1974). This synthetic ability is less well documented for microorganisms from cheese than for those of beverages, but the numerous esters formed during cheese ripening and their participation in aroma formation or in the development of certain defects, clearly show the existence of a certain activity, as do the results of cheesemaking with yeasts (Tsugo & Chang, 1970).

Oxidative degradation

Oxidative degradation is at the origin of the formation of $C(n-1)$ methylketones and the resulting secondary alcohols from Cn fatty acids (mainly from C-6:0 to C-12:0); these components are characteristic of the aroma of blue-vein cheeses (80–600 μmols methylketones/100%g fat, mainly 2-heptanone and 2-nonanone (Schwartz *et al.*, 1963)), but also of that of Camembert-type cheeses (25–60 μmols/100% g fat, mainly 2-nonanone and 2-heptanone (Schwartz & Parks, 1963)).

This metabolism, common to certain moulds of the genera *Aspergillus* and *Penicillum,* uses the pathway of oxidative degradation of fatty acids (Fig. 10). It has been extensively studied in *P. roqueforti* (see review of Kinsella & Hwang, 1976). A thiohydrolase activity has been demonstrated, as has a system responsible for the β-ketoacyl decarboxylase activity, which has been partly purified (Hwang *et al.*, 1976). These activities are high enough to favour the formation of methyl ketones, especially in the presence of a low

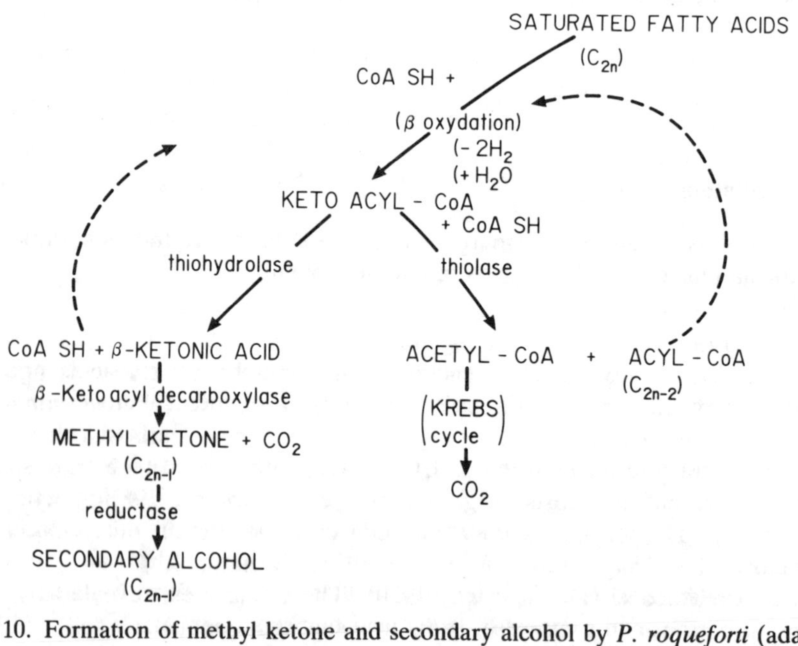

FIG. 10. Formation of methyl ketone and secondary alcohol by *P. roqueforti* (adapted from Kinsella & Wang, (1976)).

endogenous CO_2 content; they are observed in resting spores, in germinating spores or in mould mycelia. This observation has also been made with *P. camemberti* (Lamberet *et al.*, 1982).

In cheeses, the extent of methyl ketone formation increases with the level of lipolysis (Kinsella & Hwang, 1976), the medium-chain fatty acid concentrations in cheese being apparently insufficient to exert a prolonged inhibitory effect on oxidation. The relative proportions of the different methyl ketones depend on the properties of the enzyme system, the relative proportions of free fatty acid precursors (cow or ewe milk, lipolysis specificity) and the rate of disappearance (catabolism and/or formation of secondary alcohols). It has been established that the homologous (to the methyl ketones) secondary alcohols are formed by enzymatic reduction by various microorganisms (Anderson & Day, 1965*b*). However, in curd slurries the disappearance of methyl ketones does not lead to the formation of the corresponding alcohols (King & Clegg, 1979).

The biological formation from milk fat or other fats of blue-vein aroma compounds by *P. roqueforti* (Dwivedi & Kinsella, 1974; Jolly & Kosikowski, 1975; Kanisawa & Itoh, 1984 and more recently, by *P. camemberti* (Okumura & Kinsella, 1985) has been reported in a series of patents and publications. Small amounts of linoleic and linolenic acid are also susceptible to be broken down into important aroma compounds (as 1-octen-3-ol) by action of *Penicillia* in mould ripened cheeses (Karahadian *et al.*, 1985*a*, *b*).

CONCLUSION

Production of cheese with specific organoleptic characteristics implies utilisation of microorganisms with varied enzymic activities on the major curd components. Better control of the ripening process is based on a better knowledge of the flora and of their activities, but also on the production of better performing strains. The latter can be obtained by selection of strains by well-defined criteria followed by experimental manufacture, controlled by physico-chemical and organoleptic analyses. Genetic engineering will probably soon lead to the creation of strains equipped with adapted and stable biochemical potentialities and certainly represents the route of the future. Another approach toward the better control of the ripening process involves addition of enzymes to the curd which will be responsible for the desired enzymic modifications. Numerous experiments have shown that this procedure is difficult since the addition of one only enzyme (or even of few enzymes) leads to an imbalance in the formation of aroma and flavour compounds (bitter taste in the case of proteases, rancidity in the case of lipases). However, positive results have been reported in Cheddar cheese-making by the addition of protease and/or of extracts of lactic acid bacteria (Law & Wigmore, 1983). These enzymes may be incorporated into the cheese milk or during cheese salting in the case of Cheddar. Encapsulation of the enzymes within liposomes (Kirby *et al.*, 1987; Alkhalaf *et al.*, 1988) or fat capsules (Magee *et al.*, 1981) represents another possibility which should be examined further. The addition of attenuated cells (heat treatment or freeze-shocking) of lactic acid bacteria, which have lost their acidifying activities, but retained their other enzymic activities also constitutes a method to be explored (Olson *et al.*, 1986; Ardo & Pettersen, 1988; Exterkate *et al.*, 1987).

ACKNOWLEDGEMENT

The authors are very grateful to Mrs Kirsten Rérat for the translation of this manuscript into English.

REFERENCES

Accolas, J. P., Veaux, M., Vassal, L. & Mocquot, G. (1978). *Le Lait,* **58,** 118–32.
Accolas, J. P., Hemme, D., Desmazeaud, M. J., Vassal, L., Bouillanne, C. & Veaux, M. (1980). *Le Lait,* **60,** 487–524.
Adda, J., Gripon, J.-C. & Vassal, L. (1982). *Food Chem.*, **9,** 115–29.
Ahiko, K., Iwasawa, S., Ueda, M. & Miyata, N., *Reports of Research Laboratory,* Show Brand Milk Products Co. Ltd. **77,** 135, 1981.
Akpemado, K. M. & Bracquart, P. A. (1983). *Appl. Environ. Microbiol.*, **45,** 136–40.
Akuzawa, R., Ito, O. & Yokoyawa, K. (1984). *Jap. J. Zootech. Sci.,* **56,** 56–61.
Al-Hir, S. & Chandan, R. C. (1982). *J. Dairy Sci.,* **65** (Suppl. 1), 63.
Alkhalaf, W., Piard, J.-C., El Soda, M., Gripon, J.-C., Desmazeaud, M. & Vassal, L. (1988). *J. Food Sci.,* **53,** 1674–9.
Anderson, D. F. & Day, E. A. (1965*a*). *J. Dairy Sci.,* **48,** 248–9.
Anderson, D. F. & Day, E. A. (1965*b*). *J. Dairy Sci.,* **48** (Suppl.), 774.
Antilla, P. Antilla, V., Mattila, J. & Hakkarainen H. (1984). *Milchwissenchaft,* **39,** 400–4.
Ardö, Y. & Petterson, H. E. (1988). *J. Dairy Res.,* **55,** 239–45.
Auberger, B., Montals, M. & Lenoir, J. (1982). *International Dairy Congress* (Vol 1), Book, 2, p. 276.
Auberger, B., Lamberet, G. & Lenoir, J. (1985). *Science des Aliments,* **5,** 239–243.
Auclair, J. & Accolas, J. P. (1983). *Ant. Leeuvenh.,* **49,** 313–26.
Bartels, H. J., Johnson, M. E. & Olson, N. F. (1986). *Milchwissenschaft,* **42,** 83–8.
Benetaya, A., Bracquart, P. A., Le Deaut, J. Y. & Linden, G. (1986). *Le Lait,* **66,** 289–303.
Berdagué, J. L. & Grappin, R. (1988). *Le Lait,* **68,** 189–204.
Bergère, J. L. & Accolas, J. P. (1986). *Anaerobic Bacteria in Habitats Other than Man,* ed. E. M. Barnes & G. C. Mead. Blackwell Scientific Publications, Oxford, UK, pp. 373–96.
Bhowmik, T. & Marth, E. H. (1988). *J. Dairy Sci.,* **71,** 2358–65.
Biede, S. L. & Hammond, E. G. (1979). *J. Dairy Sci.,* **62,** 238–48.
Bills, D. D. & Day, E. A. (1964). *J. Dairy Sci.,* **47,** 733–8.
Blanc, B., Rüegg, M., Baer, A., Casey, M. & Lukesch, A. (1979). *Schweiz. Milch. Forschung,* **8,** 27–36.
Booth, M., Jennings, P. V., Ni Fhaolain, I. and O'Cainn, G. (1990*a*) *J. Dairy Res.,* **57,** 245–54.

Booth, M., Ni Fhaolain, I., Jennings, P. V. and O'Cainn, G. (1990*b*). *J. Dairy Res.*, **57**, 89–99.
Brandl, E. & Stahl, V. (1974). *Osterr. Milchwisch.*, **28**, 7–14.
Carini, S., Lodi, R., Braghini, T. & Fedeli, E. (1975). *L'industria del Latte*, **11**, 1–11.
Casey, M. G. & Meyer, J. (1985). *J. Dairy Sci.*, **68**, 3212–15.
Cecchi, L. & Resmini, P. (1972). *Scienza technica Lattiero-Casearia*, **23**, 389–409.
Cerning, J., Gripon, J.-C., Lamberet, G. & Lenoir, J. (1987). *Le Lait*, **67**, 3–35.
Chander, H., Chebbi, N. B. & Ranganathan, B. (1973). *Arch. Mikrobiol.*, **92**, 171–4.
Cliffe, A. J. & Law, B. A. (1979). *J. Appl. Bacteriol*, **47**, 65–73.
Cliffe, A. J. & Law, B. A. (1985). *J. Appl. Bacteriol.*, **58**, 245–50.
Cogan, T. M. (1980). *Le Lait*, **60**, 397–425.
Cogan, T. M. (1987). *J. Appl. Bacteriol.*, **63**, 551–8.
Cogan, T. M. & Daly, C. (1987). *Cheese: Chemistry, Physics and Microbiology*, ed. P. F. Fox. pp. 179–249. Elsevier Applied Science, London, UK.
Collin, J.-C., Berdagué, J.-L., Dognin-Bergeret, M. & Grappin, R. (1987). *Le Lait*, **67**, 299–318.
Collin, J.-C., Compagnone, P., Ryba, I. & Baer, A. (1988). *Le Lait*, **68**, 235–240.
Collins, L. B. & Thomas, T. D. (1974). *J. Bacteriol.*, **120**, 52–8.
Colonna, P. & Adda, J. (1976). *Le Lait*, **56**, 143–53.
Cowman, R. A. & Speck, M. L. (1967). *Appl. Microbiol.*, **15**, 851–6.
Cowman, R. A., Swaisgood, H. E. & Speck, M. L. (1967). *J. Bacteriol.*, **94**, 942–8.
Creamer, L. K. & Olson, N. (1982). *J. Food Sci.*, **47**, 631–46.
De Man, J. M. (1966). *J. Dairy Sci.*, **49**, 343–72.
De Moraes, J. & Chandan, R. C. (1982). *J. Food Sci.*, **47**, 1579–83.
Desmazeaud, M. J. (1974). *Biochimie*, **56**, 1173–81.
Desmazeaud, M. J. (1983). *Le Lait*, **63**, 267–316.
Desmazeaud, M. & Hermier, J. (1968*a*). *Ann. Biol. Anim. Biochem. Biophys.*, **8**, 419–29.
Desmazeaud, M. & Hermier, J. (1968*b*). *Ann. Biol. Anim. Biochem. Biophys.*, **8**, 565–77.
Desamzeaud, M. & Hermier, J. (1971). *Eur. J. Biochem.*, **19**, 51–5.
Desmazeaud, M. & Devoyod, J. J. (1974). *Ann. Biol. Anim. Biochem. Biophys.*, **14**, 327–41.
Desmazeaud, M. J. & Jugé, M. (1976). *Le Lait*, **56**, 241–60.
Desmazeaud, M. J. & Zevaco, C. (1976). *Ann. Biol. Anim. Biochem. Biophys.*, **16**, 851–68.
Desmazeaud, M., Gripon, J.-C., Le Bars, D. & Bergère, J.-L. (1976). *Le Lait*, **56**, 379–96.
De Vos, W. M. (1987). *FEMS Microbiol. Rev.*, **46**, 281–95.
Devoyod, J. J., Bret, G. & Auclair, J. E. (1968). *Le Lait*, **48**, 613–29.
Do Ngoc, M., Lenoir, J. & Choisy, C. (1971). *Revue Laitière Française*, **288**, 447–62.
Dovat, A. M., Reinbolt, G. N., Hammond, E. G. & Vedamathu, E. R. (1970). *J. Milk Food Technol.*, **33**, 365–72.
Dwivedi, B. K. & Kinsella, J. E. (1974). *J. Food Sci.*, **39**, 83–7 and 620–2.
Edwards, S. T. & Sandine, W. E. (1981). *J. Dairy Sci.*, **64**, 2431–8.
Eggimann, B. & Bachmann, H. (1980). *Appl. Environ. Microbiol.*, **40**, 876–82.
Eitenmiller, R. R., Vakil, J. R. & Shahani, K. M. (1970). *J. Food Sci.*, **35**, 130–3.
El Soda, M., Desmazeaud, M. J. & Bergère, J. L. (1978). *J. Dairy Res.*, **45**, 445–55.
El Soda, M., Desmazeaud, M., Abou Donia, S. & Kamal, N. (1981). *Milchwissenchaft*, **36**, 140–2.
El Soda, M., Desmazeaud, M., Abou Donia, S. & Badran, A. (1982*a*). *Milchwissenchaft*, **37**, 325–7.
El Soda, M., Zeyada, N., Desmazeaud, M. J., Mashaly, R. & Ismail, A. (1982*b*). *Sci. Alim.*, **2**, 261–73.

El Soda, M., Abd El Wahab, H., Ezzat, N., Desmazeaud, M. & Ismail, A. (1986*a*). *Le Lait*, **66**, 431–43.
El Soda, M., Desmazeaud, M. J., Le Bars, D. & Zevaco, C. (1986*b*). *J. Food Prot.*, **49**, 361–5.
El Soda, M., Fathallah, S., Ezzat, N., Demazeaud, M. & Abou Donia, S. (1986*c*). *Sci. Alim.*, **6**, 545–57.
El Soda, M., Korayen, M. & Ezzat, N. (1986*d*). *Milchwissenschaft*, **41**, 353–5.
Exterkate, F. A. (1975). *Neth. Milk Dairy J.*, **29**, 303–18.
Exterkate, F. A. (1976). *Neth. Milk Dairy J.*, **30**, 95–105.
Exterkate, F. A. (1979). *Arch. Microbiol.*, **120**, 247–54.
Exterkate, F. A. (1984). *Appl. Environ. Microbiol.*, **47**, 177–83.
Exterkate, F. A. & de Veer G. J. C. M. (1987). *System. Appl. Microbiol.*, **9**, 183–91.
Exterkate, F. A., de Veer, G. J. C. M. & Stadhouders, J. (1987). *Neth. Milk Dairy J.*, **41**, 307–20.
Ezzat, N., El Soda, M., Bouillanne C., Zevaco, C. & Blanchard P. (1985). *Milchwissenschaft*, **40**, 140–3.
Ezzat, N., Zevaco, C., El Soda, M. & Gripon, J. C. (1987). *Milchwissenschaft*, **42**, 95–7.
Ferchichi, M., Hemme, D., Nardi, M. & Pamboukjian, N. (1985). *J. Gen. Microbiol.*, **131**, 715–23.
Foissy, H. (1974). *J. Gen. Microbiol.*, **80**, 197–207.
Foissy, H. (1978). *Milchwissenschaft.*, **33**, 221–3.
Garnot, P. & Mollé, D. (1988). *J. Food Sci.*, **52**, 75–7.
Garvie, E. I. (1980). *Microbiol. Rev.*, **44**, 106–39.
Geis, A., Bockelmann, W. & Teuber, M. (1985). *Appl. Microbiol. Biotechnol.*, **23**, 79–84.
Geis, A., Kiefer, B. & Teuber, M. (1986). *Chem. Mikrobiol. Technol. Lebersm.*, **10**, 93–5.
Godinho, M. & Fox, P. F. (1981). *Milchwissenschaft*, **36**, 476–8.
Greenberg, R. S. & Ledford, R. A. (1979). *J. Dairy Sci.*, **62**, 368–72.
Gripon, J.-C. (1977*a*). *Ann. Biol. Anim. Biochim. Biophys.*, **17**, 283–98.
Gripon, J.-C. (1977*b*). *Biochimie*, **59**, 679–86.
Gripon, J.-C. (1987). *Cheese*: *Chemistry, Physics and Microbiology*, ed. P. F. Fox. Elsevier Applied Science, London, UK pp. 121–49.
Gripon, J.-C. & Hermier, J. (1974). *Biochimie*, **56**, 1323–32.
Gripon, J.-C. & Debest, B. (1976). *Le Lait*, **56**, 423–38.
Gripon, J.-C., Desmazeaud, M., Le Bars, D. & Bergère, J.-L. (1975). *Le Lait*, **55**, 502–16.
Gripon, J.-C., Auberger, B. & Lenoir, J. (1980). *Int. J. Biochem.*, **12**, 451–5.
Guéguen, M. & Lenoir, J. (1976). *Le Lait*, **56**, 439–48.
Harper, W. J., Carmona De Catril, A. & Chen, J. L. (1980). *Milchwissenschaft*, **35**, 129–32.
Harvey, R. J. & Collins, E. B. (1962). *J. Bacteriol.*, **83**, 1005–9.
Hemme, D., Nardi, M. & Jette, D. (1980). *Le Lait*, **60**, 595–618.
Hemme, D., Nardi, M. & Wahl, D. (1981). *Le Lait*, **61**, 1–18.
Hemme, D., Bouillanne, C., Metro, F. & Desmazeaud, M. J. (1982). *Sci. Aliment.*, **2**, 113–23.
Hickey, M. W., Hillier, A. J. & Jago, G. R. (1983). *Aust. J. Dairy Technol.*, **38**, 118–23.
Hickey, M. W., Hillier, A. J. & Jago, G. R. (1986). *Appl. Environ. Microbiol.*, **51**, 825–31.
Hosono, A. & Tokita, F. (1970). *Jap. J. Zootech. Sci.*, **41**, 519–27.
Hosono, A., Elliott, J. A. & McGugan, W. A. (1974). *J. Dairy Sci.*, **57**, 535–9.
Hugenholtz, J., Exterkate, F. & Konings, W. N. (1984). *Appl. Environ. Microbiol.*, **48**, 1105–10.

Hugenholtz, J., Van Sinderen, D., Kok, J. & Konings, W. N. (1987). *Appl. Environ. Microbiol.*, **53**, 853–9.
Hutkins, R. W. & Morris, H. A. (1987). *J. Food Prot.*, **50**, 876–84.
Hwang, D. H., Lee, Y. J. & Kinsella, J. E. (1976). *Int. J. Biochem.*, **7**, 165–71.
Ichishima, E., Takeuchi, M., Yamamota, K., Sano, Y. & Kikuchi, T. (1978). *Curr. Microbiol.*, **1**, 95–8.
Ichishima, E., Kawai, Y., Takeuchi, M. & Akiko, K. (1986). *Curr. Microbiol.*, **13**, 47–9.
Ismail, A. A. & Hansen, K. (1972). *Milchwissenschaft* **27**, 556–9.
Iyer, M., Richardson, T., Amundson, C. H. & Tripp, R. C. (1967). *J. Dairy Sci.*, **50**, 385.
Jensen, F. (1974). *Int. Dairy Congr.*, **1F**, 276.
Jensen, R. G. (1974). *Lipids*, **9**, 149–57.
Jensen, R. G., Sampugna, J., Quinn, J. G., Carpenter, D. L. & Marks, T. A. (1965). *J. Amer. Oil Chem. Soc.*, **42**, 1029–34.
Jensen, R. G., Dejong, F. A., & Clark, R. M. (1983). *Lipids*, **18**, 239–52.
Jolly, R. C. & Kosikowski, F. V. (1975). *J. Food Sci.*, **40**, 285–7.
Joosten, H. M. L. J. & Northolt, M. D. (1987). *Neth. Milk Dairy J.*, **41**, 259–80.
Joosten, H. M. L. J. & Stadhouders, J. (1987). *Neth. Milk Dairy J.*, **41**, 247–58.
Kaminogawa, S., Ninomiya, T. & Yamauchi, K. (1984). *J. Dairy Sci.*, **67**, 2483–92.
Kanisawa, T. & Itoh, H. (1984). *Sip. Shokuhin Kogyo Gakkaishi*, **31**, 477–82 and 483–7.
Karahadian, C., Josephson, D. B. and Lindsay, R. C. (1985*a*). *J. Agric. Food Chem.*, **33**, 339–43.
Karahadian, C., Josephson, D. B. and Lindsay, R. C. (1985*b*) *J. Dairy Sci.*, **68**, 1865–77.
Kashket, E. R. (1987). *FEMS Microbiol. Rev.*, **46**, 233–44.
Kempler, G. M. & McKay, L. L. (1981). *J. Dairy Sci.*, **64**, 1527–39.
Khalid, N. M. and North, E. H. (1990). *Appl. Environ. Microbiol.*, **56**, 381–88.
Kiefer-Partsch, B., Bockelmann, A., Geis, A. & Teuber, M. (1989). *Appl. Microbiol. Biotechnol.*, **31**, 75–8.
King, R. D. & Clegg, G. H. (1979). *J. Sci. Food Agric.*, **30**, 197–202.
Kinsella, J. E. & Hwang, D. H. (1976). *Crit. Rev. Food Sci. Nutr.*, **8**, 191–228.
Kirby, C. J., Brooker, B. E. & Law, B. A. (1987). *Int. J. Food Sci. Technol.*, **22**, 355–75.
Kiwaki, M., Ikemura, H., Shimizu-Kadota, M. & Hirashima, A. (1989). *Mol. Microbiol.*, **3**(3) 359–69.
Kleter, G. (1976). *Neth. Milk Dairy J.*, **30**, 254–70.
Kok, J. & Venema, G. (1988). *Biochimie*, **70**, 475–88.
Kok, J., Leenhouts, K. J., Haandrikman, A. J., Ledeboer, A. M. & Venema, G. (1988). *Appl. Environ. Microbiol.*, **54**, 231–8.
Kolstad, J. & Law, B. A. (1985). *J. Appl. Bact.*, **58**, 449–56.
Konings, W. N. & Otto, R. (1983). *Ant. Leeuwenh.*, **49**, 247–57.
Konings, W. N., Poolman, B. & Driessen, A. J. M. (1989). *CRC Critical Reviews in Microbiology*, **16**, 419–76.
Kowalewska, J., Zelazowska, H., Babuchowski, A., Hammond, E. G., Glutz, B. A. & Ross, F. (1985). *J. Dairy Sci.*, **68**, 2165–71.
Kuzdzal, W. & Kuzdzal-Savoie, S. (1966). *Int. Dairy Congr.*, **D2**, 335–42.
Kuzdzal-Savoie, S. (1968). *Qualitas Plantarum et Materiac Vegetabiles*, **16**, 312–19.
Kuzdzal-Savoie, S. & Kuzdzal, W. (1971). *Tech. Lait.*, **717**, 11–14.
Laan, H., Smid, E. J., de Leij, L., Schwander, E. & Konings, W. N. (1988). *Appl. Environ. Microbiol.*, **54**, 2250–6.
Lamberet, G. & Lenoir, J. (1976*a*). *Le Lait*, **56**, 119–34.
Lamberet, G. & Lenoir, J. (1976*b*). *Le Lait*, **56**, 622–44.
Lamberet, G. & Lopez, M. (1982). *International Dairy Congress* Vol. 1, Book 1, pp. 499–500.

Lamberet, G. & Menassa, A. (1983*a*). *J. Dairy Res.*, **50,** 459–68.
Lamberet, G. & Menassa, A. (1983*b*). *Le Lait*, **63,** 333–44.
Lamberet, G., Auberger, B., Canteri, G. & Lenoir, J. (1982). *Revue Laitière Française*, **406,** 13–19.
Langler, J. E. & Day, E. A. (1966). *J. Dairy Sci.*, **49,** 91–3.
Langsrud, T., Reinbold, G. W. & Hammond, E. G. (1977). *J. Dairy Sci.*, **60,** 16–23.
Langsrud, T., Reinbold, G. W. & Hammond, E. G. (1978). *J. Dairy Sci.*, **61,** 303–8.
Lavanchy, P., Buehlmann, C. & Blanc, B. (1979). *Schweiz. Milchw. Forschung*, **8,** 9–14.
Law, B. A. (1977). *J. Dairy Res.*, **44,** 309–17.
Law, B. A. (1978). *J. Gen. Microbiol.*, **105,** 113–18.
Law, B. A. (1979). *J. Appl. Bacteriol.*, **46,** 455–63.
Law, B. A. (1981). *Dairy Sci. Abstr.*, **43,** 143–54.
Law, B. A. (1984). *Progress in Industrial Microbiology*, ed. M. E. Bashell. Elsevier, Amsterdam, pp. 245–83.
Law, B. A., & Sharpe, M. E. (1978). *J. Dairy Res.*, **45,** 267–75.
Law, B. A. & Kolstad, J. (1983). *Ant. Leeuwenh.*, **49,** 225–45.
Law, B. A. & Wigmore, A. (1983). *J. Dairy Res.*, **50,** 519–26.
Law, B. A., Sezgin, E. & Sharpe, M. E. (1976). *J. Dairy Res.*, **43,** 291–300.
Lawrence, R. C., Fyer, T. F. & Reiter, B. (1967). *Nature*, **213,** 1264–5.
Leach, F. R. & Snell, E. E. (1960). *J. Biol. Chem.*, **235,** 3523–31.
Lee, H. J., Olson, N. F. & Lund, D. B. (1980). *J. Dairy Sci.*, **63,** 513–22.
Lee, C. W., Lucas, S. & Desmazeaud, M. J. (1985). *FEMS Microbiol. Lett.*, **26,** 201–5.
Lenoir, J. (1962). *CR Acad. Agric.*, **48,** 392–9.
Lenoir, J. (1963). *Le Lait*, **43,** 154–65.
Lenoir, J. (1984). *IDF Bulletin*, **171,** 3–20.
Lenoir, J. & Auberger, B. (1977). *Le Lait*, **57,** 471–91.
Lenoir, J., Auberger, B. & Gripon, J.-C. (1979). *Le Lait*, **59,** 244–268.
Lobyreva, L. B. & Marchenkova, A. I. (1980).*Mikrobiol.*, **49,** 294–30.
Lobyreva, L. B. & Marchenkova, A. I. (1981). *Mikrobiol.*, **50,** 90–5.
Lowrie, R. J. & Lawrence, R. C. (1972). *NZ J. Dairy Sci. Technol.*, **7,** 51–3.
Magee, E. L., Olson, N. F. & Lindsay, R. C. (1981). *J. Dairy Sci.*, **64,** 616–21.
Manning, D. J. (1979). *J. Dairy Res.*, **46,** 531–7.
Manning, D. J. & Robinson, H. M. (1973). *J. Dairy Sci.*, **40,** 63–75.
Manning, D. J., Chapman, H. R. & Hooking, Z. D. (1986). *J. Dairy Res.*, **43,** 313–20.
Mayo, B., Kok, J., Venerna, K., Bockelmann, M., Teuber, M., Rainke, H. & Vernerna, G. (1991). *Appl. Environ. Microbiol.*, **57,** 38–44.
Marshall, V. M. E. & Law, B. A. (1983). In *Advances in Microbiology and Biochemistry of Cheese and Fermented Milk*, ed. F. L. Davies & B. A. Law. pp. 67–98. Elsevier Applied Science Publisher, London, UK.
Matthews, D. M. & Payne, J. W. (1975). In *Peptide Transport in Protein Nutrition*, ed. D. M. Matthews & J. W. Payne. North Holland and American Elsevier, Amsterdam, pp. 1–60.
McGugan, W. A., Emmons, D. B. & Larmond, E. (1979). *J. Dairy Sci.*, **62,** 398–403.
McKay, L. L. (1983). *Ant. Leeuwenh*, **49,** 259–74.
Mellerick, D. & Cogan, T. M. (1981). *J. Dairy Res.*, **48,** 497–502.
Menassa, A. & Lamberet, G. (1982). *Lait*, **62,** 32–43.
Meyer, J. & Jordi, R. (1987). *J. Dairy Sci.*, **70,** 738–45.
Mills, O. E. & Thomas, T. D. (1980). *NZ J. Dairy Sci. Technol.*, **15,** 131.
Mills, O. E. & Thomas, T. D. (1981). *NZJ. Dairy Sci. Technol.*, **15,** 43–55.
Modler, H. W., Brunner, J. R. & Stine, G. M. (1974). *J. Dairy Sci.*, **57,** 528–34.
Monnet, V. Le Bars, D. & Gripon, J.-C. (1986). *FEMS Microbiol. Lett.*, **36,** 127–31.
Monnet, V., Le Bars, D. & Gripon, J.-C. (1987*a*) *Le Lait*, **67,** 51–61.
Monnet, V., Le Bars, D. & Gripon, J.-C. (1987*b*). *J. Dairy Res.*, **54,** 247–55.

Monnet, V., Bockelmann, W., Gripon, J.-C. & Teuber, M. (1989). *Appl. Microbiol. Biotechnol.* **31,** 112–18.
Montet, D., Ratomahenina, R., Pina, M., Graille, J. & Galzy, P. (1985). *Fat Sci. Technol.,* **87,** 181–5.
Morelli, l., Vescovo, M., Cocconcelli, P. S. & Bottazzi, V. (1986). *Can. J. Microbiol.,* **32,** 758–60.
Morishita, T., Deguchi, Y., Yajima, M., Sakurai, T. & Yura, T. (1981). *J. Bact.,* **148,** 64–71.
Motai, H., Ichishima, E. & Yoshoda, F. (1966). *Nature,* **210,** 308.
Mou, L., Sullivan, J. J. & Jago, G. R. (1975). *J. Dairy Res.,* **42,** 147–55.
Mourgues, R., Bergère, J.-L. & Vassal, L. (1978). *Technique Laitière,* **918,** 11–15.
Muderhwa, J. M., Ratomahenina, R., Pina, M., Graille, J. & Galzy, P. (1986). *Applied Microbiol. Biotech.,* **23,** 348–54.
Muset, G., Monnet, V. & Gripon, J.-C. (1989). *J. Dairy Res.,* **56,** 765–78.
Nardi, M., Chopin, M. C., Chopin, A., Cals, M. M. and Gripon, J.-C. (1991). *Appl. Environ. Microbiol.,* **57,** 45–50.
Nath, K. R. & Ledford, R. A. (1972). *J. Dairy Sci.,* **55,** 1424–7.
Neviani, E., Mucchetti, G., Giraffa, G. & Carini, S. (1984). *Scienza Tec. Latt. Casear.,* **35,** 439–50.
Nielsen, E. W. & Edelsten, D. (1974). XIX *Int. Dairy Congr.,* **1F,** 274–5.
Niki, T., Yoshioka , Y. & Ahiko, K. (1966). *Int. Dairy Congr.,* **D2,** 531–7.
Noomen, A. (1978). *Neth. Milk Dairy J.,* **32,** 49–68.
Noomen, A. (1983). *Neth. Milk Dairy J.,* **37,** 229–32.
Ohmiya, K. & Sato, Y. (1975). *Appl. Microbiol.,* **30,** 738–45.
O'Keeffe, R. B., Fox, P. F. & Daly, C. (1976). *J. Dairy Res.,* **43,** 97–107.
Okumura, J. & Kinsella, J. E. (1985). *J. Dairy Sci.,* **68,** 11–15.
Ordonez, J. A. & Ortiz-Apodaca, M. J. (1977). *Milchwissenschaft,* **32,** 531–3.
Ota, Y., Yoshioka, K., Minoda, Y. & Yamada, K. (1973). *Agr. Biol. Chem.,* **37,** 2879–83.
Oterholm, A., Ordal, Z. J. & Witter, L. D. (1968). *Appl. Microbiol.,* **16,** 524–7.
Otto, R., Brink, B., Veldkamp, H. & Konings, W. N. (1983). *FEMS Microbiol. Lett.,* **16,** 69–74.
Papon, M. & Talon, R. (1988). *J. Appl. Bacteriol.,* **64,** 107–15.
Paquet, J. & Gripon, J.-C. (1980). *Milchwissenchaft,* **35,** 72–4.
Payne, J. W. (1980). In *Microorganisms and Nitrogen Sources,* ed. J. W. Payne. John Wiley & Sons, New York, USA, pp. 211–56.
Rabier, D. & Desmazeaud, M. J. (1973). *Biochimie,* **55,** 389–404.
Reiter, B. & Oram, J. D. (1962). *J. Dairy Res.,* **29,** 63–77.
Reiter, B., Fryer, T. F., Pickering, A., Chapman, H. R., Lawrence, R. C. & Sharpe, M. E. (1967). *J. Dairy Res.,* **34,** 257–72.
Rice, G. H., Stewart, F. H. C., Hillier, A. J. & Jago, G. R. (1978). *J. Dairy Res.,* **45,** 93–107.
Roger, S., Degas, C. & Gripon, J.-C. (1988). *Food Chem.* **28,** 129–40.
Schermers, F. H., Duffus, J. H. & McLeod, A. M. (1976). *J. Int. Bromo.,* **82,** 170–4.
Schmidt, J.-L. (1982). *International Dairy Congress,* (Vol 1), Book 2, p. 365.
Schmidt, J. L. & Lamberet, G. (1981). *Adv. Biotechnol.,* **2,** 485–90.
Schousboe, I. (1976). *Biochim. Biophys. Acta,* **450,** 165–74.
Schwartz, D. P. & Parks, O. W. (1963). *J. Dairy Sci.,* **46,** 1136.
Schwartz, D. P., Parks, O. W. & Boyd, E. N. (1963). *J. Dairy Sci.,* **46,** 1422–3.
Searles, M. A., Argyle, P. J. Chandan, R. C. & Gordon, J. F. (1970). *Int. Dairy Congr.,* **1F,** 114.
Sen, N. P. (1969). *J. Food Sci.,* **34,** 22–6.
Sharpe, M. E. (1979). *J. Soc. Dairy Technol.,* **32,** 9–18.
Sørhaug, T. (1981). *Milchwissenchaft,* **36,** 137–9.

Sørhaug, T. & Ordal, Z. J. (1974). *Appl. Microbiol.*, **25,** 607–8.
Stadhouders, J. & Veringa, H. A. (1973). *Neth. Milk Dairy J.*, **27,** 77–91.
Steffen, C., Flueckiger, E., Bosset, J. O. & Ruegg, M. (1987). *Cheese: Chemistry, Physics and Microbiology*, ed. P. F. Fox, Elsevier Applied Science, London, UK, pp. 93–120.
Stepaniak, L., Cierpikowska, A. & Habaj, B. (1974). *Int. Dairy Congr.*, **1F,** 150.
Tahoun, M. K. (1987). *Fat Sci. Technol.*, **89,** 318–20.
Thomas, T. D. (1976). *Appl. Environ. Microbiol.*, **32,** 474–8.
Thomas, T. D. & Pritchard, G. G. (1987). *FEMS Microbiol. Rev.*, **46,** 245–68.
Thomas, T. D., Turner, K. W. & Crow, V. L. (1980). *J. Bacteriol.*, **144,** 672–82.
Thompson, J. (1978). *J. Bacteriol.*, **136,** 465–76.
Thompson, J. (1979). *J. Bacteriol.*, **140,** 774–85.
Thompson, J. (1980). *J. Bacteriol.*, **144,** 683–91.
Thompson, J. (1987). *FEMS Microbiol. Rev.*, **46,** 221–31.
Thompson, J. & Thomas, T. D. (1977). *J. Bacteriol.*, **130,** 583–95.
Tinson, W., Hillier, A. J. & Jago, G. R. (1982). *Aust. J. Dairy Technol.*, **37,** 8–13.
Torgersen, P. & Sørhaug, P. (1978). *Int. Dairy Congr.*, **F,** 484.
Trieu-Cuot, P. & Gripon, J.-C. (1981). *Neth. Milk Dairy J.*, **35,** 353–7.
Trieu-Cuot, P. & Gripon, J.-C. (1982). *J. Dairy Res.*, **49,** 501–10.
Trieu-Cuot, P. & Gripon, J.-C. (1983). *Le Lait,* **63,** 116–28.
Trieu-Cuot, P., Archieri-Hazé, M. J. & Gripon, J.-C. (1982*a*). *J. Dairy Res.*, **49,** 487–500.
Trieu-Cuot, P., Archieri-Hazé, M. J. & Gripon, J.-C. (1982*b*). *Le Lait,* **62,** 234–49.
Tsugo, T. & Matsuoka, H. (1962). *Int. Dairy Cong.*, **4,** 385–92.
Tsugo, T. & Chang, J. E. (1970). *Int. Dairy Congr.*, 1*E*, 339-47.
Tsujisaka, Y., Iwai, M. & Tominaga, Y. (1973). *Agr. Biol. Chem.*, **37,** 1457–64.
Turner, K. W., Morris, H. A. & Martley, F. G. (1983). *NZ J. Dairy Sci. Technol.*, **18,** 117–23.
Umemoto, Y. & Sato, Y. (1978). *Agric, Biol. Chem.*, **42,** 221–5.
Vanbelle, M., Vervak,W. & Foulon, M. (1978). *Lait,* **58,** 248–60.
Vassal, L., Monnet, V., Le Bars, D., Roux, C. & Gripon, J.-C. (1986). *Le Lait,* **66,** 341–51.
Verger, R. (1985). In *Hydrolases et Dépolymérases d'Intérêt Industriel,* ed. C. Costes, Gauthier-Villars, France, pp. 313–29.
Vescovo, M. & Bottazzi, V. (1979). *Scienza Tec. Latt. Casear.*, **30,** 434–43.
Visser, F. M. W. (1977*a*). *Neth. Milk Dairy J.*, **31,** 210–39.
Visser, F. M. W. (1977*b*). *Neth. Milk Dairy J.*, **31,** 188–209.
Visser, S., Exterkate, F. A., Slangen, C. J. & De Veer G. J. C. M. (1986). *Appl. Environ. Microbiol.*, **52,** 1162–6.
Visser, S., Slangew, C. J., Extertake, F. A. & De Veer, J. C. M. (1988). *Appl. Microbiol. Biotechnol.*, **29,** 61–6.
Vos, P., van Asseldonk, M., van Jeveren, F., Siezen, R., Simons, G. & de Vos, W. N. (1989). *J. Bacteriol.*, **172,** 2795–802.
Walstra, P., Noomen, A. & Geurts, T. J. (1987). *Cheese: Chemistry, Physics and Microbiology,* ed. P. F. Fox., Elsevier Applied Science, London, UK, pp. 45–92.
Woo, A. H. & Lindsay, R. C. (1984). *J. Dairy Sci.*, **67,** 960–8.
Woo, A. H., Kollodge, S. & Lindsay, R. C. (1984). *J. Dairy Sci.*, **67,** 874–8.
Yan, T. R., Azuma, N., Kaminogawa, S. & Yamauchi, K. (1987). *Eur. J. Biochem.*, **163,** 259–65.
Zevaco, C., Hermier, J. & Gripon, J.-C. (1973). *Biochimie,* **55,** 1353–60.
Zevaco, C., Monnet, V. & Gripon, J.-C. (1990). *J. Appl. Bacteriol.*, **68,** 357–66.

4

Microbial Enzymes in the Spoilage of Milk and Dairy Products

TERJE SØRHAUG & LESZEK STEPANIAK
Department of Dairy and Food Industries, Agricultural University of Norway, 1432 Ås-NLH, Norway

INTRODUCTION

With progress in the production and processing of milk and dairy products, the list of the most important spoilage bacteria has been limited to

(1) heat-labile psychrotrophs isolated with high frequency from cold-stored milk,
(2) psychrotrophic thermoduric bacteria, especially aerobic sporeformers,
(3) anaerobic sporeformers,
(Muir *et al.*, 1978; Cousin, 1982; Mottar, 1984, 1989; Suhren, 1989).

As evidenced by data from different countries (Muir *et al.*, 1978, 1986; Flückiger, 1981; Ewings *et al.*, 1984; Mottar, 1984; Hubble & Manners, 1985; Suhren, 1989), the microbial load of raw milk before processing is quite frequently in the range of 10^4–10^6 cfu/ml. The processing of milk containing 10^7 cfu/ml should be considered as very rare. Current microbial standards (Hubble & Manners, 1985) for grade A milk are more stringent than those of three decades ago (Foster *et al.*, 1957), and they are met by the vast majority of suppliers. The storage period for raw milk before processing and cold storage of pasteurised products have increased (Cousin, 1982); thus, the significance of psychrotrophic microbes has also increased. Part of this microflora may be true psychrophiles (Muir & Phillips, 1984; Griffiths & Phillips, 1988*a*).

Pseudomonas spp. are nowadays most often quoted among the spoilage microflora of both raw cold-stored milk and raw meat. The emphasis on biochemical activities of the microorganisms responsible for spoilage of the two product groups is, however, different (Dainty, 1982; Greer, 1989). Accumulation and action of proteolytic and lipolytic enzymes are of great importance in spoilage or potential quality deterioration of milk and dairy products. These enzymes are generally good indices of keeping quality. In meats exposed to air, microbial turnover of carbohydrates, free amino acids and nucleotides lead to the formation of end products which accumulate to levels offensive for our senses before noticeable proteolysis or lipolysis occurs.

The importance of proteolysis in the formation of bitter peptides, destabilisation of casein micelles with consequences for textural changes such as gelation and viscosity (Dainty, 1982; Andrews, 1986; Fairbairn & Law, 1986*a*; Cousin, 1989; Mottar, 1989) and the importance of lipases for flavour defects in dairy products (Stead, 1986; Cousin, 1989; Mottar, 1989) are fully recognised. What may remain disputable is the extent of the contribution by the microbial proteinases compared to the action of indigenous milk enzymes on the keeping quality of UHT-sterilised milk (Grufferty & Fox, 1988; Hill, 1988).

Pseudomonads are highly evolved microorganisms, able to degrade more than 100 organic compounds (Ribbons & Williams, 1982). Many strains of *Pseudomonas* spp., when growing in milk at 4°C, can produce enough proteinases to hydrolyse all available casein to soluble peptides (Fain, 1982). The genus is also known to be strongly lipolytic (Rapp, 1978). A simple relationship between psychrotrophic counts and enzymatic spoilage has not been found (Bishop & White, 1986; Skura, 1989). In any case, proteolysis, which is the first enzymatic reaction in a chain involving a number of enzymes in a complex process of biodegradation of proteins, also called putrefaction (Eskin *et al.*, 1971), is of consequence for the quality of milk products. A similar consideration applies to lipolysis caused by both microbial and milk lipases.

Cooling of raw milk to 4°C is a realistic compromise for economical operation of the dairy industry and to fulfil demands for high quality perishable dairy products with a relatively long shelf life. Cold storage allows, however, selective development of psychrotrophic microflora.

TECHNOLOGICAL CRITERIA FOR UNDESIRABLE PROPERTIES OF MICROBIAL ENZYMES

With freshly drawn milk as reference, all milk processed in the modern dairy industry is altered through both physical and enzymatic actions (Downey, 1980; Fox, 1981; Antila, 1982; Humbert *et al.*, 1983; Hsien-Yeh-Hsu, 1984; Keogh & Pettingill, 1984; Kang & Frank, 1988; Kosikowski, 1988). These changes start just after milking and proceed to detectable levels even without the contribution of exogenous microbial enzymes (Downey, 1980; Humbert *et al.*, 1983; Kang & Frank, 1988). In fact, both milk lipase and proteinases, particularly plasmin, can cause discernible changes in milk. Somatic cells in milk, with their pool of enzymes, mainly proteinases, also vary (Keogh & Pettingill, 1984; Kosikowski, 1988). In terms of these enzymes, their inhibitors and activators, milk is a variable material for processing.

However, the hydrolytic enzymes of psychrotrophic bacteria when they appear, interfere in a more decisive way to the detriment of the milk product quality. For instance, the enzymes of *Pseudomonas* are extremely heat resistant and only a small fraction of their activity is destroyed by pasteurisation or UHT-treatment of milk (Fox, 1981; Fairbairn & Law, 1986*a*; Stead, 1986; Kroll, 1989; Mottar, 1989).

The ability to release bitter peptides from casein is a common feature of most microbial proteinases. Bitterness is a prevalent flavour defect connected with the activity of proteinases from *Pseudomonas* spp. (Fairbairn & Law, 1986*a*; Cousin, 1989; Mottar, 1989). Chymosin-like activity makes these enzymes potent in the destabilisation of casein micelles (Andrews, 1986; Cousin, 1989). *Pseudomonas* lipases are also more active than many other lipases in the development of hydrolytic rancidity (Stead, 1986; Ren *et al.*, 1988). Enhanced lipolysis owing to enzymatic damage of the milk fat globule membrane has also been demonstrated (Marshall, 1982; Alkanhal *et al.*, 1985).

FACTORS AFFECTING COLONISATION, MICROBIAL GROWTH AND ENZYME PRODUCTION

Origin of microbial enzymes

Sources of contamination have been reviewed (Cousin, 1982; Suhren, 1989). Secretion of heat-stable enzymes by contaminating psychrotrophs in raw milk is especially important because these enzymes, directly or indirectly, may influence the quality of all dairy products.

As concluded by recent reviewers (Fairbairn & Law, 1986*a*; Stead, 1986; Cousin, 1989; Mottar, 1989), extracellular proteinases and lipases from psychrotrophic pseudomonads are recognised as primary microbial spoilage enzymes. The significance of microbial extracellular phospholipases, glycosidases and phosphatases is indicated but supported by a relatively modest amount of data. Production of different cell-associated hydrolases by *Pseudomonas fluorescens* has been confirmed (McKellar, 1986). Intracellular enzymes are considered as less important in spoilage both because of their lower heat stability and because they contribute little to the total enzyme activity (Singh *et al.*, 1973; Guamis *et al.*, 1987*a*; Shamsuzzaman & McKellar, 1987). On the other hand, a report (Keogh & Pettingill, 1984) on the significance of intracellular proteinases on the stability of UHT milk produced from raw milk of relatively good microbial quality indicated that the significance of intracellular enzymes on the total spoilage capacity may have been underestimated. Sporulation of *Bacillus* spp. is preceded by accumulation of intracellular proteinase (Strongin *et al.*, 1979). The variability in milk composition may be significant enough to influence the number and levels of intracellular and extracellular enzymes involved in nitrogen metabolism (Fairbairn & Law, 1986*a*) as well as the germination activity of spores (Cousin, 1982).

Pseudomonas spp. can grow and produce proteinases and lipases in very diluted milk or in very simple media (Patel *et al.*, 1983*a*; Rowe & Gilmour, 1983; Birkeland *et al.*, 1985; Stepaniak *et al.*, 1987*a*). After attaining maximum population density, *P. fluorescens* will survive unchanged in milk culture for a very long time (Stead, 1987). Lack of a true stationary phase for this microorganism has also been reported (Rowe & Gilmour, 1982). Some of these observations may substantiate the suspicion expressed that dilute milk residues in blind loops of cleaning systems and generally in poorly rinsed lines and containers contribute rather large amounts of enzymes as well as fresh inocula of psychrotrophic bacteria (Kielwein, 1982; Gray & Sørhaug, 1983; Stead, 1987).

Factors affecting colonisation of microflora in dairy products

Temperature is the main external factor controlling microbial growth in milk. The practical importance of slowing down growth in raw milk by lowering the temperature from 4 to 2°C is apparent from the work of Muir and Phillips (1984) and Griffiths and Phillips (1988*a*, *b*). Carbon dioxide or nitrogen purged into milk may become an auxiliary method reinforcing the effect of low temperature in controlling bacterial growth and enzyme production (Murray *et al.*, 1983; Fairbairn & Law, 1986*a*). Additionally, the lactoperoxidase system may be activated and enhanced by addition of small amounts of thiocyanate and hydrogen peroxide (Björck, 1982; Kosikowski, 1988). In raw milk cooled immediately after milking and stored at 4–5°C, very slow growth occurs during the first 2–4 days, although large differences are observed between farm tanks (Cousins *et al.*, 1977; Cousin, 1982; Mottar, 1984; Suhren, 1989), presumably due to variations in antimicrobial components.

The mean generation time found for psychrotrophic bacteria in various samples of raw milk at 4°C was 7·5 h with a range of 4·5–11·5 h (Muir & Phillips, 1984). The theoretical lowest temperature for growth (T_0) was 265·2 K (−7·8°C) for *Pseudomonas* spp., 268·0°K (−5°C) for *Alteromonas putrefaciens* and 273·3°K (−0·3°C) for *Serratia marcescens* (Stannard *et al.*, 1985). The microflora of pasteurised milk samples had T_0 values in the range 260·3–272·6°K (−12·7–−0·4°C) (Griffiths & Phillips, 1988*a*). The variation of T_0 was related to the shelf life of pasteurised milk (Griffiths & Phillips, 1988*a*, *b*). Generation times at 3–5°C of single cultures of *P. fluorescens* growing in milk or other media were 5–12 h, while different species of psychrotrophic *Bacillus* had generation times at 3–5°C ranging from 5·7 to 14 h (Shehata *et al.*, 1971; Gray & Sørhaug, 1983; Suhren, 1989). Psychrotrophic *Pseudomonas* and *Bacillus* spp., as well as psychrotrophic lactose-fermenting microflora, could grow at 2°C or lower temperatures (Shehata *et al.*, 1971; Gray & Sørhaug, 1983; Richard, 1983; Suhren, 1989). None of the *Flavobacterium*/*Cytophaga* isolates from milk could grow at 2°C (Richard, 1983). At temperatures below 10°C, outgrowth of pseudomonads by psychrotrophic *Enterobacter, Bacillus, Lactobacillus* and psychrotrophic yeasts is unlikely (Langeveld & Cuperus,

1980; Fleet & Mian, 1987; Suhren, 1989). In mildly sterilised milk or other media, no lag phase for some strains of *P. fluorescens* was observed during low temperature growth (Rowe & Gilmour, 1983; Stepaniak *et al.*, 1987*a*), although the heat and diluent stress-sensitive microorganism (Gray & Sørhaug, 1983) was inoculated from a culture prepared at room temperature. Growth of *P. fluorescens* P1 in milk sterilised at 121°C for 15 min was retarded, most likely due to the inhibitory action of Maillard reaction products (Stepaniak *et al.*, 1987*a*).

Candida, Torulopsis, Rhodotorula and *Kluyveromyces* were the fastest growing yeast isolates from cold-stored yogurt (Suriyarachchi & Fleet, 1981; Fleet & Mian, 1987) and quarg (Engel *et al.*, 1987).

Pseudomonas spp. are especially sensitive to low water activity and pH. However, along with yeasts and enterobacteria, many psychrotrophic pseudomonads can grow in cottage cheese (Stone & Naff, 1967; Brocklehurst & Lund, 1985; Herdlevær, 1988); quite rapid growth of *P. fluorescens* and yeasts was observed in some samples (Brocklehurst & Lund, 1985). Yeasts and moulds are among the main spoilage microorganisms in yogurt, cottage cheese and quarg. Low levels and no proliferation of psychrotrophic bacteria were observed in buttermilk (Cousin, 1982). Although a decline in psychrotrophic count was observed in Cheddar cheese after manufacture, large numbers of such organisms have been counted in ripened cheese (Najim, 1985; Johnston *et al.*, 1988). *Clostridium tyrobutyricum* and possibly other clostridia are the spoilage microorganisms most likely to develop during Dutch- and Swiss-cheese ripening. Also, processed cheeses may be spoiled by *C. tyrobutyricum, C. sporogenes* and other clostridia (Foster *et al.*, 1957; Bergère & Accolas, 1986). Some growth of psychrotrophic aerobic spore formers was observed in Cheddar cheese over long periods (Johnston *et al.*, 1988). In mould-ripened cheeses, lactobacilli and yeasts can grow to substantial numbers. The roles of yeasts and micrococci in the ripening of some cheeses, as seen through their enzymatic systems, may be both positive and negative (Lenoir, 1984; Bhowmik & Marth, 1988). Salt-tolerant yeasts, brevibacteria and moulds are regularly found in the natural flora of surface-ripened cheeses (Lenoir, 1984). As a result of the uneven distribution of salt in water droplets in butter, not only yeasts and moulds but also *Pseudomonas* spp. can grow even at 0°C (Thomas & Druce, 1971).

Composition of psychrotrophic microflora

Cold-stored, raw milk

The microflora of raw cold-stored milk with counts of 10^4 cfu/ml or higher is dominated by Gram-negative, non-lactose-fermenting psychrotrophs (Richard, 1983; Ewings *et al.*, 1984; Kroll *et al.*, 1984; Shelley *et al.*, 1986; Richard *et al.*, 1988; Suhren, 1989). More than 70% of the microflora in milks of various microbiological quality may consist of *Pseudomonas* spp., with *P. fluorescens* prevailing. Biovar I and III of *P. fluorescens* comprised 39 and 45%, respectively, of the *P. fluorescens* isolated from pre-cooled milk (Kroll *et al.*, 1984). *P. fluorescens* and occasionally *P. fragi* were also the predominant psychrotrophs in cold-stored caprine milk in Australia (Cox & Mac Rae, 1989). *Alcaligenes* occasionally dominated in milk containing $>10^5$ cfu/ml (Ewings *et al.*, 1984). *Acinetobacter, Aeromonas, Flavobacterium, Chromobacterium, Serratia, Vibrio,* and Enterobacteria were isolated at much lower frequencies than pseudomonads (Muir *et al.*, 1979; Fain, 1982; Richard, 1983; Ewings *et al.*, 1984; Richard *et al.*, 1988; Fox *et al.*, 1989; Suhren, 1989). In fresh, low-count milk, the mesophilic microflora usually dominate (Cousin, 1982; Suhren, 1989). Among the isolates of lactose-fermenting psychrotrophs, coliforms comprised *c.* 95% and *Aermonas c.* 5% (Richard *et al.*, 1988).

In farm or creamery raw milk the thermoduric microflora comprised about 1% of the total flora (Phillips *et al.*, 1981*a*; Schröder *et al.*, 1982; McKinnon & Pettipher, 1983; Muir *et al.*, 1986). In spite of considerable seasonal variations (Johnston & Bruce, 1982; McKinnon & Pettipher, 1983; Schröder, 1984; Phillips & Griffiths, 1986), quite a large portion of the thermoduric microflora is represented by sporeformers. However, psychrotrophic sporeformers, mainly *B. cereus,* represent only a small portion of the total spore count in milk from farm tanks (Johnston & Bruce, 1982; McKinnon & Pettipher, 1983; Phillips & Griffiths, 1986).

Pasteurised milk and cream

The composition of the microflora of pasteurised cold-stored milk and cream is less predictable than that of cold raw milk. Such microflora is composed of psychrotrophic thermoduric bacteria from raw milk and psychrotrophic thermoduric and non-thermoduric bacteria arising from recontamination (Schröder *et al.*, 1982; Schröder, 1984; Phillips & Griffiths, 1986). Post-pasteurisation contamination by sporeformers shows some seasonal variation (Phillips & Griffiths, 1986). The concentration of germination stimulants in milk is not constant (Cousin, 1982; Phillips & Griffiths, 1986). Recontamination with Gram-negative rods is often reported as negligible but even a low level of these bacteria very seriously affects the keeping quality of cold-stored, pasteurised milk (Schröder *et al.*, 1982; Schröder, 1984; Bishop & White, 1985*a*; Griffiths & Phillips, 1988*a, b*). *Pseudomonas* spp. are very likely to dominate in pasteurised milk stored <6°C at the time of spoilage (Griffiths & Phillips, 1988*a*). The storage temperature for pasteurised milk or cream is often close to 8°C and therefore quite favourable for the growth of sporeforming psychrotrophs (Griffiths *et al.*, 1981*a*; Brown *et al.*, 1984). Sporeformers and other thermoduric bacteria were most frequently responsible for spoilage when pasteurised milk was stored at 11°C; however, both at 5 and 11°C, samples spoiled by Gram-negative rods, spores, corynebacteria or *Micrococcus* were also found (Schröder *et al.*, 1982).

The predominant microorganisms in cream stored at 6°C were *Pseudomonas* (40%), *Enterobacter* (13%) and *Bacillus* (12%); storage at 10°C resulted in increased populations of *Bacillus, Enterobacter* and *Lactobacillus* (Phillips *et al.*, 1981*b*).

Capacity of psychrotrophic isolates of dairy origin to degrade proteins and fats

Zones of hydrolysis around colonies, and around wells containing extracellular enzymes, in agar media containing appropriate substrates have been used to evaluate lipolytic, proteolytic or phospholipolytic activities of microflora isolated from milk and dairy products (Muir *et al.*, 1979; Phillips *et al.*, 1981*b*; Alkanhal *et al.*, 1985; Kwan & Skura, 1985; O'Connor *et al.*, 1986; Stepaniak *et al.*, 1987*b*). Alternatively, commercial kits with substrates specific for different enzymes have been used for semi-quantitative determination of various hydrolytic activities (Lenoir, 1984; McKellar, 1986).

A number of different methods developed specifically to detect microbial hydrolases in milk have been reviewed (Fairbairn, 1989; Skura, 1989; Stead, 1989). Large variations in proteolytic and lipolytic activities or the proportionality of these activities were found within and between strains of the main psychrotrophic genera of dairy isolates (Muir *et al.*, 1979; Griffiths *et al.*, 1981*a*; McKellar, 1981; Phillips *et al.*, 1981*b*; Fain, 1982; Richard, 1983; Ewings *et al.*, 1984; O'Connor *et al.*, 1986). Consequently, a variable proportion of the psychrotrophs in milk is lipolytic and/or proteolytic (Fain, 1982; Ewings *et al.*, 1984; Mottar, 1984).

A larger number of fluorescent *Pseudomonas* isolates than non-fluorescent pseudomonads show the capacity to degrade milk proteins to a completely soluble hydrolysate. Some isolates proved to be less proteolytic by forming insoluble *para*-casein (Fain, 1982).

Among different genera isolated from cold-stored milk or spoiled cream, fluorescent *Pseudomonas* spp. and *Alcaligenes* spp. are most frequently both lipolytic and proteolytic (Muir *et al.*, 1979; Griffiths *et al.*, 1981*a*; Phillips *et al.*, 1981*b*). These genera also include the lowest number of hydrolytically-inert isolates (Muir *et al.*, 1979; Phillips *et al.*, 1981*b*), or isolates which secrete heat-labile proteinases (Ewings *et al.*, 1984). Strains of *P. fragi* isolated both from milk (Shelley *et al.*, 1986) and meat (Dainty, 1982) were more lipolytic than strains of *P. fluorescens*. The lipolytic activity of *Pseudomonas, Aeromonas* and *Acinetobacter* strains was higher than that of *Flavobacterium* isolates (Kalogridou-Vassiliadou, 1984). Relatively frequently, psychrotrophic *Bacillus* isolates possessed only proteolytic activity (Griffiths *et al.*, 1981*b*; Phillips *et al.*, 1981*b*). Some species of enterobacteria which caused measurable lipolysis and proteolysis in milk were unable to produce zones of hydrolysis in skim milk agar or Tween 80 agar (Juven *et al.*, 1981). Kwan and Skura (1985) found that, in contrast to isolates of *P. fluorescens* from milk, non-fluorescent *P. fragi*-like isolates did not hydrolyse Tween 80, or gelatin at 22°C and did not produce phospholipase C. However, both *P. fluorescens* and *P. fragi* isolates hydrolysed milk fat and milk proteins at 4 and 22°C. The *P. fragi* isolates did not hydrolyse milk proteins at 32°C.

Domination by *Pseudomonas* in raw cold-stored milk and spoiled pasteurised milk or cream in different geographical regions, competitive growth compared to other psychrotrophic bacteria, combined with high lipolytic and proteolytic activities and heat stability of secreted proteinases and lipases, are good reasons why extracellular proteinases and lipases from *Pseudomonas*, especially *P. fluorescens*, are studied most often. Phospholipases A and C are also common for *Pseudomonas*, but these enzymes have been less well characterised (Fox *et al.*, 1989; Stepaniak & Sørhaug, 1989). Much less studied are the hydrolase activities of psychrotrophic thermoduric bacteria.

Although direct comparison is not comprehensive, *Enterobacter*, *Micrococcus*, *Lactobacillus* and corynebacteria can be considered as biochemically weaker than pseudomonads (Cousin & Marth, 1977*a*; Muir *et al.*, 1979; Phillips *et al.*, 1981*b*; Ledford *et al.*, 1983; Richard, 1983). *Micrococcus* and corynebacteria isolated from Camembert cheese showed strain-variable but rather comparable proteolytic, lipolytic and aminopeptidase activities (Lenoir, 1984).

The significance of coliforms should not be overlooked. Even at a population as low as 150 cfu/ml, these microorganisms influenced the quality of raw or pasteurised milk, possibly through stimulation of the growth of *Pseudomonas* (Hankin & Shields, 1983; Ledford *et al.*, 1983). A correlation was found between the initial, low, number of coliforms and the keeping quality of raw cows' and goats' milk but not with *Pseudomonas* count (Hankin & Shields, 1983). In mixed culture with *Pseudomonas*, these bacteria significantly lowered the pH of milk at total counts of about 10^6 cfu/ml (Fain, 1982).

High levels of tyramine and histamine in cheeses are objectionable because of health aspects (Sharpe & Pettipher, 1983). Different enterococci can produce significant amounts of amines in cheese (Sharpe & Pettipher, 1983). Lactic acid bacteria used as starters, if not properly selected for their proteolytic, lactose- and citrate-fermenting activities, may be responsible for a number of defects in fermented dairy products (Sharpe & Pettipher, 1983).

Due to a unique structure, peptide bonds involving proline are often resistant to peptidases of broad specificity; however, *Escherichia coli*, *Flavobacterium* and *Micrococcus* produce proline exopeptidase activity (Walter *et al.*, 1980; Bhowmik & Marth, 1988). Some *Flavobacterium* spp. isolated from soil expressed a unique proline-specific endopeptidase activity (Walter *et al.*, 1980).

Yeasts express a variety of biochemical activities such as lipolysis, proteolysis, and pectinolysis. Some species can ferment lactose or utilise lactic acid to various degrees. Yeast enzymes are involved in hydrolysis, esterification, condensation, amination, reduction, decarboxylation, deamination and oxidation (Peppler, 1977). Proteolytic yeasts have been isolated from yogurt (Suriyarachchi & Fleet, 1981). Isolates from soft mould-ripened cheeses show both aminopeptidase and carboxypeptidase activities (Lenoir, 1984).

Threshold of microflora needed to develop flavour defects, proteolysis or lipolysis

Poor correlations between the threshold level of psychrotrophic population in raw milk and flavour and structural defects in dairy products have been widely experienced (Baker, 1983; Bishop & White, 1985*a*, 1986; Kocak & Zadow, 1985; Skura, 1989). Similar ranges of flavour defects were produced in milk or cream by psychrotrophic Gram-negative and Gram-positive bacteria (Shehata *et al.*, 1971; Griffiths *et al.*, 1981*a*; Coghill, 1982; Schröder *et al.*, 1982).

Flavour defects and measurable proteolysis or lipolysis in heat-treated milk usually developed when single strain cultures of psychrotrophic *Pseudomonas*, Enterobacteria, or *Bacillus* grew to levels of 5×10^6–1×10^7 cfu/ml (Shehata *et al.*, 1971; Juven *et al.*, 1981; Stepaniak *et al.*, 1982; Gray & Sørhaug, 1983; Birkeland *et al.*, 1985; Fairbairn & Law, 1986*a*; Shelley *et al.*, 1986; Stepaniak *et al.*, 1987*a*, *c*). Lower figures for population thresholds of single bacterial strains are rather exceptional (Gray & Sørhaug, 1983, Mottar, 1989). Reduced flavour scores in raw or pasteurised milk were most often noted when the contaminating microflora grew to or above 10^6 cfu/ml (Muir *et al.*, 1978; Gebre-Egziabher *et al.*, 1980; Schröder *et al.*, 1982; Muir & Phillips, 1984; Mottar, 1989). Populations above 10^8 cfu/ml were necessary for *P. fluorescens* AFT 36 cultured in milk at 4°C to cause detectable changes in alcohol- and heat-stability (Stepaniak *et al.*, 1982). Probably due to lowering of the pH, some enterobacteria incubated at 7°C caused changes in heat- and alcohol-stability at counts of $1{\cdot}4 \times 10^7$–2×10^8 cfu/ml (Juven *et al.*, 1981). Psychrotrophic populations of 1×10^7–3×10^7 cfu/ml in pasteurised milk were established as the limit for acceptable taste and keeping quality (Baker, 1983; Brown *et al.*, 1984; Schröder, 1984; Griffiths & Phillips, 1988*a*, *b*).

Production of volatiles in milk incubated with psychrotrophic *Pseudomonas*, *Alcaligenes*, or *Bacillus* (Pierami & Stevenson, 1976) or with natural flora (Urbach & Milne, 1988) was also noticed at populations around 10^6 cfu/ml. Threshold levels for some yeasts to cause sensoric changes in quarg varied from only $2{\cdot}4 \times 10^4$ cfu/ml to $4{\cdot}2 \times 10^6$ cfu/ml (Engel *et al.*, 1987).

The above data emphasise the lack of simple relationships between microbial population levels and spoilage or the start of spoilage. Some bacteria with a strong potential for hydrolytic spoilage, e.g. *Pseudomonas*, secrete the enzymes at rather high population levels. This point will be dealt with in more detail below.

Secretion of extracellular hydrolases in milk

The environmental and nutritional factors, and the regulatory and secretory mechanisms controlling the production of extracellular proteinases and lipases by common psychrotrophs, in particular *P. fluorescens*, have recently been extensively reviewed by McKellar (1989). Earlier reviews have treated microbial proteinases generally (Law, 1980), proteinases from psychrotrophs

(Fairbairn & Law, 1986*a*) or lipases from psychrotrophs (Stead, 1986). This section will focus primarily on factors of practical importance in controlling the secretion of hydrolases in milk.

The evidence gained so far points to energy levels in cells as the primary factor regulating the synthesis of proteinases and lipases. Decreased energy levels resulting from nutrient limitation, increased cell density and iron or oxygen limitation enhance the synthesis of extracellular enzymes (McKellar, 1989). Consistent with this assumption, the most intensive secretion of proteinase, lipase and phospholipase C by *P. fluorescens,* and by many psychrotrophic, mesophilic and thermophilic bacteria, occurs at the end of log phase of growth (Andersson, 1980; Law, 1980; Rowe & Gilmour, 1982, 1986; Dring & Fox, 1983; Griffiths & Phillips, 1984; Christen *et al.*, 1986; Fairbairn & Law, 1986*a*; Bucky *et al.*, 1987*a*; Alur *et al.*, 1988; Jeong & Frank, 1988; Vagias, 1988; McKellar, 1989).

P. fragi ATCC 4973 incubated at 21°C produced detectable amounts of proteinase after 24 h in liquid culture but after only 4 h when cultivated on the surface of an agar medium. Extracellular vesicles were 'blebbed' off the cell walls of the surface culture. The greatest number of vesicles coincided with maximum proteinase production (Myhara & Skura, 1989).

Other factors which also influence the secretion of extracellular enzymes, i.e. temperature, oxygen and nitrogen source, appear to interact in ways which make it difficult to comprehend the contribution by each factor. Most of the evidence indicates that *P. fluorescens* proteinases and lipases are at least partially constitutive or both constitutive and inducible enzymes (Fairbairn & Law, 1986*a*; McKellar, 1989). Although data on the induction of proteinases by organic nitrogen compounds are the subject of some confusion, evidence for the stimulation of synthesis by casein or other proteins and some amino acids or both is supported by independent studies with *P. fluorescens,* (McKellar, 1989) and *Aeromonas hydrophila* (Alur *et al.*, 1988). Collagenase production by a strain of *Vibrio alginolyticus* and *Acinetobacter* was inducible in media containing collagen or its high molecular weight fragments (Fox *et al.*, 1989). The role of triglycerides and fatty acid in the induction of lipases is more confusing (Sugiura, 1984; Vagias, 1988; Fox *et al.*, 1989; McKellar, 1989). Some polysaccharides, such as pectin B and gum arabic, enhanced secretion of lipase by *Serratia marcescens* (Sugiura, 1984).

A requirement for calcium in the post-secretional formation of active proteinase molecules was demonstrated for many, but not all, strains of *Pseudomonas* spp. (McKellar, 1989). Catabolite repression by adding glucose to milk was demonstrated for both proteinase and lipase from *Pseudomonas* (Patel *et al.*, 1983*a*; Griffiths & Phillips, 1984; McKellar, 1989). Skim milk is an excellent medium for the production of proteinases (Birkeland *et al.*, 1985; Stepaniak *et al.*, 1987*a*; McKellar, 1989), lipases (Christen *et al.*, 1986; Stead, 1987; Vagias, 1988; McKellar, 1989) and phospholipase C (Stepaniak *et al.*, 1987*b*; Vagias, 1988) by *Pseudomonas* and *Acinetobacter*. Some proteolytic (Gray & Sørhaug, 1983; Griffiths & Phillips, 1984; Kroll & Klostermeyer,

1984*a*; Birkeland *et al.*, 1985; Drew & Manners, 1985; Stepaniak *et al.*, 1987*a*; Jeong & Frank, 1988; Vagias, 1988) and lipolytic (Andersson, 1980; Dring & Fox, 1983; Griffiths & Phillips, 1984; Drew & Manners, 1985; Stepaniak *et al.*, 1987*a*; Vagias, 1988) activities were detectable in milk cultures of *Pseudomonas* spp. before the end of the log phase of growth.

The thresholds for detection of the activity of bacterial enzymes reflect both the sensitivity of the method used and the variable ability of strains to secrete enzymes prior to the main production at the end of the log phase of growth (Kroll & Klostermeyer, 1984*a*; Grieve & Kitchen, 1985; Stead, 1986; Kahn & Firstenberg-Eden, 1987).

After storage periods ranging from days to weeks, hydrolysis of milk proteins or fat, or changes in flavour, were detected in UHT and pasteurised milks manufactured from raw milk exposed before processing to the growth of the natural flora or of added cultures. Results from such studies which assume long incubation periods with very small amounts of enzyme provide indirect evidence for the presence of heat-stable enzymes in raw milk at about two to three log cycles below the maximum population that can be attained by pure cultures of *Pseudomonas* (Mottar, 1981; Antila, 1982; Blankenagel, 1982; Kielwein, 1982; Marshall, 1982; Muir & Phillips, 1984; Stead, 1986; Bucky *et al.*, 1987*b*, 1988; Dousset *et al.*, 1988). The evidence is much stronger for populations of *c.* 5×10^6 than for those of 1×10^6 cfu/ml.

Muir *et al.* (1978) and Muir and Phillips (1984), using their own results and data from the literature describing the relationship between psychrotrophic count and the onset of lipolysis/proteolysis or flavour defects in a number of dairy products, established a psychrotrophic count in raw milk of 5×10^6 cfu/ml as potentially affecting the quality and shelf life of the products. Combining this critical value with the range of mean generation time of raw milk microflora, they calculated probabilities for safe storage. To assure good quality, milk with initial counts greater than 2×10^3 cfu/ml should not be stored at 4°C for longer than 72 h (Muir & Phillips, 1984). Humbert *et al.* (1983) estimated that in raw milk with an average total count of $3{\cdot}9 \times 10^5$ cfu/ml and an average psychrotrophic count of $0{\cdot}3 \times 10^5$ cfu/ml, 9% of the variability in total proteolytic activity can be explained by the presence of microbial enzymes.

Proteolysis was caused by markedly lower populations of the natural milk psychrotrophic microflora than by populations of pure culture of *P. fluorescens* (Grieve & Kitchen, 1985). The natural flora of raw milk reached the decelerating phase of growth at populations of *c.* 10^7 cfu/ml, while single cultures of *P. fluorescens* grew exponentially to densities of *c.* 10^9 cfu/ml (Grieve & Kitchen, 1985).

The temperature for maximal production of proteinase is usually above that for cold storage of milk, but below the maximum growth temperature (McKellar, 1989). Three different strains of *P. fluorescens* produced more lipase at 4–8°C than at higher temperatures (Andersson, 1980; Bucky *et al.*, 1987*a*; Roussis *et al.*, 1988*a*). It has been generalised that the relatively lower number of cells in psychrotrophic cultures at low temperatures is compensated for by a higher output of enzyme or enzyme activity per cell (Fairbairn & Law, 1986*a*).

In continuously shaken milk cultures, both maximum population and maximum production of *Pseudomonas* proteinases were obtained, but a more variable effect of aeration was observed for the production of lipases (Stead, 1986; Roussis *et al.*, 1988*a*; McKellar, 1989). Limited, periodic aeration was found to have little effect on the growth of raw milk microflora at 6°C but it suppressed proteinase production; such conditions stimulated the activity of indigenous milk lipoprotein lipase (Griffiths & Phillips, 1984).

Enforced reduction of oxygen tension resulted in the premature production of both proteinase and lipase by *P. fluorescens* (Rowe & Gilmour, 1982). Earlier production of proteinase, but not lipase, was observed for cultures of this microorganism in diluted milk (Birkeland *et al.*, 1985; Stepaniak *et al.*, 1987*a*). Nitrogen and carbon dioxide atmospheres inhibited growth, and the relative inhibitory effect on proteinase and lipase secretion was greater than that on growth (Murray *et al.*, 1983; Rowe & Gilmour, 1986).

Growth rates of *Pseudomonas* spp. at 4°C and proteolysis were significantly higher in skim milk fortified to 10 or 12% total solids than in controls (Jeong & Frank, 1988). Concentration of milk by reverse osmosis had little effect on the growth of *P. fluorescens* or natural microflora but markedly reduced lipolysis and proteolysis at low temperatures (Tayfour *et al.*, 1982; Drew & Manners, 1985).

In concluding this discussion on sources of microbial enzymes in milk, the following points appear to be important:

(1) In raw milk, variable proportions of the microflora have high or low biochemical activity. The capacity of the microbes to produce high levels of enzymes is not always reflected in their ability to secrete enzymes in the early growth phase.
(2) The numbers of bacteria in raw milk and in pure cultures are poorly correlated with the secretion of enzymes. Large differences between strains should be expected not only in total enzyme production but also in specific production per cell during the early stages of growth.
(3) Raw milk contains fluctuating levels of enzyme inducers, inhibitors or growth promoters.
(4) Environmental factors such as temperature and soluble oxygen have variable effects on the secretion of microbial enzymes.

Effect of proteolysis and lipolysis on bacterial growth

Considering the activity of starters in cheese milk, both stimulatory and inhibitory effects have been attributed to the growth of psychrotrophs in cold-stored milk. Stimulation was connected to proteolysis, while inhibition was attributed to lipolysis (Cousin, 1982; Kalogridou-Vassiliadou & Alichanidis, 1984).

Culture filtrates from proteolytic *Pseudomonas* and *Bacillus* strains added to milk stimulated the growth of lactic acid bacteria (Magdoub *et al.*, 1980; Cousin, 1982). Pre-pasteurisation growth of *Pseudomonas* stimulated outgrowth of *Bacillus cereus* spores (Cousin, 1982).

Enterococcus faecium could grow and cause acid coagulation in evaporated milk only when cultured together with *Bacillus subtilis* or a proteolytic *Pseudomonas* sp. (McKellar & Nichols-Nelson, 1984). Addition of casein digest or *B. subtilis* proteinase also promoted growth and acid production by *E. faecium* (McKellar & Nichols-Nelson, 1984). Clinical isolates of *Listeria monocytogenes* grew faster in milk pre-incubated at 10°C with three strains of *P. fluorescens* or a strain of *P. fragi* than in controls. The stimulatory effect of pseudomonads on the pathogen, probably due to proteolysis, was strain-dependent (Marshall & Schmidt, 1988).

Growth of *Pseudomonas* spp. was not influenced by the addition of different fatty acids at a concentration of 100 ppm. Lactic streptococci, lactobacilli and *B. cereus* were inhibited most effectively by capric, lauric, linoleic and linolenic acids (Ergün, 1985).

CHARACTERISATION OF HYDROLASES FROM PSYCHROTROPHS

General

Proteinases and lipases from *P. fluorescens*, *P. fragi*, unspecified pseudomonads, *Alcaligenes*, *Aeromonas*, *Serratia*, *Acinetobacter*, *Flavobacterium*, *Micrococcus*, *Bacillus*, *Brevibacterium* and some other potentially psychrotrophic bacteria, isolated mostly from foods, have recently been reviewed encyclopaedically (Fox *et al.*, 1989). That and other recent reviews (Fairbairn & Law, 1986*a*; Stead, 1986; Stepaniak & Sørhaug, 1989) are dominated by the large volume of information on proteinases and lipases from pseudomonads.

Generally, among bacterial proteinases, a metalloproteinase from *Bacillus thermoproteolyticus* (thermolysin), and serine proteinases from *B. amyloliquefaciens* or *B. licheniformis* (subtilisins) have been thoroughly characterised (Aunstrup, 1980; Ward, 1983). *Bacillus* spp. are used industrially for the production of serine proteinases and α-amylases. *Bacillus* spp. are also known to produce mixtures of metallo- and serine proteinases (Aunstrup, 1980; Ward, 1983). *B. cereus* proteinase has been well characterised (Holmquist, 1977). Proteinase from *Clostridium sporogenes*, isolated from processed cheese, has also been investigated (Sinha & Sinha, 1986).

B. cereus phospholipase C has been studied in detail, and its properties have been reviewed (Bosch, 1982; Fox *et al.*, 1989, Stepaniak & Sørhaug, 1989). The crystallographic structure of the enzyme was elucidated and its gene was sequenced (Hough *et al.*, 1988). Only exceptionally (Yan *et al.*, 1985; Fox *et al.*, 1989), however, have studies of *Bacillus* spp. enzymes been concerned with possible psychrotrophicity of the producing strains.

Proteinases, exopeptidases and lipases and the role of the enzymes from moulds used for flavour formation in cheese ripening have been reviewed (Lenoir, 1984; Seth & Robinson, 1988). *Aspergillus oryzae* can secrete mixtures of aspartic, metallo- and serine proteinases plus carboxypeptidase, while *Streptomyces griseus* produces neutral and alkaline proteinases and amino- and carboxypeptidase (Adler-Nissen, 1986).

Niches for spoilage of dairy products by moulds are rather limited to surface growth on cheeses or yogurt. Presumably, before drastic spoilage in the form of visible growth is manifested, some spoilage enzymes are secreted. However, we can not quote any fungal proteinase or lipase specifically identified with spoilage at an early stage of fungal growth.

Proteinases

Table 1 summarises selected properties of proteinases from psychrotrophic bacteria. *P. fluorescens* proteinases, the most numerous group, also appear to include the enzymes with the greatest homology. The majority of strains characterised produced one principal heat-stable, neutral metalloproteinase composed of one peptide chain. Most of the enzymes were larger (mol. wt 38 000–48 000) than metalloproteinases from *Bacillus* spp. (mol. wt 34 000–37 000, Holmquist, 1977; Aunstrup, 1980) and contained six to eight atoms of calcium per molecule compared with four atoms in thermolysin (Ward, 1983; Stepaniak & Sørhaug, 1989). With one exception, *P. fluorescens* proteinases did not contain cyst(e)ine residues (Fox *et al.*, 1989). Significantly, high similarity indices of amino acid composition or sequence homology were found for a large group of these enzymes (Fairbairn & Law, 1986*b*; Mitchell *et al.*, 1986; Patel *et al.*, 1986: Vagias, 1988).

Although the information is not so complete as in the case of thermolysin or the zinc-containing elastase from *P. aeruginosa* (Morihara, 1974; Stepaniak & Sørhaug, 1989), specificity studies revealed many differences, both from thermolysin and also within the group of purified enzymes from *P. fluorescens* (Richardson, 1981; Yan *et al.*, 1985; Mitchell *et al.*, 1986; Fox *et al.*, 1989; Stepaniak & Sørhaug, 1989). Thermolysin, the metalloproteinases from *B. subtilis* and *B. cereus,* as well as the metalloproteinase from *Aeromonas proteolytica* selectively hydrolysed peptide bonds with a hydrophobic amino acid located at position P_1' of such typical substrates as furylacryloyl–glycine–leucinamide or furylacryloyl–alanyl–phenylalanine (Morihara, 1974; Holmquist, 1977; Stepaniak & Sørhaug, 1989). These substrates were not hydrolysed by proteinases from seven strains of *P. fluorescens* (Richardson, 1981; Mitchell *et al.*, 1986) or two proteinases from *Serratia marcescens* (Mitchell *et al.*, 1986).

Metal chelator-sensitive, immunologically unrelated proteinases from *P. fluorescens* or *Pseudomonas* spp. were not inhibited by phosphoramidon, a potent inhibitor of many neutral metalloproteinases (Stepaniak & Sørhaug,

1989). Weak inhibition was found, however, for six proteinases from *P. fluorescens* and two from *S. marcescens* (Mitchell *et al.*, 1986). The metalloproteinase character of some purified proteinases from *P. fluorescens* was confirmed by metal analysis and resistance to class-specific inhibitors other than metal chelators. Some of these enzymes hydrolysed typical trypsin substrates (Diermayr & Klostermeyer, 1984; Yan *et al.*, 1985; Patel *et al.*, 1986), while others did not have trypsin-like activity (Richardson, 1981; Mitchell *et al.*, 1986) but could show both endo- and exopeptidase activity (Mitchell *et al.*, 1986). The ability of *Pseudomonas* enzymes to hydrolyse bonds involving proline was also evident (Mitchell *et al.*, 1986; Fox *et al.*, 1989).

Cbz-D-phenylalaninetriethylenetetramine Sepharose has been used as an efficient ligand for the purification of proteinase from some *P. fluorescens* strains (Richardson, 1981; Patel *et al.*, 1983*b*), but in order to interact with the proteinase from *P. fluorescens* strain 112 the ligand amino acid had to be arginine (Diermayr, 1986). Biovariant-dependent difference in specificity of proteinases from *P. fluorescens* on peptide substrates was postulated (Diermayr & Klostermeyer, 1984). Proteinases from *P. fluorescens* and *S. marcescens* investigated by Mitchell *et al.* (1986) hydrolysed gramicidin, and this cyclopeptide antibiotic was used as a ligand for the purification of a proteinase from *P. fluorescens* AH-70 (Azcona *et al.*, 1988).

Production of serine proteinases or serine metalloproteinases appears to be quite common for *Aeromonas* (Table 1). Two heat-labile proteinases from *Aeromonas hydrophila*, a strain characterised by Pansare *et al.* (1986), were quite sensitive to both metal chelators and phenylmethylsulphonylfluoride. The strain also produced a thermophilic heat-stable aminopeptidase. A proteinase from *A. hydrophila* A4, with calcium-dependent heat stability, was very sensitive to metal chelators but not to diisopropylfluorophosphate (Alichanidis, 1988).

Many species of *Pseudomonas* secreted two to five proteinases. Some of the multiple enzyme systems were composed of neutral and alkaline, immunologically-unrelated metalloproteinases while others comprised both metallo- and serine proteinases. Sensitivity both to metal chelators and serine proteinase specific inhibitors was also very common among these enzymes (Fox *et al.*, 1989; Stepaniak & Sørhaug, 1989). Immunological unrelatedness within metal chelator-sensitive proteinases both from *P. fluorescens* and *Pseudomonas* spp. strains was demonstrated (Symons *et al.*, 1985; Vagias, 1988; Stepaniak & Sørhaug, 1989). The proteinases of a large group of strains of *P. fluorescens* originating from distant geographical regions were, however, immunologically related (Vagias, 1988). The appearance of several distinct proteolytic components in old cultures of *P. fluorescens* 113 (Diermayr, 1986) and the presence of multiple cell wall proteinases in *Streptococcus cremoris* (Thomas & Pritchard, 1987), were explained by autoproteolytic modification of a single proteinase. Multiple proteinase systems, composed of immunologically related or unrelated enzymes are also quite common for *Serratia* spp. (Fox *et al.*, 1989).

TABLE 1
Some Properties of Proteinases from Psychrotrophic Bacteria (Adapted from Data Compiled by Fox *et al.*, 1989)

Total Number of Strains	Pseudomonas fluorescens *27*	*Other Classified* Pseudomonas[a] *8*	*Unclassified* Pseudomonas *22*	Aeromonas *spp.*[b] *12*	Serratia *spp.*[c] *16*	Bacillus *spp.*[d] *3*	*Different G^+, G^- Psychrotrophs*[e] *19*
Proteinases							
Isolation of single or principal enzyme[f]	(26)[g]	(8)	(6)	(10)	(12)	(2)	(17)
Sensitive to metal chelators	(17)	(4)	(3)	(3)	(8)	(2)	(8)
Serine/metallo-enzyme	(1)	(2)	(1)	(1)	—	—	(2)
Serine class only	—	—	—	(3)	—	—	(4)
Multiproteinase system	(1)	—	(5)	(2)	(4)	(1)	(1)
>1 sensitive to metal chelators	—	—	(2)	—	(1)	—	—
>1 serine/metallo	—	—	(2)	(2)	(2)	—	—
>1 serine class only	—	—	—	—	—	—	—
Metalloenzyme + serine enzyme	—	—	—	—	(1)	—	—
pH optimum >8·0	(7)	(3)	(8)	(5)	(10)	—	—
pH optimum <6·0	—	—	—	—	(2)	—	—
Temperature optimum >55°C	—	(1)	(1)	(1)	—	—	(3)
Molecular weight range × 1000[h]	23–48 (23)	37–74 (6)	15–120 (10)	11–88 (12)	20–78 (16)	26–420 (3)	28–105 (18)
Molecular weight most often	38–48 (19)	35–46 (3)	41–51 (6)	35–71 (10)	38–56 (8)	26–38 (3)	29–72 (7)
p*I* range	5·1–8·8 (13)	4·5–11 (4)	3·2–6·25 (3)	3–3·5 (2)	4·4–7·3 (10)	—	5·2–8·9 (7)
p*I* most often	5·7–7·3 (6)	—		—	5·3–6·0 (6)	—	6·2–7·7 (4)
Sugar content (%)[i]	0·03–6·0 (8)	8·5 (1)	—	—	—	—	0·13–6·7 (3)

Sugar not found	(1)	—	—	—	(2)	—	—
Zn/Ca atoms/molecule	<1–1/<1–16 (13)	1/4 (1)	—	1–2/0–2 (3)	1/7 (6)	—	—
Heat stability[i]							
stable >65°C	(21)	(2)	(8)	(3)	(2)	—	—
labile >65°C	(4)	(3)	(7)	(9)	(12)	(3)	(11)
stable >65°C but labile at 55°C	(10)	—	(5)	(1)	—	—	—
require Ca^{2+} for stability >65°C	(3)	—	—	(1)	(1)	—	—
do not require Ca^{2+} for stability >65°C	(2)	(1)	(3)	—	—	—	—
Elastolytic activity	—	(1)	—	—	—	—	(2)
Collagenase activity	—	(1)	—	—	—	—	(4)
Aminopeptidases							
Heat stability							
stable	(1)	—	—	(2)	—	—	—
labile	(1)	—	—	(1)	—	—	—
Temperature optimum >55°C	—	—	—	(2)	—	—	(1)

[a] *P. fragi* (3), *P. marinoglutinosa* (2), *P. maltophilia* (2), *P. perolens* (1).
[b] *A. proteolytica* (1), *A. salmonicida* (5), *A. hydrophila* (4), *Aeromonas* sp. (1).
[c] *S. marcescens* (12), *S. piscatorum* (1), *S. liquefaciens* (1), *Serratia* sp. (2).
[d] *B. subtilis* (1), *B. coagulans* (1), *Bacillus,* sp. (1).
[e] *Acinetobacter* sp. (1), *Arthrobacter* spp. (2), *Flavobacterium* spp. (4), *Achromobacter* spp. (2), *Esherichia freundii* (1), *Enterobacter aerogenes* (1), *Alteromonas* sp. (1), *Acinetobacter* (1), *Vibrio* spp. (3), *Micrococcus* spp. (2), *Brevibacterium linens* (1).
[f] Often with minor fractions in addition.
[g] Number of enzymes reviewed is given in parantheses.
[h] By different techniques.
[i] Also from Alichanidis, 1988; Vagias, 1988.

Proteinases from most *Pseudomonas* spp. showed maximum activity at around 45°C; however, more than 15% of maximal activity was often expressed at 4°C. Substrate-dependent temperature optima were observed (Fox *et al.*, 1989). For many *Pseudomonas* proteinases, even those of alkaline character, i.e. with pH optima for casein hydrolysis above pH 8, marked activity was observed at pH 5·5 (Fairbairn & Law, 1986*a*; Vagias, 1988; Fox *et al.*, 1989). Broad pH-activity characteristics and activity at low temperatures were also observed for proteinases from *Aeromonas* spp., *Alcaligenes* sp. and *Serratia marcescens* (Zaborniak & Knaut, 1984; Alichanidis, 1988; Fox *et al.*, 1989).

Purified proteinases from psychrotrophs with pH optima <6·0 have only rarely been found (Table 1; Fox *et al.*, 1989). A raw milk isolate of *Acinetobacter lwoffi* apparently secreted multiple proteinases with secondary activity towards casein at pH 5·5. Extracellular proteinases from this strain and from *P. putida* showed a secondary activity peak with an optimum at *c.* 10°C (Akuzawa *et al.*, 1986).

Action on proteins

Secondary structure parameters such as β-sheet, β-turn and random structure are important for the milk clotting activity of a group of aspartic proteinases (Aishima *et al.*, 1987). Small variations in tertiary structure are important for elastolytic activity (Morihara, 1974). This indicates that for certain specific proteolytic functions, the specificity towards small peptide substrates has only limited predictive value in determining enzyme behaviour on complex proteins. Crude preparations of proteinases from non-dairy psychrotrophic isolates showed variable, complex patterns of activity on casein, egg albumen, gelatin and sacroplasmic proteins (Sikes & Maxcy, 1979). Some of the multiple proteinases produced by a mutant of *P. fragi* were elastolytic (Fox *et al.*, 1989). Collagenase activities were determined for proteinases from *P. marinoglutinosa* and *P. perolens* (Fox *et al.*, 1989). Elastase from *Flavobacterium elastolyticum* and collagenase from *Acinetobacter* had no activity on casein, while collagenases from *Vibrio alginolyticus* with or without caseinolytic activity have been reviewed (Fox *et al.*, 1989).

A metalloproteinase from *P. fluorescens* LY 13 showed little activity on blood serum albumin (BSA), haemoglobin or cytochrome C (Yan *et al.*, 1985), while a serine proteinase from *Flavobacterium arborescens* NRRL B-11, 022 readily hydrolysed both casein and BSA (Boguslawski *et al.*, 1983). Proteinases from *Pseudomonas* spp. were generally weakly active on undenaturated whey proteins and other globular proteins (Cousin, 1989; Fox *et al.*, 1989). Examples of relatively higher activity on β-lactoglobulin than on α-lactalbumin or vice versa can be quoted (Azcona *et al.*, 1988; Cousin, 1989; Stepaniak *et al.*, 1989). β-Lactoglobulin was only slightly digested by the purified proteinase from *P. fluorescens* AH-70, whereas the activity on α-lactalbumin was comparable to that measured on caseins (Azcona *et al.*, 1988). Proteinases from six strains of *P. fluorescens* and one of *Serratia marcescens* investigated by Mitchell and Ewings (1985) hydrolysed neither α-lactalbumin nor β-lactoglobulin. At a high concentration, a neutral proteinase from *Bacillus subtilis* hydrolysed whey proteins and immunoglobulins from milk; β-lactoglobulin was hydrolysed faster than α-lactalbumin (El Mayda *et al.*, 1986).

The maximum degree of hydrolysis of casein during incubation with strains of *P. aeruginosa, B. megaterium, B. cereus* and *S. marcescens* was in the range 64–86% of the peptide bonds (Vuillemard & Amiot, 1985). About 44% of the peptides released from buttermilk by free or immobilised cells of *S. marcescens* with both endo- and exopeptidase activity contained less than 10 amino acids/peptide (Vuillemard *et al.*, 1988).

Within and between species, differences in specificity on milk proteins of crude or purified preparations of proteinases from *Pseudomonas, Alcaligenes, Acinetobacter* and *Moraxella* were found by simple procedures, e.g. comparison of casein agar zone characteristics for precipitation of *para*-casein and total solubilisation of casein (Fain, 1982), or by comparing milk clotting and proteolytic activities (Jackman *et al.*, 1985). *P. fluorescens* T16 was not effective in catalysing the enzymatic step of milk clotting at 0°C (Jackman *et al.*, 1985). In cold-stored raw milk, the increase in concentration of peptides with characteristics similar to the glycomacropeptide released by chymosin was correlated with the growth of psychrotrophic bacteria (Mottar *et al.*, 1985).

Figure 1 illustrates that the specificities of two *Pseudomonas* proteinases are wider than that of plasmin. Differences are apparent in the relative activity on both whole casein and κ-casein. The two strains also produced different patterns of peptides when incubated in milk at 4°C (Stepaniak *et al.*, 1987*c*).

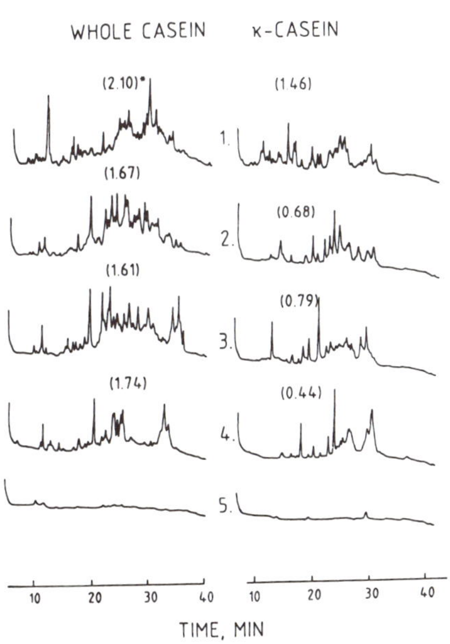

FIG. 1. Profiles of peptides soluble in 2% trichloroacetic acid released after limited proteolysis by four different proteinases (from Stepaniak, *et al.*, 1989). Enzymes were incubated at 37°C in 1% solutions of whole casein or κ-casein in synthetic milk serum ultrafiltrate, pH 6·6. The amounts of enzymes were adjusted to give comparable degrees of proteolysis of whole casein after 2 h incubation. Peptides were separated by reversed-phase HPLC and monitored at 210 nm. Proteolysis measured as the increase of free amino groups, (μmol/ml) assay method from McKellar (1981). 1 = proteinase P1 from *P. fluorescens* P1; 2 = proteinase AFT 21 III from *Pseudomonas* sp AFT 21; 3 = thermolysin (from *B. thermoproteolyticus*); 4 = bovine plasmin; 5 = unincubated controls with proteinase P1.

The broader activity of proteinases from four strains of *P. fluorescens* compared to those of trypsin, chymotrypsin or *Staphylococcus aureus* proteinase was demonstrated by separation of peptides produced from oxidised ribonuclease A; the study also confirmed differences between strains of *P. fluorescens* (Mitchell & Marshall, 1989*a*). Differences in the electrophoretic patterns of peptides released from β-casein and dephosphorylated β-casein were demonstrated for intracellular proteinases from *P. fluorescens* and *E. coli* (Gallmann & Puhan, 1982). Electrophoretic patterns of milk proteins were also different after incubation with extra- and intracellular crude preparations of enzymes from *P. fluorescens, Flavobacterium* sp. and *Cytophaga* sp. (Guamis *et al.*, 1987*b*). Strains of *P. fluorescens* subsp. *haemolytica, P. aeruginosa, Aeromonas hydrophila, B. cereus, E. coli* and *Streptococcus faecalis* subsp. *liquefaciens* incubated in milk released variable amounts of peptides which were likely to be authentic β-casomorphins or β-casomorphin precursors (Hamel *et al.*, 1985).

The preferences of some proteinases towards particular caseins can be influenced by the micellar status of casein, or by other assay conditions (Zaborniak & Knaut, 1985; Grufferty & Fox, 1988; Cousin, 1989). Based on data collected in Table 2, it can be concluded, however, that the majority of proteinases from psychrotrophs are relatively more active towards α-casein and κ-casein than cell-associated proteinases from lactic acid bacteria. High activity of metalloproteinases from psychrotrophs towards κ-casein distinguished them from serine milk alkaline proteinase and leucocyte proteinase. When present together, these enzymes should cause broad hydrolysis of casein.

Lipases

Several *Pseudomonas* lipases occur in large aggregates of varying molecular weight (Stead, 1986; Fox *et al.*, 1989; Stepaniak & Sørhaug, 1989). The formation of self aggregates and aggregates with other components is common for many bacterial and fungal lipases (Stepaniak & Sørhaug, 1989). The size of the aggregates may be influenced by the type of cultivation medium (Bucky *et al.*, 1987*a*). The molecular weight of lipase from *P. fragi* IFO 3458 calculated from its gene nucleotide sequence was 14 643 (Kugimiya *et al.*, 1986). Large differences for molecular weights of the monomeric forms of lipases from *Pseudomonas* spp. evaluated by SDS gel electrophoresis are apparent (Fox *et al.*, 1989). The lipases from *P. fluorescens* AFT29 (Dring & Fox, 1983) and *P. fluorescens* P1 had molecular weights similar to that from *P. fragi* IFO 3458. The molecular weight of the lipase subunit from *P. fluorescens* MC50 determined by SDS gel electrophoresis was 55 000 (Bozoğlu *et al.*, 1984).

Multiple lipases producing different patterns of fatty acids upon hydrolysis of butter fat were demonstrated for *P. fluorescens* 27 (Ren *et al.*, 1988). Lipase from *P. fluorescens* AFT 36 was separated by gel filtration into three molecular weight fractions which on comparison of relative specificity towards four

TABLE 2
Degradation of Caseins by Selected Proteinases

Source of Proteinase	*Substrate*[a]	*Casein Fractions, Relative Rate of Hydrolysis*	*References*
Pseudomonas fluorescens (5 strains)	M	$\kappa > \beta > \alpha_s$	Mitchell & Ewings (1985)
P. fluorescens OM227	M	$\kappa = \beta = \alpha_s$	Mitchell & Ewings (1985)
P. fluorescens AFT 36	M	$\kappa > \alpha_{s1} > \beta > \alpha_{s2}$	Stepaniak *et al.* (1982)
P. fluorescens LY 13	F	$\alpha_s > \beta > \kappa$	Yan *et al.* (1985)
P. fluorescens T 18	F	$\kappa = \alpha_s > \beta = \gamma$	Patel *et al.* (1983*a*)
P. fluorescens T 20	F	$\alpha_s > \beta > \gamma > \kappa$	Patel *et al.* (1983*a*)
Serratia marcescens (2 separate enzymes)	M	$\kappa > \beta > \alpha_s$	Mitchell & Ewings (1985)
S. marcescens	C	$\kappa = \beta > \alpha_s$	Grieve & Kitchen (1985)
Aeromonas sp. B31	M	$\kappa > \beta > \alpha_{s1}$	Richardson & Te Whaiti (1978)
A. hydrophila A4	F	$\alpha_{s1} = \alpha_{s2} = \beta$	Alichanidis (1988)
Alcaligenes sp.	M	$\kappa > \beta > \alpha_s$	Zaborniak & Knaut (1985)
Alcaligenes sp.	C	$\beta > \kappa > \alpha_s$	Zaborniak & Knaut (1985)
Micrococcus (9 strains, 4 species)	M	$\beta > \alpha_s$	Bhowmik & Marth (1988)
Bacillus subtilis LY 11	F	$\beta > \alpha_s > \kappa$	Yan *et al.* (1985)
Bacillus sp. LY 10	F	$\beta > \kappa > \alpha_s$	Yan *et al.* (1985)
Streptococcus cremoris (most strains)	C	$\beta \gg \alpha_s = \kappa$	Thomas & Pritchard (1987)
Lactobacillus bulgaricus	C	$\beta > \alpha_s = \kappa$	Chandan *et al.* (1982)
Bovine plasma (plasmin)	C	$\beta = \alpha_s \ggg \kappa$	Grieve & Kitchen (1985)
Leucocytes	C	$\alpha_s > \beta \gg \kappa$	Grieve & Kitchen (1985)

[a] M = milk, F = casein fractions, C = whole casein.

substrates, indicated the existence of a subunit which undergoes varying degrees of self association (Fox & Stepaniak, 1983).

With some exceptions, lipases from different genera of psychrotrophs were to variable extents inhibited by EDTA. The lipase from *P. fluorescens* GR 83 appeared to be a serine enzyme which also required calcium for catalytic activity (Roussis *et al.*, 1988*a*). The activities of EDTA-treated lipases GR 83 and MC50 were restored substantially by Ca^{2+}. Very little or no restoration of activity by Zn^{2+} was found for the two *P. fluorescens* enzymes (Bozoğlu *et al.*, 1984; Roussis *et al.*, 1988*a*). The presence of serine in the catalytic site is common for *P. fragi* lipases (Kugimiya *et al.*, 1986) and for some other microbial and non-microbial lipases (Stepaniak & Sørhaug, 1989). Involvement of histidine in the catalytic mechanisms of lipases from *Pseudomonas* and *Chromobacterium* was also suggested (Sugiura, 1984). Zn^{2+}, Mn^{2+}, Co^{2+}, Cu^{2+}, Hg^{2+} or Fe^{3+} were, to a variable extent, inhibitory for most lipases from psychrotrophs (Sugiura, 1984; Stead, 1986; Fox *et al.*, 1989; Stepaniak & Sørhaug, 1989). The effect of thiol reagents also varied (Stead, 1986; Fox *et* *al.*, 1989). A concentration of 2·18 M sodium chloride in a butter fat emulsion or cream was inhibitory for lipases from most isolates of *P. fluorescens, Moraxella,* and *Acinetobacter* (Fitz-Gerald & Deeth, 1983). Salt was thought to influence the enzyme–substrate interaction rather than the enzyme or substrate separately.

Variable effects of different emulsifiers, detergents or other stimulators on particular enzymes from psychrotrophs have also been widely experienced (Adams & Brawley, 1981; Fitz-Gerald & Deeth, 1983; Sugiura, 1984; McKellar & Cholette, 1986; Stead, 1986). Careful optimisation of all these factors is essential for maximum activity (McKellar & Cholette, 1986; Stead, 1986). In complex systems like milk or cream, lipases from psychrotrophs can be inhibited by milk proteins (McKellar & Cholette, 1986). Milk fat inhibited the activity of *P. fluorescens* lipases assayed by the agar diffusion method with tributyrin (Stepaniak *et al.*, 1987*a*) or when *β*-naphthyl caprylate was used as substrate (McKellar & Cholette, 1986). Optimal assay conditions for crude preparations of *P. fluorescens* lipases with cream as substrate have been established (Fitz-Gerald & Deeth, 1983).

Common for lipases from different psychrotrophic genera is activity over broad ranges of pH and temperature, with *c.* 20% or more of maximal activity expressed at temperatures of 4–7°C and at pH 5 (Dring & Fox, 1983; Fox & Stepaniak, 1983; Bozoğlu *et al.*, 1984; Stead, 1986; Fox *et al.*, 1989). Substrate-dependent temperature and pH optima were observed when lipases from common psychrotrophs were used to hydrolyse tributyrin, triolein, and homogenised and non-homogenised cream (Fitz-Gerald & Deeth, 1983).

Specificity

Lipases from *Pseudomonas* spp. and most other psychrotrophs effectively hydrolyse a range of synthetic and natural triglycerides, including butter fat in non-homogenised cream (Adams & Brawley, 1981; Fitz-Gerald & Deeth, 1983; Stead, 1986; Fox *et al.*, 1989). However, available data (Stead, 1986; Fox *et al.*, 1989) are not consistent enough to rank these enzymes according to preference of hydrolysis among simple triglycerides. *P. fluorescens* GR 83 most actively hydrolysed triglycerides with C – 8:0 or C – 10:0 fatty acids; activity towards tributyrin was *c.* 40% lower (Roussis *et al.*, 1988*a*). For *P. fragi* 3 lipase, the order of hydrolysis of synthetic triglycerides was C-12:0 > C-10:0 > C-8:0 > C-16:0 > C-4:0 > C-6:0 > C-18:0 (Lu & Liska, 1969). Preferences for hydrolysis of simple triglycerides by lipase from *P. fluorescens* or *Candida cylindracea* were influenced by temperature (Sugiura & Isobe 1975). Two thirds of 19 psychrotrophic isolates, including 10 of *P. fluorescens,* five of *Acinetobacter* spp. and two of *Moraxella* spp., hydrolysed tributyrin more effectively than butter fat (Fitz-Gerald & Deeth, 1983). Psychrotrophs showing tributyrinase activity, but incapable of hydrolysing butter fat, were also isolated (Muir *et al.*, 1979).

Preferential release of C-4–C-8 fatty acids from butter fat makes enzymes potent in the development of rancidity (McNeill *et al.*, 1986; Stead, 1986). Preferential release of short chain fatty acids by lipases from psychrotrophic bacteria in butter was suggested (Woo & Lindsay, 1984; McNeill *et al.*, 1986; Stead, 1986; Ren *et al.*, 1988). The specificity of these lipases is likely to be different from that of lipases produced by psychrotrophic yeasts such as *Cryptococcus albidus* or *Candida cylindracea* (Fitz-Gerald & Deeth, 1983; Woo & Lindsay, 1984). *Geotrichum candidum* lipase is known for its unique specificity for oleic acid (Macrae, 1983; Stead, 1986).

Since C-4:0, C-6:0 and C-8:0 fatty acids are primarily located at the *sn*-3 position of butter fat triglycerides, their preferential release is indicative of positional specificity (McNeill *et al.*, 1986). Lipase from *P. fluorescens* 27 favoured the release of short-chain fatty acids. The enzyme also showed some preference for C-14:0 acid, which is esterified primarily at the *sn*-2 position (Ren *et al.*, 1988). In a whole milk culture of *Pseudomonas* P46, butyric acid accumulated to remarkably higher molar concentration than for any of the C-6:0–C-12:0 free fatty acids (Bucky *et al.*, 1987*b*). Lipases from *Chromobacterium viscosum, Candida cylindracea, Rhizopus arrhizus* and lipase from porcine pancreas added to butter could be identified by analysis of specificity differences (Woo & Lindsay, 1983).

Positional specificity was not exhibited by lipoprotein lipases of some *Pseudomonas* spp. and *P. fluorescens* or by *Chromobacterium* lipase (Sugiura *et al.*, 1978; Sugiura, 1984), and in this respect these enzymes are different from lipase from *P. fragi* which had *sn*-1–*sn*-3 positional specificity (Kugimiya *et al.*, 1986).

Lipases from *P. fluorescens, Acinetobacter, Moraxella* and *Cryptococcus* actively hydrolysed the lipoprotein fraction isolated from butter (Stead, 1986). A lipase from a strain of *P. fluorescens* was found to be among those of microbial origin with the highest lipoproteolytic activity (Stepaniak & Sørhaug, 1989). Lacking positional specificity, it rapidly degraded serum lipoproteins to glycerol and free fatty acids (Sugiura, 1984; Stepaniak & Sørhaug, 1989). *P. fluorescens* lipoprotein lipase, together with phospholipase C from *B. cereus* and glycerol dehydrogenase, have been used for enzymatic determination of serum phospholipids (Sugiura *et al.*, 1978). On the other hand, the *P. fluorescens* lipase studied by Bengtsson and Olivecrona (1980) did not require albumin as a fatty acid acceptor for effective hydrolysis of triolein, which is a characteristic feature of lipoprotein lipase.

An additional controversy about the lipoprotein lipase character of lipases from psychrotrophs arises from a study by Fitz-Gerald and Deeth (1983). They found that addition of blood serum to cream increased the activity of lipases from *Acinetobacter* spp. and *P. aeruginosa,* but inhibited the activities of lipases from *P. fluorescens* or *Moraxella* strains. Milk lipoprotein lipase was inhibited by the protein fraction extracted from the milk fat globule membrane (Sundheim & Bengtsson-Olivecrona, 1987). Some unique specificities of lipases from *Pseudomonas* spp. towards esters other than triacylglycerols have been reviewed (Sugiura, 1984; Stead, 1986; Stepaniak & Sørhaug, 1989).

Phospholipase C

Production of both phospholipase(s) A and phospholipase C is common for *P. fluorescens* (Deeth & Fitz-Gerald, 1983; Fox *et al.*, 1989). Phospholipase C production was confirmed for strains of *Acinetobacter, Alcaligenes, Flavobacterium, Enterobacter, Bacillus, Clostridium* and *Staphylococcus* (Fox *et al.*, 1976; Deeth & Fitz-Gerald, 1983; Fox *et al.*, 1989). A *Bacillus cereus* strain produced both phospholipase C and sphingomyelinase (Fox *et al.*, 1989). Two phospholipases C from *P. aeruginosa* have been isolated (Fox *et al.*, 1989). *Serratia* spp. produced phospholipases A with no evidence of phospholipase C (Fox *et al.*, 1989).

Bacterial phospholipases C were usually very sensitive to metal chelators (Sonoki & Ikezawa, 1975; Berka & Vasil, 1982; Bosch, 1982; Stepaniak *et al.*, 1987*b*; Fox *et al.*, 1989; Stepaniak & Sørhaug, 1989). *B. cereus* phospholipase C appears to contain one histidine residue that is essential for catalytic activity and four that may coordinate zinc atoms in the structure (Little, 1977).

Compared with proteinase P1 and lipase P1, phospholipase C from *P. fluorescens* P1, with a temperature optimum at 55°C and a pH optimum at 7·3, was relatively less active at temperatures <7°C and pH<6 (Stepaniak *et al.*, 1987*b*). Phospholipases C from *P. aureofaciens, B. cereus, Clostridium perfringens* and *Clostridium novyi* showed differences in relative activities towards phospholipids from bacterial membranes (Taguchi & Ikezawa, 1977). The enzymes from *P. fluorescens* (Doi & Nojima, 1971) and from *P. aureofaciens* (Sonoki & Ikezawa, 1975) hydrolysed phosphatidylethanolamine more rapidly than phosphatidylcholine. The opposite order of preference for the two substrates was established for the enzyme from *B. cereus* (Sonoki & Ikezawa, 1975) and for heat-labile haemolytic phospholipase C from *P. aeruginosa* (Berka & Vasil, 1982).

Phospholipases C from *C. perfringens* (Kurioka & Matsuda, 1976) and *P. fluorescens* P1 (Stepaniak *et al.*, 1987*b*) hydrolysed *p*-nitrophenylphosphorylcholine. This substrate was not hydrolysed by an EDTA-insensitive acid phospholipase C from *Tetrahymena thermophila* (Florin-Christensen *et al.*, 1986).

Glycosidases

Marin and Marshall (1983) characterised glycosidase activities toward *p*-nitrophenyl-glycosides (PNP-glycosides) in the cell-free filtrate from a culture of *P. fluorescens* 26. Temperature and pH optima for the activities were as follows: 15°C and 6·0 with PNP-β-D-fucoside, 25°C and 6·0 with PNP-β-D-mannoside, 25°C and 5·5 with PNP-β-D-glucoside, 30°C and 6·5 with PNP-β-D-galactoside, 45°C and 7·0 with PNP-*N*-acetyl-β-D-glucosaminide, and 45°C and 6·0 with PNP-*N*-acetyl-β-D-galactosaminide. The crude glycosidase preparation hydrolysed glycoproteins in fat-extracted and dialysed skim milk and buttermilk or in fat globule membrane. Sialic acid was hydrolysed well, but hexosamines were hydrolysed poorly (Marin *et al.*, 1984).

Heat stability of enzymes from psychrotrophs

Psychrotrophs producing both heat-stable lipase and proteinases, heat-stable proteinases but heat-labile lipases or vice versa have been isolated from raw milk (Christen *et al.*, 1986). A larger proportion of Gram-negative and Gram-positive psychrotrophic isolates produced heat-stable proteinases compared to non-psychrotrophs (Ewings *et al.*, 1984).

Griffiths *et al.* (1981*b*) found that the heat stability of proteinases in the cell-free culture supernatants of psychrotrophic strains of seven different species of fluorescent *Pseudomonas,* four species of non-fluorescent *Pseudomonas,* four species of *Enterobacter* (including two *Serratia*), one *Alcaligenes,* one *Acinetobacter,* one *Flavobacterium,* one *Bacillus,* and one *Moraxella* was high enough to retain 20–60% of the original activity after heating at 140°C for 5 s. Lipases from the same genera were also heat-stable, but on average somewhat less so than the proteinases. Lipases from *Micrococcus freudenreichii* and *B. subtilis* appear to be resistant to pasteurisation temperatures (Singh *et al.*, 1973). Heat-labile proteinases from *Pseudomonas* have been reported occasionally (Ewings *et al.*, 1984; Fox *et al.*, 1989).

The proteinase from *Pseudomonas* sp. MC60 was among the first heat-stable enzymes from *Pseudomonas* isolates from milk to be purified and characterised (Barach *et al.*, 1976; Barach & Adams, 1977; Barach *et al.*, 1978). Its properties have been reviewed (Fox *et al.*, 1989). At 149°C, the enzyme was about 4000 times more heat-stable than spores of *B. stearothermophilus,* i.e. with a D value = 90 s and a Z value of 32·5°C. Several purified proteinases and lipases from *Pseudomonas* spp. have heat stability in the UHT region and at pH 6·6, similar to, or somewhat lower or higher than, that of proteinase MC60 or thermolysin (Barach & Adams, 1977; Fairbairn & Law, 1986*a*; Fox *et al.*, 1989; Kroll, 1989). Purified proteinases from many *Pseudomonas* spp. or *P. fluorescens* strains were heat-stable when Ca^{2+} were present in the enzyme solution; some of these enzymes were stable also in buffers without calcium (Barach & Adams, 1977; Stepaniak & Fox, 1983; Patel *et al.*, 1986; Diermayr *et al.*, 1987). Calcium was also required for high temperature stability of a proteinase from *Aeromonas hydrophila* (Alichanidis, 1988). Differences in the affinity of proteinases from *Pseudomonas* for calcium has been suggested to explain the variable effect of calcium on stability at high temperatures (Kroll, 1989). Calcium had little effect on the stability of *P. fluorescens* lipases at ultra-high temperatures (Fox & Stepaniak, 1983; Swaisgood & Bozoğlu, 1984; Kroll, 1989).

The heat stability of *Pseudomonas* lipases and proteinases in milk or casein solutions is substantially higher than in buffer systems (Kroll, 1989). Triton X-100 enhanced the thermostability at 71·5°C of a lipase from *P. fluorescens*

38. Involvement of inter- or intramolecular hydrophobic bonding in the stabilisation of the enzyme has been proposed (Bucky *et al.*, 1987*a*). A lipase in the culture supernatant of *Geotrichum candidum* was resistant to heating for 10 min at 80°C (Hassanien *et al.*, 1985).

Crude phospholipase C from four species of fluorescent *Pseudomonas* showed 31–57% residual activity after UHT heat treatment, but the enzyme from *A. hydrophila* was completely inactivated after HTST or UHT treatment (Griffiths, 1983). Partially purified phospholipase C from *P. fluorescens* P1 was heat-stable between 55 and 150°C, and its stability at high temperatures was similar to that of proteinase or lipase from the same strain (Stepaniak *et al.*, 1987*b*; Vagias, 1988). Purified phospholipase C from *P. fluorescens* A-3-1 was very unstable at pH 6·8, but it retained some activity at pH 8·7 after heating at 70°C for 5 min (Doi & Nojima, 1971).

The residual activity of proteinases from *P. fluorescens* or *Pseudomonas* spp. was higher after heating followed by cold storage for 24 h than immediately after heating at 90°C for 10 min, indicating renaturation of the enzymes (Patel *et al.*, 1986).

Bacterial isolates from Mexican-style cheeses included some which produced pasteurisation-sensitive or pasteurisation-resistant alkaline phosphatases (Pratt-Lowe *et al.*, 1988). *Streptococcus liquefaciens* produced the, relatively, most heat resistant enzyme, while phosphatases from *P. fluorescens* and *Pseudomonas* spp. were sensitive to heating at 63°C for 30 min.

Knight and Fryer (1989) reported the production of phosphatases by the natural flora of raw milk and by a psychrotrophic bacterium isolated from the milk. The bacteria were killed by pasteurisation but the phosphatases were resistant to this heat treatment. Phosphatase activity resistant to pasteurisation was detectable in raw milk after 2 days of cold storage, i.e. at $1{\cdot}2 \times 10^7$ cfu/ml or greater.

All six glycosidase activities in the culture supernatant from *P. fluorescens* 26 were heat-labile under the conditions used for the pasteurisation of milk (Marin & Marshall, 1983).

Inactivation at sub-pasteurisation temperatures

Proteinases and lipases from psychrotrophs studied by Griffiths *et al.* (1981*b*) were to various extents inactivated by heating at 55°C for 1 h. Earlier, Barach *et al.* (1976) found that the phenomenon of low temperature inactivation (LTI) at around 55°C is common for many heat-stable proteinases from psychrotrophs isolated from milk.

A characteristic discontinuity in the Arrhenius plot in the temperature range of 50–90°C for proteinases and lipases was apparent for several purified enzymes from *Pseudomonas* spp. or *P. fluorescens* (Barach *et al.*, 1978; Dring & Fox, 1983; Kroll & Klostermeyer, 1984*b*; Swaisgood & Bozoğlu, 1984; Diermayr *et al.*, 1987; Kroll, 1989). Exceptions were proteinases from *P. fluorescens* M5 (Marshall & Marstiller, 1981) and from *Pseudomonas* sp. B-25

(Malik & Mathur, 1984); the former was unstable at 40°C, while the latter was quite stable at 55°C and was active at 70°C.

Calculated *D* values at 55°C based on data for the heat stability of *Pseudomonas* proteinases or lipases >100°C are several hours, but the experimental values were in the order of seconds or minutes (Swaisgood & Bozoğlu, 1984; Diermayr *et al.*, 1987; Kroll, 1989).

Kroll and Klostermeyer (1984*b*) established that for the proteinase from *P. fluorescens* 112 the inactivation rate in milk under UHT conditions followed first-order kinetics, while at around 55°C the reaction kinetics were close to second-order. At a low enzyme concentration, first-order inactivation was observed for lipase MC50 both at 50–90°C and 100–150°C (Swaisgood & Bozoğlu, 1984). At 50–60°C, calcium had no significant effect on the inactivation rate of two proteinases from *P. fluorescens* (Stepaniak & Fox, 1983; Kroll & Klostermeyer, 1984*b*), but it partially stabilised lipases from *P. fluorescens* AFT 36, MC50 and GR 83 against the LTI (Fox & Stepaniak, 1983; Swaisgood & Bozoğlu, 1984; Roussis *et al.*, 1988*a*). The protective effect of calcium was not confirmed for *P. fluorescens* AFT 29 lipase (Dring & Fox, 1983). A lipase from *P. fluorescens* 27 was inactivated at 65°C only in the presence of histidine or histamine and magnesium chloride (Christen & Marshall, 1985). pH values within the range 6–10 had a greater effect on the LTI (at 50°C) of proteinase than of lipase from *P. fluorescens* 27 (Christen & Marshall, 1985). Considerably slower rates of LTI in the presence of milk proteins were observed for *Pseudomonas* proteinases but not for lipases (Kroll, 1989). It has been experienced for proteinases from *P. fluorescens* AFT 36 and 112 that heating to or above 90°C, followed by direct cooling to the zone of LTI, protected the residual active enzyme from self-digestion. Cooling the enzyme to 0–20°C after high temperature treatment rendered the enzyme sensitive to self-digestion at 55–60°C (Stepaniak & Fox, 1983; Diermayr *et al.*, 1987). Some proteinases from pseudomonads and other psychrotrophs apparently regain a conformation sensitive to autolysis after being held around 55°C directly after high temperature treatment (Griffiths *et al.*, 1981*b*; Kroll, 1989). Kroll (1989) suggested that some differences in the temperature required for most effective LTI can be related to the origin of the proteinases, i.e. from different biovariants of *P. fluorescens*.

Diermayr *et al.* (1987) and Kroll (1989) proposed a model to explain the inactivation of proteinase from *P. fluorescens* 112, biovar. I at high temperatures and the inactivation by self-digestion at 55°C. The model assumes intermolecular rather than intramolecular self-digestion of partially unfolded molecules. Barach *et al.* (1978) showed that LTI of proteinase from *Pseudomonas* sp. MC60 present in milk at low concentration is caused by aggregation of the enzyme with casein micelles. The effects of thermisation at 55°C of cultured milk (pH *c.* 4·5) on the inactivation of proteinase from *P. fluorescens* P1 and on the distribution of the enzyme between quarg and whey were studied using an enzyme-linked immunosorbent assay. The results indicated that the enzyme added at a concentration of 5 or 50 ng/ml of milk was inactivated by a combination of self-digestion and aggregation with casein (Zakrzewski *et al.*, 1990; Stepaniak *et al.*, 1991).

Swaisgood and Bozoğlu (1984) proposed a model for LTI of lipase from *P. fluorescens* MC50. The model suggests that inactivation occurs at two distinct stages of molecular unfolding. The first stage occurs at a low (60°C) temperature. Kroll (1989) concluded that LTI of lipases may also occur by attack on the unfolded lipase molecule by accompanying proteinases.

Bucky *et al.* (1987*b, c*, 1988) have proposed a modified UHT treatment of milk. First, the milk is heated at 140°C for 5 s, then it is cooled to 60°C and held at this temperature for 5 min. The system was effective for inactivation of both lipases and proteinases in milk incubated with selected *Pseudomonas* or with natural psychrotrophic flora. The effectiveness of the process was demonstrated by showing that the rate of proteolysis and lipolysis in UHT sterilised milk produced by the dual treatment can be reduced by up to 25-fold compared with the traditional method.

Plasmin was also partially inactivated when subjected to LTI, but the effect was considered of no practical significance (Driessen, 1983; Hill, 1988).

RELATIONSHIP BETWEEN PROTEOLYSIS, LIPOLYSIS AND FLAVOUR

Proteolysis

Peptides and amino acids are known to cause a variety of taste sensations. Hydrophobicity has been found essential for bitter taste, and casein, with relatively high hydrophobicity, is known to yield a large amount of bitter peptides (Minamiura *et al.*, 1972; Fox, 1981; Law, 1982; Adler-Nissen, 1986; Andrews, 1986). Bitter peptides were isolated from cheeses and from hydrolysates of different casein fractions, but β-casein, especially its very hydrophobic C-terminal region, is the most probable source of bitter peptides (Minamiura *et al.*, 1972; Law, 1982; Hsien-Yeh-Hsu, 1984). It is very likely that γ-caseins *per se* are astringent (Harwalkar *et al.*, 1989).

Many aspects of the bitterness of peptides have been discussed in detail by Adler-Nissen (1986). The review also reveals that there is no simple correlation between enzyme specificity and bitterness. Considering that internally-located hydrophobic amino acids contribute more to the bitterness of peptides than terminal ones, metalloproteinases, with their specificities for peptide bonds comprised of hydrophobic amino acid, should contribute less to bitterness than trypsin which prefers non-hydrophobic amino acids. However, the overall effect, which also depends on the relative length of peptides released and on other factors, reduces the value of such predictions considerably. Appropriate exopeptidases may reduce bitterness substantially by further degradation of bitter peptides (Minamiura *et al.*, 1972; Adler-Nissen, 1986).

The minimum level of proteolysis needed by enzymes from psychrotrophs to develop bitterness in milk depended on enzyme origin (McKellar, 1981; Mitchell & Ewings, 1985). Proteolysis corresponding to the accumulation of 0·289–0·746 μmols of free amino groups per ml of pasteurised or UHT-treated milk containing proteinases from either of three strains of *P. fluorescens* was sufficient to cause off-flavours (McKellar, 1981). This indicates differences in the ability to accumulate bitter peptides. With purified proteinase preparations from six *P. fluorescens* strains and one *Serratia marcescens* strain added to UHT milk, the following stages in flavour development were noted: fresh > lack of freshness > slightly stale > unclean or mild off-flavour > slightly bitter > bitter (Mitchell & Ewings, 1985). Flavour changes were detectable in UHT milk stored at 37°C for 72–192 h after addition of 0·01–0·0001% of cell-free supernatants of fully grown proteolytic cultures of *P. fluorescens* (Rollema *et al.*, 1989). Bitterness in UHT milk has also been attributed to the activity of plasmin (Driessen, 1983; Hill, 1988).

For pasteurised milk incubated with *P. fluorescens* P27 at 7°C, an impedance value was a better predictive parameter of shelf life than proteinase activity, psychrotrophic count or lipopolysaccharide (Gram-negative bacteria) content (Bishop & White, 1985*a*). All these measurements were, however, significantly correlated with shelf life of the product. A linear relationship between flavour score and proteinase activity in skim and whole milk was established (Janzen *et al.*, 1982*a*). A significant correlation was also found between proteolysis and bitterness in cold-stored double-pasteurised cream (Griffiths *et al.*, 1981*a*). Mitchell *et al.* (1989) found that the shelf life of pasteurized milk inoculated after laboratory pasteurisation with 21 and 130 ng/ml of purified proteinase from *P. fluorescens* OM82 and kept at 7°C is 13 and 3 days respectively. At the expiry dates the total colony count was less than 10^4 cfu/ml. However, when *P. fluorescens* OM82 grew in milk at the same temperature from a level of 8×10^4 cfu/ml the shelf life of the milk expired after 6 d at a psychrotrophic population of 9×10^8 cfu/ml and a concentration of secreted proteinase of *c.* 1·6 ng/ml. The reasons for reduced release of bitter peptides during incubation with pure proteinase were suggested to be its concentration-dependent inhibition by substrate or by products of hydrolysis, or alternatively participation of other metabolites absent in the pure enzyme preparation. Bitterness in milk also developed as a result of growth of an extracellular proteinase- and lipase-deficient mutant of *P. fluorescens* (Torrie *et al.*, 1983).

Flavour defects developed on addition of a lipase-free proteinase preparation from *P. fluorescens* P26 to milk, cottage cheese and Cheddar cheese. With cottage cheese, the frequency distribution of bitter/astringent and other off-flavours in proteinase-treated milk was different from that in milk inoculated with the corresponding microorganism (White & Marshall, 1973). In an experiment with *P. fluorescens* P27 added to cream dressing, proteolysis, impedance value and lipopolysaccharides, with correlation coefficients of −0·87, +0·90 and −0·81, respectively, were better indices of the potential shelf life of cottage cheese than plate count ($r = -0{\cdot}61$) (Bishop & White, 1985*b*).

Quarg produced from milk containing 5 ng/ml of proteinase from *P. fluorescens* P1 had, due to bitterness, significantly reduced flavour scores ($P < 0{\cdot}001$) after 3 weeks of storage at 7°C. The defect was eliminated by thermisation of cultured milk before separation of quarg (Zakrzewski *et al.*, 1990).

Lipolysis

Lipolysis and resulting rancidity in cream, butter, cheeses or UHT-sterilised milk developed on addition of *Pseudomonas, Acinetobacter* or yeast lipases (Deeth *et al.*, 1979; Andersson *et al.*, 1981; Woo & Lindsay, 1983; Christen *et al.*, 1986; Stead, 1986; Mottar, 1989). Rancidity also occurred during cold storage after pre-heat-treatment of milk with added lipolytic strains of pseudomonads or with natural microflora (Mottar, 1981; Stead, 1986; Bucky *et al.*, 1988; Mottar, 1989). Lipolysis by lipases from psychrotrophs was not well correlated with flavour defects in double-pasteurised cream; preceding spontaneous lipolysis was an interfering factor (Griffiths *et al.*, 1981*b*). Hill (1988) concluded that lipolysed flavour in UHT milk occurs less frequently than bitterness caused by proteolysis.

Butter made from sweet cream (pH 6·6) without added salt is relatively more susceptible to lipolysis than to autoxidation but copper-catalysed autoxidation is generally the prime cause of flavour impairment in salted, ripened cream butter (Downey, 1980).

Lipolytic rancidity in butter, also described as butyric or lipase flavour (Fitz-Gerald & Deeth, 1983), is best correlated with the content of short-chain free fatty acids (McNeill *et al.*, 1986; Stead, 1986). Acid degree value (ADV), a measure of total free fatty acids and the content of long-chain free fatty acids are poorly correlated with the rancidity of butter (Deeth *et al.*, 1979; McNeill *et al.*, 1986). The value corresponding to deterioration of flavour may vary according to the specificity of the lipase responsible for lipolysis (Deeth *et al.*, 1979). Flavour threshold values of free fatty acids expressed as ADV, are in the ranges of 0·8–2·74 meq/100 g fat for milk and 0·75–2·5 meq/100 g fat for butter (Deeth & Fitz-Gerald, 1983). Short-chain fatty acids are detected first by smell (Deeth *et al.*, 1979), and at much lower weight concentrations than are C-16:0 or C-18:1 fatty acids (Scanlan *et al.*, 1965). Rancid and soapy flavours in pasteurised, homogenised milk could be perceived after addition of 46·1 mg/kg of C-4:0, 30·4 mg/kg of C-6:0 or 22·5 mg/kg of C-8:0 fatty acid (Scanlan *et al.*, 1965). On the basis of these findings, 0·25 μmol/ml of short-chain free fatty acids in UHT milk was regarded as the cut-off value for prediction of shelf life (Bucky *et al.*, 1988). Rancidity in butter was detected when the concentrations of C-4:0 and C-6:0 free fatty acids in butter were 35 and 26 mg/kg, respectively (McNeill *et al.*, 1986).

C-10:0–C-12:0 fatty acids are responsible for most of the unclean and soapy flavours of milk (Deeth & Fitz-Gerald, 1983; Stead, 1986). The flavour of milk with added caproic or caprylic acids at concentrations ranging from 6·25 to

100 μmol/100 g has been described as 'cowy' and 'goaty' (Al-Shabibi *et al.*, 1964). Some incomplete glycerides may contribute to bitterness (Deeth & Fitz-Gerald, 1983). Free unsaturated fatty acids are most susceptible to oxidation to aldehydes and ketones resulting in off-flavours described as cardboardy, oxidized or metallic (Stead, 1986). Addition of lipase from *P. fluorescens* at 0·3 units/ml (one unit defined as the amount of enzyme which produced 1·0 micro-equivalent of fatty acids in 1 h at 37°C, pH 7·4) to milk before sterilisation resulted in the development of rancidity in UHT milk after 5–8 days of storage at 8°C (Andersson *et al.*, 1981). The appearance of rancidity in butter was related to the presence of heat-stable lipases from psychrotrophs because the product made from milk stored cold for longer periods was more susceptible to rancidity development (McNeill *et al.*, 1986).

The spoilage of cream or butter due to lipolysis may be promoted by some affinity of lipase from psychrotrophs for the cream phase of milk (Stead, 1986, 1989).

Relationship between flavour and some other metabolic activities

Pseudomonas spp. secrete many enzymes (Ribbons & Williams, 1982) which are very likely to be responsible for many hydrolytic and oxido-reductive enzymatic reactions that occur simultaneously with or prior to proteolysis and lipolysis. The concentrations of lactate and glucose in raw milk were reduced during incubation at 7°C with *P. fluorescens* NCDO 2085 before the bacterial count reached 10^8 cfu/ml and before the production of detectable levels of proteinase (Sanjose *et al.*, 1987).

Preferential release of butyric acid from milk fat by *Pseudomonas* lipases (Bucky *et al.*, 1987*b*; Ren *et al.*, 1988) indicates spoilage potential due to both rancidity and possible formation of esters. *P. fragi* is an especially active producer of volatile esters (Reddy *et al.*, 1969) responsible for the 'fruity' odours sometimes noted in spoiled milk, cheeses and meats (Cousin, 1982; Dainty, 1982; Greer, 1989). Addition of ethanol to a milk culture stimulated the production of esters by *P. fragi* (Reddy *et al.*, 1969). The patterns of volatile compounds were found to be characteristic for some psychrotrophs (Cousin, 1982).

Differences between *P. fluorescens* strains in the capacity to produce propan-2-ol in milk have been observed (Urbach & Milne, 1988). Marked increases in the concentration of ethanol in pasteurised milk stored at 4, 7 or 10°C corresponded to the onset of off-flavour development and plate counts in the order of 10^6–10^7 cfu/ml (Urbach & Milne, 1988). Dehydrogenase activities were shown for *P. fluorescens*, *P. fragi*, *P. putrefaciens* and a marine *Pseudomonas* sp. (Keenan *et al.*, 1967). At 21°C all these species reduced added acetaldehyde, propionaldehyde and butyraldehyde to the corresponding alcohols. No consistent relationship was observed between psychrotrophic contamination and the reduction of diacetyl in refrigerated buttermilk, but species of *Pseudomonas*, *Enterobacter*, *Acinetobacter*, *Escherichia* and yeasts isolated from this product reduced 25–75% of the diacetyl within 24 h at 21°C (Wang & Frank, 1981). Volatile sulphides in milk arise primarily as a result of thermal processing rather than microbial action (Schwimmer, 1981). Methanethiol contributes positively to the overall aroma of some surface-ripened cheeses, but it is responsible for putrid-like aroma in spoiled fish and meat (Lindsay & Rippe, 1986). Methioninase from *P. putida* has been partially characterised (Lindsay & Rippe, 1986).

INTERACTION OF ENZYMES

The rapid decrease in lipolytic activity observed in some cultures of *Pseudomonas fluorescens, Geotrichum candidum* and *Acinetobacter* (Fox & Stepaniak, 1983; Hassanien *et al.*, 1985; Christen *et al.*, 1986) may be explained by the hydrolysis of lipase by proteinase. However, incubation of purified lipase with purified proteinase, both from *P. fluorescens* P1, did not prove the assumption (Stepaniak *et al.*, 1987*b*; Vagias, 1988). Also, a mixture of lipase and proteinase produced by *P. fluorescens* GR 83 lost only 16% of its lipolytic activity on incubation; modification of lipase by limited proteolysis was suggested (Roussis *et al.*, 1988*a*). Subtilisin, however, quickly inactivated a lipase from *P. fluorescens* (Andersson, 1980).

Cousin (1989) has reviewed possible consequences of synergistic and antagonistic action of bacterial hydrolases, including the significance of *Pseudomonas* proteinases, phospolipases and glycosidases in enhancing lipolysis through damaging the fat globule membrane, and the responsibility of phospholipase C from *Bacillus cereus* for bittiness defects. Enhanced lipolysis during cheese ripening may also be influenced by the activities of phospholipases from psychrotrophs (Najim, 1985).

Removal of the carbohydrate moieties of κ-casein may cause loss of its micelle-stabilising property (Marin & Marshall, 1983; Hsien-Yeh-Hsu, 1984). The release of the carbohydrate moiety may also eliminate steric hindrance, facilitating the access of proteinases to the modified substrate (Marin & Marshall, 1983). Adverse effects of dephosphorylation of β-casein on curd firmness and increased rennet coagulation time have been shown (Jimenez-Flores & Richardson, 1988). A phosphoryl group in the hydrophilic region of β-casein is probably responsible for the curd tension (Yun *et al.*, 1982). Direct evidence on the influence of glycosidases or (protein) phosphatases from psychrotrophs is lacking. Variable results were obtained regarding both curd tension and coagulation time for milk pre-cultured with psychrotrophic *Pseudomonas* (Cousin, 1982; Fairbairn & Law, 1986*a*).

Addition of a partially purified proteinase preparation from *P. fluorescens* W 11 to a solution of bovine xanthine oxidase resulted in an immediate and marked increase in xanthine oxidase activity and an immediate decrease in xanthine dehydrogenase activity towards nitro-blue tetrazolium (NBT). Activation of xanthine oxidase was also possible with preparations containing heat-inactivated proteinase. Activation by a heat-stable, non-dialyzable activator present in the preparation was suggested. The decrease in NBT-dehydrogenase activity could be caused by a heat-labile inhibitor (Laohaviranit, 1982).

Gelation of UHT milk

Independent data have been obtained to demonstrate that gelation of UHT milk can occur either as a result of milk alkaline proteinase (plasmin) action and/or through hydrolysis by heat-stable proteinases secreted by psychrotrophic bacteria (Fox, 1981; Driessen, 1983; Suhren, 1983; Andrews, 1986; Grufferty & Fox, 1988; Hill, 1988; Mottar, 1989). The UHT process is insufficient to inactivate either type of enzyme (Snoeren *et al.*, 1979; Driessen, 1983). A variety of physico-chemical processes also occur during the long-term storage of UHT milk, which must be considered (Hill, 1988). The extent of proteolysis in UHT milk stored at 40°C was somewhat faster than in milks stored at 25 or 30°C (Kocak & Zadow, 1985), while at 2, 10, 37, 40 and 50°C gelation was retarded compared with the 20–30°C temperature range (Kocak & Zadow, 1985; Manji *et al.*, 1986). These observations support the significance of physico-chemical factors. Gelation induced merely by physico-chemical changes is likely to occur in concentrated UHT sterilised milk (de Koning *et al.*, 1985).

There is strong evidence that purified *Pseudomonas* proteinases added to milk before UHT sterilisation or the growth of natural or added psychrotrophic microorganisms to high numbers before UHT treatment may cause gelation after short storage periods, e.g. <6 weeks (Driessen, 1976; Law *et al.*, 1977; Mitchell & Ewings, 1985; Dousset *et al.*, 1988; Hill, 1988, Table 3). Adverse effect on stability at ambient temperatures can be expected upon addition of less than 1 ng/ml of heat-stable proteinase from *P. fluorescens* or *Serratia marcescens* (Richardson & Newstead, 1979; Mitchell & Ewings, 1985). Concentrations of 1 ng/ml are far below the lowest value reported for plasmin concentration in milk (Rollema *et al.*, 1983).

A convincing correlation, $r = 0{\cdot}86$, was found between gelation time and the amount of proteinase (six different) from *P. fluorescens* and (two different) from *S. marcescens*. However, no significant relationship was established between the extent of proteolysis, as measured by non-casein nitrogen, caused by the eight enzymes and gelation time (Mitchell & Ewings, 1985). Poor correlation between the extent of proteolysis and gelation of UHT milk was also demonstrated by other investigators (Kocak & Zadow, 1985; Hill, 1988).

Electron micrographs of UHT milk, gelled after addition of proteinase from *P. fluorescens* NCDO 2085, showed clumping of casein micelles into a network indistinguishable from that of milk gelled by chymosin (Law, 1982; Andrews, 1986). On the other hand, only 5–20% of the κ-casein was degraded in some gelled UHT milks, compared with 80–90% in rennet-coagulated milks (Mitchell & Ewings, 1985; Hill, 1988). Differences in the specificity of proteinases from psychrotrophs, referred to earlier, may partially explain the lack of correlation between proteolysis and time for onset of gelation.

Plasmin, despite its low activity in milk towards κ-casein (Grieve & Kitchen, 1985; Grufferty & Fox, 1988), was also demonstrated to promote gelation. Koning *et al.* (1985) found that after addition of selective plasmin inhibitors, gelation and proteolysis were not observed even after 9 months of storage at 20°C. Kohlmann *et al.* (1988) found that addition of plasmin at 0·3 or 1·5 mg protein/litre or trypsin at concentrations of 1·5 or 7·5 mg protein/litre did not cause gelation. They showed, however, that gelation was initiated by plasminogen-derived activity.

It is reasonable to conclude that because of the common practice of cold storage of raw milk and because of the natural wide variation of milk alkaline proteinase activity, the process of milk gelation is strongly influenced by the interaction of indigenous and exogenous proteinases. Limited proteolysis may be considered as an initiation step of the gelation process which develops over long periods (Hill, 1988). It will be difficult to positively identify the major factor(s) responsible for gelation unless specific and sensitive methods are developed to quantify particular enzymes involved. Immunological methods show promise in this respect.

Cheese ripening

Proteinases from *P. fluorescens* and *P. aeruginosa* have been used as milk coagulants for Cheedar cheese production. These cheeses developed a soft texture and a bitter and unclean flavour (Juffs, 1974).

Several reports, reviewed by Law (1979), Cousin (1982) and Fairbairn and Law (1986*a*), indicated that good quality cheese could be made from cold-stored milk. Law *et al.* (1976, 1979) produced Cheddar cheeses from milk pre-incubated with selected proteolytic strains of *Pseudomonas* spp. and *Acinetobacter* or with lipolytic strains of *P. fluorescens* and *P. fragi*. The proteolytic microflora that grew before pasteurisation to *c.* 10^7 cfu/ml caused no defects attributable to proteolysis (Law *et al.*, 1979). The lipolytic bacteria were responsible for the development of rancidity (Law *et al.*, 1976; Table 3). Mottar (1989) compiled reports showing that rancidity occurred also in Swiss-type, Teleme and Camembert cheeses manufactured from milk of high psychrotrophic counts.

Feta cheeses were manufactured from milk pre-cultured with *P. fluorescens* GR 83 to psychrotrophic populations of $8{\cdot}6 \times 10^7$ cfu/ml. Proteolysis and lipolysis in these cheeses were higher than in the controls, but no flavour or texture defects followed (Roussis *et al.*, 1988*b*). However, some reports discussed below and data in Table 3 indicate negative effects on taste or structure by the psychrotrophs or their enzymes when cheeses were made from milk of reasonable microbial quality.

In a study by White and Marshall (1973), *P. fluorescens* P26 was added ($4{\cdot}3 \times 10^4$ cfu/ml) to pasteurised milk and incubated for 12 h at 4°C before the

TABLE 3

Relationship Between Psychrotrophic Microflora in Cold Stored Milk and Quality of Some Dairy Products. Growth of Recontaminating Microflora was in all Cases Insignificant

Product	*Microflora*[a]	*Count* ($\times 10^5$ *cfu/ml*)	*Effect*	*Reference*
Pasteurised milk	NF	10–100	Bitter flavours	Patel & Blankenagel (1972)
	NF	3·0	Inferior flavour as compared with that made of fresh raw milk with count *c.* $0{\cdot}5 \times 10^5$ cfu/ml	Antila (1982)
	NF	3–100	Bitter flavour after 2 weeks at 7°C	Blankenagel (1982)
	NF[b]	0·05	Keeping quality 30 days	Kielwein (1982)
		3–11	Keeping quality 11–14 days	
		23–85	Keeping quality 3–11 days	
	Pschr	23	Reduced flavour scores after 7 days of storage	Marshall (1982)
UHT Milk	NF + Pf	8	Stable at least 20 weeks	Law *et al.* (1977)
		80	Unstable after 8–10 weeks	
	P	200	Rancidity expected after 1 week	Bucky *et al.* (1987*a*)
	NF	60	Rancidity expected after 70 days calculated from rate of free fatty acid formation	Bucky *et al.* (1988)
		36–60	Off-flavours due to proteolysis expected after 20–60 days calculated from rate of proteolysis	
	NF	<1	No defects >100 days	Dousset *et al.* (1988)
		8	Slight bitterness and proteolysis	
	NF + Pf	40	Unstable after 2 weeks	
Milk powder	NF	20	Some reduction of heat stability of reconstituted powders	Muir *et al.* (1986)
Freeze-dried milk	NF + Pf	100	Decrease in dispersibility, increase in foaming capacity	Burlingame-Frey & Marth (1984)
	NF + Lb	100	Increase in dispersibility, decrease in foaming capacity	

(*continued*)

TABLE 3—*contd.*

Product	*Microflora*[a]	*Count* (×10^5 *cfu/ml*)	*Effect*	*Reference*
Cheeses	Al	36–38	Increased lipolysis in Dutch cheeses	Driessen & Stadhouders (1971)
	NF + GNR	240–350	Rancidity after 4 months in Cheddar	Law *et al.* (1976)
	Pf	30–50	Rancidity after 6–8 months in Cheddar	Law (1979)
	NF	310–2500	Flavour defects after 4 months (mainly rancidity) in Cheddar	Cousin & Marth (1977*b*)
	NF,P	0·2	Inferior quality of Raclette cheeses	Glättli (1982)
Cottage cheese	NF or P	1–700	Significant correlation between cfu/ml in raw milk and bitter taste	Herdlevær (1988)
	P[c]	680	Unclean, fruity flavours, reduced yield	Cousin & Marth, (1977*c*)
	P[d]	190		
	Fl	450		
Yogurt	P[c]	680	No defects	Cousin & Marth, (1977*c*)
	P[d]	190	Bitter, unclean, fruity flavours	
	Fl	450		
Butter	NF	7	*C.* 50% higher level of FFA as compared with control milk with 2 × 10^4 psychrotrophs/ml	Antila (1982)

[a] NF = natural microflora, Pschr = unspecified psychrotrophs, Pf = *P. fluorescens*, P = *Pseudomonas* sp., Lb = *lactobacillus*, Al = *Alcaligenes*, GNR = Gram-negative rods, Fl = *Flavobacterium* sp.
[b] Raw milk was not cold stored.
[c,d] Different strains.

manufacture of Cheddar cheeses. Alternatively, proteinase from this microorganism was added at a concentration 0·94 units/ml (One unit of enzyme released the equivalent of 1 μg of tyrosine per ml skim milk at 21°C in 48 h). Increased proteolysis in both experimental cheeses was paralleled by reduced flavour scores due to bitterness and acid taste. Tye *et al.* (1988), however, successfully used the proteolytic, cell-free supernatant from *P. fluorescens* T25 as an agent to accelerate the ripening of Cheddar cheese. The preparation which was added to milk prior to renneting with porcine pepsin enhanced Cheddar flavour and neither off-flavours nor bitterness was observed.

Manufacture of Cheddar cheeses from grade A milk inoculated with 10^4 or 10^5 cfu/ml of *P. fluorescens* 27 or *P. fluorescens* 103 and stored at 7°C for 2 days before processing led to the development of a number of defects. Adverse effects were also demonstrated when these microorganisms were added at similar population densities to pasteurised milk just before cheese manufacture. Negative effects of coliforms inoculated at levels of 10^3 cfu/ml to the raw or pasteurised milk were also shown. The most frequent defect attributed to the two *P. fluorescens* strains was rancidity (Najim, 1985). Apart from rancidity, soapy, yeasty, butyric and fruity flavours have also been identified as a result of excessive lipolytic activity in cheeses due to bacterial enzymes (Stead, 1986).

Reduced yields were also observed from milks containing 2×10^5–1×10^6 psychrotrophs/ml (Ellis & Marth, 1984; Fairbairn & Law, 1986*a*; Mottar, 1989).

Rancidity in cheese as the most probable flavour defect due to enzymatic activity of psychrotrophs has some rational support. Experience with accelerated cheese ripening indicates that the right enzyme cocktail should be composed of proteinase and exopeptidases (Law, 1988). Addition of proteinases alone may result in an increase in the concentration of protein breakdown products, but at reasonable levels these will cause no serious consequence for taste but rather for the structure of the cheese. Addition of lipases as flavour promoters has been rather unsuccessful except in cheeses like Italian, Roquefort and Feta in which lipolytic flavour is appreciated (Law, 1988). It is unlikely that lipases from psychrotrophs will produce the balance of fatty acids required for flavour in Cheddar and Swiss-type cheeses. The negative role of lipases is also emphasised by the fact that, like milk alkaline proteinase, the major part of lipases from psychrotrophs may be transmitted to cheese curd (Stead, 1986; Grufferty & Fox, 1988). On the contrary, the proportion of *Pseudomonas* proteinases in milk that partitions into cheese curd is reported to be as low as that for chymosin (Åshammer, 1985; Fox *et al.*, 1989).

Considering the effect of pH on the activity of enzymes participating in proteolysis, the low pH of many cheese varieties favours proteinases from psychrotrophs and milk acid proteinase rather than milk alkaline proteinase (Fox, 1981; Hsien-Yeh-Hsu, 1984; Fox *et al.*, 1989).

PHYSICO-CHEMICAL FACTORS AFFECTING PROTEOLYSIS AND LIPOLYSIS IN MILK

Micellar status of casein

pH influences colloidal calcium phosphate and therefore the integrity of casein micelles and their susceptibility to proteolysis. Therefore, small variations in pH may influence proteolysis in milk to a greater extent than that of non-micellar casein substrates (Hsien-Yeh-Hsu, 1984; Hsien-Yeh-Hsu & Shipe, 1986). The small natural variation in the pH of bulk milk was thus correlated with total proteolytic activity (Humbert *et al.*, 1983).

Proteolysis by indigenous milk proteinases increased in calcium-free or colloidal calcium phosphate-free milks. The effect of calcium removal was greater for skim milk than for whole milk (Hsien-Yeh-Hsu, 1984).

Proteolysis in skim milk and whole milk

Janzen *et al.* (1982*a*) found higher proteinase activity at 4·5 or 7°C in skim milk than in whole milk with a similar psychrotrophic count. Consequently, the shelf life of the skim milk was significantly shorter. Hsien-Yeh-Hsu (1984) measured higher proteolytic activity of both indigenous milk proteinase and *Pseudomonas* proteinase in skim milk than in whole milk; more γ-casein was formed during a 1–4 day incubation period at 5°C in skim milk than in whole milk. Hydrophobic association of β- and κ-caseins to fat globules was suggested to explain the reduced proteolysis in whole milk (Hsien-Yeh-Hsu, 1984). Contrary to these findings, Brown *et al.* (1984) found no differences between the shelf lives of commercial pasteurised skim milks and milks containing fat. McKellar (1981) reported that crude proteinases from *P. fluorescens* were twice as active in UHT sterilised milk containing 2% fat than in laboratory pasteurised skim milk. This indicates that the conditions of thermal processing may be a more important factor than fat content.

Effect of polysaccharides, surfactants and other additives

Anionically charged carrageenans, which are commonly used as stabilisers, are effective inhibitors of both indigenous milk lipoprotein lipase and *P. fluorescens* lipases. Inhibition was not selective (Stern *et al.*, 1988; Stepaniak & Sørhaug, 1989). Most (93·7%) of milk lipoprotein lipase and 87·6% of *P. fluorescens* lipase was inhibited in whole milk by λ-carrageenan which was the most effective of a number of gums tested (Stern *et al.*, 1988). These gums were less effective inhibitors of proteolysis in raw whole milk (Hsien-Yeh-Hsu, 1984; Hsien-Yeh-Hsu & Shipe, 1986), but λ-carrageenan was again the most potent inhibitor. Involvement of the carrageenans in the stabilisation of casein micelles has been suggested (Hsien-Yeh-Hsu, 1984; Hsien-Yeh-Hsu & Shipe, 1986).

Treatment with oil-soluble surfactants, such as Tween 20, increased proteolysis in raw whole milk. Release of milk fat globule membrane-bound proteinase, proteinase associated with casein micelles, as well as some denaturation of proteins and dissociation of micelles caused by surfactants were proposed as factors responsible for stimulation (Hsien-Yeh-Hsu, 1984; Hsien-Yeh-Hsu & Shipe, 1986).

Addition of sucrose to 15% inhibited proteolysis by enzymes of yogurt microflora (Slocum *et al.*, 1988). The extent of inhibition was dependent on the type of yogurt culture. The stability at 58°C of rulactine, a proteinase from *Micrococcus caseolyticus,* was increased by the addition of polyhydric alcohols and disaccharides (Yoovidhya *et al.*, 1986).

Further research on food-grade enzyme inhibitors and their use in foods, as suggested by Schwimmer (1981), should be conducted.

Physical treatments

Activation of milk lipoprotein lipase by physical treatment is an acknowledged phenomenon (Deeth & Fitz-Gerald, 1983). Activation of proteolysis in raw milk by agitation in a Waring blender has been demonstrated (Hsien-Yeh-Hsu, 1984; Hsien-Yeh-Hsu & Shipe, 1986). Proteolysis is stimulated also by freezing and thawing. Homogenisation, however, had an inhibitory effect on proteolysis (Hsien-Yeh-Hsu, 1984).

Enzyme activity in extreme environments

Activity in partially frozen systems is common for enzymes of both microbial and animal origin. Compared with lipoxygenases and lipases, proteinases are generally less active in such systems (Fennema, 1975). Water activity thresholds are also lower for lipases and phospholipases than for proteinases (Drapron, 1985).

Measurable activity was found for lipase from *P. fluorescens* in milk powder of $a_w \sim 0{\cdot}2$. The enzyme-catalysed lipolysis in an olive oil emulsion continued at −30°C during the first 3 months of storage (Andersson, 1980). Activity at sub-freezing temperatures is common for *Pseudomonas* and for many other microbial lipases (Alford & Pierce, 1961; Stead, 1986). Lipolysis in butter with added *Pseudomonas* spp. lipase proceeded at −10°C (Stead, 1986).

The activation energy for the hydrolysis of casein by the proteinase from *P. fluorescens* T16 was 13·4 kJ/mol (Patel *et al.*, 1983*b*), and those for the hydrolysis of azocasein by four proteinases from *P. fluorescens* strains were in the range 18·8–58·6 kJ/mol (Mitchell & Marshall, 1989*b*). The lipase from *P. fluorescens* MC50 had an activation energy of 4·9 kJ/mol for hydrolysis of tributyrin (Bozoğlu *et al.*, 1984). The activation energies for glycosidases from *P. fluorescens* 26 for hydrolysis of four PNP-glycosides were in the range 25·1–106·4 kJ/mol (Marin & Marshall, 1983). Most of these values are low, compared to 62·4 kJ/mol obtained for the proteinase from *Lactobacillus murinus* (Strasser de Saad *et al.*, 1988), 37·5 kJ/mol for rulactase from *M. caseolyticus* (Yoovidhya *et al.*, 1986), and 50·2 kJ/mol for trypsin (Fairbairn & Law, 1986*a*). The results suggest that *Pseudomonas* hydrolases are well adapted for the psychrotrophic temperature range.

Proteinase from *Alcaligenes* retained *c.* 15% of its maximal activity at −5°C (Zaborniak & Knaut, 1984). Proteolytic enzymes from *Pseudomonas* spp. with temperature optima at or above 55°C have rarely been reported (Table 1). Some residual activity of phospholipase C from *P. fluorescens* P1 was observed at 65°C (Stepaniak *et al.*, 1987*b*). A temperature optimum at 70°C for lipase from *P. mephitica* var. *lipolytica* has been reported (Kosugi & Kamibayashi, 1971).

Hydrolysis of α-, β- and κ-caseins by an industrial preparation of *B. subtilis* neutral proteinase was inhibited up to 20% by 10% sodium chloride, while hydrolysis of whey proteins was inhibited by *c.* 50% (El Mayda *et al.*, 1985). A halophilic metalloproteinase from a marine *Pseudomonas* sp. with maximal activity at 18% sodium chloride has been isolated (van Qua *et al.*, 1981). Hydrolysis of β-casein by chymosin is inhibited strongly by 10% sodium chloride (Fox & Walley, 1971). The inhibitory effect was interpreted as arising from changes in substrate conformation and modification of the enzyme molecule. Degradation of casein and soy protein by thermolysin was not affected by pressure. Degradation of β-lactoglobulin occurred only at a high pressure (2000 atm), while α-lactalbumin was resistant to proteolysis at both normal and high pressure conditions (Hayashi *et al.*, 1987).

Development on storage of 'gluey' and stale off-flavour in dry casein preparations was related to non-enzymatic browning reaction. Proteolysis was not implicated (Ramshaw & Dunstone, 1969).

EFFECT OF MICROBIAL HEAT-STABLE ENZYMES ON THE SHELF LIFE OF DAIRY PRODUCTS

Data in Table 3 show the relationships between the microbial quality of raw milk and the quality of dairy products. Post heat-treatment contamination was minimized. Thus, heat-stable enzymes from the raw milk and from milk residues in poorly-washed equipment and, more exceptionally, spores, will account for quality deterioration during storage of the products.

Antila (1982) made the point that today all dairy products are made from stored milk, and therefore there are very few references to commercial products manufactured from completely fresh milk. Pasteurised milk made experimentally from fresh milk had superior keeping quality to milk made from raw cold-stored milk of good microbial quality (Antila, 1982). Janzen *et al.*, (1982*b*) have formulated a predictive equation to calculate flavour scores and shelf life of laboratory pasteurised milk from the age of raw milk stored at 4·5°C before pasteurisation.

Lack of a significant correlation between time for gelation of UHT milk and pre-heat treatment count of psychrotrophic proteolytic microorganisms in the range 7×10^3–$3{\cdot}2 \times 10^7$ cfu/ml was reported by Kocak and Zadow (1985). Mottar (1984) found, however, a significant impact of initial microbial quality and storage time of raw milk on the development of off-flavours and shelf life of UHT milk.

Non-enzymatic oxidative changes may, however, prevail as flavour defects in pasteurised milk (Allen & Joseph, 1985). Pre-processing lipolysis by indigenous enzymes and post-pasteurisation non-microbial enzyme activity also contribute to quality deterioration (Allen & Joseph, 1985). Nomenclature for off-flavours of microbial and non-microbial origin in milk has been compiled (Shipe *et al.*, 1978).

DEFECTS RELATED TO POST-PASTEURISATION BACTERIAL GROWTH

Milk and cream

It has been estimated that 20–25% of the problems associated with the shelf life of pasteurised milk may be caused by the thermoduric microflora (Cousin, 1982). Levels of post-pasteurisation contamination with psychrotrophic Gram-negative rods varied from <1/100 ml to 10^3 cfu/ml (Schröder, 1984; Griffiths & Phillips, 1988*a,b*). Recontamination with Gram-negative rods is considered the most serious cause of poor keeping quality because these bacteria grow fast at low temperatures (Schröder, 1984; Griffiths & Phillips, 1988*a,b*). A shorter shelf life of pasteurised milk was observed during the summer–autumn period than in winter–spring. This was explained by a combination of factors: a higher incidence of recontamination with psychrotrophic spores during the former period and faster germination and outgrowth of recontaminating spores compared with those originating from raw milk (Phillips & Griffiths, 1986). At 5·6°C, pasteurised milk was spoiled by both proteolytic and acid-producing microbes, but at 1·7°C only by proteolytic psychrotrophs (Hankin *et al.*, 1977). Psychrotrophic *Bacillus circulans* has the capability to ferment lactose (Cousin, 1982). By raising the storage temperature from 5 to 11°C, the average shelf life of commercial and laboratory pasteurised milk was reduced from 13 to 5 days and 28 to 6 days, respectively (Schröder *et al.*, 1982). Bittiness, and fruity and bitter flavours, were found in milk stored at 5 or 11°C, most frequently accompanied by psychrotrophic counts around 1×10^6 cfu/ml (Schröder *et al.*, 1982). Off-flavours and bitterness encountered in milks at low counts, $<10^5$ cfu/ml (Blankenagel, 1982; Janzen *et al.*, 1982*a,b*; Marshall, 1982), indicate participation of heat-stable microbial enzymes originating from raw milk. The occurrence of rancidity in commercially pasteurised milks, concomitant with increased free fatty acid levels, has also been reported. Bitterness can mask other flavours (Blankenagel, 1982; Baker, 1983).

Phillips *et al.* (1981*a*) estimated that over 80% of the spoilage of cream is caused mainly by post-pasteurisation contamination. Effective cooling is stressed to avoid spoilage by thermoduric psychrotrophs (Phillips *et al.*, 1981*a*). Spoilage of cream due to proteolysis or both lipolysis and proteolysis is common; spoilage due to acid production is also fairly common (Griffiths *et al.*, 1981*a*). Bitterness was associated with the growth of *Bacillus* spp. (Griffiths *et al.*, 1981*a*; Phillips *et al.*, 1981*b*). Flavour defects in cream occurred at 6 or 10°C when post-pasteurisation growth exceeded 1×10^7 cfu/ml (Griffiths *et al.*, 1981*a*); however, defects have also been observed at lower populations (Cousin, 1982; Stead, 1986).

Cheeses

High numbers of non-starter *Lactobacillus casei* were associated with a soft (pasty) body in Mozzarella cheese; the defect could not be explained by comparing levels of TCA-soluble peptides with those in normal cheeses (Hull *et al.*, 1983). Heterofermentative lactobacilli were responsible for open structure in Canadian Cheddar and Oka cheeses (Laleye *et al.*, 1987). Fermented flavour in Cheddar cheeses developed after 6 months of maturation; elevated amounts of ethanol, ethyl acetate and ethyl butyrate occurred in cheeses which contained high numbers of *Candida* spp. (Horwood *et al.*, 1987).

Cottage Cheeses

In cottage cheese, the most common defects originating from psychrotrophic growth are slime formation, off-odours and off-flavours. Bitter and fruity flavours are also quite common (Cousin, 1982; Bishop & White, 1985*b*; Brocklehurst & Lund, 1985; Herdlevær, 1988). *Pseudomonas* spp., as well as yeasts and enterobacteria, were responsible for spoilage at 7°C (Brocklehurst & Lund, 1985). A very short shelf life was observed for cheeses with Gram-negative counts above 10^6 cfu/ml (Bishop & White, 1985*b*; Brocklehurst & Lund, 1985).

Butter

Post-pasteurisation contamination and growth of psychrotrophic yeasts, and more commonly of *Pseudomonas,* cause hydrolytic rancidity and breakdown of proteins responsible for cheesy or putrid flavours or surface taints. The latter defect is usually attributed to *P. putrefaciens* (Thomas & Druce, 1971; Cousin, 1982). Partial reduction of diacetyl may occur due to the activity of diacetyl reductase produced by some pseudomonads (Cousin, 1982).

Yogurt

Moulds and yeasts were isolated in high numbers from yogurt, especially fruit yogurt (García & Fernández, 1984). Driessen *et al.* (1983) found *B. cereus* responsible for the spoilage of yogurt fermented at 31°C. *B. cereus* could grow down to pH 5·7, especially in aerated milk. Populations of *B. cereus* of *c.* 1×10^6 cfu/ml caused off-flavours, lumpy sructure and separation of whey in yogurt after overnight incubation.

FINAL REMARKS

Hydrolases from many genera of psychrotrophic bacteria, especially *Pseudomonas* spp., are recognised as the major group of microbial enzymes involved in the spoilage of dairy products. These enzymes are generally heat-stable and express their activities over wide ranges of pH and temperature. The proteinases are well characterised whereas knowledge of phospholipases and glycosidases vary considerably. References in this field, many in recent reviews (Fairbairn & Law, 1986*a*; Stead, 1986; Cousin, 1989; Fox *et al.*, 1989; McKellar, 1989; Stepaniak & Sørhaug, 1989), are numerous.

Investigations on the impact of these enzymes on the quality of dairy products are more limited. The lack of sensitive and specific enzyme assays reduces the precision of such studies. The role of bacterial enzymes is also difficult to elucidate because they regularly contribute in addition to the activities of indigenous enzymes or in concert with the activities of starter bacteria. Moreover, due to specificity divergence within microbial enzymes, no simple relationships are encountered between taste or structural changes and proteolysis or lipolysis.

Awareness is growing that total bacterial or psychrotrophic counts are poor predictive indices of storage stability. With the increasing demand for longer shelf life, the role of enzymatic spoilage without parallel bacterial growth is increasing.

ACKNOWLEDGEMENTS

Support from the Norwegian Dairies Association and the Agricultural Research Council of Norway to our related research greatly stimulated the writing of this chapter.

REFERENCES

Adams, D. M. & Brawley, T. G. (1981). *J. Food Sci.*, **46,** 677–80.
Adler-Nissen, J. (1986). *Enzymic Hydrolysis of Food Proteins.* Elsevier Applied Science Publishers, London, UK.
Aishima, T., Yada, R. Y., Smith, A. M. & Nakai, S. (1987). *J. Food Biochem.*, **11,** 121–32.
Akuzawa, R., Yoshida, S. & Yokoyama, K. (1986). *Bull. Nippon Veter. Zootech. Coll.*, **35,** 202–6.
Alford, J. A. & Pierce, D. A. (1961). *J. Food. Sci.*, **26,** 518–24.
Alichanidis, E. (1988). *J. Dairy Res.*, **55,** 97–107.
Alkanhal, H. A., Frank, J. F. & Christen, G. L. (1985). *J. Dairy Sci.*, **68,** 3162–70.
Allen, J. C. & Joseph, G. (1985). *J. Dairy Res.*, **52,** 469–87.
Al-Shabibi, M. M. A., Langner, E. H., Tobias, J. & Tuckey, S. L. (1964). *J. Dairy Sci.*, **47,** 295–6.
Alur, M. D., Nerkar, D. P. & Venugopal, V. (1988). *J. Food Sci.*, **53,** 243–7.
Andersson, R. E. (1980). *Appl. Environ. Microbiol.*, **39,** 36–40.
Andersson, R. E., Danielsson, G., Hedlund, C. B. & Svensson, S. G. (1981). *J. Dairy Sci.*, **64,** 375–9.
Andrews, A. T. (1986). Report 1985–86, Food Research Institute, Reading, UK, pp. 85–94.
Antila, V. (1982). *Kieler Milchwirtsch. Forsch.*, **34,** 170–3.
Åshammer, L. (1985). Thesis, Agricultural University of Norway, Ås, Norway.
Aunstrup, K. (1980). In *Microbial Enzymes and Bioconversions, Economic Microbiology* (Vol. 5), ed. A. H. Rose. Academic Press, London, UK, pp. 50–114.
Azcona, J. I., Martin, R., Asensio, M. A., Hernández, P. E. & Sanz, B. (1988). *J. Dairy Res.*, **55,** 217–26.
Baker, S. K. (1983). *Aust. J. Dairy Technol.*, **38,** 124–7.

Barach, J. T. & Adams, D. M. (1977). *Biochim. Biophys. Acta,* **485,** 417–23.
Barach, J. T., Adams, D. M. & Speck, M. L. (1976). *J. Dairy Sci.,* **59,** 391–5.
Barach, J. T., Adams, D. M. & Speck, M. L. (1978). *J. Dairy Sci.,* **61,** 523–8.
Bengtsson, G. & Olivecrona, T. (1980). *Eur. J. Biochem.,* **106,** 557–62.
Bergère, J.-L. & Accolas, J.-P. (1986). In *Anaerobic Bacteria in Habitats Other than Man,* ed. E. M. Barnes & G. C. Mead. Blackwell Scientific Publications, Oxford, UK, pp. 373–96.
Berka, R. M. & Vasil, M. L. (1982). *J. Bacteriol.,* **152,** 239–45.
Bhowmik, T. & Marth, E. H. (1988). *J. Dairy Sci.,* **71,** 2358–65.
Birkeland, S.-E., Stepaniak, L. & Sørhaug, T. (1985). *Appl. Environ. Microbiol.,* **49,** 382–7.
Bishop, J. R. & White, C. H. (1985*a*). *J. Food Prot.,* **48,** 663–7.
Bishop, J. R. & White, C. H. (1985*b*). *J. Food Prot.,* **48,** 1054–7.
Bishop, J. R. & White, C. H. (1986). *J. Food Prot.,* **49,** 739–53.
Björck, L. (1982). *Kieler Milchwirtsch. Forsch.,* **34,** 5–11.
Blankenagel, G. (1982). *Kieler Milchwirtsch. Forsch.,* **34,**158–62.
Boguslawski, G., Shultz, J. L. & Yehle, C. O. (1983). *Anal. Biochem.,* **132,** 41–9.
Bosch, H. van den (1982). In *Phospholipids,* ed. J. N. Hawthorne & G. B. Ansell. Elsevier Biomedical Press, Amsterdam, The Netherlands, pp. 313–57.
Bozoğlu, F., Swaisgood, H. E. & Adams, D. M. Jr (1984). *J. Agric. Food Chem.,* **32,** 2–6.
Brocklehurst, T. F. & Lund, B. M. (1985). *Food Microbiol.,* **2,** 207–33.
Brown, J. V., Ranjith, H. M. P. & Prentice, G. A. (1984). *J. Soc. Dairy Technol.,* **37,** 132–5.
Bucky, A. R., Robinson, D. S. & Hayes, P. R. (1987*a*). *Food Chem.,* **23,** 159–73.
Bucky, A. R., Hayes, P. R. & Robinson, D. S. (1987*b*). *J. Dairy Res.,* **54,** 275–82.
Bucky, A. R., Robinson, D. S. & Hayes, P. R. (1987*c*). *Int. J. Food Sci. Technol.,* **22,** 35–40.
Bucky, A. R., Hayes, P. R. & Robinson, D. S. (1988). *J. Dairy Res.,* **55,** 373–80.
Burlingame-Frey, J. P. & Marth, E. H. (1984). *J. Food Prot.,* **47,** 288–92.
Chandan, R. C., Argyle, P. J. & Mathison, G. E. (1982). *J. Dairy Sci.,* **65,** 1408–13.
Christen, G. L. & Marshall, R. T. (1985). *J. Dairy Sci.,* **68,** 594–604.
Christen, G. L., Wang, W.-C. & Ren, T. J. (1986). *J. Dairy Sci.,* **69,** 2769–78.
Coghill, D. (1982). *Aust. J. Dairy Technol.,* **37,** 147–8.
Cousin, M. A. (1982). *J. Food Prot.,* **45,** 172–207.
Cousin, M. A. (1989). In *Enzymes of Psychrotrophs in Raw Food,* ed. R. C. McKellar. CRC Press, Boca Raton, FL, USA, pp. 205–25.
Cousin, M. A. & Marth, E. H. (1977*a*). *Milchwissenschaft,* **32,** 337–41.
Cousin, M. A. & Marth, E. H. (1977*c*) *Cultured Dairy Products J.,* **12**(1), 15–18, 30.
Cousin, M. A. & Marth, E. H. (1977*b*). *J. Dairy Sci.,* **60,** 1048–56.
Cousins, C. M., Sharpe, E. M. & Law, B. A. (1977). *Dairy Ind. Int.,* **42**(7), 12–17.
Cox, J. M. & Mac Rae, I. C. (1989). *J. Appl. Bacteriol.,* **66,** 137–52.
Dainty, R. H. (1982). *Food Chem.,* **9,** 103–13.
Deeth, H. C. & Fitz-Gerald, C. H. (1983). In *Developments in Dairy Chemistry, 2—Lipids,* ed. P. F. Fox. Applied Science Publishers, London, UK, pp. 195–239.
Deeth, H. C., Fitz-Gerald, C. H. & Wood, A. F. (1979). *Aust. J. Dairy Technol.,* **34,** 146–9.
Diermayr, P. (1986). PhD thesis, Technischen Universität, Munich, Germany.
Diermayr, P. & Klostermeyer, H. (1984). *Hoppe-Seyler's Z. Physiol. Chem.,* **365,** 1345–50.
Diermayr, P., Kroll, S. & Klostermeyer, H. (1987). *J. Dairy Res.,* **54,** 51–60.
Doi, O. & Nojima, S. (1971). *Biochim. Biophys. Acta,* **248,** 234–44.
Dousset, X., Demaimay, M., Ravaud, C., Levesque, A., Pinet, X. & Kergo, Y. (1988). *Le Lait,* **68,** 143–56.

Downey, W. K. (1980). *J. Dairy Res.*, **47**, 237–52.
Drapron, R. (1985). In *NATO Advanced Science Institutes*, (Series E No. 90), ed. D. Simatos & J. L. Multon. Martinus Nijhoff Publishers, Dordrecht, The Netherlands, pp. 171–90.
Drew, P. G. & Manners, J. G. (1985). *Aust. J. Dairy Technol.*, **40**, 108–12.
Driessen, F. M. (1976). *Zuivelzicht*, **68**, 514–15.
Driessen, F. M. (1983). PhD thesis, Agricultural University, Wageningen, The Netherlands.
Driessen, F. M. & Stadhouders, J. (1971). *Neth. Milk Dairy J.*, **25**, 141–4.
Driessen, F. M., Kingma, F. & Stadhouders, J. (1983). *Zuivelzicht*, **75**, 499–502.
Dring, R. & Fox, P. F. (1983). *Ir. J. Food Sci. Technol.*, **7**, 157–71.
Ellis, B. & Marth, E. H. (1984). *J. Food Prot.*, **47**, 713–16.
El Mayda, E., Pâquet, D. & Ramet, J. P. (1985). *J. Food Sci.*, **50**, 1745–6.
El Mayda, E., Pâquet, D., Ramet, J. P. & Linden, G. (1986). *J. Dairy Sci.*, **69**, 305–10.
Engel, G., Einhoff, K., Prokopek, D. & Teuber, M. (1987). *Milchwissenschaft*, **42**, 428–30.
Ergün, Ö. (1985). *Dairy Sci. Abstr.*, **47**, 393.
Eskin, N. A. M., Henderson, H. M. & Townsend, R. J. (1971). *Biochemistry of Foods*, Academic Press, New York, U.S.A.
Ewings, K. N., O'Connor, R. E. & Mitchell, G. E. (1984). *Aust. J. Dairy Technol.*, **39**, 65–8.
Fain, A. R. Jr (1982). PhD thesis, University of Georgia, Athens, GA, USA.
Fairbairn, D. J. (1989). In *Enzymes of Psychrotrophs in Raw Food*, ed. R. C. McKellar. CRC Press, Boca Raton, FL, USA, pp. 189–204.
Fairbairn, D. J. & Law, B. A. (1986*a*). *J. Dairy Res.*, **53**, 139–77.
Fairbairn, D. & Law, B. A. (1986*b*). *J. Dairy Res.*, **53**, 457–66.
Fennema, O. (1975). In *Water Relations of Foods.*, ed. R. B. Duckworth. Academic Press, London, UK, pp. 397–413.
Fitz-Gerald, C. H. & Deeth, H. C. (1983). *Aust. J. Dairy Technol.*, **38**, 97–103.
Fleet, G. H. & Mian, M. A. (1987). *Int. J. Food Microbiol.*, **4**, 148–55.
Florin-Christensen, J., Florin-Christensen, M., Rasmussen, L. & Knudsen, J. (1986). *Comp. Biochem. Physiol.*, **85B**, 143–8.
Flückiger, E. (1981). *Kieler Milchwirtsch. Forsch.*, **33**, 347–56.
Foster, E., Nelson, F. E., Speck, M. L., Doetsch, R. N. & Olson, J. C. Jr. (1957). *Dairy Microbiology*, Prentice-Hall, Englewood Cliffs, New Jersey, USA.
Fox, P. F. (1981). *Neth. Milk Dairy J.*, **35**, 233–53.
Fox, P. F. & Stepaniak, L. (1983). *J. Dairy Res.*, **50**, 77–89.
Fox, P. F. & Walley, B. F. (1971). *J. Dairy Res.*, **38**, 165–70.
Fox, C. W., Chrisope, G. L. & Marshall, R. T. (1976). *J. Dairy Sci.*, **59**, 1857–64.
Fox, P. F., Power, P. & Cogan, T. M. (1989). In *Enzymes of Psychrotrophs in Raw Food*, ed. R. C. McKellar. CRC Press, Boca Raton, FL, USA, pp. 57–120.
Gallmann, P. & Puhan, Z. (1982). *Milchwissenschaft*, **37**, 396–400.
García, M. A. & Fernández, S. G. (1984). *J. Food Prot.*, **47**, 629–36.
Gebre-Egziabher, A., Humbert, E. S. & Blankenagel, G. (1980). *J. Food Prot.*, **43**, 709–12.
Glättli, H. (1982). *Kieler Milchwirtsch. Forsch.*, **34**, 164–70.
Gray, R. J. H. & Sørhaug, T. (1983). In *Food Microbiology, Economic Microbiology* (Vol. 8), ed. A. H. Rose. Academic Press, London, UK, pp. 1–45.
Greer, G. G. (1989). In *Enzymes of Psychrotrophs in Raw Food*, ed. R. C. McKellar. CRC Press, Boca Raton, FL, USA, pp. 267–92.
Grieve, P. A. & Kitchen, B. J. (1985). *J. Dairy Res.*, **52**, 101–12.
Griffiths, M. W. (1983). *J. Food Technol.*, **18**, 495–505.
Griffiths, M. W. & Phillips, J. D. (1984). *J. Food Prot.*, **47**, 697–702.

Griffiths, M. W. & Phillips, J. D. (1988*a*). *J. Appl. Bacteriol.*, **65,** 269–78.
Griffiths, M. W. & Phillips, J. D. (1988*b*). *J. Soc. Dairy Technol.*, **41,** 97–102.
Griffiths, M. W., Phillips, J. D. & Muir, D. D. (1981*a*). *J. Soc. Dairy Technol.*, **34,** 142–6.
Griffiths, M. W., Phillips, J. D. & Muir, D. D. (1981*b*). *J. Appl. Bacteriol.*, **50,** 289–303.
Grufferty, M. B. & Fox, P. F. (1988). *J. Dairy Res.*, **55,** 609–30.
Guamis, B., Huerta, T. & Garay, E. (1987*a*). *Milchwissenschaft,* **42,** 651–3.
Guamis, B., Huerta, T. & Garay, E. (1987*b*). *Milchwissenschaft,* **42,** 89–91.
Hamel, U., Kielwein, G. & Teschemacher, H. (1985). *J. Dairy Res.*, **52,** 139–48.
Hankin, L. & Shields, D. (1983). *J. Food Prot.*, **46,** 873–7.
Hankin, L., Dillman, W. F. & Stephens, G. R. (1977). *J. Food Prot.*, **40,** 848–53.
Harwalkar, V. R., Boutin-Muma, B., Cholette, H., McKellar, R. C., Emmons, D. B. & Klassen, G. (1989). *J. Dairy Res.*, **56,** 367–73.
Hassanien, F. R., Shehata, M. I. & Ragab, M. (1985). *Die Nährung,* **29,** 269–74.
Hayashi, R., Kawamura, Y. & Kunugi, S. (1987). *J. Food Sci.*, **52,** 1107–8.
Herdlevær, M. (1988). Thesis, Agriculture University of Norway, Ås, Norway.
Hill, A. R. (1988). *Food Technol.*, **41,** 92–7.
Holmquist, B. (1977). *Biochem.*, **16,** 4591–4.
Horwood, J. F., Stark, W. & Hull, R. R. (1987). *Aust. J. Dairy Technol.*, **42,** 25–6.
Hough, E., Hansen, L. K., Birknes, B., Jynge, K., Hordvik, A., Little, C., Dodson, E. J. & Derewenda, Z. (1988). In *Proceedings of the 3rd Nordic Symposium on Gene Technology,* Hardanger, 15–19 June, Laboratory of Biotechnology, University of Bergen, Norway, p. 115.
Hsien-Yeh-Hsu (1984). PhD thesis, Cornell University, Ithaca, USA.
Hsien-Yeh-Hsu & Shipe, W. F. (1986). *J. Dairy Sci.*, **69,** 1491–7.
Hubble, I. B. & Manners, J. G. (1985). *Aust. J. Dairy Technol.*, **40,** 143–51.
Hull, R. R., Roberts, A. W. & Mayes, J. J. (1983). *Aust. J. Dairy Technol.*, **38,** 78–80.
Humbert, G., Brun-Bellut, J., Mora, T. & Alais, C. (1983). *J. Dairy Sci.*, **66,** 2271–7.
Jackman, D. M., Patel, T. R. & Haard, N. F. (1985). *J. Food Sci.*, **50,** 602–4, 609.
Janzen, J. J., Bishop, J. R. & Bodine, A. B. (1982*a*). *J. Dairy Sci.*, **65,** 2237–40.
Janzen, J. J., Bishop, J. R., Bodine, A. B. & Caldwell, C. A. (1982*b*). *J. Dairy Sci.*, **65,** 2233–6.
Jeong, D. K. & Frank, J. F. (1988). *J. Food Prot.*, **51,** 643–7.
Jimenez-Flores, R. & Richardson, T. (1988). *J. Dairy Sci.*, **71,** 2640–54.
Johnston, D. W. & Bruce, J. (1982). *J. Appl. Bacteriol.*, **52,** 333–7.
Johnston, D. E., Murphy, R. J., Gilmour, A., McGuiggan, J. T. M. & Mullan, W. M. A. (1988). *Milchwissenschaft,* **43,** 211–15.
Juffs, H. S. (1974). *Aust. J. Dairy Technol.*, **29,** 74–8.
Juven, B. J., Gordin, S., Rosenthal, I. & Laufer, A. (1981). *J. Dairy Sci.*, **64,** 1781–4.
Kahn, P. & Firstenberg-Eden, R. (1987). *J. Dairy Sci.*, **70,** 1544–70.
Kalogridou-Vassiliadou, D. (1984). *Milchwissenschaft,* **39,** 601–3.
Kalogridou-Vassiliadou, D. & Alichanidis, E. (1984). *J. Dairy Res.*, **51,** 629–36.
Kang, Y.-J. & Frank, J. F. (1988). *J. Dairy Sci.*, **71,** 1484–91.
Keenan, T. W., Bills, D. D. & Lindsay, R. C. (1967). *Appl. Microbiol.*, **15,** 1216–18.
Keogh, B. P. & Pettingill, G. (1984). *J. Food Prot.*, **47,** 105–7.
Kielwein, G. (1982). *Kieler Milchwirtsch. Forsch.*, **34,** 174–7.
Knight, A. H. & Fryer, S. M. (1989). *J. Soc. Dairy Technol.*, **42,** 81–6.
Kocak, H. R. & Zadow, J. G. (1985). *Aust. J. Dairy Technol.*, **40,** 14–21.
Kohlmann, K. L., Nielsen, S. S. & Ladisch, M. R. (1988). *J. Dairy Sci.*, **71,** 1728–39.
Koning, P. J. de, Kaper, J., Rollema, H. S. & Driessen, F. M. (1985). *Neth. Milk Dairy J.*, **39,** 71–87.
Kosikowski, F. V. (1988). *J. Dairy Sci.*, **71,** 557–73.
Kosugi, Y. & Kamibayashi, A. (1971). *J. Ferment. Technol.*, **49,** 968–80.

Kroll, S. (1989). In *Enzymes of Psychrotrophs in Raw Food,* ed. R. C. McKellar. CRC Press, Boca Raton, FL, USA, pp. 121–52.
Kroll, S. & Klostermeyer, H. (1984*a*). *Z. Lebensm. Unters. Forsch.,* **178,** 9–14.
Kroll, S. & Klostermeyer, H. (1984*b*). *Z. Lebensm. Unters. Forsch.,* **179,** 288–95.
Kroll, S., Kellerer, B., Busse, M. & Klostermeyer, H. (1984). *Milchwissenschaft,* **39,** 538–40.
Kugimiya, W., Otani, Y., Hashimoto, Y. & Takagi, Y. (1986). *Biochem. Biophys. Res. Commun.,* **141,** 185–90.
Kurioka, S. & Matsuda, M. (1976). *Anal. Biochem.,* **75,** 281–9.
Kwan, K. K. & Skura, B. J. (1985). *J. Dairy Sci.,* **68,** 1902–9.
Laleye, L. C., Simard, R. E., Lee, B.-H., Holley, R. A. & Giroux, R. N. (1987). *J. Food Prot.,* **50,** 1009–12.
Langeveld, L. P. M. & Cuperus, F. (1980). *Neth. Milk Dairy, J.,* **34,** 106–25.
Laohaviranit, L. (1982). PhD thesis, University of Missouri, Columbia, MO, USA.
Law, B. A. (1979). *J. Dairy Res.,* **46,** 573–88.
Law, B. A. (1980). In *Microorganisms and Nitrogen Sources,* ed. J. W. Payne. J. Wiley & Sons, New York, pp. 381–409, 783–6.
Law, B. A. (1982). In *Food Proteins,* ed., P. F. Fox & J. J. Condon. Applied Science Publishers, London, UK, pp. 307–28.
Law, B. A. (1988). In *Proceedings of the 8th International Biotechnology Symposium,* ed. G. Durand, L. Bobichon & J. Florent. Sociéte Française de Microbiologie, Paris, France pp. 843–54.
Law, B. A., Sharpe, M. E. & Chapman, H. R. (1976). *J. Dairy Res.,* **43,** 459–68.
Law, B. A., Andrews, A. T. & Sharpe, M. E. (1977). *J. Dairy Res.,* **44,** 145–8.
Law, B. A., Andrews, A. T., Cliffe, A. J., Sharpe, M. E. & Chapman, H. R. (1979). *J. Dairy Res.,* **46,** 497–509.
Ledford, R. A., Senyk, G. F., Shipe, W. F., Katsides, E. & Wolff, E. T. (1983). *J. Dairy Sci.,* **66,** 1611–15.
Lenoir, J. (1984). *Int. Dairy Fed. Bull.,* Document no. 171, 3–20.
Lindsay, R. C. & Rippe, J. K. (1986). In *Biogeneration of Selected Aromas,* ed. T. H. Parliment & R. Croteau. ACS Symposium Series no. 317, American Chemical Society, Washington, DC, USA, pp. 286–308.
Little, C. (1977). *Biochem. J.,* **167,** 399–404.
Lu, J. Y. & Liska, B. J. (1969). *Appl. Microbiol.,* **18,** 108–13.
Macrae, A. R. (1983). In *Microbial Enzymes and Biotechnology,* ed. W. M. Fogarty. Applied Science Publishers, London, UK, pp. 225–50.
Magdoub, M. N. I., Shehata, A. E., Fayed, E. O. & Hoffi, A. A. (1980). *Milchwissenschaft,* **35,** 474–6.
Malik, R. K. & Mathur, D. K. (1984). *J. Dairy Sci.,* **67,** 522–30.
Manji, B., Kakuda, Y. & Arnott, D. R. (1986). *J. Dairy Sci.,* **69,** 2994–3001.
Marin, A. & Marshall, R. T. (1983). *J. Food Prot.,* **46,** 676–80.
Marin, A., Mawhinney, T. P. & Marshall, R. T. (1984). *J. Dairy Sci.,* **67,** 52–9.
Marshall, R. T. (1982). *Kieler Milchwirtschaft. Forsch.,* **34,** 149–57.
Marshall, R. T. & Marstiller, J. K. (1981). *J. Dairy Sci.,* **64,** 1545–50.
Marshall, D. L. & Schmidt, R. H. (1988). *J. Food Prot.,* **51,** 277–82.
McKellar, R. C. (1981). *J. Dairy Sci.,* **64,** 2138–45.
McKellar, R. C. (1986). *J. Dairy Sci.,* **69,** 658–64.
McKellar, R. C. (1989). In *Enzymes of Psychrotrophs in Raw Food,* ed. R. C. McKellar. CRC Press, Boca Raton, FL, USA, pp. 153–72.
McKellar, R. C. & Cholette, H. (1986). *J. Dairy Res.,* **53,** 301–12.
McKellar, R. C. & Nichols-Nelson, D. (1984). *J. Food Prot.,* **47,** 853–5, 860.
McKinnon, C. H. & Pettipher, G. L. (1983). *J. Dairy Res.,* **50,** 163–70.
McNeill, G. P., O'Donoghue, A. & Connolly, J. F. (1986). *Ir. J. Food Sci. Technol.,* **10,** 1–10.

Minamiura, N., Matsumura, Y., Fukumoto, J. & Yamamoto, T. (1972). *Agric. Biol. Chem.*, **36**, 588–95.
Mitchell, G. E. & Ewings, K. N. (1985). *NZ J. Dairy Sci. Technol.*, **20**, 65–76.
Mitchell, S. L. & Marshall, R. T. (1989*a*). *J. Dairy Sci.*, **72**, 875–82.
Mitchell, S. L. & Marshall, R. T. (1989*b*). *J. Dairy Sci.*, **72**, 864–74.
Mitchell, G. E., Ewings, K. N. & Bartley, J. P. (1986). *J. Dairy Res.*, **53**, 97–115.
Mitchell, G. E., Symons, M. H. & Ewings, K. N. (1989). *Aust. J. Dairy Technol.*, **44**, 70–3.
Morihara, K. (1974). *Adv. Enzymol.*, **41**, 179–243.
Mottar, J. (1981). *Milchwissenschaft*, **36**, 87–91.
Mottar, J. (1984). *Le Lait*, **64**, 29–45.
Mottar, J. (1989). In *Enzymes of Psychrotrophs in Raw Food*, ed. R. C. McKellar. CRC Press, Boca Raton, FL, USA, pp. 227–43.
Mottar, J., Van Renterghem, R. & De Vilder, J. (1985). *Milchwissenschaft*, **40**, 717–21.
Muir, D. D. & Phillips, J. D. (1984). *Milchwissenschaft*, **39**, 7–11.
Muir, D. D., Kelly, M. E. & Phillips, J. D. (1978). *J. Soc. Dairy Technol.*, **31**, 203–8.
Muir, D. D., Phillips, J. D. & Dalgleish, D. G. (1979). *J. Soc. Dairy Technol.*, **32**, 19–23.
Muir, D. D., Griffiths, M. W., Phillips, J. D., Sweetsur, A. W. M. & West, I. G. (1986). *J. Soc. Dairy Technol.*, **39**, 115–18.
Murray, S. K., Kwan, K. K. H., Skura, B. J. & McKellar, R. C. (1983). *J. Food Sci.*, **48**, 1166–9.
Myhara, R. M. & Skura, B. J. (1989). *J. Food Sci.*, **54**, 686–90, 694.
Najim, N. H. (1985). PhD thesis, Louisiana State University, Baton Rouge, LA, USA.
O'Connor, R. E., Ewings, K. N., Hayward, A. C. & O'Rukke, P. K. (1986). *J. Appl. Bacteriol.*, **61**, 25–38.
Pansare, A. C., Lewis, N. F. & Venugopal, V. (1986). *Agric. Biol. Chem.*, **50**, 1743–9.
Patel, G. B. & Blankenagel, G. (1972). *J. Milk Food Technol.*, **35**, 203–6.
Patel, T. R., Bartlett, F. M. & Hamid, J. (1983*a*). *J. Food Prot.*, **46**, 90–4.
Patel, T. R., Jackman, D. M. & Bartlett, F. M. (1983*b*). *Appl. Environ. Microbiol.*, **46**, 333–7.
Patel, T. R., Jackman, D. M., Williams, G. J. & Bartlett, F. M. (1986). *J. Food Prot.*, **49**, 183–8.
Peppler, H. J. (1977). *Food Technol.*, **30**, 62–5.
Phillips, J. D. & Griffiths, M. W. (1986). *J. Appl. Bacteriol.*, **61**, 275–85.
Phillips, J. D., Griffiths, M. W. & Muir, D. D. (1981*a*). *J. Soc. Dairy Technol.*, **34**, 109–13.
Phillips, J. D., Griffiths, M. W. & Muir, D. D. (1981*b*). *J. Soc. Dairy Technol.*, **34**, 113–18.
Pierami, R. M. & Stevenson, K. E. (1976). *J. Dairy Sci.*, **59**, 1010–15.
Pratt-Lowe, E. L., Geiger, R. M., Richardson, T. & Barret, E. L. (1988). *J. Dairy Sci.*, **71**, 17–23.
Qua, D. van, Simidu, U. & Taga, N. (1981). *Can. J. Microbiol.*, **27**, 505–10.
Ramshaw, E. H. & Dunstone, E. A. (1969). *J. Dairy Res.*, **36**, 203–13.
Rapp, M. (1978). *Milchwissenschaft*, **33**, 493–6.
Reddy, M. C., Bills, D. D. & Lindsay, R. C. (1969). *Appl. Microbiol.*, **17**, 779–82.
Ren, T. J., Frank, J. F. & Christen, G. L. (1988). *J. Dairy Sci.*, **71**, 1432–8.
Ribbons, W. D. & Williams, P. A. (1982). *Basic Life Sci.*, **19**, 211–32.
Richard, J. (1983). *Le Lait*, **63**, 148–70.
Richard, J., Piton, C. & Gratadoux, J. J. (1988). *Le Lait*, **68**, 311–22.
Richardson, B. C. (1981). *NZ J. Dairy Sci. Technol.*, **16**, 195–207.
Richardson, B. C. & Newstead, D. F. (1979). *NZ J. Dairy Sci. Technol.*, **14**, 273–9.
Richardson, B. C. & Te Whaiti, I. (1978). *NZ J. Dairy Sci. Technol.*, **13**, 172–6.

Rollema, H. S., Visser, S. & Poll, J. K. (1983). *Milchwissenschaft,* **38,** 214–17.
Rollema, H. S., McKellar, R. C., Sørhaug, T., Suhren, G., Zadow, J. G., Law, B. A., Poll, J. K., Stepaniak, L. & Vagias, G. (1989). *Milchwissenschaft,* **44,** 491–6.
Roussis, I. G., Karabalis, I., Papadopoulou, C. & Drainas, C. (1988*a*). *Lebensm.-Wiss. Technol.,* **21,** 188–94.
Roussis, I. G., Voudouris, E. K. & Kontominas, M. G. (1988*b*). *Lebensm.-Wiss. Technol.,* **21,** 131–6.
Rowe, M. T. & Gilmour, A. (1982). *Milchwissenschaft,* **37,** 597–9.
Rowe, M. T. & Gilmour, A. (1983). *Milchwissenschaft,* **38,** 705–7.
Rowe, M. T. & Gilmour, A. (1986). In Milk the vital force, Posters presented at the XXII Int. Dairy Congr., Haag, ed. Int. Dairy Federation, D. Reidel Publishing Co., Dordrecht, p. 164.
Sanjose, C., Fernandes, L. & Palacios, P. (1987). *J. Food Prot.,* **50,** 1004–8.
Scanlan, R. A., Sather, L. A. & Day, E. A. (1965). *J. Dairy Sci.,* **48,** 1582–4.
Schröder, M. J. A. (1984). *J. Dairy Res.,* **51,** 59–67.
Schröder, M. J. A., Cousins, C. M. & McKinnon, C. (1982). *J. Dairy Res.,* **49,** 619–30.
Schwimmer, S. (1981). *Source Book of Food Enzymology,* AVI Publishing, Westport, CT, USA.
Seth, R. J. & Robinson, R. K. (1988). In *Developments in Food Microbiology–4*, ed. R. K. Robinson. Elsevier Applied Science, London, UK, pp. 23–46.
Shamsuzzaman, K. & McKellar, R. C. (1987). *J. Dairy Res.,* **54,** 283–93.
Sharpe, M. E. & Pettipher, G. L. (1983). In *Food Microbiology, Economic Microbiology* (Vol. 8), ed. A. H. Rose. Academic Press, London, UK, pp. 200–41.
Shehata, T. E., Duran, A. & Collins, E. B. (1971). *J. Dairy Sci.,* **54,** 1579–82.
Shelley, A. W., Deeth, H. C. & McRae, I. C. (1986). *J. Appl. Bacteriol.,* **61,** 395–400.
Shipe, W. F., Basette, R., Deane, D. D., Dunkley, W. L., Hammond, E. G., Harper, W. J., Kleyn, D. H., Morgan, M. E., Nelson, J. H. & Scanlan, R. A. (1978). *J. Dairy Sci.,* **61,** 855–69.
Sikes, A. & Maxcy, R. B. (1979). *J. Food Sci.,* **44,** 1228–31.
Singh, A., Srinivasan, R. A. & Dudani, A. T. (1973). *Milchwissenschaft,* **28,** 164–6.
Sinha, P. R. & Sinha, R. N. (1986). *Indian J. Dairy Sci.,* **39,** 410–14.
Skura, B. J. (1989). In *Enzymes of Psychrotrophs in Raw Food,* ed. R. C. McKellar. CRC Press, Boca Raton, FL, USA, pp. 245–63.
Slocum, S. A., Jasinski, E. M., Anantheswaran, R. C. & Kilara, A. (1988). *J. Dairy Sci.,* **71,** 589–95.
Snoeren, T. H. M., van der Spek, C. A., Dekker, R. & Both, P. (1979). *Neth. Milk Dairy J.,* **33,** 31–9.
Sonoki, S. & Ikezawa, H. (1975). *Biochim. Biophys. Acta,* **403,** 412–24.
Stannard, C. J., Williams, A. P. & Gibbs, P. A. (1985). *Food Microbiol.,* **2,** 115–22.
Stead, D. (1986). *J. Dairy Res.,* **53,** 481–505.
Stead, D. (1987). *J. Dairy Res.,* **54,** 535–43.
Stead, D. (1989). In *Enzymes of Psychrotrophs in Raw Food,* ed. R. C. McKellar. CRC Press, Boca Raton, FL, USA, pp. 173–87.
Stepaniak, L. & Fox, P. F. (1983). *J. Dairy Res.,* **50,** 171–84.
Stepaniak, L. & Sørhaug, T. (1989). In *Enzymes of Psychrotrophs in Raw Food,* ed. R. C. McKellar. CRC Press, Boca Raton, FL, USA, pp. 35–55.
Stepaniak, L., Fox, P. F. & Daly, C. (1982). *Ir. J. Food Sci. Technol.,* **6,** 135–46.
Stepaniak, L., Birkeland, S.-E., Vagias, G. & Sørhaug, T. (1987*a*). *Milchwissenschaft,* **42,** 168–72.
Stepaniak, L., Birkeland, S.-E., Sørhaug, T. & Vagias, G. (1987*b*). *Milchwissenschaft,* **42,** 75–9.
Stepaniak, L., Sørhaug, T., Tingstad, H. & Vagias, G. (1987*c*). In *Proc. Euro Food Chem. IV,* eds. W. Baltes, P. Baardseth, R. Norang & K. Søyland, Norwegian Food Res. Inst., 1432 Ås-NLH, Norway, vol. 2, pp. 382–6.

Stepaniak, L., Zakrzewski, E. & Sørhaug, T. (1989). *Italian J. Food Sci.*, **1**(4), 5–12.
Stepaniak, L., Zakrzewski, E. & Sørhaug, T. (1991). *Milchwissenschaft*, **46**, 139–42.
Stern, K. K., Foegeding, E. A. & Hansen, A. P. (1988). *J. Dairy Sci.*, **71**, 41–5.
Stone, W. K. & Naff, D. M. (1967). *J. Dairy Sci.*, **50**, 1497–500.
Strasser de Saad, A. M., Manca de Nadra, M. C., Pesce de Ruiz Holgado, A. & Oliver, G. (1988). *Folia Microbiol.*, **33**, 96–100.
Strongin, A. Y., Abramov, Z. T., Yaroslavtseva, N. G., Baratova, L. A., Shaginyan, K. A., Belyanowa, L. P. & Stepanow, V. M. (1979). *J. Bacteriol.*, **137**, 1017–19.
Sugiura, M. (1984). In *Lipases*, ed. B. Borgström & H. L. Brockman, Elsevier, Amsterdam, The Netherlands, pp. 505–22.
Sugiura, M. & Isobe, M. (1975). *Chem. Pharm. Bull.*, **23**, 681–3.
Sugiura, M., Oikawa, T. & Hirano, K. (1978). *Clinica Chimica Acta*, **89**, 447–53.
Suhren, G. (1983). *Int. Dairy Fed. Bul.*, Document no. 157, 17–25.
Suhren, G. (1989). In *Enzymes of Psychrotrophs in Raw Food*, ed. R. C. McKellar. CRC Press, Boca Raton, FL, USA, pp. 3–34.
Sundheim, G. & Bengtsson-Olivecrona, G. (1987). *J. Dairy Sci.*, **70**, 1815–21.
Suriyarachchi, V. R. & Fleet, G. H. (1981). *Appl. Environ. Microbiol.*, **42**, 574–9.
Swaisgood, H. E. & Bozoğlu, F. (1984). *J. Agric. Food Chem.*, **32**, 7–10.
Symons, M. H., Clements, R. S., Mitchell, G. E. & Ewings, K. N. (1985). *NZ J. Dairy Sci. Technol.*, **20**, 173–8.
Taguchi, R. & Ikezawa, H. (1977). *J. Biochem.*, **82**, 1225–30.
Tayfour, A., Milliere, J. B. & Veillet-Poncet, L. (1982). *Milchwissenschaft*, **37**, 720–3.
Thomas, S. B. & Druce, R. G. (1971). *Dairy Ind. Intern.*, **36**, 75–80, 145–50.
Thomas, T. D. & Pritchard, G. G. (1987). *FEMS Microbiol. Rev.*, **46**, 245–68.
Torrie, J. P., Cholette, H., Froehlich, D. A. & McKellar, R. C. (1983). *J. Dairy Res.*, **50**, 365–74.
Tye, T. M., Haard, N. F. & Patel, T. R. (1988). *Can. Inst. Food Sci. Technol. J.*, **21**, 373–377.
Urbach, G. & Milne, T. (1988). *Aust. J. Dairy Technol.*, **43**, 53–8.
Vagias, G. (1988). PhD thesis, Agricultural University of Norway, Ås, Norway.
Vuillemard, J. C. & Amiot, J. (1985). *Microbiologie–Aliments–Nutrition*, **3**, 333–43.
Vuillemard, J. C., Goulet, J., Amiot, J. & Vijayalakshmi, M. A. & Terre, S. (1988). *Enzymes Microb. Technol.*, **10**, 2–8.
Walter, R., Simmons, W. H. & Yoshimoto, T. (1980). *Molec. Cell. Biochem.*, **30**, 111–27.
Wang, J. J. & Frank, J. F. (1981). *J. Dairy Sci.*, **64**, 2154–60.
Ward, O. P. (1983). In *Microbial Enzymes and Biotechnology*, ed. W. M. Fogarty, Applied Science Publishers, London, UK, pp. 251–317.
White, C. H. & Marshall, R. T. (1973). *J. Dairy Sci.*, **56**, 849–53.
Woo, A. H. & Lindsay, R. C. (1983). *J. Dairy Sci.*, **66**, 2070–5.
Woo, A. H. & Lindsay, R. C. (1984). *J. Dairy Sci.*, **67**, 1194–8.
Yan, L., Langlois, B. E., O'Leary, J. & Hicks, C. L. (1985). *J. Dairy Sci.*, **68**, 1323–36.
Yoovidhya, T., Combes, D. & Monsan, P. (1986). *Biotechnol. Lett.*, **8**, 333–8.
Yun, S.-E., Ohmiya, K. & Shimizu, S. (1982). *Agric. Biol. Chem.*, **46**, 1505–11.
Zaborniak, A. & Knaut, T. (1984). *Archiv. Lebensmittelhygiene*, **35**, 41–3.
Zaborniak, A. & Knaut, T. (1985). *Deutsche Molkerei-Z.*, **106**, 368–70.
Zakrzewski, E., Stepaniak, L., Abrahamsen, R. K. & Sørhaug, T. (1990). In Posters and Brief Communications of the XXIII International Dairy Congress, Montreal, 8–12 October, p. 470.

5

Exogenous Enzymes in Dairy Technology

P. Fox & M. B. Grufferty
Department of Food Chemistry, University College, Cork, Republic of Ireland

INTRODUCTION

The use of rennets (crude proteinase preparations) in cheese manufacture is among the earliest applications of exogenous enzymes in food processing, dating from at least 6000 BC. Although rapid growth has occurred in the use of amylases and glucose isomerase in recent years, the use of rennets is probably still the largest single application of enzymes in food technology, possibly of all industrial enzymes. In addition to rennets, proteinases have found some new applications in dairy technology in recent years, e.g. acceleration of cheese ripening, modification of functional properties and preparation of dietetic products.

In comparison to rennets, other applications of exogenous enzymes in dairy technology are relatively minor. Principal among these are lipases and β-galactosidases which have important growing applications. Enzymes with minor or potential applications include glucose oxidase, lysozyme, catalase, glucose isomerase, lactoperoxidase, superoxide dismutase and sulphydryl oxidase.

PROTEINASES

Cheese manufacture

The distinguishing feature of mammals, of which there are ~4300 species, is the secretion of a fluid, milk, by specialised glands on the female for the nutrition of the neonate. For most species, milk is the sole source of nutrients for some period after birth, and, therefore, milk must be able to supply all the nutritional requirements of the neonate. The nutritional value of milk was recognised by early man who domesticated goats, sheep and cattle about 10 000 years ago and contrived to share the mother's milk with its off-spring.

Milk is a very perishable commodity but, by accident or design, man developed techniques for converting the principal constituents of milk into more stable products. Although milk was probably preserved by dehydration in prehistoric times, preservation by fermentation (fermented milks, cheeses and butter) was much more widespread and probably the only significant preservation technique until the development of sterilisation in the late 19th century.

While fermentation alone extends the shelf life of milk, such products are relatively unstable. However, by combining fermentation with concentration/dehydration and the addition of salt, relatively stable products can be produced. In butter manufacture, concentration of the fat is accomplished by destabilising the lipid emulsion in milk or cream which leads to clumping of the fat; further dehydration may be accomplished by heating, as in ghee manufacture.

The manufacture of all cheese varieties essentially involves concentrating the protein and fat of milk six- to twelve-fold, depending on the variety. Concentration is achieved by (i) coagulating the principal milk protein group, casein (if present, the milk fat is occluded in the coagulum), (ii) cutting or breaking the coagulum and inducing it to synerese under the influence of heat and acid, (iii) separation of curd and whey, (iv) acidification, pressing and salting of dewheyed curd. Coagulation of the casein is induced by:

(1) Limited proteolysis by a crude proteinase (rennet), which is exploited in the manufacture of the vast majority of ripened and some fresh cheeses.
(2) Isoelectric precipitation at ~pH 4·6, used mainly for fresh cheeses, usually by in-situ production of lactic acid by a culture (starter) of lactic acid bacteria and less frequently by direct acidification with pre-formed acid, usually hydrochloric acid, or acidogen, usually gluconic acid-δ-lactone.
(3) Acid plus heat, i.e. acidification to ~pH 5·2 with acid whey, acid milk, citrus juice, vinegar or acetic acid at 80–90°C, e.g. Ricotta, Questo Blanco.

Recent developments in ultrafiltration permit concentration of the total colloidal phase of milk (i.e. fat and total protein) to the level present in cheese, i.e. to pre-cheese. Rennet and starter are added to the concentrate to modify texture and initiate flavour development. Ultrafiltration is now used commercially for the manufacture of several cheese varieties.

In this review, we are concerned only with rennet-coagulated cheeses, in which the primary role of rennet is to induce coagulation of the casein but it also plays a key, probably an essential, role during ripening.

Enzymatic coagulation of milk

It has been known since the work of Hammersten in the 1880s that the rennet coagulation of milk involves proteolysis with the formation of '*para*-casein' and non-protein nitrogen. It is now well established that coagulation is a two-stage process, the first (primary) phase involving the enzymatic production of *para*-casein and TCA-soluble peptides ((glyco) macropeptides), the secondary phase involving the calcium-induced gelation of *para*-casein at ~30°C. Proteolysis is essentially complete before the onset of coagulation.

The milk protein system

Since the enzymatic coagulation of milk involves the specific modification of the milk protein system, an understanding of this system is essential for the thorough understanding of the coagulation process. A detailed description of milk proteins is beyond the scope of this chapter and the reader is referred to Fox (1982, 1989*a*, 1992). The key features of milk proteins in relation to rennet coagulation are summarised in the following.

Bovine milk contains two markedly different groups of proteins: the caseins, i.e. phosphoproteins insoluble at pH 4·6 and 20°C, and the whey (or non-casein) proteins, in the approximate ratio of 4:1. Casein consists of four principal proteins, α_{s1}-, α_{s2}-, β- and κ-, in the approximate ratio of 40:10:35:12, and several minor proteins, most of which originate via proteolysis of the primary caseins by alkaline milk proteinase, plasmin. These include γ^1-, γ^2- and γ^3-caseins and proteose peptones five and eight fast, all derived from β-casein, and at least 30 unidentified peptides.

All the caseins are phosphorylated; α_{s1}-, α_{s2}-, β- and κ-caseins contain 8–9, 10–13, 4–5 and 1–2 mol P/mol, respectively. Because of their high phosphate contents, α_{s1}-, α_{s2}- and β-caseins bind Ca^2 strongly and precipitate at $[Ca^{2+}] > 6$ mM. However, κ-casein, which predominantly has one phosphate residue, binds calcium weakly and is soluble at high $[Ca^{2+}]$. It also reacts hydrophobically with α_{s1}-, α_{s2}- and β-caseins and can stabilise up to 10 times its weight of the calcium-sensitive caseins against precipitation by forming coarse colloidal aggregates—micelles.

In milk, >95% of the casein exists naturally as micelles, which consist, on a dry weight basis, of ~94% protein and 6% low molecular weight species, mainly calcium and phosphate with some magnesium and citrate, collectively called colloidal calcium phosphate (CCP). The micelles are spherical, 50–600 nm (mean ~ 100 nm) in diameter, with molecular weights of $\sim 10^8$ Da, i.e. a typical micelle contains ~5000 monomers (mol. wt 20–24 kDa). The micelles typically bind ~2 g water/g protein.

Although the structure of the micelles is not yet fully established, there is fairly general agreement that they consist of sub-micelles (spherical particles, mol. wt $\sim 5 \times 10^6$ Da) held together by CCP bridges, hydrophobic interactions and hydrogen bonds. The micelles dissociate when CCP is removed, e.g. by calcium-chelators or acidification, when the pH is increased to above 9, or on addition of SDS or urea. The precise structure of the sub-micelles is unknown. In most hypotheses it is envisaged that the calcium-sensitive α_s- and β-caseins interact hydrophobically to form the core of the sub-micelles with κ-casein having a predominantly surface location. The *N*-terminal two-thirds of κ-casein is hydrophobic and reacts hydrophobically with the core proteins, leaving the hydrophilic *C*-terminal regions projecting into the surrounding environment. It is proposed that sub-micelles contain variable amounts of surface κ-casein and aggregate such that κ-casein-rich sub-micelles predominate at the surface of the micelles with the κ-casein-deficient sub-micelles buried within. The micelles are stabilised by a zeta potential of ~−20 mV and by steric factors caused by the protruding *C*-terminal segments of κ-casein which give the micelles a 'hairy' appearance and which prevent close approach.

Primary phase of rennet action

The micelle-stabilising properties of κ-casein are lost during the primary phase of rennet action (Waugh & von Hippel, 1956) when it is the only protein hydrolysed to a significant extent (Wake, 1959). The cleavage site is Phe-105–Met-106 (Delfour *et al.*, 1965), which is many times more susceptible to hydrolysis by acid proteinases (which include all commercial rennets) than any other bond in the milk protein system.

The unique sensitivity of the Phe–Met bond has aroused interest. The dipeptide, H.Phe–Met.OH, is not hydrolysed nor are tri- or tetrapeptides containing a Phe–Met bond. However, this bond is hydrolysed in the pentapeptide, H.Ser–Leu–Phe–Met–Ala–OMe (Hill, 1968, 1969) and reversing the positions of serine and leucine to give the correct sequence for κ-casein, i.e. Leu–Ser–Phe–Met–Ala, increases the susceptibility of the Phe–Met bond to chymosin (Schattenkerk & Kerling, 1973). The length of the peptide and the sequence around the cleavage site are important determinants of enzyme–substrate interaction. Ser-104 is particularly important (Hill, 1968, 1969) and its replacement by glycine (Visser *et al.*, 1977) or alanine (Raymond *et al.*, 1972) in the above pentapeptide renders the Phe–Met bond very resistant to hydrolysis by chymosin but not by pepsins (Raymond & Bricas, 1979); even substituting D-serine for L-serine markedly reduces its sensitivity (Raymond & Bricas, 1979). Extension of the above pentapeptide from the *N*- and/or *C*-terminal to reproduce the sequence of κ-casein increases the efficiency of hydrolysis of the Phe–Met bond by chymosin (Visser *et al.*, 1976, 1977). At pH 4·7, these studies showed that relative to the pentapeptide H.Ser–Phe–Met–Ala–Ile.OH, i.e. the sequence 104–108 of κ-casein, extending the peptide towards the *C*-terminal by three residues, to give Ser-104–Lys-111, caused a six-fold increase in the catalytic ratio, k_{cat}/K_m, while addition of Leu-103 to the pentapeptide increased the ratio 600-fold (Table 1). Addition of His-102 and Pro-101 (i.e. Pro-101–Ile-108) increased k_{cat}/K_m a further five-fold. Hill and Hocking (1978) reported that the sequence His-98–Glu-129 of κ-casein includes all the residues necessary to render the Phe–Met bond as susceptible to hydrolysis by chymosin as it is in intact κ-casein. In fact, the peptide representing the sequence His-98–Lys-111 includes all the necessary determinants (Visser *et al.*, 1980). This peptide is hydrolysed ~66 000 faster than the parent pentapeptide, with a k_{cat}/K_m of ~2 $M^{-1}\,s^{-1}$, which is very close

TABLE 1
Kinetic Parameters for Hydrolysis of κ-Casein Peptides by Chymosin at pH 4·7

Peptide	*Sequence*	κ_{cat} *(per s)*	K_m (mM)	κ_{cat}/K_m *(per s/mM)*
S.F.M.A.I.	104–108	0·33	8·5	0·038
S.F.M.A.I.P.	104–109	1·05	9·2	0·114
S.F.M.A.I.P.P.	104–100	1·57	6·8	0·231
S.F.M.A.I.P.P.K.	104–111	0·75	3·2	0·239
L.S.F.M.A.I.	103–108	18·3	0·85	21·6
L.S.F.M.A.I.P.	103–109	38·1	0·69	55·1
L.S.F.M.A.I.P.P.	103–110	43·3	0·41	105·1
L.S.F.M.A.I.P.P.K.	103–111	33·6	0·43	78·3
L.S.F.M.A.I.P.P.K.K.	103–112	30·2	0·46	65·3
H.L.S.F.M.A.I.	102–108	16·0	0·52	30·8
P.H.L.S.F.M.A.I.	101–108	33·5	0·34	100·2
H.P.H.P.H.L.S.F.M.A.I.P.P.K.	98–111	66·2	0·026	2 509
H.P.H.P.H.L.S.F.M.A.I.P.P.K.[a]	98–111	46·2	0·029	1 621
κ-Casein[b]		2–20	0·001–0·005	200–2 000
L.S.F. (NO_2)Nle A.L.OMe		12	0·95	12·7

[a] pH 6·6.
[b] pH 6·4.

to that for intact κ-casein. κ-Casein and the peptide His-98–Lys-111 are also readily hydrolysed at pH 6·6 but smaller peptides are not.

The two residues, Phe–Met, are not intrinsically essential for chymosin action. Replacement of phenylaninine by phenylalanine (NO_2) or cyclohexylalanine reduces k_{cat}/K_m by ~3- and ~50-fold, respectively (Visser *et al.*, 1977). Oxidation of Met-106 reduces k_{cat}/K_m by ~10-fold but substitution of isoleucine for methionine increases this ratio by ~3-fold. Although porcine or human κ-caseins do not possess a Phe–Met bond [isoleucine replaces methionine in both (Chobert *et al.*, 1976; Brignon *et al.*, 1985; Fiat *et al.*, 1977), they are readily hydrolysed by calf chymosin, although more slowly than bovine κ-casein; in contrast, porcine milk is coagulated more effectively than bovine milk by porcine chymosin (Foltmann, 1981*a*), indicating that unidentified subtle structural features influence chymosin action. Thus, the sequence around the Phe–Met bond, rather than the bond itself, contains the important determinants of hydrolysis. The particularly important residues are Ser-104, Leu-103, Ile-108, at least one of the three histidines (residues 98, 100, 102), some or all of the four prolines (residues 99, 101, 109, 110) and Lys-111.

Based on the results of studies on peptides representing the sequence His-98–Lys-112 of κ-casein, Visser *et al.* (1987) suggest that the sequence Leu-103–Ile-108, which probably exists as an extended β-structure (Loucheux-Lefebvre *et al.*, 1978; Raap *et al.*, 1983), fits into the active site cleft of acid proteinases. The hydrophobic residues, Phe-105, Met-106, Leu-103 and Ile-108, are directed to hydrophobic pockets along the active site cleft while the hydroxyl group of Ser-104 forms a hydrogen bridge with some counterpart on the enzyme. The sequences 98–102 and 109–111 form β-turns around the edges of the active site cleft in the enzyme–substrate complex; this conformation is stabilised by proline residues at positions 99, 101, 109 and 110. One or more of the three histidine residues, 98, 100, 102, are probably involved in electrostatic bonding between enzyme and substrate but Lys-112 appears not to be important in enzyme–substrate binding as long as Lys-111, which participates in electrostatic interactions, is present (Visser *et al.*, 1987). The importance of electrostatic interactions in chymosin–substrate complex formation is indicated by the inhibitory effect of low concentrations of sodium chloride on the enzymatic phase (Visser *et al.*, 1980).

Small peptides, mimicking or identical to the sequence of κ-casein around the Phe–Met bond, are very useful substrates for determining the activity of rennets in absolute units, i.e. independent of variations in the non-enzymatic phase of clotting of different milks and form the principle of methods for quantifying the proportions of acid proteinases in commercial rennets (e.g. Raymond *et al.*, 1973; Schattenkerk & Kerling, 1973; Salesse & Garnier, 1976; Koning *et al.*, 1978; Martin *et al.*, 1981; Visser & Rollema, 1986).

Although the hydrolysis of κ-casein by acid proteinases other than calf chymosin has not been studied thoroughly, it is generally assumed that all commonly-used rennets hydrolyse the Phe–Met bond. However, the aggregation characteristics of micelles altered by rennet substitutes differ from those of chymosin-treated micelles, suggesting differences in the extent and/or specificity of κ-casein hydrolysis (Lawrence & Creamer, 1969). Furthermore, the common rennets have markedly different specific activities on synthetic κ-casein-related peptides (Salesse & Garnier, 1976; Koning *et al.*, 1978; Martin *et al.*, 1981).

Reviews on the enzymatic coagulation of milk include Dalgleish (1982, 1987), Fox (1984, 1987, 1988); and Fox and Mulvihill (1990).

Rennets

The rennets used to coagulate milk are crude preparations of selected proteinases. Many proteinases will coagulate milk but most are too proteolytic relative to their milk clotting activity; consequently, they hydrolyse the coagulum too quickly, causing reduced cheese yield and/or defective cheese with a propensity to bitterness. Gastric proteinases from calves, kids or lambs have been used traditionally as rennets, with very few exceptions. The principal proteinase in rennets from young ruminants is chymosin, but it is replaced by pepsin as the animal ages.

Owing to increasing world cheese production (~4% per annum over the past 20 years), concomitant with a reduced supply of calf rennet (due to a decrease in calf numbers and a tendency to slaughter more mature calves), the supply of veal rennet has been inadequate for many years. This has led to an increased price for calf rennet and to a search for rennet substitutes. In spite of the availability of numerous milk coagulants, only six rennet substitutes have been found to be more or less acceptable: bovine, porcine and chicken pepsins and the acid proteinases from *Mucor miehei, M. pusillus* and *Endothia parasitica.*

Chicken pepsin is the least suitable of these and is used widely only in Israel, and, to a lesser extent, in Czechoslovakia. Bovine pepsin gives generally satisfactory results with respect to cheese yield and quality; many commercial 'calf rennets' contain a substantial proportion (perhaps 50%) of bovine pepsin. Porcine pepsin is very sensitive to pH > 6·6 and is extensively denatured during cheesemaking; consequently, it contributes little to proteolysis during cheese ripening. A 50:50 mixture of porcine pepsin and calf rennet gives generally acceptable results. Although the proteolytic specificities of the three commonly used fungal rennets are considerably different from that of calf chymosin, they generally yield acceptable cheese and are widely used in the US but 'calf rennet' is still predominant in Europe, Australia and New Zealand. The extensive literature on rennet substitutes has been reviewed (e.g. Sardinas, 1972; Ernstrom, 1974; Nelson, 1975; Sternberg, 1976; Green, 1977; Phelan, 1986).

Chymosin and all the commercially successful rennet substitutes are acid proteinases, the molecular and catalytic aspects of which have been reviewed (Foltmann, 1981*a, b,* 1987). Acid proteinases have relatively narrow specificities with a preference for peptide bonds to which hydrophobic residues supply the carboxyl group; this narrow specificity is significant for successful cheese manufacture. However, some acid proteinases are too active even under the prevailing relatively unfavourable conditions and are not suitable as rennets.

Although they are relatively cheap, rennets represent the largest single industrial usage of enzymes with a world market of ~25×10^6 litres of standard rennet per annum (worth ~£100 million). Therefore, rennets are attractive to industrial enzymologists and biotechnologists. The gene for calf chymosin has been cloned in selected bacteria and yeasts (e.g. Smith *et al.*, 1985; Cullen *et al.*, 1987; Foltmann, 1987; Hicks *et al.*, 1988). Microbial chymosins are now commercially available. Large-scale cheesemaking trials have given very satisfactory results. Cheddar cheese made using chymosin produced by *Kluyveromyces lactis* (Gist Brocades) was indistinguishable organoleptically or chemically from controls made using calf rennet (Bines *et al.*, 1989; O'Sullivan & Fox, 1991). Hicks *et al.* (1988) also report chymosin from *Escherichia coli* K12 (Genencor) to be satisfactory for Cheddar.

Since most (>90%) of the rennet added to cheese milk is lost in the whey, immobilisation would extend its catalytic life. Continuous coagulation of milk using immobilised rennets would also be technologically attractive. Several rennets have been immobilised but their efficiency as milk coagulants has been questioned. There is fairly general support for the view that properly immobilised enzymes can not coagulate milk owing to inaccessibility of the Phe–Met bond of κ-casein and that the apparent coagulating activity of immobilised rennets is due to leaching of enzyme from the support. Even if enzymatically effective, immobilised rennets would experience other problems, e.g. hygiene, especially if used in columns. Immobilised rennets have not been commercialised (see Fox, 1981; Carlson *et al.*, 1986; Dalgleish, 1987 for reviews).

Factors affecting the hydrolysis of κ-casein

pH. The pH optimum for chymosin and bovine pepsin on the small Phe–Met containing peptides is ~4·7 (Hill, 1968; Raymond & Bricas, 1979; Koning *et al.*, 1978) and 5·3–5·5 on the sequence His-98–Lys-111 of κ-casein (Visser *et al.*, 1987) or on κ-casein itself (Humme, 1972). The pH optimum for the first stage of rennet action in milk is ~6·0 (Hooydonk *et al.*, 1986). Milk for most cheese varieties is renneted at ~pH 6·5.

Ionic strength. Increasing ionic strength (0·01–0·11) reduces the rate of hydrolysis of the His-98–Lys-112 peptide, the effect becoming more marked as the reaction pH is increased (Visser *et al.*, 1980). At 1 mM, sodium chloride, calcium chloride and magnesium chloride stimulate the hydrolysis of κ-casein in isolated form and in sodium caseinate (Kato *et al.*, 1970).

Temperature. The optimum temperature for the coagulation of milk by calf rennet at pH 6·6 is ~45°C (Fox, 1969). The temperature coefficient for the hydrolysis of κ-casein in solutions of sodium caseinate is ~1·8, the activation energy is ~42 kJ/mol and the activation entropy is ~−163·8 J/°C mol (Nitschmann & Bohren, 1955); generally similar values were reported by Garnier (1963) for the hydrolysis of isolated κ-casein by chymosin.

Degree of glycosylation. The efficiency of κ-casein as a substrate for chymosin decreases with the degree of glycosylation (Doi *et al.*, 1979; Vreeman *et al.*, 1986). At pH 6·6, k_{cat} decreased from ~43 s^{-1} for carbohydrate-free κ-casein to ~25/s for κ-casein containing 6 mol *N*-acetyl neuraminic acid (NANA)/mol (Vreeman *et al.*, 1986). However, K_m was optimal for the κ-casein component containing 3 mol NANA/mol. Polymerisation (aggregation) markedly increased K_m with little effect on k_{cat}.

Heat treatment of milk. Rennet coagulation is adversely affected by heat treatments at temperatures >65°C and is prevented by very severe heat treatments (>90°C for 10 min). Although changes in salts equilibria are contributory factors, complexation of κ-casein with β-lactoglobulin and/or α-lactalbumin is primarily responsible for the increased rennet coagulation time of heated milk; both the primary and secondary phases are adversely affected (Kannan & Jenness, 1961; Morrissey, 1969; Sawyer, 1969; Hindle & Wheelock, 1970; Wilson & Wheelock, 1972; Wilson *et al.*, 1974; Shalabi & Wheelock, 1976, 1977). The adverse effects of heating can be reversed by acidification before or after heating or by addition of calcium chloride (Banks & Muir, 1985; Marshall, 1986; Hooydonk *et al.*, 1987; Singh *et al.*, 1988).

Secondary (non-enzymatic) phase of coagulation

Hydrolysis of κ-casein during the primary phase of rennet action causes the release of highly-charged, hydrophilic peptides, representing the *C*-terminal segments of κ-casein, from the casein micelles, thereby reducing their zeta potential from $-10/-20\,mV$ to $-5/-7\,mV$ and removing the protruding peptides from their surfaces, thus reducing the intermicellar repulsive forces (electrostatic and steric) and micelle stability. When ~85% of the total κ-casein has been hydrolysed, the casein micelles begin to aggregate or gel but an individual micelle can not participate in gelation until ~97% of its κ-casein has been hydrolysed. Reducing the pH or increasing the temperature from the normal values (~6·6 and ~31°C, respectively) permits coagulation at a lower degree of κ-casein hydrolysis.

The coagulation of rennet-altered micelles, the strength of the resulting gel and the syneresis of the gel when cut or broken will not be discussed here. These phenomena, which together with salting of the cheese curd, are critically important in controlling cheese composition and quality and have been reviewed by Dalgleish (1982, 1987), Fox and Mulvihill (1990), Green and Grandison (1987), Walstra *et al.* (1987) and Guinee and Fox (1987).

Proteolysis during ripening

Acid-coagulated cheeses are usually consumed fresh but the vast majority of rennet-coagulated cheeses are ripened (matured) for periods ranging from ~4 weeks to >2 years; the duration of ripening is inversely related to the moisture content of the cheese. During ripening, a multitude of microbiological, chemical and biochemical events occur, as a result of which the principal constituents of the cheese, i.e. the proteins, lipids and residual lactose, are transformed to primary, and later to secondary, products. Among the principal compounds present in most cheese varieties are: peptides, amino acids, amines, acids, thiols, thioesters (from proteins); fatty acids, methyl ketones, lactones and esters (from lipids); organic acids (lactic acid, acetic, propionic), carbon dioxide, esters and alcohols (from lactose). In the right combinations, these compounds are responsible for the characteristic flavour of the various cheese varieties.

Significance of proteolysis

While lipolysis and glycolysis are important in all, and critical in some, cheese varieties, e.g. Blue cheeses, hard Italian varieties and Swiss-type cheeses, proteolysis is essential in all varieties, especially in internal- and surface–bacterially ripened cheeses in which it is probably the principal biochemical event during ripening. Proteolysis contributes to cheese ripening in at least four ways: (i) a direct contribution to flavour, or off-flavour, e.g. bitterness, or indirectly via catabolism of amino acids to amines, acids, thiols, thioesters, etc., (ii) facilitates the release of sapid compounds during mastication, (iii) changes in pH via the formation of ammonia, (iv) changes in texture owing to breakdown of the protein network, increase in pH and greater water binding by the newly-formed amino and carboxyl groups. A high correlation exists between the intensity of Cheddar cheese flavour and the concentration of free amino acids.

Considerable information is available on the level and type of proteolysis in the principal cheese groups; reviews include Castberg and Morris (1976), Desmazeaud and Gripon (1977), Grappin *et al.* (1985), Rank *et al.* (1985), Law (1987) and Fox (1989*b*).

Proteolytic agents in cheese

Four, and in some varieties five, agents are involved in the ripening of cheese: (i) rennet or rennet substitute, (ii) indigenous milk enzymes, particularly in raw milk cheeses, (iii) starter bacteria and their enzymes, released on cell lysis, (iv) non-starter bacteria, which either survive pasteurisation of the cheese milk or gain access to the pasteurised milk or curd during manufacture, (v) secondary inocula, e.g. propionic acid bacteria, *Brevibacterium linens,* yeasts and moulds (*Penicillium roqueforti* and *P. candidum*), and their enzymes are of major importance in some varieties.

Techniques have been developed which permit quantitation of the contribution of each of these five agents to the primary aspects of cheese ripening (see Fox, 1989*b*). Here, we are concerned only with the contribution of the coagulant; the significance of starter and other bacterial enzymes in cheese ripening is discussed in Chapter 3.

Contribution of coagulant to proteolysis in cheese

Only ~6% of the rennet added to cheese milk is normally retained in the curd. However, the amount retained is influenced by the type of rennet, e.g. porcine pepsin is extensively denatured during cheesemaking, pH (low pH favours retention of chymosin but not pepsins or microbial rennets (Holmes *et al.*, 1977; Creamer *et al.*, 1985) and reduces denaturation) and cooking temperature, e.g. very little, if any, coagulant survives the cooking conditions used for Swiss-type cheeses (Matheson, 1981).

The proteolytic specificity of calf chymosin and the principal rennet substitutes on α_{s1}- and β-caseins is fairly well established and these findings can, largely, be extended to cheese (see Fox, 1989*b* for references).

β-Casein in solution is sequentially hydrolysed at bonds 192–193, 189–190, 163–164 and 139–140 to yield the peptides β-I^1, β I^{11}, β-II and β-III, respectively; bonds 165–166, 167–168 may also be hydrolysed to yield peptides indistinguishable electrophoretically from β-II. At low pH (2–3), bond 127–128 is also hydrolysed to yield β-IV. The hydrolysis of β-casein by chymosin is strongly inhibited by 5%, and completely by 10%, sodium chloride; presumably, the effect is related to water activity, A_w, but the influence of varying A_w *per se* on the proteolysis of individual caseins has not been studied.

β-Casein is quite resistant to proteolysis in internal bacterially-ripened cheeses; such proteolysis as does occur is probably due to plasmin and/or bacterial proteinases and peptidases since the β-peptides produced by rennet, i.e. β-I, β-II, are not formed. Undoubtedly, sodium chloride in cheese has an inhibitory effect but even in the absence of sodium chloride, the extent of β-casein hydrolysis in Cheddar cheese is slight.

α_{s1}-Casein in solution has several chymosin-susceptible bonds, hydrolysis of which is dependent on pH and sodium chloride concentration. In contrast to β-casein, sodium chloride up to 5% stimulates the hydrolysis of α_{s1}-casein and significant proteolysis occurs at 20%. α_{s1}-Casein is readily hydrolysed in cheese, initially to α_{s1}-I (f24–199) and later to α_{s1}-V (f30–199), α_{s1}-VII (f56–179) and small amounts of α_{s1}-II (f24–169). In Cheddar and Dutch-type cheeses, α_{s1}-casein is completely degraded to α_{s1}-I and some further products by the end of ripening. In mould-ripened cheeses, α_{s1}-casein is completely degraded to at least α_{s1}-I prior to the mould-ripening phase and very extensive degradation occurs thereafter due to the action of fungal proteinases and peptidases (see Gripon, 1987). α_{s2}-Casein and *para*-κ-casein appear to be very resistant to rennets and remain largely intact in bacterially-ripened cheeses.

Significance of secondary coagulant proteolysis

Studies on aseptic curd and cheese with controlled microflora have shown that the coagulant is responsible for the level of proteolysis detected by gel electrophoresis and for most of the nitrogen soluble in water or at pH 4·6; however, little TCA- or PTA-soluble nitrogen is produced by the coagulant (O'Keeffe *et al.*, 1978). Proteolysis by rennet is believed (Jong, 1976; Creamer *et al.*, 1982) to be responsible for the softening of cheese texture during early ripening via the hydrolysis of α_{s1}-casein to α_{s1}-I which is sufficient to break the continuous protein matrix. Additional proteolysis by both coagulant and bacterial proteinases further modifies the texture. Even in surface mould-ripened cheese, and probably in smear cheeses, coagulant is considered to be essential for textural changes, e.g. in Camembert, although the very marked increase in pH (to 7) caused by the catabolism of organic acids and the production of ammonia (by deamination of amino acids) is also essential (Noomen, 1983; Lenoir, 1984).

In Swiss and hard Italian cheeses little, if any, coagulant survives the cooking process (Matheson, 1981) and plasmin plays a significant role in primary proteolysis. Plasmin activity in these cheeses is much higher than in Cheddar or Cheshire cheese due to either the higher pH at whey drainage (Lawrence *et al.*, 1984) or to the higher cooking temperature which appears to activate plasminogen or inactivate plasmin inhibitors (Farkye & Fox, 1990).

The secondary proteolytic action of the coagulant influences flavour in three ways.

(1) Some rennet-produced peptides probably have a positive influence on flavour but excessive or unbalanced proteolysis, e.g. due to too much or excessively proteolytic rennet (or to certain starters) or unsuitable environmental conditions, e.g. too much moisture or too little sodium chloride, leads to bitterness.

(2) Rennet-produced peptides serve as substrates for microbial proteinases and peptidases which produce small peptides and amino acids. These contribute at least to background flavour, and perhaps to bitterness if the activity of such enzymes is excessive. Catabolism of amino acids by microbial enzymes and perhaps alterations via chemical mechanisms, leads to a range of sapid compounds—amines, acids, ammonia, thiols—which are major contributors to characteristic cheese flavours.
(3) Alterations in cheese texture appear to influence the release of flavourful and aromatic compounds, arising from proteolysis, lipolysis, glycolysis and secondary metabolic changes, from cheese during mastication and this may be the most significant contribution of proteolysis to cheese flavour (McGugan *et al.*, 1979).

Acceleration of cheese ripening

The original objective of cheese manufacture was to conserve the principal nutrients in milk, i.e. lipids and proteins; this was achieved by a combination of acidification, dehydration, low E_h and salting. When cheese production was totally or primarily for the purpose of conserving milk constituents, then the more stable the product, i.e. the less change, the better. However, storage stability, while still important, is no longer the primary objective of cheese manufacture. Indeed, because ripening is quite an expensive process (~£1/t day), acceleration of ripening, especially of low-moisture, slow-ripening varieties, is desirable, at least under certain circumstances, provided that the whole process can be maintained balanced.

Objective of accelerating ripening

Fermentation of the residual lactose in cheese curd and its constituent monosaccharides to lactic acid occurs so quickly that is it not necessary to accelerate glycolysis. With some exceptions, notably Blue and hard Italian varieties, lipolysis in cheese is quite limited and is, usually, not rate-determining in the ripening of most cheese varieties. The use of lipases in blue and Italian varieties is discussed later.

To date, research on the acceleration of cheese ripening has concentrated on proteolysis in Cheddar, with only a few reports on other varieties. Reviews include Law (1978, 1980, 1982, 1983, 1984, 1986, 1987), Moskowitz and Noelck (1987) and Fox (1988/89).

Methods for accelerating cheese ripening

The methods available to accelerate ripening can be categorised into four groups, the principal advantages and limitations of which are summarised in Table 2. These methods aim to accelerate ripening either by increasing the level(s) of key enzymes or by making the conditions more favourable for the activity of 'indigenous' enzymes. Only exogenous enzymes will be considered here; the interested reader is referred to, e.g. Fox, (1988/89), for a critical discussion on the alternative methods.

TABLE 2
Major Methods for Accelerating Cheese Ripening

Method	*Advantages*	*Limitations*
Elevated temperature	No legal barriers; technically simple; no cost (perhaps saving).	Non-specific action; increased risk of microbial spoilage.
Exogenous enzymes	Low cost; specific action; choice of flavour options.	Limited choice of useful enzymes; possible legal barriers; difficult to ensure uniform incorporation of enzymes; risk of over-ripening.
Modified starters	Easy to incorporate; natural enzyme balance retained.	Technically complex; uneconomic at present.
Cheese slurries	Very rapid flavour development	High risk of microbial spoilage; final product requires processing.

Exogenous enzymes

Since cheese maturation involves the action of indigenous, endogenous and exogenous enzymes rather than viable microorganisms, it seems logical to augment those enzymes known to play definite, identifiable roles in ripening. Law (1978, 1980, 1983, 1984, 1986, 1987) is of the view that enzyme addition offers a more specific alternative to elevated temperatures for accelerating flavour-producing reactions during cheese ripening but identifies the following problems: enzyme addition is not legally permitted in all countries; the range of commercially available useful enzymes is limited; some enzymes are difficult to control, both in terms of their addition to, and their action in, cheese. There is, in addition, a cost factor.

Cheese ripening is a very complex biochemical phenomenon involving the orchestrated action of many enzymes; although some of these reactions are fairly well characterised, many are scarcely recognised. It appears unlikely that acceleration of only one of these many reactions will lead to the balanced acceleration of cheese ripening, unless that reaction is clearly rate-limiting. Studies to date on the use of exogenous enzymes to accelerate ripening have concentrated on proteinases with less emphasis on lipases and peptidases; little attention has been focussed on the myriad of secondary reactions. The implication/assumption is that proteolysis is the limiting reaction in cheese ripening, which may not be totally true.

Since the level of proteolysis in hard and semi-hard cheeses is relatively limited, a proteinase with limited specificity is indicated. Since, as discussed above, the coagulant is responsible for the primary proteolysis in most cheeses, a higher level of rennet might be expected to accelerate ripening. However, although a four-fold increase in the level of rennet used in the manufacture of Dutch cheese accelerated the rate of formation of (water) soluble nitrogen, the formation of amino acid nitrogen was not affected (Stadhouders, 1960). It was concluded that sufficient peptides are produced in Dutch cheese by the normal level of rennet and increasing it does not stimulate amino acid production and flavour development. The high level of rennet led to bitterness. Increased levels of both starter and rennet may accelerate ripening but this has not been reported.

As previously discussed, the proportion of added chymosin retained in cheese curd increases as the pH at whey drainage decreases. Creamer *et al.* (1985) showed that the flavour score of mature Cheddar cheese decreased as the pH at drainage decreased, especially when calf rennet was used; bitterness was the most common defect. Thus, it appears that attempts to accelerate cheese ripening through the use of higher levels of rennet are unlikely to be successful.

The earliest reports on the use of exogenous enzymes (combinations of commercially available acid and neutral proteinases, lipases, decarboxylases and lactases) to accelerate the ripening of Cheddar cheese appear to be those of Iwasaki and Kosikowski (1973), Kosikowski and Iwasaki (1975), and Kosikowski (1976). Microbial acid proteinases caused pronounced bitterness but certain neutral proteinases and peptidases gave marked increases in flavour intensity after 1 month at 20°C, although an over-ripe flavour and free fluid were evident after 1 month at 32°C. Incorporation of up to 20% of the enzyme-treated cheese in processed cheese gave a marked increase in Cheddar flavour.

Sood and Kosikowski (1979*a*) described the production of good quality, medium sharp Cheddar in 3 months at 10°C through addition of combinations of selected proteinases and lipases. Up to 60% of enzyme-treated (fungal lipases and proteinases) ultrafiltration retentate could be successfully incorporated into processed cheese (Sood & Kosikowski, 1979*b*).

Law (1980) described the results of Cheddar cheesemaking trials in which a proteinase–peptidase preparation from a *Pseudomonas* culture was incorporated into the curd at salting. The intensity of Cheddar flavour in 8-week-old cheeses with a low level of enzyme addition was equivalent to that in 16-week-old controls but on further ripening, the difference decreased and had disappeared at 22 weeks. Characteristic Cheddar flavour was masked by bitterness and other off-flavours when larger amounts of enzyme were added.

Law and Wigmore (1982*a*,*b*) compared the influence of acid, neutral and alkaline proteinases on proteolysis and flavour development in Cheddar. An acid proteinase from *Aspergillus oryzae* was very proteolytic in cheese and always caused bitterness over a 100-fold range of enzyme concentrations. A low level of Neutrase (*Bacillus subtilis*) enhanced flavour development but higher levels caused bitterness. Subtilisin and Ponase (*Streptomyces griseus*) produced very bitter cheese. Neutrase was considered to be the most promising of the enzymes tested, possibly because it is unstable in cheese and hence activity is somewhat limited. These studies showed that while it was possible to increase the concentration of small peptides and amino acids in 2-month-old cheese by up to 400%, increases greater than ~150% gave bitter cheese, probably due to an imbalance between the endo- and exopeptidases which Law and Wigmore (1983) attempted to correct this by combining a preparation of intracellular enzymes from starter streptococci with Neutrase. This combination gave better results than Neutrase alone, but flavour intensity did not increase *pro rata* with proteolysis, suggesting that subsequent amino acid transformations to sapid compounds were rate-limiting and were not catalysed by the streptococcal enzymes. This enzyme preparation is produced commercially by Imperial Biotechnology, London, UK, and marketed as 'Accelase'. Its use in large-scale commercial cheesemaking trials was described by Fullbrook (1987).

Frick *et al.* (1984) reported that the neutral proteinase, Rhozyme PII (*A. oryzae*), added to Colby curd at salting accelerated ripening without bitter flavour development. However, Fedrick *et al.* (1986*a*) reported that the lowest level of this enzyme that gave detectable flavour advancement also caused bitterness that intensified with increasing level of added enzyme. Law and Wigmore (1982*a,b*) also found that the acid proteinase of *A. oryzae* was unsatisfactory.

A comparative study on proteolysis and textural changes in granular Cheddar cheeses supplemented with Neutrase, calf lipase, Neutrase plus calf lipase or NaturAge (a culture–enzyme mixture from Miles Laboratories, Elkhart, IN) was undertaken by Lin *et al.* (1987). TCA-insoluble nitrogen increased rapidly in all proteinase-supplemented cheeses but free amino acid levels increased more slowly. Textural changes reflected gross proteolysis better than the formation of free amino acids; the flavours of the cheeses were not reported.

The combined influence of Neutrase, lac^- prt^- mutant starter and elevated temperature (15°C) was studied by Fedrick *et al.* (1986*b*). Storage at 15°C was the most effective single treatment, reducing ripening time by >50%. Neutrase alone gave ~25% reduction. A slightly bitter flavour was noted in the Neutrase-treated cheeses but did not significantly affect panel preferences or grade until late in ripening. Best results were obtained with a combination of Neutrase and mutant starter at 15°C.

A proteinase (*Penicillium candidum*)–peptidase (*Streptococcus lactis* or *Lactobacillus casei*) preparation for accelerating the ripening of Dutch, Tilsit or Lowicki-type cheese was described by Kalinowski *et al.* (1979, 1982). The preparation, when added to the cheese milk, accelerated proteolysis in cheese and approximately halved the ripening time.

Addition of a cell-free extract of *Lb. casei* to Cheddar curd accelerated proteolysis but the cheeses were bitter (El Soda *et al.*, 1981). In a later study (El Soda *et al.*, 1982), cell-free extracts from *Lb. lactis, Lb. helveticus* and *Lb. bulgaricus* accelerated, in the order given, the formation of soluble nitrogen and volatile fatty acids in Cheddar which developed marked bitterness after 2 months. A crude cell-free homogenate of *S. lactis,* containing mainly aminopeptidase activity, produced a high level of free amino acid in a model cheese system (Gripon *et al.*, 1977).

β-Galactosidase

The manufacturing time for Cheddar cheese was reported (Thompson & Brower, 1976; Marschke & Dulley, 1978) to be reduced, the flavour, body and texture of the resulting cheese improved and ripening accelerated by ~50% when the cheesemilk was treated with β-galactosidase. Apparently, the accelerating effect of β-galactosidase treatment was due to contamination of the β-galactosidase preparation (Maxilact from *Kluyveromyces lactis*) used with a proteinase which accelerated proteolysis during ripening (Hemme *et al.*, 1979; Marschke *et al.*, 1980; Grieve *et al.* 1983). However, Gooda *et al.* (1983) claim that stimulation of starter and secondary flora, and not a contaminating proteinase, is responsible for the improved flavour development in Cheddar cheese made from Maxilact-treated milk; however, microorganisms were not enumerated. More recent studies, all reporting the accelerating effect of β-galactosidase treatment of cheese-milk on the ripening of cheese made therefrom, include Farahat *et al.* (1985) (Ras cheese), Omer and El-Shibiny (1985) (Domiati), Kanawija and Singh (1987) (Cheddar), El-Zayat (1987) (Domiati) and Fernandez-Garcia *et al.* (1988) (Manchego). We are not aware of studies on the use of the Maxilact proteinase alone to accelerate cheese ripening although Hemme *et al.* (1979) used a heat-treated Maxilact preparation in which β-galactosidase, but not proteinase, was inactivated.

Method of enzyme addition

The method of adding exogenous enzymes to cheese has posed some problems. Essentially, three methods are available (i) direct addition to cheesemilk, (ii) direct addition to cheese curd and (iii) addition of encapsulated enzymes to cheese milk. Each method has advantages and limitations.

Direct addition of enzymes to cheese milk ensures uniform distribution of enzyme throughout the curd but poses at least three problems: (i) proteinases are usually water soluble and most of the added enzyme is lost in the whey, (ii) enzyme-contaminated whey must be heat-treated as soon as possible if the whey is to be used as a source of functional food proteins; the choice of enzyme is limited to those that are inactivated at temperatures below those that cause thermal denaturation of whey proteins, (iii) according to Law and King (1985), addition of sufficient Neutrase to milk to give the level of enzyme in the curd used by Law and Wigmore (1982*a*) caused a 25–80% reduction in rennet coagulation time, yielded a soft curd and at least 20% of the β-casein was hydrolysed at pressing; reduced cheese yields may be expected.

With the exception of FlavorAge (90% of the proteinase–peptidase in which and lost in the whey (Freund, 1986)) the exogenous soluble enzymes used in the studies reviewed were added to the curd with the salt. Since the diffusion coefficient of large molecules like proteinases is very low, this method is applicable only to Cheddar-type cheeses which are salted as chips at the end of manufacture and is not applicable to surface-salted (brine or dry) cheeses, which represent the majority of varieties. Even with Cheddar-type cheeses, the added enzyme will probably be concentrated at the surface of chips, which may be quite large. Uneven mixing of the salt–enzyme mixture with the curd may lead to 'hot-spots' where excessive proteolysis and lipolysis, with concomitant off-flavours, may occur.

Enzyme encapsulation offers the possibility of overcoming the problems outlined above. The microcapsules, being sufficiently large, are occluded in the curd; the main problem is to achieve the release of the enzymes after curd formation. Extensive studies by Olson and co-workers (Magee & Olson, 1981*a,b*; Magee *et al.*, 1981; Braun *et al.*, 1982; Kim & Olson, 1985; Braun & Olson, 1986*a,b*) showed that it is possible to encapsulate multi-enzyme systems, including cofactors, in milk fat and that these enzymes are active in model systems.

Law and King (1985) entrapped Neutrase in multi-lamellar visicles prepared from phosphatidylcholine and cholesterol. When the microcapsules, which ranged in diameter from ~0·2–4·5 μm, were added to milk they were retained in cheese curd. Apparently, the microcapsules ruptured in the cheese, due either to osmotic shock on salting or to phospholipase action, releasing active proteinase. Law (1986, 1987) reported that up to 90% of the visicles were retained in cheese curd.

Cheese slurries

Perhaps the most marked acceleration of ripening has been achieved using slurry systems, introduced by Kristoffersen *et al.* (1967). Cheddar cheese slurried in ~3% sodium chloride to 40% solids developed full flavour in 4–5 days when reduced glutathione was included. The process was refined by Singh and Kristoffersen (1970, 1971*b*, 1972). Cheddar cheese acquired the flavour characteristics of Brick cheese on extended ripening (7 days) and a Romano flavour could be developed by addition of lipase (Kristoffersen *et al.*, 1967). The slurry method was adopted for the rapid ripening of Feta by Zerfiridis and Kristoffersen (1970) and for Swiss cheese by Singh and Kristoffersen (1971*b*).

Ripened Cheddar cheese slurries have been successfully incorporated into processed cheese at levels up to 20% (Sutherland, 1975). Slurries of ripened cheese have also been added to cheese milk or curd to increase the rate of flavour development and reduce ripening time (von Bockelman & Lodin, 1974; Dulley, 1976; Abdel Baky *et al.*, 1982*a,b*); introduction of a high number of lactobacilli may be responsible for the acceleration of flavour development.

Another variant of the slurry system has been developed by Novo Industries A/S (Bagsvaerd, Denmark) for the production of 'Enzyme-Modified Cheese (EMC)'. Medium-aged cheese is mixed with water, 'emulsified', homogenised and pasteurised (to control the indigenous microflora). After cooling, exogenous enzyme, e.g. 'Palatase' (a *Mucor miehei* lipase) and/or proteinase is added at the required level and the mixture incubated at 40°C for 12–96 h. The mixture is then repasteurised (66–72°C for 4–8 min) to yield an EMC paste that may be used as such in processed cheese, soups, dips, dressings, snack foods, etc., or spray dried. EMC has 5–20 times the flavour impact of the mild

Cheddar from which it is made and may be used at 2–3% of a processed cheese blend.

A similar approach is adopted by the Miles Marschall Company (Madison, WI) to prepare the Marstar range of EMC products (Cheddar, Romano, Swiss) for use in processed cheeses or cheese-containing recipes (Talbott & McCord, 1981). Other methods for the preparation of EMC have been reviewed by Kilara (1985*a*) and Lee *et al.* (1987).

OTHER APPLICATIONS OF PROTEINASES

Dietary-products

Extensive protein hydrolysates for use in soups, gravies, flavourings and dietetic foods are generally prepared from soy proteins, gluten, milk proteins, meat scrap or fish protein by acid hydrolysis. Neutralisation results in a high salt content which is acceptable for certain applications but may be unsuitable for dietetic foods and food supplements. Furthermore, acid hydrolysis causes total or partial destruction of some amino acids. Enzymatic hydrolysis is a viable alternative (Adler-Nissen, 1986) but bitterness due to hydrophobic peptides is frequently encountered. Caseins are strongly hydrophobic and yield very bitter hydrolysates but bitterness may be eliminated, or at least reduced to an acceptable level, by one of several treatments (see Fox *et al.*, 1982; Adler-Nissen, 1986).

There is increasing interest in the production of casein-derived peptides with special nutritional or physiological properties; some of the possibilities were reviewed by Maubois (1984). Apart from the interest in casein hydrolysates for the nutrition of patients with digestive problems, e.g. cystic fibrosis or following abdominal surgery (see Adler-Nissen (1986) for references), the interest of nutritionists in casein hydrolysates has focussed on phosphopeptides derived from casein which stimulate the absorption of calcium (Sato *et al.*, 1986). Such peptides are resistant to proteolysis due to their high density of negative charges; they have been detected in the small intestine of the rat and may pass intact through the intestinal wall. Stimulation of calcium absorption by phosphopeptides is proportional to the degree of phosphorylation. Iron is also bound by phosphoserine residues although the influence of this binding on the bioavailability of iron is ambiguous. Casein phosphopeptide–iron complexes have been proposed as useful dietary supplements (Meisel & Frister, 1989). Methods for the production of caseinophosphopeptides for nutritional and/or medical applications have been developed (Brule *et al.*, 1982; Juillerat *et al.*, 1989).

The serine phosphate groups of casein may have specific metabolic functions (Williams & Sanger, 1959). One of the phosphate groups of β-casein is released faster than the others (West & Dalgleish, 1976) and one may be transferred to ADP by a specific AMP-independent casein kinase to form

ATP, suggesting that it may be a type of high-energy phosphate (Cochet *et al.*, 1983). The structure and function of the phosphorylated residues of casein were reviewed by West (1986).

Physiologically active peptides from milk proteins

Peptides with various types of physiological activity have been isolated from casein hydrolysates; at least some of these peptides are produced *in vivo* and may play a physiological role.

Caseinomorphins

The existence of several endogenous opioid peptides (endorphins) has been demonstrated in various mammalian tissues (Teschemacher, 1978). Although these peptides range in size from 5 to 31 residues, their *N*-terminal sequences are invariably H–Tyr–Gly–Gly–Phe. Peptides with opioid activity have also been identified in enzymatic digests of proteins derived from foods; these are called exorphins.

Peptides with opioid activity were detected at considerable concentrations in certain batches of baby foods, casein digests and cows' milk by Brantl and Teschemacher (1979). These peptides were isolated from enzymatic digests of casein and characterised as a family of peptides containing 4–7 amino acids with a common *N*-terminal sequence, H–Tyr–Pro–Phe–Pro–, and 0–3 additional residues (glycine, proline and isoleucine) (Brantl *et al.*, 1979; Henschen *et al.*, 1979). The peptides, which represent residues 60–63/6 of β-casein, were called β-caseinomorphins 4–7, respectively, and were found to possess μ-type opiate receptor activity. Obviously, since their amino acid sequences differ from those of the endorphins, the caseinomorphins (CM) represent a new class of opioid peptides. β-CM-5 is the most effective of the β-CMs, which are 300–4000 times less effective than morphine (Brantl *et al.*, 1981). A new β-caseinomorphin, β-CM-11 (β-CN f60–70) was isolated from the duodenal chyme of minipigs (Meisel, 1986; Meisel & Frister, 1989). It is less effective than β-CM-5, but it may be hydrolysed to smaller, more active β-CM by peptidases in the brush border. Various physiological effects of β-CMs, which are very resistant to enzymes of the gastrointestinal tract and appear in the juice of the small intestine following ingestion of milk were investigated by Brantl's group (see Brantl & Neubert, 1986; Hautefeuille *et al.*, 1986). Teschemacher *et al.* (1986) found no evidence for β-CM in human plasma after ingestion of cows' milk and milk products. However, Teschemacher (1987) reported the occurrence of β-CM precursor in the plasma of new born infants, and of pregnant, lactating and non-lactating females.

The sequence 51–57 of human β-casein, Tyr–Pro–Phe–Val–Glu–Pro–Ile, corresponds to bovine β-casein 60–66 (i.e. Tyr–Pro–Phe–Pro–Gly–Pro–Ile). The opioid activity of synthetic human β-CMs-4 and 5 was considerably less than that of bovine β-CM-4 and 5 (Brantl, 1984). Several synthetic peptides with sequences corresponding to segments of human β-casein 51–57 had opioid activity (Yoshikawa *et al.*, 1986). Synthetic peptides corresponding to human β-casein 59–63 and 41–44 also had weak opioid activity, as did the peptides Tyr–Gly–Leu–Phe, which occurs in α-lactalbumin, and Tyr–Leu–Leu–Phe, which occurs β-lactoglobulin (see also Chiba & Yoshikawa, 1986).

Two exorphins, representing residues 90–95 and 90–96 (Arg–Tyr–Leu–Gly–Tyr–Leu(–Glu)), have also been isolated from peptic hydrolysates of α-casein (Zioudrou *et al.*, 1979). The *N*-terminal arginine appears to be important for activity (Loukas *et al.*, 1983). Obviously, the α-casein-derived exorphins have considerably different structures from the β-caseinomorphins.

The sequences 25–34, 35–41 and 57–60 of bovine κ-casein have opioid antagonistic activity, especially when in the form of methyl esters, e.g. Ser–Arg–Tyr–Pro–Ser–Tyr–OCH_3; these peptides were called casoxins (Chiba & Yoshikawa, 1986; Chiba *et al.*, 1989). Three peptides with opioid antagonistic properties were isolated from peptic digests of lactotransferrin: Tyr–Leu–Gly–Ser–Gly–Tyr (lactoferroxin A), Arg–Tyr–Tyr–Gly–Tyr (lactoferroxin B), Lys–Tyr–Leu–Gly–Pro–Gln–Tyr (lactoferroxin C) (Yoshikawa *et al.*, 1988). The activity of the three peptides was markedly increased by methylation of the α-carboxyl group.

Thus, some milk products contain pre-formed peptides with opioid properties while all the major milk proteins contain sequences which, when liberated by gastrointestinal proteinases, possess opioid activity. These peptides are very resistant to proteolysis by gastrointestinal proteinases and can be absorbed intact from the intestine. They have been shown to have physiological activity *in vitro* but their activity *in vivo* is as yet uncertain.

Immunomodulating peptides

Migliore-Samour *et al.* (1989) reported that enzymatic digests of bovine and human caseins contain immunomodulating peptides which stimulate the phagocytic activity of human macrophages *in vitro* and exert a protective effect *in vivo* in mice against *Klebsiella pneumoniae* infection.

DNA-synthesis-stimulating peptides

Azuma *et al.* (1989) isolated two peptides for hydrolysates of human β-casein (f1–18 and f101–117) and one peptide (f177–183) from bovine β-casein that stimulated DNA synthesis in BALB/C 3T3 cells.

Platelet-modifying peptides

The undecapeptide, Met–Ala–Ile–Pro–Pro–Lys–Lys–Asn–Gln–Asp–Lys (residues 106–116 of bovine κ-casein) inhibited the aggregation of ADP-treated platelets; its behaviour was similar to that of a structurally related *C*-terminal dodecapeptide (residues 400–411) from human fibrinogen γ-chain (Jolles *et al.*, 1986). This peptide was produced from the caseino(glyco) macropeptide, residues 106–169 of κ-casein, produced by the action of chymosin on κ-casein or whole casein. Shorter peptides, κf106–112 and κf113–116 also inhibited platelet aggregation but were less effective than the undecapeptide. Fiat *et al.* (1989) reported the isolation of a peptide with similar properties from a hydrolysate of lactotransferrin.

Modification of protein functionality

Milk proteins are among the principal functional proteins used in food products (Fox, 1989*a*). In general, milk proteins process very good functional properties but suffer some limitations, notably the insolubility of casein in the pH range 3·0–5·5. The functional properties of milk proteins may be improved by limited proteolysis (see Kilara, 1985*b*; Nakai & Li-Chan (1989) for a review of the enzymatic modification of milk proteins). An acid-soluble casein, free of off-flavour and suitable for incorporation into beverages and other acid foods, was prepared by limited proteolysis (Hagget, 1974). The antigenicity of casein is destroyed by proteolysis and the hydrolysate is suitable for use in milk protein-based foods for infants allergic to cows' milk (Takase *et al.*, 1979). Controlled proteolysis improves the meltability of directly-acidified cheese but excessive proteolysis causes bitterness (Lazaridis *et al.*, 1981). Casein solutions are very viscous at relatively low concentrations (>20%, w/v), which increases the cost of drying caseinates and causes some processing problems. Hooker *et al.* (1982) investigated the potential of limited proteolysis by Alcalase to reduce the viscosity of sodium caseinate; he concluded that owing to the high concentration-dependence of the viscosity of caseinate, an increase of only 1% protein was possible following an acceptable degree of proteolysis and this was too low to be commercially useful.

The surface activity of sodium caseinate was increased considerably by treatment with plasmin, apparently owing to the formation of γ^2- and γ^3-caseins (βf106–209, βf108–209) which were very surface active (Wilson *et al.*, 1989). The proteolytic specificity of trypsin is similar to that of plasmin and its action on casein might be expected to have a similar effect on surface activity. Pepsin selectively cleaves α_{s1}-casein between residues 23 and 24. The peptide α_{s1} f24–199 (α_{s1}-I) is less hydrophobic than α_{s1}-casein and has inferior emulsifying properties (Shimizu *et al.*, 1984). The emulsifying properties of the peptide, α_{s1} (f1–23) were very poor > pH 7 (Shimizu *et al.*, 1986). Preliminary work on the surface activity of β-casein peptides (f1–25, f1–28, f193–209) and κ-casein f106–169 (glycomacropeptide) has been published (Lee *et al.*, 1987); the emulsifying properties of the individual peptides were poor but a mixture of β-casein f193–209 and κ-casein f106–169 had good emulsifying properties.

Limited proteolysis of lactalbumin, which is insoluble and has very poor functional properties, yields a product that is almost completely soluble > pH 6 (Monti & Jost, 1978); although the product was slightly bitter, it appears promising as a food ingredient. A hydrolyzate with ~70% solubility at pH 6–7 was produced from lactalbumin by treatment with Alcalase (Ennis & Harper,

1986). The hydrolysate had good emulsifying properties but did not whip or form gels on heating. Almost complete solubilisation of lactalbumin was obtained by treatment with Alcalase to a degree of hydrolysis of 11–12%; solubility decreased on more extensive hydrolysis (Smith *et al.*, 1989). Various applications of hydrolysed lactalbumin in foods, therapeutic and cosmetic products have been proposed (see Smith *et al.*, 1989). Solubilisation of heat-denatured whey proteins by hydrolysis with papain or Protease A2 (*B. licheniformis*) was also reported (Saint-Paul *et al.*, 1984). Limited proteolysis of whey protein concentrate decreased its emulsifying capacity, increased its specific foam volume but decreased foam stability (Kuelher & Stine, 1974). The heat stability of whey protein concentrate is considerably improved by limited hydrolysis without concomitant impairment of other functional properties or off-flavour development (Hildalgo & Gamper, 1977).

The plastein reaction

Exploitation of the plastein reaction to modify the nutritional and functional properties of proteins was initiated in about 1970 by a group at Tokyo University whose work has been reviewed, e.g. by Eriksen and Fagerson (1976); Fujimaki *et al.* (1977) and Watanabe and Arai (1988) [see Chapter 20 for more recent work]. The Tokyo group considers the mechanism of plastein formation to be via transpeptidation, possibly with some peptide condensation. However, Sukan and Andrews (1982*a*,*b*) consider hydrophobic interactions, with possible contributions from ionic attractions, as the most likely mechanism.

Most of the work on plastein formation has been on soya protein and to a lesser extent on single cell proteins, zein and ovalbumin. A comprehensive study on plastein formation from casein or skim milk powder was undertaken by Sukan and Andrews (1982*a*,*b*). Optimum conditions for plastein formation were a 20–35% (w/v) peptic hydrolysate with peptides in the molecular weight range 400–800, pH 4–5, 37°C for 24 h (or 50°C for 4–6 h). The same proteinase should be used for both protein hydrolysis and plastein synthesis (to avoid over-hydrolysis). Under ideal conditions, plastein yield never exceeded 27% of the weight of the peptide material taken and the authors concluded that at present the plastein reaction is of nothing more than curiosity value. Plastein yields from individual caseins were in the order $\beta > \alpha_s > \kappa$. The plastein formed thixotropic gels which were completely dissolved and the plasteins dissociated by chotropic agents, indicating that covalent bonds were not formed during plastein formation. Plastein yields of ~37% were reported by Sciancalepore and Longone (1988) who used α-chymotrypsin on the 450–1450 mol. wt fraction of a peptic hydrolysate of whey protein.

As far as we are aware, the functional properties of casein or whey protein plasteins, other than gel characteristics (Sukan & Andrews, 1982*b*), have not been reported. Since casein has good biological and functional properties, little advantage can be expected from its modification via the plastein reaction but lactalbumin, which has very poor functionality, should benefit.

LIPASES

Lipases have a number of relatively low-volume applications in dairy processing which have been reviewed, e.g. by Neelakanten *et al.* (1971), Seitz (1974), Arnold *et al.* (1975), Shahani *et al.* (1976), Huang and Dooley (1976) and Kilara (1985*a*). Considerable information (see e.g. Brockerhoff & Jensen, 1974; Shahani, 1975; Borgstron & Brochman, 1984) has been accumulated on the characteristics of lipolytic systems which makes it possible to select a lipase best suited for a specific purpose.

The principal application of lipases in dairy technology is in cheese manufacture, particularly hard Italian varieties. The characteristic 'piquant' flavour of these cheeses is due primarily to short-chain fatty acids resulting from the action of lipase(s) in the rennet paste traditionally used in their manufacture. Rennet pastes are prepared from the stomachs of calves, lambs or kids slaughtered after suckling; the stomachs and contents are held for about 60 days prior to maceration. Because of possible risks to public health, the use of rennet pastes, which have proteolytic and lipolytic activities, is not permitted in some countries. However, attempts to produce cheeses using rennet extract, which is prepared from washed vells (stomachs) and does not contain a lipase, were unsuccessful. The lipase in rennet paste is of oral origin, being secreted by a gland at the base of the tongue, which is stimulated by suckling; the secreted lipase is washed into the stomach with the ingested milk. The physiological significance of oral lipase, which is secreted by several species and is generally referred to as pre-gastric esterase (PGE), has been extensively investigated and the considerable literature has been comprehensively reviewed by Nelson *et al.* (1977). PGE shows a high specificity for short chain fatty acids, especially butyric acid, esterified on the sn-3 position of glycerol, although some interspecies differences in specificity have been reported. Most PGEs are maximally active at 32–42°C, pH 4·8–5·5 and in 0·5 M sodium chloride. Calf, kid and lamb PGEs, partially purified from commercial extracts, have molecular weights of 172, 168 and 150 kDa, respectively (Lee *et al.*, 1980), i.e. considerably larger than pancreatic or milk lipases.

A liquid extract of calf oral tissue as an alternative source of PGE for Italian cheese manufacture was patented by Farnham in 1950. Extensive studies (see Harper, 1957) showed that PGE, either as rennet paste or extract, was essential for the development of the characteristic flavour of Romano and Provolone cheeses; PGE from calf, kid or lamb, which are now commercially available, gave satisfactory results, although slight differences in specificity render one or other more suitable for particular applications. Although PGE extract is now widely used in the US and elsewhere in the manufacture of hard Italian cheese varieties, connoisseurs of Italian cheese claim that rennet paste gives superior results, and it is cheaper. The superiority of rennet paste may be due to the presence of enzymes in addition to chymosin and PGE. Such enzymes could be of microbial origin since bacterial growth probably occurs during ageing of the stomachs prior to maceration. A second lipase, termed gastric lipase, has been identified in an extract of cleaned gastric tissue and partially characterised (Richardson *et al.*, 1971); a combination of calf gastric lipase and goat PGE gave Cheddar and Provolone cheese a superior quality to that made with PGE alone. Traditional rennet paste would be expected to contain both PGE and gastric lipase. However, Nelson *et al.* (1977) expressed reservations on the occurrence of a gastric lipase distinct from PGE.

Mucor miehei secretes a lipase that is reported to give satisfactory results in Italian cheese manufacture (Peppler *et al.*, 1976). The enzyme has been characterised (Moskowitz *et al.*, 1977) and is commercially available as 'Piccantase' prepared by Societe Rapidase, France. The lipases secreted by selected strains of *Penicillium roqueforti* and *P. candidum* are considered to be potentially useful for the manufacture of Italian and other cheese varieties (Kornacki *et al.*, 1979).

Extensive lipolysis also occurs in Blue cheese varieties and in addition to making a direct contribution to flavour, the free fatty acids serve as substrates for fungal enzyme systems in the biosynthesis of methyl ketones which are the principal contributors to the typical flavour of Blue cheese (Kinsella & Hwang, 1977; Gripon, 1987). *P. roqueforti* lipase predominates the ripening of Blue cheeses, perhaps with a contribution from indigenous milk lipase. However, Blue cheese ripening may be accelerated and quality improved by the use of added lipases (Arnold *et al.*, 1975; Jolly & Kosikowski, 1975*a*; Anon, 1976). In the study by Jolly and Kosikowski (1975*a*), *Aspergillus* lipases, which also contained proteinases, gave best results and reduced ripening times by >50%. All the normal parameters of ripening were elevated in the enzyme-treated cheeses. Apparently, the oxidative system of *P. roqueforti* was stimulated by the elevated levels of free fatty acids. Proteolysis in these enzyme-supplemented cheeses was studied further by Jolly and Kosikowski (1978). This method is patented (Kosikowski and Jolly, 1976).

Blue cheese is a popular ingredient for salad dressings and cheese dips. High-quality natural cheese is not normally required for these applications and there is considerable interest in the production of cheaper substitutes. A 'Blue cheese product' prepared by mixing freeze- or spray-dried skim milk ultrafiltration retentate with cream that had been incubated (20°C for 40 h) with fungal lipase (*Aspergillus oryzae*), *P. roqueforti* spores and a lactic acid culture is described by Jolly and Kosikowski (1975*b*). Blue cheese flavour has also been developed in milk fat by culturing with *P. roqueforti* or perhaps by treatment with the isolated lipolytic system of *P. roqueforti* (Nelson, 1970*a,b*; Dwivedi & Kinsella, 1974*a,b*; Jolly & Kosikowski, 1975*c*). The production of a fast-ripening particulate Blue cheese was described by Harte and Stine (1977).

Although, Cheddar-type and Dutch-type cheeses undergo little lipolysis during ripening, it has been claimed that addition of rennet paste (Babel & Hammer, 1945) or gastric lipase (Chaudhari & Richardson, 1971; Richardson *et al.*, 1971) improves the flavour of Cheddar cheese, especially that made from pasteurised milk. Several patents have been issued for the use of lipases to improve the flavour of 'American' or 'processed American' cheeses (see Arnold *et al.*, 1975; Nelson *et al.*, 1977). The enzyme mixtures used by Iwasaki and Kosikowski (1973), Kosikowski and Iwasaki (1975), and Sood and Kosikowski (1979*a,b*) to accelerate Cheddar cheese ripening contained lipases. Law and Wigmore (1985) reported that the addition of PGE or *M. miehei* lipase, with or without proteinase (Neutrase), to Cheddar cheese curd did not enhance flavour development; the lower lipase level used had a slight, but not significant, detrimental effect on the flavour quality of some cheeses due to slight rancidity while the higher level of lipase used invariably caused marked rancidity which reduced the quality score significantly without increasing perceived intensity of cheese flavour.

The FlavorAge enzyme system produced by Chr Hansen Laboratory (Milwaukee, WI) contains an unique lipase from *A. oryzae* which has an exceptionally high specificity for the C-6–C-8 acids (Arbige *et al.*, 1986) Another interesting feature of this enzyme is its association characteristics: it forms micelles, ~0·2 μm in diameter, in aqueous media as a result of which ~94% of the enzyme added to milk is recovered in the cheese curds. The formation of short-chain fatty acids paralleled flavour intensity in Cheddar cheese containing this enzyme. In contrast to the free fatty acid profile caused by calf PGE, which liberated high concentrations of C-4:0, the free fatty acid profile in the cheese containing *A. oryzae* lipase was similar to that in the control cheese except that the level of free fatty acids was much higher (Arbige *et al.*, 1986).

Frick *et al.* (1984) compared the fatty acid profiles in Colby cheese to which Genencor *A. oryzae* lipase/proteinase or Miles 600 lipase plus proteinase (e.g. *A. oryzae* Rhozyme PII) were added. *A. oryzae* lipase released longer chain free fatty acids than did the Miles 600 lipase. The latter produced a Romano-type flavour while *A. oryzae* lipase produced a flavour more typical of an aged Cheddar at similar enzyme activities, as determined by free fatty acid release.

Addition of an unspecified lipase to Samsoe yielded a cheese with a flavour that closely resembled that of Greek Kasseri cheese (Jensen, 1970). Feta cheese produced from cow's milk with a 1:1 blend of *Streptococcus lactis* and *Lactobacillus casei* and a blend of kid and lamb PGEs developed the body, flavour and texture of authentic Feta cheese (Efthymiou & Mattick, 1964). Addition of unspecified microbial lipases to ultrafiltration retentate yielded Feta-type cheese with typical flavour (Mohmoud & Kosikowski, 1980). The flavour of Egyptian Cephalotypre Ras cheese was improved by addition of PGE or lipases from *Mucuor miehei* or *M. pusillus* (El Shibiny *et al.*, 1978). Addition of low levels of PGE improved and accelerated (~50% reduction in ripening time) flavour development in Domiati cheese but prolonged ripening led to rancid off-flavours in the enzyme-treated cheeses (El Neshawy *et al.*, 1982). The flavour of Latin America White cheese was improved by addition of low levels of PGE (Torres & Chandan, 1981).

Lipases are used to hydrolyse milk fat for a variety of uses in the confectionery, candy, chocolate, sauce and snack food industries; the partially hydrolysed fat imparts a greater intensity of butter-like flavour to the products and delays staling, presumably through the emulsifying influence of the di- and monoglycerides. Table 3, from Nelson *et al.* (1977), lists some of the numerous applications claimed for PGE. Other applications are discussed by Nelson (1972), Seitz (1974), Arnold *et al.* (1975), Shahani *et al.* (1976) and Kilara (1985*a*).

An important new application of lipases is in the trans/interesterification of fatty acids on triglycerides to modify melting characteristics and hence their rheological properties.

TABLE 3
Application of PGE Modified Milk Fat Products (Modified From Nelson *et al.*, 1977)

Bakery and related products
Cake and cookie mixes
Cheese-cake mixes
Chemically leavened bakery products
Dinner rolls
Pancake mixes
Sweet doughs
Butter flavour
Butter flavour on turkey roasts
Butter sauce
Popcorn oils
Cheese-flavoured foods
Cheese dips
Cheese flavours, Italian
Cheese sauces
Confectionery products
Chocolate syrups and beverages
Coconut
Compound coatings
Cremes cream centres
Fudges
Milk chocolate
Peanut butter
Toffee, caramel, butterstotch
Dairy alternate products
Coffee whiteners
Filled and imitation dairy products
Margarines
Egg formulations
Salad dressings
Sauces
Butter sauce
Pizza sauce
Sauces and gravies
Soups
Snack items

The treatment of milk with phospholipases C and D is reported (Shipe *et al.*, 1975) to increase stability to oxidative rancidity, presumably by modification of the fat globule membrane.

β-GALACTOSIDASE

Lactose is by far the dominant carbohydrate in milks which are, in turn, the only significant natural sources of lactose. Bovine milk contains ~4·8% lactose while human milk contains ~7%. Lactose is an important source of energy for the new-born mammal but when very energy-dense milks are required (mammals in aquatic or polar environments), its lipid rather than its lactose content is increased; in fact there is a fairly good inverse relationship between fat and lactose contents.

Among sugars lactose possesses many fairly unique properties: low solubility (~18 g/100 ml water at 20°C); a marked tendency to form supersaturated solutions which are difficult to crystallise; when crystallisation does occur, the crystals are hard and sharp and unless kept to dimensions <20 μm cause a sandy texture in foods; crystallisation is complicated by its mutarotation characteristics since α- and β-lactose differ considerably in solubility and degree of hydration; it has low sweetness (16% as sweet as sucrose at 1%, w/v); owing to its crystallisation and mutarotation characteristics, it is hygroscopic and may cause 'caking' of dairy powders; it has a strong tendency to absorb flavours and odours. Unless properly controlled, these characteristics may cause defects in concentrated, dehydrated and frozen dairy products. However, these same characteristics may be exploited to render lactose an interesting and useful food additive, e.g. as a free-flowing agent, an agglomerating (instantising) agent, an additive to stabilise colour, flavour and texture, especially when concomitant sweetness is undesirable, as a reducing sugar in products in which Maillard browning is desirable as a source of colour and flavour. The chemistry, properties, problems, modifications and applications of lactose and lactose derivatives have been reviewed (Nickerson, 1974; Shukla, 1975; Fox, 1980, 1985; Hobman, 1984; Zadow, 1984; Modler, 1985; IDF, 1987*a*).

It is claimed that lactose promotes the intestinal absorption of calcium and phosphorus and hence should be nutritionally beneficial, especially in infant nutrition (see Morrissey, 1985). However, lactose is involved in two enzyme-deficiency diseases: lactase deficiency and galactosaemia. There are in fact two forms of galactosaemia, both arising from the congenital deficiency of an enzyme in the Leloir pathway for galactose metabolism (see Flynn, 1985). Classical galactosaemia is due to a deficiency of galactose-1-phosphate:uridyl transferase; ingested galactose (e.g. from lactose) is phosphorylated to galactose 1-phosphate which is not metabolised further, leading to the accumulation of galactose and galactose-1-phosphate. Galactosaemic infants appear normal at birth but develop various symptoms, including mental retardation, unless put on a galactose-free diet within 2–3 months. The second form, galactokinase-deficient galactosaemia, results in the failure to phosphorylate galactose, some of which is metabolised to galactitol which accumulates in the eye, causing cataracts.

Disaccharides must be hydrolysed to monosaccharides in the intestine prior to absorption, in the case of lactose by β-galactosidase. The vast majority of infants secrete adequate levels of β-galactosidase in the brush border of the small intestine to hydrolyse ingested lactose; however a small minority of infants secrete an inadequate level of β-galactosidase. The level of intestinal β-galactosidase reaches a maximum shortly after birth and declines thereafter to a low level. With the exception of north-western Europeans and a few African tribes, the level of intestinal β-galactosidase becomes so low post-weaning as to render the subject incapable of hydrolysing ingested lactose at an adequate rate, leading to lactose intolerance. The unhydrolysed lactose passes to the large intestine where it leads to flatulence, cramps, diarrhoea and possibly death. Thus, only a minority of the world's population can consume large quantities of lactose-containing foods with impunity. The subject of lactose intolerance has been reviewed extensively (e.g. Dahlqvist, 1984; Paige & Davies, 1985).

The third feature of the lactose problem is the development of economic outlets for the large quantities ($\sim 4 \times 10^9$ kg per annum) available from cheese and casein wheys. The technology for lactose production is well developed but only ~10% of the potentially available lactose is recovered as such; although lactose has a number of special food applications, the market appears to be static (see Fox, 1980). Increasing concern about environmental pollution has focussed attention on more complete utilisation of whey for which there are many options (see Coton, 1980; Fox, 1980; Hobman, 1984; Zadow, 1984; Modler, 1985; IDF, 1987*a*).

The dairy industry has developed methods for controlling the physico-chemical problems posed by lactose and lactose intolerance may not be too serious if lactose-containing products are introduced gradually to the diet. However, all the various problems posed by lactose (technological, nutritional, utilisation) may be solved via hydrolysis by β-galactosidase (lactase) to the less problematic sugars, glucose and galactose.

β-Galactosidase (β-D-galactoside galactohydrolase; EC 3.2.1.23) typically catalyses the hydrolysis of lactose to its component monosaccharides, glucose and galactose. The use of β-galactosidase in dairy technology has been considered as one of the most promising applications of exogenous enzymes in food processing. However, although many interesting applications have been demonstrated, the use of β-galactosidase has not yet become commercially successful for economic reasons. The voluminous literature on the preparation, properties and uses of β-galactosidases has been the subject of several reviews (including Shukla, 1975; Finocchiaro *et al.*, 1980; Mahoney, 1985; Gekas & Lopez-Leiva, 1985; Harju, 1987). It is not intended comprehensively to review again the literature on β-galactosidase; only more recent publications are reviewed.

Although β-galactosidase is widely distributed in plant, animal and microbial sources (Shukla, 1975; Mahoney, 1985; Gekas & Lopez-Leiva, 1985), only the enzymes from *Aspergillus niger, A. oryzae, Kluyveromyces lactis* (*K marxianus*), *K. fragilis, Bacillus stearothermophilus* and *Escherichia coli* are commercially available. β-Galactosidases from the first five species listed currently have GRAS (generally recognised as safe) status but β-galactosidase from *E. coli* is not permitted for food use and is of interest in the food industry only for the enzymatic determination of lactose. β-Galactosidases from several sources have been isolated and characterised; the important properties have been summarised by Mahoney (1985) and Gekas and Lopez-Leiva (1985). In general, β-galactosidases of fungal origin have acid pH optima (2·5–4·5) and are therefore best suited for use in acid wheys while yeast and bacterial β-galactosidases have pH optima in the range 6–7·5 and are suitable for use in milk or rennet wheys. Isolation of β-galactosidases with high thermal stability has been reported recently. Heat stable enzymes are attractive because they can be used at high temperatures at which microbial growth is slow or absent. The heat stability of *K. lactis* β-galactosidase is 20–50 times greater in milk than in buffer systems due to the combined effects of casein and lactose (Mahoney & Wilder, 1988). However, the activity of *K. lactis* β-galactosidase in whey, ultrafiltration permeate and skim milk decreased in the order given (Bernal & Jelen, 1985). The β-galactosidase from *Streptococcus salivarius* subsp. *thermophilus* is also much more stable at 60°C in milk (half-life of 146 min) than in buffers (Chang & Mahoney, 1989).

β-Galactosidases from several sources have been immobilised by encapsulation, entrapment in fibres, gels or semi-permeable membranes, adsorption or covalent attachment by a variety of techniques on various supports, e.g. porous glass, collagen, cellulose derivatives, various resins (see Shukla, 1975; Finocchiario *et al.*, 1980; Mahoney, 1985; Gekas & Lopez-Leiva, 1985).

The principal applications of β-galactosidase in dairy technology are:

(1) production of low-lactose milk and dairy products for β-galactosidase-deficient patients,
(2) acceleration of yoghurt and cheese production,
(3) modification of dairy products for use in ice-cream, baked goods, yoghurt, etc,
(4) pre-treatment of milk for freezing,
(5) production of syrups and sweeteners for food applications.

In spite of the widespread incidence of β-galactosidase deficiency among non-Caucasians, such subjects adjust to lactose-containing diets if introduced gradually and the response to lactose is moderated by other components in the diet; researchers are divided as to the desirability of including milk in the diets of lactose-intolerant subjects (see Paige & Davis, 1985). However, it appears to be generally agreed that treatment with β-galactosidase would enhance the nutritional value of dairy products in such cases and render protein-rich dairy products suitable for supplementation of nutritionally deficient diets. Recent studies on this subject include Barillas and Solomons (1987). Lactose-hydrolysed milk is commercially available in the US but is more expensive than normal milk; β-galactosidase is also available in powder or liquid form for home use (see Mahoney, 1985). Direct addition of β-galactosidase to milk at mealtime, 'enzyme replacement therapy', has also given satisfactory results (Rosado *et al.*, 1984; Solomons *et al.*, 1985); *K. lactis* β-galactosidase gave better results than the *A. niger* enzyme. A promising method of reducing treatment cost is the addition of very low levels of soluble lactase to UHT milk which during prolonged storage induces extensive hydrolysis (Dahlqvist *et al.*, 1977). This approach has been commercialised by the Tetra-Pak Company (Lund, Sweden) (Rausing, 1982). The increased sweetness of lactose-hydrolysed milk does not appear to be a problem and may actually be preferred by some individuals (see Paige & Davis, 1985).

β-Galactosidase can act as a transferase resulting in the synthesis of several oligosaccharides, the range and concentration of which appear to vary with the source of the β-galactosidase and duration of treatment; many of the oligosaccharides are β-(1→6) galactosides which are not hydrolysed in the small intestine but pass into the large intestine where they are acted on by bacteria leading to intestinal disturbances, mainly flatulence (Huber *et al.*, 1976; Asp *et al.*, 1980; Burvall *et al.*, 1980). An HPLC method for the analysis of oligosaccharides in lactose hydrolysates was described by Jeon and Mantha (1985).

Hydrolysis of lactose in milk for yoghurt has been suggested as a means of increasing the sweetness of yoghurt without a concomitant increase in calories; it is reported to reduce fermentation time (see Mahoney, 1985). Although Gunther and Burger (1982) reported that lactose-hydrolysed yoghurt had superior texture and consistency, with less wheying off, than controls, Broome *et al.* (1983) found that all but one commercial lactase preparation tested caused off-flavours in yoghurt; even the preparation that did not cause off-flavours was considered of doubtful benefit in most cases. Off-flavours in yoghurt prepared from lactase-treated milk were also reported by Dariani *et al.* (1982). Treatment of skim milk with a commercial β-galactosidase preparation did not stimulate acid production by thermophilic *lactobacilli* but did stimulate *Bacillus thermophilus*, probably through the action of a contaminating proteinase (Hemme *et al.*, 1979). Yoghurt and other cultured milks are well tolerated by lactose-intolerant subjects, apparently due to the secretion of β-galactosidase by the thermophilic culture used (see Savaiano & Levitt, 1987; Martini *et al.*, 1987). However, Newcomer *et al.* (1983) reported that lactase-deficient patients with irritable bowel syndrome were as intolerant to acidophilus milk as to regular milk.

It is also claimed that the manufacturing time for Cheddar and other cheeses is reduced by pre-treatment of cheese milk with β-galactosidase and, possibly more significantly, the quality of the cheese is improved and ripening is accelerated. However, acceleration of cheese ripening by lactase has been attributed to a proteinase present in the commercial lactase preparations used rather than to the action of β-galactosidase (see the section on proteinases (p. 234), and Mahoney, 1985).

Hydrolysis of lactose improves the functionality of milk powder in bakery products: the glucose moiety, being fermentable by baker's yeast, increases loaf volume while the non-fermentable galactose contributes to flavour and crust colour through Maillard browning. Frozen storage of milk is attractive under certain circumstances but the caseinate system is destabilised unless certain precautions are observed; since the destabilisation is due, indirectly, to the consequences of lactose crystallisation, pre-hydrolysis of lactose using β-galactosidase provides a freeze-stable product (see Muir, 1984; Mahoney, 1985).

Dulche de Leche, a sweetened concentrated dairy product, is popular in Latin America as a dessert or spread. Lactose crystallisation is a major problem in this highly concentrated product. Growth of *K. lactis* in milk for the preparation of Dulche de Leche has been recommended for the control of lactose crystallisation (Sabioni *et al.*, 1984*a*). About 50% of the lactose was hydrolysed in 12 h (at $5{\cdot}2 \times 10^7$ cells/ml) or 100% in 24 h. When the delactosed milk was mixed in various proportions with normal milk and concentrated, no lactose crystallisation occurred even when one part of lactose-hydrolysed milk was mixed with two parts of normal milk and no significant changes in flavour were noted. The use of permeabilised cells of *K. lactis* for the same application was found to be very satisfactory (Sabioni *et al.*, 1984*b*). Presumably, isolated β-galactosidase could be used successfully in the preparation of Dulche de Leche.

Lactose crystallisation may also be a problem in conventional sweetened condensed milk; although the problem can be controlled by appropriate manufacturing steps, pre-hydrolysis with β-galactosidase offers an alternative solution.

β-Galactosidase was used to prevent lactose crystallisation in a whey retentate–buttermilk powder spread (Patocka & Jelen, 1988); acid (fungal) β-galactosidase was preferable to neutral (yeast) enzyme and hydrolysis of 30% of the lactose was sufficient. Lactose-hydrolysed whey appears to be suitable for incorporation into ice-cream in which up to 25% of the skim milk solids may be replaced by hydrolysed whey syrup without adverse effects on quality; 50% of the sucrose may also be replaced by whey syrup (see Mahoney, 1985; Harju, 1987; Martinez & Speckman, 1988*a*). Lactose-hydrolysed whey may be fed in larger amounts than normal whey to animals, especially pigs, (see Harju, 1987).

Probably the principal commercial interest in β-galactosidase is for the production of syrups as a profitable outlet for lactose. Glucose–galactose syrups are ~70% as sweet as sucrose and about four times sweeter than lactose. The hydrolysis of lactose to glucose and galactose by strong mineral acids has been known for a long time and there has been renewed interest in the process, using free acid or ion-exchange resins, as a means of producing glucose–galactose syrups. Production costs are reported to be lower than for enzymatic hydrolysis but acid hydrolysis is applicable only to purified lactose solutions or perhaps ultrafiltration permeates (see McBean, 1979; Gekas & Lopez-Leiva, 1985).

The technological feasibility of producing glucose–galactose syrups by the action of β-galactosidase (soluble or immobilised) is now well established (see Mahoney, 1985; Gekas & Lopez-Leiva, 1985). Recent studies on the immobilisation of β-galactosidase include Mann and Thompkinson (1988) and Roland (1986). Reactor design (Fauqueux *et al.*, 1984; Huffman & Harper, 1985) and cheaper methods of immobilisation appear to attract much attention. A low cost system using bone charcoal was reported by Daniels (1985). Permeabilised (heat and solvent) cell of *K. fragilis,* immobilised on a highly porous diatomite carrier hydrolysed the lactose in whey very effectively (Martinez & Speckman, 1988*b*). The performance of β-galactosidase from *A. oryzae,* immobilised on chitosan-coated silica gel, was reported by Griethuysen-Dilber *et al.* (1988); the study was performed at pH 6·0 (to avoid protein fouling at the optimum pH for the enzyme, 4–5). Although both the activity and stability of the enzyme were lower at pH 6 than pH 3–4, the performance at the higher pH was considered satisfactory. Immobilisation of *A. oryzae* β-galactosidase in a macroporous beaded carrier, chemically similar to Plexiglas, was reported by Sprossler and Plainer (1983); the beads were resistant to compression and enzyme leakage was very low; the estimated cost of producing lactose-hydrolysed whey was ~6 cents/kg. Soluble enzyme from *K. lactis* for the hydrolysis of concentrated (up to 30% lactose) sweet whey permeate was used by Giec and Kosikowski (1983) and Chiu and Kosikowski (1985).

Production of β-galactosidase from *K. fragilis* grown in whey and the permeabilisation of harvested cells by treatment with acetone or ethanol was reported by Barbosa *et al.* (1985). β-Galactosidase from *Fusarium oxysporum* var. *lini* was purified and characterised by Brandao *et al.* (1987). The enzyme, which has pH and temperature optima at 5·0 and 55°C, respectively, has a momomeric molecular weight of 36 300 (SDS–PAGE) but normally exists as a hexamer (mol. wt. 224 000).

Numerous applications have been reported for glucose–galactose syrups (in addition to the use of lactose-hydrolysed whey) (see Mahoney, 1985; Gekas & Lopez-Leiva, 1985; Harju, 1987). Although effective systems for the production of glucose–galactose syrups and other lactose-hydrolysed products are available and continued research will undoubtedly lead to improved systems, the cost of such syrups *vis-à-vis* alternatives, mainly glucose syrups from starch, remains a problem. Greenberg and Mahoney (1981) maintain that enzymatically hydrolysed lactose would be commercially viable in Europe but not in the US where corn syrups are cheaper. The economics of various semi-commercial systems for the production of glucose–galactose syrups are discussed by Mahoney (1985) and Gekas and Lopez-Leiva (1985). However, other considerations may render glucose–galactose syrups economical in comparison to alternative methods for whey disposal/utilisation.

The sweetness of glucose–galactose syrups may be increased by isomerising the glucose to fructose (which is about twice as sweet as glucose) using glucose isomerase which is now widely used in the commercial production of high-fructose syrups. Such a scheme has been patented (Weetall & Yaverbaum, 1974) and a glucose–fructose–galactose–lactose syrup with a sweetness equal to sucrose (both at 10% solution) has been prepared by treating a glucose-fortified lactose hydrolysate with glucose isomerase (Poutanen *et al.*, 1978). Conversion of lactose to lactulose offers another avenue for the production of novel sugar mixtures. Lactulose is a disaccharide consisting of galactose and fructose which can be produced from lactose by mild alkaline treatment. At least some β-galactosidases can hydrolyse lactulose, e.g. Maxilact 20 000 (*K. lactis*) hydrolysed lactulose at ~40% its rate on lactose (Ramos *et al.*, 1982).

Undoubtedly, β-galactosidase has numerous applications in the dairy industry. However, in spite of very considerable research and demonstrated technical feasibility in pilot-scale experiments, β-galactosidase is not yet exploited commercially on a significant scale.

LYSOZYME

Lysozymes (muramidase, EC 3.2.1.17), a group of enzymes which cause lysis of certain bacteria by hydrolysing cell-wall polysaccharides, are widely distributed in animal tissues and secretions; some bacteria and bacteriophage secrete similar enzymes. Egg white is a particularly rich source of lysozyme which is the best-characterised of these enzymes; it is the usual source of commercial lysozyme. The molecular and biological properties of lysozyme and its use in food preservation and as a pharmaceutical have been very comprehensively reviewed by Proctor and Cunningham (1988).

Lysozyme was isolated from human milk by Jolles and Jolles (1961) who failed to detect it in bovine milk. However, the latter contains a low level of lysozyme which was isolated and characterised by Chandan *et al.* (1965). Human milk contains ~3000 times as much lysozyme as bovine milk (Parry *et al.*, 1969); mare's milk is also a very rich source (Jauregui-Adell, 1975). Various aspects of indigenous milk lysozyme are reviewed in Chapter 2.

In view of its anti-bacterial activity, the large difference in the lysozyme content of human and bovine milks may have significance in infant nutrition. It is generally accepted that breast-fed babies thrive better and suffer fewer enteric disorders than bottle-fed babies, especially in underdeveloped countries where hygiene is poor. There are many compositional differences between human and bovine milks, especially with respect to anti-bacterial factors which could account for the advantages of breast-feeding (Blanc, 1981; Lonnerdal, 1985; Atkinson & Lonnerdal, 1989) but lysozyme may be a significant contributor. It is claimed (see Haneberg & Finne, 1974; Proctor & Cunningham, 1988) that supplementation of baby-food formulations based on cows' milk with egg-white lysozyme gives beneficial results, especially with premature babies. However, Haneberg and Finne (1974) could not substantiate such claims and attributed the benefits of breast-feeding to other factors; they failed to detect lysozyme in the faeces of most infants fed formulae fortified with egg white lysozyme while the faeces of breast-fed babies were rich in lysozyme. Lysozyme is reported to increase the digestibility of casein (Kisza *et al.*, 1974), possibly due to the increased rennet coagulability of milk on lysozyme addition (Green, 1973; Green & Marshall, 1977; Pierre, 1983; Niki, 1985; Thapon & Brule, 1986) which may result in the formation of smaller protein flocs, such as formed in human milk.

Most current interest in the application of lysozyme in dairy products is focussed on its use in Dutch, Swiss, Italian and other cheese varieties to prevent late gas blowing and/or off-flavours caused by the growth of *Clostridium tyrobutyricum*. Contamination of cheesemilk with *Clostridia* can be reduced by good hygienic practices and populations may be further reduced by bactofugation. However, in most countries it is normal practice to add nitrate as a further precaution. The use of nitrate in foods is under scrutiny because of its involvement in nitrosamine formation and many countries have reduced permitted levels or prohibited its use.

Pulay (1966) was probably the first to demonstrate the potential of lysozyme to control the butyric acid fermentation in cheese. This was confirmed by Wasserfall (1977) and Wasserfall and Prokopek (1978). Since then the application of lysozyme in many cheese varieties to improve flavour and prevent gas blowing has been widely investigated. The results of these studies have been reviewed by Proctor and Cunningham (1988) and IDF (1987*b*, 1988). The latter conclude that 'the published information indicates that for some cheese types, lysozyme is a suitable substance for the control of late blowing provided the number of clostridial spores is low. For these cheese types, which may be considered less sensitive to late blowing, the level of lysozyme already permitted by regulatory authorities in a number of countries appears to be satisfactory until further evidence is obtained. Lysozyme appears to be of value in the control of late blowing in countries which prohibit nitrate.

Published information indicates that in other cheese types which are very sensitive to late blowing, such as Gouda, lysozyme used at the current normal addition under normal manufacturing and storage conditions is less effective than the usual amount of nitrate. In this case, lysozyme can not be considered a suitable alternative to nitrate at present.

More information will become available for the various cheese types about the critical number of spores in the raw milk to cause defects when lysozyme is used. Combinations of lysozyme addition and other control measures can then be evaluated further' (IDF 1987*b*, 1988).

New information is being produced and some recent publications include Neviani *et al.* (1988*a,b*), Mayr (1988), Stadhouders *et al.* (1987), and Goudedranche *et al.* (1986).

For Italian varieties (Provolone, Grana and Parmesan) lysozyme and formaldehyde (a traditional bacteriocidal agent for Italian cheese-milk) act synergistically (Battistotti *et al.*, 1984; Bottazzi *et al.*, 1984; Carini & Lodi, 1984; Losi & Chiavari, 1984).

Lysozyme appears to be quite effective against *Listeria monocytogenes* and other bacteria involved in food spoilage and food borne disease (Hughey & Johnson, 1987). Considering the widespread attention now focussed on *Listeria* in dairy products, especially cheeses, it is likely that this application of lysozyme will be the subject of research in the future. The preservative effects of lysozyme in other foods were reviewed by Proctor and Cunningham (1988). It is possible that some of these may be applicable to dairy products.

GLUCOSE OXIDASE

Glucose oxidase (GO) catalyses the oxidation of glucose to gluconic acid (via glucono-δ-lactone) according to the following reactions:

$$\text{glucose} \xrightarrow{\text{GO}} \text{glucono-}\delta\text{-lactone} + FADH_2$$

$$\text{glucono-}\delta\text{-lactone} \xrightarrow[\text{spontaneous}]{\text{lactonase or}} \text{gluconic acid}$$

$$FADH_2 \xrightarrow{O_2,\ \text{GO}} FAD + H_2O_2$$

$$H_2O_2 \xrightarrow{\text{catalase}} H_2O + \tfrac{1}{2}O_2$$

The hydrogen peroxide formed is normally reduced by catalase present as a contaminant in commercial preparations of GO (from *P. notatum, P. glaucum* or *Aspergillus niger*) or added separately. GO, which has a pH optimum ~5·5, is highly specific for D-glucose and may be used to assay specifically for D-glucose in the presence of other sugars.

In the food industry, GO has four principal applications (Scott, 1975; Sharma & Messing, 1980; Nijpels & Ramet, 1982):

(1) *Removal of residual, trace levels of glucose.* This application, which is particularly useful for the treatment of egg white prior to dehydration (although alternative procedures using yeast fermentation are more commonly used), is of little, if any, significance in dairy technology.

(2) *Removal of trace levels of oxygen.* Traces of oxygen in wines and fruit juices cause discolouration and/or oxidation of ascorbic acid. Chemical reducing agents may be used to scavenge oxygen but enzymatic treatment with GO may be preferred. GO has been proposed as an antioxidant system for high-fat products such as mayonnaise, butter and whole milk powder but it does not appear to be used commercially for this purpose, probably because of cost *vis à vis* chemical antioxidants (if permitted) and the relative effectiveness of inert gas flushing of canned milk powder.

(3) *Generation of hydrogen peroxide* in situ. The hydrogen peroxide generated by glucose oxidase has a direct bacteriocidal effect (which appears to be a useful side-effect of GO applied to egg products) but its bacteriocidal properties can be much more effectively exploited as a component of the lactoperoxidase/hydrogen peroxide/thiocyanate system (Reiter, 1978, 1985; Björck, 1980; Chapter 2). Two components of this system occur naturally in milk: lactoperoxidase is present at ~ 30 mg/litre, while thiocyanate, produced in the rumen by hydrolysis of thioglucosides from brassicae plants, varies from 0·017 to 0·26 mM. Hydrogen peroxide does not occur naturally in bacteria-free milk but it can be generated metabolically by catalase-negative bacteria, added as an exogenous supply (which is usually preferred) or produced *in situ* from sodium percarbonate or by the action of xanthine oxidase on added hypoxanthine (Björck & Claesson, 1979; Collins *et al.*, 1988) or by the action of glucose oxidase on glucose, either added (Björck *et al.*, 1979; Beumer *et al.*, 1985; Sandholm *et al.*, 1988) or produced *in situ* from lactose by β-galactosidase (Björck & Rosen, 1976). Activation of the lactoperoxidase–hydrogen peroxide–thiocyanate system suppressed the growth of psychrotrophs in milk stored at 5°C for up to 5 days and has given promising results as a milk preservative in tropical regions where refrigeration is lacking (see Chapter 2). It would appear that in such applications the use of exogenous hydrogen peroxide is the simplest and most appropriate.

(4) *Production of acid* in situ. Direct acidification of dairy products, particularly cottage and Mozzarella cheese, is now fairly common (Fox, 1978). Acidification is normally performed by addition of acid or acidogen (usually glucono-δ-lactone) or by a combination of acid and acidogen. *In situ* production of gluconic acid from added glucose or from glucose produced *in situ* from lactose or added sucrose has been proposed (Rand & Hourigan, 1975). Sharma and Messing (1980) reported on the possible production of gluconic acid from glucose using immobilised glucose oxidase. However, it is doubtful whether immobilised glucose oxidase could be applied to the acidification of milk because of the high probability of fouling by precipitated protein even if low temperatures, which would lead to less efficient casein precipitation, were used. Production of lactobionic acid from lactose by lactose dehydrogenase has also been proposed for the direct acidification of dairy and other foods (Wright & Rand, 1973).

SUPEROXIDE DIMUTASE

Superoxide dimutase (SOD) catalyses the reduction of superoxide anions, $O_2^{\bar{\cdot}}$, according to the following:

$$2O_2^{\bar{\cdot}} + 2H^+ \rightarrow H_2O_2 + O_2$$

The hydrogen peroxide formed may be reduced by catalase, peroxidase or a suitable reducing agent. The biological function of SOD was discovered by McCord and Fridovich (1969). The enzyme occurs widely in tissues where it is believed to play a major antioxidant role in scavenging superoxide radicals which are produced through the action of several enzymes, e.g. xanthine oxidase and peroxidases.

Milk contains a low level of indigenous SOD which has been isolated and characterised (Hill, 1975; Hicks *et al.*, 1975; Korycka-Dahl *et al.*, 1979); the milk enzyme appears to be identical to SOD from bovine erythrocytes. The level of indigenous SOD in milk, which is entirely in the skim phase, shows considerable variation among individual cows, with stage of lactation and mastitic infection (Holbrook & Hicks, 1978). It may play a role in the oxidative stability of milk but attempts to correlate stability with SOD activity have been inconclusive (Hicks, 1980), presumably due to the interaction of various pro- and antioxidants. Kankare and Antila (1982) found that the tendency of fat in milk to oxidise was increased by increases in xanthine oxidase and slightly reduced by increases in SOD activity. Studies on indigenous milk SOD are reviewed in Chapter 2. A low level of exogenous SOD, coupled with catalase, is a very effective antioxidant in dairy products (Hill, 1979; Hicks, 1980; Hill *et al.*, 1977) and it has been suggested that treatment of milk with SOD may be effective in preserving the flavour of UHT milk which is prone to lipid oxidation. Valenzuela *et al.* (1981) found a combination of SOD and catalase to be a more effective antioxidant than BHA (butylated hydroxyanisole). However, SOD appeared to be ineffective as an antioxidant in the presence of 0·1 ppm Cu^{2+} (Hill *et al.*, 1977; Allen & Wrieden, 1982), apparently because Cu^{2+} effectively competes with SOD for $O_2^{\cdot-}$ and converts it to lipid-reactive species such as $^{\cdot}OH$. Obviously, the commercial feasibility of SOD as an antioxidant depends on cost, particularly *vis à vis* chemical antioxidants, if permitted. As far as is known, SOD is not used commercially as an antioxidant in the dairy industry.

SULPHYDRYL OXIDASE

An enzyme capable of oxidising sulphydryl groups to disulphides was first demonstrated in bovine milk by Kiermeier and Petz (1967):

$$2RSH + O_2 \rightarrow RSSR + H_2O$$

The enzyme, which has molecular and catalytic properties distinctly different from thioloxidase (EC 1.8.3.2), glulathione: protein disulphide oxido-reductase (EC 1.8.4.2) and protein disulphide isomerase (EC 5.3.4.1), has been isolated and well characterised (see Swaisgood, 1980; Kitchen, 1985; Chapter 2). The enzyme has also been demonstrated in human milk (Isaacs *et al.*, 1984).

It has been proposed (Swaisgood & Horton, 1980) that sulphydryl oxidase has a physiological role in the formation of disulphide bonds *in vivo* to give proteins the correct three dimensional structure, e.g. it can reactivate reduced ribonuclease (Janolino & Swaisgood, 1975), link IgM and J-chain (Roth & Koshland, 1981) and may be involved in the synthesis of sIgA in the mammary gland (Isaacs *et al.*, 1984). Industrially, the potential of the enzyme lies in its ability to ameliorate the cooked flavour of UHT-treated milk (Swaisgood, 1980). The milk enzyme occurs exclusively in the serum (whey) from which it may be easily isolated by exploiting its marked tendency to aggregate, thus facilitating its isolation by chromatography on porous glass CPG-3000 (Janolino *et al.*, 1981). The enzyme has been immobilised on porous glass and tetania and its effectiveness in ameliorating the cooked flavour of UHT-treated milk has been demonstrated on a pilot-scale using immobilised enzyme columns (see Swaisgood, 1980). For industrial-scale use, an adequate supply of the enzyme from whey may be a limitation but production of the enzyme by genetically engineered microorganisms might be feasible.

CATALASE

Hydrogen peroxide is used for the cold-sterilisation of milk in regions lacking refrigeration and perhaps in some developed countries also, e.g. it may be used to treat cheese milk in the US. In underdeveloped regions, treatment with hydrogen peroxide is performed at ambient temperature and excess hydrogen peroxide is reduced by indigenous milk catalase or by chemical interaction with milk proteins in which it causes some physico-chemical changes, principally oxidation of methionine, with adverse effects on cheese quality (Fox & Kosikowski, 1967). Side-effects can be reduced by short exposure to hydrogen peroxide at ~65°C, after which the residual hydrogen peroxide is reduced by added catalase, usually from beef liver or *Aspergillus niger*. Sweet potato appears to be a good source of catalase for treatment of milk (Moser, 1988). Most of the recent interest in the use of hydrogen peroxide as a preservative for milk has focussed on its use as a component of the lactoperoxidase hydrogen peroxide–thiocyanate system in which low concentrations of hydrogen peroxide are required. However at 400 ppm, hydrogen peroxide alone was a more effective long-term bactericidal agent than 8·5 ppm hydrogen peroxide in the lactoperoxidase system (Kaman & Kroger, 1984). There was no significant difference in the quality of cultured milk made from either lactoperoxidase activated or hydrogen peroxide-treated milk; the activity of lactic starter was significantly lower in the former but the rennet coagulability of the latter was extended.

There is interest in using immobilised catalase reactors for milk pasteurisation or for glucose oxidase–catalase reactions. Although catalase may be readily immobilised, it is inactivated rapidly on exposure to hydrogen peroxide (Richardson & Olson, 1974; Chu *et al.*, 1975). Hydrogen peroxide appears to be an effective agent for inactivation of aflotoxin M_1 (Applebaum & Marth, 1982). As discussed above, catalase, usually as a contaminant, is necessary for many of the applications of glucose oxidase and also to optimise the antioxidative properties of SOD.

LACTOPEROXIDASE

Bovine milk is rich in lactoperoxidase (~30 μg/ml) which is distinct from the myloperoxidase of leucocytes and salivary peroxidase. Lactoperoxidase has been purified by several investigators and well characterised (see Fox & Morrissey, 1981; Kitchen, 1985; see also Chapter 2).

The most important physiological and technological feature of lactoperoxidase is its ability in the presence of hydrogen peroxide and thiocyanate to inhibit the growth of several bacteria. Since lactoperoxidase is relatively heat resistant, there is adequate lactoperoxidase activity even in milk that had been moderately severely heat-treated. Thiocyanate occurs in many animal tissues and fluids. It is produced endogenously during the detoxification of thiosulphates and metabolic products of sulphur amino acids and cyanide and from foods containing thioglucosides. Cows on pastures containing clover (rich in $^-$CN) and non-grasses (e.g. cruciferrae containing thioglucosides) yield milk containing higher concentrations of thiocyanide (0·02–0·25 mM) than cows on winter feed or lay pastures. Saliva contains high levels of thiocyanate which is also secreted by gastric mucosal cells. Milk does not contain indigenous hydrogen peroxide but it may be produced *in situ* by catalase-negative bacteria, generated enzymatically (e.g. by added glucose oxidase or indigenous xanthine oxidase if suitable substrates, i.e. glucose, xanthine or hypoxanthine, are added). Thus, bovine milk possesses an effective antibacterial which probably affects the intestinal microflora of calves (and presumably other species). Lactoperoxidase also appears to protect the mammary gland against mastitic infection, especially during the dry period. The antibacterial significance of lactoperoxidase *in vivo* has been reviewed by Reiter (1985) and in references therein. There has been controversy as to whether or not human milk contains lactoperoxidase but recent evidence (Gothefors & Marklund, 1981; Langbakk & Flatmark, 1984) suggests the presence of a low level of lactoperoxidase (<0·1 μg/ml) as well as myeloperoxidase and eosinophil peroxidase.

The lactoperoxidase system may be exploited *in vitro* to extend the shelf life of milk under conditions where refrigeration and pasteurisation facilities are lacking (see Reiter, 1985; Chapter 2).

While most interest in lactoperoxidase has focussed on the indigenous enzyme, it may also acquire significance as an exogenous enzyme. Techniques have been developed for the commercial-scale purification of lactoperoxidase from milk. Large-scale trials have shown that addition of purified lactoperoxidase to calf milk replacers markedly reduced the incidence of diarrhoea and increased weight gain. Although the antibacterial properties of human milk do not depend on lactoperoxidase, addition of it to bovine milk-based infant formulae may have beneficial effects (see Reiter, 1985; Chapter 2).

β-LACTAMASE

Residues of penicillin in milk, resulting from treatment of mastitis-infected cows with antibiotics, inhibit or prevent the growth of lactic acid bacteria used in the production of cheese and fermented milks. They may also evoke an allergic response in susceptible consumers and create conditions for the selection of penicillin-resistant pathogens. The usual step taken to prevent contamination of the milk supply is to withhold milk from treated cows for an adequate period after treatment; obviously, this results in economic losses to farmers.

Procedures developed for the removal of β-lactam-type antibiotics from milk include passage through a column of activated charcoal (Charm, 1980) or ultrafiltration–diafiltration (Kosikowski & Jimenez-Flores, 1985). While these methods may be suitable under certain circumstances, they have obvious limitations. An alternative approach is the inactivation of penicillin by β-lactamase which catalyses the following reaction:

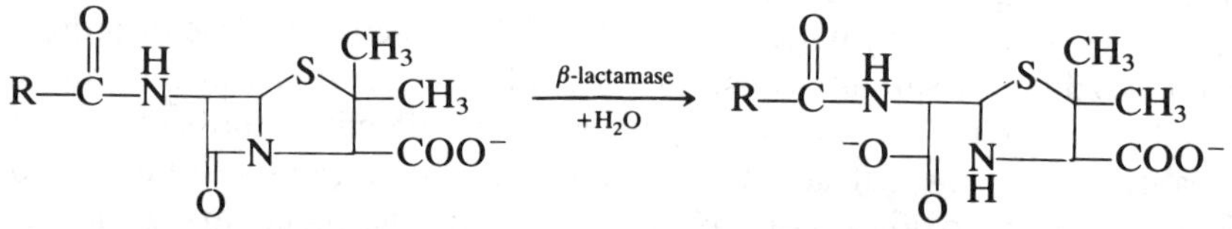

Penicillin   Penicilloic acid

Korycka-Dahl *et al.* (1985) demonstrated the ability of β-lactamase from *Bacillus cereus* to inactivate penicillin G such that Cheddar or Swiss cheeses of normal quality could be produced from contaminated milk. The enzyme was active at 4°C so that penicillin in milk could be inactivated during cold storage on the farm or at the factory. The enzyme lost only 50% of its activity following heating at 63°C for 30 min and hence would retain considerable activity in pasteurised milk, which may be illegal. These problems may be solved by using immobilised β-lactamase, as developed by Lee and Richardson (1987). This procedure appears promising and awaits further development.

An alternative approach to solving the problems caused by penicillin in fermented products is the selection of β-lactamase-producing mutants of lactic acid bacteria, as reported by Khosravi and Sandine (1987) for *Streptococcus cremoris*.

REFERENCES

Abdel Baky, A. A., El Fak, A. M., Rabia, A. M. & El Neshewy, A. A. (1982*a*). *J. Food Protect.*, **45**, 894–7.

Abdel Baky, A. A., El Neshewy, A. A., Rabie, A. H. M. & Farahat, S. M. (1982*b*). *J. Dairy Res.*, **49**, 337–41.

Adler-Nissen, J. (1986). *Enzymatic Hydrolysis of Food Proteins*. Elsevier Applied Science Publishers, London, UK.

Allen, J. C. & Wrieden, W. L. (1982). *J. Dairy Res.*, **49**, 249–63.

Anon. (1976). Technical Bulletin, Dairyland Food Laboratories, Waukesha, WI, USA.

Applebaum, R. S. & Marth, E. H. (1982). *J. Food Protect.*, **45**, 557–60.

Arbige, M. V., Freund, P. R., Silver, S. C. & Zelro, J. T. (1986). *Food Technol.*, **40**(4), 91–8.

Arnold, R. S., Shahani, K. M. & Dwivedi, B. K. (1975). *J. Dairy Sci.*, **58**, 1127–43.

Asp, N.-G., Burvall, A., Dahlqvist, A., Hallgren, P. & Lunblad, A. (1980). *Food Chem.*, **5**, 147–53.

Atkinson, S. A. L. & Lonnerdal, B. (1989). *Protein and Non-protein Nitrogen in Human Milk*. CRC Press, Boca Raton, FL, USA.

Azuma, N., Nagaune, S.-I., Kaminogawa, S. & Yamauchi, K. (1989). *J. Dairy Res.*, **56**, 552–3.

Babel, F. J. & Hammer, B. W. (1945). *J. Dairy Sci.*, **28**, 201–8.

Banks, J. M. & Muir, D. D. (1985). *Milchwissenschaft*, **40**, 209–12.

Barbosa, M. F. S., Silva, D. O., Pinheiro, A. J. R., Guimarães, W. V. & Borges, A. C. (1985). *J. Dairy Sci.*, **68**, 1618–23.

Barillas, C. & Solomons, N. W. (1987). *Pediatrics*, **79**, 766–72.

Battistotti, B., Bottazzi, V., Scolari, C. L. & Chiusa, P. (1984). *Sci. Tecnica Lattiero-Caseria*, **35**, 193–200.

Bernal, V. & Jelen, P. (1985). *Can. Inst. Food Sci. Technol. J.*, **18**, 97–9.

Beumer, R. R., Noomen, A., Marijs, J. A. & Kampelmacher, E. H. (1985). *Neth. Milk Dairy J.*, **39**, 107–14.

Bines, V. G., Young, P. & Law, B. A. (1989). *J. Dairy Res.*, **56**, 657–64.

Björck, L. (1980). International Dairy Federation, Doc. 126, pp. 5–7.

Björck, L. & Claessen, O. (1979). *J. Dairy Sci.*, **62**, 1211–15.

Björck, L. & Rosen, C.-G. (1976). *Biotechnol. Bioengng.*, **18**, 1463–72.

Björck, L., Rosen, C.-G., Marshall, V. & Reiter, B. (1979). *Appl. Microbiol.*, **30**, 199–204.

Blanc, B. (1981). *Wld Rev. Nutr. Diet.*, **36**, 1–89.

Borgstron, B. & Brochman, H. L. (eds) (1984). *Lipases*. Elsevier, Amsterdam, The Netherlands.

Bottazzi, V., Battistotti, B., Chiusa, P. & Scolari, G. (1984). *Sci. Technica Lattiero-Caseria*, **35**, 377–83.

Brandao, R. L., Nicoli, J. R. & Figueiredo, A. F. S. (1987). *J. Dairy Sci.*, **70**, 1331–7.

Brantl, V. (1984). *Eur. J. Pharmacol.*, **106**, 213–14.

Brantl, V. & Neubert, K. (1986). *Trends Pharmacol. Sci.*, **7**, 6–7.

Brantl, V. & Teschemacher, H. (1979). *Naunyn-Schmiedeberg's Arch. Pharmacol.*, **306**, 301–4.

Brantl, V., Teschemacher, H., Henschen, A. & Lottspeich, F. (1979). *Hoppej-Seyler's Z. Physiol. Chem.*, **360**, 1211–16.

Brantl, V., Teschemacher, H., Bläsig, J., Henschen, A. & Lottspeich, F. (1981). *Life Sci.*, **28**, 1903–9.

Braun, S. D. & Olson, N. F. (1986*a*). *J. Dairy Sci.*, **69**, 1202–8.

Braun, S. D. & Olson, N. F. (1986*b*). *J. Dairy Sci.*, **69**, 1209–18.

Braun, S. D., Olson, N. F. & Lindsay, R. C. (1982). *J. Food Sci.*, **47**, 1803–67.
Brignon, G., Chtourou, A. & Ribadeau Dumas, B. (1985). *FEBS Lett.*, **188**, 48–54.
Brockerhoff, H. & Jensen, R. G. (1974). *Lipolytic Enzymes*. Academic Press, New York, USA.
Broome, M. C., Roginski, H. & Hickey, M. W. (1983). *Aust. J. Dairy Technol.*, **38**, 35–7.
Brule, G., Roger, L., Fauquant, J. & Piot, M. (1982). US Patent 4, 358,456.
Burvall, A., Asp, N.-G. & Dahlqvist, A., (1980). *Food Chem.*, **5**, 189–94.
Carlson, A., Hill, G. C. & Olson, N. F. (1986). *Enzyme Microb. Technol.*, **8**, 642–50.
Carini, S. & Lodi, R. (1984). *Sci. Tecnica Lattiero-Casearia*, **35**, 418–22.
Castberg, H. B. & Morris, H. A. (1976). *Milchwissenschaft*, **31**, 85–90.
Chandan, R. C., Parry, R. M. Jr & Shahani, K. M. (1965). *Biochim. Biophys. Acta*, **110**, 389–98.
Charm, S. E. (1980). US Patent 4, 238, 521.
Chang, B.-S. & Mahoney, R. R. (1989). *J. Dairy Res.*, **56**, 117–27.
Chaudhari, R. V. & Richardson, G. H. (1971). *J. Dairy Sci.*, **54**, 467–71.
Chiba, H. & Yoshikawa, M. (1986). In *Protein Tailoring for Food and Medical Uses*, ed. R. E. Feeney & J. R. Whitaker. Marcel Dekker, New York, pp. 123–53.
Chiba, H., Tani, F. & Yoshikawa, M. (1989). *J. Dairy Res.*, **56**, 363–6.
Chiu, C. P. & Kosikowski, F. V. (1985). *J. Dairy Sci.*, **68**, 16–22.
Chobert, J.-M., Mercier, J.-C., Bahy, C. & Haze, G. (1976). *FEBS Lett.*, **72**, 173–8.
Chu, H. D., Leeder, J. G. & Gilbert, S. G. (1975). *J. Food Sci.*, **40**, 641–3.
Cochet, C., Feige, J.-J. & Chambaz, E. M. (1983). *Biochim. Biophys. Acta*, **774**, 147–54.
Collins, R. A., Parsons, K. R., Field, T. R. & Bramley, A. J. (1988). *J. Dairy Res.*, **55**, 25–32.
Coton, S. G. (1980). *J. Soc. Dairy Technol.*, **33**, 89–94.
Creamer, L. K., Zoerb, H. F., Olson, N. F. & Richardson, T. (1982). *J. Dairy Sci.*, **65**, 902–6.
Creamer, L. K., Lawrence, R. C. & Gilles, J. (1985). *NZJ. Dairy Sci. Technol.*, **20**, 185–203.
Cullen, D., Gray, G. L., Wilson, L. J., Hayenga, K. J., Lamsa, J. H., Rey, M. W., Norton, S. & Berka, R. M. (1987). *Biotechnology*, **5**, 369–76.
Dahlqvist, A. (1984). *Nutr. Abstr. Rev.*, **54**, 649–58.
Dahlqvist, A., Asp, N.-G., Burvall, A. & Rausing, H. (1977). *J. Dairy Res.*, **44**, 541–8.
Dalgleish, D. G. (1982). *Developments in Dairy Chemistry–1*, ed. P. F. Fox. Applied Science Publishers, London, UK, pp. 157–87.
Dalgleish, D. G. (1987). In *Cheese: Chemistry, Physics and Microbiology* (Vol. 1), ed. P. F. Fox. Elsevier Applied Science, London, UK, pp. 63–96.
Daniels, M. J. (1985). *Food Technol.*, **39**(10), 68–70.
Dariani, D. N., Frank, J. F. & Lowenstein, M. (1982). *Cult. Dairy Prod. J.*, **17**(2), 18–19.
Delfour, A., Jolles, J., Alais, C. & Jolles, P. (1965). *Biochem. Biophys. Res. Commun.*, **19**, 452–5.
Desmazeaud, M. J. & Gripon, J.-C. (1977). *Milchwissenschaft*, **32**, 731–4.
Doi, H., Kawaguchi, V., Ibuki, F. & Kanamori, M. (1979). *J. Nutr. Sci. Vitaminol.*, **25**, 33–41.
Dulley, J. R. (1976). *Aust. J. Dairy Technol.*, **31**, 143–8.
Dwivedi, B. K. & Kinsella, J. E. (1974*a*). *J. Food Sci.*, **39**, 83–7.
Dwivedi, B. K. & Kinsella, J. E. (1974*b*). *J. Food Sci.*, **39**, 620–2.
Efthymiou, C. C. & Mattick, J. F. (1964). *J. Dairy Sci.*, **47**, 593–8.
El Neshawy, A. A., Abdel Baky, A. A. & Farahat, S. M. (1982). *Dairy Ind. Intern.*, **47**(2), 29, 31.

El Shibiny, S., Soliman, M. A., El-Bagoury, E., Gad, A. & Abd El Salam, M. H. (1978). *J. Dairy Res.*, **45,** 497–500.
El Soda, M., Desmazeaud, M. J., Abou Donia, S. & Kamal, N. (1981). *Milchwissenschaft,* **36,** 140–2.
El Soda, M., Desmazeaud, M. J., Abou Donia, S. & Badran, A. (1982). *Milchwissenschaft,* **37,** 325–7.
El-Zayat, A. I. (1987). *Nahrung,* **31,** 27–37.
Ennis, B. M. & Harper, W. J. (1986). *NZJ. Dairy Sci. Technol.*, **21,** 205–16.
Eriksen, S. & Fagerson, I. S. (1976). *J. Food Sci.*, **41,** 490–3.
Ernstrom, C. A. (1974). In *Fundamentals of Dairy Chemistry* (2nd edn), ed. B. H. Webb, A. H. Johnson, & J. A. Alford. AVI Publishing Co. Inc., Westport, CT, USA, pp. 662–771.
Farahat, S. M., Rabie, A. M., Abdel-Baky, A. A., El-Neshawy, A. A. & Mobasher, S. (1985). *Nahrung,* **29,** 247–54.
Farnham, M. G. (1950). Cheese modifying enzyme proudcts. US Patent 2,531,329.
Farkye, N. Y. & Fox, P. F. (1990). *J. Dairy Res.*, **57,** 413–18.
Fauqueux, P. F., Flaschel, E. & Renken, A. (1984). *Chimica,* **38,** 262–9.
Fedrick, I. A., Aston, J. W., Nottingham, S. M. & Dulley, J. R. (1986*a*). *NZJ. Dairy Sci. Technol.*, **21,** 9–19.
Fedrick, I. A., Cromie, S. J., Dulley, J. R. & Giles, J. E. (1986*b*). *NZJ. Dairy Sci. Technol.*, **21,** 191–203.
Fernandez-Garcia, E., Ramos, M., Polo, C., Juarez, M. & Olano, A. (1988). *Food Chem.*, **28,** 63–80.
Fiat, A.-M., Jolles, J. & Jolles, P. (1977). *C.R. Acad. Sci. Paris* (Ser. D.), **284,** 393–5.
Fiat, A.-M., Levy-Toledano, S., Caen, J. P. & Jolles, P. (1989). *J. Dairy Res.*, **56,** 351–5.
Finnocchiaro, T., Olson, N. F. & Richardson, T. (1980). *Adv. Biochem. Engng,* **15,** 71–88.
Flynn, A. (1985). In *Developments in Dairy Chemistry—3,* ed. P. F. Fox. Elsevier Applied Science Publishers, London, UK, pp. 133–41.
Foltmann, B. (1981*a*). *Neth. Milk Dairy J.*, **35,** 223–31.
Foltmann, B. (1981*b*). In *Essays in Biochemistry* (Vol. 17), ed. P. N. Campbell & R. D. Marshall. Academic Press, London, UK.
Foltmann, B. (1987). In *Cheese: Chemistry, Physics and Microbiology* (Vol. 1), ed. P. F. Fox. Elsevier Applied Science, London, UK, pp. 33–61.
Fox, P. F. (1969). *J. Dairy Res.*, **36,** 427–33.
Fox, P. F. (1978). *Dairy Sci. Abstr.*, **40,** 727.
Fox, P. F. (1980). IDF Document 125, International Dairy Federation, Brussels, Belgium, pp. 22–59.
Fox, P. F. (1981). In *Proteinases and their Inhibitors: Structure, Function and Applied Aspects,* ed. V. Turk & Lj. Vitale. Pergamon Press, Oxford, UK, pp. 245–67.
Fox, P. F. (1982). *Developments in Dairy Chemistry—1—Proteins.* Applied Science Publishers, London, UK.
Fox, P. F. (1984). In *Developments in Food Proteins—3,* ed. B. F. J. Hudson. Elsevier Applied Science, London, UK, pp. 69–112.
Fox, P. F. (ed.) (1985). *Developments in Dairy Chemistry—3.* Elsevier Applied Publishers, London, UK.
Fox, P. F. (1987). *Proc. 22nd Intern. Dairy Congr.*, (the Hague), D. Reidel Pub. Co., Dordrecht, Holland, pp. 61–73.
Fox, P. F. (1988). *Biotechnol. Appl. Biochem.*, **10,** 522–35.
Fox, P. F. (1988/89). *Food Biotechnol.*, **2,** 133–85.
Fox, P. F. (ed.) (1989*a*). *Developments in Dairy Chemistry—4.* Elsevier Applied Science Publishers, London, UK.
Fox, P. F. (1989*b*). *J. Dairy Sci.*, **71,** 1379–400.

Fox, P. F. (1992). *Advanced Dairy Chemistry—1—Proteins*. Elsevier Applied Science Publishers, London, UK (in press).
Fox, P. F. & Kosikowski, F. V. (1967). *J. Dairy Sci.*, **50,** 1183–8.
Fox, P. F. & Morrissey, P. A. (1981). In *Enzyme and Food Processing,* ed. G. G. Birch, N. N. Blakebrough & K. J. Parker. Elsevier Applied Publishers, London, UK, pp. 213–38.
Fox, P. F., Morrissey, P. A. & Mulvihill, D. M. (1982). In *Developments in Food Proteins—1,* ed. B. J. F. Hudson. Applied Science Publishers, London, UK, pp. 1–60.
Fox, P. F. & Mulvihill, D. M. (1990). In *Food Gels,* ed. P. Harris. Elsevier Applied Science Publishers, London, UK, pp. 121–73.
Freund, P. R. (1986). A unique system for enhancing Cheddar cheese aging. 7th Biennial Cheese Industry Conference, Logan.
Frick, C. M., Hicks, C. L. & O'Leary, J. (1984). *J. Dairy Sci.,* **67** (suppl. 1), 89.
Fujimaki, M., Arai, S. & Yamashita, M. (1977). In *Food Proteins: Improvement through chemical and enzymatic modification,* ed. R. E. Feeney & J. R. Whitaker. American Chemical Society, Washington, DC, USA, pp. 156–84.
Fullbrook, P. (1987). Biotechnology and related developments as they apply to cheese and other dairy products. Society Dairy Technology, Spring Conference, April 26–29.
Garnier, J. (1963). *Biochim. Biophys. Acta.,* **66,** 366–77.
Gekas, V. & Lopez-Leiva, M. (1985). *Process Biochem.,* **20**(2), 2–12.
Giec, A. & Kosikowski, F. V. (1983). *J. Dairy Sci.,* **66,** 396–9.
Gooda, E., Bednarki, W. & Poznanski, S. (1983). *Milchwissenschaft,* **38,** 83–6.
Gothefors, L. & Marklund, S. (1981). *Infect. Immun.,* **11,** 1210–15.
Goudedranche, H., Ducruet, P., Vachot, J. C., Pannetier, R. & Maubois, J. L. (1986). *Le Lait,* **66,** 189–206.
Grappin, R., Rank, T. C. & Olson, N. F. (1985). *J. Dairy Sci.,* **68,** 531–40.
Green, M. L. (1973). *Neth. Milk Dairy J.,* **27,** 278–85.
Green, M. L. (1977). *J. Dairy Res.,* **44,** 159–88.
Green, M. L. & Grandison, A. S. (1987). In *Cheese: Chemistry, Physics and Microbiology.* (Vol. 1), ed. P. F. Fox. Elsevier Applied Science, London, UK.
Green, M. L. & Marshall, R. J. (1977). *J. Dairy Res.,* **44,** 521–31.
Greenberg, N. A. & Mahoney, R. R. (1981). *Process Biochem.,* **16**(2), 2–8, 49.
Griethuysen-Dilber, E. van, Flaschel, E. & Renken, A. (1988). *Process Biochem.,* **23,** 55–9.
Grieve, P. A., Kitchen, B. J., Dulley, J. R. & Bartley, J. (1983). *J. Dairy Res.,* **50,** 469–80.
Gripon, J.-C. (1987). In *Cheese: Chemistry, Physics and Microbiology* (Vol. 2), ed. P. F. Fox. Elsevier Applied Science Publishers, London, UK, pp. 121–49.
Gripon, J.-C., Desmazeaud, M. J., Le Bars, D. & Bergere, J.-L. (1977). *J. Dairy Sci.,* **60,** 1532–8.
Guinee, T. P. & Fox, P. F. (1987). In *Cheese: Chemistry, Physics and Microbiology* (Vol. 2), ed. P. F. Fox. Elsevier Applied Science Publishers, London, UK, pp. 251–97.
Gunther, E. & Burger, E. (1982). In *Uses of Enzymes in Food Technology,* ed. P. Dupuy. Technique et Documentation Lavoisier, Paris, France, pp. 243–8.
Hagget, T. O. R. (1974). *Proc. 19th Int. Dairy Congr.,* (Delhi) Thompson Press (India) Ltd, New Delhi. *IE,* 339.
Haneberg, B. & Finne, P. (1974). *Acta Paediatr. Scand.,* **63,** 588–94.
Harju, M. (1987). IDF Bulletin 212, International Dairy Federation, Brussels, Belgium, pp. 50–5.
Harper, W. J. (1957). *J. Dairy Sci.,* **40,** 556–63.
Harte, B. R. & Stine, C. M. (1977). *J. Dairy Sci.,* **60,** 1266–72.

Hautefeuille, M., Brantl, V., Dumontier, A. M. & Desjeux, J.-F. (1986). *Amer. J. Physiol.*, **250**, 692–7.

Hemme, D., Vassal, L., Foyen, H. & Auclair, J. (1979). *Le Lait*, **59**, 597–614.

Henschen, A., Lottspeich, F., Brantl, V. & Teschemacher, H. (1979). *Hoppe-Seyler's Z. Physiol. Chem.*, **360**, 1217–24.

Hicks, C. L. (1980). *J. Dairy Sci.*, **63**, 1199–204.

Hicks, C. L., Korycka-Dahl, M. & Richardson, T. (1975). *J. Dairy Sci.*, **58**, 796 (Abstr.).

Hicks, C. L., O'Leary, J. & Bucy, J. (1988). *J. Dairy Sci.*, **71**, 1127–31.

Hildalgo, J. & Gamper, E. (1977). *J. Dairy Sci.*, **60**, 1515–18.

Hill, R. D. (1968). *Biochem. Biophys. Res. Commun.*, **33**, 659–63.

Hill, R. D. (1969). *J. Dairy Res.*, **36**, 409–15.

Hill, R. D. (1975). *Aust. J. Dairy Technol.*, **30**, 26–8.

Hill, R. D. (1979). *CSIRO Food Res. Quart.*, **39**, 33–7.

Hill, R. D. & Hocking, V. M. (1978). *NZJ Dairy Sci. Technol.*, **13**, 195–201.

Hill, R. D., van Leeuwen, V. & Wilkinson, R. A. (1977). *NZJ. Dairy Sci. Technol.*, **12**, 69–77.

Hindle, E. J. & Wheelock, J. V. (1970). *J. Dairy Res.*, **37**, 389–96.

Hobman, P. G. (1984). *J. Dairy Sci.*, **67**, 2630–53.

Holbrook, J. & Hicks, C. L. (1978). *J. Dairy Sci.*, **61**, 1072–7.

Holmes, D. G., Deursch, J. W. & Ernstrom, C. A. (1977). *J. Dairy Sci.*, **60**, 862–9.

Hooker, P. H., Munro, P. A. & O'Meara, G. M. (1982). *NZJ. Dairy Sci. Technol.* **17**, 35–40.

Hooydonk, A. C. M. van, Boerrigter, I. J. & Hagendoorn, H. G. (1986). *Neth. Milk Dairy J.*, **40**, 297–313.

Hooydonk, A. C. M. van, Koster, P. G. de & Boerrigter, I. J. (1987). *Neth. Milk Dairy J.*, **41**, 3–18.

Huang, H. T. & Dooley, J. C. (1976). *Biotechnol. Bioengng*, **18**, 909–19.

Huber, R. E., Kurz, G. & Wallenfels, K. (1976). *Biochem.*, **15**, 1994–2001.

Huffman, L. M. & Harper, W. J. (1985). *NZJ. Dairy Sci. Technol.*, **20**, 57–63.

Hughey, V. L. & Johnson, E. A. (1987). *Appl. Environ. Microbiol.*, **53**, 2165–70.

Humme, H. E. (1972). *Neth. Milk Dairy J.*, **26**, 180–5.

IDF (1987*a*). IDF Document 212, International Dairy Federation, Brussels, Belgium.

IDF (1987*b*). IDF Bulletin 216, International Dairy Federation, Brussels, Belgium, pp. 2–16.

IDF (1988). A-Doc. 112, International Dairy Federation, Brussels, Belgium, pp. 40–3.

Isaacs, C. E., Pascal, T., Wright, C. E. & Gaull, G. E. (1984). *Pediatr. Res.*, **18**, 532–5.

Iwasaki, T. & Kosikowski, F. V. (1973). *J. Dairy Sci.*, **56**, 623–4.

Janolino, V. G. & Swaisgood, H. E. (1975). *J. Biol. Chem.*, **250**, 2532–8.

Janolino, V. G., Clare, D. A. & Swaisgood, H. E. (1981). *Biochim. Biophys. Acta*, **658**, 406–9.

Jauregui-Adell, J. (1975). *J. Dairy Sci.*, **58**, 835–8.

Jensen, F. (1970). *Proc. 18th Intern. Dairy Congr.*, (Sydney). Oswald Ziegler Publications, Sydney. *IE*, 364 (Abstr.).

Jeon, I. J. & Mantha, V. R. (1985). *J. Dairy Sci.*, **68**, 581–8.

Jolles, P. & Jolles, J. (1961). *Nature*, **192**, 1187–8.

Jolles, P., Levy-Toledano, S., Fiat, A.-M., Soria, C., Gillessen, D., Thomaidis, A., Dunn, F. W. & Caen, J. P. (1986). *Eur. J. Biochem.*, **158**, 379–82.

Jolly, R. C. & Kosikowski, F. V. (1975*a*). *J. Dairy Sci.*, **58**, 846–52.

Jolly, R. C. & Kosikowski, F. V. (1975*b*). *J. Dairy Sci.*, **58**, 1272–5.

Jolly, R. C. & Kosikowski, F. V. (1975*c*). *J. Food Sci.*, **40**, 285–7.

Jolly, R. C. & Kosikowski, F. V. (1978). *J. Dairy Sci.*, **61**, 536–41.

Jong, L. de (1976). *Neth. Milk Dairy J.*, **30**, 242–53.

Juillerat, M. A., Baechler, R., Berrocal, R., Chanton, S., Scherz, J.-C. & Jost, R. (1989). *J. Dairy Res.*, **56,** 603–11.
Kalinowski, L., Frackiewicz, E., Janiszewska, L., Pawlik, A. & Kikolska, D. (1979). US Patent 4, 158, 607.
Kalinowski, L., Frackiewicz, E. & Janiszewska, L. (1982). *Proc. XXI Intern. Dairy Congr.*, Mir Publishers, Moscow. Book 2, pp. 500 (Abstr.).
Kaman, D. M. & Kroger, M. (1984). *Milchwissenschaft,* **39,** 658–61.
Kanawija, S. K. & Singh, S. (1987). *Indian Dairyman,* **39,** 467–70.
Kankare, V. & Antila, V. (1982). *Meijeritietiecttinen Aikakausjirga,* **40**(2), 32–40.
Kannan, A. & Jenness, R. (1961). *J. Dairy Sci.*, **44,** 808–22.
Kato, I., Mikawa, K., Kim, Y. K. & Yasui, T. (1970). *Men. Fac. Agric. Hokkaido University,* **7,** 477–81.
Khosravi, L. & Sandine, W. E. (1987). *J. Dairy Sci.*, **70,** 1529–43.
Kiermeier, F. & Petz, E. (1967). *Z. Lebensm. unters Forsch.*, **132,** 342–52.
Kilara, A. (1985*a*). *Process. Biochem.*, **20**(2), 35–45.
Kilara, A. (1985*b*). *Process. Biochem.*, **20**(5), 149–57.
Kim, S. C. & Olson, N. F. (1985). *J. Microencapsulation,* **2,** 197–206.
Kinsella, J. E. & Hwang, D. H. (1977). *CRC Crit. Rev. Food Sci. Nutr.*, **8,** 191–228.
Kisza, J., Panfil, H. & Ziajka, S. (1974). *Proc. 19th Intern. Dairy Congr.*, Thompson Press (India) Ltd, New Delhi *IE,* 565–6.
Kitchen, B. J. (1985). In *Developments in Dairy Chemistry—3,* ed. P. F. Fox. Elsevier Applied Science Publishers, London, UK, pp. 239–79.
Koning, P. J. de, Rooijen, P. J. van & Visser, S. (1978). *Neth. Milk Dairy J.*, **32,** 232–44.
Kornacki, K., Stepaniak, L., Adamiec, I., Grabska, J. & Wrona, K. (1979). *Milchwissenschaft,* **34,** 340–3.
Korycka-Dahl, M., Richardson, T. & Hicks, C. L. (1979). *J. Food Prot.*, **42,** 867–71.
Korycka-Dahl, M., Richardson, T. & Bradley, R. L. (1985). *J. Dairy Sci.*, **68,** 1910–16.
Kosikowski, F. V. (1976). US Patent 3,975,544.
Kosikowski, F. V. & Iwasaki, T. (1975). *J. Dairy Sci.*, **58,** 963–70.
Kosikowski, F. V. & Jimenez-Flores, R. (1985). *J. Dairy Sci.*, **68,** 3224–33.
Kosikowski, F. V. & Jolly, R. C. (1976). US Patent 3,973,042.
Kristoffersen, T., Mikolajcik, E. M. & Gould, I. A. (1967). *J. Dairy Sci.*, **50,** 292–7.
Kuehler, C. A. & Stine, C. M. (1974). *J. Food Sci.*, **39,** 379–82.
Langbakk, B. & Flatmark, T. (1984). *FEBS Lett.*, **174,** 300–3.
Law, B. A. (1978). IDF Document 108, International Dairy Federation, Brussels, Belgium, pp. 40–8.
Law, B. A. (1980). *Dairy Ind. Intern.*, **45**(5), 15, 17, 19, 20, 22, 48.
Law, B. A. (1982). In *Economics Microbiology* (Vol. 7), Academic Press, London, UK, pp. 147–98.
Law, B. A. (1983). IDF Document 157, International Dairy Federation, Brussels, Belgium, pp. 33–9.
Law, B. A. (1984). In *Advances in the Microbiology and Biochemistry of Cheese and Fermented Milk,* ed. F. L. Davies & B. A. Law. Elsevier Applied Science Publishers, London, UK, pp. 209–28.
Law, B. A. (1986). Annual Report, Food Research Institute, Reading, UK, pp. 111–32.
Law, B. A. (1987). In *Cheese: Chemistry, Physics and Microbiology* (Vol. 1), ed. P. F. Fox. Elsevier Applied Science, London, UK, pp. 365–92.
Law, B. A. & King, J. S. (1985). *J. Dairy Res.*, **52,** 183–8.
Law, B. A. & Wigmore, A. (1982*a*). *J. Dairy Res.*, **49,** 137–46.
Law, B. A. & Wigmore, A. (1982*b*). *J. Soc. Dairy Technol.*, **35,** 75–6.
Law, B. A. & Wigmore, A. (1983). *J. Dairy Res.*, **50,** 519–25.
Law, B. A. & Wigmore, A. S. (1985). *J. Soc. Dairy Technol.*, **38,** 86–8.

Lawrence, R. C. & Creamer, L. K. (1969). *J. Dairy Res.*, **36,** 11–20.
Lawrence, R. C., Heap, H. A. & Creamer, L. K. (1984). *J. Dairy Sci.*, **67,** 1632–45.
Lazaridis, H. N., Rosenau, J. R. & Mahoney, R. R. (1981). *J. Food Sci.*, **46,** 332–9.
Lee, M.-Z. P. & Richardson, T. (1987). *J. Dairy Sci.*, **70,** 2032–9.
Lee, H. J., Olson, N. F. & Ryan, D. S. (1980). *J. Dairy Sci.*, **63,** 1834–8.
Lee, C. R., Lin, C. F. & Melachouris, N. (1986*a*). US Patent 4,575,594.
Lee, S. W., Shimizu, M., Kaminogawa, S. & Yamauchi, K. (1986*b*). *Dairy Sci. Abstr.*, **49,** No. 562.
Lenoir, J. (1984). IDF Bulletin 171, pp. 3–20.
Lin, J. C. C., Jeon, I. J., Roberts, H. A. & Milliken, G. A. (1987). *J. Food Sci.*, **52,** 620–5.
Lonnerdal, B. (1985). *Amer. J. Clin. Nutr.*, **42,** 1299–317.
Losi, G. & Chiavari, C. (1984). *Sci. Tecnica Lattiero-Casearia,* **35,** 427–33.
Loucheux-Lefebvre, M. H., Aubert, J.-P. & Jolles, P. (1978). *Biophys. J.*, **23,** 323–6.
Loukas, S., Varoucha, D., Zioudrou, C., Streaty, R. A. & Klee, W. A. (1983). *Biochem.*, **22,** 4567–73.
Magee, E. L. Jr & Olson, N. F. (1981*a*). *J. Dairy Sci.*, **64,** 600–10.
Magee, E. L. Jr & Olson, N. F. (1981*b*). *J. Dairy Sci.*, **64,** 611–15.
Magee, E. L. Jr, Olson, N. F. & Lindsay, R. C. (1981). *J. Dairy Sci.*, **64,** 616–21.
Mahoney, R. R. (1985). In *Developments in Dairy Chemistry—3,* ed. P. F. Fox. Elsevier Applied Science Publishers, London, UK, pp. 69–109.
Mahoney, R. R. & Wilder, T. (1988). *J. Dairy Res.*, **55,** 423–33.
Mann, S. & Thompkinson, D. K. (1988). *Aust. J. Dairy Technol.*, **43,** 47–9.
Marschke, R. J. & Dulley, J. R. (1978). *Aust. J. Dairy Technol.*, **33,** 139–42.
Marschke, R. J., Nickerson, D. E. J., Jarrett, W. D. & Dulley, J. R. (1980). *Aust. J. Dairy Technol.*, **35,** 84–8.
Marshall, R. J. (1986). *J. Dairy Res.*, **53,** 313–22.
Martin, P., Collin, J.-C., Garnot, P., Ribadeau-Dumas, B. & Mocquot, G. (1981). *J. Dairy Res.*, **48,** 447–56.
Martinez, S. B. & Speckman, R. A. (1988*a*). *J. Dairy Sci.*, **71,** 893–900.
Martinez, S. B. & Speckman, R. A. (1988*b*). *J. Dairy Sci.*, **71** (Suppl.), 90 (Abstr.).
Martini, M. C., Bollweg, G. L., Levitt, M. D. & Savaiano, D. A. (1987). *Amer. J. Clin. Nutr.*, **45,** 432–6.
Matheson, A. R. (1981). *NZJ Dairy Sci. and Technol.*, **15,** 33-41.
Maubois, J. L. (1984). *Le Lait,* **64,** 485–95.
Mayr, W. (1988). *Deutsche Molk Z.*, **109,** 483–5.
McBean, R. D. (1979). *NZJ Dairy Sci. Technol.*, **14,** 113–18.
McCord, J. M. & Fridovich, I. (1969). *J. Biol. Chem.*, **244,** 6049–55.
McGugan, W. A., Emmons, D. B. & Larmond, E. (1979). *J. Dairy Sci.*, **62,** 398–403.
Meisel, H. (1986). *FEBS Lett.*, **196,** 223–7.
Meisel, H. & Frister, H. (1989). *J. Dairy Res.*, **56,** 343–9.
Migliore-Samour, D., Floch, F. & Jolles, P. (1989). *J. Dairy Res.*, **56,** 357–62.
Modler, H. W. (1985). *J. Dairy Sci.*, **68,** 2206–14.
Mohmoud, M. M. & Kosikowski, F. V. (1980). *J. Dairy Sci.*, **63** (Suppl. 1), 47 (Abstr.).
Monti, J. C. & Jost, R. (1978). *J. Dairy Sci.*, **61,** 1233–7.
Morrissey, P. A. (1969). *J. Dairy Res.*, **36,** 333–41.
Morrissey, P. A. (1985). In *Developments in Dairy Chemistry—3,* ed. P. F. Fox. Elsevier Applied Science Publishers, London, UK, pp. 1–34.
Moser, P. (1988). *Schweizerische Milchwertschaft. Forsch.* **17**(2), 32–5.
Moskowitz, G. T. & Noelck, S. S. (1987). *J. Dairy Sci.*, **70,** 1761–9.
Moskowitz, G. T., Shen, T., West, I. R., Gassaigne, R. & Feldman, L. I. (1977). *J. Dairy Sci.*, **60,** 1260–5.
Muir, D. D. (1984). *J. Dairy Res.*, **51,** 649–64.

Nakai, S. & Li-Chan, E. (1989). In *Developments in Dairy Chemistry—4,* ed. P. F. Fox. Elsevier Applied Science Publishers, London, UK, pp. 347–76.
Neelakanten, S., Shahani, K. M. & Arnold, R. G. (1971). *Food Prod. Dev.,* **5**(7), 52–8.
Nelson, J. H. (1970*a*). *Food Prod. Dev.,* **4**(1), 52, 54, 56.
Nelson, J. H. (1970*b*). *J. Agric. Food Chem.,* **18,** 567–9.
Nelson, J. H. (1972). *J. Amer. Oil Chem. Soc.,* **49,** 559–62.
Nelson, J. H. (1975). *J. Dairy Sci.,* **58,** 1739–50.
Nelson, J. H., Jensen, R. G. & Pitas, R. E. (1977). *J. Dairy Sci.,* **60,** 327–62.
Neviani, E., Carminati, D., Giraffa, G., Carini, S. & Lanzanova, M. (1988*a*). *Ind. Latte,* **24,** 3–26.
Neviani, E., Carminati, D., Giraffa, G. & Carini, S. (1988*b*). *Ind. Latte,* **24,** 27–34.
Newcomer, A. D., Park, H. S., O'Brien, P. C. & McGill, D. B. (1983). *Amer. J. Clin. Nutr.,* **38,** 257–63.
Nickerson, T. A. (1974). In *Fundamentals of Dairy Chemistry* (2nd edn), ed. B. H. Webb, A. H. Johnson & J. A. Alford. AVI Publishing Co. Inc., Westport, CT, USA, pp. 273–324.
Nijpels, H. H. & Ramet, J. P. (1982). *Proc. 21st Intern. Dairy Congr., 1,* Book 2, 511 (Abstr.).
Niki, R. (1985). *Jap. J. Zootech. Sci.,* **56,** 611–12.
Nitschmann, Hs. & Bohren, H. U. (1955). *Helv. Chim. Acta,* **38,** 1953–63.
Noomen, A. (1983). *Neth. Milk Dairy J.,* **37,** 229–232.
O'Keeffe, A. M., Fox, P. F. & Daly, C. (1978). *J. Dairy Sci.,* **45,** 465–77.
Omer, M. M. & El-Shibiny, S. (1985). *Egyptian J. Dairy Sci.,* **13,** 33–9.
O'Sullivan, M. & Fox, P. F. (1991) *Food Biotechnol.* **5,** 19–32.
Paige, D. M. & Davis, L. R. (1985). *Developments in Dairy Chemistry—3,* ed. P. F. Fox. Elsevier Applied Science Publishers, London, UK, pp. 111–32.
Parry, R. M. Jr, Chandan, R. C. & Shahani, K. M. (1969). *Arch. Biochem. Biophys.,* **103,** 59–65.
Patocka, J. & Jelen, P. (1988). *J. Food Sci.,* **53,** 1370–2.
Peppler, H. J., Dooley, J. G. & Huang, H. T. (1976). *J. Dairy Sci.,* **59,** 859–62.
Phelan, J. A. (1986). PhD thesis, National University of Ireland, Republic of Ireland.
Pierre, A. (1983). *Le Lait,* **63,** 217–29.
Poutanen, K., Linko, Y. Y. & Linko, P. (1978). *Milchwissenschaft,* **33,** 435–8.
Proctor, V. A. & Cunningham, F. E. (1988). *CRC Crit. Rev. Food Sci. Nutr.,* **26,** 359–95.
Pulay, G. (1966). *Proc. 17th Intern. Dairy Congr.,* Allgauer Struckerei und Verlagsantalt GmbH, Kempten, Germany. *D,* 641–6.
Raap, J., Kerling, K. E. T., Vreeman, H. E. & Visser, S. (1983). *Arch. Biochem. Biophys.,* **221,** 117–24.
Ramos, M., Olano, A. & Martinez-Castro, I. (1982). *Proc. 21st Intern. Dairy Congr.,* (Moscow) Mir Publishers, Moscow. *1* Book 2, 516–17.
Rand, A. G. Jr & Hourigan, J. A. (1975). *J. Dairy Sci.,* **58,** 1144–50.
Rank, T. C., Grappin, R. & Olson, N. F. (1985). *J. Dairy Sci.,* **68,** 801–5.
Rausing, H. A. (1982). French Patent, FR 2 500 269 A1.
Raymond, M. N. & Bricas, E. (1979). *J. Dairy Sci.,* **62,** 1719–25.
Raymond, M. N., Garnier, J., Bricas, E., Cilianu, S., Blasnic, M., Chaix, A. & Lefrancier, P. (1972). *Biochemie,* **54,** 145–54.
Raymond, M. N., Bricas, E., Salesse, R., Garnier, J., Garnot, P. & Ribadeau Dumas, B. (1973). *J. Dairy Sci.,* **56,** 419–22.
Reiter, B. (1978). *J. Dairy Res.,* **45,** 131–47.
Reiter, B. (1985). In *Developments in Dairy Chemistry—3,* ed. P. F. Fox. Elsevier Applied Science Publishers, London, UK, pp. 281–336.
Richardson, T. & Olson, N. F. (1974). In *Immobilized Enzymes in Food and Microbiol*

Processes, ed. A. C. Olson & C. C. Looney. Plenum Press, New York, USA, pp. 19–40.
Richardson, G. H., Nelson, J. H. & Farnham, M. G. (1971). *J. Dairy Sci.*, **54,** 643–7.
Roland, J. F. (1986). US Patent 4,585,738.
Rosado, J. L., Solomons, N. W., Lisker, R. & Bourges, H. (1984). *Gastroenterology*, **87,** 1072–82.
Roth, R. A. & Koshland, M. E. (1981). *J. Biol. Chem.*, **256,** 4633–9.
Sabioni, J. G., Silva, D. O., Pinheiro, A. J. R., Borges, A. C. & Chaves, J. B. C. (1984*a*). *J. Dairy Sci.*, **67,** 1694–8.
Sabioni, J. G., Pinheiro, A. J. R., Silva, D. O., Chaves, J. B. P. & Borges, A. C. (1984*b*). *J. Dairy Sci.*, **67,** 2210–15.
Saint-Paul, F., Humbert, G., Paquet, D. & Linden, G. (1984). *Sci. Aliments*, **4,** 259–64.
Salesse, R. & Garnier, J. (1976). *J. Dairy Sci.*, **59,** 1215–21.
Sandholm, M., Ali-Vehmas, T., Kaartinen, L. & Junnila, M. (1988). *J. Veter. Med.*, **35,** 346–52.
Sardinas, J. L. (1972). *Adv. Appl. Microbiol.*, **15,** 39–73.
Sato, R., Noguchi, T. & Naito, H. (1986). *J. Nutr. Sci. Vitaminol.*, **32,** 67–76.
Savaiano, D. A. & Levitt, M. D. (1987). *J. Dairy Sci.*, **70,** 397–406.
Sawyer, W. H. (1969). *J. Dairy Sci.*, **52,** 1347–55.
Schattenkerk, C. & Kerling, K. E. T. (1973). *Neth. Milk Dairy J.*, **27,** 286–7.
Sciancalepore, V. & Longone, V. (1988). *J. Dairy Res.*, **55,** 547–50.
Scott, D. (1975). In *Enzymes in Food Processing* (2nd edn), ed. G. Reed. Academic Press, New York, USA, pp. 219–54.
Seitz, E. W. (1974). *J. Amer. Oil Chem. Soc.*, **51,** 12–16.
Shahani, K. M. (1975). In *Enzymes in Food Processing* (2nd edn). ed. G. Reed. Academic Press, New York, USA, pp. 181–217.
Shahani, K. M., Arnold, R. G., Kilara, A. & Dwivedi, B. K. (1976). *Biotechnol. Bioengng.*, **18,** 891–905.
Shalabi, S. I. & Wheelock, J. V. (1976). *J. Dairy Res.*, **43,** 331–5.
Shalabi, S. I. & Wheelock, J. V. (1977). *J. Dairy Res.*, **44,** 351–5.
Sharma, B. P. & Messing, R. A. (1980). In *Immobilized Enzymes for Food Processing*, ed. W. H. Pitcher Jr. CRC Press Inc., Boca Raton, FL, USA, pp. 185–209.
Shimizu, M., Lee, S. W., Kaminogawa, S. & Yamauchi, K. (1984). *J. Food Sci.*, **49,** 1117–20.
Shimizu, M., Lee, S. W., Kaminogawa, S. & Yamauchi, K. (1986). *J. Food Sci.*, **51,** 1248–52.
Shipe, W. F., Lee, E. C. & Senyk, G. F. (1975). *J. Dairy Sci.*, **58,** 1123–6.
Shukla, T. P. (1975). *CRC Crit. Rev. Food Technol.*, **5,** 325–56.
Singh, S. & Kristoffersen, T. (1970). *J. Dairy Sci.*, **53,** 533–6.
Singh, S. & Kristoffersen, T. (1971*a*). *J. Dairy Sci.*, **54,** 1589–94.
Singh, S. & Kristoffersen, T. (1971*b*). *J. Dairy Sci.*, **54,** 349–54.
Singh, S. & Kristoffersen, T. (1972). *J. Dairy Sci.*, **55,** 744–9.
Singh, H., Shalabi, S. I., Fox, P. F., Flynn, A. & Barry, A. (1988). *J. Dairy Res.*, **55,** 205–15.
Smith, R. A., Duncan, M. J. & Moir, D. T. (1985). *Science*, **229,** 1219–24.
Smith, C. G., Munro, P. A., Otter, D. E. & Brauer, R. M. (1989). *J. Dairy Res.*, **56,** 87–95.
Solomons, N. W., Guerrero, A. M. & Torun, B. (1985). *Amer. J. Clin. Nutr.*, **41,** 222–7.
Sood, V. K. & Kosikowski, F. V. (1979*a*). *J. Dairy Sci.*, **62,** 1865–72.
Sood, V. K. & Kosikowski, F. V. (1979*b*). *J. Dairy Sci.*, **62,** 1713–8.
Sprossler, B. & Plainer, H. (1983). *Food Technol.*, **37**(10), 93–5.
Stadhouders, J. (1960). *Neth. Milk Dairy J.*, **14,** 83–110.

Stadhouders, J., Stegink, H., van den Berg, M. G. & van Ginkel, W. (1987). *Voedingsmiddelen-technologie,* **20**(25), 17–20 (*Dairy Sci. Abstr.*, **50,** (1988) 5762.)
Sternberg, M. (1976). *Adv. Appl. Microbiol.*, **20,** 135–57.
Sukan, G. & Andrews, A. T. (1982*a*). *J. Dairy Res.*, **49,** 265–78.
Sukan, G. & Andrews, A. T. (1982*b*). *J. Dairy Res.*, **49,** 279–92.
Sutherland, B. J. (1975). *Aust. J. Dairy Technol.*, **30,** 138–42.
Swaisgood, H. E. (1980). *Enzyme Microb. Technol.*, **2,** 265–72.
Swaisgood, H. E. & Horton, H. R. (1980). In *Sulfur in Biology*. (Ciba Foundation Symposium 72), Excerpta Medica, Amsterdam, The Netherlands, pp. 205–22.
Takase, M., Fukuwatari, Y., Kawase, K., Kiyosawa, I., Ogasa, K., Suzuki, S. & Kuroume, T. (1979). *J. Dairy Sci.*, **62,** 1570–7.
Talbott, L. L. & McCord, C. (1981). Paper presented at the 2nd Biennial Marschall International Cheese Conference, Madison, WI, 15–18 Sept.
Teschemacher, H. (1978). In *Developments in Opiate Research,* ed. A. Hertz. Marcel Dekker, New York, USA, pp. 67–151.
Teschemacher, H. (1987). In *Human Lactation, 3,* ed. A. S. Goldman, S. A. Atkinson & L. A. Hanson. Plenum Press, New York, USA, pp. 213–35.
Teschemacher, H., Umbach, M., Hamel, U., Praetorius, K., Amnert-Hilger, G., Brandl, V., Lottspeich, F. & Henschen, A. (1986). *J. Dairy Res.*, **53,** 135–8.
Thapon, J. L. & Brule, G. (1986). *Le Lait,* **66,** 19–30.
Thompson, M. P. & Brower, D. P. (1976). *Cultured Dairy Prod. J.*, **11**(1), 22–3.
Torres, N. & Chandan, R. C. (1981). *J. Dairy Sci.*, **64,** 2161–9.
Valenzuela, A., Adarmes, H. & Guerra, R. (1981). *Alimentos,* **6,** 5–11.
Visser, S. & Rollema, H. S. (1986). *Anal. Biochem.*, **153,** 235–41.
Visser, S., Rooijen, S. J. van, Schattenkerk, C. & Kerling, K. E. T. (1976). *Biochim. Biophys. Acta,* **438,** 265–72.
Visser, S., Rooijen, P. J. van, Schattenkerk, C. & Kerling, K. E. T. (1977). *Biochim. Biophys. Acta.*, **481**, 171–6.
Visser, S., Rooijen, P. J. van & Slangen, C. J. (1980). *Eur. J. Biochem.*, **108,** 415–21.
Visser, S., Slangen, C. J. & Rooijen, P. J. van (1987). *Biochem. J.*, **224,** 553–8.
Von Bockelman, I. & Lodin, L. O. (1974). *Proc. XIX Intern. Dairy Congr.*, (New Delhi). Thompson Press (India) Ltd, New Delhi. *IE,* 441 (Abstr.).
Vreeman, H. J., Visser, S., Slangen, C. J. & van Riel, J. A. M. (1986). *Biochem. J.*, **240,** 87–97.
Wake, R. G. (1959). *Aust. J. Biol. Sci.*, **12,** 479–89.
Walstra, P., Dijk, H. J. M. van & Geurts, T. J. (1987). In *Cheese: Chemistry, Physics and Microbiology* (Vol. 1), ed. P. F. Fox. Elsevier Applied Science, London, UK, pp. 135–77.
Wasserfall, F. (1977). *Milchwissenschaft,* **32,** 654–7.
Wasserfall, F. & Prokopek, D. (1978). *Milchwissenschaft,* **33,** 288–91.
Watanabe, M. & Arai, S. (1988). In *Developments in Food Proteins—6,* ed. B. J. F. Hudson. Elsevier Applied Science Publishers, London, UK, pp. 179–217.
Waugh, D. F. & Hippel, P. H. von (1956). *J. Amer. Chem. Soc.*, **78,** 4567–82.
Weetall, H. H. & Yaverbaum, S. (1974). US Patent 3,852,469.
West, D. W. (1986). *J. Dairy Res.*, **53,** 333–52.
West, D. W. & Dalgleish, D. G. (1976). *Biochim. Biophys. Acta,* **438,** 169–75.
Williams, J. & Sanger, F. (1959). *Biochim. Biophys. Acta,* **33,** 294–6.
Wilson, M., Mulvihill, D. M., Donnelly, W. J. & Gill, B. P. (1989). *J. Dairy Res.*, **56,** 487–94.
Wilson, G. A. & Wheelock, J. V. (1972). *J. Dairy Res.*, **39,** 413–19.
Wilson, G. A., Wheelock, J. V. & Kirk, A. (1974). *J. Dairy Res.*, **41,** 37–44.
Wright, D. G. & Rand, A. G. Jr (1973). *J. Food Sci.*, **38,** 1132–5.

Yoshikawa, M., Tani, F. & Chiba, H. (1988). In *Peptide Chemistry 1987*, ed. T. Shiba. Osaka Protein Research Foundation, Osaka, Japan, pp. 473–6.
Yoshikawa, M., Tani, F., Yoshimura, T. & Chiba, H. (1986). *Agric. Biol. Chem.*, **50,** 2419–21.
Zadow, J. G. (1984). *J. Dairy Sci.*, **67,** 2654–79.
Zerfiridis, G. & Kristoffersen, T. (1970). *Proc. 18th Intern. Dairy Congr.*, (Sydney), Oswald Ziegler Publications, Sydney. *IE*, 351 (Abstr.).
Zioudrou, C., Streaty, R. A. & Klee, W. A. (1979). *J. Biol. Chem.*, **254**, 2446–9.

6

The Role of Ethylene in Fruit Ripening and Senescence

Jeffrey F. D. Dean* & Autar K. Mattoo

Plant Hormone Laboratory and Plant Molecular Biology Laboratory, Beltsville Agricultural Research Center (W), ARS, US Department of Agriculture, Beltsville, Maryland 20705, USA

INTRODUCTION

Every fruit displays a unique assortment of characteristics that its consumer considers attractive and indicative of ripeness. The overall characteristics of ripening generally involve changes in color, texture, and flavor as a result of specific alterations in the patterns of respiration, sugar and aromatic content, cell wall metabolism, and the turnover of pigments. Because these metabolic changes occur in concert during a very brief, determinate period in fruit development, it is thought that a fundamental signal may trigger these varied responses. The need to influence the degree or timing of these metabolic shifts in commodity crops has long been the basis of a search for this 'ripening controller'.

The gaseous phytohormone, ethylene, has been a focus in this hunt ever since Cousins (1910) suggested that a gas produced by oranges accelerated the ripening of bananas, and Gane (1934) demonstrated ethylene production by apples. Ethylene has since been shown to be produced by all higher plants tested, and is recognized as an effector of numerous growth processes in addition to ripening, including germination, abscission, and senescence (Lieberman, 1979; Lürssen, 1985; Reid, 1985; Mattoo & Aharoni, 1988). Production of ethylene in response to environmental stresses and disease is an early and rapid event common to many plant tissues (Hislop *et al.*, 1973; Yang & Hoffman, 1984; Kader, 1985; Ecker & Davis, 1987).

The gaseous nature of ethylene under physiological conditions holds many advantages for its use as a hormone in organisms that lack a pressurized vasculature or central nervous system, but still require an ability to respond quickly to environmental stimuli. The ability of ethylene to diffuse readily through intercellular spaces makes it an appropriate messenger for coordinat-

* Present address: Department of Biochemistry, University of Georgia, Athens, GA 30602, USA.

ing responses of plants to wounding, as well as for the ripening of various differentiated tissues (Brecht, 1987; Reid, 1987). Also, the rapid diffusion of ethylene out of the tissue precludes any need for elaborate mechanisms to render it inactive once the requirement for stimulation has passed. The ability of exogenously applied ethylene to diffuse into tissues and bring about metabolic changes, together with the advent of its rapid assay, have facilitated research on the biological effects of ethylene (Abeles, 1973; Lürssen, 1985; Lougheed *et al.*, 1987; Reid, 1987; Mattoo & Aharoni, 1988, 1991). The responsiveness of harvested fruit to a diffusible gas has encouraged the use of ethylene and ethylene-releasing compounds to influence fruit quality at market (Lürssen, 1985; Watada, 1986; Lougheed *et al.*, 1987).

Although ripening encompasses a wide array of metabolic changes, many of which are specific to each type of fruit, all fruits may be classified as either climacteric or non-climacteric according to their respective patterns of ethylene production and respiration (Biale & Young, 1981). Climacteric fruits, such as apple, avocado, banana, peach and tomato show a transient increase in ethylene production and respiration near the end of the cell enlargement stage of development. This 'respiratory climacteric' (Rhodes, 1980) coincides with the initial structural and anatomical changes considered indicative of ripening in these fruit. Despite much study, a precise cause-and-effect relationship between ethylene production and respiration remains to be defined. In most cases, ethylene synthesis precedes or coincides with the onset of respiration (Biale & Young, 1981). Application of ethylene to climacteric fruit stimulates production of endogenous ethylene (auto-enhancement), leading to an advancement of the climacteric (Tucker & Grierson, 1987). In contrast, non-climacteric fruit, such as citrus, cucumber and strawberry, show a general decline in both respiration and ethylene production during ripening. However, when exposed to exogenous ethylene, non-climacteric fruits undergo a respiratory increase in proportion to the concentration of ethylene. Subsequent removal of ethylene allows the fruit to return to its original level of respiration. It is interesting to note that immature 'climacteric fruit' also display increased respiration when exposed to exogenous ethylene (Eaks, 1980). Thus, it appears that the primary difference between climacteric and non-climacteric fruits is a developmentally dependent capacity for auto-enhancement of ethylene production that develops in mature climacteric-type fruit (Tucker & Grierson, 1987).

In this chapter, our aim has been to cover the current conception of ethylene biosynthesis and its effects on the enzymes affecting ripening and post-harvest characteristics of both climacteric and non-climacteric fruits. The effects of ethylene as related to other agricultural uses have been reviewed elsewhere (Lürssen, 1985; Lougheed *et al.*, 1987, 1991).

THE BIOSYNTHESIS OF ETHYLENE

Early attempts to identify intermediates of the ethylene biosynthetic pathway involved model systems capable of generating ethylene from substrates such as linoleic acid, propanol or methionine upon exposure to free radicals and transition metal catalysts (Lieberman, 1979). It was subsequently shown that apple tissue slices, when incubated with [3,4-^{14}C]-methionine, produced radiolabelled ethylene (Lieberman *et al.*, 1966). Similar results were later obtained with several other plant tissues (Lieberman, 1979). A breakthrough was made in 1979 when 1-aminocyclopropane-1-carboxylic acid (ACC) was identified as the immediate precursor of ethylene (Adams & Yang, 1979; Lürrsen *et al.*, 1979) in the following sequence: methionine→*S*-adenosylmethionine (SAM)→ACC→ethylene (Fig. 1). A variety of plant tissues convert ACC to ethylene (Yang & Hoffman, 1984). However, severe stress conditions can elicit ethylene production in the absence of ACC via free radical reactions with lipids and other cellular components (Sandmann & Böger, 1980; Diolez *et al.*, 1986; Mattoo *et al.*, 1986).

ACC synthase

Cell-free extracts from tomato fruit were shown to catalyze the formation of ACC from SAM (Böller *et al.*, 1979; Konze & Kende, 1979). The enzyme catalyzing this reaction, ACC synthase (EC 4.4.1.4), is considered a rate-

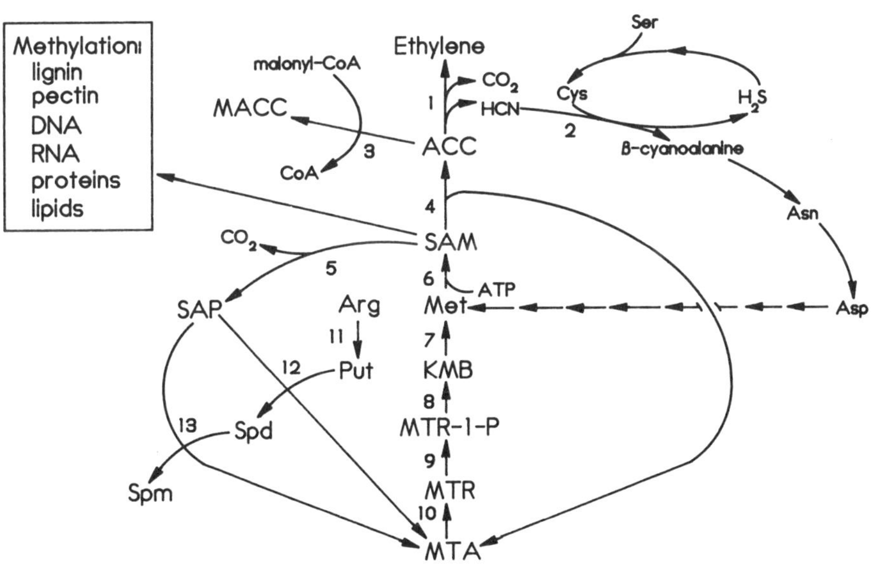

FIG. 1. Schematic representation of the ethylene biosynthetic pathway depicting inter-linkage with methionine recycling, polyamine biosynthesis and cyanide detoxification. The numbered reactions are catalyzed by the following enzymes: (1) EFE, (2) β-cyanoalanine synthase, (3) ACC–*N*-malonyltransferase, (4) ACC synthase, (5) SAM decarboxylase, (6) SAM synthetase, (7) 2-keto-4-methylthiobutyrate (KMB) transaminase, (8) unidentified (probably multiple steps), (9) methylthioribose (MTR) kinase, (10) 5′-methylthioadenosine (MTA) nucleosidase, (11) arginine decarboxylase, (12) spermidine (Spd) synthase, (13) spermine (Spm) synthase. Put is for putrescine.

limiting factor for ethylene biosynthesis in most plant tissues (Yang & Hoffman, 1984; Mansour *et al.*, 1986), and has thus been the focus of considerable research.

ACC synthase constitutes *c.* 0·0002% of the total soluble protein in light-red tomato fruit (Kende *et al.*, 1985). The enzyme has been purified from mung beans and squash (Nakajima *et al.*, 1988; Tsai *et al.*, 1988), and also identified in tomato using monoclonal antibodies (Bleeker *et al.*, 1986, 1988*b*; Mehta *et al.*, 1988*a*) or affinity-labeling (Privalle & Graham, 1987; Mehta *et al.*, 1988*b*). Some confusion exists regarding the reported subunit molecular weights for this enzyme (Mattoo *et al.*, 1989*a*). This confusion may have resulted from the characterization of different isoforms of ACC synthase. Mehta *et al.*, (1988*a*) identified three separate ACC synthase activities by isoelectric focussing. In addition, Nakagawa *et al.* (1988) showed that antibodies specific for the wound-induced ACC synthase from squash do not cross-react with the auxin-induced ACC synthase. Thus, it now appears that differentially regulated, multiple forms of ACC synthase exist. This may explain in part why experiments to block transcription of ACC synthase worked in some tissues (Acaster & Kende, 1983) but not in others (Chappell *et al.*, 1984). A more detailed understanding of this enzyme should soon be forthcoming now that a cDNA encoding an ACC synthase from squash has been isolated (Sato & Theologis, 1989).

Although calcium ions do not affect tomato ACC synthase activity via calmodulin (Mattoo *et al.*, 1982/83), they appear to stimulate ACC production either by increased synthesis of the enzyme or through some post-translational event (Ben-Arie *et al.*, 1982; Burns & Evenson, 1986). Increases in ACC synthase activity have been noted in ripening fruit (Böller *et al.*, 1979), wounded tissues (Böller & Kende, 1980; Hyodo & Nishino, 1981; Kende & Böller, 1981; Mattoo & Anderson, 1984), senescent flowers (Mor *et al.*, 1985), auxin-treated seedlings (Jones & Kende, 1979; Yu & Yang, 1979; Yoshii & Imaseki, 1981), and galactose-infiltrated tomato fruit (Gross, 1983).

ACC synthase activity is also regulated by ethylene in a tissue-specific manner. Auxin-induced mung bean and citrus peel ACC synthases are repressed by ethylene (Riov & Yang, 1982; Yoshii & Imaseki, 1982), whereas the enzyme from mature pre-climacteric fruits increases rapidly with increased ethylene synthesis (Liu *et al.*, 1985*a*; Sitrit *et al.*, 1986).

The ACC synthases from wounded winter squash (Nakajima & Imaseki, 1986) and mung bean hypocotyls (Satoh & Esashi, 1986) are rapidly inactivated by their substrate, SAM, under conditions conducive to the catalytic production of ACC. Radiolabelling experiments suggest that this inactivation results from the covalent attachment of a fragment of SAM to the enzyme (Satoh & Yang, 1988); thus, catalysis for these ACC synthases may be considered a 'suicide' event. Such a regulatory mechanism seems metabolically expensive, but it may allow for tight control of ACC production under conditions requiring transient bursts of ethylene.

Ethylene-forming enzyme (EFE)

The terminal step in ethylene production involves oxidation of ACC to ethylene, carbon dioxide and cyanide, catalyzed by the ethylene-forming enzyme (EFE). The enzyme requires oxygen for activity (Adams & Yang, 1979; Mattoo *et al.*, 1989*b*), is inhibited by CO (Yu & Yang, 1979) and salicylic acid (Romani *et al.*, 1989), and is feedback-regulated by carbon dioxide (Kao & Yang, 1982; Bufler, 1986). This enzyme activity seems constitutive in most plant tissues; however, exceptions include pre-climacteric fruits (Yang & Hoffman, 1984). Application of ethylene to many plant tissues enhances EFE activity (Riov & Yang, 1982; Chalutz *et al.*, 1984; Gupta & Anderson, 1989), but the ability of climacteric fruits to respond in a similar manner only develops upon maturation (Liu *et al.*, 1985*a*; Bufler, 1986). This developmentally acquired ethylene enhancement of EFE activity, in combination with ethylene-stimulated ACC synthase activity, is probably the initial trigger for the auto-enhancement cascade leading to accelerated ripening (Abeles, 1973). Unlike ACC synthase, ethylene enhancement of EFE activity in avocado does not require de novo protein synthesis (Sitrit *et al.*, 1986). In non-climacteric fruits, the enhancement of EFE activity in response to ethylene has not been observed.

Cyanide released from ACC during ethylene synthesis is scavenged by β-cyanoalanine synthase (EC 4.4.1.9) to form β-cyanoalanine from cysteine, and does not appear to inhibit respiration in tissues producing copious ethylene (Yip & Yang, 1988). Ethylene stimulates β-cyanoalanine synthase activity in several plant tissues (Goudey *et al.*, 1989).

Purification of EFE has proven even more intractable than ACC synthase. The enzyme appears to be a highly structured membrane protein whose activity is sensitive to membrane perturbants, including detergents, ionophores and cold shock (Mattoo & Aharoni, 1988). Stereospecificity of EFE activity has been demonstrated in membrane vesicles (Mitchell *et al.*, 1988) and isolated vacuoles (Guy & Kende, 1984). Removal of vacuoles from petunia protoplasts resulted in a 94% decrease in protoplastic EFE activity, and under appropriate conditions, the isolated vacuoles were capable of converting ACC to ethylene at rates approaching those of normal protoplasts (Erdmann *et al.*, 1989). Reformation of the central vacuole in evacuolated petunia protoplasts required de novo protein synthesis, and led to the light-dependent revovery of EFE activity. Experiments with the ACC analog, α-aminoisobutyric acid (α-AIB), suggest that the potential for active transport of ACC into tomato pericarp vacuoles increases immediately prior to the climacteric and then falls during later stages of ripening in parallel with ethylene production (Saftner & Baker, 1987). This work appears to further support a tonoplast localization for EFE.

However, other work, using isolated vacuoles from pea and bean protoplasts, suggests that protoplasts may contain <5% of the total EFE activity of intact tissues, and that significant levels of activity may be associated with the plasmalemma (Porter *et al.*, 1986). Inhibition of EFE activity in apple tissue by high levels (100 mM) of calcium suggests that as much as 40% of the activity may associate with the plasmalemma, where it is probably localized on the inner surface of the membrane (Cheverry *et al.*, 1988). Previous work also suggested an association between ethylene production and the cell wall plasmalemma, and demonstrated a stabilization of ethylene production rate by the addition of Ca^{2+} or Mg^{2+} (Mattoo & Lieberman, 1977; Anderson *et al.*, 1979; Lieberman & Wang, 1982). The low recovery rates for EFE activity in all of these systems may reflect a requirement of EFE for a transmembrane ion potential (Mayne & Kende, 1986) or the release of sequestered flavonoid or proteinaceous inhibitors of the enzyme during tissue homogenization (Sakai & Imaseki, 1973; Shih *et al.*, 1989).

ACC conjugation

Plants further control ethylene production by limiting the free ACC pool. ACC is condensed with malonyl-CoA (Kionka & Amrhein, 1984) to form 1-(malonylamino)cyclopropane-1-carboxylic acid (MACC), which is subsequently sequestered in the vacuole. This malonylation of ACC appears to render it biologically inactive (Yang & Hoffman, 1984), and thus appears to be a control step in maintaining low ethylene levels in immature fruit (Sitrit *et al.*, 1986), particularly under conditions where ACC is imported from other parts of the plant. Conjugation of ACC may also be important for protecting tissues containing limited EFE activity from the toxic effects of ACC (Philosoph-Hadas & Aharoni, 1987). High levels of MACC accumulate in the skin of apples during the climacteric and have been suggested to be important for the active lipid synthesis required for cuticular wax deposition in this tissue (Mansour *et al.*, 1986). An enzyme capable of releasing ACC from MACC is inducible in tobacco and watercress (Jiao *et al.*, 1986); thus, MACC may also serve as a storage form of ACC under appropriate conditions. The ACC-*N*-malonyltransferase appears to provide additional feedback control of ethylene production as this enzyme activity has been demonstrated to increase in response to ethylene (Liu *et al.*, 1985*b*).

Methionine cycling

ACC synthase catalyzes the stoichiometric synthesis of ACC and 5′-methylthioadenosine (MTA) from SAM. MTA is then recycled to methionine (Wang *et al.*, 1982; Yung *et al.*, 1982; Giovanelli *et al.*, 1983) via the pathway shown in Fig. 1. The MTA nucleosidase and methylthioribose (MTR) kinase enzymes are under developmental control (Adams & Yang, 1977; Guranowski *et al.*, 1981; Kushad *et al.*, 1982, 1985, 1988), but do not seem to be coordinately expressed. Very little information is available regarding the regulation of the remaining enzymes. However, a SAM synthetase (EC 2.5.1.6) gene from *Arabidopsis thaliana* has been cloned, and gene expression studies indicate that transcripts for this gene exist primarily in the vascular tissues during normal growth (Peleman *et al.*, 1989). Analysis of genomic DNA indicates that a second gene exists in *Arabidopsis* and further work will be necessary to determine whether either of these genes has a regulatory function for ethylene biosynthesis in fruit.

Polyamine biosynthesis

SAM and MTA are also substrate and product, respectively, for another important pathway in cells, that of polyamine biosynthesis (see Fig. 1). Polyamines are modulators of numerous processes in fruits and other plant tissues (Smith, 1985; Galston & Kaur-Sawhney, 1987), including ethylene biosynthesis (Apelbaum *et al.*, 1981; Ben-Arie *et al.*, 1982; Even-Chen *et al.*, 1982). The first step to polyamine biosynthesis is the decarboxylation of SAM to form the propylamine donor, *S*-(5′-adenosyl)-3-methylmercaptopropylamine (SAP) (Icekson *et al.*, 1985). The initial polyamine carbon skeleton, putrescine (1,4-diaminobutane), derived from arginine (Apelbaum *et al.*, 1985; Smith, 1985), is the acceptor for one (spermine) or two (spermidine) propylamine groups from SAP. MTA is released after transfer of the propylamine group and is recycled to methionine by the pathway previously described.

Polyamines suppress ethylene production by inhibiting both ACC synthase (Even-Chen *et al.*, 1982; Mattoo & Anderson, 1984) and EFE (Apelbaum *et al.*, 1981; Suttle, 1981). Stabilization of cellular membranes by polyamines has been implicated in the inhibition of EFE activity (Ben-Arie *et al.*, 1982), but more recent studies suggest that polyamines restrict ACC uptake into tomato pericarp vacuoles, and thereby limit ethylene production by EFE (Saftner, 1989). Conversely, ethylene may suppress polyamine biosynthesis by inhibiting arginine decarboxylase and SAM decarboxylase (Icekson *et al.*, 1986). This tight coupling and reciprocal control of polyamine and ethylene production by mutual inhibition of synthesis suggests that the two pathways normally do not compete for the same substrates (Kushad *et al.*, 1988).

Ethylene receptors

The most vexing problem in ethylene biology, as with many other plant hormones, centers on perception. The molecular mechanism whereby a plant cell responds to the presence of ethylene is poorly understood and until recently, two very different proposals vied to explain the mechanism of ethylene action.

One hypothesis suggested that the oxidation products of ethylene metabolism were required for ethylene action (Beyer, 1975). Ethylene incorporation and oxidation were both blocked by Ag^+ or carbon dioxide in parallel with inhibition of ethylene action by the same affectors (Beyer, 1979). However, the low rate of ethylene metabolism at physiological concentrations of ethylene, and the inability to saturate the reaction (even at 10–15 times the concentration for saturating physiological responses) undermined this hypothesis (Smith & Hall, 1984). In a careful study, it was demonstrated that while four different cyclic olefins at very low concentrations could block ethylene metabolism, only two of them, viz. 2,5-norbornadiene and 1,3-cyclohexadiene, were capable of specifically blocking ethylene action (Raskin & Beyer, 1989). Thus, ethylene metabolism is not a prerequisite for ethylene action, disproving the first hypothesis.

The second hypothesis calls for the binding of ethylene to a receptor or receptors that in turn mediate ethylene action. Some information regarding ethylene-binding proteins has been collected. A kinetic study of ethylene displacement in tobacco leaves suggests about 4000 binding sites per cell (Sisler, 1979). In-situ properties, such as a requirement for oxygen and inhibition by Ag^+, suggest that ethylene binding may involve a metal ion (Burg & Burg, 1967; Lieberman, 1979; Thompson *et al.*, 1983; Goren *et al.*, 1984). Some progress has been made on the purification of an ethylene binding protein (Sisler 1982, 1984), and the data point to its localization in the endoplasmic reticulum (Evans *et al.*, 1982*a*,*b*; Hall *et al.*, 1984). The numbers of these ethylene binding proteins do not appear to vary greatly during fruit ripening (Blankenship & Sisler, 1989), suggesting that the number of ethylene receptors is not limiting the climacteric response in immature climacteric fruit. However, ethylene binding may be modulated by cellular metabolites, such as short-chain fatty acids, which have been shown to increase the sensitivity of the ethylene recognition system during pollination-induced senescence of petunia flowers (Whitehead & Halevy, 1989). A mutant of *Arabidopsis* displaying characteristics of impaired ethylene sensitivity and binding, such as might be associated with an altered receptor, has been isolated (Bleeker *et al.*, 1988*a*). Cloning of the affected gene may eventually yield the tool for solving problems concerning the ethylene receptor.

Tissue-specific regulation of ethylene production

Because of their bulky nature, fruit are often viewed as being composed of a single homogeneous tissue. Consequently, biochemical and enzymological analyses are often performed without regard for cotyledons, seeds, locules, pericarp, cortex, skin or other tissues that constitute the specific fruit. Mansour *et al.* (1986) have demonstrated a two-fold difference in ACC and ethylene levels between the inner cortex and the skin in apples. Similar developmental changes and differences have been found in both avocado (Sitrit *et al.*, 1987) and tomato (Brecht, 1987; Tieman & Handa, 1989). These studies generally suggest that ethylene production initially begins in the region of the seeds. Given that the primary function of a fruit is to provide the seed with some form of survival or distributive advantage, it seems logical that the seed should be able to signal the fruit when to begin its maturation. However, at present there is no evidence to support this idea.

EFFECTS OF ETHYLENE DURING FRUIT RIPENING

Because ripening tissues undergo a series of catabolic events leading to breakdown and degradation of a variety of cellular structures, ripening was long construed as just an early stage of senescence (Sacher, 1973). More recent observations of orchestrated changes in protein synthesis (Baker *et al.*, 1985) and gene transcription (Grierson, 1985; Slater *et al.*, 1985) have led to consideration of the ripening process as a later stage of tissue differentiation. These changes in gene expression seem to be required for ripening to proceed, and probably account for features of ripening not associated with senescence in other tissues, such as pigment synthesis and shifts in cell wall metabolism (Grierson, 1987). Nevertheless, ripening still comprises certain catabolic processes common to senescence, and hastens the imminent demise of many fruit tissues (Brady, 1987). It would be surprising indeed if ripening did not depend upon the interplay of controlled differentiation and senescent catabolism. Some of the changes associated with fruit ripening are listed in Table 1.

Although levels of the well-defined endogenous plant growth regulators change with the onset of ripening (Lieberman, 1979; Rhodes, 1980; Ludford, 1987), only ethylene can be considered to have a direct regulatory role in

TABLE 1
Changes Associated with Fruit Ripening

- Increased ethylene synthesis, ACC content, ACC synthase and EFE activities
- Decreased polyamine biosynthesis
- Increased chlorophyll breakdown and dismantling of the photosynthetic apparatus
- β-Carotene and lycopene accumulation in osmiophilic granules in plastids
- Increased anthocyanin production
- Increased respiration rate, phosphofructokinase activity, and fructose-2,6-bisphosphate content
- Depolymerization of starch
- Increased acid invertase activity
- Decreased malate and increased citrate levels
- Increased malic enzyme activity
- Changed pattern of polypeptide synthesis
- Changed pattern of mRNA synthesis
- Decreased phospholipid content
- Altered membrane permeability to ions and small molecules
- Reduced galactan, araban, and polyuronide content in cell walls
- Partial depolymerization of polyuronides; loss of electron density in the middle-lamella
- Appearance of endopolygalacturonase activity and accumulation of endopolygalacturonase protein
- Increase in β-1:4-glucanase (cellulase) activity
- Changes in pectin methylesterase activity and redistribution of activity between molecular forms
- Calcium mobilization
- Production of various volatiles and aromatics

ripening, particularly in the climacteric fruit (Brady, 1987; Tucker & Grierson, 1987). Several well-documented observations support this proposition. First is the fact that ethylene production coincides with or preceeds other changes characterizing the climacteric (Sawamura *et al.*, 1978; Su *et al.*, 1984, Christofferson, 1987). Secondly, ripening is accelerated by ethylene addition (Abeles, 1973; McGlasson *et al.*, 1978) and slowed upon removal of ethylene by scrubbers or hypobaric storage (Lougheed *et al.*, 1987; see Mattoo *et al.*, 1989*b*). Thirdly, ripening is inhibited by application of ethylene-synthesis blockers such as aminoethoxyvinylglycine (AVG) and by treatments interfering with ethylene recognition (Ag^+, norbornadiene, CO_2) (Yang & Hoffman, 1984). Fourthly, there exist mutant lines with apparent defects in ethylene synthesis or recognition whose fruit fail to ripen (Lieberman, 1979; Roberts *et al.*, 1987). And finally, ethylene has been shown to stimulate the accumulation of specific mRNA transcripts in ripening tomato (Grierson, 1985; Maunders *et al.*, 1987), while blockage of ethylene perception prevents accumulation of these same mRNAs (Davies *et al.*, 1988).

By the broad criteria mentioned above, ethylene may be categorized as a ripening regulator, but it does not regulate each specific biochemical reaction of the ripening process. Using controlled atmosphere (CA) techniques to study tomato ripening, Jeffery *et al.* (1984) have classified biochemical changes into three groups: those not stimulated by ethylene, those enhanced by ethylene, and those intimately dependent on ethylene. Changes that are not affected by ethylene, including shifts in organic acid and hexose concentrations, occur immediately after the fruit is removed from the vine. Chlorophyll degradation, likewise, begins at harvest, but is accelerated by the addition of ethylene. The synthesis of lycopene, polygalacturonase and invertase takes place only after endogenous ethylene synthesis begins. Enzymes known to be responsive to ethylene are listed in Table 2. The judicial use of CA, blockers of ethylene synthesis and action, and ripening mutants to define which enzymes are regulated directly by ethylene has greatly increased our understanding of the biochemical and morphological changes that constitute ripening.

Respiration

Nearly all plant tissues increase their respiration upon treatment with ethylene (Rhodes, 1980) and the close temporal association between ethylene production and the respiratory climacteric led many to believe that ethylene was the cause of increased respiration. However, some fruit appear to initiate the respiratory climacteric prior to significant increases in ethylene synthesis (Biale & Young, 1981). Goodenough (1986) suggested that this is the case for tomato; however, blockage of ethylene perception in tomato by treatment with Ag^+ arrests climacteric respiration while ethylene production continues (Atta-Aly *et al.*, 1987). Consequently, claims describing an uncoupling of the respiratory climacteric from increased ethylene production should be scrutinized carefully.

TABLE 2
Enzymes Influenced by Ethylene Exposure

Enzyme	*Tissue*	*Effect*	*Reference*
ACC: *N*–malonyltransferase	Tomato	+	Liu *et al.* (1985*b*)
ACC synthase	Orange	+	Riov & Yang (1982)
	Preclimacteric fruit	+	Liu *et al.* (1985*a*)
		+	Sitrit *et al.* (1986)
Acid phosphatase	Mango	+	Mattoo *et al.* (1968)
	Banana	+	Kanellis *et al.* (1989*a*)
Alcohol dehydrogenase	Apple	+	Hulme *et al.* (1971)
α–Amylase			
	Banana	+	Mao & Kinsella (1981)
	Mango	+	Mattoo *et al.* (1975)
β-Amylase	Banana	+	Terra *et al.* (1983)
Arginine decarboxylase	Pea	−	Palavan *et al.* (1984)
		−	Apelbaum *et al.* (1985)
	Tomato	−	Kushad *et al.* (1988)
Catalase	Mango	+	Mattoo *et al.* (1968)
	Sweet potato	−	Esaka *et al.* (1983)
Cellulase	Tomato	+	Babbitt *et al.* (1973)
	Avocado	+	Pesis *et al.* (1978)
		+	Awad & Young (1979)
		+	Tucker & Laties (1984)
		+	Christofferson *et al.* (1984)
Chlorophyllase	Calmondin	+	Barmore (1975)
	Orange	+	Shimokawa *et al.* (1978)
	Tomato	+	Jeffery *et al.* (1984)
β-Cyanoalanine synthase	Several	+	Goudey *et al.* (1989)
		+	Yip & Yang (1988)
*Endo*polygalacturonase	Several	+	Sawamura *et al.* (1978)
		+	Brady *et al.* (1982)
		+	Tucker & Grierson (1982)
		+	Saltveit & McFeeters (1980)
		+	Dellapenna *et al.* (1986)
EFE	Several	+	Riov & Yang (1982)
		+	Chalutz *et al.* (1984)
EFE	Tomato	+	Liu *et al.* (1985*a*)
	Apple	+	Bufler (1986)
Fructose-6-phosphate, 2-kinase	Banana	+	Stitt *et al.* (1986)
β-1,3-Glucanase	Tomato, peach	+	Hinton & Pressey (1980)
Invertase	Banana	+	Terra *et al.* (1983)
		+	Iki *et al.* (1978)
	Tomato	+	Jeffery *et al.* (1984)
Isoperoxidase	Asparagus	+	Haard *et al.* (1974)
Lysine decarboxylase	Pea	+	Icekson *et al.* (1986)
Malic enzyme	Apple	+	Hulme *et al.* (1971)
MTA nucleosidase	Tomato	+?	Kushad *et al.* (1988)
MTR kinase	Tomato	+?	Kushad *et al.* (1988)

TABLE 2—*contd.*

Enzyme	*Tissue*	*Effect*	*Reference*
Ornithine decarboxylase	Pea	−	Palavan *et al.* (1984)
	Tomato	−	Bakanashvili *et al.* (1987)
Pectinmethylesterase	Tomato	+/−	Medina *et al.* (1981)
	Papaya	+	Paull & Chen (1983)
Peroxidase	Sweet potato	+	Imaseki (1970)
	Potato	+	Gahagan *et al.* (1968)
		+	Matsuno & Uritani (1972)
	Mango	+	Mattoo *et al.* (1968)
Phenylalanine ammonia-lyase	Apple	+	Blankenship & Unrath (1988)
	Grapefruit	+	Riov *et al.* (1969)
Polyphenol oxidase	Potato	+	Stahmann *et al.* (1966)
Protein kinase	Jerusalem artichoke	+	Koritsas (1988)
Pyruvate decarboxylase	Apple	+	Hulme *et al.* (1971)
SAM decarboxylase	Pea	−	Icekson *et al.* (1986)
Sucrose synthetase	Banana	+	Terra *et al.* (1983)
	Mango	+	Ghai & Modi (1972)
Xylanase	Papaya	+	Paull & Chen (1983)

Climacteric respiration is induced in two phases (Brady & O'Connell, 1976; Tucker & Laties, 1984), with the first phase involving altered protein synthesis and the second comprising expression of new genes. It has been suggested that the mechanism for increased respiration may involve increased electron transport through the cyanide-insensitive alternate oxidation pathway in mitochondria (Solomos, 1977). Such a situation could cause production of excess lipid peroxides that could, in turn, lead to ethylene production without the involvement of ACC (Brennan & Frenkel, 1977; Mattoo *et al.*, 1986; Wang & Yang, 1987). Thus, increased respiration could be responsible in part for enhanced ethylene evolution, particularly if mechanisms for controlling peroxide levels are impaired at the climacteric (Rabinowitch *et al.*, 1982). However, lipid breakdown has been shown to actually decrease in potato slices after induction of the alternate oxidation pathway by ethylene (Laties, 1982). Identification of the proteins that constitute the alternate oxidation pathway (Elthon *et al.*, 1989) should facilitate investigations into their synthesis and turnover during the respiratory climacteric. Also needed are more detailed measurements throughout the ripening sequence of peroxide levels and the activities of enzyme systems responsible for eliminating peroxides.

The cellular demands driving climacteric respiration are still unclear—both increased protein turnover and membrane leakiness have been invoked as possible sinks for the increased energy output from respiration (Brady, 1987). Fitting with these possibilities is a proposal suggesting that the respiratory rise is a homeostatic response by semi-autonomous mitochondria to senescence-driven biochemical changes in other parts of the cell (Romani, 1987).

Carbohydrate and organic acid metabolism

Increased respiration demands increased substrates for mitochondrial oxidation. These may be provided through the glycolytic pathway after starch breakdown in the amyloplasts and transport of triose-phosphates or hexoses to the cytosol (Beaudry *et al.*, 1987). Starch degradation has been shown to begin immediately upon harvest, and does not seem to require the synthesis of ethylene (Jeffery *et al.*, 1984). Alternatively, glycolytic intermediates may be provided by hydrolysis of sucrose, and synthesis of the enzyme responsible for this hydrolysis, β-fructofuranoside fructohydrolase or invertase (EC 3.2.1.26), is under the control of ethylene (Jeffery *et al.*, 1984; Tucker & Brady, 1987).

Treatment of carrot root with ethylene leads to increased content of glycolytic intermediates, and a doubling in the level of fructose-2,6-bisphosphate (F26BP) (Stitt *et al.*, 1986). The increase in F26BP seems to follow from increased activity of fructose-6-phosphate, 2-kinase (EC 2.7.1). F26BP, a key regulator of the glycolytic pathway, inhibits gluconeogenesis and sucrose synthesis at fructose-2,6-bisphosphatase (F26BPase) (EC 3.1.3), while stimulating glycolysis by modulating pyrophosphate-dependent phosphofructokinase activity (PP_i-PFK) (EC 2.7.1.90) (Huber, 1986). A transient rise in F26BP coincides with an increase in ethylene production in ripening banana, and is preceeded by the respiratory climacteric (Beaudry *et al.*, 1987). In these studies, increased activity of fructose-1,6-bisphosphatase (EC 3.1.3.11) and ATP-dependent phosphofructokinase (ATP-PFK) (EC 2.7.1.56) concomitant with the respiratory rise was observed along with an overall decline in PP_i-PFK activity. In-vivo ^{31}P-NMR spectroscopy of freshly-harvested avocado demonstrated increased levels of ATP and F26BP, leading to a speculation that increased glycolytic carbon flow may be a result of F26BP-mediated activation of PP_i-PFK (Bennett *et al.*, 1987). These results appear to support a role for ethylene in modulating F26BP regulation of glycolytic carbon flow, although some differences at specific biochemical steps, e.g. total PP_i-PFK activity levels, exist from fruit to fruit. In contrast, CA storage of pears in air + 10% CO_2, a treatment that reduces both respiration and ethylene evolution, led to declines in protein and fructose-1,6-bisphosphate contents, and ATP-PFK and PP_i-PFK activities, while fructose-6-phosphate and F26BP contents increased as compared to pears stored in air alone (Kerbel *et al.*, 1988). Although, in this case, reduced level of phosphofructokinase may reflect a decreased rate of protein synthesis in elevated CO_2 atmospheres, the increase in F26BP content without a concomitant increase in ethylene production suggests a less direct regulatory relationship between ethylene and F26BP.

A source of carbohydrate substrate for glycolysis and climacteric respiration, other than starch breakdown, may be provided by cell wall hydrolysis during fruit softening (see below). Although it is uncertain whether hexoses released from the cell wall are actually used in this manner, infiltration of tomato (Kim *et al.*, 1987) and pear (Watkins & Frenkel, 1987) fruit with hexoses corresponding to ones turned over during cell wall metabolism (Gross, 1983) leads to profound, but opposite, effects on respiration and ethylene evolution. In the case of tomato, including the ripening mutants *rin* (ripening inhibitor) and *nor* (non-ripening), infiltration with galactose, mannose or galacturonic acid stimulated a transient increase in ACC synthase activity and, subsequently, ethylene production. In contrast, pears infiltrated with mannose displayed decreased respiration, ethylene production and ripening. The changes in mannose-treated pears were ascribed to mannose-6-phosphate toxicity and a decrease in intracellular *ortho*phosphate. Obviously, different fruit have very different capacities for dealing with free hexoses, and further studies as to the pertinence of these observations *in vivo* are warranted.

Oxidation of malate plays an important part in the respiratory climacteric of some fruit (Moreau & Romani, 1982; Goodenough, 1986). However, Jeffery *et al.* (1984) demonstrated that changes in the activities of the enzymes involved in malate and citrate metabolism occurred even in the absence of ethylene. The switch to malate metabolism in tomato fruit has been suggested to stimulate respiration which in turn spurs ethylene production and thus initiates the respiratory climacteric (Goodenough, 1986).

Membrane permeability and calcium

Exposure of both climacteric (Ferguson & Watkins, 1981) and non-climacteric (Elkashif & Huber, 1988) fruit tissues to ethylene results in enhanced electrolyte leakage. This effect is similar to that seen in membranes from senescent or ethylene-treated flowers in which the lipid bilayers transform to a gel phase (McKersie & Thompson, 1979; Thompson *et al.*, 1982) and become increasingly rigidified (Borochov *et al.*, 1978). The increased rigidification and electrolyte leakage is closely correlated with increased lipid peroxidation (Pauls & Thompson, 1980; Dhindsa *et al.*, 1981), possibly as a result of increased lipoxygenase and lipase activities (Thompson *et al.*, 1987). Similar rigidification of the tomato plasmalemma was shown to increase concurrently with the climacteric rise in ethylene production; however, plastid membranes from these same fruit showed no sign of increased rigidification (Legge *et al.*, 1986). Ethylene treatment of mitochondria isolated from apples did not affect normal membrane fluidity, but did reduce the normal increase in fluidity of the inner membrane in response to succinate (Djebar & Moreau, 1988). Although enzyme compartmentalization (Mattoo & Vickery, 1977) and ultrastructural studies (Platt-Aloia & Thomson, 1981; Crookes & Griersen, 1983) indicate that the membranes in fruit cells remain largely intact during ripening, ethylene-induced damage to the membranes resulting in influx of ions such as Ca^{2+} and P_i could lead to major effects on glycolysis and protein phosphorylation.

As mentioned previously, calcium affects several dimensions of ethylene production, some of which may occur through membrane interactions (Ben-Arie *et al.*, 1982). Calcium interacts directly with membranes to reduce deterioration, increase fluidity and control ion transport (Paliyath *et al.*, 1984;

Poovaiah, 1986, 1987). High concentrations of calcium in the incubation buffer can inhibit membrane peroxidation and protein leakage, and stabilize EFE activity in apple slices (Lieberman & Wang, 1982). Post-harvest infiltration of calcium into apple (Paliyath & Poovaiah, 1984; Conway & Sams, 1987) and avocado (Eaks, 1985) has been shown to reduce respiration and ethylene production, maintain fruit firmness and lengthen shelf-life. Thus, the observation that the total calcium content of cell walls from mature green tomatoes drops prior to the increase in ethylene production may be important for more than just softening during fruit ripening (Rigney & Wills, 1981).

A report of an increased ratio of soluble calcium: bound calcium during tomato ripening is suggestive of a role for this cation in the initiation of ethylene biosynthesis (Rigney & Wills, 1981); however, measurements of soluble versus wall-bound calcium in ripening fruits must be scrutinized carefully due to the ability of accumulated malate and citrate to mobilize this cation (Brady *et al.*, 1985, 1987). An increase in free apoplastic calcium would lead to increased influx of calcium into the cells, which in turn would be expected to influence the cellular metabolism (respiratory climacteric) to maintain homeostatic calcium gradients. Influx of calcium into fruit cells promotes ethylene synthesis by increasing the endogenous level of ACC synthase activity (Burns & Evenson, 1986), and may also influence translocation of ACC to the EFE (Cheverry *et al.*, 1988; Saftner, 1989). Calcium influx could also lead to an increase in membrane lipid peroxidation, a process that increases in some senescent tissues, by activation of calcium-dependent lipase and lipoxygenase enzymes (Leshem, 1984; Thompson *et al.*, 1987). The resultant ionic leakiness of such damaged membranes could then lead to further calcium influx.

Significant decreases in the calcium: calmodulin phosphorylation of membrane and soluble proteins in apple (Paliyath & Poovaiah, 1985) and tomato (Raghothama *et al.*, 1985; Poovaiah, 1986) occur during the climacteric. Although ethylene has a demonstrable effect on protein phosphorylation during xylem-cell differentiation in Jerusalem artichoke explants (Koritsas, 1988), the relationship between developmentally regulated protein phosphorylation and ethylene biosynthesis in fruit tissue is unknown.

Cell wall metabolism and softening

Fruit softening is one of the primary morphological indices for ripeness, and as such, is the focus of concern for both the producer and consumer. Although intracellular changes, such as starch degradation (banana), can lead to fruit softening, alterations in cell wall metabolism are thought to be the primary cause of fruit softening. Details of cell wall structural changes and enzymology have been presented elsewhere (Huber, 1983*a*; Dey & Brinson, 1984; Fry, 1985). There are also surprising parallels to be found in comparing ethylene-induced cell wall changes during fruit softening with those induced during abscission (Sexton & Roberts, 1982).

An increase in soluble pectins is an early sign of fruit softening, and consequently the pectolytic enzymes, endopolygalacturonase (EC 3.2.1.15), exopolygalacturonase (EC 3.2.1.67) and oligo-D-galactosiduronate hydrolase (EC 3.2.1.82), are considered to be the controlling factors for this process. Polygalacturonases (PGs) are closely correlated with softening in fruits, but whereas fruits containing high levels of endo-PG, such as tomato, avocado and freestone peach, demonstrate marked softening when ripe (Hobson, 1965; Pressey & Avants, 1978; Awad & Young, 1979), fruits containing only exo-PG, such as clingstone peach and apple, remain much firmer (Bartley, 1978; Pressey & Avants, 1978). An exception to this pattern is strawberry which softens without evident PG activity (Huber, 1983*a*).

The endo-PG from tomato is probably the best studied of all ripening-induced proteins. Based on studies of *nor* tomato mutants, endo-PG was once considered to be the primary factor controlling ripening (Tigchelaar *et al.*, 1978; Poovaiah & Nukaya, 1979). However, endo-PG activity was later shown to be controlled by ethylene in both climacteric (Sawamura *et al.*, 1978; Brady *et al.*, 1982; Tucker & Grierson, 1982) and non-climacteric (Saltveit & McFeeters, 1980) fruits. The gene for endo-PG from tomato has been isolated, and synthesis of endo-PG mRNA has been shown to be under the control of ethylene (Dellapenna *et al.*, 1986; Grierson *et al.*, 1986; Lincoln *et al.*, 1987; Sheehy *et al.*, 1987; Biggs & Handa, 1989).

The endo-PG protein is initially synthesized as a larger protein with a leader sequence that is removed during its transit through the membrane (Sato *et al.*, 1984; Dellapenna *et al.*, 1986). The protein subsequently undergoes glycosylation prior to localization in the cell wall (Zainon & Brady, 1982). Endo-PG activity can be separated into at least two isoenzymic forms (Pressey & Avants, 1973; Tucker *et al.*, 1980). The isoenzymes share a common catalytic polypeptide, but one form aggregates with another protein of unknown function (Tucker *et al.*, 1981; Pressey, 1984). The high molecular weight aggregate form, endo-PG1, is secreted from the cell coincident with initiation of the climacteric. Later in ripening the smaller isoenzyme, endo-PG2, rapidly accumulates to levels far exceeding *endo*-PG1 (Tucker *et al.*, 1980; Brady *et al.*, 1982). It is not known whether ethylene affects the relative ratios of the different endo-PG isozymes in the ripening tomato (Tucker *et al.*, 1980; Brady *et al.*, 1982; Zainon & Brady, 1982). However, the much reduced level of ethylene production in *Nr* (never-ripe) (Tucker *et al.*, 1980) and *alcobaca* (Mutschler & Guttieri, 1987) tomato mutants suggests that failure to synthesize endo-PG2, the smaller, more-active isoform(s) that usually becomes predominant during the climacteric, can prevent an important coupling between ethylene biosynthesis and cell wall degradation in tomato. The observation that treatment of mature green tomatoes with methyl jasmonate uncouples endo-PG activity from ethylene production and chlorophyll degradation (Saniewski *et al.*, 1987) warrants further investigation, particularly in regard to differential effects on the endo-PG isozymes.

The role of endo-PG in ripening was further elucidated by elegant experiments using anti-sense endo-PG mRNA to block expression of polygalacturonase in transgenic plants. Expression of an endo-PG anti-sense gene in transgenic tomato reduced endo-PG activity by 90%, but without affecting fruit softening (Smith *et al.*, 1988). A different approach introduced a chimeric endo-PG gene into *rin* tomato fruit, a mutant that does not produce endo-PG even when treated with ethylene. Although the chimeric endo-PG released polyuronides in amounts comparable to those seen in the wild-type tomato, this was not sufficient for the induction of fruit softening, lycopene accumulation or ethylene production (Giovannoni *et al.*, 1989). The lack of ethylene induction in transgenic tomato expressing all of the endo-PG isoenzymes, appears to contradict work showing that infiltration of tomato fruit with a purified tomato endo-PG induces ethylene biosynthesis (Baldwin & Pressey, 1988). However, given the breadth of defects in *rin* development, it cannot be said that pectic fragments released during normal tomato ripening do not influence ethylene synthesis. Induction of ethylene synthesis in tomato and other tissues has also been demonstrated with pectic fragments released autolytically from cell walls (Brecht & Huber, 1988), or by digestion with fungal pectinases (Tong *et al.*, 1986; Baldwin & Biggs, 1988).

The role of pectin methylesterase (PME) (EC 3.1.1.11) in polyuronide degradation and fruit softening is somewhat tenuous, as this enzyme is often found in large quantities in immature fruit and other plant tissues (Hobson, 1981). However, the possibility that PME activity might influence pectin solubility during ripening by modifying the ability of pectin to bind calcium remains tantalizing (Huber, 1983*a*). PME activity in banana declines after the initiation of ripening, and although transfer to a low (2·5%) oxygen atmosphere slowed this decline, addition of ethylene to this atmosphere failed to reverse the PME activity to its original high rates (Kanellis *et al.*, 1989*a*). At least two isoforms of the enzyme exist in ripening tomato fruit (Tucker *et al.*, 1982), and there is one report that ethylene affects the relative distribution of activity between isoforms (Medina *et al.*, 1981). However, Sawamura *et al.* (1978) reported that PME activity in tomato showed no correlation with ripening and ethylene levels. This discrepancy may result from different extraction protocols used as this enzyme is reportedly quite sensitive to phenolic compounds and cations. Similar contradictory reports on the changes in PME activity of other fruits make it difficult to draw any conclusions regarding the involvement of PME in ethylene biosynthesis or ripening (Huber, 1983*a*).

The levels of another predominant cell wall hydrolase, β-1,4-glucanase or C_x-type cellulase (EC 3.2.1.4), often increase in the ripening fruit (Huber, 1983*a*). Babbitt *et al.* (1973) demonstrated that cellulase levels in mature green tomato rise with the onset of the respiratory climacteric, and precede the increase in endo-PG activity. However, any correlation between ethylene, cellulase and softening in tomato is still somewhat uncertain as *rin* mutants are able to accumulate near normal levels of cellulase, yet do not soften (Buescher & Tigchelaar, 1975). On the other hand, avocado cellulase is definitely synthesized in response to ethylene application (Pesis *et al.*, 1978; Awad & Young, 1979; Tucker & Laties, 1984), and the protein undergoes a series of glycosylations prior to secretion from the cell (Bennett & Christofferson, 1986). It has not been determined whether secretion of avocado cellulase is regulated by ethylene as was shown for the ethylene-induced cellulase responsible for bean leaf abscission (Abeles & Leather, 1971). A clone of the avocado cellulase gene has been used to show that primary regulation in response to ethylene is pre-translational (Christofferson *et al.*, 1984), and that cellulase might comprise a small gene family in avocado (Tucker *et al.*, 1987). Accumulation of avocado cellulase mRNA transcripts appears to be biphasic (Christofferson, 1987). This two-phase induction mechanism resembles those previously discussed for ethylene-induced climacteric respiration and abscission (Sexton & Roberts, 1982; Tucker & Laties, 1984). Christofferson (1987) has suggested that the initial phase may represent the time necessary to overcome effects of a 'ripening inhibitor' (Sawamura *et al.*, 1978; Yang & Hoffman, 1984). The increase in avocado cellulase activity initiated by ethylene treatment is arrested upon transfer of the fruit to a 2·5% oxygen atmosphere, and addition of further ethylene to this atmosphere does not seem to reverse this effect (Kanellis *et al.*, 1989*b*). This decrease in cellulase activity under conditions of low oxygen tension appears to result from decreased transcription of the gene (Kanellis *et al.*, 1989*c*).

Even if cellulase is primarily responsible for softening in avocado, there remains a mystery as to which cell wall polysaccharide is its substrate. This form of cellulase is ineffective against the crystalline cellulose that composes a majority of the cell wall (Hatfield & Nevins, 1986). One intriguing suggestion by these authors and others (Hayashi *et al.*, 1984) is that the enzyme could be active against xyloglucan. A very discrete fragment of this polysaccharide has a demonstrated anti-auxin activity (York *et al.*, 1984; McDougall & Fry, 1988, 1989). Thus, if auxin plays the role of the 'ripening inhibitor', as has been suggested for avocado (Ludford, 1987), then xyloglucan fragments could act as feedback activators for ripening. Such a mechanism might explain in part the decrease in molecular weight of hemicellulosic polymers rich in glucose and xylose during tomato ripening (Huber, 1983*b*). As ethylene inhibits the deposition of xyloglucan and cellulose in the cell wall (Hayashi & Maclachlan, 1984), this process could play an important role in fruit softening.

It is becoming quite clear that the fruit cell wall is not an inert, static structure passively awaiting degradation by hydrolytic enzymes. There is good evidence that cell wall synthesis continues throughout ripening (Labavitch, 1981; Mitcham *et al.*, 1989), and that qualitative shifts in de novo synthesis do occur, particularly in the hemicellulosic fractions (Tong & Gross, 1988; Mitcham *et al.*, 1989).

Pigmentation

Blanching, resulting from chlorophyll degradation, has long been recognized as an effect of ethylene on many fruit and vegetables (Lougheed *et al.*, 1987).

Chlorophyll degradation in tomato proceeds slowly when ethylene levels are kept low. However, chlorophyllase (EC 3.1.1.14) activity increases when higher ethylene concentrations are applied (Jeffery *et al.*, 1984). A tomato which is defective in chlorophyll breakdown has been reported (Tucker & Grierson, 1987). This variety, *greenflesh,* produces red–brown fruit.

Ethylene-induced chlorophyllase from orange peel has been purified to homogeneity (Shimokawa, 1982). Chlorophyll degradation in ethylene-treated oranges was shown to result in the accumulation of α-tocopherol (vitamin E) (Rise *et al.*, 1989), presumably by release of the α-tocopherol precursor, phytol, during degradation (Amir-Shapira *et al.*, 1987). The accumulation of α-tocopherol, a potent free-radical scavenger, may represent a homeostatic response by the plastids to neutralize free-radicals generated by the partly-dismantled photosynthetic apparatus so as to avoid excessive lipid peroxidation. A similar photoprotective function could possibly be ascribed to β-carotene and lycopene, the other major pigments that accumulate in the chromoplasts during ripening (Spurgeon & Porter, 1980).

Phytol and α-tocopherol share a common precursor in the isoprenoid biosynthetic pathway with lycopene and β-carotene; thus, a build-up of intermediates in the phytol pathway might result in the accumulation of carotenoids. Although lycopene is the presumed precursor for β-carotene synthesis, there is evidence that lycopene produced during tomato ripening is derived from another pathway (Goodwin, 1980). Ethylene has been shown to be required for tomato lycopene synthesis in that CA storage (Jeffery *et al.*, 1984), ethylene synthesis inhibitors (AVG, AOA (aminoxyacetic acid, α-AIB)) (Edwards *et al.*, 1983) and ethylene perception blockers (silver thiosulfate) (Atta-Aly *et al.*, 1987) inhibit lycopene synthesis. Methyl jasmonate enhances ethylene synthesis and chlorophyll degradation, but blocks lycopene synthesis in tomato (Saniewski *et al.*, 1987).

Anthocyanin accumulation is responsible for the color in apple, cherry, berries (Tucker & Grierson, 1987) and olive (Rugini *et al.*, 1982). The biosynthesis of the anthocyanin, cyanidin 3-glycoside, in apple skin is associated with ethylene production (Chalmers & Faragher, 1977; Swindeman & Dilley, 1985). Although the activity of phenylalanine ammonia-lyase (PAL) (EC 4.3.1.5), the enzyme primarily responsible for regulating production of the phenylpropanoid precursors of anthocyanins, is stimulated by ethylene, increased PAL activity could not be correlated with anthocyanin accumulation (Faragher & Chalmers, 1977; Blankenship & Unrath, 1988). Thus, the primary regulatory steps in anthocyanin biosynthesis are as yet unknown.

Stress ethylene in post-harvest physiology

Stress, wounding, and infection can all stimulate ethylene biosynthesis, and in mature, pre-climacteric fruits this stress ethylene can initiate the ripening process (MacLeod *et al.*, 1976; Yang & Pratt, 1978). Some work suggests that stress ethylene can also promote the degradation of climacteric fruit tissues by pathogens (Hislop *et al.*, 1973; El-Kazzaz *et al.*, 1983). Environmental stimuli can also induce ethylene production in non-climacteric fruit, often resulting in decreased market quality through tissue softening or accumulation of unpalatable secondary metabolites, or phytoalexins (Kader, 1985; Watada, 1986; Elkashif *et al.*, 1989).

Little is known about the mechanisms whereby mechanical or physical stresses elicit ethylene production, but rapid progress is being made towards understanding disease-induced ethylene biosynthesis. Pectic fragments from the plant cell wall (Tong *et al.*, 1986; Baldwin & Biggs, 1988; Brecht & Huber, 1988) and glucan fragments from fungal cell wall (Chappell *et al.*, 1984; Toppan & Esquerre-Tugaye, 1984) released by the action of hydrolases can act singly or synergistically to stimulate ethylene production (Davis *et al.*, 1986), presumably by binding to specific receptors on the plant cell plasmalemma (Schmidt & Ebel, 1987). Injection of orange peel or infiltration of tomato with endo-PGs or pectin lyases (EC 4.2.2.3) purified from a mixture of *Aspergillus* pectolytic enzymes (Baldwin & Biggs, 1988; Baldwin & Pressey, 1989) or a purified *endo*-xylanase (EC 3.2.1.8) from *Trichoderma viride* (J. F. D. Dean, K. C. Gross & J. D. Anderson, unpublished observations) also stimulate ethylene production in these fruits. However, in the case of the endo-xylanase, it has not been shown that the enzyme generates heat-stable oligosaccharide fragments capable of stimulating ethylene synthesis on their own (Anderson *et al.*, 1989; Dean *et al.*, 1989; Fuchs *et al.*, 1989).

The oligosaccharide fragments are somehow recognized by the cell causing induction of ACC synthase activity and ethylene production (Chappell *et al.*, 1984) through an undefined signal transduction mechanism. Ethylene, in turn, induces transcription of defensive genes resulting in phytoalexin accumulation and synthesis of cell wall proteins (Ecker & Davis, 1987). Also, the production of several hydrolases that can degrade fungal cell walls are induced (Böller, 1985). Two of these hydrolases, chitinase (EC 3.2.1.14) and β-1,3-glucanase (EC 3.2.1.39), are sequestered primarily in the vacuole, suggesting that their release in high concentrations upon lysis of the plant cell is more effective at inhibiting fungal infection than their steady secretion at the site of infection (Mauch & Staehelin, 1989; Bulke *et al.*, 1989).

General mechanisms of ethylene-induced gene expression

As previously discussed, climacteric fruits are distinguished from non-climacteric fruit by auto-enhancement of ethylene production when exposed to exogenous ethylene. As non-climacteric fruit can still respond to exogenous ethylene and may produce ethylene in response to stress conditions, McMurchie *et al.* (1972) have postulated two control systems for ethylene biosynthesis. System I is responsible for controlling synthesis of background and stress-related ethylene, and is found in both climacteric and non-climacteric fruits. System II controls the auto-enhanced production of ethylene, and is found only in climacteric fruit. In this proposal of a two-receptor scheme, the system I receptor is responsive to the concentration of ethylene, and promotes EFE and ACC:malonyltransferase activities (Yang, 1987). The system II receptor is temporally sensitive to ethylene, and is responsible for inducing ACC synthase activity. Temporal sensitivity is imparted to the system II receptor via a 'ripening inhibitor'. The inhibitor either suppresses receptor II synthesis or interferes directly with ethylene binding, but the interference is relieved through the action of system I ethylene. Removal of the inhibitor by the action of system I allows auto-enhancement to take over, and the ripening 'climacteric' commences.

Studies using cDNAs to mRNAs from ethylene-treated green or ripe tomatoes to probe RNA from fruit at various stages of development have demonstrated responses that seem to parallel the physiological effects of ethylene (Lincoln *et al.*, 1987). Four gene products, including one resembling a proteinase inhibitor (Graham *et al.*, 1985), responded immediately to ethylene application (system I); whereas, a PGase gene product only increased with the onset of ripening (system II). However, the puzzle gets more complex as these four genes are rapidly transcribed in response to ethylene in green fruit, but are differentially transcribed during fruit development. Two of these genes show increased sensitivity to baseline levels of ethylene and start transcription with the onset of the climacteric, while the remaining two genes do not show significant transcription until the climacteric when ethylene synthesis is substantially increased. The data partly contradict the hypothesis that increased sensitivity to ethylene is primarily responsible for ripening (Trewavas, 1982); increased sensitivity may initiate ripening, but increased levels of ethylene are mandatory for complete ripening. Further work has demonstrated that ethylene induction of these genes exhibits dose-response transcription curves that saturate with kinetics similar to those observed for certain animal hormones in which the hormone–receptor complex affects gene transcription by direct interaction with the genome (Lincoln & Fischer, 1988*a*). Expression of the same genes in ethylene-treated *rin* tomato may suggest further regulation of the gene products at the post-transcriptional level (Lincoln & Fischer, 1988*b*). One of the genes that is responsive to ethylene, and whose expression is activated with the onset of ripening, appears to be flanked by nucelotide sequences capable of interacting with a DNA-binding factor that is produced during ripening (Deikman & Fischer, 1988).

These studies (Deikman & Fischer, 1988; Lincoln & Fischer, 1988*a,b*) suggest that while transcription of ethylene-induced genes is regulated by both *trans*-acting proteins and *cis*-acting DNA sequences, ethylene also affects post-transcriptional processes. Biggs and Handa (1989) working specifically with endo-PG, likewise, point out the importance of mRNA stability, and post-translational processes such as protein modification and stability, and the production of specific inhibitors of enzyme activity in the control of ripening. Thus, the techniques of molecular biology are being used to more precisely elucidate some aspects of ethylene involvement in fruit development and ripening.

PROSPECTS FOR FUTURE RESEARCH

As is presently the case for many of the life sciences, our understanding of fruit biology is rapidly accelerating with the help of new ideas, new biological models and new technology. The identification of DNA sequences that confer both constitutive and developmentally dependent ethylene sensitivity to genes will not only provide probes with which to identify other ethylene-regulated genes, but will also allow precise control of new genes introduced into transgenic plants. Isolation of the *trans*-acting protein factors that recognize ethylene-sensitive regulatory sequences should enable study of their tissue-specific and developmental regulation, and help determine whether binding of these factors is a function of direct recognition of ethylene or some post-translational modification process stimulated by ethylene. Nucleic acid and immunological probes to both ethylene biosynthetic and ethylene-responsive genes, and gene products, are needed to define precisely which tissues are responsible for initiating the process of fruit development. Such work is already leading to a new understanding of the mechanisms regulating SAM synthetase, the enzyme that catalyzes the synthesis of SAM, the immediate precursor of ACC (Peleman *et al.*, 1989).

The advent of modern molecular biology notwithstanding, there are many purely biochemical and physiological questions that need to be addressed with respect to fruit development and ethylene. Cell culture techniques have been developed in an effort to overcome the limitations to studying the interplay of ethylene and respiration in bulky fruit tissues (Brady & Romani, 1988). NMR spectroscopy of ^{13}C-, ^{15}N-, ^{31}P-, and ^{35}Cl-labeled metabolites and ions in cell culture (Roberts, 1984, 1987; Ojalvo *et al.*, 1987), tissue explants (Lang & Martin, 1986; Chang & Roberts, 1989), seedlings (Roberts & Testa, 1988) and whole fruit (Bennett *et al.*, 1987) promise to answer questions regarding compartmentation and pathway fluxes of ethylene precursors. However, a similar, non-invasive technique for quantifying the relative concentrations of wall-bound, cytosolic, and vacuolar Ca^{2+} levels needs to be developed. And a much more detailed analysis of the interaction of Ca^{2+} with enzymes involved in ethylene synthesis and action needs to be undertaken.

The knowledge that specific oligosaccharides of either plant or fungal origin can induce ethylene biosynthesis has profound implications for models of pathogenesis, as well as fruit ripening. Identification and purification of specific pectic fragments capable of eliciting ethylene production (Tong *et al.*, 1986; Baldwin & Biggs, 1988; Brecht & Huber, 1988) should enable identification of receptors (Schmidt & Ebel, 1987) important for coordination of cell wall degradation and ethylene synthesis during fruit ripening. Identification of these receptors might facilitate discovery of specific, reversible blocking agents capable of controlling post-harvest ripening in a more precise manner than is currently available.

Perhaps the biggest challenge in unravelling the function of ethylene in plant biology is the integration of biochemical and cell biology information with actual morphological and physiological processes underway in the intact plant. Borrowing from environmental chemists studying air pollutants, Woltering *et al.* (1988) have demonstrated that laser-based photo-acoustic detection of ethylene can easily provide a 100-fold improvement in sensitivity over current gas chromatographic methods. This would allow real-time detection of ethylene emanating from whole plants or even single fruit maintained on the plant. One question that could be answered using such a system is whether the rhythmical production of ethylene that has been shown in detached cotyledons (Rikin *et al.*, 1985) extends to entire plants, and whether such a rhythmicity is the basis for the observation of cycles of ethylene produced in the leaves of olive trees followed by cycles of ethylene production in olive fruit (Sanchez-Raya, 1981). Such a coordination in signal transmission between the tree and fruit might provide the trigger that actually initiates ethylene synthesis during fruit ripening on the vine.

The isolation of new mutant genetic lines has provided new models of plant responses to pathogens (Mauch & Staehelin, 1989), and may provide the key to isolating the ethylene receptor (Bleeker *et al.*, 1988*a*). Tomato ripening mutants have provided numerous advancements in our understanding of the role of ethylene in tomato ripening (Roberts *et al.*, 1987), but vast repositories of genetic variation remain untapped (Rick, 1987). Large variations in ethylene biosynthetic capability have been documented in several tomato cultivars (Mattoo & Anderson, 1984); however, the specific cultivars used in particular experiments are not always reported. More thorough investigation of cultivar-specific differences and wider study of the wild forerunners of agronomic crops could provide answers to some of the intractable questions regarding the regulation of ethylene biosynthesis and its relationship to fruit ripening.

ACKNOWLEDGEMENTS

The authors are indebted to Dr J. D. Anderson for many years of support and encouragement in the study of ethylene. Thanks also to Drs K. C. Gross, B. Parsons, and M. Tucker for suggestions and critical reading of the original manuscript. This work was supported in part by USDA Competitive Grant #88-37261-3680.

REFERENCES

Abeles, F. B. (1973). *Ethylene in Plant Biology*. Academic Press, New York, USA.
Abeles, F. B. & Leather, G. R. (1971). *Planta*, **97**, 87–91.
Acaster, M. A. & Kende, H. (1983). *Plant Physiol.*, **72**, 139–45.
Adams, D. O. & Yang, S. F. (1977). *Plant Physiol.*, **60**, 892–6.
Adams, D. O. & Yang, S. F. (1979). *Proc. Natl Acad. Sci. USA*, **76**, 170–4.

Amir-Shapira, D., Goldschmidt, E. E. & Altman, A. (1987). *Proc. Natl Acad. Sci. USA,* **84,** 1901–5.

Anderson, J. D., Lieberman, M. & Stewart, R. N. (1979). *Plant Physiol.,* **63,** 931–5.

Anderson, J. D., Dean, J. F. D., Gamble, H. R. & Mattoo, A. K. (1989). In *Biochemical and Physiological Aspects of Ethylene Production in Higher and Lower Plants,* eds. H. Clijsters, M. De Proft, R. Marcelle & M. Van Poucke. Kluwer Academic Publishers, Dordrecht, The Netherlands, pp. 49–56.

Apelbaum, A., Burgoon, A. C., Anderson, J. D., Lieberman, M., Ben-Arie, R. & Mattoo, A. K. (1981). *Plant Physiol.,* **68,** 453–6.

Apelbaum, A., Goldlust, A. & Icekson, I. (1985). *Plant Physiol.,* **79,** 635–40.

Atta-Aly, M. A., Saltveit, M. E. & Hobson, G. E. (1987). *Plant Physiol.,* **83,** 44–8.

Awad, M. & Young, R. E. (1979). *Plant Physiol.,* **64,** 306–8.

Babbitt, J. K., Powers, M. J. & Patterson, M. E. (1973). *J. Amer. Soc. Hort. Sci.,* **98,** 77–81.

Bakanashvili, M., Barkai-Golan, R., Kopeliovitch, E. & Apelbaum, A. (1987). *Physiol. Molec. Plant Pathol.,* **31,** 41–50.

Baker, J. E., Anderson, J. D. & Hruschka, W. R. (1985). *J. Plant Physiol.,* **120,** 167–79.

Baldwin, E. A. & Biggs, R. H. (1988). *Physiol. Plant.,* **73,** 58–64.

Baldwin, E. A. & Pressey, R. (1988). *J. Amer. Soc. Hort. Sci.,* **113,** 92–5.

Baldwin, E. A. & Pressey, R. (1989). *Plant Physiol.,* **90,** 191–6.

Barmore, C. R. (1975). *Hortsci.,* **10,** 595–9.

Bartley, I. M. (1978). *Phytochem.,* **17,** 213–16.

Beaudry, R. M., Paz, N., Black, C. C. & Kays, S. J. (1987). *Plant Physiol.,* **85,** 277–82.

Ben-Arie, R., Lurie, S. & Mattoo, A. K. (1982). *Plant Sci. Lett.,* **24,** 239–47.

Bennett, A. B. & Christofferson, R. E. (1986). *Plant Physiol.,* **81,** 830–35.

Bennett, A. B., Smith, G. M. & Nichols, B. G. (1987). *Plant Physiol.,* **83,** 973–6.

Beyer, E. M. (1975). *Plant Physiol.,* **56,** 273–8.

Beyer, E. M. (1979). *Plant Physiol.,* **63,** 169–73.

Biale, J. B. & Young, R. E. (1981). In *Recent Advances in the Biochemistry of Fruits and Vegetables,* ed. J. Friend & M. J. C. Rhodes. Academic Press, New York, USA, pp. 1–39.

Biggs, M. S. & Handa, A. K. (1989). *Plant Physiol.,* **89,** 117–25.

Blankenship, S. M. & Sisler, E. C. (1989). *J. Plant Growth Regul.,* **8,** 37–44.

Blankenship, S. M. & Unrath, C. R. (1988). *Phytochem.,* **27,** 1001–3.

Bleeker, A. B., Kenyon, W. H., Somerville, S. C. & Kende, H. (1986). *Proc. Natl. Acad. Sci. USA,* **83,** 7755–9.

Bleeker, A. B., Estelle, M. A., Somerville, C. & Kende, H. (1988*a*). *Science,* **241,** 1086–9.

Bleeker, A. B., Robinson, G. & Kende, H. (1988*b*). *Planta,* **173,** 385–90.

Böller, T. (1985). In *Cellular and Molecular Biology of Plant Stress,* ed. J. L. Key & T. Kosuge. Alan R. Liss, New York, USA, pp. 247–62.

Böller, T. & Kende, H. (1980). *Nature,* **286,** 259–60.

Böller, T., Herner, R. C. & Kende, H. (1979). *Planta,* **145,** 293–303.

Borochov, A., Halevy, A. H., Borochov, H. & Shinitzsky, M. (1978). *Plant Physiol.,* **61,** 812–15.

Brady, C. J. (1987). *Ann. Rev. Plant Physiol.,* **38,** 155–78.

Brady, C. J. & O'Connell, P. B. H. (1976). *Aust. J. Plant Physiol.,* **3,** 301–10.

Brady, C. J. & Romani, R. J. (1988). *Plant Physiol.,* **87,** 571–6.

Brady, C. J., MacAlpine, G., McGlasson, W. B. & Ureda, Y. (1982). *Aust. J. Plant Physiol.,* **9,** 171–8.

Brady, C. J., McGlasson, W. B., Pearson, J. A., Meldrum, S. K. & Kopeliovitch, E. (1985). *J. Amer. Soc. Hort. Sci.,* **110,** 254–8.

Brady, C., McGlasson, B. & Speirs, J. (1987). In *Tomato Biotechnology,* ed. D. J. Nevins & R. A. Jones. Alan R. Liss, New York, USA, pp. 270–88.
Brecht, J. K. (1987). *Hortsci.,* **22,** 476–9.
Brecht, J. K. & Huber, D. J. (1988). *Plant Physiol.,* **88,** 1037–41.
Brennan, T. & Frenkel, C. (1977). *Plant Physiol.,* **59,** 411–6.
Buescher, R. W. & Tigchelaar, E. C. (1975). *Hortsci.,* **10,** 624–5.
Bufler, G. (1986). *Plant Physiol.,* **80,** 539–43.
Bulke, M. van den, Bauw, G., Castresana, C., Montague van, M. & Vandekerckhove, J. (1989). *Proc. Natl Acad. Sci. USA,* **86,** 2673–7.
Burg, S. P. & Burg, E. A. (1967). *Plant Physiol.,* **42,** 144–52.
Burns, J. K. & Evensen, K. B. (1986). *Physiol. Plant.,* **66,** 609–15.
Chalmers, D. J. & Faragher, J. D. (1977). *Aust. J. Plant Physiol.,* **4,** 111–17.
Chalutz, E., Mattoo, A. K., Solomos, T. & Anderson, J. D. (1984). *Plant Physiol.,* **74,** 99–103.
Chang, K. & Roberts, J. K. M. (1989). *Plant Physiol.,* **89,** 197–203.
Chappell, J., Hahlbrock, K. & Boller, T. (1984). *Planta,* **161,** 475–80.
Cheverry, J. L., Pouliquen, J., Le Guyader, H. & Marcellin, P. (1988). *Physiol. Plant,* **74,** 53–7.
Christofferson, R. E. (1987). In *Plant Senescence: Its Biochemistry and Physiology,* ed. W. W. Thomson, E. A. Nothnagel & R. C. Huffaker. American Society of Plant Physiologists, Rockville, MD, USA, pp. 89–97.
Christofferson, R. E., Tucker, M. L. & Laties, G. G. (1984). *Plant Mol. Biol.,* **3,** 385–91.
Conway, W. S. & Sams, C. E. (1987). *J. Amer. Soc. Hort. Sci.,* **112,** 300–3.
Cousins, H. H. (1910). Annual Report of the Department of Agriculture, Jamaica, p. 7.
Crookes, P. R. & Grierson, D. (1983). *Plant Physiol.,* **72,** 1088–93.
Davis, K. R., Darvill, A. G. & Albersheim, P. (1986). *Plant Mol. Biol.,* **6,** 23–32.
Davies, K. M., Hobson, G. E. & Grierson, D. (1988). *Plant Cell Environ.,* **11,** 729–38.
Dean, J. F. D., Gamble, H. R. & Anderson, J. D. (1989). *Phtyopathol.,* **79,** 1071–8.
Deikman, J. & Fischer, R. L. (1988). *EMBO J.,* **7,** 3315–20.
Dellapenna, D., Alexander, D. C. & Bennett, A. B. (1986). *Proc. Natl Acad. Sci. USA,* **83,** 6420–4.
Dey, P. M. & Brinson, K. (1984). *Adv. Carbohydr. Chem. Biochem.,* **42,** 265–382.
Dhindsa, R. J., Dhindsa, P. P. & Thorpe, T. A. (1981). *J. Exper. Bot.,* **32,** 93–101.
Diolez, P., Davy De Virille, J., Latche, A., Moreau, F., Pech, J. C. & Reid, M. (1986). *Plant Sci.,* **43,** 13–17.
Djebar, M.-R. & Moreau, F. (1988). *C. R. Acad. Sci. Paris,* **307,** 741–4.
Eaks, I. L. (1980). *J. Amer. Soc. Hort. Sci.,* **105,** 744–7.
Eaks, I. L. (1985). *J. Amer. Soc. Hort. Sci.,* **110,** 145–8.
Ecker, J. R. & Davis, R. W. (1987). *Proc. Natl Acad. Sci. USA,* **84,** 5202–6.
Edwards, J. I., Saltveit, M. E. & Henderson, W. R. (1983). *J. Amer. Soc. Hort. Sci.,* **108,** 512–14.
Elkashif, M. E. & Huber, D. J. (1988). *J. Amer. Soc. Hort. Sci.,* **113,** 378–81.
Elkashif, M. E., Huber, D. J. & Brecht, J. K. (1989). *J. Amer. Soc. Hort. Sci.,* **114,** 81–5.
El-Kazzaz, M. K., Sommer, N. F. & Kader, A. A. (1983). *Phytopathol.,* **73,** 998–1001.
Elthon, T. E., Nickels, R. L. & McIntosh, L. (1989). *Plant Physiol.,* **89,** 1311–17.
Erdmann, H., Griesbach, R. J., Lawson, R. H. & Mattoo, A. K. (1989). *Planta,* **179,** 196–202.
Esaka, M., Takeshi, T. & Asahi, T. (1983). *Plant Cell Physiol.,* **24,** 625–33.
Evans, D. E., Bengochea, T., Cairns, A. J., Dodds, J. H. & Hall, M. A. (1982*a*). *Plant Cell Environ.,* **5,** 101–7.

Evans, D. E., Dodds, J. H., Lloyd, P. C., apGwynn, I. & Hall, M. A. (1982*b*). *Planta,* **154,** 48–52.
Even-Chen, Z., Mattoo, A. K. & Goren, R. (1982). *Plant Physiol.,* **69,** 385–8.
Faragher, J. D. & Chalmers, D. J. (1977). *Aust. J. Plant Physiol.,* **4,** 133–40.
Ferguson, I. B. & Watkins, C. B. (1981). *Aust. J. Plant Physiol.,* **8,** 155–64.
Fry, S. C. (1985). In *Oxford Surveys of Plant Molecular & Cell Biology* (vol. 2), ed. B. J. Miflin. Clarendon, Oxford, England, pp. 1–42.
Fuchs, Y., Saxena, A., Gamble, H. R. & Anderson, J. D. (1989). *Plant Physiol.,* **89,** 138–43.
Gahagan, H. E., Holm, R. E. & Abeles, F. B. (1968). *Physiol. Plant.,* **21,** 1270–8.
Galston, A. W. & Kaur-Sawhney, R. (1987). In *Plant Hormones and Their Role in Plant Growth and Development,* ed. P. J. Davies. Martinus Nijhoff Publishers, Dordrecht, The Netherlands, pp. 280–95.
Gane, R. (1934). *Nature,* **134,** 1008.
Ghai, G. & Modi, V. V. (1972). *Biochem. Biophys. Res. Comm.,* **41,** 1088–94.
Giovanelli, J., Datko, A. H., Mudd, S. H. & Thompson, G. A. (1983). *Plant Physiol.,* **71,** 319–326.
Giovannoni, J. J., DellaPenna, D., Bennett, A. B. & Fischer, R. L. (1989). *Plant Cell,* **1,** 53–63.
Goodenough, P. W. (1986). *Plant Growth Regul.,* **4,** 125–37.
Goodwin, T. W. (1980). *The Biochemistry of the Carotenoids* (Vol. 1). Chapman & Hall, London, UK.
Goren, R., Mattoo, A. K. & Anderson, J. D. (1984). *J. Plant Physiol.,* **117,** 243–8.
Goudey, J. S., Tittle, F. L. & Spencer, M. S. (1989). *Plant Physiol.,* **89,** 1306–10.
Graham, J. S., Pearce, G., Merryweather, J., Titani, K., Ericsson, L. & Ryan, C. A. (1985). *J. Biol. Chem.,* **260,** 6555–60.
Grierson, D. (1985). *CRC Crit. Rev. Plant Sci.,* **3,** 113–32.
Grierson, D. (1987). *Hortsci.,* **22,** 859–62.
Grierson, D., Tucker, G. A., Keen, J., Ray, J., Bird, C. R. & Schuch, W. (1986). *Nucl. Acid Res.,* **14,** 8595–603.
Gross, K. C. (1983). *Phytochem.,* **22,** 1137–9.
Gupta, K. & Anderson, J. D. (1989). *Plant Cell Physiol.,* **30,** 345–9.
Guranowki, A. B., Chiang, P. K. & Cantoni, G. L. (1981). *Eur. J. Biochem.,* **114,** 293–9.
Guy, M. & Kende, H. (1984). *Planta,* **160,** 281–7.
Haard, N. F., Sharma, S. C., Wolfe, R. & Frenkel, C. (1974). *J. Food Sci.,* **39,** 452–6.
Hall, M. A., Smith, A. R., Thomas, C. J. R. & Howorth, C. J. (1984). In *Ethylene: Biochemical, Physiological, and Applied Aspects,* ed. Y. Fuchs & E. Chalutz. Martinus Nijhoff/Junk Publishers, The Hague, The Netherlands, pp. 55–63.
Hatfield, R. & Nevins, D. J. (1986). *Plant Cell Physiol.,* **27,** 541–52.
Hayashi, T. & Maclachlan, G. (1984). *Plant Physiol.,* **76,** 739–42
Hayashi, T., Wong, Y. & Maclachlan, G. (1984). *Plant Physiol.,* **75,** 605–10.
Hinton, D. M. & Pressey, R. (1980). *J. Amer. Soc. Hort. Sci.,* **105,** 499–502.
Hislop, E. C., Hoad, G. V. & Archer, S. A. (1973). In *Fungal Pathogenicity and the Plant's Response,* ed. R. J. W. Byrde & C. V. Cutting. Academic Press, London, UK, pp. 87–113.
Hobson, G. E. (1965), *J. Hort. Sci.,* **40,** 66–72.
Hobson, G. E. (1981). In *Recent Advances in the Biochemistry of Fruits and Vegetables,* ed. J. Friend & M. J. C. Rhodes. Academic Press, New York, USA, pp. 123–32.
Huber, D. J. (1983*a*). *Hortic. Rev.,* **5,** 169–219.
Huber, D. J. (1983*b*). *J. Amer. Soc. Hort. Sci.,* **108,** 405–9.
Huber, S. C. (1986). *Ann. Rev. Plant Physiol.,* **37,** 233–46.
Hulme, A. C., Rhodes, M. J. C. & Wooltorton, L. S. C. (1971). *Phytochem.,* **10,** 1315–23.

Hyodo, H. & Nishino, T. (1981). *Plant Physiol.*, **67,** 421–3.
Icekson, I., Goldlust, A. & Apelbaum, A. (1985). *J. Plant Physiol.*, **119,** 335–45.
Icekson, I., Bakhanashvili, M. & Apelbaum, A. (1986). *Plant Physiol.*, **82,** 607–9.
Iki, K., Sekiguchi, K., Kurata, K., Tada, T., Nakagawa, H., Oguva, N. & Tukehana, H. (1978). *Phytochem.*, **17,** 311–12.
Imaseki, H. (1970). *Plant Physiol.*, **46,** 172–4.
Jeffery, D., Smith, C., Goodenough, P., Prosser, I. & Grierson, D. (1984). *Plant Physiol.*, **74,** 32–8.
Jiao, X.-Z., Philosoph-Hadas, S., Su, L.-Y. & Yang, S. F. (1986). *Plant Physiol.*, **81,** 637–41.
Jones, J. F. & Kende, H. (1979). *Planta,* **146,** 649–56.
Kader, A. A. (1985). *Hortsci.*, **20,** 54–7.
Kanellis, A. K., Solomos, T. & Mattoo, A. K. (1989*a*). *Plant Physiol.*, **90,** 251–8.
Kanellis, A. K., Solomos, T. & Mattoo, A. K. (1989*b*). *Plant Physiol.*, **90,** 259–6.
Kanellis, A. K., Solomos, T., Mehta, A. M. & Mattoo, A. K. (1989*c*). *Plant Cell Physiol.*, **30,** 817–23.
Kao, C. H. & Yang, S. F. (1982). *Planta,* **155,** 261–6.
Kauss, H. (1987). *Ann. Rev. Plant Physiol.*, **38,** 47–72.
Kende, H. & Böller, T. (1981). *Planta,* **151,** 476–81.
Kende, H., Bleeker, A. B., Kenyon, W. H. & Mayne, R. G. (1985). In *Plant Growth Substances 1985*, ed. M. Bopp. Springer-Verlag, Berlin, Germany, pp. 120–8.
Kerbel, E. L., Kader, A. A. & Romani, R. J. (1988). *Plant Physiol.*, **86,** 1205–9.
Kim, J., Gross, K. C. & Solomos, T. (1987). *Plant Physiol.*, **85,** 804–7.
Kionka, C. & Amrhein, N. (1984). *Planta,* **162,** 226–35.
Konze, J. R. & Kende, H. (1979). *Plant Physiol.*, **63,** 507–10.
Koritsas, V. M. (1988). *J. Exper. Bot.*, **39,** 375–86.
Kushad, M. M., Richardson, D. G. & Ferro, A. J. (1982). *Biochem. Biophys. Res. Commun.*, **108,** 167–73.
Kushad, M. M., Richardson, D. G. & Ferro, A. J. (1985). *Plant Physiol.*, **79,** 525–9.
Kushad, M. M., Yelenosky, G. & Knight, R. (1988). *Plant Physiol.*, **87,** 463–7.
Labavitch, J. M. (1981). *Ann. Rev. Plant Physiol.*, **32,** 385–406.
Lang, G. A. & Martin, G. C. (1986). *J. Amer. Soc. Hort. Sci.*, **111,** 577–81.
Laties, G. G. (1982). *Ann. Rev. Plant Physiol.*, **33,** 519–55.
Legge, R. L., Cheng, K.-H., Lepock, J. R. & Thompson, J. E. (1986). *Plant Physiol.*, **81,** 954–9.
Leshem, Y. Y. (1984). In *Structure, Function, and Metabolism of Plant Lipids,* ed. P.-A. Siegenthaler & W. Eichenberger. Elsevier Science Publishers B. V., Amsterdam, The Netherlands, pp. 181–8.
Lieberman, M. (1979). *Ann. Rev. Plant Physiol.*, **30,** 533–91.
Lieberman, M. & Wang, S. Y. (1982). *Plant Physiol.*, **69,** 1150–5.
Lieberman, M., Kunishi, A. T., Mapson, L. W. & Wardale, D. A. (1966). *Plant Physiol.*, **41,** 376–82.
Lincoln, J. E. & Fischer, R. L. (1988*a*). *Mol. Gen. Genet.*, **212,** 71–5.
Lincoln, J. E. & Fischer, R. L. (1988*b*). *Plant Physiol.*, **88,** 370–4.
Lincoln, J. E., Cordes, S., Read, E. & Fischer, R. L. (1987). *Proc. Natl Acad. Sci. USA,* **84,** 2793–7.
Liu, Y., Hoffman, N. E. & Yang, S. F. (1985*a*). *Plant Physiol.*, **77,** 407–11.
Liu, Y., Su, L.-Y. & Yang, S. F. (1985*b*). *Plant Physiol.*, **77,** 891–5.
Lougheed, E. C., Murr, D. P. & Toivonen, P. M. A. (1987). In *Postharvest Physiology of Vegetables,* ed. J. Weichmann, Marcel Dekker Publishers, New York, USA, pp. 255–76.
Ludford, P. M. (1987). In *Plant Hormones and Their Role in Plant Growth and Development,* ed. P. J. Davies. Martinus Nijhoff Publishers, Dordrecht, The Netherlands, pp. 257–79.

Lürssen, K. (1985). *Adv. Agric. Biotechnol.*, **16,** 307–51.
Lürssen, K., Naumann, K. & Schroder, R. (1979). *Z. Pflanzenphysiol.*, **92,** 285–94.
MacLeod, R. F., Kader, A. A. & Morris, L. L. (1976). *Hortsci.*, **11,** 604–9.
Mansour, R., Latche, A., Vaillant, V., Pech, J.-C. & Reid, M. S. (1986). *Physiol. Plant.*, **66,** 495–502.
Mao, W. W. & Kinsella, J. E. (1981). *J. Food Sci.*, **46,** 1400–3.
Matsuno, H. & Uritani, I. (1972). *Plant Cell Physiol.*, **13,** 1091–101.
Mattoo, A. K. & Aharoni, N. (1988). In *Senescence and Aging in Plants,* ed. L. D. Nooden & A. C. Leopold. Academic Press, New York, USA, pp. 241–80.
Mattoo, A. K. & Anderson, J. D. (1984). In *Ethylene: Biochemical, Physiological, and Applied Aspects,* ed. Y. Fuchs & E. Chalutz. Martinus Nijhoff/Dr W. Junk Publishers, The Hauge, The Netherlands, pp. 139–47.
Mattoo, A. K. & Lieberman, M. (1977). *Plant Physiol.*, **60,** 794–9.
Mattoo, A. K. & Vickery, R. S. (1977). *Plant Physiol.*, **60,** 496–8.
Mattoo, A. K. & Suttle, J. C. (1991). *The Plant Hormone Ethylene.* CRC Publ., Inc, Boca Raton, Florida, USA.
Mattoo, A. K., Modi, V. V. & Reddy, V. V. (1968). *Ind. J. Biochem.*, **5,** 111–14.
Mattoo, A. K., Murata, T., Pantastico, E. B., Chachin, K., Ogata, K. & Phan, C. T. (1975). In *Postharvest Physiology, Handling and Utilization of Tropical and Subtropical Fruits and Vegetables,* ed. E. B. Pantastico. AVI Publishing Co., Inc., Westport, CT, USA, pp. 103–27.
Mattoo, A. K., Adams, D. O., Patterson, G. W. & Lieberman, M. (1982/83). *Plant Sci. Lett.*, **28,** 173–9.
Mattoo, A. K., Baker, J. E. & Moline, H. E. (1986). *J. Plant Physiol.*, **123,** 193–202.
Mattoo, A. K., Mehta, A. M., Dean, J. F. D. & Anderson, J. D. (1989*a*). In *Cell Separation in Plants,* NATO ASI Series, vol. H35, eds D. J. Osborn & M. J. Jackson. Springer-Verlag, Berlin, pp. 39–49.
Mattoo, A. K., Kanellis, A. K., Solomos, T. & Mehta, A. M. (1989*b*). In *Proceedings of the Symposium on the Physiology of Fruit Drop Ripening, Storage and Post-harvest Processing of Fruits,* ed. R. Jona. Università degli Studii di Torino, Turin, Italy.
Mauch, F. & Staehelin, L. A. (1989). *Plant Cell,* **1,** 447–57.
Maunders, M. J., Holdsworth, M. J., Slater, A., Knapp, J. E., Bird, C. R., Schuch, W. & Grierson, D. (1987). *Plant Cell Environ.*, **10,** 177–84.
Mayne, R. G. & Kende, H. (1986). *Planta,* **167,** 159–65.
McDougall, G. J. & Fry, S. C. (1988). *Planta,* **175,** 412–16.
McDougall, G. J. & Fry, S. C. (1989). *Plant Physiol.*, **89,** 883–7.
McGlasson, W. B., Wade, N. L. & Adato, I. (1978). In *Phytohormones and Related Compounds* (Vol. 2), ed. D. S. Lethem, P. B. Goodwin & T. J. V. Higgins. Elsevier, Amsterdam, The Netherlands, pp. 447–93.
McKersie, B. D. & Thompson, J. E. (1979). *Biochim. Biophys. Acta,* **550,** 48–58.
McMurchie, E. J., McGlasson, W. B. & Eaks, I. L. (1972). *Nature,* **237,** 235–6.
Medina, P. V. L., Dostali, H. C. & Medina, R. M. T. (1981). *Rev. Ceres,* **28,** 435–43.
Mehta, A. M., Jordan, R. L., Anderson, J. D. & Mattoo, A. K. (1988*a*). *Proc. Natl Acad. Sci. USA,* **85,** 8810–14.
Mehta, A. M., Anderson, J. D. & Mattoo, A. K. (1988*b*). *FASEB J.*, **2,** A1024 (Abstr.).
Mitcham, E. J., Gross, K. C. & Ng, T. J. (1989). *Plant Physiol.*, **89,** 477–81.
Mitchell, T., Porter, A. J. & John, P. (1988). *New Phytol.*, **109,** 313–16.
Mor, Y., Halevy, A. H., Spiegelstein, H. & Mayak, S. (1985). *Physiol. Plant,* **65,** 196–202.
Moreau, F. & Romani, R. (1982). *Plant Physiol.*, **70,** 1385–90.
Mutschler, M. A. & Guttieri, M. (1987). In *Tomato Biotechnology,* ed. D. J. Nevins & R. A. Jones. Alan R. Liss, New York, USA, pp. 289–97.

Nakagawa, N., Nakajima, N. & Imaseki, H. (1988). *Plant Cell Physiol.*, **29,** 1255–9.
Nakajima, N. & Imaseki, H. (1986). *Plant Cell Physiol.*, **27,** 969–80.
Nakajima, N., Nakagawa N. & Imaseki, H. (1988). *Plant Cell Physiol.*, **29,** 989–98.
Ojalvo, I., Rokem, J. S., Navon, G. & Goldberg, I. (1987). *Plant Physiol.*, **85,** 716–19.
Palavan, N., Goren, R. & Galston, A. W. (1984). *Plant Cell Physiol.*, **25,** 541–6.
Paliyath, G. & Poovaiah, B. W. (1984). *Proc. Natl Acad. Sci. USA*, **81,** 2065–9.
Paliyath, G. & Poovaiah, B. W. (1985). *Plant Cell Physiol.*, **26,** 977–86.
Paliyath, G., Poovaiah, B. W., Munske, G. R. & Magnuson, J. A. (1984). *Plant Cell Physiol.*, **25,** 1083–7.
Paull, R. E. & Chen, N. J. (1983). *Plant Physiol.*, **72,** 382–5.
Pauls, K. P. & Thompson, J. E. (1980). *Nature*, **283,** 504–6.
Peleman, J., Boerjan, W., Engler, G., Serunick, J., Botterman, J., Alliotte, T., Van Montague, M. & Inze, D. (1989). *Plant Cell*, **1,** 81–93.
Pesis, E., Fuchs, Y. & Zauberman, G. (1978). *Plant Physiol.*, **61,** 416–19.
Philosoph-Hadas, S. & Aharoni, N. (1987). *Plant Physiol.*, **83,** 8–11.
Platt-Aloia, K. A. & Thomson, W. W. (1981). *Ann. Bot.* **48,** 451–65.
Poovaiah, B. W. (1986). *Food Technol.*, **40,** 86–9.
Poovaiah, B. W. (1987). In *Plant Senescence: Its Biochemistry and Physiology*, ed. W. W. Thomson, E. A. Nothnagel & R. C. Huffaker. The American Society of Plant Physiologists, Rockville, MD, USA, pp. 182–9.
Poovaiah, B. W. & Nukaya, A. (1979). *Plant Physiol.*, **64,** 543–7.
Porter, A. J. R., Borlakoglu, J. T. & John, P. (1986). *J. Plant Physiol.*, **125,** 207–16.
Pressey, R. (1984). *Eur. J. Biochem.*, **144,** 212–13.
Pressey, R. & Avants, J. K. (1973). *Biochim. Biophys. Acta*, **309,** 363–9.
Pressey, R. & Avants, J. K. (1978). *J. Food Sci.*, **43,** 1415–17.
Privalle, L. S. & Graham, J. S. (1987). *Arch. Biochem. Biophys.*, **253,** 333–40.
Rabinowitch, H. D., Sklan, D. & Budowski, P. (1982). *Physiol. Plant*, **54,** 369–74.
Raghothama, K. G., Veluthambi, K. & Poovaiah, B. W. (1985). *Plant Cell Physiol.*, **26,** 1565–72.
Raskin, I. & Beyer, E. M. (1989). *Plant Physiol.*, **90,** 1–5.
Reid, M. S. (1985). *Hortsci.*, **20,** 45–50.
Reid, M. S. (1987). In *Plant Hormones and Their Role in Plant Growth and Development*, ed. P. J. Davies. Martinus Nijhoff Publishers, Dordrecht, The Netherlands, pp. 257–79.
Rhodes, M. J. C. (1980). In *The Biochemistry of Plants* (Vol. 2), ed. D. D. Davies. Academic Press, New York, USA, pp. 419–62.
Rick, C. M. (1987). In *Tomato Biotechnology*, ed. D. J. Nevins & R. A. Jones. Alan R. Liss, New York, USA, pp. 17–26.
Rigney, C. J. & Wills, R. B. H. (1981). *Hortsci.*, **16,** 550–1.
Rikin, A., Chalutz, E. & Anderson, J. D. (1985). *Planta*, **163,** 227–31.
Riov, J. & Yang, S. F. (1982). *Plant Physiol.*, **70,** 136–41.
Riov, J., Monselise, S. P. & Kahan, R. S. (1969). *Plant Physiol.*, **44,** 631–5.
Rise, M., Cojocaru, M., Gottlieb, H. E. & Goldschmidt, E. E. (1989). *Plant Physiol.*, **89,** 1028–30.
Roberts, J. A., Grierson, D., Tucker, G. A. (1987). In *Hormone Action in Plant Development: A Critical Appraisal*, ed. G. V. Hoad, J. R. Leuton, M. B. Jackson & R. K. Atkin. Butterworths, London, UK, pp. 107–18.
Roberts, J. K. M. (1984). *Ann. Rev. Plant Physiol.*, **35,** 375–86.
Roberts, J. K. M. (1987). In *The Biochemistry of Plants* (Vol. 13) ed. D. D. Davies. Academic Press, New York, USA, pp. 181–227.
Roberts, J. K. M. & Testa, M. P. (1988). *Plant Physiol.*, **86,** 1127–30.
Romani, R. J. (1987). *Hortsci.*, **22,** 865–8.
Romani, R. J., Hess, B. M. & Leslie, C. A. (1989). *J. Plant Growth Regul.*, **8,** 63–9.
Rugini, E., Bongi, G. & Fontanazza, G. (1982). *J. Amer. Soc. Hort. Sci.*, **107,** 835.

Sacher, J. A. (1973). *Ann. Rev. Plant Physiol.*, **24,** 197–224.
Saftner, R. (1989). *Physiol. Plant,* **75,** 485–91.
Saftner, R. & Baker, J. E. (1987). *Plant Physiol.*, **84,** 311–17.
Sakai, S. & Imaseki, H. (1973). *Planta,* **113,** 115–28.
Saltveit, M. E. & McFeeters, R. F. (1980). *Plant Physiol.*, **66,** 1019–23.
Sanchez-Raya, A. J. (1981). *Fruit Sci. Rep.*, **8,** 97–100.
Sandmann, G. & Böger, P. (1980). *Plant Physiol.*, **66,** 797–800.
Saniewski, M., Urbanek, H., Czapski, J. (1987). *J. Plant Physiol.*, **127,** 177–81.
Sato, T. & Theologis, A. (1989). *Proc. Natl Acad. Sci. USA,* **86,** 6621–4.
Sato, T., Sakiku, K., Nakagawa, H. & Ogura, N. (1984). *Plant Cell Physiol.*, **25,** 1069–71.
Satoh, S. & Esashi, Y. (1986). *Plant Cell Physiol.*, **27,** 285–91.
Satoh, S. & Yang, S. F. (1988). *Plant Physiol.*, **88,** 109–14.
Sawamura, M., Knegt, E. & Bruisnima, J. (1978). *Plant Cell Physiol.*, **19,** 1061–9.
Schmidt, W. E. & Ebel, J. (1987). *Proc. Natl Acad. Sci. USA,* **84,** 4117–21.
Sexton, R. & Roberts, J. A. (1982). *Ann. Rev. Plant Physiol.*, **33,** 133–62.
Sheehy, R. E., Pearson, J., Brady, C. J. & Hiatt, W. R. (1987). *Mol. Gen. Genet.*, **208,** 30–6.
Shih, C. Y., Dumbroff, E. B. & Thompson, J. E. (1989). *Plant Physiol.*, **89,** 1053–9.
Shimokawa, K. (1982). *Phytochem.*, **21,** 543–5.
Shimokawa, K., Shimada, S. & Yaed, K. (1978). *Sci. Hortic.*, **8,** 129–35.
Sisler, E. C. (1979). *Plant Physiol.*, **64,** 538–42.
Sisler, E. C. (1982). *J. Plant Growth Regul.*, **1,** 211–18.
Sisler, E. C. (1984). In *Ethylene: Biochemical, Physiological, and Applied Aspects,* ed. Y. Fuchs & E. Chalutz. Martinus Nijhoff/Junk Publishers, The Hague, The Netherlands, pp. 45–54.
Sitrit, Y., Riov, J. & Blumenfeld, A. (1986). *Plant Physiol.*, **81,** 130–5.
Sitrit, Y., Blumenfeld, A. & Riov, J. (1987). *Physiol. Plant,* **69,** 504–10.
Slater, A., Maunders, M. J., Edwards, K., Schuch, W. & Grierson, D. (1985). *Plant Mol. Biol.*, **5,** 137–47.
Smith, A. R. & Hall, M. A. (1984). In *The Biosynthesis and Metabolism of Plant Hormones,* ed. A. Crozier & J. R. Hillman. Cambridge University Press, New York, USA, pp. 201–29.
Smith, C. J. S., Watson, C. F., Ray, J., Bird, C. R., Morris, P. C., Schuch, W. & Grierson, D. (1988). *Nature,* **334,** 724–6.
Smith, T. A. (1985). *Ann. Rev. Plant Physiol.*, **36,** 117–43.
Solomos, T. (1977). *Ann. Rev. Plant Physiol.*, **28,** 279–97.
Spurgeon, S. L. & Porter, J. W. (1980). In *The Biochemistry of Plants: A Comprehensive Treatise* (Vol. 4), ed. P. K. Stumpf, Academic Press, New York, USA, pp. 419–83.
Stahmann, M. A., Clare, B. G. & Woodbury, W. (1966). *Plant Physiol.*, **41,** 1505–12.
Stitt, M., Cseke, C. & Buchanan, B. (1986). *Plant Physiol.*, **80,** 246–8.
Su, L.-Y., McKeon, T., Grierson, D., Cantwell, M. & Yang, S. F. (1984). *Hortsci.*, **19,** 576–8.
Suttle, J. C. (1981). *Phytochem.*, **20,** 1477–80.
Swindeman, A. M. & Dilley, D. R. (1985). *Hortsci.*, **20,** 608.
Terra, N. N., Garcia, E. & Lajolo, F. M. (1983). *J. Food Sci.*, **48,** 1097–100.
Thompson, J. E., Mayak, S., Shinitzsky, M. & Halevy, A. H. (1982). *Plant Physiol.*, **69,** 859–63.
Thompson, J. E., Paliyath, G., Brown, J. H. & Duxbury, C. L. (1987). In *Plant Senescence: Its Biochemistry and Physiology,* ed. W. W. Thomson, E. A. Nothnagel & R. C. Huffaker. The American Society of Plant Physiologists, Rockville, MD, USA, pp. 146–55.
Thompson, J. S., Harlow, R. L. & Whitney, J. F. (1983). *J. Amer. Chem. Soc.*, **105,** 3522–7.

Tieman, D. M. & Handa, A. K. (1989). *Plant Physiol.*, **90,** 17–20.

Tigchellaar, E. C., McGlasson, W. B. & Buescher, R. W. (1978). *Hortsci.*, **13,** 508–13.

Tong, C. B. & Gross, K. C. (1988). *Physiol. Plant,* **74,** 365–70.

Tong, C. B., Labavitch, J. M. & Yang, S. F. (1986). *Plant Physiol.*, **81,** 929–30.

Toppan, A. & Esquerre-Tugaye, M.-T. (1984). *Plant Physiol.*, **75,** 1133–8.

Trewavas, A. J. (1982). *Physiol. Plant,* **55,** 60–72.

Tsai, D.-S., Arteca, R. N., Bachmann, J. M. & Phillips, A. T. (1988). *Arch. Biochem. Biophys.*, **264,** 632–40.

Tucker, G. A. & Brady, C. J. (1987). *J. Plant Physiol.*, **127,** 165–9.

Tucker, G. A. & Grierson, D. (1982). *Planta,* **155,** 64–7.

Tucker, G. A. & Grierson, D. (1987). In *The Biochemistry of Plants* (Vol. 12), ed. D. D. Davies. Academic Press, New York, USA, pp. 265–318.

Tucker, G. A., Robertson, N. G. & Grierson, D. (1980). *Eur. J. Biochem.*, **112,** 119–24.

Tucker, G. A., Robertson, N. G. & Grierson, D. (1981). *Eur. J. Biochem.*, **115,** 87–90.

Tucker, G. A., Robertson, N. G. & Grierson, D. (1982). *J. Sci. Food Agric.*, **33,** 396–400.

Tucker, M. L. & Laties, G. G. (1984). *Plant Physiol.*, **74,** 307–15.

Tucker, M. L., Durbin, M. L., Clegg, M. T. & Lewis, L. N. (1987). *Plant Mol. Biol.*, **9,** 197–203.

Wang, S. Y., Adams, D. O. & Lieberman, M. (1982). *Plant Physiol.*, **70,** 117–21.

Wang, T.-T. & Yang, S. F. (1987). *Planta,* **170,** 190–6.

Watada, A. E. (1986). *Food Technol.*, **40,** 82–5.

Watkins, C. B. & Frenkel, C. (1987). *Plant Physiol.*, **85,** 56–61.

Whitehead, C. S. & Halevy, A. H. (1989). *Plant Growth Regul.*, **8,** 41–54.

Woltering, E. J., Harren, F. & Boerrigter, H. A. M. (1988). *Plant Physiol.*, **88,** 506–10.

Yang, S. F. (1987). In *Plant Senescence: Its Biochemistry and Physiology,* ed. W. W. Thomson, E. A. Nothnagel & R. C. Huffaker. American Society of Plant Physiologists, Rockville, MD, USA, pp. 156–66.

Yang, S. F. & Hoffman, N. E. (1984). *Ann. Rev. Plant Physiol.*, **35,** 155–89.

Yang, S. F. & Pratt, H. K. (1978). In *Biochemistry of Wounded Plant Tissues,* ed. G. Kahl. Walter de Gruyter, Berlin, Germany, pp. 595–622.

Yip, W.-K. & Yang, S. F. (1988). *Plant Physiol.*, **88,** 473–6.

York, W. S., Darvill, A. G. & Albersheim, P. (1984). *Plant Physiol.*, **75,** 295–7.

Yoshii, H. & Imaseki, H. (1981). *Plant Cell Physiol.*, **22,** 369–79.

Yoshii, H. & Imaseki, H. (1982). *Plant Cell Physiol.*, **23,** 639–49.

Yu, Y. B. & Yang, S. F. (1979). *Plant Physiol.*, **64,** 1074–7.

Yung, K. H., Yang, S. F. & Schlenk, F. (1982). *Biochem. Biophys. Res. Commun.*, **104,** 771–7.

Zainon, M. A. & Brady, C. J. (1982). *Aust. J. Plant Physiol.*, **9,** 155–69.

7

The Significance of Endogenous and Exogenous Pectic Enzymes in Fruit and Vegetable Processing

W. PILNIK & A. G. J. VORAGEN

Department of Food Science, Agricultural University, PO Box 8129, 6700 EV Wageningen, The Netherlands

INTRODUCTION

Pectic enzymes occur in higher plants and are produced by microorganisms. Their substrate is the group of pectic substances which occur as structural polysaccharides in the middle lamella and the primary cell walls of higher plants and are prominent in parenchymous tissues. Endogenous pectic enzymes can therefore produce important textural changes in fruits and vegetables during ripening, storage and processing. Microbial pectic enzymes serve functions in plant pathology and fermented foods but are also produced industrially as processing aids in the food industry. Pectic enzymes purified to well-defined activities are used in pectin analysis, research on the structure of pectins and in cell wall studies. The subject has been extensively reviewed in the last decade by Rombouts and Pilnik (1979, 1980), Pilnik and Rombouts (1979*a,b*, 1981), Gierschner (1981), Rombouts (1981), Kilara (1982), Pilnik (1982), Voragen and Pilnik (1989), Pilnik and Voragen (1989).

PECTIC SUBSTANCES AND PECTIC ENZYMES

Pectic substances (pectins, pectin) are polymeric molecules with a rhamno-galacturonan backbone. This backbone consists of 'smooth' α-D-1,4-galacturonan regions which are interrupted to a small extent by insertion of 1,2-linked α-L-rhamnosyl residues and highly branched regions with an almost alternating rhamno-galacturonan chain (Aspinall, 1980; Neukom *et al.*, 1980; McNeil *et al.*, 1984; Selvendran 1985). Side chains composed of neutral sugars are attached by glycosidic linkages to the C-3 and/or C-4 atoms of the rhamnose units and C-2 and/or C-3 atom of the galacturonic acid units, giving the rhamno-galacturonan part of the pectin backbone the character of a 'hairy region' (de Vries *et al.*, 1986; Fig. 1). The predominant sugars, D-galactose and

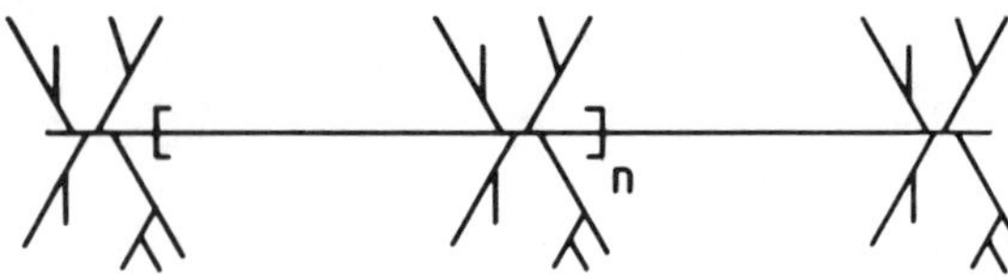

FIG. 1. Schematic structure of apple pectin; rhamnogalacturonan backbone with regions rich in neutral sugar side chains (hairy regions) (de Vries *et al.*, 1986).

L-arabinose, are present in complex chains of considerable length (McNeil *et al.*, 1984), whereas some xylose occurs as monomeric or oligomeric side chains. Commercial apple or citrus pectins, extracted for use as gelling agents, have a degree of polymerization usually not exceeding 500 but pectin fractions from fruits may have much higher molecular weights. Pectins extracted from most fruits under mild conditions usually have a degree of esterification (DE, % of galacturonic acid monomers which are methyl-esterified) of over 70 (Doesburg, 1965). Acetic acid may occur as a substituent of the hydroxyl groups of galacturonic acid at the C-2 and/or C-3 atoms. Acetyl groups are prominent in sugar beet pectins which also carry feruloyl groups linked as esters to terminal galactose or arabinose units of side chains (Fry, 1982; Rombouts & Thibault, 1986*a*).

Pectic enzymes are classified according to their mode of attack on the galacturonan part of the pectin molecule. Relating to the scheme in Fig. 2, one can distinguish between pectin-methylesterases (PE; EC 3.1.11.1) which di-esterify pectins to low methoxyl pectins or pectic acids and pectin depolymerases which split the glycosidic linkages between galacturonosyl (methylester) residues. Polygalacturonases (PG; EC 3.2.1.15 and 3.2.1.67) split glycosidic linkages next to free carboxyl groups by hydrolysis; pectate lyases

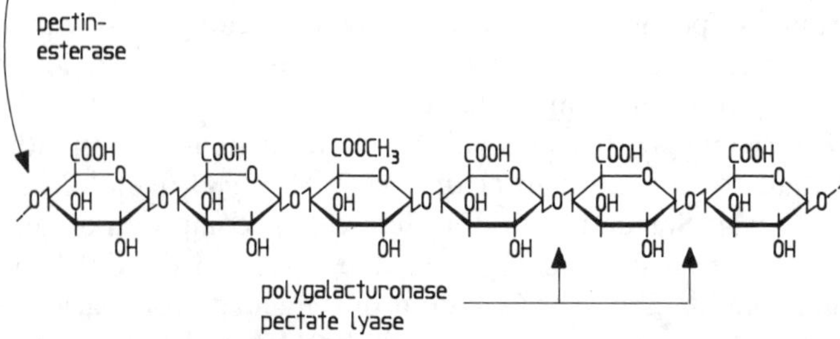

FIG. 2. Fragment of a pectin molecule and points of attack of pectic enzymes.

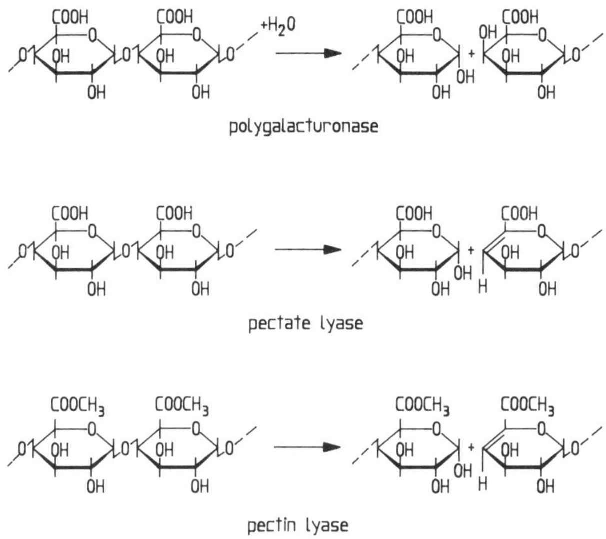

FIG. 3. Splitting of glycosidic bonds in pectin by hydrolysis (polygalacturonase) and by β-elimination (pectate lyase and pectin lyase).

(pectic acid lyases, PAL; EC 4.2.2.2 and 4.2.2.9) split glycosidic linkages next to free carboxyl groups by β-elimination (Fig. 3). Both endo- and exo-types of PGs and PALs are known. The endo-types (EC 3.2.1.15, EC 4.2.2.2) split the pectin chain at random. Exo-PGs (EC 3.2.1.67) split off monomers or dimers from the non-reducing end of the chain, while exo-PALs (EC 4.2.2.9) split off unsaturated dimers from the reducing end. For PGs and PALs, pectates and low-methoxyl pectins are, therefore, the preferred substrates; they have very low activity on highly methylated pectins. Such pectins are degraded by pectin lyases (PL; EC 4.2.2.10) of which only the endo-type has been described. Highly esterified pectins are, of course, also degraded by a combination of PE with PG or PAL (Neukom, 1969; Fogarty & Ward, 1974; McMillan & Sheiman, 1974; Kulp, 1975; Rexova-Benkova & Marcovic, 1976; Rombouts & Pilnik, 1980). The existence of a polymethylgalacturonase for hydrolytic degradation of high ester pectin has never been proved and has not been found by careful screening of commercial preparations and microorganisms in the authors' laboratory.

Occurrence and microbial production of pectic enzymes

Table 1 shows the presence of PEs and PGs in higher plants. These enzymes are also produced by microorganisms, which in addition produce PAL and PL (Table 2; Rombouts & Pilnik, 1980).

TABLE 1
Occurrence of Pectin Esterase and Polygalacturonase in Higher Plants (Various authors)

Plants	*PE*	*PG*		
		Endo	*Exo*	*Not specified*
Apple	+		+	
Apricot	(+)			+
Avocado		+		
Banana	+	+		
Berries	+			+
Citrus				
lime	+			
orange	+			
grapefruit	+		+	
mandarin	+			
Cherries	+	+		
Currant				+
Grapes	+			+
Mango	+	+		
Papaya	+	+	+	
Passion fruit	+	+	+	
Peach		+	+	
Pear	+	+	+	
Plums	+			
Beans	+			
Carrots	+		+	
Cauliflower	+			
Cucumber	+	+		
Leek	+			
Onions	+			
Pea	+			
Potato	+		+	
Radish	+			+
Tomato	+	+	+	

ASSAY METHODS

An extended discussion by many authors on methods used for pectic enzymes from specific microorganisms and plant cell cultures has recently been edited by Wood and Kellogg (1988). This is a more general treatment of the subject.

Pectinesterase activities

Pectinesterase activities produce free carboxyl groups and free methanol. The increase in free carboxyl groups is easily monitored by automatic titration. At pH values below 4·5 or when assessing PE action in a natural substrate not suited for automatic titration, the increase in free methanol can be measured

TABLE 2
Occurrence of Pectic Enzymes in Microorganisms (Rombouts & Pilnik, 1980)

Organism	*PE*[a]	*PG*		*PAL*		*PL*
		Endo	*Exo*	*Endo*	*Exo*	*Endo*
Fungi						
Aspergillus niger	+ +[b]	+ +	+ +			+
Penicillium	+	+				+
Fusarium	+	+		+		
Rhizopus	+	+ +				
Sclerotinia	+	+				+
Collecotrichum		+	+			
Yeasts						
Kluyveromyces		+ +				
Bacteria						
Bacillus polymyxa				+ + +		
Clostridium	+				+	
Erwinia		+	+	+ + +	+	+
Pseudomonas				+ + +		
Arthrobacter				+		

[a] PE, PG, see Table 7. PAL, pectate lyase; PL, pectin lyase.
[b] + + + means produced in high levels, no data means activity absent.

by GLC (McFeeters & Armstrong, 1984) or HPLC. Headspace GLC after conversion of methanol to methyl nitrite is convenient and sensitive (Bartolome & Hoff, 1972; Litchman & Upton, 1972). Even more convenient, although less sensitive, is HPLC performed, for example on an alkaline isopropanol extract with an ion-exchange column (Voragen *et al.*, 1986*a*). One unit of PE is defined as the amount of enzyme which liberates one micro-mole of free carboxyl groups or methanol per minute under specified assay conditions. Random or blockwise distribution of free carboxyl groups obtained by various PEs (discussed below) can be distinguished by high performance ion exchange chromatography of the substrate (Schols *et al.*, 1989) or by measuring the affinity for Ca^{2+} which is much stronger in the case of blockwise distribution (Kohn *et al.*, 1968). Semi-quantitative tests are sensitive and conveniently done by observing the pH-shift in a neutralized pectin solution or gelling when Ca^{2+} has been added (Pilnik & Rothschild, 1960). Such tests are useful, e.g. to check for (undesirable) residual PE activity in citrus juices after pasteurization (IFU, 1985).

Lyase activities

Lyase activities can be measured by monitoring the increase in light absorbance at 232 nm caused by the formation of the double bond (Fig. 3) at the new non-reducing ends of the molecules. Differentiation between PALs and PLs is possible by choosing low methoxyl or high methoxyl pectins, respectively, as substrates. Care must be taken to use clear pectin solutions since any change in turbidity by pectin degradation would falsify the values measured. Both exo- and endo-activities, initial velocity and extent of depolymerization are measured in this way. One unit of lyase liberates one micromole of unsaturated products per minute under specified assay conditions. The specific extinctions at 232 nm for PAL: 4600 mol/cm (Macmillan & Phaff, 1966) and for PL: 5500 mol/cm (Edstrom & Phaff, 1964).

Polygalacturonase activities

Polygalacturonase activities can be determined by measuring the increase in reducing end groups. The test of Nelson–Somogyi (Spiro, 1966) or that of Milner and Avigad (1967) are preferentially used, with galacturonic acid as standard. These tests measure endo- and exo-activities as well as the activity of oligopolygalacturonases. One unit of activity is defined as the quantity of enzyme that liberates one micromole of reducing groups per minute under specified conditions. The substrate used is low methoxyl pectin or pectate which also responds to PAL, the activity of which also results in an increase in non-reducing end groups. PG, however, does not form unsaturated products and can thus be differentiated. Furthermore, the pH optima differ widely (Table 3) and PAL has an absolute requirement for Ca^{2+}. It is recommended

TABLE 3
Properties of Pectic Enzymes from Some Plants and Food Grade Microorganisms

Enzyme Type/Source	*Mol. wt*	*Iso-electric Point*	*Specific Activity (units/mg)*	*Optimum pH*	*K_m value (mg/ml)*	*Ref.*
PE						
Orange						
I[a]	36 000	10·0	694	7·6	0·083	Versteeg *et al.* (1980)
II[a]	36 200	11·0	762	8·8	0·0046	Versteeg *et al.* (1980)
III[a]	54 000	10·2	≥180	8	0·041	Versteeg *et al.* (1980)
A. niger	39 000	3·9	406	4·5	3	Baron *et al.* (1980)
PG						
Tomato						
I[a]	100 000	8·6		4·5		Pressey (1986)
II[a]	44 000	9·4		4·5		Pressey (1986)
Kluyveromyces fragilis	46 000–48 000	6·1	89	4·0		[b]
A. aculeatus	47 000	4·7–4·8	67	4·0–5·0		[b]
A. niger[a]	42 000–49 000	6·0–6·4	275	4·0		[b]
PE						
A. niger I	35 400	3·65	17	6	5·0	van Houdenhoven (1975)
A. niger II	33 100	3·75	44	6	0·9	van Houdenhoven (1975)
PAL						
B. subtilis	33 000	9·85		8·5		Chesson & Codner (1978)

[a] Multiple molecular forms (iso-enzymes).
[b] Unpublished results of the authors' laboratory.

that viscosity measurements should be combined with end-group (PG) and light-absorbance (lyases) measurements. This allows the calculation of glycosidic linkages split when the specific viscosity has reached half the original value. For endo-activities, cleavage of only a few percent of the glycosidic bonds is necessary whereas values as high as 40% are required for exo-activities. Information about the presence of exo- and endo-activities and the breakdown patterns in general can be obtained by gel chromatography of oligomers and by peak shifts in elution diagrams from size exclusion chromatography for which versatile HPLC methods are now available (Schols *et al.*, 1990). A reliable qualitative test for complete pectin degradation, e.g. in fruit juice clarification, is the absence of flocculation upon addition of two volumes of alcohol.

PREPARATIONS OF PECTIC ENZYMES FOR THE LABORATORY

Pectinesterase with only weak exo-polygalacturonase activity is obtained from orange pulp by extraction with a borax-acetate buffer (Krop, 1974). Polygalacturonase is found as the only pectic activity in liquid growth medium of *Kluyveromyces* (Phaff, 1966). Pectate lyases as single pectic activities are

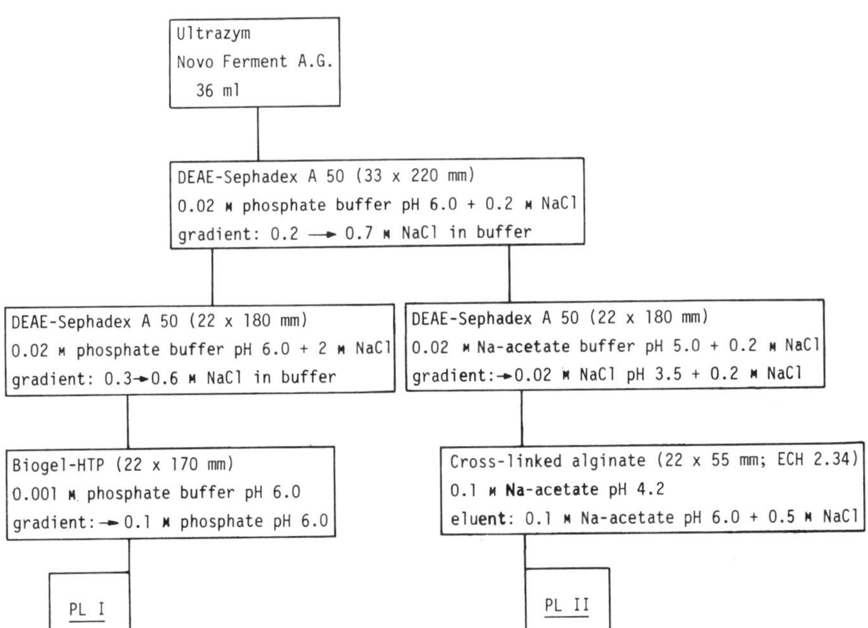

FIG. 4. Purification scheme for pectin lyase from a technical pectinase preparation of fungal origin. (Adapted from van Houdenhoven, 1975).

isolated from liquid cultures of certain strains of *Pseudomonas fluorescence* (Pilnik *et al.*, 1973; Rombouts *et al.*, 1978). There is no simple source of pectin lyase. Separation and clean-up of these activities is achieved by chromatographic methods. The usual steps are as follows:

(1) desalting by gel filtration (Biogel P100),
(2) protein separation by anion exchange (DEAE Biogel A), and/or cation exchange (CM Biogel A),
(3) separation of pectic enzymes by cross-linked alginate, and
(4) final cleaning-up step by FPLC.

Cross-linked alginate works by a combination of affinity and electrostatic effects and replaces the crosslinked pectate (Versteeg *et al.*, 1978; Rombouts *et al.*, 1982*a*). It is less problematic to prepare and has a greater capacity. In the authors' laboratory, Biogel is used for its stability towards glycosidases. Detailed procedures have been published for PG (Phaff, 1966; Versteeg, 1979), for PAL (Rombouts *et al.*, 1978), for PL (van Houdenhoven, 1975), and for PE (Rexova-Benkova *et al.*, 1977*a*; Versteeg *et al.*, 1978; Rombouts *et al.*, 1979; Versteeg, 1979). Figure 4 gives the flow diagram for the purification of pectin lyases I and II from a commercial preparation (van Houdenhoven, 1975).

TECHNOLOGICALLY IMPORTANT PROPERTIES OF PECTIC ENZYMES

Some molecular and kinetic properties are given in Table 3.

Pectinesterases

Plant PEs (Solms & Deuel, 1955; Miller & Macmillan, 1971) are thought to attack either at the non-reducing end or next to a free carboxyl group and then proceed along the molecule by a single-chain mechanism, creating blocks of unesterified galacturonic acid which are extremely calcium sensitive (Kohn *et al.*, 1968). Irregularities in the galacturonan chain, such as acetylated monomers, ester groups transformed into amides or reduced to primary alcohol, or the occurrence of hairy regions, inhibit PE activity (Solms & Deuel, 1955). PE is highly specific for the methylester of polygalacturonic acid. Other esters are attacked only very slowly (MacDonnell *et al.*, 1950; Manabe, 1973) and the methylester of polymannuronic acid not at all (MacDonnell *et al.*, 1950). The rate of pectin de-esterification depends on chain length; trimethyl trigalacturonate is not attacked at all (McCready & Seegmiller, 1954). Fungal PEs differ from plant PEs by obeying a multichain mechanism, removing methoxyl groups at random (Ishii *et al.*, 1979*a*). Further properties of PEs will be discussed below in the section on technological significance.

Polygalacturonases

Activity measurements by viscosity on methylesters and glycolesters of pectic acid show a rapid decrease in the rate and degree of hydrolysis with increasing esterification (Pilnik *et al.*, 1973). Rexova-Benkova *et al.* (1977*b*) have shown that acetyl groups decrease the extent of degradation by reducing the affinity for the substrate molecules through the blocking of binding sites. The limitation of degradation by the presence of acetyl groups was confirmed by Rombouts and Thibault (1986*b*) using sugar beet pectin as substrate (Table 4). The limiting effect of methoxyl groups was shown in the same experiment. On the other hand, Koller and Neukom (1969) have isolated a fungal PG which was fully active on 70% acetylated pectin. Also, the mode of attack is known to differ for PGs from different origins. An endo-enzyme may hydrolyze randomly one bond in a single enzyme–substrate encounter, followed by complete dissociation of enzyme and products. Further such scissions follow, resulting in the gradual appearance of oligomers from the depolymerization of higher oligogalacturonides. This pattern is observed with endo-PG from *Kluyveromyces fragilis* (Phaff, 1966). It is characterized as multichain attack. In the case of single chain-multiple attack, found for example with PG from *Colletotrichum lindemuthianum* (English *et al.*, 1972), a single, random hydrolytic scission is followed by a number of non-random attacks on one of the products, resulting in rapid liberation of oligogalacturonates. Many multiple forms and isoenzymes of PG have been described (Pilnik & Rombouts, 1981). Points of difference are dependence on cations and action pattern on oligomers. Particularly relevant to their industrial application is possible inhibition by vegetable extracts which is noted for fungal PGs

TABLE 4
Degradation Limits[a] of Sugar-Beet Pectins Before and After Enzymic Demethylation and Saponification (Rombouts & Thibault, 1986*a*, *b*)

Pretreatment and Mode of Degradation	*Pectins Water-soluble*	*Oxalate-soluble*	*Acid-soluble*	*Alkali-soluble*
No pretreatment				
β-Elimination	7·2	4·9	4·6	—
Endo-PL	8·6	4·3	4·1	—
Endo-PG	1·2	3·7	1·7	26·9
Endo-PAL	≤1·0	4·8	1·3	23·7
After pectin esterase treatment				
Endo-PG	19·1	22·4	16·3	—
Endo-PAL	11·8	13·0	9·9	—
After saponification				
Endo-PFG	37·5	36·7	30·8	29·3
Endo-PAL	33·4	34·5	31·0	26·4

[a] Percent of galacturonic acid bonds broken.

(Albersheim & Anderson, 1971), but not for PG from tomatoes (Bock *et al.*, 1972).

The best substrate for endo-pectin-lyases, measured at pH ≥ 7, is completely esterified pectin as indicated by degradation limits and affinity (per mol). At lower pH values the affinity for less highly esterified pectins increases and a marked stimulation by Ca^{2+} and other cations is observed. This makes PL a useful enzyme for fruit processing (Voragen, 1972; Ishii & Yokotsuka, 1975). These enzymes need the methylester groups and are inactive on glycol esters and on amidated pectates (Pilnik *et al.*, 1973, 1974).

Endo-pectate-lyases

Endo-pectate-lyases on the other hand, do not differentiate between the methylester and the glycolester of pectic acids. Surprisingly, pectate is not the best substrate for bacterial PALs. They show maximal activity (initial velocity and degree of degradation) on low methoxyl pectins. Figure 5 shows that the DE optima for PALs from *Arthrobacter* 547, *Arthrobacter* 370 and *Bacillus polymyxa* are 21, 44 and 26%, respectively (Pilnik *et al.*, 1973). This and their high pH optima make them interesting for vegetable processing Changes produced by PALs are conveniently studied since, due to an absolute requirement for Ca^{2+}, enzyme action can be stopped by the addition of a chelating agent (Rombouts, 1972).

Rhamno-galacturonase

Recently, an enzyme has been described (Schols *et al.*, 1990) which splits the galacturonic acid–rhamnose glycosidic linkage in hairy regions of apple pectins with strongly enhanced activity when these have been de-esterified and arabinose has been removed by acid hydrolysis (modified hairy regions). The

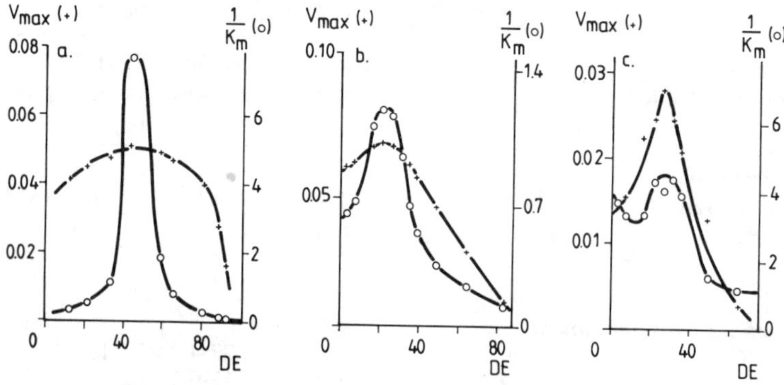

FIG. 5. Influence of DE of the substrate on V_{max} and $1/K_m$ of pectate lyases of (a) *Arthrobacter* 370, (b) *Arthrobacter* 547, and (c) *Bacillus polymyxa* measured by spectrophotometry (Pilnik *et al.*, 1973).

TABLE 5
Specific Enzyme Activities of Multi-enzyme Complexes in Technical Fungal Pectinase Preparations (Expressed as International Units per g of Enzyme Preparation) (Voragen *et al.*, 1986*b*)

Enzyme Activity	*Substrate*	*A*	*B*	*C*
PG	Polygalacturonic acid	1 982	3 314	1 878
PL	Pectin DE 90	43	53	74
PE	Pectin DE 65	548	448	227
Combined pectolytic	Pectin DE 75	198	290	274
C-*x*-cellulase	CMC	998	180	1 228
C-1-cellulase	Avicel	99	1	22
Arabanase: linear arabinan	1-5-α-L-arabinan	9	10	16
Arabanase: branched arabinan	1,3;1,2;1,5-α-L-arabinan	14	14	16
α-L-Arabinofuranosidase	PNP-Arabinofuranoside	35	37	333
Galactomannanase	Galactomannan	3	4	9
Mannanase	1,4-β-D-Mannan	7	0	0
Galactanase	1,4-β-D-Galactan	11	58	91
Xylanase	1,4-β-D-Xylan	4	0	2

enzyme was found in a commercial pectinase preparation and must surely be grouped with pectic enzymes. The end products are oligomers with alternating galacturonic acid and rhamnose units, with rhamnose forming the non-reducing end. The availability of these compounds made it possible to determine the anomeric form of the rhamnose units as α, using advanced NMR techniques (Colquhoun *et al.*, 1990).

Commercial pectinases

Commercial pectinases are fungal preparations, mainly from *Aspergillus* species, produced by a number of companies (Gist-Brocades, NOVO, Rohm, Rohm & Haas, Biocon, Finnsugar, Genecor, Miles Laboratories). They are usually mixtures of PE, PG and PL activities, hemicellulases and endo-β-glucanases (C-*x*-cellulases). The activities of three different commercial preparations are shown in Table 5 (Voragen *et al.*, 1986*b*). The activities found all originate from the mould used for the production of the pectic enzymes with the exception of the C-1-cellulase (cellobiohydrolase) which is added for technological purposes. The arabinanases play a special role in fruit juice processing (see below). Preparations containing PG as sole pectic activity are on the market and experimental quantities of fungal PE are available. Only one pectinase has so far been found to contain rhamno-galacturonase activity.

Genetic technology will surely be used to produce desirable single activity preparations which will open new possibilities for use as such or in optimal combinations.

TECHNOLOGICAL SIGNIFICANCE OF ENDOGENOUS ENZYMES

Pectinesterase

It was due to the methanol content of distillates from fermented fruit pulps that pectin was recognized as methylester by von Fellenberg (1913). Endogenous PE liberates this methanol during fermentation and distilleries pasteurize fruit pulp before fermentation to ensure that the methanol content of products does not exceed legal limits (Tanner, 1970). The presence of PE in citrus fruit is the cause of one of the most intensively studied problems in food technology: i.e. cloud loss in orange juice. If PE is not inhibited directly after juice extraction by heat inactivation or by freezing, the native pectin will be de-esterified and will be coagulated by the Ca^{2+} in the juice (Pilnik, 1958; Joslyn & Pilnik, 1961; Krop, 1974; Versteeg, 1979). On standing, the juice then separates into a clear supernatant and a layer of sediment. In the case of concentrated juice, the formation of a calcium pectate gel occurs from which no juice can be reconstituted (Rouse & Atkins, 1953; Kew & Veldhuis, 1961). These are serious quality defects. Citrus juices have a very heat sensitive flavour so that alternatives to the expensive frozen product are sought. Chelation of Ca^{2+} can prevent cloud loss by PE action but this creates legal problems. Inhibition of PE by polyphenols has been patented (Kew & Veldhuis, 1961) but this also creates problems of taste and of product identity. PE is known to display end-product inhibition, but the addition of pectic acid clarifies the juice, thus creating the effect one wants to prevent (Krop & Pilnik, 1974*a*; Baker, 1976). However, it has been established that by degrading pectic acid, a degree of polymerization can be found (8–10) which has the inhibitory effect of the high-polymer preparations but does not coagulate with Ca^{2+} (Termote *et al.*, 1977). The addition of such preparations will not prevent but will delay self-clarification, probably sufficiently long to be useful for short-term distribution of fresh chilled juice. Another possibility is the addition of exogenous enzymes (Fig. 6): PG which breaks down the low ester pectins formed before calcium coagulation occurs (Krop & Pilnik, 1974*b*) or PL (not readily available) which degrades the juice pectin to low molecular products which are

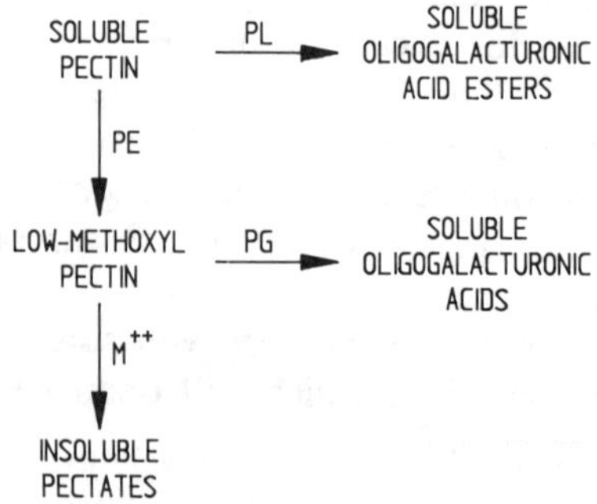

FIG. 6. Prevention of calcium coagulation of PE de-esterified orange serum pectin by PL or PG (Baker, 1980).

not calcium-sensitive (Baker, 1980), even if de-esterified and in fact might also serve the function of PE inhibitor. Such pectin depolymerizing action by exogenous enzymes has also been considered for lowering juice viscosity, thus facilitating the concentration of orange juice to higher Brix values than ordinarily possible (Crandall *et al.*, 1986). It should be pointed out that the use of multistage evaporators which allow optimized heat treatment and aroma recovery combined with subsequent frozen storage and shipment of the concentrated juice, have largely overcome the PE inactivation–heat damage problem. However, the problem still exists in developing countries in the manufacture of citrus and other cloudy juices which have strong PE activity and a heat-sensitive flavour, such as guava and mango. The general trend for minimally heat-processed or fresh products may reintroduce the PE problem, even in the modern citrus industry.

The intensive research on citrus juice cloud and gelling problems has resulted in better knowledge of citrus PEs. Versteeg (1979) and Versteeg *et al.* (1978) have isolated various multiple forms of PEs from oranges and shown that they differ in affinity for pectins and pectates (Table 6; Rombouts *et al.*, 1982*b*). They also have different heat stabilities (Versteeg *et al.*, 1980) and it is understandable that they play different roles in cloud loss phenomena. One form, characterized by its high molecular weight, represents about 5% of the total activity and has various outstanding properties. It is heat resistant (Fig. 7) and therefore is the main activity responsible for cloud loss after insufficient heat treatment of pasteurized juice. It is also active at low temperatures and at low pH (Versteeg, 1979) and must have been the important activity in the now almost extinct process of self-clarification for clear lemon and lime juices which were obtained by letting the juices stand in tanks under sulphur dioxide preservation. Such clear juices are still important commercial products but are now produced with commercial pectinases.

Of the two other iso-enzymes in Table 6, only form I has clarifying power. PE II has no clarifying power; the amount of free methanol generated is less than that from the clarifying enzymes. Table 6 shows the extremely high

TABLE 6
General Properties of Pectinesterases Isolated from Navel Oranges (Rombouts *et al.*, 1982*b*)

	PE I	*HMW-PE*	*PE II*
Specific activity (μ/mg protein)	694	⩾180	762
Mol. wt	36 200	54 000	36 200
Isoelectric point	10·05	10·2	⩾11
pH optimum	7·6	8·0	8·0
K_m value (mg pectin/ml)	0·083	0·041	0·0046
K_i value (mg pectate/ml)	0·416	0·010	0·0016
V_{max} (mmols/mg min)	0·564	—	0·561
Turnover number (mmol/mmol s)	340	—	338

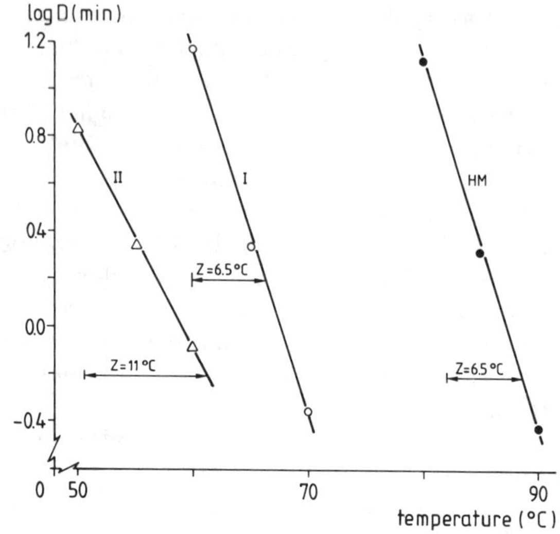

FIG. 7. Thermal destruction curves of pectinesterases I and II and of the high-molecular-weight pectinesterase (HM) in orange juice of pH 4·0. (Versteeg, 1980) (D = optical density).

inhibition (K_i) by pectate which probably stops the activity at a certain point of de-esterification. This effect is surely enhanced by a blockwise distribution of liberated acid groups. Figure 8 shows that reactivation occurs on addition of PG breaking down pectate blocks and indeed Versteeg (1979) has also noted that the rate at which free methanol is formed in PE-active orange juice is enhanced by PG. Such kinetic differences between PEs can also explain the incomplete inhibition by oligogalacturonates. There is one PE in orange juice which does not cause self-clarification, even after liberating the same amount of methanol as do the clarifying PEs (Versteeg, 1979). A different mode of attack must be supposed which produces low-ester pectins with reduced calcium sensitivity, probably by random distribution of free carboxyl groups. This mechanism, usually connected with fungal PEs, has also been postulated for a PE from *A. japonicus* (Ishii *et al.*, 1979*a*) and its production patented (Ishii *et al.*, 1979*b*) for the manufacture of low-ester pectin for use in low sugar jams. Such pectins are usually manufactured by an acid process (May, 1989). Fungal pectinesterases have also been developed for clarification in the French cider industry (Baron & Drilleau, 1982). The two orange iso-enzymes with orange juice clarifying properties are found in all component parts of orange fruit. All orange cultivars and citrus species tested so far contain the 12 PEs found in the authors' institute (Rombouts *et al.*, 1982*b*). The different affinities to pectate have been used for separating PEs (Rombouts *et al.*, 1979). It should be noted that PEs are inhibited by high sugar concentrations so that their activity may not be observed in a 60° Brix fruit juice concentrate but will manifest itself upon reconstitution.

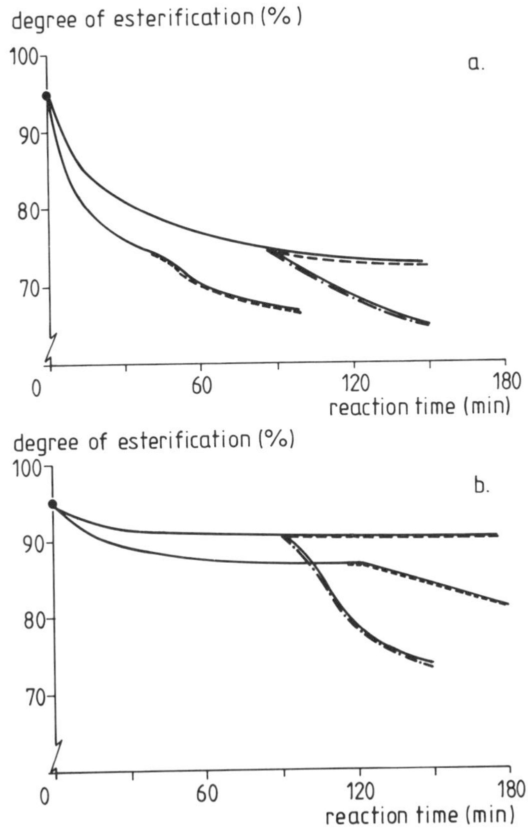

FIG. 8. Stimulation of pectinesterase by PG (– · – · –) and by PL (– – – – –). The stimulation is stronger for PG and stronger in experiment b, in which the esterase used was more strongly inhibited by end-product than in experiment a. Obviously PL (——) which can not degrade the inhibiting de-esterified regions, has no effect (Versteeg, 1979).

The phenomenon of calcium coagulation of PE de-esterified pectins is exploited when drying citrus peels for cattle feed after juice extraction. A slurry of calcium hydroxide is added to the peels while grinding them to bring the pH to neutrality or higher. The high pH, and also the calcium ions, activate the PE; rapid de-esterification and calcium pectate coagulation occurs. Bound liquid phase is released and easily pressed out, so that only a fraction of the original water content needs to be removed by expensive thermal drying. The press liquor is similar in composition to the juice and is concentrated to molasses which is also used as cattle feed (Kesterson & Braddock, 1976). This so-called liming process is probably the second biggest use of a native enzyme in the food and feed industry, the biggest being the use of native amylases in the cereal processing industries (malt, beer, bread). Of course, for the production of dried citrus peels as raw material for pectin extraction, any PE action must be prevented by immediate blanching of the peels. If this is not done, the extracted pectin, even if only slightly de-esterified, will have blocks of free galacturonic acid and be too calcium sensitive to be used as a gelling agent (May, 1989).

Endogenous PE is also exploited for protecting and improving the texture and firmness of several processed fruits and vegetables; these include apple slices (Wiley & Lee, 1970), canned tomatoes (Hsu *et al.*, 1965), cauliflower (Hoogzand & Doesburg, 1961), carrots (Lee *et al.*, 1979), potatoes (Bartolome & Hoff, 1972) and peas (Steinbuch, 1976). A low-temperature–long-time blanching process will activate PE, causing a partial de-esterification of the pectins which then react with (added) Ca^{2+}, resulting in stronger intercellular cohesion.

Pectinesterase and polygalacturonase

These two enzymes are abundantly present together in tomatoes, a fact which has a great impact on tomato processing. Immediate heat inactivation of these highly active enzymes is necessary in order to obtain the highly viscous juice expected by the consumer or the high consistency concentrated juice (paste) used for sauces, soups, ketchup and similar products (McColloch & Kertesz, 1949; Gould, 1974). Such juices are characterized as 'hot break' to indicate the heat treatment in special equipment in which tomatoes are crushed directly into circulating hot tomato pulp. Obviously, heat treatment adds to the flavour characteristics of processed tomato products. In cases where tomato ingredients are used for colour and flavour only and consistency is provided by other ingredients, such as starch, a more easily handled thin 'cold break' juice is the starting material for paste production; a holding time is introduced between crushing and heat treatment to ensure break-down of the pectins by combined PE/PG action. A process has been patented to macerate vegetables in fresh unpasteurized tomato pulp for the preparation of vegetable juices (Bock *et al.*, 1983). Lately, there have been press reports about field trials with 'anti-sense' tomatoes that possess, among other advantages, increased total solids content, firmness and viscosity of juice. These tomatoes have been engineered with a gene that significantly reduces PG (Calgene and Campbell Soup Company).

Pectic activities from microbial infections

These are, of course, of exogenous nature but the consequences of their presence makes them similar to endogenous activities (Rombouts & Pilnik, 1980). Yeast PG and bacterial PAL in cooperation with endogenous PE are involved in the softening of cucumbers and olives in brine (Bock *et al.*, 1970). This quality defect can be avoided by inhibitors, e.g. polyphenols leaching out from vine leaves into the brine (Bell *et al.*, 1962) as practised in some Mediterranean regions. Recirculated brine must in any case be pasteurized.

Byssochlamys fulva, a heat resistant mould, produces enzymes which cause decay of strawberries in syrup or jam (Put & Kruiswijk, 1964; Eckhardt & Ahrens, 1977). A highly resistant PG from *Rhizopus* spp. is responsible for texture break-down in canned apricots (Luh *et al.*, 1978). Mould infection has been reported to cause clarification of aseptically filled orange juice. It is not clear whether this is caused by PE or PG or both, but an increased amount of methanol was found in the clarified juices (Nussinovitch & Rosen, 1989). Yeast PG can cause pectin break-down in fermenting apple pomace which then loses its value as raw material for the pectin industry (Sulc & Ristic, 1954).

Desirable aspects of pectolysis by microorganisms are coffee (Castelein & Pilnik, 1976) and cocoa fermentation. During fermentation, the mucilage layer surrounding these beans is decomposed and washing of the beans before drying is greatly facilitated. Noble rot or *Botrytis cinerea* rot of over-ripe grapes is often considered desirable. The skin of the grapes is pierced; as a result of water evaporation, higher sugar concentrations are obtained, yielding a wine with higher alcohol content. It has a special flavour from metabolites of the mould (Drawert & Kreft, 1977).

USE OF EXOGENOUS PECTIC ENZYMES AS PROCESSING AIDS

Fruit juice clarification

This is the oldest and still the largest use of pectinases, applied mainly to deciduous juices and grape juice (Kertesz, 1930; Mehlitz, 1930). Enzyme treatment is necessary because raw press juice is viscous owing to dissolved pectin and has a persistent cloud of cell wall fragments and complexes of cell wall fragments with cytoplasmic proteins. Addition of pectinase lowers the viscosity and causes cloud particles to aggregate to larger units ('break') which sediment and are easily removed by centrifugation and/or (ultra)filtration. Indeed, in the early days, pectinase preparations were known as filtration enzymes. Thirty-five years after this use of enzymes was introduced the mechanism was explained by Japanese workers for apples, pears and grapes. The cloud particles are essentially composed of a protein nucleus carrying an outward positive charge which is coated by negatively charged pectin molecules. Partial degradation of this pectin coat results in aggregation of oppositely charged particles (Fig. 9; Yamasaki *et al.*, 1964). The reduction of viscosity of the juice is, of course, due to depolymerization of the dissolved pectin (Endo, 1965*a*; Yamasaki *et al.*, 1967). Careful experiments with purified enzymes have shown that this effect is reached either by a combination of PE and PG (Endo, 1965*b*) or by PL alone in the case of apple juice which contains highly esterified pectin (≥80%; Ishii & Yokotsuka, 1972), whereas in the case of grape juice, which contains pectins with a lower DE (44–65%), PL alone does not perform as well (Ishii & Yokotsuka, 1973). The clarification of juices

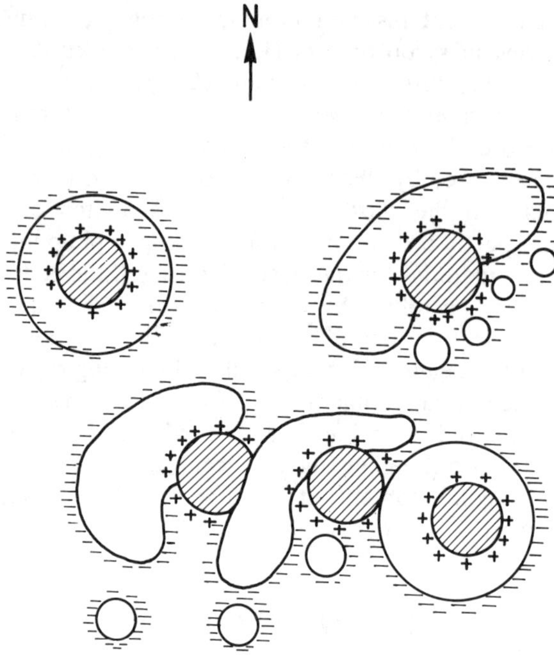

FIG. 9. Aggregation of cloud particles (Yamasaki *et al.*, 1964).

by pectin degradation is also important in the manufacture of high Brix concentrates to avoid gelling and the development of haze. A hot clarification process that saves enzyme and time (Grampp, 1977) and continuous clarification processes have been described (Richard, 1974; Wucherpfennig & Otto, 1988). The advent of membrane processes has created the possibility for ultrafiltration of enzyme-treated juices, a process which is increasingly used in the juice industry (Milner, 1984; Moslang, 1984).

Juice extraction

Pulping soft fruits with a high content of water soluble pectin, e.g. blackcurrants, results in a semi-gelled mass which is very difficult to press. Treatment of such pulps with pectinases yields thin free-run juice and a pulp with good pressing characteristics (Beltman & Pilnik, 1971). This pulp enzyme-treatment process is almost as old as enzymatic clarification (Charley, 1969). In the early 1970s, a start was made to apply it to apples with poor pressing characteristics due to the use of certain cultivars (Golden Delicious) and the necessity for storage (Vos & Pilnik, 1973). An important aspect in the flow diagram (Fig. 10) is the aeration of the pulp in an oxidation tank. Phenolics are oxidized by endogenous polyphenoloxidase and subsequently polymerize to high-molecular compounds which are unable to inhibit the added pectinase. At the same time the danger of a polyphenol haze in the finished product is diminished

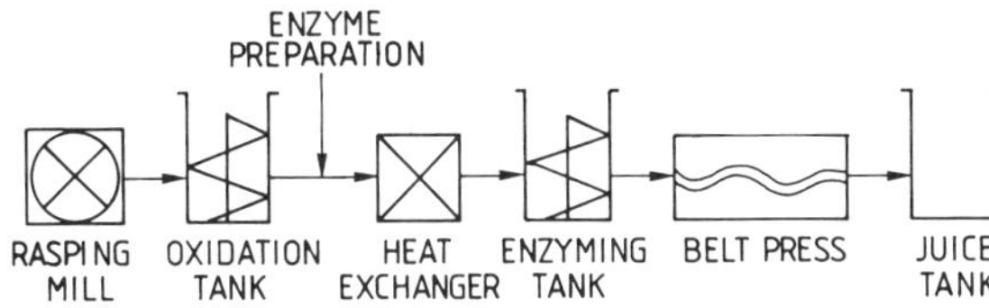

FIG. 10. Flowsheet of juice extraction from apple pulp treated with enzymes (de Vos & Pilnik, 1973).

(van Buren & Way, 1978). An alternative is the addition of polyvinylpolypyrrolidone which binds the phenolic substances. Enzyme preparations that perform well in juice clarification are also suitable for the enzymic treatment of pulp. In the case of apples, it has been shown that any combination of enzymes which depolymerizes highly esterified pectin (De $\geq$ 90) can be used successfully (Voordouw *et al.*, 1974). The good pressability of enzyme-treated pulp is shown in Table 7. High yields are always connected with enzyme treatment (Bielig *et al.*, 1971; Hak *et al.*, 1973). In the case of red fruit, the release of anthocyanins into the juice by enzymatic destruction of cell wall is a further important aspect of the pulp enzyme-treatment process. It has also attracted the interest of red wine manufacturers. It permits the attainment of good yields of red grape juice, especially in combination with a mild heat treatment, which can be fermented directly, as is the case for white wines. This is a simpler technique than the classical fermentation 'on the skins' which is

TABLE 7
Effect of Enzyme Treatment and Storage on Apple Juice Yields

Variety	*Storage*	*Enzyme Treatment*	*Press Yield (%)*
Treatment with 0·04% Pectinol flüssig® at 45°C, for 10 min (Bielig *et al.*, 1971)			
Horneburger	+	−	66
Horneburger	+	+	86
Golden Delicious	+	−	59
Golden Delicious	+	+	73
Golden Delicious	−	−	71
Golden Delicious	−	+	83
Jonathan	−	−	61
Jonathan	−	+	88
Treatment with 0·004% Ultrazym 100® at 40°C, for 20 min, after pre-oxidation (Hak *et al.*, 1973)			
Golden Delicious	+	+	78
Golden Delicious	+	−	52
Golden Delicious	+	+	75
Golden Delicious	+	−	62
Karmijn	+	+	82
Karmijn	+	+	74

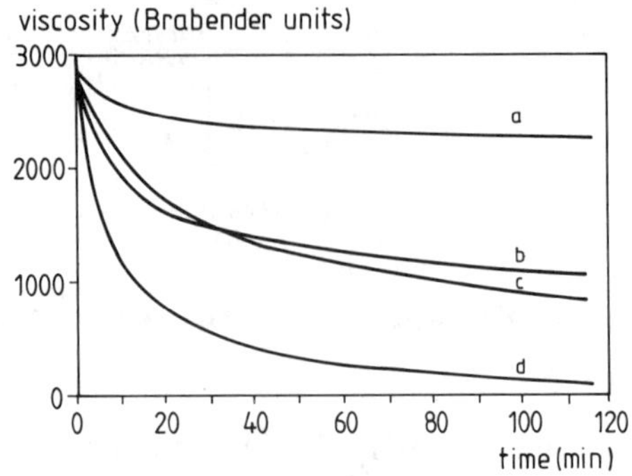

FIG. 11. Viscosity decrease of stirred apple pulp at 25°C: (a) no enzyme added, (b) cellulase added (C-1 + C-*x*), (c) pectinase added, (d) cellulase and pectinase added (Pilnik *et al.*, 1975).

necessary for red wine to obtain the desired colour. In the USA, the extraction of red grape juice from pectin-rich Concord grapes (*Vitis labrusca*) by the enzymic process is widely used (Neubeck, 1975). Enzyme treatment of pulp of olives, of palm fruit and of coconut to increase oil yield has also been described (Neubeck, 1975).

Liquefaction

Enzymatic treatment of pulp can be carried through to almost complete liquefaction of fruits and vegetables if an exo-β-glucanase (C-1-cellulase, cellobiohydrolase) is used together with pectinase preparations which usually contain fungal endo-β-glucanase (C-*x*-cellulase). Figure 11 (Pilnik *et al.*, 1975) shows the viscosity decrease of stirred apple pulp during treatment with pectinase, C-1-cellulase and a mixture of the two enzyme preparations. For the mixture, a synergistic effect is seen. The low viscosity values reached correspond to complete liquefaction and at that stage microscopic examination shows that cell walls have disappeared. The liquefaction juices are almost clear (papaya, cucumber), cloudy (apples, peaches) or pulpy (carrots), depending on the accessibility of the cell wall compounds for the enzymes (lignin). The liquefaction products can then be clarified further by usual techniques,

TABLE 8
Neutral Sugars and Galacturonides Released from Apple Alcohol-insoluble Solids by Pure Enzymes (Percentage Recovery, Corrected for Control) (Voragen *et al.*, 1980)

Enzyme Treatment	*Neutral Sugars*							*Anhydro-galact-uronic Acid*	*Total*
	Rhamnose	*Arabinose*	*Galactose*	*Mannose*	*Xylose*	*Galact-uronic Acid*	*Total*		
PE (citrus)	—	—	—	—	—	—	—	—	0·1
PG	4	9	5	—	4	—	3	21	8
PL	19	39	27	—	20	4	17	57	28
PE + PG	25	52	34	—	13	—	18	75	33
C-1	9	3	12	26	20	22	16	5	13
C-1 + PE + PG	58	89	69	84	71	79	78	82	80

preference being given increasingly to ultrafiltration. The action of the enzymes can be followed chemically in enzyme-treated cell wall material by comparing the amounts of galacturonic acid and neutral sugars in alcohol-insoluble residues with the amounts of the control sample (Voragen *et al.*, 1980). Table 8, which shows results of experiments on apple cell wall material with purified enzymes, reveals that pectinase activity (PE + PG) alone releases 75% of the pectic material, including the pectin sugars arabinose, galactose, xylose and rhamnose. No glucose from cellulose was released. Cellulase alone had little influence on pectin and solubilized only 22% of the cellulose. Combined cellulase–pectinase activities released 80% of the polysaccharides. This is the synergistic effect observed in pulp viscosity experiments (Fig. 11). A similar effect has been found for grapefruit segment membranes (Ben-Shalom, 1986) when treated enzymatically. The breakdown products increase the soluble solids content of the juices. In this way high yields of juice and solids are obtained (Voragen *et al.*, 1980). This is especially interesting for manufacturers of concentrated juices (Pilnik, 1988). Liquefaction is also interesting for vegetables and fruits (Dongowski & Bock, 1977; Steinbuch & van Deelen, 1986; Pilnik, 1988) from which pressing yields no juice or for which no presses have been developed (mango, guava, bananas, etc.) (Schreier *et al.*, 1985). Therefore, much interest is shown in this new, simple, low capital, easy maintenance technology in developing countries (Fig. 12). Enzymic liquefaction is attractive in all countries as costly waste disposal is virtually eliminated. In industrialized countries socio-economic factors are increasing the necessity to process stored apples throughout the year and this situation also generates interest in the process. Furthermore, legal restraints with regard to the definition of fruit juice and the use of *Trichoderma viride* cellulase are being overcome.

Maceration

Maceration is the process by which organized tissue is transformed into a suspension of intact cells (Grampp, 1972; Bock *et al.*, 1983) resulting in pulpy

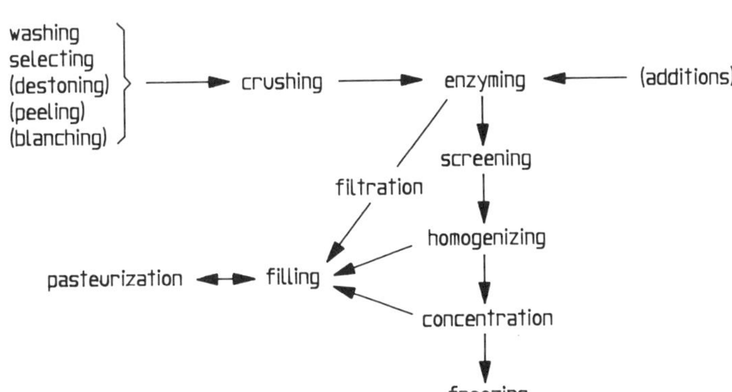

FIG. 12. Flow diagram for liquefied fruit and vegetable juices.

products used as base material for pulpy juices and nectars, as baby foods and as ingredients for dairy products like puddings and yoghurts. The enzymic process for the manufacture of such products must be contrasted to the mechanical process in which, inevitably, many cells are disrupted with negative consequences for quality due to action of endogenous enzymes damaging flavour, colour and ascorbic acid and in which finisher operation to remove coarse tissue results in mechanical losses of nutritive constituents. Enzymatic degradation of pectin after a mild mechanical treatment greatly improves product properties if carried out so as to leave as many cells as possible intact. This is achieved by so-called macerases, enzyme preparations with only one specific depolymerizing activity which acts in the first place on the middle-lamella pectin which is responsible for cell cohesion. Both the use of purified PG (Zetelaki-Horvath & Gatai, 1977*a,b*; Struebi *et al.,* 1978; Zetelaki-Horvath & Vas, 1980) or of purified PL (Ishii & Yokotsuka, 1971; Ishii & Yokotsuka, 1975) have been described and the effectiveness of these macerases on various substrates has been compared (Ishii, 1976).

For maceration of vegetables, bacterial PAL is interesting because of its high pH optimum (Rombouts *et al.,* 1978; Wegener & Henniger, 1987). The restricted pectin degradation solubilizes rather than degrades the middle lamella pectin which may help the creamy mouth-feel. The process is often used for carrots for baby foods. Whether maceration or cell wall breakdown has occurred can be checked by microscopy and by the absence of enzymatic browning. A very interesting use of enzymic maceration is pulp for the production of dried instant potato mash. It prevents starch leaking out of the cells, thus avoiding the quality defect of glueiness of the reconstituted product. Potatoes are blanched in order to gelatinize starch and to inactivate endogenous PE which together with the added PLG would extend the possibilities for pectin degradation and turn the maceration process into pulp enzyming technology. Inactivation of endogenous PE is important for the maceration of many products (Sulc & Vujicic, 1973; Dongowski & Bock, 1980). It is interesting to reflect that most uses of exogenous pectinases for fruit juice extraction and clarification are based on the destruction of pectin and other desirable dietary fibre components of fruits. Macerated products with only limited pectin degradation may, therefore, in future gain an increasing appeal for the consumer and promote the production of purified PGs and lyases. The use for pulpy products also means that immobilization of these enzymes has little interest.

Product specific uses of pectinases

Pulp wash in the citrus industry is the product obtained by counter-current water extraction of citrus pulp after de-juicing. It is added back to the juice before concentration, but is often too viscous for this purpose because of dissolved pectin. The viscosity is easily and efficiently reduced by addition of pectinase. There are also claims of an increased solids yield (Braddock &

Kesterson, 1979). Also in the area of citrus, a process has been patented (Bruemmer, 1981) to facilitate the manufacture and improve the quality of citrus fruit segments. Citrus fruit is heated to a core temperature of 20–40°C, the peel surface is scored to barely penetrate albedo without penetrating juice sections. Gas bubbles in albedo are then replaced with pectinase solution and fruit incubated to permit removal of peel and other membrane from intact juice segments.

Apricot nectar stabilization

Apricot nectar exhibits a yet unexplained phenomenon of cloud loss. The nectar is made by mixing finely screened pasteurized apricot pulp with water, sugar and acid to contain from 30 to 50% apricot ingredient. On standing, the slowly sedimenting pulp particles form a gel which contracts leaving clear syneresis fluid as supernatant (Siliha & Pilnik, 1985). This separation can be avoided by grinding the pulp so finely that at least one third of the cells are broken open (Weiss & Sämann, 1972). This is difficult to control during production, but cells can be broken open by the action of pectolytic enzymes (Siliha & Pilnik, 1985). For this purpose, an arabinan-rich pectin fraction must

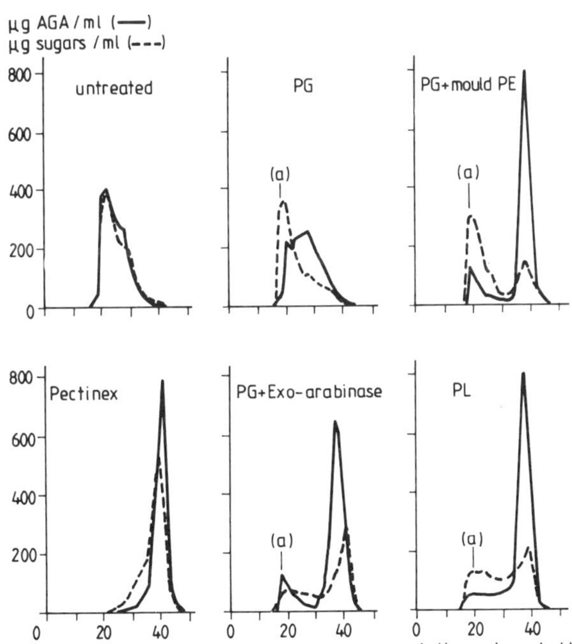

FIG. 13. Gel filtration chromatography of the hydrochloric acid-soluble apricot pectin fraction degraded by various enzyme systems on a Sephacryl S-300 column. AGA: Anhydrogalacturonic acid, Pectinex is a wide spectrum commercial pectinase, PG activity alone hardly degrades the pectin and does not prevent cloud loss (Siliha, 1985).

be degraded (Siliha, 1985). Pure PG or commercial preparations containing mainly PG are not suitable. The cell walls become thinner but do not break. If PE or an *exo*-arabanase is added, the enzyme preparation becomes a cloud-stabilizing enzyme, capable of degrading the extracted arabinan-rich pectin and of breaking open cells in the apricot pulp from which a cloud-stable product can then be made. Figure 13 shows the elution patterns of gel filtration chromatography of the araban-rich pectin after treatment with various enzyme systems. Only treatments with cloud-stabilizing enzymes move the uronide peak into the included volume, indicating depolymerization (Siliha & Pilnik, 1985).

QUALITY ASPECTS OF CLEAR FRUIT JUICES, OBTAINED BY ENZYME TECHNOLOGY

The primary quality requirement of the consumer for sparkling clarity is achieved in all processes (juice clarification, enzyme treatment of pulp, liquefaction) provided pectin is sufficiently degraded (alcohol flocculation test). Nevertheless, haze may appear in concentrated apple and pear liquefaction juices and sometimes in concentrated juices from pulp enzyming treatments (see below).

Uronide content of apple juices

Table 9 (Kolkman, 1983) shows the total uronide content of juices obtained in a laboratory study on the composition of juices prepared with the use of enymes. The uronide content is seen to increase with the intensity of pectic enzyme treatment. Free methanol from PE action is also present in quantities corresponding to *c.* 10% of the uronide content, i.e. from 60 to 450 mg/litre. Concern about free methanol is repeatedly expressed but does not take into consideration the fact that ester-bound methanol also enters the metabolism; 450 mg/litre would be about 100 mg/kg above the range found in all kinds of beverages (Wucherpfennig *et al.*, 1989). Of course, in modern fruit juice

TABLE 9
Uronide Content of Apple Juice (12° Brix) (Kolkman, 1983)

Methods of Preparation	*Uronide Content (mg/ml)*
Pressing + clarification	0·58
Water extraction	1·31
Methanol extraction	0·35
Enzymic pulp treatment	2·51
Enzymic pulp liquefaction	4·32

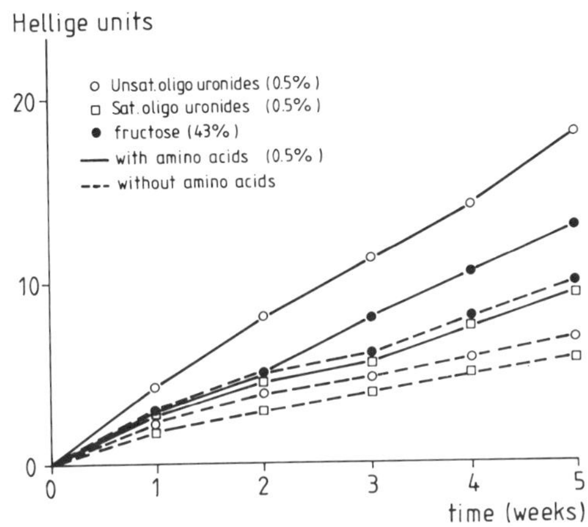

FIG. 14. Unsaturated oligogalacturonides as precursors for non-enzymic browning reactions.

technology volatiles are stripped from the juice and kept and traded separately and mixed for use. This means that in reconstituted juices, methanol content can be controlled. Depending on the PL action of the enzyme preparations used, 20–40% of the uronides are present as unsaturated oligomers (Table 9). These enzyme-treated juices undergo browning more rapidly than traditional juices. Figure 14 presents the results of model experiments from which it is seen that the unsaturated oligomer uronides are powerful browning precursors (Voragen *et al.*, 1988). The models for concentrated apple juice were made up as follows: indicated quantities of oligo-uronides and fructose were brought to 70° Brix with sorbitol and to pH 3·5 with malic acid. In some experiments, 0·5% of an amino acid mixture (asparagine, aspartic acid and glutamic acid) were also added. Storage temperature was 50°C. At high temperatures, unsaturated oligomers have been seen to split into saturated oligomers with one unit less than the original compound. The unsaturated monomer reacts further to a number of compounds among which the known browning precursors 2-furoic acid and 2-formyl-2-furoic acid have been identified. These findings are surely of interest in view of the large shipments of concentrated apple and pear juice under non-refrigerated conditions.

NON-PECTIC ENZYMES INVOLVED IN PECTIC ENZYME TECHNOLOGY

Cellulases

Cellulases have been shown to be necessary for liquefaction processes (see above; Voragen *et al.*, 1986*c*). Cellulolytic enzymes are produced by a large number of microorganisms such as fungi, actinomycetes and bacteria. Cellulases with high activity on crystalline cellulose are excreted by several fungi. Cellulases from *Trichoderma* spp. (*T. viride, T. reesi* and *T. koningii*), *Fusarium solani, Penicillium funiculosum, Sporotrichum pulverulentum* and *Talaromyces emersonii* have been investigated thoroughly (Wood, 1985; Beldman, 1986). Commercially available cellulases are derived from *Trichoderma* spp. The degradation of crystalline cellulose requires the participation of a set of enzymes (Fig. 15). With respect to hydrolysis cellulases can be divided into endo- and exo-acting enzymes. The presence of an exo-β-glucanase, sometimes referred to as C-1-cellulase, is typical of cellulase preparations which are able to degrade highly ordered cellulose. The enzyme releases cellobiose from the non-reducing chain-end and is therefore called a cellobiohydrolase (1,4-β-D-glucan cellobiohydrolase, EC 3.2.1.91). This enzyme works in cooperation with an endo-β-glucanase (endo-1,4-β-D-glucan glucanohydrolase, EC 3.2.1.4), also known as C-*x*-cellulase, and a β-glucosidase, also called cellobiase (β-D-glucoside glucohydrolase, EC 3.2.1.21). Enzymatic hydrolysis of amorphous cellulose, leading to cellobiose, can be obtained by a sequential attack of endo-β-glucanase and cellobiohydrolase via cello-oligosaccharides. However, degradation of crystalline cellulose is thought to occur at the surface of the cellulose fibrils, where chain cleavage by an endo-β-glucanase is immediately followed by the catalytic action of a cellobiohydrolase. The ability of these cellulolytic enzymes to adsorb to their substrate is a prerequisite for the latter type of hydrolysis and is thought to be the driving force for synergistic effects, which are often observed during hydrolysis with mixtures of these enzymes. Cellobiose, a competitive inhibitor

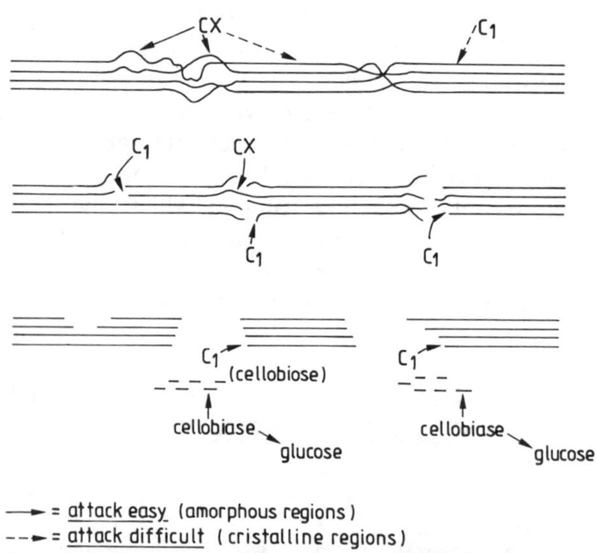

FIG. 15. Points of attack of C-1 cellulase (cellobiohydrolase) and C-*x* cellulases on cellulose micelles.

of cellulases, is hydrolyzed by cellobiase to glucose, a less inhibitory product. Some cellulase systems contain a glucohydrolase (1,4-β-D-glucan glucohydrolase, EC 3.2.1.74), which forms glucose directly from the polymeric substrate.

Hemicellulases

Hemicellulases are contained in commercial fungal pectinase preparations (Table 5; Voragen *et al.*, 1986*a*). Experiments on apple pulp with cleaned-up galactanase and arabanase (Voragen *et al.*, 1982) did not show an enhancing effect on pulp enzyming or liquefying action although there is release of arabans of various structures and degrees of polymerization including oligo- and monomers from the hemicellulose substrates containing arabinose present in (fruit) cell walls. When ultrafiltration is used to clarify fruit juices there may be a benefit in obtaining low-molecular degradation products which pass through the membrane and raise the soluble solids level in the juice. Table 10 shows the concentration and composition of high-molecular-weight fragments present in apple juices prepared by conventional pressing, pulp enzyme treatment or enzymatic liquefaction using four commercial enzyme preparations which differed in the level and type of pectic and (hemi)-cellulolytic activities (Veld, 1986). The polysaccharide fractions were isolated as freeze-dried dialysis

TABLE 10
Amounts and Characteristics of Polysaccharides Isolated from Apple Juice Prepared by Different Processes and Different Enzymes. Analysis of Dialysis Retentates (Veld, 1986)

Analysis	*Pressing*	*Pulp Enzyme Treatment*			*Liquefaction*	
		A[a]	*B*[a]	*C*[a]	*C*[a]	*D*[a]
Polysaccharides content (mg/kg pulp)	155	1075	1860	1075	3000	4300
Sugar composition (mol%)						
Rhamnose	3	5	3	5	4	4
Arabinose	21	38	6	52	43	54
Xylose	2	5	2	5	6	8
Galactose	19	9	4	7	6	8
Glucose	2	1	1	1	1	0
Anh.galacturonic acid	53	42	84	30	40	26
DM[b]	52	72	76	31	59	39
DA[c]	15	18	6	13	28	29

[a] Different enzyme preparations used. A, B and D were experimental preparations from Gist-Brocades, Delft, The Netherlands: A, was a pectinase; B, a pectinase rich in hemi-cellulase activity, and D, a pectinase–cellulase mix. C, was ultra SP-L from Novo Ferment AG, Dittingen, Switzerland.
[b] Degree of methylation.
[c] Degree of acetylation calculated on anhydrogalacturonic acid content.

retentates from raw juices clarified with bentonite, gelatin and Klar-Sol 30. The effect of enzyme treatment is unequivocal; the largest amounts of polysaccharides were released in the liquefaction process. The amounts and composition of the polysaccharide fractions also depended on the technical preparations used; distinct variations in arabinose, galactose, galacturonic acid and xylose contents were found. Also, variations in the degree of methylation were observed; fragments with increased contents of acetyl groups accumulated in the liquefaction juices. Juices from deciduous fruit with a high amount of non-dialysable polysaccharides, a high proportion of which is arabinose (Table 10), are in danger of haze formation under conditions of chilled storage in concentrated form, a quality defect observed when pulp enzyming processes were introduced. Analysis of haze isolated from apple juice (Table 11; Voragen *et al.*, 1982) showed it to consist mainly of arabinose of which 88·4% occurred in unbranched α-1,5-linked chains. Confirming results were also reported by Churms *et al.* (1983). These phenomena of different arabinan release and eventual haze formation are explained by the different arabinanase activities found in fungal preparations (Kaji, 1984). An arabinanase that releases arabinose as the only product from polymeric and oligomeric substrates (including arabinosides) is called α-L-arabinofuranosidase, ((1, 5)(1, 3)(1, 2)-α-L-arabinofuranoside arabinofuranhydrolase, EC 3.2.1.55). This type of enzyme was isolated from *Aspergillus niger* (Rombouts *et al.*, 1988) and *Cortitium rolfsi* (Komae *et al.*, 1982). Another type of arabinofuranosidase is active only on low molecular weight substrates and does not act on arabinose-containing polymers. Sources for this enzyme are *A. niger* (Rombouts *et al.*, 1988) and *Streptomyces purpurascens* (Kaji & Yoshihara, 1970; Komae *et al.*, 1982). There are also arabinanases which hydrolyse

TABLE 11
Sugar Composition of Haze Material and Glycosidic Linkage Composition for Arabinosyl Residues (Voragen *et al.*, 1982)

	Amount of Sugar (g/100 g haze)	*Glycosidic Linkages[a] Arabinosyl Units (%)*	
Rhamnose	1·2	4·2%	1-linked, (terminal furanose units)
Arabinose	90·0		
Xylose	0·3	0·2%	1-linked, (terminal pyranose units)
Mannose	0·7		
Galactose	1·6	2·3%	1,3-linked, (unbranched furanose units)
Glucose	0·1		
Galacturonic acid	3·9	88·4%	1,5-linked, (unbranched furanose units)
Total	97·8	3%	1,3,5-linked (branched furanose units)
		1%	1,2,5-linked (branched furanose units)

[a] Derived from methylation studies.

1,5-α-L-arabinan by an *endo*-mechanism, resulting in a mixture of products, comprising the monomer and/or oligomers. This *endo*-1,5-α-L-arabinanase (1,5-α-L-arabinan 1,5-α-L-arabinohydrolase, EC 3.2.1.99) was also purified from *A. niger* (Rombouts *et al.*, 1988). Figure 16 shows the points of attack of these enzymes on branched arabinans as found in apple juice obtained by pulp enzyming or liquefaction (Voragen *et al.*, 1986*c*). The arabinofuranosidase II obviously serves as a debranching enzyme, having a preference for 1,3 links, but it can also remove arabinose from the α-1,5-non-reducing end of the α-1,5-arabinan backbone. The de-branched arabinan chain is able to retrograde and then precipitates as haze. Obviously, enzyme manufacturers wish to enrich their preparation with endo-arabinanase to degrade the main chain before retrogradation occurs. There is a combined action (Voragen *et al.*, 1987) of endo-arabinanase and arabinofuranosidase similar to the combined action of PG and PE. The arabinofuranosidase I, working on low molecular

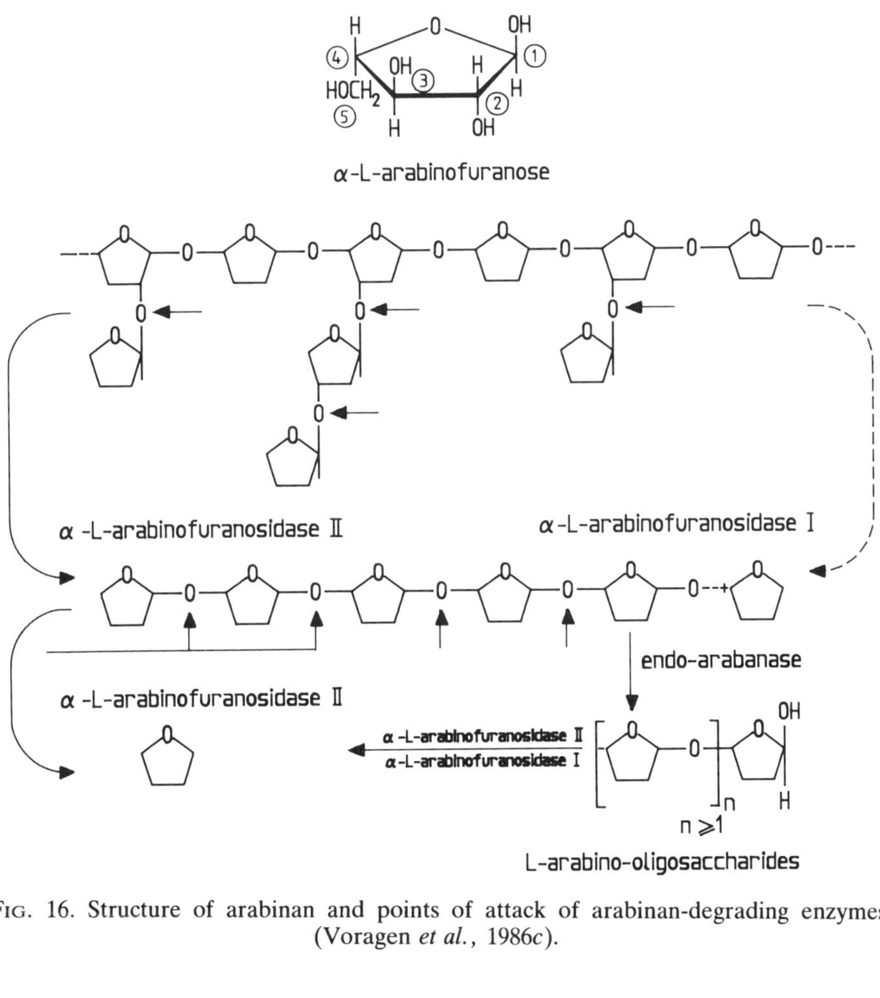

FIG. 16. Structure of arabinan and points of attack of arabinan-degrading enzymes (Voragen *et al.*, 1986*c*).

weight substrates, appears to serve no technological function. Unfortunately, there are no simple tests for arabinans, like the alcohol test for pectin or the iodine test for starch. Detection of the presence of the (partly) debranched arabinan chain involves separation by gel filtration and degradation by pure endo-arabinase (Endresz *et al.*, 1986; Ducroo, 1987).

REFERENCES

Albersheim, P. & Anderson, A. J. (1971). *Proc. Nat. Acad. Sci. USA,* **68,** 1815–19.

Aspinall, G. O. (1980). In *The Biochemistry of Plants* (Vol. III), ed. J. Preis. Academic Press, New York, USA, pp. 473–500.

Baker, R. A. (1976). *J. Food Sci.,* **41,** 1198–200.

Baker, R. A. (1980). In *Citrus Nutrition and Quality,* ed. S. N. Nagy & J. A. Attaway. ACS Symposium Series 143, American Chemical Society, Washington, DC, USA, pp. 109–28.

Baron, A. & Drilleau, J. F. (1982). In *Use of Enzymes in Food Technology,* ed. P. Dupuy. Technique et Documentation Lavoisier, Paris, France, p. 471–6.

Baron, A., Rombouts, F. M., Drilleau, J. F. & Pilnik, W. (1980). *Lebensm. Wiss. m.-Technol.,* **13,** 330–3.

Bartolome, L. G. & Hoff, J. E. (1972). *J. Agric. Food Chem.,* **20,** 262–6.

Beldman, G. (1986). PhD thesis, Agricultural University, Wageningen, The Netherlands.

Bell, T. A., Etchells, J. L., Williams, C. F. & Porter, W. L. (1962). *The Botanical Gazette,* **123**(3), 220–3.

Beltman, H. & Pilnik, W. (1971). *Confructa,* **16**(1), 4–9.

Ben-Shalom, N. (1986). *J. Food Sci.,* **51**(3), 720–1, 730.

Bielig, H. J., Wolff, J. & Balcke, K. J. (1971). *Flüssiges Obst.,* **38,** 408, 410–12, 414.

Bock, W., Krause, M. & Dongowski, G. (1970). *Ernährungsforschung,* **15,** 403–15.

Bock, W., Dongowski, G. & Krause, M. (1972). *Nahrung,* **16,** 787–99.

Bock, W., Krause, M. & Dongowski, G. (1983). Verfahren zur Herstellung von Pflanzenmazeraten, GDR Patent 84317.

Braddock, R. J. & Kesterson, J. W. (1979). *Food Technol.,* **33**(11), 78–83.

Bruemmer, J. H. (1981). US Patent 4 284 651.

Buren, J. P. van & Way, R. D. (1978). *J. Food Sci.,* **43,** 1235–7.

Castelein, J. M. & Pilnik, W. (1976). *Lebensm. Wiss. Technol.,* **9**(5), 277–83.

Charley, V. L. S. (1969). *Chem. Ind.,* 635–41.

Chesson, A. & Codner, R. C. (1978). *J. Appl. Bacteriol.,* **44,** 347–64.

Churms, S. C., Merrifield, E. H., Stephen, A. M., Walwylor, D. R., Poson, A., van der Merke, K. J., Spies, H. S. C. & Losta, N. (1983). *Carbohydr. Res.,* **113,** 339–44.

Colquhoun, I. J., de Ruiter, G. A., Schols, H. A. & Voragen, A. G. J. (1990). Carbohydr. Res., **206,** 131–4.

Crandall, P. G., Chen, C. S., Marcy, J. E. & Martin, F. G. (1986). *J. Food Sci.,* **51,** 1017–20.

Doesburg, J. J. (1965). Pectic Substances in Fresh and Preserved Fruits and Vegetables. IBVT Communication No. 25, Wageningen, The Netherlands.

Dongowski, G. & Bock, W. (1977). *Die Lebensmittelindustrie,* **24,** 33.

Dongowski, G. & Bock, W. (1980). *Berichte der Wiss.-Technol. Komm. der IFU,* **XVI,** 275–88.

Drawert, F. & Kreft, M. (1977). *Chem. Mikrob. Technol. Lebensm.,* **5,** 105–12.

Ducroo, P. (1987). *Flüssiges Obst,* **5,** 265–9.

Eckhardt, C. & Ahrens, E. (1977). *Chem. Mikrob. Technol. Lebensm.*, **5**, 71–5.
Edstrom, R. D. & Phaff, H. J. (1964). *J. Biol. Chem.*, **239**, 2403–7.
Endo, A. (1965*a*). *Agric. Biol. Chem.*, **29**, 229–33.
Endo, A. (1965*b*). *Agric. Biol. Chem.*, **29**, 29–136.
Endresz, H.-U., Tableros, M., Omran, H. & Gierschner, K. (1986). *Proc. Intern. Fruit Juice Symp.*, The Hague, The Netherlands, pp. 293–306. Juris Druck + Verlag, Zurich, Switzerland.
English, P. D., Maglothin, A., Keegstra, K. & Albersheim, P. (1972). *Plant Physiol.*, **49**, 293–7.
Fellenberg, T. von (1913). *Mitt. Lebensm. Hyg.*, **4**, 122, 273.
Fogarty, W. M. & Ward, O. P. (1974). In *Progress in Industrial Microbiology*, ed. D. J. D. Hockenhull. Churchill Livingstone, Edinburgh, UK, **13**, 59–119.
Fry, S. C. (1982). *Biochem. J.*, **203**, 493–504.
Gierschner, K. (1981). *Gordian*, **81**, 171–210.
Gould, W. A. (1974). *Tomato Production, Processing and Quality Evaluation*, AVI, Westport, CT, USA.
Grampp, E. (1972). *Dechema-Monographien*, **70**, 175–86.
Grampp, E. (1977). *Food Technol.*, **31**(11), 38–41.
Hak, J., Pilnik, W., Voragen, A. G. J. & de Vos, L. (1973). MSc thesis, Agricultural University, Wageningen, The Netherlands.
Hoogzand, C. & Doesburg, J. J. (1961). *Food Technol.*, **15**, 160–3.
Houdenhoven, F. E. A., van (1975). PhD thesis, Agricultural University, Wageningen, The Netherlands.
Hsu, C. P., Deshpande, S. N. & Desrosier, N. W. (1965). *J. Food Sci.*, **30**, 583–8.
IFU (1985). Analysis Method No. 46, Intern. Fed of Fruit Juice Producers. Juris Druck + Verlag, Zurich, Switzerland.
Ishii, Sh. & Yokotsuka, T. (1971). *Agric. Biol. Chem.*, **35**, 1157–9.
Ishii, Sh. & Yokotsuka, T. (1972). *J. Agric. Food Chem.*, **20**, 787–91.
Ishii, Sh. & Yokotsuka, T. (1973). *J. Agric. Food Chem.*, **21**, 269–72.
Ishii, Sh. & Yokotsuka, T. (1975). *Agric. Biol. Chem.*, **39**, 313–21.
Ishii, Sh. (1976). *Phytopathol.*, **66**, 281–9.
Ishii, Sh., Kiho, K., Sugiyama, S. & Sugimoto, H. (1979*a*): *J. Food Sci.*, **44**, 611–14.
Ishii, S., Kiho, K., Sugiyama, S. & Sugimoto, H. (1979*b*). FRG Patent Offenlegungsschrift 2843351.
Joslyn, M. A. & Pilnik, W. (1961). In *The Orange: Its Biochemistry and Physiology*, ed. W. B. Sinclair. Univ. of California, USA, p. 373–435.
Kaji, A. (1984). *Adv. Carbohydr. Chem. Biochem.*, **42**, 383–94.
Kaji, A. & Yoshihara, O. (1970). *Agric. Biol. Chem.*, **34**, 1249–53.
Kertesz, Z. I. (1930). New York State Agric. Exper. Sta., Bull. 689. US Patent 1.932.833.
Kesterson, J. W. & Braddock, R. J. (1976). *Bull.* (technical), **784**, Agric. Exper. Sta., Univ. of Florida, Gainsville, USA.
Kew, T. J. & Veldhuis, M. K. (1961). US Patent 2,995,448.
Kilara, A. (1982). *Process Biochem.*, **17**, 35–41.
Kohn, R., Furda, I. & Kopec, Z. (1968). *Collection Czechoslov. Chem. Commun.* **33**, 264–9.
Kolkman, J. R. (1983). MSc thesis, Agricultural University, Wageningen, The Netherlands.
Koller, A. & Neukom, H. (1969). *Eur. J. Biochem.*, **7**, 485–9.
Komae, K., Kaji, A. & Sato, M. (1982). *Agric. Biol. Chem.*, **46**, 1899–905.
Krop, J. J. P. (1974). PhD thesis, Agricultural University, Wageningen, The Netherlands.
Krop, J. J. P. & Pilnik, W. (1974*a*). *Lebensm.-Wiss. Technol.*, **7**(1), 62–3.
Krop, J. J. P. & Pilnik, W. (1974*b*). *Lebensm.-Wiss. Technol.*, **7**(2), 121–4.

Kulp, K. (1975). In *Enzymes in Food Processing,* ed. G. Reed. Academic Press, London, UK, pp. 53–122.

Lee, C. Y., Bourne, M. C. & Buren, J. P., van (1979). *J. Food Sci.,* **44,** 615–16.

Litchman, M. A. & Upton, R. P. (1972). *Anal. Chem.,* **44,** 1495–7.

Luh, B. S., Ozbilgin, S. & Liu, Y. K. (1978). *J. Food Sci.,* **43,** 713–20.

MacDonnell, L. R., Jang, R., Jansen, E. F. & Lineweaver, H. (1950). *Arch. Biochem.,* **28,** 260–73.

Macmillan, J. D. & Phaff, H. J. (1966). *Methods in Enzymology,* **8,** 632–5.

Manabe, M. (1973). *J. Agric. Chem. Soc. Japan,* **47,** 385–90.

May, D. (1989). *Carbohydr. Polym.,* **12,** 79–99.

McColloch, R. J. & Kertesz, Z. I. (1949). *Food Technol.,* **3** (3) 94–6.

McCready, R. M. & Seegmiller, C. G. (1954). *Arch. Biochem. Biophys.,* **50,** 440–50.

McFeeters, R. F. & Armstrong, S. A. (1984). *Anal. Biochem.,* **139,** 212–17.

McMillan, J. D. & Sheiman, M. I. (1974). In *Food Related Enzymes,* ed. J. R. Whitaker. American Chemical Society, Washington, DC, USA, pp. 101–30.

McNeil, M., Darvill, A. G., Frey, S. C. & Albersheim, P. (1984). *Ann. Rev. Biochem.,* **53,** 625–63.

Mehlitz, A. (1930). *Konservenindustrie,* **17,** 306, 321.

Miller, L. & Macmillan, J. D. (1971). *Biochem.,* **10,** 570–6.

Milner, B. A. (1984). New York State Agric. Exper. Sta. Special Report 54.

Milner, Y. & Avigad, G. (1967). *Carbohydr. Res.,* **4,** 359–61.

Moslang, H. (1984). *Confructa,* **28,** 219–24.

Neubeck, C. E. (1975). In *Enzymes in Food Processing* (2nd edn), ed. G. Reed. Academic Press, New York, USA, p. 397.

Neukom, H. (1969). Pektinspaltende Enzyme (Suppl. 8), zur *Z. für Ernährungswissenschaft,* **8,** 91–6.

Neukom, H., Amado, R. & Pfister, M. (1980). *Lebensm.-Wiss Technol.,* **13,** 1–6.

Nussinovitch, A. & Rosen, B. (1989). *Lebensm.-Wiss. Technol.,* **22,** 60–4.

Phaff, H. J. (1966). *Methods in Enzymology,* **8,** 636–41.

Pilnik, W. (1958). In *Berichte Wiss.-Techn. Komm. Intern. Fruchtsaft Union.* Juris Verlag, Zürich, Switzerland, pp. 203–28.

Pilnik, W. (1982). In *Use of Enzymes in Food Technology,* ed. P. Dupuy. Technique et Documentation Lavoisier, Paris, France, pp. 425–50.

Pilnik, W. (1988). *Proc. X. Intern. Congress of Fruit Juices,* Orlando, Florida, USA, pp. 159–79. Kündig Druck A.G., Zug, Switzerland.

Pilnik, W. & Rombouts, F. M. (1979*a*). In *Polysaccharides in Food,* ed. J. M. V. Blanshard & J. R. Mitchell. Butterworths, London, UK, pp. 109–26.

Pilnik, W. & Rombouts, F. M. (1979*b*). In *Proc. 5th Intern. Congress of Food Science and Technology,* ed. H. Chiba, M. Fujimaki, K. Iwai, H. Mitsuda & Y. Morita. Elsevier Scientific Publ. Co., Amsterdam, The Netherlands, pp. 269–77.

Pilnik, W. & Rombouts, F. M. (1981). In *Enzymes and Food Processing,* ed. G. G. Birch, N. Blakebrough & K. J. Parker. Applied Science Publishers, London, UK, pp. 105–28.

Pilnik, W. & Rothschild, G. (1960). *Die Fruchtsaftindustrie,* **5,** 131–8.

Pilnik, W. & Voragen, A. G. J. (1989). In *Quality Factors of Fruits and Vegetables, Chemistry and Technology,* ed. J. J. Jen. ACS Symposium Series 405. USA, pp. 250–69.

Pilnik, W., Rombouts, F. M. & Voragen, A. G. J. (1973). *Chem. Microbiol. Technol.,* **2,** 122–8.

Pilnik, W., Voragen, A. G. J. & Rombouts, F. M. (1974). *Lebensm.-Wiss. Technol.,* **7**(6), 353–5.

Pilnik, W., Voragen, A. G. J. & Vos, L. de (1975). *Flüssiges Obst.,* **42**(11), 448–51.

Pressey, R. (1986). In *Recent Advances in the Chemistry and Function of Pectins,* ed. M. Fishman & J. J. Jen. ACS Symposium Series 310. American Chemical Society, Washington, DC, USA, pp. 157–74.

Put, H. M. C. & Kruiswijk, J. T. (1964). *J. Appl. Bacteriol.*, **27,** 53–8.
Rexová-Benková, L. & Marcovic, O. (1976). *Adv. Carbohydr. Chem. Biochem.*, **33,** 323–85.
Rexová-Benková, L., Markovic, O. & Foglietti, M. J. (1977*a*). *Collect. Czech. Chem. Commun.*, **42,** 1736–41.
Rexová-Benková, L., Mracková, M., Luknár, D. & Kohn, R. (1977*b*). *Collect. Czech. Chem. Commun.*, **42,** 3204–13.
Richard, J. P. (1974). US Patent 3,795,521.
Rombouts, F. M. (1972). PhD thesis, Agricultural University, Wageningen, The Netherlands (Agric Res. Report 779).
Rombouts, F. M. (1981). In *Adv. Biotechnol. Proc. VIth Int. Ferm. Symp. and Vth Int. Symp. on Yeasts.* Pergamon Canada Ltd, Willowdale, Ontario, Canada, pp. 585–92.
Rombouts, F. M. & Pilnik, W. (1979). Proc. Congrès Intenational Microbiologie et Industrie Alimentaire, October 1979, Paris. APRIA, Paris, France.
Rombouts, F. M. & Pilnik, W. (1980). In *Economic Microbiology,* (Vol. 5), ed. A. H. Rose. Academic Press, London, UK, p. 227–82.
Rombouts, F. M. & Pilnik, W. (1984). Enzymes for structural analysis of plant polysaccharides. *Proc. Intern. Workshop on Plant Polysaccharides, Structure and Function,* Nantes, 9–11 July 1984. Organized by the INRA, Lab. de Biochimie et Technologie des Glucides, Nantes, France.
Rombouts, F. M. & Thibault, J. F. (1986*a*). *Carbohydr. Res.*, **154,** 177–87.
Rombouts, F. M. & Thibault, J. F. (1986*b*). *Carbohydr. Res.*, **154,** 189–203.
Rombouts, F. M., Spaansen, C. H., Visser, J. & Pilnik, W. (1978). *J. Food Biochem.*, **2,** 1–22.
Rombouts, F. M., Wissenburg, A. K. & Pilnik, W. (1979). *J. Chromatogr.*, **168,** 151–61.
Rombouts, F. M., Geraeds, C. C. J. M., Visser, J. & Pilnik, W. (1982*a*). In *Affinity Chromatography and Related Techniques,* ed. T. C. J. Gribnau, J. Visser & R. J. V. Nivard. Elsevier, Amsterdam, The Netherlands, pp. 255–60.
Rombouts, F. M., Versteeg, C., Karman, A. M. & Pilnik, W. (1982*b*). In *Use of Enzymes in Food Technology,* Technique et Documentation Lavoisier, Paris, France, pp. 483–7.
Rombouts, F. M., Voragen, A. G. J., Searle-van Leeuwen, M. F., Geraeds, C. C. J. M., Schols, H. A. & Pilnik, W. (1988). *Carbohydr. Polym.*, **9,** 25–47.
Rouse, A. H. & Atkins, C. D. (1953). *Food Technol.*, **7,** 221–3.
Schols, H. A., Reitsma, J. C. E., Voragen, A. G. J. & Pilnik, W. (1989). *Food Hydrocolloids,* **6**(2), 115–21.
Schols, H. A., Geraeds, C. C. J. M., Searle-van Leeuwen, M. F., Kormelink, F. J. M. & Voragen, A. G. J. (1990). *Carbohydr. Res.*, **206,** 105–15.
Schreier, R., Kiltsteiner-Eberle, R. & Idstein, H. (1985). *Flüssiges Obst,* **52,** 365–70.
Selvendran, R. R. (1985). *J. Cell. Sci.*, (Suppl. 2), 51–88.
Siliha, H. A. I. (1985). PhD thesis, Agricultural University, Wageningen, The Netherlands.
Siliha, H. A. I. & Pilnik, W. (1985). *Berichte der Wiss.-Techn. Kommission der IFU,* **XVIII,** 325–34.
Solms, J. & Deuel, H. (1955). *Helvetica Chimica Acta,* **38,** 321–9.
Spiro, R. G. (1966). *Methods in Enzymology,* **8,** 3–26.
Steinbuch, E. (1976). *J. Food Technol.*, **11,** 313–6.
Steinbuch, E. & Deelen, W. van (1986). *Confructa,* **30,** 15–23.
Struebi, P., Escher, F. E. & Neukom, H. (1978). *J. Food Sci.*, **43,** 260–3.
Sulc, D. & Ristic, M. (1954). *Z. Lebensmittelunters.u.-Forschung,* **98,** 430–4.
Sulc, D. & Vujicic, B. (1973). *Flüssiges Obst.*, **40,** 79–83, 130–7.
Tanner, H. (1970). *Schweiz. Z. Obst Weinbau,* **106,** 625–9.
Termote, F., Rombouts, F. M. & Pilnik, W. (1977). *J. Food Biochem.*, **1**(1), 15–34.

Veld, P., In't (1986). MSc thesis, Agricultural University, Wageningen, The Netherlands.
Versteeg, C. (1979). PhD thesis, Agricultural University, Wageningen, The Netherlands.
Versteeg, C., Rombouts, F. M. & Pilnik, W. (1978). *Lebensm.-Wiss.u.-Technol.,* **11,** 267–74.
Versteeg, C., Rombouts, F. M., Spaansen, C. H. & Pilnik, W. (1980). *J. Food Sci.,* **45,** 969–71, 998.
Voordouw, G., Voragen, A. G. J. & Pilnik, W. (1974). *Flüssiges Obst,* **41**(7), 282–4.
Voragen, A. G. J. (1972). PhD thesis, Agricultural University, Wageningen, The Netherlands.
Voragen, A. G. J. & Pilnik, W. (1989). In *Biocatalysis in Agricultural Biotechnology.* ed. J. R. Whitaker & Ph. E. Sonnet. ACS Symposium Series 389, American Chemical Society, Washington, DC, USA, pp. 93–115.
Voragen, A. G. J., Heutink, R. & Pilnik, W. (1980). *J. Appl. Biochem.,* **2,** 452–68.
Voragen, A. G. J., Geerst, F. & Pilnik, W. (1982). In *Use of Enzymes in Food Technology,* ed. P. Dupuy. Technique et Documentation Lavoisier, Paris, France, pp. 497–502.
Voragen, A. G. J., Schols, H. A. & Pilnik, W. (1986*a*). *Food Hydrocolloids,* **1,** 65–70.
Voragen, A. G. J., Wolters, H., Verdonschot-Kroef, T., Rombouts, F. M. & Pilnik, W. (1986*b*). International Fruit Juice Symposium, The Hague (NL), May 1986, pp. 453–62. Juris Druck + Verlag, Zurich, Switzerland.
Voragen, A. G. J., Schols, H. A., Siliha, H. A. I. & Pilnik, W. (1986*c*). In *Chemistry and Function of Pectins,* ed. M. L. Fishman & J. J. Jen. ACS Symposium Series 310. American Chemical Society, Washington, DC, USA, pp. 230–47.
Voragen, A. G. J., Rombouts, F. M., Searle-van Leeuwen, M. F., Schols, H. A. & Pilnik, W. (1987). *Food Hydrocolloids,* **1,** 423–37.
Voragen, A. G. J., Schols, H. A. & Pilnik, W. (1988). *Z. Lebensm. Unters. Forsch.,* **187,** 315–20.
Vos, L. de & Pilnik, W. (1973). *Process Biochem.,* **8**(8), 18–19.
Vries, J. A. de, Voragen, A. G. J., Rombouts, F. M. & Pilnik, W. (1986). In *Chemistry and Function of Pectins,* ed. M. L. Fishman & J. J. Jen. ACS Symposium Series 310, American Chemical Society, Washington, DC, USA, pp. 38–48.
Wegener, Ch. & Henninger, H. (1987). *Nahrung,* **31,** 297–303.
Weiss, J. & Sämann, H. (1972). *Mitt. Rebe Wein Obstbau und Früchteverwertung,* **22,** 177–84.
Wiley, R. C. & Lee, Y. S. (1970). *Food Technol.,* **24,** 1168–70.
Willaman, J. J. & Kertesz, Z. I. (1933). US Patent 1,932,833.
Wood, T. M. (1985). *Biochem. Soc. Trans.,* **13,** 407–10.
Wood, W. A. & Kellogg, S. T. (1988). *Methods in Enzymology,* **161,** 329–99.
Wucherpfennig, K. & Otto, K. (1988). *Flüssiges Obst,* **55**(3), 112–16.
Wucherpfennig, K., Dietrich, H. & Bechtel, J. (1989). *Flüssiges Obst,* **83,** 348–54.
Yamasaki, M., Yasui, T. & Arima, K. (1964). *Agric. Biol. Chem.,* **28,** 779–87.
Yamasaki, M., Kato, A., Chu, S. Y. & Arima, K. (1967). *Agric. Biol. Chem.,* **31,** 552.
Zetelaki-Horváth, K. & Gátai, K. (1977*a*). *Acta Alimentaria,* **6**(3), 227–40.
Zetelaki-Horváth, K. & Gátai, K. (1977*b*). *Acta Alimentaria,* **6,** 335–76.
Zetelaki-Horváth, K. & Vas, K. (1980). *Berichte der Wiss.-Techn. Kommission der IFU,* **XVI,** 291–308.

8

Significance of Lipoxygenase in Fruits and Vegetables

T. P. O'CONNOR[a] & N. M. O'BRIEN[b]

[a]*Department of Food Chemistry*, [b]*Department of Nutrition, University College, Cork, Republic of Ireland*

INTRODUCTION

Lipoxygenase, formerly known as lipoxidase or carotene oxidase (EC 1.13.11.12), is an iron-containing dioxygenase which catalyses the oxidation of polyunsaturated fatty acids containing *cis,cis*-1,4-pentadiene units to the corresponding conjugated *cis,trans* dienoic monohydroperoxides.

Theorell *et al.* (1947) were the first to isolate and purify the enzyme, from soy-beans. Lipoxygenase is widely distributed in nature, being found in essentially all higher plants and animals. This review is confined to plant lipoxygenases. In most sources, it exists as several isozymes (e.g. soy-bean contains three isozymes: lipoxygenase-1, -2 and -3) which can differ significantly in properties such as optimum pH, substrate specificity, end-products produced, heat stability and the ability to participate in cooxidation reactions. Two major categories of plant lipoxygenases exist: Type-1 lipoxygenases (e.g. soy-bean lipoxygenase-1), which have been reported in relatively few plants, have pH optima of 9 and little tendency to participate in co-oxidation reactions; Type 2 lipoxygenases (e.g. soy-bean lipoxygenase-2 and -3) occur widely in plants, have pH optima of 6·5 and a strong tendency to participate in co-oxidation reactions. Thus, in experimental studies involving lipoxygenase, it is critical to completely separate different isozymes and correctly identify them to ensure correct interpretation of results (Faubion & Hoseney, 1981).

Lipoxygenase isozymes have molecular weights between 95 and 100 kDa and contain one mole of iron per mole of enzyme. Although the exact subcellular location of lipoxygenase has not been fully established, there is good evidence that the isozymes are located both in organelles and in the cytoplasm of cells (Galliard, 1975, Lebedev *et al.*, 1978). Grossman *et al.* (1972*a*) reported lipoxygenase activity associated with mitochondria and chloroplasts, in addition to the particle-free supernatant from alfalfa leaf tissue. Studies involving immunofluorescence labelling of lipoxygenase in soy-beans have shown that the enzyme is randomly distributed in the cell (Vernooy-Gerritsen *et al.*, 1983*a*).

The enzyme is of interest to the food scientist primarily due to its role in the genesis of flavour and aroma compounds in plant products. Lipoxygenase-produced flavour and aroma compounds are desirable in many foods but may also give rise to off-flavours, particularly in soy-bean products.

Various aspects of lipoxygenase have been reviewed by Grosch (1972), Axelrod (1974), Eskin *et al.* (1977), Veldink *et al.* (1977), Gardner (1980), Nicolas and Drapron (1981), Vliegenthart and Veldink (1982), Klein *et al.* (1985*a*) and Mack *et al.* (1987).

SOURCES OF LIPOXYGENASE

Lipoxygenases are found in both plant and animal tissue and can be considered to be ubiquitous among eukaryotic organisms (Hildebrand, 1989). Lipoxygenase occurs widely in plant tissues, not only in seeds and legumes but also in the green parts and fruits. A comprehensive list of plant sources in which lipoxygenase has been identified is given in Table 1.

Apart from soy-bean lipoxygenases, which have been well characterized, relatively few lipoxygenases have been studied in detail. Lists of plants in which lipoxygenase has been identified have been published by Rhee and Watts (1966*a*), Holden (1970), Pinsky *et al.* (1971), Axelrod (1974), Eskin *et al.* (1977), Galliard and Chan (1980) and Sekiya *et al.* (1983). Lipoxygenase activity in 14 different legumes under identical assay conditions was reported by Chang and McCurdy (1985). They noted a high level of activity in soy-bean, lentil and cowpea, a medium level of activity in adsuki bean, faba bean, field pea, black bean, great northern bean, navy bean, pinto bean and kidney bean, and low activity in chickpea, lima bean and mung bean. Lipoxygenase has also been reported in mushrooms (*Agaricus bisporus*) (de Lumen *et al.*, 1978) and in lower forms of plants such as *Chlorella pyrenoidosa* (Zimmerman & Vick, 1973).

ISOLATION AND PURIFICATION OF LIPOXYGENASE

All lipoxygenases, except soy-bean lipoxygenase-1, are extraordinarily labile and, therefore, their purification is difficult (Schewe *et al.*, 1987). The enzyme exists in different isomeric forms in soy-beans, peas, other legumes, wheat, rice, alfalfa, barley and potato (Veldink *et al.*, 1972; Pinsky *et al.*, 1973; Axelrod, 1974; Wallace & Wheeler, 1979; Yoon & Klein, 1979; Ida *et al.*, 1983).

Theorell *et al.* (1947) first prepared crystalline lipoxygenase from soy-beans. Since then, virtually every protein purification method has been used to isolate the enzyme, ranging from conventional techniques such as ammonium sulphate

TABLE 1
Plant Sources of Lipoxygenase

Common Name	*Genus and Species*	*Reference*
Adsuki bean	*Phaseolus angularis*	Chang & McCurdy (1985)
Alfalfa	*Medicago sativa*	Grossman *et al.* (1969, 1974); Ben Aziz *et al.* (1971*a*); Wheeler & Finley (1981); Sekiya *et al.* (1983)
Apple	*Malus sylvestris*	Wooltorton *et al.* (1965); Rhee & Watts (1966*a*); Pinsky *et al.* (1971); Kim & Grosch (1979); Marcel *et al.* (1980)
Artichoke	*Cynara scolymus*	Pinsky *et al.* (1971)
Asparagus	*Asparagus officinalis*	Rhee & Watts (1966*a*); Ganthavorn & Powers (1989)
Avocado	*Persea americana*	Pinsky *et al.* (1971); Prusky *et al.* (1985); Marcus *et al.* (1988)
Banana	*Musa paradisaca*	Rhee & Watts (1966*a*); Pinsky *et al.* (1971)
Barley	*Hordeum distichum*	Yabuuchi & Amaha (1975); Baxter (1982); Schwarz & Pyler (1984)
Beetroot	*Beta vulgaris*	Baardseth & Slinde (1987)
Black bean	*Phaseolus vulgaris*	Chang & McCurdy (1985)
Blackeye pea	*Vigna sinensis*	Rhee & Watts (1966*a*)
Broad bean	*Vicia faba*	Pinsky *et al.* (1971); Beaux *et al.* (1973); Al-Obaidy & Siddiqi (1981)
Broccoli	*Brassica oleracea* var. *botrytis* f. *asparagoides*	Baardseth & Slinde (1987)
Burdock	*Arctium lappa*	Sekiya *et al.* (1983)
Bush bean	*Phaseolus vulgaris*	Hurt & Axelrod (1977)
Cabbage	*Brassica oleracea* var. *capitata*	Rhee & Watts (1966*a*); Sekiya *et al.* (1983); Baardseth & Slinde (1987)
Cape Marigold	*Dimorphotheca sinuata*	Gardner *et al.* (1973)
Carnation	*Dianthus caryophyllus*	Lynch *et al.* (1985)
Carrot	*Daucus carota*	Rhee & Watts (1966*a*); Pinsky *et al.* (1971)
Cauliflower	*Brassica oleracea* var. *botrytis*	Rhee & Watts (1966*a*); O'Reilly *et al.* (1969); Pinsky *et al.* (1971)

TABLE 1—*Contd*

Common Name	*Genus and Species*	*Reference*
Celeriac	*Apium graveoleus* var. *rapaceum*	Baardseth & Slinde (1987)
Celery	*Apium graveolens*	Rhee & Watts (1966*a*); Pinsky *et al.* (1971)
Chickpea	*Cicer arietium*	Chang & McCurdy (1985); Borthakur & Ramadoss (1986); Sosulski & Gadan (1988)
Chinese cabbage	*Brassica napus*	Sekiya *et al.* (1983)
Corn (maize)	*Zea mays*	Fritz & Beevers (1955); Rhee & Watts (1966*a*); Sekiya *et al.* (1983)
Cowpea	*Vigna anguiculata*	Den & Mendoza (1982); Chang & McCurdy (1985); N'Doye *et al.* (1985)
Cucumber	*Cucumis sativus*	Galliard & Phillips (1976); Wardale & Lambert (1980)
Eggplant	*Solanum melongena*	Rhee & Watts (1966*a*); Pinsky *et al.* (1971); Grossman *et al.* (1972*b*); Sredni & Grossman (1980); Sekiya *et al.* (1983)
Flax	*Linum usitatissimum*	Zimmerman & Vick (1970*a*); Grosch *et al.* (1976)
Garden pea	*Pisum sativum*	Siddiqi & Tappel (1956); Rhee & Watts (1966*a*); Eriksson & Svensson (1970); Pinsky *et al.* (1971); Arens *et al.* (1973); Haydar & Hadziyev (1973); Anstis & Friend (1974*a,b*); Spaapen *et al.* (1977); Reynolds & Klein (1982); Chang & McCurdy (1985); Chen & Whitaker (1986)
Garlic	*Allium sativum*	Pinsky *et al.* (1971)
Ginko	*Ginko bilboa*	Sekiya *et al.* (1983)
Gooseberry	*Ribes grossularia*	Kim & Grosch (1978)
Grape	*Vitis vinifera*	Cayrel *et al.* (1983); Zamora *et al.* (1985)
Grapefruit	*Citrus paradisi*	Pinsky *et al.* (1971)
Great northern bean	*Phaeolus vulgaris*	Chang & McCurdy (1985)
Green bean	*Phaeolus vulgaricus* var. *cascade*	Rhee & Watts (1966*a*); Pinsky *et al.* (1971); Adams & Ongley (1989)
Haricot bean	*Phaseolus vulgaris*	Kermasha & Metche (1986)

Holly	*Ilexintegra thunb*	Sekiya *et al.* (1983)
Japanese persimmon	*Diospyros kaki*	Sekiya *et al.* (1983)
Kale	*Brassica oleracea* var. *acephala*	Baardseth & Slinde (1987)
Kidney bean	*Phaseolus vulgaris*	Pinsky *et al.* (1971); Sekiya *et al.* (1983); Chang & McCurdy (1985)
Kohlrabi	*Brassica caulorapa*	Pinsky *et al.* (1971)
Leek	*Allium porrum*	Baardseth & Slinde (1987)
Lentil	*Lens culinaris*	Chang & McCurdy (1985)
Lettuce	*Lactuca sativa*	Pinsky *et al.* (1971); Sekiya *et al.* (1983); Baardseth & Slinde (1987)
Lima bean	*Phaseolus limensis*	Rhee & Watts (1966*a*); Pinsky *et al.* (1971); Chang & McCurdy (1985)
Lupin	*Lupinus albus*	Olias & Valle (1988)
Mulberry	*Morus bombycis*	Sekiya *et al.* (1983)
Mung bean	*Phaseolus aureus*	Chang & McCurdy (1985)
Navy bean	*Phaseolus vulgaris*	Chang & McCurdy (1985)
Oats	*Avena sativa*	Lascorz & Drapron (1987)
Onion	*Allium cepa*	Rhee & Watts (1966*a*); Pinsky *et al.* (1971);
Orange	*Citrus sinensis*	Rhee & Watts (1966*a*)
Parsley	*Petroselinum crispum*	Pinsky *et al.* (1971)
Peanut	*Arachis hypogaea*	Dillard *et al.* (1961); Triplett (1985)
Pear	*Pyrus communis*	Rhee & Watts (1966*a*); Kim & Grosch (1978)
Pepper	*Capsicum annuum*	Rhee & Watts (1966*a*); Pinsky *et al.* (1971); Daood & Biacs (1986)
Pinto bean	*Phaseolus vulgaris*	Sekiya *et al.* (1982); McCurdy *et al.* (1983); Chang & McCurdy (1985)
Potato	*Solanum tuberosum*	Rhee & Watts (1966*a*); Galliard & Phillips (1971); Pinsky *et al.* (1971, 1973); Sekiya *et al.* (1977, 1983); Mulliez *et al.* (1987);
Pumpkin	*Cucurbita maxima*	Pinsky *et al.* (1971); Sekiya *et al.* (1983); Hidaka *et al.* (1986)
Radish	*Raphanus sativus*	Rhee & Watts (1966*a*); Pinsky *et al.* (1971)

TABLE 1—*Contd*

Common Name	*Genus and Species*	*Reference*
Rhubarb	*Rheum rhaponticum*	Rhee & Watts (1966*a*)
Rice	*Oryza sativa*	Shastry & Rao (1975); Yamamoto *et al.* (1980); Sekhar & Reddy (1982) Sekiya *et al.* (1983); Ida *et al.* (1983); Jhon & Lee (1985); Ohta *et al.* (1986*a,b*)
Snap bean	*Phaseolus vulgaris*	Baardseth & Slinde (1987)
Soy-bean	*Glycine max*	Theorell *et al.* (1947); Pinsky *et al.* (1971); Chan (1973); Sekiya *et al.* (1983); Alsoe & Adler-Nissen (1988)
Spinach	*Spinacia oleracea*	Pinsky *et al.* (1971); Sekiya *et al.* (1983)
Squash	*Cucurbita pepo* var. *melopepo*	Rhee & Watts (1966*a*); Pinsky *et al.* (1971)
Strawberry	*Fragaria vesca*	Pinsky *et al.* (1971); Kim & Grosch (1978)
Sunflower	*Helianthus annuus*	Sekiya *et al.* (1983); Leoni *et al.* (1985)
Sweet gum	*Liquidambar styraciflua*	Sekiya *et al.* (1983)
Sweet potato	*Ipomoca batatas*	Rhee & Watts (1966*a*); Sekiya *et al.* (1983)
Tangarine	*Citrus reticulata*	Pinsky *et al.* (1971)
Tea	*Thea sinensis*	Takeo & Tsushida (1980); Hatanaka *et al.* (1982)
Tobacco	*Nicotiana tabacum*	Sekiya *et al.* (1983)
Tomato	*Lycopersicon esculentum*	Rhee & Watts (1966*a*); Pinsky *et al.* (1971); Bonnet & Crouzet (1977); Sekiya *et al.* (1983); Zamora *et al.* (1987); Daood & Biacs (1988)
Turnip	*Brassica rapa*	Rhee & Watts (1966*a*)
Watermelon	*Citrullus lanatus*	Vick & Zimmerman (1976); Sekiya *et al.* (1983)
Wheat	*Triticum vulgare*	Blain & Todd (1955); Guss *et al.* (1967)
Wheat (Durum)	*Triticum durum*	McDonald (1979)
Wheat germ	*Triticum aestivum*	Wallace & Wheeler (1975, 1979)
White clover	*Trifolium repens*	Sekiya *et al.* (1983)
Winged bean	*Psophocarpus tetragonolobus*	Den *et al.* (1982*a,b,c*); Mtebe & Gordon (1987); King & Puwastien (1987)
Zucchini	*Curcurbita pepo* var. *medullosa*	Rhee & Watts (1966*a*)

precipitation, gel filtration, ion-exchange and hydroxyapatite chromatography to more recently developed chromatographic techniques (Mitsuda *et al.*, 1967*a,b*; Allen, 1968; Catsimpoolas, 1969; Christopher *et al.*, 1970; Eriksson & Svensson, 1970; Yamamoto *et al.*, 1970; Chang *et al.*, 1971; Grossman *et al.*, 1972*b,c*; Verhue & Francke, 1972; Finazzi-Agro *et al.*, 1973; Beaux *et al.*, 1973; Anstis & Friend, 1974*b*; Arens *et al.*, 1973; Wallace & Wheeler, 1975; Sekiya *et al.*, 1978; Ramadoss & Axelrod, 1982; Vernooy-Gerritsen *et al.*, 1982; Funk *et al.*, 1985).

Isoelectric focusing techniques have been used successfully by Catsimpoolas (1969) and Funk *et al.* (1985) to achieve a high degree of purification. Soy-bean lipoxygenase-1 has been separated from lipoxygenase-2 by affinity chromatography on a column containing polyclonal antibodies directed against lipoxygenase-1 covalently coupled to Sepharose 4B (Vernooy-Gerritsen *et al.*, 1982). High performance liquid chromatography is capable of resolving lipoxygenase-1, -2 and -3 in crude extracts of ground soy-beans (Ramadoss & Axelrod, 1982).

ASSAY OF LIPOXYGENASES

This topic has been comprehensively reviewed by Grossman and Zakut (1979), Galliard and Chan (1980), Nicolas and Drapron (1981) and Schewe *et al.* (1987). An early assay for lipoxygenase involved the bleaching of carotene in a coupled reaction with oxidized unsaturated fatty acids or esters (Sumner & Sumner, 1940; Balls *et al.*, 1943; Cosby & Sumner, 1945). Ben-Aziz *et al.* (1971*b*) used Tween 80 to maintain linoleic acid and carotene in aqueous dispersion, thus facilitating the direct spectrophotometric assay of lipoxygenase activity. These coupled procedures are not ideal for routine assays because of the limited range over which activity is proportional to enzyme activity, the instability of the substrate mixture, and the poor correlation between the carotene-oxidizing activity of different lipoxygenases and their peroxidizing activity (Axelrod, 1974).

Surrey (1964) developed a spectrophotometric method for lipoxygenase assay based on the absorption at 234 nm of the hydroperoxides formed when the *cis,cis*-1,4-pentadiene-containing substrate is oxidized. This method is useful for pure lipoxygenase preparations but is not applicable if, as occurs with crude extracts of cereals, the hydroperoxides are further converted to products which do not absorb at 234 nm (Galliard, 1983). A further disadvantage of this method is the difficulty of assaying enzyme activity in turbid solutions. The procedures of Surrey (1964) and Ben Aziz *et al.* (1970, 1971*b*) use linoleic acid (C-18:2) as an assay substrate. However, its solubility in aqueous media is pH dependent. The earlier practice of increasing solubility by raising the pH to 9·0 has been discontinued since some lipoxygenase isozymes have pH optima near neutrality (Mack *et al.*, 1987). Its dispersibility can be improved by adding an emulsifier, e.g. Tween 20, to the reaction mixture.

Allen (1969) used a water-soluble substrate, octadeca-9,12-dienylsulphate, instead of linoleic acid in the spectrophotometric procedure with improved results. However, since soybean lipoxygenase-2 and -3 have low activity on this substrate it is rarely used (Bild *et al.*, 1977).

Enzyme activity can also be assayed by measuring oxygen uptake. Irvine and Anderson (1953) used a Warburg manometer to determine O_2 uptake by linoleic acid in an aqueous system containing Triton X100. However, this method is not very precise and suffers the further disadvantage that lipoxygenase is highly sensitive to denaturation on shaking.

The disadvantages of the manomeric method are avoided by use of an oxygen electrode to determine O_2 uptake. This polarographic method is particularly useful for kinetic studies and is generally the method of choice for lipoxygenase assay in crude extracts (Galliard, 1983). A potential disadvantage of this method is the uptake of O_2 due to factors other than lipoxygenase activity in crude extracts (Nicolas & Drapron, 1981).

Theoretically, lipoxygenase can also be assayed by quantifying the peroxides formed. The most commonly used procedure involves the oxidation of $Fe(CNS)_2$ to the coloured $Fe(CNS)_3$ which can be measured spectrophotometrically at 480 nm (Koch *et al.*, 1958). However, the method is not very precise and is rarely used.

A number of other colorimetric assays of lipoxygenase have been reported (Guss *et al.*, 1967; Hale *et al.*, 1969; Grossman *et al.*, 1971*a*; Verhue & Francke, 1972; de Lumen & Kazeniac, 1976). While these have not been very useful as assays of lipoxygenase activity due to their generally non-quantitative nature, they can be used to locate lipoxygenase isozymes in gel electrophoretograms.

Other proposed methods of lipoxygenase assay include a method utilizing paper impregnated with linoleic acid and *N,N*-dimethyl-*p*-phenylenediamine (de Lumen & Kazeniac, 1979) and an apparatus developed by Johns *et al.* (1973, 1974) which is capable of simultaneously measuring lipoxygenase activity by polarography and pentane production. The amount of pentane generated is proportional to the quantity of hydroperoxides formed. A sensitive assay for lipoxygenase exploits the ability of lipoxygenase to co-oxidize luminol (5-amino-2,3-dihydronaphthalazine-1,4-dione) (Lilius & Laasko, 1982).

Triplett (1985) briefly described a 'Western blot' procedure sensitive enough to detect nanogram quantities of peanut lipoxygenase and suggested that the technique would be valuable for rapidly screening peanut samples for lipoxygenase activity.

MECHANISM OF LIPOXYGENASE REACTION

Lipoxygenase was reported by Chan (1973) and Roza and Francke (1973) to contain one atom of non-haem iron per molecule. Pistorius and Axelrod (1974) confirmed this for purified soy-bean lipoxygenase-1, -2 and -3, and subsequently it was confirmed by several other investigators (de Groot *et al.*, 1975*a*; Spaapen *et al.*, 1977; Hurt & Axelrod, 1977; Diel & Stan, 1978). The presence of one iron atom per molecule is believed to be a general property of lipoxygenases and removal of iron results in the loss of enzymic activity (Klein *et al.*, 1985*a*).

There has been considerable work on the role of Ca^{2+} in stimulating lipoxygenase activity (Koch *et al.*, 1959; Koch, 1968; Holman *et al.*, 1969; Christopher *et al.*, 1972; Verhue & Francke, 1972; Dreesen *et al.*, 1982). It is thought that Ca^{2+} may influence the physical state of the fatty acid substrate in solution, particularly at low pH (Zimmerman & Snyder, 1974). Ca^{2+} have also been reported to inhibit lipoxygenase activity (Den *et al.*, 1982*b*).

Matsuda *et al.* (1978) reported that cobalt is required for activity of a lipoxygenase isolated from *Fusarium oxysporum*.

Since lipoxygenases do not contain a haem prosthetic group or an iron–sulphur cluster, the iron must be bound directly to functional groups of the enzyme. Iron appears to be involved in electron transfer during the incorporation of O_2 into unsaturated fatty acids containing *cis,cis*-1,4-pentadiene systems. Lipoxygenase must be in the oxidized (Fe^{3+}) form for oxidation reactions to proceed (Hildebrand, 1989). The oxidized form of lipoxygenase then can catalyse the stereospecific removal of hydrogen from the C-11 methylene group of linoleic (C-18:2) or linolenic (C-18:3) acid (Fig. 1, step 1).

Removal of the 11-pro-*S*-hydrogen leads to the formation of a C-13 radical and lipoxygenase is reduced to the Fe^{2+} form (Fig. 1, step 2). Under aerobic conditions, the lipoxygenase–fatty-acid radical complex usually reacts with O_2, forming a lipid peroxy radical (Fig. 1, step 3). Under anaerobic conditions, step 3 is impossible and alternative reactions occur (step 3a), releasing fatty acid radicals from the enzyme. This leads to the so-called lipohydroperoxidase reactions of lipoxygenase which yield fatty-acid dimers and oxodienoic acids (Hildebrand, 1989). Step 3a may also occur under aerobic conditions for some forms of lipoxygenase (e.g. soy-bean lipoxygenase-3 and pea lipoxygenase) and the released fatty acid radicals can react with O_2, yielding peroxy radicals (Hildebrand, 1989) which can participate in co-oxidation reactions with carotenoids and chlorophylls (Schewe *et al.*, 1986). Step 4 involves the reduction of the fatty-acid hydroperoxy radical to a hydroperoxide with the simultaneous re-oxidation of lipoxygenase to the Fe^{3+} state and the release of the fatty-acid hydroperoxide. Lipoxygenases from different sources tend to generate different relative amounts of the 9- and 13-hydroperoxides. The peroxidation specificity of lipoxygenase from different sources has been reviewed by Grosch (1982). Thus, the key feature of all lipoxygenase-catalysed reactions in plants and animals is the cleavage of a sigma bond with the formation of intermediate radicals (Hildebrand, 1989).

Absorption and fluorescence spectroscopy studies (Egmond *et al.*, 1975; de Groot *et al.*, 1975*b*) have shown the existence of three distinct forms of lipoxygenase: the ‘native’ Fe^{2+} form, the ‘yellow’ Fe^{3+} form and the ‘purple’ form which is probably a complex of the ‘yellow’ form and the hydroperoxide product.

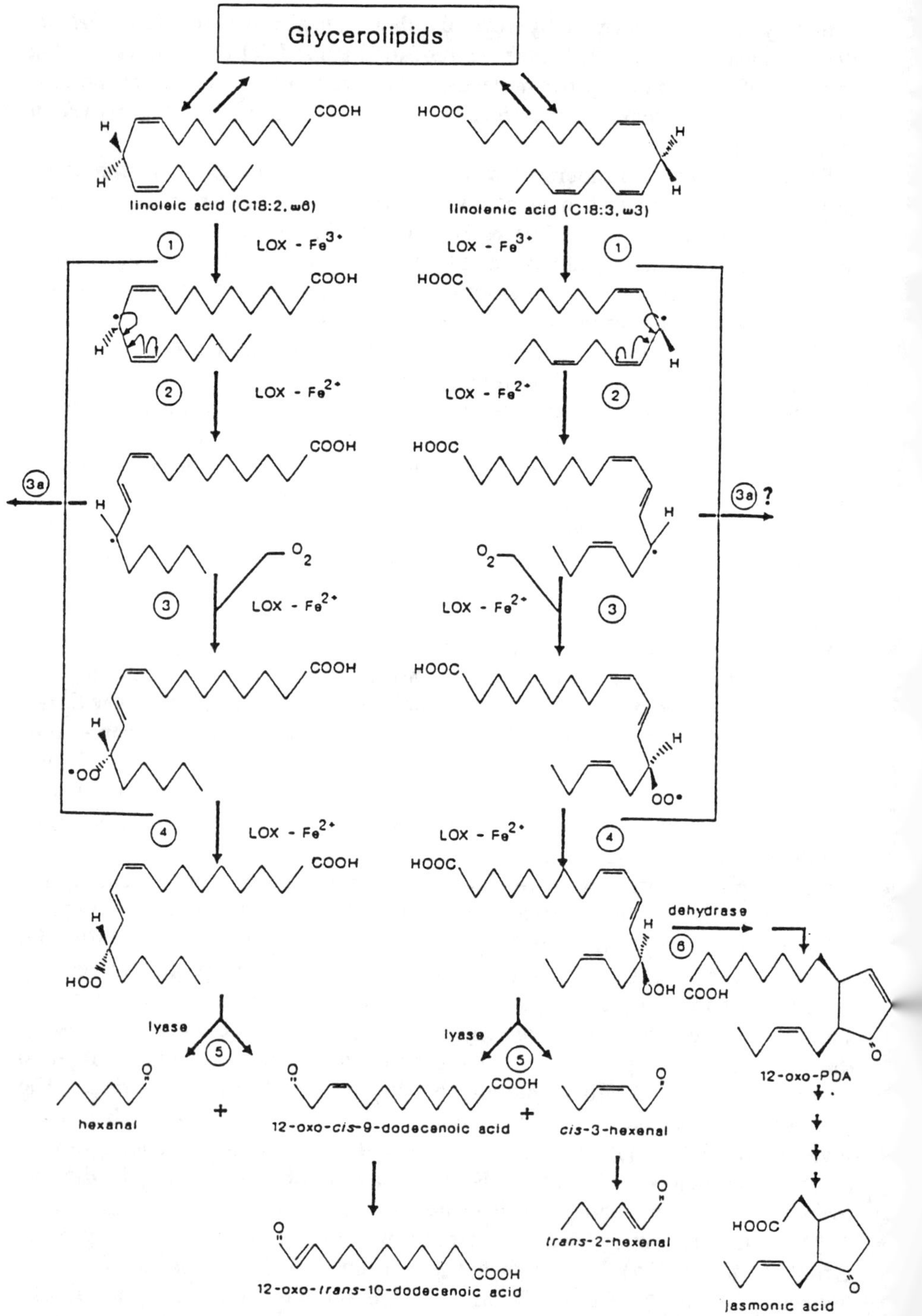

Glycerolipids
linoleic acid (C18:2, ω6)
linolenic acid (C18:3, ω3)
LOX - Fe3+
LOX - Fe2+
O2
3a
3a ?
dehydrase
6
lyase
5
hexanal
12-oxo-cis-9-dodecenoic acid
cis-3-hexenal
trans-2-hexenal
12-oxo-trans-10-dodecenoic acid
12-oxo-PDA
jasmonic acid

Two major metabolic pathways are available for the fatty-acid hydroperoxides produced via lipoxygenase action in higher plant tissues (Vick & Zimmerman, 1987*a*). The first pathway (Fig. 1, step 5) involves cleavage by hydroperoxide lyase of 13-hydroperoxy linoleic or linolenic acids into the 12-carbon compound, 12-oxo-*cis*-9-dodecenoic acid, and the 6-carbon aldehydes, hexenal or *cis*-3-hexenal. Cleavage by hydroperoxide lyase of 9-hydroperoxy fatty acids results in the formation of 9-carbon oxo fatty acids and aldehydes. The volatile products from the 9-hydroperoxides of linoleic and linolenic acids are *cis*-3-nonenal and *cis,cis*-3,6-nonadienal, respectively. The 12-oxo-*cis*-9-dodecenoic acid and *cis*-3-hexenal undergo isomerization to the more stable 12-oxo-*trans*-10-dodecenoic acid and *trans*-2-hexenal (Hatanaka *et al.*, 1987; Vick & Zimmerman, 1987*b*). The second pathway for metabolism of hydroperoxy fatty acids in plants involves hydroperoxide dehydrase (Fig. 1, step 6). The hydroperoxy fatty acids are converted to allene oxides which can undergo hydrolysis with the formation of ketols (Vick & Zimmerman, 1987*b*). Alternatively, the allene oxides may undergo rearrangement and cyclization, resulting in the formation of 12-oxo-*cis,cis*-10,15-phytodienoic acid which can be converted to jasmonic acid (Vick & Zimmerman, 1987*b*), which promotes senescence in plants (Dathe *et al.*, 1981).

An interesting feature of the kinetics of lipoxygenase action is that it is affected by enzyme-product interactions. A specific interaction between lipoxygenase and its hydroperoxide product was first noted by Haining and Axelrod (1958) in kinetic experiments. Dioxygenation of linoleic acid exhibits a marked lag period at the start of the reaction which can be abolished by adding product hydroperoxides (Smith & Lands, 1972; Sekiya *et al.*, 1977). It is important to note that the effect of the hydroperoxide is very specific: only

---

FIG. 1. The lipoxygenase (LOX) pathway. The initial event is thought to be the release of free fatty acids by lipases, but it is not known that this is always necessary. Linoleic and linolenic acids are the principal substrates for LOX. The first step of the LOX-catalyzed reaction is the stereo-specific removal of hydrogen from the C-11 methylene group (step 1). Removal of the 11-pro-*S*-hydrogen results in a rearrangement to form a free radical at C-13 (step 2) and LOX is reduced to the Fe^{2+} form. Under aerobic conditions, the LOX-fatty acid radical complex subsequently reacts with O_2, forming a lipid peroxy radical (step 3). However, step 3 cannot take place under anaerobic conditions and thus the fatty acid radical is released from the LOX and alternative reactions occur for some LOX isozymes (step 3a), at least for the linoleoyl radical and presumably for the linolenyl radical. The fatty acid radicals (formed after steps 1 and 2) and peroxy fatty acid radicals (formed at step 3) are normally bound to the LOX (intermediates bound to the LOX are enclosed in brackets). The final step of the LOX-catalyzed reaction is the reduction of the fatty acid peroxy radical to a hydroperoxide and the re-oxidation of LOX to the Fe^{3+} form and release of the fatty acid product from the enzyme (step 4). The fatty acid hydroperoxides are subsequently metabolized by hydroperoxide lyase to C-6 and C-12 products (step 5) (or C-9 products with some other LOX/lyase systems) or by hydroperoxide dehydrase C (at least in the case of hydroperoxy-linolenic acid) to 12-oxo-*cis,cis*-10,15-phytodienoic acid (12-oxo-PDA) which can be converted ultimately to jasmonic acid. (From Hildebrand (1989), reproduced by permission of the University of Lund, Sweden.)

the hydroperoxide actually produced by the particular lipoxygenase isozyme under investigation will exert this change in kinetic behaviour (Funk *et al.*, 1981). The hydroperoxide may exert its effect by oxidizing Fe^{2+} to Fe^{3+}, hence activating the enzyme.

All plant lipoxygenases studied to date readily oxidize free linoleic and linolenic acids. Some legume lipoxygenases are also capable of rapidly oxidizing methyl linoleate and trilinolein (Dillard *et al.*, 1960; Christopher *et al.*, 1970). An exhaustive study on the action of soy-bean lipoxygenase-1 on various C-18-dienoic acids having the pentadiene structure in different positions on the molecule indicated that linoleic acid (9-*cis*, 12-*cis*) was the best substrate followed by the 13-*cis*,16-*cis* isomer which was half as good (Holman *et al.*, 1969). Lipoxygenase-1 exhibited some activity on all isomers of C-18:2 from 5-*cis*,8-*cis* to 13-*cis*,16-*cis*.

INHIBITION AND INACTIVATION OF LIPOXYGENASE

Because of the nature of the enzyme, it is not surprising that lipoxygenase activity in model systems is inhibited by various antioxidants (Holman, 1947; Tappel *et al.*, 1952, 1953; Yasumoto *et al.*, 1970; Wheeler & Finley, 1981; Takahama, 1985). However, the ability of antioxidants to control lipoxygenase activity is limited by the difficulty of introducing them into plant tissue prior to disruption of the cells (Eskin *et al.*, 1977). Originally, the effect of inhibitors was attributed solely to their antioxidant properties. However, since iron is a cofactor for the enzyme, it is possible that inhibitors may also act by coordinating to iron in the enzyme.

Yasumoto *et al.* (1970) reported that in order to obtain 50% inhibition of lipoxygenase, the following concentrations of antioxidants were required: pyrocatechol, 1×10^{-3} M; homocatechol, $5{\cdot}5 \times 10^{-4}$ M; propyl gallate, $3{\cdot}2 \times 10^{-4}$ M; nordihydroguaiaretic acid, 1×10^{-5} M; resorcinol, $1{\cdot}7 \times 10^{-2}$ M; phloroglucinol, $6{\cdot}3 \times 10^{-3}$ M and hydroquinone, $5{\cdot}2 \times 10^{-3}$ M. Rhee and Watts (1966*b*) reported that butylated hydroxyanisole, propyl gallate, quercetin, gallic acid and turnip green extract inhibit lipoxygenase. Takahama (1985) also reported inhibition by quercetin while ferulic acid, isolated from lucerne extracts (Ben Aziz *et al.*, 1968) and various flavonoids and related compounds (King & Klein, 1987) are also effective inhibitors.

Adams (1989) has shown that cyanide is a weak, competitive inhibitor of green-bean lipoxygenase; it was suggested that the nitrogen of HCN forms a coordination bond with the iron at the active site of lipoxygenase, thus preventing interaction with the fatty acid substrate. However, earlier work on the inhibition of lipoxygenase by cyanide gave conflicting results. Some investigators found little or no effect (Hale *et al.*, 1969; Galliard & Phillips, 1971; Grossman *et al.*, 1972*b*; Jadhav *et al.*, 1972; Eskin & Henderson, 1974; Bonnet & Crouzet, 1977) while others reported significant inhibition (Chan, 1973; Haydar & Hadziyev, 1973; Flick *et al.*, 1975; de Lumen *et al.*, 1978; Den *et al.*, 1982*b*; Den & Mendoza, 1982; Hidaka *et al.*, 1986). Adams (1989) suggests that all of the above studies suffered from unsuitable methodology.

Hydrogen peroxide is a very potent, irreversible inhibitor of soy-bean lipoxygenase-1 (Mitsuda *et al.*, 1967*a*; Egmond *et al.*, 1975). A chemical reaction between hydrogen peroxide (H_2O_2) and the enzyme, causing structural alterations in the coordination sphere of iron, is thought to be responsible for the inhibition. The inactivation of lipoxygenase by thiols such as cysteine, 2-mercaptoethanol and glutathione has been attributed (Veldink *et al.*, 1977) to H_2O_2 since the auto-oxidation of thiols in the presence of trace amounts of metals (Cu^{2+}, Mn^{2+}, Fe^{3+}) yields H_2O_2. Veldink *et al.* (1977) state that small amounts of these metals are found even in the purest enzyme preparations.

A further class of inhibitors has been reported by Downing *et al.* (1970, 1972). These include eicosa-5,8,11,14-tetraenoic acid and octadec-9,12-dienoic acid which irreversibly inhibit soy-bean lipoxygenase at very low concentrations. The proposed mechanism involves hydrogen abstraction from the ω-8 carbon and a subsequent irreversible reaction of the allene product with the enzyme.

Lipoxygenase-produced off-flavours are a significant potential problem in soy-bean products and considerable research effort has been devoted to optimizing the conditions necessary for the inactivation of lipoxygenase in such products. Methods investigated include heating, pH adjustment and addition of antioxidants (Wilkens *et al.*, 1967; Mustakas *et al.*, 1969; Baker & Mustakas, 1973; Nelson *et al.*, 1976). However, the heat treatments used by these investigators reduced protein solubility and adversely affected the functional properties of soy products. Brown *et al.* (1982) succeeded in inactivating 99% of the lipoxygenase activity while retaining over 70% protein solubility by adjusting soy-bean cotyledon moisture to 16·3% with pH 9·8 buffer and then heating with steam for 10 s. Ediriweera *et al.* (1987) investigated the effects of various processing conditions such as soaking, moderate heat, pH and shearing action on the inactivation of soy-bean lipoxygenase. Lipoxygenase activity was reduced by homogenization together with moderate heating with good retention of protein solubility. The resistance of the lipoxygenase isozymes to inactivation by homogenization and heating was in the order L1 > L2 > L3. Total inactivation of soy-bean lipoxygenase was achieved by soaking the beans at pH 8·5 at 50°C for at least 3–4 h, followed by homogenization in water at 60°C for 15–30 min. The mechanism by which homogenization increased the susceptibility of lipoxygenase to inactivation is unknown. The shearing action may cause molecular distortions that result in the breaking of molecular bonds or stretching without breaking, thus causing inactivation (Charm & Wong, 1981).

Studies on lipoxygenase isozymes at different degrees of purification indicate that thermal inactivation follows first-order kinetics (Farkas & Goldblith, 1962; Svensson & Eriksson, 1972; Svensson, 1977; Borhan & Snyder, 1979).

Park *et al.* (1988) studied the thermal inactivation kinetics of two purified lipoxygenase isozymes from potatoes. Inactivation of the two isozymes individually followed first-order kinetics but thermal inactivation of a mixture of the isozymes was best described by second-order kinetics. It is known that enzymes in their natural environment, where substrates are often present, are generally better protected against thermal denaturation than the same enzymes in a purified state (Alsoe & Adler-Nissen, 1988).

Farkas and Goldblith (1962) reported that pea solids had a protective effect for soy-bean lipoxygenase against thermal inactivation, and Svennson and Eriksson (1972) reported similar findings for pea lipoxygenase. However, Borhan and Snyder (1979) concluded that thermal inactivation of soy-bean lipoxygenase-1 and -2 was largely unaffected by the presence of soy-bean solids. Half-lives at 69°C were 25 min and 0·7 min for purified lipoxygenase-1 and -2, respectively, and 15 min and 0·8 min, respectively, for crude preparations of lipoxygenase-1 and -2. In conflict with these findings, however, Alsoe and Adler-Nissen (1988) reported a considerable reduction in the rate of denaturation of lipoxygenase when increasing concentrations of soy-bean solids were present in the assay system. They suggest that the major factor responsible for stabilization is probably the binding of the enzyme to its natural substrates, linoleic and linolenic acids.

Lipoxygenase in soy-beans may also be inactivated by soaking them in ethanolic solutions at 25°C for 24 h (Eldridge *et al.*, 1977). Borhan and Snyder (1979) investigated the effect of a combined heat/ethanol treatment on soy-bean lipoxygenase. They found that heating for 6 h at 50°C in 15% ethanol was sufficient to destroy lipoxygenase and maintain adequate protein solubility; only 50% of the soy-bean trypsin inhibitor was destroyed. However, later work by the same group suggested that lipoxygenase may not be completely inactivated by the above conditions (Ashraf & Snyder, 1981).

The effect of microwave heating on lipoxygenase stability has been investigated in recent studies (Esaka *et al.*, 1986, 1987; Wang & Toledo, 1987). Wang and Toledo (1987) demonstrated that the duration of microwave heating needed for lipoxygenase inactivation in soy-beans is shorter and the retention of protein solubility is higher than with conventional heat processing; they suggested that the industrial applicability of the method warranted further investigation. Esaka *et al.* (1987) reported that lipoxygenase was completely inactivated in winged beans by microwave heating for 3 min with 0·5 kW power at 2450 MHz. Soaking of the seeds prior to microwave heating considerably decreased the heating time needed to inactivate the enzyme.

Acidification to pH 3·85 in combination with heat treatment is effective in inactivating soy-bean lipoxygenase as determined by the inability of a sensory panel to detect off-flavours (Kon *et al.*, 1970). The acidic slurry can be neutralized with sodium hydroxide but the resulting salt levels may be undesirable (Wolf, 1975).

Peroxidase is generally used as an index of adequate blanching in the frozen vegetables industry. However, a considerable body of work indicates that there is not a good correlation between residual peroxidase activity and the keeping quality of frozen foods (Burnette, 1977). Several workers have suggested that inactivation of lipoxygenase is a better index of adequate heat treatment than peroxidase inactivation (Lee & Wagenknecht, 1958; Eskin *et al.*, 1977; Schwimmer, 1981). According to Chen and Whitaker (1986), lipoxygenase is more important than peroxidase in the flavour deterioration of frozen green peas.

The milk protein, casein, is reported (Laakso & Lilius, 1982) to inhibit lipoxygenase activity in a model system as assayed by the chemiluminescence method of Lilius and Laakso (1982). The authors suggested that the hydrophobic nature of lipoxygenase (Allen *et al.*, 1977) might cause its spontaneous association with casein and that this 'immobilized' enzyme is inactive. The authors concluded that casein offers an alternative to the use of antioxidants and heat treatments in the stabilization of essential fatty acids in plant materials.

LIPOXYGENASE ISOZYMES

Soy-beans have been studied extensively for the presence of isozymes of lipoxygenase. There is general agreement that there are three isozymes in soy-beans (Mack *et al.*, 1987); a fourth isozyme was reported by Christopher *et al.* (1972) but many workers believe that it is in fact isozyme-3. The three isozymes in soy-beans are monomeric globular proteins with molecular weights ranging from 94 to 97 kDa and isoelectric points ranging from 5·7 to 6·3 (Hildebrand *et al.*, 1988). Each isozyme contains a single atom of non-haem iron (Chan, 1973; Pistorius & Axelrod, 1974) which is essential for enzyme activity (de Groot *et al.*, 1975*a*; Pistorius *et al.*, 1976).

Soy-bean lipoxygenase-1 has an optimum pH of 9·5 and an isoelectric point of 5·68 and is quite different from lipoxygenase-2 and -3 which have very similar isoelectric points: 6·25 and 6·15, respectively, but differ in their optimum pH. Lipoxygenase-2 has a sharp pH optimum around 6·5 while lipoxygenase-3 is quite active over the pH range 4·5–9·0. Ca^{2+} stimulate the activity of lipoxygenase-2 but not that of the other two isozymes (Koch *et al.*, 1971; Dreesen *et al.*, 1982). Collectively, lipoxygenase-2 and -3 are sometimes referred to as type 2 lipoxygenase while lipoxygenase-1 is referred to as type 1 lipoxygenase.

The amino acid composition of the three isozymes was reported by Holman *et al.* (1950), Stevens *et al.* (1970) and Pistorius (1974); the amino acid compositions of lipoxygenase-1 and -3 are very similar and somewhat different from that of lipoxygenase-2. There is controversy regarding the number of cysteine and cystine residues in the molecule with reported values of four (Mitsuda *et al.*, 1967*b*), five (Spaapen *et al.*, 1980), eight (Stevens *et al.*, 1970), nine (Diel and Stan, 1978) or twelve (Axelrod, 1974). Shibata *et al.* (1987), who reported the complete primary structure of soy-bean lipoxygenase-1, found only four cysteine residues.

There is also a lack of clarity regarding the precise structure of the active site of lipoxygenase. Downing *et al.* (1970) and Smith and Lands (1972) suggested that a histidine residue is involved in the active site of lipoxygenase based on the finding that certain acetylenic compounds irreversibly inhibit soy-bean lipoxygenase. Their hypothesis was that the acetylenic acid acted as a substrate for the enzyme, forming a highly reactive allene hydroperoxide that could undergo a covalent reaction with histidine residues at the active site. However, Kühn *et al.* (1984) presented data which contradicted the above hypothesis.

Grossman *et al.* (1971*b*) suggested that both tryptophan and tyrosine are involved in the active site since tetranitromethane and *n*-bromosuccinimide significantly inhibited lipoxygenase activity. Finazzi-Agro *et al.* (1973) suggested that type-1 lipoxygenase has a large hydrophobic site that contains 20 tryptophan residues as well as the iron atom. Klein *et al.* (1985*b*) provided evidence for two tryptophan residues in the active site of lipoxygenase-1: when both residues are oxidized the enzyme loses its ability to oxidize linoleic acid and the ability to co-oxidize β-carotene and chlorophyll is reduced greatly. Zakut *et al.* (1976) demonstrated the involvement of a methionine residue in the active site of soy-bean lipoxygenase. Iodoacetate (^{14}C labelled) alkylated a methionyl residue, resulting in total inhibition of the enzyme. Rapoport *et al.* (1984) confirmed the essential role of methionine in the active site in studies involving a mammalian lipoxygenase.

Lipoxygenase-3 is the most abundant isozyme in mature soy-beans on a protein basis; lipoxygenase-1 is almost as abundant, with lipoxygenase-2 least abundant. However, lipoxygenase-2 has the highest specific activity, so that on an activity basis there are similar amounts of the three isozymes in soy-beans (Hildebrand *et al.*, 1988). The three lipoxygenase isozymes in soy-beans have slightly different substrate–product specificities. Lipoxygenase-1 is most active on linoleic acid, while lipoxygenase-2 is more active on arachidonic acid than on linoleic acid and lipoxygenase-2 and -3 are somewhat more active on methyl linoleate than on linoleic acid (Bild *et al.*, 1977; Axelrod *et al.*, 1981). Furthermore, soy-bean lipoxygenase-1 will react with the water-soluble linoleyl sulphate while isozyme-2 and -3 show only limited activity on this substrate (Bild *et al.*, 1977).

The isozymes also differ in the products produced from linoleic acid. Lipoxygenase-1 produces the 13-hydroperoxide almost exclusively (95%), while lipoxygenase-2 produces approximately equal amounts of the 13- and 9-hydroperoxides and lipoxygenase-3 produces more 9-hydroperoxide (65%) than 13-hydroperoxide (35%) (Axelrod *et al.*, 1981). Gardner (1980), Nicolas and Drapron (1981) and Grosch (1982) comprehensively reviewed the ratio of 13- to 9-hydroperoxides produced by lipoxygenase isozymes from a wide variety of plant sources other than soy-bean. The data indicate that type-2 lipoxygenases are much more widespread in plants than type-1. Lipoxygenase-1 activity has been reported in pea seeds, but at much lower levels than lipoxygenase type-2 activity (Yoon & Klein, 1979; Reynolds & Klein, 1982).

A recent development in the study of lipoxygenase isozymes is the raising of polyclonal antibodies against purified soy-bean isozymes-1, -2 and -3 (Vernooy-Gerritsen *et al.*, 1982, 1983*b*; Yabuuchi *et al.*, 1982; Peterman & Siedow, 1985*a*). Polyclonal antibodies have been used to screen soy-bean seed collections to identify varieties lacking a particular isozyme (Hildebrand & Hymowitz, 1981; Kitamura *et al.*, 1983; Kitamura, 1984). The technique has also been used to compare lipoxygenases from soy-bean seedlings, other legumes and non-legumes to mature soy-bean lipoxygenases (Vernooy-Gerritsen *et al.*, 1982; Peterman & Siedow, 1985*a*). However, the immunological relatedness of lipoxygenases from a wide range of plant species has yet to be elucidated (Mack *et al.*, 1987).

Soy-bean seed collections have been systematically screened to identify varieties that lack each of the lipoxygenase isozymes. Varieties have been identified that lack a particular isozyme and do not produce the isozyme protein as determined by immunological analysis and the absence of the protein from gel electrophoretograms and fractions from ion-exchange columns (Hildebrand & Hymowitz, 1981; Kitamura *et al.*, 1983; Kitamura, 1984).

Genetic studies indicate that soy-bean lipoxygenase-1 and -3 are inherited independently of each other (Kitamura *et al.*, 1983). Genetic crosses have also indicated that soy-bean lipoxygenase-2 is inherited independently of lipoxygenase-3 but is closely linked to the gene for lipoxygenase-1 (Davies & Nielsen, 1986). No naturally occurring soy-bean variety lacking more than one isozyme has been identified to date (Mack *et al.*, 1987). However, using appropriate crosses, soy-bean seeds have been developed that lack either isozymes-1 and -3 or isozymes-2 and -3. Plants grown from seeds lacking two lipoxygenase isozymes showed no obvious deleterious effects when grown under glasshouse conditions (Kitamura, 1984).

Soy-bean varieties lacking some or all of the lipoxygenase isozymes should theoretically be less susceptible to lipoxygenase-mediated oxidation and the production of undesirable off-flavours. However, a recent study (Frankel *et al.*, 1988) involving soy-beans lacking lipoxygenase-1, found no significant differences in flavour quality or stability of the oil compared to that from conventional soy-beans. The authors concluded that factors other than lipoxygenase-1 appear to affect the quality of soybean oil. Engeseth *et al.* (1987) reached the same conclusion based on the results of a similar type of study. Promising results were obtained by Davies *et al.* (1987) with respect to flavour improvement of full-fat soy-flour and unblanched soy-milk by genetic removal of lipoxygenase-2; the soy-milk received lower scores for beany, rancid and oily flavour and aroma than the control. Thiobarbituric acid numbers, a measure of lipid oxidation, were lower in soy-flour from lines lacking lipoxygenase-2. However, genetic removal of lipoxygenase-1 or -3 did not result in improved flavour scores or lower TBA (thiobarbituric acid) values in this study.

The results of Davies *et al.* (1987) supported earlier work by Matoba *et al.* (1985*a*), which showed that *n*-hexanal (one of the major compounds responsible for soy-bean off-flavour) was produced at very low levels in homogenates prepared from soybean genetically deficient in lipoxygenase-2. Homogenates prepared from lipoxygenase-1 deficient, lipoxygenase-3 deficient and lipoxygenase-1 and -3 deficient soybeans generated significant amounts of hexanal.

Hildebrand *et al.* (1988) also produced data indicating that lipoxygenase-2 is the most effective enzyme in hexanal formation and therefore has the greatest impact on the flavour of soy-beans. However, their work also indicated that lipoxygenase-1 is quite effective in hexanal formation. Addition of sodium linoleate to the seed extracts resulted in the production of larger amounts of hexanal which is consistent with the observation that lipoxygenase works better on free fatty acids than on esterified fatty acids. The production of hexanal in soy-beans *in vivo* probably requires the action of three enzymes: lipase, lipoxygenase and hydroperoxide lyase (Vick & Zimmerman, 1987*a,b*). It remains to be seen, however, whether lipoxygenase is essential for the survival of cultivated soybeans and, thus, whether a totally lipoxygenase-free soy-bean, if developed, will prove to be a viable solution to overcoming the commercial problems associated with soybean lipoxygenase (Mack *et al.*, 1987).

ROLE OF LIPOXYGENASE IN THE BIOGENESIS OF FLAVOUR

A very large number of different types of compounds, volatile and non-volatile, contribute to the aroma of fruit and vegetables. Some aroma compounds are present in the intact product, being produced during the normal metabolism of the plant at concentrations above the odour thresholds and positively influence the aroma. However, many of the volatile aroma compounds are produced only when the raw fruit or vegetable is subjected to chewing, cutting or processing that allow mixing of enzymes and potential substrates which are normally compartmentalized. Only those aroma and flavour compounds resulting from lipoxygenase action are considered in this review.

Tressl *et al.* (1970*a*) and Tressl and Drawert (1973) have proposed that the lipoxygenase-catalysed formation of hydroperoxides from linoleic and linolenic acids, followed by a cleavage of the hydroperoxides by a lyase, is the pathway involved in production of many different volatile carbonyl compounds important in the flavour of fruits and vegetables. Galliard *et al.* (1976), Galliard and Phillips (1976), Galliard *et al.* (1977), Galliard and Matthew (1977), Matthew and Galliard (1978), Phillips and Galliard (1978) and Phillips *et al.* (1979) provided evidence to support the above hypothesis by showing that at least three enzymes (lipoxygenase, hydroperoxide lyase and a *cis*-3,*trans*-2-enal isomerase) are involved in the formation of hexanal and *cis*-3,*trans*-2-nonenal from linoleic acid. Further support for the involvement of lipoxygenase and hydroperoxide lyase in the biogenesis of volatiles has been provided in watermelon seedlings (Vick & Zimmerman, 1976), tea chloroplasts (Hatanaka *et al.*, 1979), alfalfa and cucumber seeds (Sekiya *et al.*, 1979), pears (Kim & Grosch, 1981) and apples and tomatoes (Schreier & Lorenz, 1982). The differences in the type and proportions of volatile compounds produced have an impact on the distinctive flavour and aroma of a particular fruit or vegetable and are due to the nature and properties of the enzymes involved in the oxidative degradation of linoleic and linolenic acids.

Tomatoes

The characteristic flavour and aroma of tomato are produced by a large group of volatile compounds including aldehydes, ketones and alcohols (Yu *et al.*, 1967; Dalal *et al.*, 1968). The principal volatiles produced from the breakdown of fatty acids in tomatoes are hexanal and *cis*-hex-3-enal (Kazeniac & Hall, 1970; Jadhav *et al.*, 1972; Stone *et al.*, 1975). Tomato lipoxygenase, which has been purified and characterized (Zamora *et al.*, 1987), forms 9-hydroperoxides predominantly. However, only the minor 13-hydroperoxides are susceptible to the action of the tomato hydroperoxide lyase and are converted to the C-6 aldehydes (Galliard & Matthew, 1977; Schreier & Lorenz, 1982). The metabolic fate of the 9-hydroperoxides is unknown. An isomerase can transform *cis*-hex-3-enal to *trans*-hex-2-enal, although the level of this isomerase in tomatoes is low (Schreier, 1981). Alcohol dehydrogenase can then reduce these aldehydes to the corresponding alcohols.

Schreier *et al.* (1977) reported that the formation of hexanal in tomato pulp was reduced by 50–70% by heating. They concluded that only a part of the hexanal resulted from the heat-labile lipoxygenase action while the remainder was formed by the action of some heat-stable factor. Furthermore, the occurrence in tomatoes of a wide variety of carbonyl compounds with structures known to be generated during auto-oxidation or light-induced oxidation suggests that non-specific oxidation pathways are involved in the development of flavour volatiles (Schormüller & Kochman, 1969; Schreier *et al.*, 1977; Tressl *et al.*, 1981).

Bananas

Goldstein and Wick (1969) reported that the lipid content of bananas is 0·2–0·5% of the fresh weight with the major fatty acids being palmitic, oleic and linoleic acids. A three-fold decrease in the linoleic acid content was observed on ripening, suggesting a possible correlation between the oxidation of this fatty acid by lipoxygenase and genesis of volatiles. The role of unsaturated fatty acids as precursors of volatile compounds in bananas was confirmed by Tressl and Drawert (1973). Incubation of ^{14}C-linoleic and ^{14}C-linolenic acids with banana homogenates resulted in the incorporation of the radioactive label into volatile compounds. The concentrations of 16- and 18-carbon fatty acids decrease following the climacteric rise in respiration which is also the stage of maximum formation of aroma compounds (Tressl *et al.*, 1970*b*).

The spectrum of volatile compounds isolated from bananas depends on the stage of ripeness of the fruit. Green bananas contained *trans*-2-nonenal, *trans*-2,*cis*-6-nonadienal and 9-oxonanoic acid. However, bananas treated with ethylene and stored for four days at 15°C contained hexanal, *trans*-2-hexenal and 12-oxo-*trans*-10-dodecenoic acid (Tressl & Drawert, 1973). Hultin and Proctor (1961) had earlier identified 2-hexenal as the principal carbonyl in banana volatiles, reaching its maximum concentration in the ripe fruit. *Trans*-2-hexenal is also present in banana volatiles (Issenberg & Wick, 1963; Tressl & Jennings, 1972).

Apples

Apples have a low lipid content (e.g. 947 mg of lipid per kg of Cox's Orange Pippin apple pulp) (Galliard, 1968). Lipase activity apparently increases with the respiration climacteric and this could increase the levels of free fatty acids (Rhodes & Wooltorton, 1967). Lipoxygenase has been extracted from Golden Delicious apples and partly purified (Kim & Grosch, 1979). Feys *et al.* (1980) noted lipoxygenase activity in Schone von Boskoop apples, with highest activity in the core and peel. Hexanal and *trans*-hex-2-enal are the principal products of the action of lipoxygenase on polyunsaturated fatty acids in apples; lesser amounts of hexan-1-ol, *trans*-hex-2-en-1-ol, *cis*-hex-3-enal and *cis*-hex-3-en-1-ol are also formed (Flath *et al.*, 1967; Drawert *et al.*, 1973; Schreier *et al.*, 1978).

Cucumber and melon

The principal aroma compounds in cucumber are C-9 volatiles although C-6 volatiles are also present. Studies using ^{14}C linoleic and linolenic acids showed that these are the precursors of the C-9 compounds (Grosch & Schwarz, 1971). These volatiles are not found in intact cucumber tissue but cutting or chewing of the tissue induces the production of C-9 carbonyl compounds, the most important of which is *trans*-2, *cis*-6-nonadienal which provides most of the characteristic aroma of cucumber (Forss *et al.*, 1962; Fleming *et al.*, 1968). In melons, nona-*cis,cis*-3,6-dienol and nona-*cis*-6-enal are reported to be the key aroma volatiles (Kemp *et al.*, 1972, 1974).

Galliard and Phillips (1976) and Galliard *et al.* (1976) showed that at least three enzymes are involved in the genesis of volatiles from fatty-acid precursors in cucumber tissue: lipoxygenase, hydroperoxide lyase and *cis*-3, *trans*-2-enal isomerase. The membrane-bound hydroperoxide lyase from cucumber has been partially purified (Phillips & Galliard, 1978) and appears to be similar to the enzyme found in watermelon seedlings (Vick & Zimmerman, 1976). It has been reported (Wardale & Lambert, 1980) that cucumber peel contains twice as much lipoxygenase activity as the flesh.

Tea

The flavour and aroma of tea are due to a wide range of compounds; lipoxygenase activity contributes to the overall flavour. It has been known for a considerable time that macerated, fresh tea leaves produce large quantities of C-6 aldehydes in the presence of oxygen (Nye & Spoehr, 1943; Major *et al.*, 1963; Wright and Fishwick, 1979). Several authors have shown that hexanal, *cis*-3-hexenal, *trans*-2-hexenal and *cis*-3-hexen-1-ol are generated from linoleic and linolenic acids in macerated tea leaves (Gonzales *et al.*, 1972; Saijyo & Takeo, 1972; Hatanaka & Harada, 1973). An enzyme system which degrades linoleic acid to hexanal and linolenic acid to *cis*-3-hexenal and, ultimately, to *trans*-2-hexenal was reported in tea leaf chloroplasts (Hatanaka *et al.*, 1976; Sekiya *et al.*, 1976). These workers were unable to isolate the enzymes at that time but the same research group (Hatanaka *et al.*, 1979) subsequently detected hydroperoxylinoleic acid in tea and the 13-hydroperoxide appeared to be the predominant form (Kajiwara *et al.*, 1980). A hydroperoxide lyase was partially purified from tea chloroplasts by Hatanaka *et al.* (1982). The current hypothesis is that C-6 volatile compounds are generated in tea chloroplasts by an enzyme system consisting of lipoxygenase and hydroperoxide lyase (Hatanaka *et al.*, 1982). The activity of lipoxygenase in tea leaves increases after plucking (Takeo & Tsushida, 1980) and there is a further increase on subsequent dehydration. Lipoxygenase activity is higher in leaves used for making high quality black tea than in those used for lower quality products.

OFF-FLAVOUR PRODUCTION

Off-flavour production is a potential problem in legumes such as soy-beans (Wilkens *et al.*, 1967; Arai *et al.*, 1970; Mattick & Hand, 1969; Rackis *et al.*, 1970, 1972; Kalbrenner *et al.*, 1971; Hildebrand & Kito, 1984), peas (Siddiqi & Tappel, 1956; Lee & Wagenknecht, 1958; Bengtsson & Bosund, 1966; Murray *et al.*, 1968; Haydar & Hadziyev, 1973; Haydar *et al.*, 1975) and lentils (Kon *et al.*, 1970). Maceration of the legumes exacerbates the problem (Mattick & Hand, 1969).

Much of the interest in lipoxygenase in soy-beans is due to its role in the production of volatile flavour and aroma compounds. Lipoxygenase represents about 1–2% of the protein in soy-beans (Axelrod *et al.*, 1981) and soy-bean oil contains approximately 55% linoleic (C-18:2) and 8% linolenic acids (C-18:3), which are substrates for lipoxygenase. Volatile degradation products of these fatty acids have been associated with grassy-beany and rancid off-flavours in soy-beans (Dutton *et al.*, 1951; Mattick & Hand, 1969; Rackis *et al.*, 1979) which reduce the consumer acceptability of soy-bean products. Flavour problems can occur in soy-bean protein as well as oil products because undesirable compounds can bind covalently to proteins during processing (Gardner, 1979; Rackis *et al.*, 1979). Heat treatment is widely used to inactivate lipoxygenase in soy-beans, but it may adversely affect the functional properties of the protein and can itself affect the flavour of soy-bean products.

The initial products of lipoxygenase activity are hydroperoxides which then undergo rapid degradation to a variety of products including several aldehydes, ketones and alcohols, many of which have low flavour thresholds (Eskin *et al.*, 1977; MacLeod & Ames, 1988). While lipoxygenase-generated volatile compounds generally cause off-flavours in legumes, the same compounds are often responsible for the desired flavour characteristics of many fruits.

Arai *et al.* (1970) reported that the major volatile compounds that affect the quality of soy-bean products are *n*-hexanal, *n*-hexanol, *n*-pentanol and *n*-heptanol. Grosch & Schwencke (1969) reported similar findings. Fujimaki *et al.* (1965) showed that hexanal is primarily responsible for the green-bean-like flavour of defatted soy-flour owing to its extremely low flavour threshold.

Lipoxygenase-2 is the isozyme primarily responsible for the generation of *n*-hexanal in soy-beans (Matoba *et al.*, 1985*a*). The same group (Matoba *et al.*, 1985*b*) has also isolated a soy-bean hydroperoxide lyase which is involved in *n*-hexanal formation.

Non-volatile compounds produced by lipoxygenase may also contribute to off-flavour in soy-beans. These arise from the action of peroxidase or hydroperoxide isomerase on lipoxygenase-generated 9- and 13-hydroperoxides giving rise to epoxy-hydroxy unsaturated fatty acids. These are then hydrolysed to produce, in the case of linoleic acid, the bitter compounds 9,12,13-trihydroxy-octadec-10-enoic acid and 9,10,13-trihydroxy-octadec-11-enoic acid (Baur *et al.*, 1977; Moll *et al.*, 1979). Evans *et al.* (1960) have also suggested that bitterness in soy-bean oil may be due in part to fatty acid dimers formed from lipoxygenase-generated hydroperoxides.

The principal approaches to the reduction of off-flavour development in soy-bean products are inactivation and inhibition of lipoxygenase and the development of cultivars deficient in lipoxygenase isozymes. These approaches were discussed earlier.

Wagenknecht (1959) attributed off-flavours in sweet corn to lipoxygenase activity while Koch *et al.* (1959) implicated the enzyme in off-flavour development in cottonseed oil.

Lipoxygenase-generated *cis*-3-hexenal, *n*-hexanal and *trans*-2-hexenal contribute to the characteristic flavour of fresh tomatoes. These compounds were reported to improve the flavour of canned tomato juice when added at low levels but when they were added at levels above the recommended ranges they resulted in the development of 'green' and rancid off-flavours (Kazeniac & Hall, 1970).

It is essential that green peas (*Pisum sativum*) are blanched and frozen rapidly after harvesting to avoid the development of lipoxygenase-generated off-flavours. While many volatile carbonyl compounds have been detected in green peas, no one compound has been specifically implicated in off-flavour development (Pattee *et al.*, 1982). However, lipoxygenase is clearly involved in off-flavour development since the addition of purified lipoxygenase to blanched peas produces the characteristic off-flavour (Williams *et al.*, 1986).

Winged beans (*Psophocarpus tetragonolobus*) are a promising agricultural crop for the humid tropics where soy-beans do not flourish. However, products prepared from winged beans often have a strong beany off-flavour. Mtebe & Gordon (1987) showed that the principal volatile components responsible for this flavour are hexanal and 2-pentyl furan. Mature winged beans possess lipoxygenase at levels comparable to soy-beans (Den *et al.*, 1982*a*). These authors purified two major isozymes of lipoxygenase from winged beans; both isozymes were very thermostable when present in crude extracts or *in situ* (Den *et al.*, 1982*b*).

Under certain conditions, potatoes can develop a cucumber-like off-flavour (Samotus & Schwimmer, 1963). The action of lipoxygenase on linolenic acid in potatoes produces the volatile *trans*-2, *cis*-6-nonadienal which is thought to be responsible for the off-flavour (Galliard, 1973).

Lipoxygenase-generated off-flavour is a potential problem in wines. Hexanal, *cis*-3-hexenal and *trans*-2-hexenal are particularly undesirable in terms of the organoleptic quality of wine since they impart a 'grassy' taste even at very low concentrations (Cordonnier & Bayonove, 1978; Cayrel *et al.*, 1983; Mesias *et al.*, 1983). These C-6 volatile compounds are formed during the crushing and pressing of grapes to release the juice for fermentation. During fermentation, however, the C-6 aldehydes are almost completely metabolized by the yeast to their corresponding C-6 alcohols. The volatile alcohols have a much higher detection threshold and do not constitute the same potential problem as the volatile aldehydes.

Cayrel *et al.* (1983) and Crouzet *et al.* (1985) showed that three enzymes are necessary to generate C-6 volatile alcohols from fatty acid precursors in grapes: lipoxygenase, hydroperoxide lyase and alcohol dehydrogenase. Grape lipoxygenase produces mainly (80%) 13-hydroperoxides and 20% 9-hydroperoxides (Cayrel *et al.*, 1983). The hydroperoxide lyase has been characterized by Valentin & Crouzet (1989). It is active only on 13-hydroperoxides and does not cleave 9-hydroperoxides, as is true also for tomato hydroperoxide lyase (Galliard & Matthew, 1977).

CO-OXIDATION REACTIONS OF LIPOXYGENASE

As outlined previously, there are two distinct groups of plant lipoxygenases: types 1 and 2. Type 1 lipoxygenase has been reported in relatively few plants and has optimum activity around pH 9 with little tendency to cause co-oxidation of other compounds such as carotenoids or chlorophyll (Weber *et al.*, 1974; Grosch *et al.*, 1977*a*; Reynolds & Klein, 1982). However, under anaerobic conditions, type 1 lipoxygenase is a powerful pigment bleaching agent (Klein *et al.*, 1984). Type 2 lipoxygenase occurs widely in plants, has optimum activity around pH 6·5 and has a strong tendency to catalyse the co-oxidation of other compounds (Grosch *et al.*, 1977*a*; Weber *et al.*, 1974). For example, the carotene oxidase activity of soy-bean is associated with the type 2 lipoxygenases only (Kies *et al.*, 1969). Substrates for co-oxidation reactions include carotene (Sumner & Sumner, 1940), bixin (Sumner & Smith, 1947), chlorophyll (Yoon & Klein, 1979), crocin (Weber *et al.*, 1974), lutein (McDonald, 1979) and dyes (Ikediobi, 1977). One technological application of the ability of lipoxygenase to co-oxidize plant pigments is in flour bleaching (Drapron & Uzzan, 1968). However, that application is outside the scope of this review.

It should be emphasized that in co-oxidation reactions both the lipoxygenase substrate (fatty acids) and the substance being co-oxidized must be present initially in the reaction system. The hydroperoxide products of fatty acids do not oxidize the co-substrate.

Carotenoids

Sumner & Sumner (1940) first noted the co-oxidation of carotenoids in a lipoxygenase system. Kies (1947) reported that the carotene bleaching capacity was susceptible to heat destruction while the lipid oxidation capacity was not. This discrepancy in activities was attributed (Kies *et al.*, 1969) to the existence of soy-bean isozymes differing in their heat lability and carotene bleaching activity. Different co-oxidation potentials of isozymes from wheat (McDonald, 1976) and peas (Arens *et al.*, 1973) have also been reported.

Several studies have clearly demonstrated that type-2 lipoxygenases are responsible for the co-oxidation reactions involving carotenes, canthaxanthin and crocin, when linoleic acid and oxygen are present (Weber *et al.*, 1973; Fischer & Grosch, 1977; Grosch *et al.*, 1977*a*). Hidaka *et al.* (1986) showed that the lipoxygenase isozyme in pumpkins involved in carotene co-oxidation was similar to soy-bean lipoxygenase-3; lipoxygenase type-1 showed little co-oxidation activity. However, Ikediobi and Snyder (1977) and Hildebrand and Hymowitz (1982) suggested that soy-bean lipoxygenase-1 did play a role in the co-oxidation of carotene and chlorophyll.

Type-1 lipoxygenase shows little co-oxidation activity under aerobic conditions, perhaps, because the peroxyl radicals formed by enzyme action are converted to hydroperoxides so rapidly that they cannot react with polyenes (Grosch *et al.*, 1976). Weber *et al.* (1974) and Weber and Grosch (1976) suggested that the high co-oxidation potential of type-2 lipoxygenases is due to the fact that a large proportion of the peroxyl radicals is not directly converted to hydroperoxides by this isozyme. The peroxyl radical would therefore abstract a hydrogen atom from the carotenoid producing a polyene radical which would then combine with oxygen to produce a carbonyl compound. Klein *et al.* (1985*a*) speculated that type-2 lipoxygenases are able to orientate carotenoids so that the peroxyl radical can attack the polyenes directly.

An alternative hypothesis for the variation in co-oxidizing potential of lipoxygenase isozymes is that the stability of the enzyme–fatty-acid radical complexes, which release free radicals, may vary (Veldink *et al.*, 1977).

There is very little information on the active site of lipoxygenase involved in co-oxidation reactions. Spaapen *et al.* (1980) reported that modification of type

1-lipoxygenase with methylmercuric halides increased the rate of crocin co-oxidation at pH 10, and they suggested that the modified type-1 enzyme behaves more like a type-2 lipoxygenase. Similar results were reported by Grossman *et al.* (1984). These results suggest that free sulphydryl groups may be involved in the active site for co-oxidation reactions.

Kanofsky and Axelrod (1986) presented evidence for the production of singlet oxygen by soy-bean lipoxygenase-3 and suggested that singlet oxygen is involved in the co-oxidation of carotenes. Lipoxygenase-3 is also the most active isozyme for carotene co-oxidation. However, they stated that the production of singlet oxygen demonstrated in their study was insufficient to explain all the destruction of β-carotene observed in co-oxidation studies. Lipoxygenase-1 produced detectable quantities of singlet oxygen only under anaerobic conditions which correlates with the finding that β-carotene is co-oxidized by this isozyme only under anaerobic conditions (Klein *et al.*, 1984). However, Kanofsky and Axelrod (1986) were of the opinion that the most likely mechanism of β-carotene co-oxidation is the one involving peroxy radicals.

Some workers believe that synergistic effects between soy-bean type-1 and type-2 lipoxygenases are necessary for optimum co-oxidation of carotene (Ramadoss *et al.*, 1978). Using methyl linoleate as substrate, lipoxygenase-2 or lipoxygenase-3 individually were not good carotene co-oxidizers. However, combinations of lipoxygenase-1 and -3, or lipoxygenase-2 and -3 were much more effective in bleaching carotene. They also found that 13-hydroperoxide in combination with lipoxygenase-3 was effective in bleaching carotene. Their hypothesis was that lipoxygenase-3 on its own is a poor co-oxidizer of carotene since it could not generate sufficient 13-hydroperoxide to promote co-oxidation. However, combination with lipoxygenase-1 or -2 overcame this problem.

However, Grosch and Laskawy (1979) reported conflicting findings. Both soy-bean lipoxygenase-2 and -3 individually co-oxidized carotene or canthaxanthine as effectively as a combination of the two isozymes and they were unable to confirm the requirement for 13-hydroperoxide for co-oxidation.

Chlorophyll

Bleaching of chlorophyll generally occurs at a slower rate than that of carotenoids in plant extracts and tissues. Wagenknecht *et al.* (1952) and Wagenknecht and Lee (1956) first attributed chlorophyll destruction to lipoxygenase. Holden (1965, 1970) reported that heat treatment inactivated, and antioxidants inhibited, the chlorophyll-bleaching activity of legume seed extracts. She suggested that bleaching of chlorophyll occurred during decomposition of the lipid hydroperoxides and was not due directly to lipoxygenase action.

Zimmerman and Vick (1970*b*) noted that lipoxygenases purified from soy-beans or flax seeds would not bleach chlorophyll although crude extracts from these sources, which also contained hydroperoxide isomerase (which isomerizes hydroperoxides to the corresponding ketohydroxy compounds), did bleach chlorophyll. Bleaching was thought to result from a redox reaction between the ketohydroxy fatty acid and chlorophyll.

A study (Abbas *et al.*, 1989) showed that type-2 lipoxygenases isolated from fresh beans had good chlorophyll bleaching properties which were stimulated by the presence of hydroperoxides. The same result was observed for the co-oxidation of canthaxanthine (Grosch & Laskawy, 1979). These findings suggest that hydroperoxides promote chlorophyll bleaching via activation of lipoxygenase (Grosch & Laskawy, 1979).

Fat-soluble vitamins

Tappel *et al.* (1953) were the first to show that α-tocopherol is destroyed during the lipoxygenase-catalysed oxidation of linoleic acid. Vitamins A (retinol) and D do not occur in plant tissue but they may be present in food products formulated from both plant and animal tissues. Thus, a recent study has investigated the co-oxidation of vitamins A, D and E by soy-bean lipoxygenase (Gordon & Barimalaa, 1989). Vitamins A, D_2, D_3 and E were all susceptible to rapid co-oxidation at neutral pH in the presence of soy-bean lipoxygenase. Retinal appeared to be more stable towards co-oxidation than retinol, which showed similar stability to β-carotene. The products from the co-oxidation of β-carotene indicate that oxidation occurs at several positions along the polyene chain (Grosch *et al.*, 1977*b*). However, since vitamin D, with only three conjugated double bonds, was oxidized at quite a fast rate it is clear that extensive conjugation is not required for co-oxidation (Gordon & Barimalaa, 1989).

POSSIBLE PHYSIOLOGICAL FUNCTIONS OF LIPOXYGENASE IN PLANTS

Despite considerable speculation on the physiological role of lipoxygenase in plant tissue, little definitive information is available. The topic has been extensively reviewed (Mack *et al.*, 1987; Vick & Zimmerman, 1987*b*; Hildebrand *et al.*, 1988).

Lipoxygenase has been postulated to be involved in

(1) plant growth and development, since it may provide substrates for other metabolically important enzymes;
(2) senescence;
(3) wound response and resistance to pests.

It has also been postulated that lipoxygenase has a role in alternative cyanide-insensitive respiration, but available evidence suggests that this is unlikely (Vick & Zimmerman, 1987*b*).

Lipoxygenase-mediated lipid peroxidation has been postulated to be responsible for the membrane damage observed during plant infection (Lupu *et al.*, 1980), wounding (Galliard, 1978) and senescence (Grossman & Leshem, 1978; Leshem *et al.*, 1979). Galliard (1978) reported that lipoxygenase activity increases in wounded or infected potato tubers and Lupu *et al.* (1980) showed that lipoxygenase activity increased by four to sevenfold in tobacco leaves infected with *Erysiphe cichoracearum*.

Lipid hydroperoxy radicals and some of the secondary oxidation products resulting from lipoxygenase action are toxic to plant pests (Hildebrand, 1989; Hildebrand *et al.* 1986). Furthermore, lipoxygenase catalyses the production of traumatin (12-oxo-*trans*-10-dodecenoic acid) in wounded plant tissue. Autooxidation of traumatin results in the formation of traumatic acid. Both traumatin and traumatic acid are involved in wound healing in plant tissues (Zimmerman & Caudron, 1979) and thus reduce the entry of plant pathogens through the wounds.

Lipoxygenase has also been implicated as playing a role in the senesence of plant tissues (Grossman & Leshem, 1978; Leshem *et al.*, 1979, 1981; Leshem & Barness, 1982; Leshem, 1984*a,b*). These studies provide evidence of a reasonable correlation between lipoxygenase activity and plant senescence. It has been speculated that the increased lipoxygenase activity during senescence results in an increased rate of free radical production with a subsequent increase in the destruction of cell membranes and proteins and the inhibition of photochemical activity in chloroplasts. However, Dhindsa *et al.* (1982), in a study utilizing salicylhydroxamic acid to inhibit lipoxygenase activity, clearly showed that the lipid peroxidation that accompanies senescence in *Avena sativa* L. and *Rumex obtusifolius* L. leaves was not mediated by lipoxygenase.

Another potential role of lipoxygenase in senescence is in the synthesis of jasmonic acid (see Fig. 1) which has been reported to promote plant cell senescence and inhibit growth (Dathe *et al.*, 1981; Vick & Zimmerman, 1984, 1987*b*). Lipoxygenase may be involved in the production of the plant hormone, ethylene (Wooltorton *et al.*, 1965; Legge & Thompson, 1983; Bousquet & Thimann, 1984). Ethylene initiates fruit ripening and regulates many aspects of plant growth, development and senescence. Although most evidence does not support a role for lipoxygenase in the synthesis of ethylene (Pirrung, 1986; Mack *et al.*, 1987; Wang & Yang, 1987), one report does indicate that the hydroperoxides produced by microsomal-associated lipoxygenase from carnations facilitate the conversion of the ethylene precursor 1-aminocyclopropane-1-carboxylic acid to ethylene *in vitro* (Lynch *et al.*, 1985). However, the authors stated that the in-vitro reaction is not reflective of the situation *in vivo*.

Several investigators have observed an increase in lipoxygenase activity in plants during periods of active growth, especially during seed germination and early seedling growth (Guss *et al.*, 1968; Oelze-Karow *et al.*, 1970; Hildebrand & Hymowitz, 1983; Peterman & Siedow, 1985*b*; Ohta *et al.*, 1986*a*). However, the role, if any, of lipoxygenase during seed germination and establishment is unknown.

REFERENCES

Abbas, J., Rouet-Mayer, M. A. & Lauriere, C. (1989). *Phytochem.*, **28,** 1019–24.
Adams, J. B. (1989). *Food Chem.*, **31,** 243–50.
Adams, J. B. & Ongley, M. H. (1989). *Food Chem.*, **31,** 57–71.
Allen, J. C. (1968). *Eur. J. Biochem.*, **4,** 201–8.
Allen, J. C., (1969). *J. Chem. Soc.*, **16,** 906–7.
Allen, J. C., Eriksson, C. & Galpin, J. R. (1977). *Eur. J. Biochem.*, **73,** 171–7.
Al-Obaidy, H. M. & Siddiqi, A. M. (1981). *J. Food Sci.*, **46,** 622–5, 629.
Alsoe, K. O. & Adler-Nissen, J. (1988). *Lebensm.-Wiss. Technol.*, **21,** 8–12.
Anstis, P. J. P. & Friend, J. (1974*a*). *Phytochem.*, **13,** 567–73.
Anstis, P. J. P. & Friend, J. (1974*b*). *Planta,* **115,** 329–35.
Arai, S., Nogushi, M., Kaji, M., Kato, H. & Fujimaki, M. (1970). *Agric. Biol. Chem.*, **34,** 1420–23.
Arens, D., Seilmeier, W., Weber, F., Kloos, G. & Grosch, W. (1973). *Biochim. Biophys. Acta,* **327,** 295–305.
Ashraf, H. R. L. & Snyder, H. E. (1981). *J. Food Sci.*, **46,** 1201–4.
Axelrod, B. (1974). In *Food Related Enzymes,* ed. J. R. Whitaker. American Chemical Society, Washington DC, USA, pp. 324–48.
Axelrod, B., Cheesbrough, T. M. & Laasko, S. (1981). In *Methods in Enzymology* (*Vol. 71*), ed. J. M. Lowenstein. Academic Press, New York, USA, pp. 441–51.
Baardseth, P. & Slinde, E. (1987). *Norwegian J. Agric. Sci.*, **1,** 111–17.
Baker, E. C. & Mustakas, G. C. (1973). *J. Amer. Oil Chem. Soc.*, **50,** 137–41.
Balls, A. K., Axelrod, B. & Kies, M. W. (1943). *J. Biol. Chem.*, **149,** 491–504.
Baur, C., Grosch, W., Wieser, H. & Jugel, H. (1977). *Z. Lebensm. Unters. Forsch.*, **164,** 171–6.
Baxter, E. D. (1982). *J. Inst. Brewing,* **88,** 390–6.
Beaux, Y., Drapron, R., Nicolas, J. & Caillat, J. M. (1973). *Biochimie,* **55,** 253–62.
Ben Aziz, A., Grossman, S., Budowski, P., Ascarelli, I. & Bondi, A. (1968). *J. Sci. Food Agric.*, **19,** 605–8.
Ben Aziz, A., Grossman, S., Ascarelli, I. & Budowski, P. (1970). *Anal. Biochem.*, **34,** 88–100.
Ben Aziz, A., Grossman, S., Budowski, P. & Ascarelli, I. (1971*a*). *Phytochem.*, **10,** 1823–30.
Ben Aziz, A., Grossman, S., Ascarelli, I. & Budowski, P. (1971*b*). *Phytochem.*, **10,** 1445–52.
Bengtsson, B. & Bosund, I. (1966). *J. Food Sci.*, **31,** 474–81.
Bild, G. S., Ramadoss, C. S. & Axelrod, B. (1977). *Lipids,* **12,** 732–5.
Blain, J. A. & Todd, J. P. (1955). *J. Sci. Food Agric.*, **6,** 471–9.
Bonnet, J. L. & Crouzet, J. (1977). *J. Food Sci.*, **42,** 625–8.
Borhan, M. & Snyder, H. E. (1979). *J. Food Sci.*, **44,** 586–90.
Borthakur, A. & Ramadoss, C. S. (1986). *J. Agric. Food Chem.*, **34,** 1016–18.
Bousquet, J. F. & Thimann, K. V. (1984). *Proc. Nat. Acad. Sci. USA,* **81,** 1724–7.
Brown, B. D., Wei, L. S., Steinberg, M. P. & Villota, A. (1982). *J. Amer. Oil Chem. Soc.*, **59,** 88–92.
Burnette, F. S. (1977). *J. Food Sci.*, **42,** 1–6.
Catsimpoolas, N. (1969). *Arch. Biochem. Biophys.*, **136,** 352–60.
Cayrel, A., Crouzet, J., Chen, H. W. S. & Price, K. R. (1983). *Amer. J. Enol. Vitic.*, **34,** 77–82.
Chan, H. W. S. (1973). *Biochim. Biophys. Acta,* **237,** 32–5.
Chang, C. C., Esselman, W. J. & Clagett, C. O. (1971). *Lipids,* **6,** 100–6.
Chang, P. R. Q. & McCurdy, A. R. (1985). *Can. Inst. Food Sci. Technol. J.*, **18,** 94–6.
Charm, S. E. & Wong, B. L. (1981). *Enzym. Microb. Technol.*, **3,** 111–18.
Chen, A. O. & Whitaker, J. R. (1986). *J. Agric. Food Chem.*, **34,** 203–11.

Christopher, J. P., Pistorius, E. & Axelrod, B. (1970). *Biochim. Biophys. Acta,* **198,** 12–19.

Christopher, J. P., Pistorius, E., Regnier, F. E. & Axelrod, B. (1972). *Biochim. Biophys. Acta,* **289,** 82–7.

Cordonnier, R. & Bayonove, C. (1978). *Parfum Cosmetiques Aromes,* **24,** 67–77.

Cosby, E. L. & Sumner, J. B. (1945). *Arch. Biochem.,* **8,** 259–64.

Crouzet, J., Nocolas, M., Molina, J. & Valentin, G. (1985). In *Progress in Flavour Research,* ed. J. Adda. Elsevier Science Publishers, Amsterdam, The Netherlands, pp. 401–8.

Dalal, K. B., Salunkhe, D. K., Olson, L. E., Do, J. Y. & Yu, M. H. (1968). *Plant Cell Physiol.,* **9,** 389–400.

Daood, H. & Biacs, P. A. (1986). *Acta Aliment.,* **15,** 307–18.

Daood, H. & Biacs, P. A. (1988). *Acta Aliment.,* **17,** 53–65.

Dathe, W., Ronsch, H., Preiss, A., Schade, W., Sembdner, G. & Schreiber, K. (1981). *Planta,* **153,** 530–5.

Davies, C. S. & Nielsen, N. C. (1986). *Crop Sci.,* **26,** 460–3.

Davies, C. S., Nielsen, S. S. & Nielsen, N. C. (1987). *J. Amer. Oil Chem. Soc.,* **64,** 1428–33.

de Groot, J. J., Veldink, G. A., Vliegenthart, J. F. G., Boldingh, J., Wever, R. & Van Gelder, B. F. (1975*a*). *Biochim. Biophys. Acta,* **377,** 71–9.

de Groot, J. J., Garssen, G. J., Veldink, G. A., Vliegenthart, J. F. G. & Boldingh, J. (1975*b*). *FEBS Lett.,* **56,** 50–4.

de Lumen, B. O. & Kazeniac, S. J. (1976). *Anal. Biochem.,* **72,** 428–32.

de Lumen, B. O. & Kazeniac, S. J. (1979). *Anal. Biochem.,* **99,** 118–20.

de Lumen, B. O., Stone, E. J., Kazeniac, S. J. & Forsythe, R. H. (1978). *J. Food Sci.,* **43,** 698–702, 708.

Den, T. van & Mendoza, E. M. T. (1982). *J. Agric Food Chem.,* **30,** 54–60.

Den, T. van, Raymundo, L. C. & Mendoza, E. M. T. (1982*a*). *Food Chem.,* **8,** 187–201.

Den, T. van, Raymundo, L. C. & Mendoza, E. M. T. (1982*b*). *Food Chem.,* **8,** 277–89.

Den, T. van, Raymundo, L. C. & Mendoza, E. M. T. (1982*c*). *Food Chem.,* **8,** 307–12.

Dhindsa, R. S., Plumb-Dhindsa, P. L. & Reid, D. M. (1982). *Physiol. Plant.,* **56,** 453–7.

Diel, E. & Stan, H. J. (1978). *Planta,* **142,** 321–8.

Dillard, M. G., Henick, A. S. & Koch, R. B. (1960). *Food Res.,* **25,** 544–53.

Dillard, M. G., Henick, A. S. & Koch, R. B. (1961). *J. Biol. Chem.,* **236,** 37–40.

Downing, D. T., Ahern, D. G. & Bachta, M. (1970). *Biochem. Biophys. Res. Commun.,* **40,** 218–23.

Downing, D. T., Barve, J., Gunstone, F. D., Jacobsberg, F. R., & Lie Ken Jie, M. (1972). *Biochim. Biophys. Acta,* **280,** 343–7.

Drapron, R. & Uzzan, A. (1968). *Ann. Nutr. Alim.,* **22,** B393–436.

Drawert, F., Tressl, R., Heimann, W., Emberger, R. & Speck, M. (1973). *Chem. Mikrobiol. Technol. Lebensm.,* **2,** 10–22.

Dreesen, T. D., Dickens, M. & Koch, R. B. (1982). *Lipids,* **17,** 964–9.

Dutton, H. J., Lancaster, C. R., Evans, C. D. & Cowan, J. C. (1951). *J. Amer. Oil Chem. Soc.,* **28,** 115–18.

Ediriweera, N., Akiyama, Y. & Saio, K. (1987). *J. Food Sci.,* **52,** 685–90.

Egmond, M. R., Finazzi-Agro, A., Fasella, P. M., Veldink, G. A. & Vliegenthart, J. F. G. (1975). *Biochim. Biophys. Acta,* **397,** 43–9.

Eldridge, A. C., Warner, K. & Wolf, W. J. (1977). *Cereal Chem.,* **54,** 1229–37.

Engeseth, N. J., Klein, B. P. & Warner, K. (1987). *J. Food Sci.,* **52,** 1015–19, 1029.

Eriksson, C. E. & Svensson, S. G. (1970). *Biochim. Biophys. Acta,* **198,** 449–59.

Esaka, M., Suzuki, K. & Kubota, K. (1986). *Agric. Biol. Chem.,* **50,** 2395–6.

Esaka, M., Suzuki, K. & Kubota, K. (1987). *J. Food Sci.,* **52,** 1738–9.

Eskin, N. A. M. & Henderson, H. M. (1974). *Phytochem.*, **13,** 2713–16.
Eskin, N. A. M., Grossman, S. & Pinsky, A. (1977). *CRC Crit. Rev. Food Sci. Nutr.*, **9,** 1–40.
Evans, C. D., Frankel, E. N., Cooney, P. M. & Moser, H. A. (1960). *J. Amer. Oil Chem. Soc.*, **37,** 452–6.
Farkas,, D. F. & Goldblith, S. (1962). *J. Food Sci.*, **27,** 262–76.
Faubion, J. M. & Hoseney, R. C. (1981). *Cereal Chem.*, **58,** 175–80.
Feys, M., Naessens, W., Tobback, P. & Maes, E. (1980). *Phytochem.*, **19,** 1009–11.
Finazzi-Agro, A., Avigliano, L., Veldink, G. A., Vliegenthart, J. F. G. & Boldingh, J. (1973). *Biochim. Biophys. Acta,* **326,** 462–70.
Fischer, K. H. & Grosch, W. (1977). *Z. Lebensm. Unters. Forsch.*, **165,** 137–9.
Flath, R. A., Black, D. R., Guadagni, D. G., McFadden, W. H. & Schultz, T. H. (1967). *J. Agric. Food Chem.*, **15,** 29–33.
Fleming, H. P., Cobb, W. Y., Etchells, J. L. & Bell, T. A. (1968). *J. Food Sci.*, **33,** 572–6.
Flick, G. J., St Angelo, A. J. & Ory, R. L. (1975). *Lipids,* **10,** 564–6.
Forss, D. A., Dunstone, E. A., Ramshaw, E. H. & Stark, W. (1962). *J. Food Sci.*, **27,** 90–3.
Frankel, E. N., Warner, K. & Klein, B. P. (1988). *J. Amer. Oil Chem. Soc.*, **65,** 147–50.
Fritz, G. & Beevers, H. (1955). *Arch. Biochem. Biophys.*, **55,** 436–46.
Fujimaki, M., Kato, S., Kirigaya, N. & Sakurai, Y. (1965). *Agric. Biol. Chem.*, **29,** 855–8.
Funk, M. O., Kim, S. H. S. & Alteneder, A. W. (1981). *Biochem. Biophys. Res. Commun.*, **98,** 922–9.
Funk, M. O., Whitney, M. A., Hausknecht, E. C. & O'Brien, E. M. (1985). *Anal. Biochem.*, **146,** 246–51.
Galliard, T. (1968). *Phytochem.*, **7,** 1915–22.
Galliard, T. (1973). *J. Sci. Food Agric.*, **24,** 617–22.
Galliard, T. (1975). In *Recent Advances in the Chemistry and Biochemistry of Plant Lipids,* ed. T. Galliard & E. I. Mercer. Academic Press, London, UK, pp. 319–57.
Galliard, T. (1978). In *Biochemistry of Wounded Plant Tissues,* ed. G. Kahl. Watter de Gruyter, Berlin, Germany, pp. 155–201.
Galliard, T. (1983). In *Lipids in Cereal Technology,* ed. P. J. Barnes. Academic Press, London, UK, pp. 403–7.
Galliard, T. & Chan, H. W. S. (1980). In *The Biochemistry of Plants,* (Vol. 4), ed. P. K. Stumpf. Academic Press, London, UK, pp. 131–61.
Galliard, T. & Matthew, J. A. (1977). *Phytochem.*, **16,** 339–43.
Galliard, T. & Phillips, D. R. (1971). *Biochem. J.*, **124,** 431–8.
Galliard, T. & Phillips, D. R. (1976). *Biochim. Biophys. Acta,* **431**, 278–87.
Galliard, T., Phillips, D. R. & Reynolds, J. (1976). *Biochim. Biophys. Acta,* **441,** 181–92.
Galliard, T., Matthew, J. A., Wright, A. J. & Fishwick, M. J. (1977). *J. Sci. Food Agric.*, **28,** 863–8.
Ganthavorn, C. & Powers, J. R. (1989). *J. Food Sci.*, **54,** 371–8.
Gardner, H. W. (1979). *J. Agric. Food Chem.*, **27,** 220–9.
Gardner, H. W. (1980). In *Autooxidation in Foods and Biological Systems,* ed. M. G. Simic & M. Karel. Plenum Press, New York, USA, pp. 447–504.
Gardner, H. W., Christianson, D. D. & Kleiman, R. (1973). *Lipids,* **8,** 271–6.
Goldstein, J. L. & Wick, E. L. (1969). *J. Food Sci.*, **34,** 482–4.
Gonzales, J. G., Coggon, P. & Sanderson, G. W. (1972). *J. Food Sci.*, **37,** 797–8.
Gordon, M. H. & Barimalaa, I. S. (1989). *Food Chem.*, **32,** 31–7.
Grosch, W. (1972). *Fette Seifen Anstrichmittel,* **74,** 375–81.
Grosch, W. (1982). In *Food Flavours, A: Introduction,* ed. I. D. Morton, & A. J. MacLeod. Elsevier Scientific Publishers, Amsterdam, The Netherlands, pp. 325–98.

Grosch, W. & Laskawy, G. (1979). *Biochim. Biophys. Acta,* **575,** 439–45.
Grosch, W. & Schwarz, J. (1971). *Lipids,* **6,** 351–62.
Grosch, W. & Schwencke, D. (1969). *Lebensm. Wissen. und Technol.,* **2,** 109–12.
Grosch, W., Laskawy, G. & Weber, F. (1976). *J. Agric. Food Chem.,* **24,** 456–9.
Grosch, W., Laskawy, G. & Kaiser, K. P. (1977*a*). *Z. Lebensm. Unters. Forsch.,* **165,** 77–81.
Grosch, W., Weber, F. & Fischer, K. H. (1977*b*). *Ann. Technol. Agric.,* **26,** 133–7.
Grossman, S. & Leshem, Y. Y. (1978). *Physiol. Plant.,* **43,** 359–62.
Grossman, S. & Zakut, R. (1979). In *Methods of Biochemical Analysis* (Vol. 25), ed. D. Glick. J. Wiley and Sons, New York, USA, pp. 303–29.
Grossman, S., Ben Aziz, A., Budowski, P., Ascarelli, I., Gertler, A., Birk, Y. & Bondi, A. (1969). *Phytochem.,* **8,** 2287–93.
Grossman, S., Pinsky, A. & Goldweitz, Z. (1971*a*). *Anal. Biochem.,* **44,** 642–4.
Grossman, S., Yaroni, S., Pinsky, A. & Wilchek, M. (1971*b*). *Proc. XII Ann. Meet. Israel Biochem. Soc.,* p. 50.
Grossman, S., Ben Aziz, A., Ascarelli, I. & Budowski, P. (1972*a*). *Phytochem.,* **11,** 509–14.
Grossman, S., Trop, M., Avtalion, R. & Pinsky, A. (1972*b*). *Lipids,* **7,** 467–73.
Grossman, S., Trop, M., Yaroni, S. & Wilchek, M. (1972*c*). *Biochim. Biophys. Acta,* **289,** 77–81.
Grossman, S., Ben Aziz, A., Ascarelli, I. & Budowski, P. (1974). *Phytochem.,* **13,** 1379–81.
Grossman, S., Klein, B. P., Cohen, B., King, D. & Pinsky, A. (1984). *Biochim. Biophys. Acta,* **793,** 72–9.
Guss, P. L., Richardson, T. & Stahman, M. A. (1967). *Cereal Chem.,* **44,** 607–10.
Guss, P. L., Macko, V., Richardson, T. & Stahmann, M. H. (1968). *Plant Cell Physiol.,* **9,** 415–22.
Haining, J. L. & Axelrod, B. (1958). *J. Biol. Chem.,* **232,** 193–202.
Hale, S. A., Richardson, T., Elbe, J. H. von & Hagedorn, D. J. (1969). *Lipids,* **4,** 209–15.
Hatanaka, A. & Harada, T. (1973). *Phytochem.,* **12,** 2341–6.
Hatanaka, A., Kajiwara, T. & Sekiya, J. (1976). *Phytochem.,* **15,** 1125–6.
Hatanaka, A., Kajiwara, T., Sekiya, J. & Fujimara, K. (1979). *Agric. Biol. Chem.,* **43,** 175–6.
Hatanaka, A., Kajiwara, T., Sekiya, J., Imoto, M. & Inouye, S. (1982). *Plant Cell Physiol.,* **23,** 91–9.
Hatanaka, A., Kajiwara, T. & Sekiya, J. (1987). *Chem. Phys. Lipids,* **44,** 341–61.
Haydar, M. & Hadziyev, D. (1973). *J. Sci. Food Agric.,* **24,** 1039–53.
Haydar, M., Steele, L. & Hadziyev, D. (1975). *J. Food Sci.,* **40,** 808–14.
Hidaka, T., Katsuki, S., Nagata, Y. & Nakatsu, S. (1986). *J. Food Biochem.,* **101,** 55–73.
Hildebrand, D. F. (1989). *Physiol. Plant.,* **76,** 249–53.
Hildebrand, D. F. & Hymowitz, T. (1981). *J. Amer. Oil Chem. Soc.,* **58,** 583–6.
Hildebrand, D. F. & Hymowitz, T. (1982). *J. Agric. Food Chem.,* **30,** 705–8.
Hildebrand, D. F. & Hymowitz, T. (1983). *Bot. Gaz.,* **144,** 212–16.
Hildebrand, D. F. & Kito, M. (1984). *J. Agric. Food Chem.,* **32,** 815–19.
Hildebrand, D. F., Rodriguez, J. G., Brown, G. C., Luu, K. J. & Volden, C. S. (1986). *J. Econ. Entomol.,* **79,** 1459–65.
Hildebrand, D. F., Hamilton-Kemp, T. R., Legg, C. S. & Bookjans, G. (1988). *Curr. Topics Plant Biochem. Physiol.,* **7,** 201–19.
Holden, M. J. (1965). *J. Sci. Food Agric.,* **16,** 312–24.
Holden, M. J. (1970). *Phytochem.,* **9,** 1771–7.
Holman, R. T. (1947). *Arch. Biochem.,* **15,** 403–13.
Holman, R. T., Panzer, F., Schweigert, B. S. & Ames, S. R. (1950). *Arch. Biochem. Biophys.,* **26,** 199–204.

Holman, R. T., Egwim, P. O. & Christie, W. W. (1969). *J. Biol. Chem.*, **244,** 1149–51.
Hultin, H. O. & Proctor, B. E. (1961). *Food Technol.*, **15,** 440–4.
Hurt, G. B. & Axelrod, B. (1977). *Plant Physiol.*, **59,** 695–700.
Ida, S., Masaki, Y. & Morita, Y. (1983). *Agric. Biol. Chem.* **47,** 637–41.
Ikediobi, C. (1977). *Agric. Biol. Chem.*, **41,** 2369–75.
Ikediobi, C. & Snyder, H. E. (1977). *J. Agric. Food Chem.*, **25,** 124–7.
Irvine, G. N. & Anderson, J. A. (1953). *Cereal Chem.*, **30,** 247–55.
Issenberg, P. & Wick, P. L. (1963). *J. Agric. Food Chem.*, **11,** 2–8.
Jadhav, S., Singh, B. & Salunkhe, D. K. (1972). *Plant Cell Physiol.*, **13,** 449–59.
Jhon, D. Y. & Lee, H. J. (1985). *Korean Biochem. J.*, **18,** 190–5.
Johns, E. B., Pattee, H. E. & Singleton, J. A. (1973). *J. Agric. Food Chem.*, **21,** 570–3.
Johns, E. B., Singleton, J. A. & Pattee, H. E. (1974). *J. Amer. Oil Chem. Soc.*, **51,** 32–3.
Kalbrenner, J. E., Elridge, A. C., Moser, H. A. & Wold, W. J. (1971). *Cereal Chem.*, **48,** 595–600.
Kajiwara, T., Nagata, N., Hatanaka, A. & Noashima, Y. (1980). *Agric. Biol. Chem.*, **44,** 437–8.
Kanofsky, J. R. & Axelrod, B. (1986). *J. Biol. Chem.*, **261,** 1099–104.
Kazeniac, S. J. & Hall, R. M. (1970). *J. Food Sci.*, **35,** 519–30.
Kemp, T. R., Knavel, D. E. & Stoltz, L. P. (1972). *Phytochem.*, **11,** 3321–2.
Kemp, T. R., Knavel, D. E., Stoltz, L. P. & Lundin, R. E. (1974). *Phytochem.*, **13,** 1167–70.
Kermasha, S. & Metche, M. (1986). *J. Food Sci.*, **51,** 1224–7.
Kies, M. W. (1947). *Fed. Proc.*, **6,** 267 (Abstr.).
Kies, M. W., Haining, J. L., Pistorius, E., Schroeder, D. H. & Axelrod, B. (1969). *Biochem. Biophys. Res. Commun.*, **36,** 312–15.
Kim, I. S. & Grosch, W. (1978). *Z. Lebensm. Unters. Forsch.*, **167,** 324–6.
Kim, I. S. & Grosch, W. (1979). *J. Agric. Food Chem.*, **27,** 243–5.
Kim, I. S. & Grosch, W. (1981). *J. Agric. Food Chem.*, **29,** 1220–5.
King, D. L. & Klein, B. P. (1987). *J. Food Sci.*, **52,** 220–1.
King, R. D. & Puwastien, P. (1987). *J. Food Sci.*, **52,** 106–8.
Kitamura, K. (1984). *J. Agric. Biol. Chem.*, **48,** 2339–46.
Kitamura, K., Davies, C. S., Kaizuma, N. & Nielsen, N. C. (1983). *Crop. Sci.*, **23,** 924–7.
Klein, B. P., Grossman, S., King, D., Cohen, B. & Pinsky, A. (1984). *Biochim. Biophys. Acta,* **793,** 72–9.
Klein, B. P., King, D. & Grossman, S. (1985*a*). *Adv. Free Radical Biol. Med.*, **1,** 309–43.
Klein, B. P., Cohen, B., Grossman, S., King, D., Malovany, H. & Pinsky, A. (1985*b*). *Phytochem.*, **24,** 1903–6.
Koch, R. B. (1968). *Arch. Biochem. Biophys.*, **125,** 303–7.
Koch, R. B., Stern, B. & Ferrari, C. G. (1958). *Arch. Biochem. Biophys.*, **78,** 165–79.
Koch, R. B., Smith, J. W. & Henick, A. S. (1959). *J. Amer. Oil Chem. Soc.*, **36,** 205–9.
Koch, R. B., Brumfiel, B. L. & Brumfiel, M. N. (1971). *J. Amer. Oil Chem. Soc.*, **48,** 532–8.
Kon, S., Wagner, J. P., Guadagni, D. G. & Horvat, R. J. (1970). *J. Food Sci.*, **35,** 343–5.
Kühn, H., Holzhütter, H. G., Schewe, T., Hiebsch, C. & Rapoport, S. M. (1984). *Eur. J. Biochem.*, **139,** 577–83.
Laasko, S. & Lilius, E. M. (1982). *J. Agric. Food Chem.*, **30,** 913–16.
Lascorz, M. & Drapron, R. (1987). *Phytochem.*, **26,** 349–51.

Lebedev, S. I., Savehenko, N. P. & Krivozucho, L. G. (1978). *Soviet Plant Physiol.*, **25,** 465–8.
Lee, F. A. & Wagenknecht, A. C. (1958). *Food Res.*, **23,** 584–9.
Legge, R. L. & Thompson, J. E. (1983). *Phytochem.*, **22,** 2161–6.
Leoni, O., Iori, R. & Palmieri, S. (1985). *J. Food Sci.*, **50,** 88–92.
Leshem, Y. Y. (1984*a*). *Can. J. Bot.*, **62,** 2943–9.
Leshem, Y. Y. (1984*b*). In *Structure, Function and Metabolism of Plant Lipids,* ed. P. A. Siegenthaler & W. Eichenberger. Elsevier Biomedical Press, Amsterdam, The Netherlands, pp. 181–8.
Leshem, Y. Y. & Barness, G. (1982). In *Biochemistry and Metabolism of Plant Lipids,* ed. J. F. G. M. Wintermans & P. J. C. Kuiper. Elsevier Biomedical Press, Amsterdam, The Netherlands, pp. 275–8.
Leshem, Y. Y., Grossman, S., Frimer, A. & Ziv, J. (1979). In *Advances in the Biochemistry and Physiology of Plant Lipids,* ed. L. A. Appelqvist & C. Lijenberg. Elsevier Biomedical Press, Amsterdam, The Netherlands, pp. 193–8.
Leshem, Y. Y., Wurzburger, J., Grossman, S. & Frimer, A. A. (1981). *Physiol. Plant.*, **53,** 9–12.
Lilius, E. M. & Laasko, S. (1982). *Anal. Biochem.*, **119,** 135–42.
Lupu, R., Grossman, S. & Cohen, Y. (1980). *Physiol. Plant Pathol.*, **16,** 241–8.
Lynch, D. V., Sridhara, S. & Thompson, J. E. (1985). *Planta,* **164,** 121–5.
Mack, A. J., Peterman, T. K. & Siedow, J. N. (1987). In *Isozymes: Current Topics in Biological and Medical Research* (Vol. 13), ed. M. C. Rattazzi, J. G. Scandalios & G. S. Whitt. Alan R. Liss, Inc., New York, USA, pp. 127–54.
MacLeod, G. & Ames, J. (1988). *CRC Crit. Rev. Food Sci. Nutr.*, **27,** 219–400.
Major, R. T., Marchini, P. & Boulton, A. J. (1963). *J. Biol. Chem.*, **238,** 1813–16.
Marcel, F., Naeseus, W., Tobback, P. & Maes, E. (1980). *Phytochem.*, **19,** 1009–11.
Marcus, L., Prusky, D. & Jacoby, B. (1988). *Phytochem.*, **27,** 323–7.
Matoba, T., Hidaka, H., Narita, H., Kitamura, K., Kaizuma, N. & Kito, M. (1985*a*). *J. Agric. Food Chem.*, **33,** 852–5.
Matoba, T., Hidaka, H., Kitamura, K., Norihiko, K. & Kito, M. (1985*b*). *J. Agric. Food Chem.*, **33,** 856–8.
Matsuda, Y., Peppu, T. & Arima, K. (1978). *Biochim. Biophys. Acta,* **530,** 439–50.
Matthew, J. A. & Galliard, T. (1978). *Phytochem.*, **17,** 1043–4.
Mattick, L. R. & Hand, D. R. (1969). *J. Agric. Food Chem.*, **17,** 15–17.
McCurdy, A. R., Nagel, C. W. & Swanson, B. G. (1983). *Can. Inst. Food Sci. Technol. J.*, **16,** 179–84.
McDonald, C. E. (1976). *J. Agric. Food Chem.*, **24,** 133. (Abstr.).
McDonald, C. E. (1979). *Cereal Chem.*, **56,** 84–9.
Mesias, J. L., Maynar, J. I., Henao, F. & Mareca, I. (1983). *Rev. Fr. Oenol.*, **90,** 51–8.
Mitsuda, H., Yamamoto, K. & Yamamoto, A. (1967*a*). *Agric. Biol. Chem.*, **31,** 853–60.
Mitsuda, H., Yamamoto, K., Yamamoto, A. & Kusano, T. (1967*b*). *Agric. Biol. Chem.*, **31,** 115–18.
Moll, C., Biermann, O. & Grosch, W. (1979). *J. Agric. Food Chem.*, **27,** 239–43.
Mtebe, K. & Gordon, M. H. (1987). *Food Chem.*, **23,** 175–82.
Mulliez, E., Leblanc, J. P., Girerd, J. J., Rigaud, M. & Chottard, J. C. (1987). *Biochim. Biophys. Acta,* **916,** 13–23.
Murray, K. F., Shipton, J., Whitfield, F. B., Kennett, B. H. & Stanley, G. (1968). *J. Food Sci.*, **33,** 290–4.
Mustakas, G. C., Albrecht, W. J., McGree, J. E., Black, L. T., Bookwalter, G. N. & Griffin, E. L. (1969). *J. Amer. Oil. Chem. Soc.*, **46,** 623–6.
N'Doye, A. S., Drapron, R. & Nicolas, J. (1985). *Sci. des Aliments,* **5,** 465–82.
Nelson, A. I., Steinberg, M. P. & Wei, L. S. (1976). *J. Food Sci.*, **41,** 57–61.
Nicolas, J. & Drapron, R. (1981). *Sci. des Aliments,* **1,** 91–168.

Nye, W. & Spoehr, H. A. (1943). *Arch. Biochem.*, **2,** 23–35.
Oelze-Karow, H., Schopfer, P. & Mohr, H. (1970). *Proc. Nat. Acad. Sci. USA,* **65,** 51–7.
Ohta, H., Ida, S., Mikami, B. & Morita, Y. (1986*a*). *Agric. Biol. Chem.*, **50,** 3165–71.
Ohta, H., Ida, S., Mikami, B. & Morita, Y. (1986*b*). *Plant Cell Physiol.*, **27,** 911–18.
Olias, J. M. & Valle, M. (1988). *J. Sci. Food Agric.*, **45,** 165–74.
O'Reilly, S., Prebble, J. & West, S. (1969). *Phytochem.*, **8,** 1675–7.
Park, K. H., Kim, Y. M. & Lee, C. W. (1988). *J. Agric. Food Chem.*, **36,** 1012–15.
Pattee, H. E., Salunkhe, D. K., Sathe, S. K. & Reddy, N. R. (1982). *CRC Crit. Rev. Food Sci. Nutr.*, **17,** 97–139.
Peterman, T. K. & Siedow, J. N. (1985*a*). *Arch. Biochem. Biophys.*, **238,** 476–83.
Peterman, T. K. & Siedow, J. N. (1985*b*). *Plant Physiol.*, **78,** 690–5.
Phillips, D. R. & Galliard, T. (1978). *Phytochem.*, **17,** 355–8.
Phillips, D. R., Matthew, J. A., Reynolds, J. & Fenwick, G. R. (1979). *Phytochem.*, **18,** 401–4.
Pinsky, A., Grossman, S. & Trop, M. (1971). *J. Food Sci.*, **36,** 571–2.
Pinsky, A., Sporn, J., Grossman, S. & Rivlin, M. (1973). *Phytochem.*, **12,** 1051–5.
Pirrung, M. C. (1986). *Biochem.*, **25,** 114–19.
Pistorius, E. K. (1974). PhD thesis, Purdue University.
Pistorius, E. K. & Axelrod, B. (1974). *J. Biol. Chem.*, **249,** 3183–6.
Pistorius, E. K., Axelrod, B. & Palmer, G. (1976). *J. Biol. Chem.*, **251,** 7144–8.
Prusky, D., Kobiler, I., Jacoby, B., Sims, J. J. & Midland, S. L. (1985). *Physiol. Plant Pathol.*, **27,** 269–79.
Rackis, J. J., Honig, D. H., Sessa, D. J. & Steggerda, F. R. (1970). *J. Agric. Food Chem.*, **18,** 977–82.
Rackis, J. J., Honig, D. H., Sessa, D. J. & Moser, H. A. (1972). *Cereal Chem.*, **49,** 586–97.
Rackis, J. J., Sessa, D. J. & Honig, D. H. (1979). *J. Amer. Oil Chem. Soc.*, **56,** 262–71.
Ramadoss, C. S. & Axelrod, B. (1982). *Anal. Biochem.*, **127,** 25–31.
Ramadoss, C. S., Pistorius, E. & Axelrod, B. (1978). *Arch. Biochem. Biophys.*, **190,** 549–52.
Rapoport, S. M., Hartel, B. & Hausdorf, G. (1984). *Eur. J. Biochem.*, **139,** 573–6.
Reynolds, P. A. & Klein, B. P. (1982). *J. Agric. Food Chem.*, **30,** 1157–63.
Rhee, K. S. & Watts, B. W. (1966*a*). *J. Food Sci.*, **31,** 664–8.
Rhee, K. S. & Watts, B. W. (1966*b*). *J. Food Sci.*, **31,** 669–74.
Rhodes, M. J. C. & Wooltorton, L. S. C. (1967). *Phytochem.*, **6,** 1–12.
Roza, M. & Francke, A. (1973). *Biochim. Biophys. Acta,* **327,** 24–31.
Saijyo, R. & Takeo, T. (1972). *Plant Cell Physiol.*, **13,** 991–8.
Samotus, B. & Schwimmer, S. (1963). *J. Food Sci.*, **28,** 1–5.
Schewe, T., Rapoport, S. M. & Kuhn, H. (1986). *Adv. Enzymol. Related Areas Mol. Biol.*, **58,** 191–272.
Schewe, T., Kühn, H. & Rapoport, S. M. (1987). In *Prostaglandins and related substances—a practical approach,* ed. R. G. Benedetto, R. G. McDonald-Gibson, S. Nigam & T. F. Slater. IRL Press, Oxford, UK, pp. 229–42.
Schormüller, J. & Kochmann, H. J. (1969). *Z. Lebensm. Unters. Forsch.*, **136,** 271–8.
Schreier, P. (1981). In *Quality in Stored and Processed Vegetables and Fruit,* ed. P. W. Goodenough & R. K. Atkins. Academic Press, New York, USA, pp. 355–71.
Schreier, P. & Lorenz, G. (1982). *Z. Naturforsch,* **37c,** 165–73.
Schreier, P., Drawert, F. & Junker, A. (1977). *Z. Lebensm. Unters. Forsch.*, **165,** 23–7.
Schreier, P., Drawert, F. & Schmid, M. (1978). *J. Sci. Food Agric.*, **29,** 728–36.
Schwarz, P. B. & Pyler, R. E. (1984). *J. Amer. Soc. Brew. Chem.*, **42,** 47–53.
Schwimmer, S. (1981). *Source Book of Food Enzymology.* AVI Publishing Co., Westport, CT, USA.

Sekhar, B. P. S. & Reddy, G. M. (1982). *J. Sci. Food Agric.*, **33,** 1160–3.
Sekiya, J., Numa, S., Kajiwara, T. & Hatanaka, A. (1976). *Agric. Biol. Chem.*, **40,** 185–90.
Sekiya, J., Aoshima, H., Kajiwara, T., Togo, T. & Hatanaka, A. (1977). *Agric. Biol. Chem.*, **41,** 827–32.
Sekiya, J., Kajiwara, T. & Hatanaka, A. (1978). *Agric. Biol. Chem.*, **42,** 677–8.
Sekiya, J., Kajiwara, T. & Hatanaka, A. (1979). *Agric. Biol. Chem.*, **43,** 969–80.
Sekiya, J., Kamiuchi, H. & Hatanaka, A. (1982). *Plant Cell Physiol.*, **23,** 631–8.
Sekiya, J., Kajiwara, T. & Munechika, T. (1983). *Phytochem.*, **22,** 1867–9.
Shastry, B. S. & Rao, M. R. R. (1975). *Cereal Chem.*, **52,** 597–603.
Shibata, D., Steczko, J., Dixon, J. E., Hermodson, M., Yazdanparast, R. & Axelrod, B. (1987). *J. Biol. Chem.*, **262,** 10080–5.
Siddiqi, A. M. & Tappel, A. L. (1956). *Arch. Biochem. Biophys.* **60,** 91–9.
Smith, W. L. & Lands, W. H. (1972). *J. Biol. Chem.*, **247,** 1038–47.
Sosulski, F. W. & Gadan, H. M. (1988). *J. Amer. Oil Chem. Soc.* **65,** 369–72.
Spaapen, L. J. M., Vliegenthart, J. F. G. & Boldingh, J. (1977). *Biochim. Biophys. Acta,* **488,** 517–20.
Spaapen, L. J. M., Verhagen, J., Veldink, G. A. & Vliegenthart, J. F. G. (1980). *Biochim. Biophys. Acta,* **618,** 153–62.
Sredni, D. & Grossman, S. (1980). *Phytochem.*, **19,** 1335–7.
Stevens, F. C., Brown, D. M. & Smith, E. L. (1970). *Arch. Biochem. Biophys.*, **136,** 413–21.
Stone, E. J., Hall, R. M. & Kazeniac, S. J. (1975). *J. Food Sci.*, **40,** 1138–41.
Sumner, J. B. & Smith, G. N. (1947). *Arch. Biochem.*, **14,** 87–92.
Sumner, J. B. & Sumner, R. J. (1940). *J. Biol. Chem.*, **134,** 531–3.
Surrey, K. (1964). *Plant Physiol.*, **39,** 65–70.
Svensson, S. (1977). In *Physical, Chemical and Biological Changes in Food Caused by Thermal Processing,* ed. T. Hoyem & O. Kvale. Applied Science Publishers Ltd, London, UK, pp. 202–17.
Svensson, S. G. & Eriksson, C. E. (1972). *Lebensm. Wissen. Technol.*, **5,** 118–23.
Takahama, U. (1985). *Phytochem.*, **24,** 1443–6.
Takeo, T. & Tsushida, T. (1980). *Phytochem.*, **19,** 2521–2.
Tappel, A. L., Boyer, P. D. & Lundberg, W. O. (1952). *J. Biol. Chem.*, **199,** 267–81.
Tappel, A. L., Lundberg, W. O. & Boyer, P. D. (1953). *Arch. Biochem. Biophys.*, **42,** 293–304.
Theorell, H., Holman, R. T. & Akeson, A. (1947). *Acta Chem. Scand.*, **1,** 571–6.
Tressl, R. & Drawert, F. (1973). *J. Agric. Food Chem.*, **21,** 560–5.
Tressl, R. & Jennings, W. G. (1972). *J. Agric. Food Chem.*, **20,** 189–92.
Tressl, R., Drawert, F. & Heimann, W. (1970*a*). *Z. Lebensm. Unters. Forsch.*, **142,** 393–7.
Tressl, R., Drawert, F., Heimann, W. & Emberger, R. (1970*b*). *Z. Lebensm. Unters. Forsch.*, **144,** 4–12.
Tressl, R., Bahri, D. & Engel, K. H. (1981). In *Quality of selected fruits and vegetables of North America,* ed. R. Teranashi & H. Barrera-Benitez. ACS Symposium Series 170, American Chemical Society, Washington DC, USA, pp. 213–32.
Triplett, B. A. (1985). *Proc. Amer. Peanut. Res. Ed. Soc.*, **17,** 69. (Abstr.).
Valentin, G. & Crouzet, J. (1989). In *Actualités Oenologiques 1989.* VI Symposium International d'Oenologie de Bordeau, Dunod, Paris.
Veldink, G. A., Garssen, J. G., Vliegenthart, J. F. G. & Boldingh, J. (1972). *Biochem. Biophys. Res. Commun.*, **47,** 22–6.
Veldink, G. A., Vliegenthart, J. F. G. & Boldingh, J. (1977). *Prog. Chem. Fats and Other Lipids,* **15,** 131–66.
Verhue, W. M. & Francke, A. (1972). *Biochim. Biophys. Acta,* **285,** 43–53.
Vernooy-Gerritsen, M., Bos, A. L. M., Veldink, G. A. & Vliegenthart, J. F. G. (1982). *Biochim. Biophys. Acta,* **708**, 330–4.

Vernooy-Gerritsen, M., Bos, A. L. M., Veldink, G. A. & Vliegenthart, J. F. G. (1983*a*). *Plant Physiol.*, **73**, 262–7.
Vernooy-Gerritsen, M., Bos, A. L. M., Veldink, G. A. & Vliegenthart, J. F. G. (1983*b*). *Biochim. Biophys. Acta*, **748**, 148–52.
Vick, B. A. & Zimmerman, D. C. (1976). *Plant Physiol.*, **57**, 780–8.
Vick, B. A. & Zimmerman, D. C. (1984). *Plant Physiol.*, **75**, 458–61.
Vick, B. A. & Zimmerman, D. C. (1987*a*). *Plant Physiol.*, **85**, 1073–8.
Vick, B. A. & Zimmerman, D. C. (1987*b*). In *Biochemistry of Plants: A Comprehensive Treatise* (Vol. 9) ed. P. K. Stumpf. Academic Press, Orlando, FL, USA, pp. 53–90.
Vliegenthart, J. F. & Veldink, G. A. (1982). In *Free Radicals in Biology* (Vol. V), ed. W. A. Pryor. Academic Press, New York, USA, pp. 29–64.
Wagenknecht, A. C. (1959). *Food Res.*, **24**, 539–47.
Wagenknecht, A. C. & Lee, F. A. (1956). *Food Res.*, **21**, 605–10.
Wagenknecht, A. C., Lee, F. A. & Boyle, F. P. (1952). *Food Res.*, **17**, 343–50.
Wallace, J. M. & Wheeler, E. L. (1975). *J. Agric. Food Chem.*, **23**, 146–50.
Wallace, J. M. & Wheeler, E. L. (1979). *Phytochem.*, **18**, 389–93.
Wang, S. H. & Toledo, M. C. F. (1987). *J. Food Sci.*, **52**, 1344–7.
Wang, T. T. & Yang, S. F. (1987). *Planta*, **170**, 190–6.
Wardale, D. A. & Lambert, E. A. (1980). *Phytochem.*, **19**, 1013–16.
Weber, F. & Grosch, W. (1976). *Z. Lebensm. Unters. Forsch.*, **161**, 223–30.
Weber, F., Arens, D. & Grosch, W. (1973). *Z. Lebensm. Unters. Forsch.*, **152**, 152–4.
Weber, F., Laskawy, G. & Grosch, W. (1974). *Z. Lebensm. Unters. Forsch.*, **155**, 142–50.
Wheeler, E. L. & Finley, J. W. (1981). *J. Agric. Food Chem.*, **29**, 912–14.
Wilkens, W. F., Mattick, L. R. & Hand, D. B. (1967). *Food Technol.*, **21**, 1630–3.
Williams, D. C., Lim, M. H., Chen, A. O., Pangborn, R. M. & Whitaker, J. R. (1986). *Food Technol.*, **40**, 130–40.
Wolf, W. J. (1975). *J. Agric. Food Chem.*, **23**, 136–41.
Wooltorton, L. S. C., Jones, J. D. & Hulme, A. C. (1965). *Nature*, **207**, 999–1001.
Wright, A. J. & Fishwick, M. J. (1979). *Phytochem.*, **18**, 1511–13.
Yabuuchi, S. & Amaha, M. (1975). *Phytochem.*, **14**, 2569–72.
Yabuuchi, S., Lister, R. M., Axelrod, B., Wilcox, J. R. & Nielsen, N. C. (1982). *Crop Sci.*, **22**, 333–7.
Yamamoto, A., Yasumoto, K. & Mitsuda, H. (1970). *Agric. Biol. Chem.*, **34**, 1169–77.
Yamamoto, A., Fujii, Y., Yasumoto, K. & Mitsuda, H. (1980). *Lipids*, **15**, 1–5.
Yasumoto, K., Yamamoto, A. & Mitsuda, H. (1970). *Agric. Biol. Chem.*, **34**, 1162–8.
Yoon, S. & Klein, B. P. (1979). *J. Agric. Food Chem.*, **27**, 955–62.
Yu, M. H., Olson, L. E. & Salunkhe, D. K. (1967). *Phytochem.*, **6**, 1457–67.
Zakut, R., Grossman, S., Pinsky, A. & Wilchek, M. (1976). *FEBS Lett.*, **71**, 107–110.
Zamora, R., Maynar, J. I. & Mesias, J. L. (1985). *Amer. J. Enol. Vitic.*, **36**, 316–18.
Zamora, R., Olias, J. M. & Mesias, J. L. (1987). *Phytochem.*, **26**, 345–7.
Zimmerman, D. C. & Caudron, C. A. (1979). *Plant Physiol.*, **63**, 536–41.
Zimmerman, D. C. & Vick, B. A. (1970*a*). *Lipids*, **5**, 392–7.
Zimmerman, D. C. & Vick, B. A. (1970*b*). *Plant Physiol.*, **46**, 445–53.
Zimmerman, D. C. & Vick, B. A. (1973). *Lipids*, **8**, 264–6.
Zimmerman, G. L. & Snyder, H. E. (1974). *J. Agric. Food Chem.*, **22**, 802–5.

9

Phenoloxidases and their Significance in Fruit and Vegetables

Alfred M. Mayer & Eitan Harel
Department of Botany, The Hebrew University of Jerusalem, Jerusalem 91904, Israel

INTRODUCTION

The phenoloxidases comprise essentially two types of enzymes, *o*-diphenol oxidase (catechol oxidase, tyrosinases, phenolase, polyphenoloxidase (PPO)) and *p*-diphenoloxidase or laccases. Both types of enzyme are phenol–oxygen oxido reductases. Initially, the enzymes were designated EC 1.10.3.1 and EC 1.10.3.2, but later the nomenclature was revised and both were called EC 1.14.18.1 monophenol monooxygenases. This has been revised yet again to give the two sequential activities (oxidation of a monophenol to a diphenol and the subsequent oxidation of the latter) distinct numbers EC 1.14.18.1 and 1.10.3.2, respectively. Laccase is now given the number 1.10.3.1. The numbers and their revision can cause a considerable amount of confusion. We shall use the terminology catechol oxidase or PPO to indicate the activity in **I** but will not

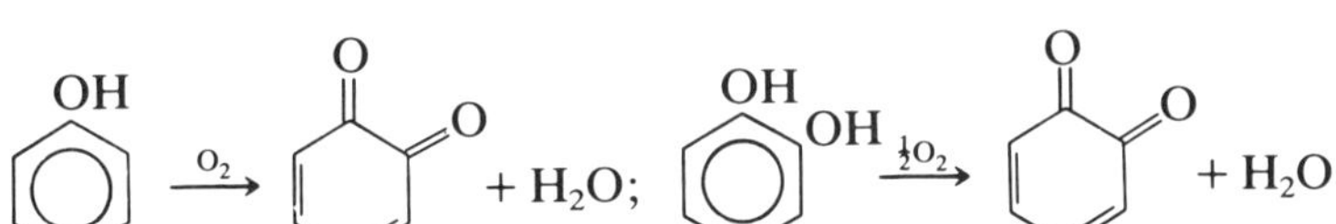

I

deal with the distinction between catecholase and cresolase activities, catalysed by the same enzyme. We will use the term laccase to describe the enzyme reaction **II**.

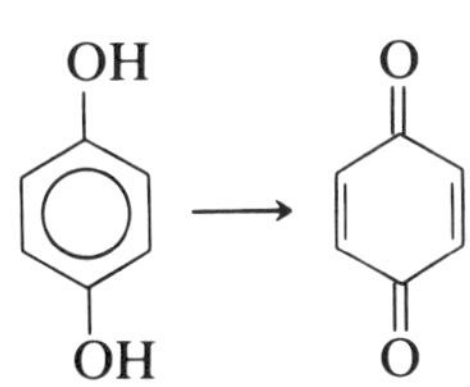

II

These two reactions are different in their properties, potential substrates, conditions under which they take place and the molecular properties of the enzymes which catalyse them.

The literature on the two enzymes has been reviewed a number of times in the last 20 years and a very large body of information has accumulated (Mason, 1955; Bonner, 1957; Brooks & Dawson, 1966; Levine 1966; Malmström, 1970; Mayer & Harel, 1979, 1981; Mayer, 1987). The PPOs are copper enzymes and the oxidation reaction involves a change in valency of the copper atoms in the molecule and abstraction of electrons from oxygen. The reaction catalysed by the enzymes results in the formation of water. In this respect, catechol oxidase and laccase are similar. However, in almost all other respects they are entirely different (Lerch, 1987; Mayer, 1987).

The basic properties of the catechol oxidases and the laccases are shown in Table 1, but it must be remembered that this table merely presents a brief summary and does not take into account the considerable variation and diversity in properties within each group.

The immediate products of the oxidation of phenols, both by PPO and laccase, are quinones. Due to the great reactivity of the quinones, they usually undergo secondary reactions, either between quinone molecules or with other substances if present in the immediate vicinity. Consequently, a variety of coloured compounds is formed, which are not the direct result of the primary oxidation reaction. The formation of such coloured compounds, often brown or black, is one of the reasons for the great interest in the PPOs in food products. Nevertheless, it must be remembered that not all dark colours in a food product are the result of PPO activity. The well-known Maillard reaction

TABLE 1
Some Molecular Properties of Polyphenol Oxidase

Property	*Catechol Oxidase*	*Laccase*
Mol. wt	35–120 kD	35–350 kD
Sub-units	Present, identical or heavy and light	Present, identical
State of Cu	Cu^+–Cu^+ (binuclear clusters)	Type 1, blue, paramagnetic Type 2, paramagnetic Type 3, paramagnetic
Substrates	Monohydroxyphenols, *o*-dihydroxyphenol	Monohydroxyphenols, *o*- and *p*-dihydroxy phenols, ascorbic acid,
Protein structure	Polypeptide	Glycopolypeptide
pH of maximal activity	Variable	Variable
Nature of oxidation step	Completely enzymatic	First step enzymatic Second step non-enzymatic
Reaction products	Quinones	Quinones

can also result in the formation of dark products during food processing as a result of the interaction of amino acids with each other or with, for example, sugars (Waller & Feather, 1983). Peroxidase can also cause the oxidation of phenolic compounds (Huystee & Chibbar, 1987), which can result in the formation of coloured substances. Consequently, the result of reactions catalysed by peroxidase and PPO at first sight appear to be similar. However, these reactions can be distinguished fairly easily, as will be discussed later.

Problems relating to the subcellular location and latency of laccase and PPO will be discussed in the section on properties. As shown in (**I**), the typical PPO reaction catalysed by catechol oxidase can be a two step reaction, starting with a monophenol or a one-step reaction, in which a dihydroxy phenol is oxidised. Whereas the latter reaction is common and has been studied in great detail, the ability to oxidise monophenols has been frequently reported to be absent in enzyme preparations from plant tissues. This is probably due to the fact that (i) this reaction is inherently less readily detected, (ii) is usually associated with a lag period and (iii) apparently the ability of the enzyme to hydroxylate monophenols is readily lost, perhaps due to changes in enzyme conformation. Hydroxylation of monophenols to diphenols has been achieved using alginate-entrapped cells, under conditions where the product is immediately removed (Pras *et al.*, 1988).

No treatment of PPOs can ignore the recurrent question of their biological function. This question remains unanswered. In most cases, in higher plants only catechol oxidases are present, but at least in some tissues laccases are also found. In fungi, very often both types of enzyme, catechol oxidase (tyrosinase) and laccase, are present. The simultaneous or sequential appearance of two enzymes apparently catalysing reactions resulting in the same final product—quinones—makes the question of biological function even more puzzling. We have reviewed theories about the function of phenol oxidases (Mayer & Harel, 1979; Mayer, 1987) and the question has also engaged others (Vaughn *et al.*, 1988). We can not offer any definite answer to this question of function. However, it seems extremely likely that different functions must be ascribed to the same enzyme present in different tissues, at different developmental stages and at different locations within the cell. Defence mechanisms in higher plants, attack mechanisms in fungi, pigment formation, oxygen buffering or free radical scavenging are all possible candidates for functions of PPOs. Enzymes, which have apparently appeared early in evolution are unlikely to be merely evolutionary relics.

The problem of function is one of several aspects related to this group of enzymes which could benefit from the application of recombinant DNA techniques. It appears that PPO is not regulated to a large extent at the transcriptional level. However, analysis of the promoters involved and the use of reporter genes in chimeric gene-constructs can permit quantitation of expression in particular cells and tissues and examination of expression of PPO genes under stress conditions, e.g. pathogen attack, heat shock or water stress. The use of promoter-specific anti-sense RNA to silence genes has recently been shown to be feasible in plants (Delauney *et al.*, 1988; Smith *et al.*, 1988). Many PPO-containing species can be transformed with *Agrobacterium* vectors, a requirement for the promoter analysis and the anti-sense RNA approaches. Even the transformation of yam has recently been reported (Kahl *et al.*, 1988). Silencing of PPO in a particular tissue could make an important contribution to solving the problem of PPO function.

The reaction mechanism of the enzymes will be discussed briefly later. Detailed reviews have appeared which pay particular attention to the subject. (Mayer & Harel, 1979; Farver & Pecht, 1984; Reinhammar, 1984; Reinhammar & Malmström, 1981; Solomon, 1981, 1988; Robb, 1984; Butt, 1985; Lerch, 1987; Mayer, 1987; Morpurgo, 1987)

DETECTION AND DETERMINATION OF PPOs

It is often desirable to get a quick idea whether or not catechol oxidases (or laccases) are present in a given tissue. As a preliminary step, simply grinding the tissue in dilute buffer and observing whether the extract darkens is probably the easiest procedure. Browning in such extracts suggests that phenol oxidases are present, but by no means proves it and absence of browning does not completely rule out the presence of the enzyme. Tissue extracts darken due to a variety of reactions, not all of which are oxidative. Failure to darken may be due to the absence of endogenous substrate or to latency. Nevertheless, darkening is a preliminary indication of the presence of enzyme. A more reliable preliminary indication can be obtained by using histochemical reactions. Incubation of thin sections of the tissue with a suitable substrate, resulting in staining, will give good preliminary evidence of the presence or absence of the enzyme. A refinement of the method has recently been described (Spruce *et al.*, 1987) whereby the section is blotted onto nitrocellulose, the imprint rinsed, buffered and then developed for enzyme activity. Coloration as the result of the enzyme reaction (red turning to black using dihydroxyphenylalanine (DOPA) in the case of PPO, pink using syringaldazine for laccase) is quite definite proof for the presence of enzyme activity, provided that peroxides are absent or have been removed using catalase. Since these peroxides, as well as endogenous inhibitors, can be removed by rinsing the blot, the method is reasonably reliable.

A much more difficult problem is the accurate estimation of enzyme activity, again for the reasons already mentioned [for discussion see, for example, Kean (1964), Mayer *et al.* (1966), Yamaguchi *et al.* (1969), Mayer & Harel (1979), Mayer (1987)]. Since the products of PPO activity are coloured, spectrophotometric assays are frequently adopted. These are rapid and convenient, but contain inherent possibilities for inaccuracies. One is that the presence of reducing agents, which are often present in the tissues (e.g. ascorbic acid), will delay colour formation. Secondly, the reaction products—quinones—are very unstable and reactive and react with each other or with other cell constituents. As a result, colour formation is never linear with time, and initial rates of reaction must be measured. The measurement of disappearance of substrate is, in many ways, to be preferred, particularly as simple and rapid means exist for measuring the disappearance of oxygen, using oxygen electrodes. Here too, initial rates must be determined, since the stoichiometry of the reaction for oxygen changes with time. Both disappearance of substrate and appearance of product can be measured, following separation of the compounds present in the reaction mixture using HPLC (Badiani *et al.*, 1983; Goodenough *et al.*, 1983; Wichers *et al.*, 1984). This method could be very reliable, but is of course rather cumbersome. The hydroxylation step during oxidation of monohydroxy phenols can be measured by following the release of ^3H from a titrated substrate (Pomeranz, 1966; Kahn & Pomeranz, 1980). The method is very sensitive, but requires the availability of a suitable titrated substrate.

The precise substrate selected for the various methods is, in some ways, irrelevant and numerous substrates have been used. Generally, it is recommended that 4-methylcatechol or 3,4-DOPA be used as substrates for catechol oxidase and *p*-quinol or syringaldazine (Harkin & Obst, 1973) for laccase. Since PPOs generally have low affinities (high K_m) for both their substrates it is important to ensure that substrate concentration does not limit the rate of reaction. A further complication of assays is the presence of inhibitors of enzyme activity in crude extracts. Accurate estimation of enzyme activity should preferably be made on partially purified enzyme preparations. Even then, difficulties exist. Since the cresolase activity (oxidation of monophenols) is readily lost, partial purification may underestimate this activity or it may be missed completely, or only appear with a considerable time lag. When a reducing agent (e.g. ascorbic acid) is added the lag period can be considerably shortened (Mayer *et al.*, 1966). Tanning reactions during enzyme extraction can cause partial inactivation of the enzymes. To avoid these, reducing agents are often added during extraction, but must be removed before assay, or, alternatively, phenolic substrates must be removed prior to assay, using insoluble polyvinylpyrrolidone.

In recent years, antibodies have been successfully raised against both PPOs and laccases. Although these are not as yet entirely specific, it is likely that refinement of their preparation will lead to antibodies sufficiently specific to quantify the amount of enzyme present in a given preparation. Some problems exist with antibodies towards PPO (Lanker *et al.*, 1988) and there are distinct differences between monoclonal and polyclonal ones.

The methods outlined above are valid, provided it is clear which enzyme is present in the tissue and that only one of them (laccase or PPO) is in fact present. It is extremely difficult to estimate the activity of both enzymes, when both are present or even to differentiate between them. Furthermore, if endogenous phenols are present in the enzyme preparation, artefacts are very readily obtained. Due to the great reactivity of the product—quinones—

coupled reactions can take place. As a result, either of the two enzymes can oxidise either an *o*- or a *p*-dihydroxy phenol provided even a small amount of the right endogenous substrate is present. Identification must, therefore, be based on additional criteria. One such criterion is selective inhibition of one type of enzyme in the presence of the other. This is generally much more difficult than it appears at first sight. Clearly, inhibitors whose action is based on copper chelation can not be used for this purpose, nor is it possible to utilise inhibitors whose action depends on —SH group binding or on their action as reducing agents. However, a number of suitable inhibitors have been described. Among these inhibitors are: phenylhydrazine (Lerner *et al.*, 1971; Joel *et al.*, 1978), carbon monoxide (Keilin & Mann, 1940; Mayer *et al.*, 1964), 2,3-naphthalenediol (Mayer *et al.*, 1964; Mayer & Harel, 1968) and cinnamic acid derivatives (Walker & McCallion, 1980), all of which inhibit catechol oxidase without affecting laccase. In contrast, cetyl trimethyl ammonium bromide (CTAB) has been reported to inhibit laccase without inhibiting catechol oxidase (Walker & McCallion, 1980). Among the more novel catechol oxidase inhibitors are salicylhydroxamic acid (SHAM), mimosine and tropolone (Kahn & Andrawis, 1985*a,b,c*; Cabanes *et al.*, 1987; Allan & Walker, 1988). The latter does not apparently inhibit laccase (Mayer, A. M., unpublished data), but mimosine and SHAM have not been examined in this respect. In crude preparations it is desirable to ensure the absence of peroxidase activity. This can be easily done by addition of catalase and alcohol to the preparation which will remove residual peroxides and as a result prevent peroxidase activity.

We will not describe in detail the techniques used to separate and purify the laccases and catechol oxidases. Much of the literature was reviewed in Mayer and Harel (1979) and Mayer (1987); see also Watson and Flurkey (1986), Ingebrigtsen and Flurkey (1988), Wichers *et al.* (1984).

Since laccases are glycoproteins, while the consensus is that catechol oxidases are not, it is easy to distinguish the two following electrophoretic separation on SDS or nondenaturing acrylamide gels. It is possible to effect separation, stain for enzyme activity using a suitable substrate and then in parallel, identify protein and glycoprotein on the gels. PPO activity cannot be reliably detected in SDS gels due to denaturation and inactivation. Laccases are much more difficult to separate by electrophoresis than are the catechol oxidases. Isoelectric focusing techniques can, however, be applied to the separation of laccases.

OCCURRENCE, DISTRIBUTION AND PROPERTIES

The two types of phenol oxidases under discussion are very widely distributed. However, while the catechol oxidase type is almost universally present in all major groups, from bacteria, via the fungi and higher plants to animals, including man, laccases appear much more sporadically. The presence of laccase has been reported in most, but not all, fungi and in certain higher plants (Mayer, 1987). The latter include peaches, *Aesculus,* the genus *Rhus,* the Burmese lacquer tree, *Melanorrhea,* and all members of the Anacardiaceae, including the mango fruit where it is restricted to the cavity of secretory ducts (Joel *et al.,* 1978). Although catechol oxidase is apparently not ubiquitous in plants (Mayer & Harel, 1979) there are few fruits and vegetables in which it has not been found when looked for. Lack of enzyme activity in a particular variety or tissue, or at a certain developmental stage, can not be taken as proof of the absence of gene(s) coding for catechol oxidase or even absence of the protein itself. The latency of catechol oxidase and the presence of endogenous inhibitors of the enzyme are well documented phenomena (Mayer & Harel, 1979; Carlberg, 1987; Mayer, 1987; Udayasekhara Rao & Deosthale, 1987). Furthermore, the level of enzyme activity changes during fruit development, ripening and storage and may drop to negligible levels in some cases (Mayer & Harel, 1981). In addition, peroxidase action can be mistaken for catechol oxidase involvement. Ways to discriminate between the two are discussed in the 'Detection and Determination' section. Thus, until screening of cDNA and genomic libraries with the proper probes along with a wider use of antibodies and in-situ hybridisation techniques will be applied to the study of PPO, no reliable and complete picture of the occurrence and distribution of the enzyme can be obtained. Such information is particularly important when evolutionary problems and possible funcion(s) of the enzyme are considered. Catechol oxidase is frequently found in thylakoids and its possible involvement in photosynthesis or reactions related to it have been discussed (Tolbert, 1973; Vaughn & Duke, 1984). In non-photosynthetic prokaryotes the enzyme is present, but it has not been reported in cyanobacteria. In fungi, it is soluble in the cytosol and not membrane-bound in organelles. Taken together, these observations leave unexplained gaps in the 'evolutionary picture' of the enzyme. If catechol oxidase was incorporated into the photosynthetic membrane after the development of oxygenic photosynthesis, its function in the latter must be reconsidered. It would be interesting to establish the extent of homology of the enzyme from higher plants to fungal tyrosinases. One of the copper sites of *Neurospora crassa* tyrosinase has recently been shown to possess 80% homology with the amino acid sequence of a similar site in *Streptomyces glaucescens* tyrosinase and even 66% homology with a copper site in an arthropod haemocyanin, an oxygen-binding copper protein (Lerch, 1987).

A survey of recent reports on catechol oxidases in fruits and vegetables and their products adds several new species to the already long list of those reported to exhibit enzyme activity (Mayer & Harel, 1979, 1981; Mayer, 1987). The recent literature abounds with more studies on catechol oxidase in deciduous fruits, grapes and wine, potatoes, avocados, olives, bananas, mangoes, tea, coffee, cocoa, and mushroom. Among the new additions, some of which are 'exotic' species, are citrus (Fujita & Tono, 1982; Santana *et al.,* 1982), sago palm (*Metroxylon saqus*) (Okamoto *et al.,* 1988), dates (Hasegawa

& Maier, 1980; Jarrah & Benjamin, 1982), palm hearts (Ferreira *et al.*, 1982), strawberries (Pilando *et al.*, 1985), Japanese persimmon (Taira *et al.*, 1987), Japanese apricot (Goto *et al.*, 1984), pomegranate (Ben-Arie & Or, 1986), kiwifruit (Park & Luh, 1985), annona (Sanchez de Medina *et al.*, 1987), guava (Augustin *et al.*, 1985), langsat (*Lansium domesticum*) (Suhaila & Leong, 1987) and starfruit (*Averrhoa carambola*). The latter is an example of contradicting reports concerning the occurrence of PPO in a particular tissue. Starfruit from Malesia was reported to contain the enzyme (Adnin *et al.*, 1986) while it was absent from a Brazilian crop (Baldini *et al.*, 1982). Many of these reports reflect the extension of the scope of cultivation and export of fruit crops and the intensification of agriculture and agricultural research in developing countries, for example the studies on catechol oxidase in yam (Ikediobi & Oti, 1983), taro (*Colocasia esculenta*) (Rhee & Iwata, 1982) and cassava (Richard, 1985) and reports on the enzyme in spice-crops: chillies (Bajaj & Kaur, 1986), black pepper (Mangalakumari *et al.*, 1983) and Amla (*Phyllanthus emblica*) for curry powder (Vijay, 1986). Catechol oxidase has recently been studied also in sweet potatoes (Walter & Schadel, 1982), eggplants (Sidhu *et al.*, 1982) and Jerusalem artichoke (Zawistowski *et al.*, 1987*a,b*). Following reports on PPO in wheat (Interesse *et al.*, 1980; Lamkin *et al.*, 1981), the enzyme was studied in fractions of flour (Marsh & Galliard, 1986). The occurrence of catechol oxidase in legume seeds as well as its latency in some species is well documented and reports on its presence in additional species continue to appear (Udayasekhara Rao & Deosthale, 1987).

Although the above cited reports widen the spectrum of catechol oxidase occurrence, regretfully little has been added to our basic knowledge of its properties, let alone to the understanding of its reaction mechanism or physiological function. No reports on the occurrence of laccase in additional species of higher plants have come to our attention.

Although fungal laccase is not exclusively extracellular (Mayer & Harel, 1979) as has been suggested (Levine, 1966), it is readily excreted by hyphae. Both laccase and tyrosinase have been suggested to be involved in colour formation during fruiting in fungi. Both are cytosolic, soluble enzymes in fungi and in contrast to the catechol oxidases, no clear connection to a subcellular organelle or a membrane fraction has been demonstrated for higher plant laccases either. It is interesting to note that sycamore (*Acer pseudoplatanum*) cells in suspension cultures secrete large amounts of laccase into the medium (Bligny & Douce, 1983).

Catechol oxidase in higher plants has been found in a variety of organs and tissues. It is often abundant in leaves, tubers, storage roots, floral parts and fruits. Its abundance in tubers and fruits at early stages of development, along with high levels of phenolic substrates, raised speculation as to its possible role in making the unripe fruit and storage organs inedible to predators (Mayer & Harel, 1981). Although there are well-established cases of cytosolic catechol oxidase in higher plants [e.g. potato tubers, sugarbeet roots, bananas (Mayer & Harel, 1979)], the enzyme is mostly membrane-bound in non-senescing tissues. In a 1979 review, it was stated that 'the weight of evidence indicates the wide occurrence of membrane-bound catechol oxidases, particularly in chloroplasts' (Mayer & Harel, 1979). The data that have accumulated since strengthen this conclusion, although only a few of the studies on the location of PPO have employed satisfactory criteria of fractionation and identification of subcellular organelles (see Vaughn & Duke, 1984 for discussion). Some (Henry *et al.*, 1981; Vaughn & Duke, 1981; Vaughn & Duke, 1984) claimed that the enzyme is located only in plastids, from where it might be solubilised and leak to the cytosol, mainly during senescence and deterioration of organelles. We believe that some of the claims about 'authentic' soluble catechol oxidases or reports on location in mitochondria can not be ignored (Mayer & Harel, 1979; Mayer, 1987, but also see Carlberg *et al.* 1985; Soederhall *et al.*, 1985; Soederhall & Soederhall, 1989). However, we agree that catechol oxidase is predominantly a thylakoid enzyme in higher plants. This fact has important implications for possible functions of PPO and for the distribution of the enzyme in organs and tissues as well as practical aspects in fruit and vegetable technology.

Only a limited number of histochemical studies have been carried out on PPO. Some involved localisation of the enzyme in organelles using oxidation of DOPA and electron microscopy (Parish, 1972; Shomer *et al.*, 1979; Vaughn & Duke, 1981) or immuno-gold labelling (Golan-Goldhirsh *et al.*, 1986). Localisation in organs and tissues could benefit from a wider use of the blotting or 'contact print' technique recently described (Spruce *et al.*, 1987) which has already been used in a study of the distribution of tyrosinase during the development of mushrooms (Moore *et al.*, 1988). An interesting observation was made by Vaughn and Duke (1981) on the distribution of PPO in leaves of a C-4 plant. C-4 plants differ from C-3 plants by having a lower compensation point for CO_2 and low rates of photorespiration as well as in their anatomy and distribution of enzymes between bundle sheath and mesophyll cells. The enzyme was detected only in the mesophyll cells of sorghum leaves, a C-4 plant, and was absent from bundle sheath ones. Since bundle sheath cells are devoid of grana stacks and low in PS II activity and oxygen evolution, the observation has implications regarding the possible function of the enzyme and the regulation of expression of its genes. The absence of PPO from plastids of stomata guard-cells is apparently a related phenomenon. On the other hand, the enzyme was detected in root plastids, potato amyloplasts, *Aegopodium* leucoplasts, etioplasts and chromoplasts. In non-photosynthesising plastids, the enzyme was usually detected in vesicles which appeared to be attached to the plastid envelope (see Vaughn & Duke (1984) for details). These observations, along with latency phenomena and the marked changes often observed in the level of the enzyme during development, suggest that the regulation of PPO expression is quite complex and might operate at several levels.

Chloroplastic catechol oxidase of higher plants is apparently coded in the nucleus (Lax *et al.*, 1984), but more work is required to establish whether this is true for all PPOs studied. The import of PPO into plastids has not yet been studied. Immunoprecipitation of poly A^+ mRNA translation products with an antibody against *Vicia faba* PPO suggests that the enzymes synthesised in the cytoplasm is identical in size with that observed in the chloroplasts (Flurkey, 1985). This is the only report of a nuclear coded plastid protein out of a large number studied, which does not have an amino terminal transit peptide which is removed following import into the plastid (Schmidt & Mishkind, 1986; Keegstra & Bauerle, 1988). The report should therefore be treated with some caution and needs confirmation. Final judgement should await cloning of PPO genes with active promoters attached, to enable in-vitro transcription and translation of the protein. Such an approach, along with information on the amino acid sequence of the protein, will open the way to answer some interesting questions regarding the enzyme, its location and function. Among these questions are (i) Is the enzyme located in grana or stroma lamellae? (ii) What is the route of its assembly? (iii) How is it folded in the thylakoid membrane? (iv) On which face of the membrane is the active site of the enzyme exposed and what ensures the anchoring of the protein in the membrane. Since fungal tyrosinases are soluble proteins, it would be interesting to compare their amino acid sequence to that of catechol oxidases from higher plants with respect to the membrane anchoring domain(s) (Kohorn & Tobin, 1989) as well as other features of the enzymes.

The levels of PPO and its substrates often change markedly during the life cycle of the plant, particularly in fruits and vegetables. The point of departure for many of the studies on fruit PPOs is technological and practical. This imposes some bias on the choice of material and the time-span of the observations performed. Food technologists are often interested in the potential for browning. The picture is therefore often incomplete and not always correct as to the pattern of PPO expression. Latency, endogenous inhibitors, leakage from organelles and mistaken identity of the enzymes involved are all factors that should be considered. Even the introduction of antibodies into such studies may lead to wrong conclusions. Purified PPOs are often 'tanned' proteins to which phenols and their oxidation products are attached. Antibodies prepared against such preparations can give erroneous results if they were raised against the phenolic epitope. The removal of phenolics and prevention of browning markedly reduced the number of 'multiple forms' of PPO observed (Smith & Montgomery, 1985).

In a previous survey (Mayer & Harel, 1981) we summarised observations by us and others on changes in PPO activity and substrates during fruit development and ripening. Observations which have been published since usually fit the general picture obtained earlier. Four points emerge from such a survey.

(1) Enzyme(s) and substrates do change, often dramatically, during fruit development.
(2) Differences in the pattern exist among species and even varieties of the same fruit (e.g. Sapis *et al.*, 1983). The differences sometimes observed

with the same variety in different years (Kidron *et al.*, 1978; Sapis *et al.*, 1983), in response to fertilisers (Drawert *et al.*, 1985) or as a result of stress (Bower & Lelyveld, 1985) probably reflect the complex effects of environmental factors on the level of PPO and its substrates, a subject we know very little about.

(3) Enzyme activity and phenolic content are usually high during the early stages of development and decrease later-on.

(4) As the fruit ages, more enzyme becomes soluble. Solubilisation (Mayer & Harel, 1981), and sometimes also activation, can occur during post-harvest ripening of certain fruits, e.g. annona (Sanchez de Medina *et al.*, 1987), guava (Mowlah & Itoo, 1982).

However, work on the subcellular location of enzymes, especially their assignment to a particular organelle during ripening, should take into account the enormous changes which take place in the properties of cell organelles during the process, such as the distribution of organelles during density gradient centrifugation and the location of marker enzymes.

The failure of recent work to add to our understanding of PPO is partly due to the 'applied' approach which guided many of these studies and partly to difficulties inherent in the properties of the enzyme and the tissues in which it is found. High levels of enzyme are usually found in tissues which are also rich in phenolic compounds. The membranous location of the enzyme in many cases requires harsh extraction measures, which involve denaturation of the protein. Tanning and denaturation often affect the multiple form pattern of the enzyme as well as the apparent molecular weight(s), specificity for substrates (especially the hydroxylation reaction), enzyme kinetics, sensitivity to inhibitors, etc. The in-vitro translation products of poly A^+ mRNA from eight plant species produced one prominent protein of 45 kD which was precipitated by antibodies against PPO (Flurkey, 1986). This was in contrast to the multiplicity pattern observed for PPO in many of these species and in spite of the fact that antibodies can also react with latent forms of the enzyme (Lanker *et al.*, 1987, 1988). Although post-translational modifications might be responsible for the formation of multiple forms, the observations described suggest that at least part of the multiplicity encountered is an artefact. New reports on the apparent molecular weight of PPO in various fruits do not help to clarify the picture of the primary structure of the protein. Mango (Katwa *et al.*, 1982), yam (Anosike & Ayaebene, 1981), cocoyam (Anosike & Ojimelukwe, 1982), and grapes (Nakamura *et al.*, 1983) were reported to possess only one form of PPO of 137, 115, 47 and 40 kD, respectively. Banana (Mowlah *et al.*, 1982), kiwifruit (Park & Luh, 1985) and DeChaunac, hybrid grapes (Lee *et al.*, 1983), on the other hand, were reported to have multiple forms (from three up to 14) of various sizes. Although multiplicity in a hybrid could result from genetic differences between the parent species, it is probably not the only reason for the multiple forms observed. In any case, the properties of the authentic, undenatured form of the enzyme in higher plants and its relation to the primary structure of PPOs from other sources, e.g. fungi, awaits reliable determination. Conclusive answers might be more easily obtained by isolation and sequencing of the respective genes and perhaps even by expressing them, by means of a suitable vector, in cells which are free of phenolic substrates for the enzyme.

Attempts to purify the enzyme continue to employ phenol binding agents such as poly vinylpyrrolidone and ion exchange resins (Smith & Montgomery, 1985) along with routine methods of protein fractionation. Hydrophobic gel filtration media, such as phenyl Sepharose CL 4B, are being increasingly used in chromatography of the enzyme (Lee *et al.*, 1983; Wisseman & Montgomery, 1985). Isoelectric focusing (Interesse *et al.*, 1981) is also occasionally used. The powerful tool of affinity chromatography has so far been employed only to a limited extent in the purification of PPO, mostly with mushroom tyrosinase (Golan-Goldhirsh *et al.*, 1986; Ingebrigtsen & Flurkey, 1988). Zawistovski *et al.* (1988*b*) used an interesting approach in the purification of PPO from Jerusalem artichoke. They employed copper immobilised by chelation to imino diacetic acid for affinity chromatography of the enzyme. The addition of inhibitors of proteolytic enzymes during the extraction and fractionation of PPO might attain wider use since it was reported to decrease the number of 'isozymes' observed (Flurkey & Jen, 1980). Although a considerable degree of purification was achieved in some cases (Hasegawa & Maier, 1980; Interesse *et al.*, 1982), only few cases of purification of PPO to homogeneity (as judged by silver staining after polyacrylamide gel electrophoresis) were reported (e.g. Wisseman & Montgomery, 1985) and these too usually consisted of more than one form of the enzyme.

Information on the catalytic and other properties of the enzyme usually comes from partially purified preparations, often containing more than one active form of the enzyme. We will cite only a few such reports. Many of the partially purified or even crude preparations lack the labile cresolase activity. This is particularly true for ripe fruits. Of interest is the requirement for NADH for the hydroxylation of *p*-coumaric acid but not for that of phloridzin by *Malus pumila* catechol oxidase and the unusually low *Km* value (0·2 μM) of the enzyme for 3-hydroxyphloridzin and caffeic acid, while the affinity for oxygen remained quite low (Goodenough *et al.*, 1983). Unusual, too, is the oxidative decarboxylation of 3,4-dihydroxymandelate to 3,4-dihydroxybenzaldehyde by mushroom tyrosinase (Sugumaran, 1986). Ito *et al.* (1984) confirmed and elaborated on earlier reports about the oxidation of tyrosyl residues in proteins by mushroom tyrosinase. Interaction of the resulting DOPA-quinone residues with cysteinyl groups produces 5-*S*-cysteinyl 3,4-DOPA attached to the protein.

Unfortunately for the efforts to prevent browning during food processing, no significant additions to our knowledge have been made in recent studies on potential inhibitors of catechol oxidases apart from the observations on mimosine and tropolone already mentioned. The inhibition by sulphur dioxide received further attention (Sayavedra-Soto & Montgomery, 1986). The inhibition of mushroom tyrosinase by relatively high concentrations of some amino acids (Kahn, 1985), which might be due to their ability to chelate copper, recalls earlier reports on inhibition of PPO by oligopeptides (Mayer & Harel, 1979). This phenomenon and its possible use in food technology has not received the attention it deserves in applied research efforts. Of the numerous reports describing latency and activation of PPO, the ones from cell suspension cultures of carrot are of particular interest (Carlberg *et al.*, 1985; Soederhall *et al.*, 1985; Carlberg, 1987). The enzyme can be purified to electrophoretic homogeneity in an inactive form. It is converted to an active form by millimolar concentrations of calcium chloride or by trypsin at lower calcium concentrations. This activation is inhibited by soyabean and carrot trypsin inhibitors. The activity of PPO in carrot cells is restricted to embryogenic cultures and correlates with their embryogenic capacity. Such cells secrete a 67 kD protein activator of PPO.

A vast gap exists between the information on the properties of fungal tyrosinase and laccases from both fungi and *Rhus* on one hand and higher plant PPOs on the other. We will briefly mention some of the new data obtained on the molecular properties, active site structure and reaction mechanism of the enzymes from fungi since no comparable data are available on PPOs from plants. Some of the detailed reviews on tyrosinase and laccase have been recorded in the introductory section.

The complete amino acid sequence of *N. crassa* tyrosinase was deduced by sequencing the complement of peptides obtained by cyanogen bromide cleavage of the purified protein. The enzyme is composed of a single 46 kD polypeptide chain, which contains an unusual thioether bridge between the cysteinyl residue at 94 and a histidyl at 96 (Lerch, 1982; Lerch *et al.*, 1982). Ruegg *et al.* (1982) went on to show that a thermostable form of the enzyme has an aspartic acid at position 201 while the thermolabile form has asparagine at this position. Since thermostability increases with ionic strength, the authors suggested that it is achieved by formation of an ionic bridge between the aspartate and another group, possibly histidine. These studies demonstrate the degree of sophistication achieved and possibilities opened by the research on fungal tyrosinases. The fungal enzyme is a monooxygenase containing a coupled copper pair, which is the site of interaction with both oxygen and the phenolic substrate. This type of a copper active site, referred to as 'type 3 copper', is found also in haemocyanin, laccase, ascorbate oxidase and ceruloplasmin. The molecular and active site structure of tyrosinase were discussed recently by Lerch (1987). He compared the amino acid sequence of tyrosinases from *S. glaucescens* and *N. crassa,* focussing on the domains which are apparently involved in the ligation of copper. The bacterial enzyme is much smaller than the fungal one, 273 as opposed to 407 amino acid residues, and the overall homology between the two is rather low (24·2%). However, two domains show more than 50% homology and hence were presumed to play an important role in the active site structure. Chemical and spectroscopic studies have suggested similarities between the structure of the active site of tyrosinase and haemocyanin. Sequence data for the two proteins were therefore compared and related to the three-dimensional structure of haemocyanin obtained by X-ray crystallography. The latter shows that the two copper ions Cu(A) and Cu(B) in the cluster are ligated by different domains of the polypeptide chain, each containing three histidyl residues. The Cu(B) site shows considerable similarity among the three proteins. However, the second copper site of haemocyanin shows only limited sequence homology to the region in *N. crassa* tyrosinase which was demonstrated to contain the Cu(A) site by photoinactivation studies. *S. glaucescens* tyrosinase completely lacks this site and appears to possess another region with two histidine residues and a high degree of sequence homology to a domain in the *N. crassa* tyrosinase. This latter region appears to be the Cu(A) site in the *S. glaucescens* enzyme, a suggestion that was supported by site-directed mutagenesis of the gene. Lerch (1987) went on to speculate that the bacterial Cu(A) and Cu(B) sites were the original ligands of the copper pair in the type 3 copper proteins. The Cu(A) site was replaced during evolution by the site noted in *N. crassa,* which shows similarity to the Cu(A) region of haemocyanin. It is suggested that the tyrosinase molecule has increased in size during evolution at both the amino and carboxy ends of the protein. The X-ray analysis of haemocyanin clearly showed that these extensions in the arthropod protein shield the binuclear copper complex in a way that enables binding of oxygen but prevents interaction with organic molecules, characteristic of tyrosinase.

Lerch (1987) also analysed the various derivatives of the binuclear cluster of copper ions of tyrosinase and haemocyanin, on which a wealth of chemical and spectroscopic data has accumulated. He summarised the properties of the five forms identified in the enzyme: met, oxy, dimer, deoxy and half met, (most of which result from treatment of tyrosinase with redox reagents or inhibitors) and their suggested structure and interaction with ligands. Based on the information about the electronic and geometric structure of the binuclear copper cluster, on the way different ligands interact with it and on rapid-kinetics studies, a scheme for the mechanism of action of tyrosinase was proposed (Fig. 1). The catecholase activity involves oxidation of two *o*-diphenols to two *o*-quinones with concomitant $4e^-$ reduction of molecular oxygen to two water molecules. It starts with the binding of one *o*-diphenol to the met-form of the enzyme which is followed by reduction of the bicupric cluster to form deoxytyrosinase. The latter forms oxytyrosinase upon the binding of oxygen. Oxytyrosinase binds the second *o*-diphenol which reduces the peroxide to water and is itself converted to *o*-quinone. In the cresolase activity, a monophenol is coordinated to one of the coppers in oxytyrosinase, followed by rearrangement to a trigonal bipyramidal intermediate, in which a deprotonated polarised peroxide is bound to the other copper. This hypothetical, highly reactive species hydroxylates the properly positioned monophenol by an electrophilic attack on the aromatic ring. The resulting *o*-diphenol is then oxidised from the equatorial position, with the formation of an *o*-quinone and deoxytyrosinase.

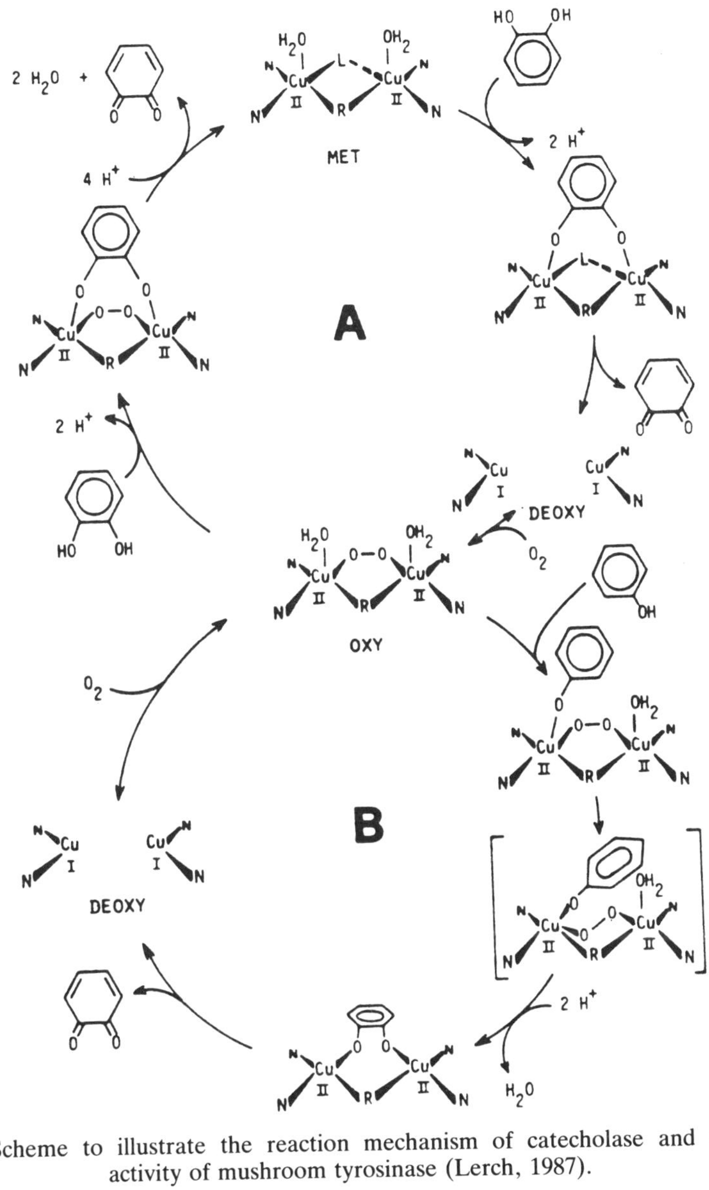

FIG. 1. Scheme to illustrate the reaction mechanism of catecholase and cresolase activity of mushroom tyrosinase (Lerch, 1987).

The laccases are glycoproteins which contain four copper ions per molecule. Together with ceruloplasmin and ascorbate oxidase, they belong to the class of 'blue' copper oxidases which catalyse the reduction of oxygen to water by a variety of substrates, such as polyphenols, aromatic polyamines and ascorbic acid (see Malmström *et al.*, 1975; Reinhammar, 1985 for reviews). Laccase has been purified from several sources: *Rhus vernicifera, Polyporus versicolor, N. crassa* and *Podospora anserina*. The laccase from *Botrytis* appears to be rather different, both in molecular weight and in the degree of glycosylation

(Marbach *et al.*, 1983, 1984). With the exception of the enzyme from *Podospora,* which has four subunits, laccases apparently consist of a single polypeptide chain of more than 500 amino acid residues. The differences in molecular weight reported for the enzyme from various sources (64–140 kD) can be ascribed to differences in the carbohydrate content which ranges from 14 to 45%. The high carbohydrate content is probably the reason for the failure to crystallise the enzyme. The four copper ions of laccase are divided into three types of copper centres based on spectral and other features (see Table 1). The two mononuclear sites—type 1 (or 'blue' copper) and type 2 ('normal' copper), and type 3 (EPR silent) or coupled binuclear site. Type 3 is associated with an absorption feature at 330 nm. Azide binding studies suggest that the type 2 and type 3 centres are very close to each other, creating a trinuclear centre, with oxygen binding and stabilisation of reduction intermediates occurring at the type 2 centre (Morpurgo, 1987; Solomon *et al.*, 1987). Some of the copper at site 2 can be reversibly removed and its removal also affects absorption due to type 3 copper (Hanna *et al.*, 1988). While oxygen oxidises deoxytyrosinase to the oxy derivative, binding as a $\mu-1,2$ bridging peroxide, it is irreversibly reduced by laccase to water molecules, using four electrons from the organic substrate. A wealth of information has been obtained on the spectroscopic, magnetic and oxidation–reduction properties of laccase. However, the data about its molecular structure are still limited and there is no consensus on the reaction mechanism of the enzyme and the way the various copper centres are involved in it (Blumberg & Peisach, 1987; Morpurgo, 1987; Solomon *et al.*, 1987; Hanna *et al.*, 1988).

The genes coding for the two allelic forms of *N. crassa* laccase have recently been isolated and sequenced (Germann *et al.*, 1988). The two forms differ in 12 out of 619 amino acid residues. A somewhat higher divergence is found in the nontranslated regions. The coding part of the two genes is interrupted by a short intron. The enzyme is synthesised as a precursor, with a 49 amino acid residue extension at the amino terminus and a 13 residue extension at the carboxy end. The amino terminal and possibly also glycosylation are believed to be involved in the secretion of the extracellular enzyme. The carboxy terminal extension is suggested to interact with the active site preventing the binding of copper ions. The resulting latent form of the enzyme is activated by proteolytic processing. It is interesting to note that *N. crassa* tyrosinase was also demonstrated to derive from a precursor having a C-terminal extension. Whether other laccases behave in the same way remains to be determined.

IMPORTANCE OF PHENOL OXIDASES IN FRUITS, VEGETABLES AND THEIR PRODUCTS

As already stated in the introduction, the physiological function of the phenol oxidases in plant and fungal tissues is still unclear. The most likely functions are (i) Action as oxygen buffers. (ii) Affording some degree of protection against pathogens, due to the great reactivity of their reaction products. (iii) Deterrence of predators, by making the tissues unpalatable due to the interaction of phenol oxidation products with cell components. (iv) Signalling, by decreases in activity, that the fruit has reached a stage of ripeness so that seeds can be distributed, by increasing palatability. The functions which have been discussed by a number of authors (Swain, 1962; Mayer & Harel, 1979, 1981; Mayer, 1987, 1989, Vaughn *et al.*, 1988) indicate the importance of these enzymes in fruits and vegetables. Two aspects must be considered in this respect. These are the function of the enzymes in the plant tissue during its growth and development on the one hand, and their importance in the actual processing or utilisation of the tissues for eventual human consumption. It is plausible that an enzyme which is an evolutionary 'veteran' is of great survival value to the species. Yet, despite this intuitive conclusion it is at present impossible to point to any clear evolutionary advantage afforded to plants or fungi due to the presence of phenol oxidases. No proof of such an advantage has ever been provided, or even sought. Most of the physiological work, in which enzyme activity has been described or followed, is deficient in this respect. Almost no attempts have been made to study the possible effect of reducing or removing enzyme activity either by selective breeding or by the techniques of molecular biology. One can probably account for this failure in the research. First, it is as yet extremely difficult to obtain phenol oxidase-free mutants. Secondly, anti-sense mRNA for PPOs or laccases have not yet been employed. Thirdly, even if the aforementioned are achieved, the physiological role and significance for survival must be established over a long period, probably several generations and in situations in which selective pressures are present. The lack of function, in a food crop, over one or two generations when such selective pressure may be absent can tell us little about the possible long-term importance of phenol oxidases in a tissue. We would suggest that this is a major area of research on the phenoloxidases which requires attention. It should be mentioned that phenoloxidase-free mutants of fungi have been developed (for a review of some of the data see Mayer & Harel, 1979; Mayer, 1987). These experiments have not produced the hoped-for clear-cut results.The reason for the failure to obtain clear-cut results is the same as that already discussed above with regards to the higher plants, i.e. the systems studied were unsuitable for the purpose of determining function.

The second aspect—the importance of phenoloxidases in the processing of food products—is much clearer. Here it should first be stressed that phenoloxidase activity is **not** always detrimental. In a number of food products such activity plays an important role in determining the quality of the product. We will mention only a few examples. PPO activity in cocoa beans is essential for product quality and colour (Holden, 1961; Roberts, 1961; Quesnel & Jugmohunsingh, 1970; Villeneuve *et al.*, 1985) and its activity is also significant in coffee (Amorim & Amorim, 1976). In vinification, the presence of phenoloxidases in the must is an important part of the fermentation process and is clearly linked to wine quality (Poux, 1966; Wisseman & Lee, 1980; Cantarelli, 1986). It is mostly detrimental but sometimes beneficial. At the same time the presence of extraneous phenoloxidase activity, laccase, originating in infecting *Botrytis cinerea,* may be extremely detrimental (Ribereau-Gayon *et al.*, 1980). The importance of tannins and flavanoids in tea leaves can hardly be overemphasised from the point of view of product quality (Roberts, 1962; Swain, 1962; Jain & Takeo, 1984; Pierpoint, 1985; Reeves *et al.*, 1987). The precise nature of these phenolics, which are significant for both colour and flavour, is determined by the presence and amount of PPO in the tea leaves and tips during the fermentation process (Sanderson 1964*a,b*; Coggon *et al.*, 1973; Robertson, 1983*a,b,c*). Equally in beer production, oxidation of phenolics in the hops and barley is responsible for some aspects of flavour and colour (Meilgaard & Peppard, 1986) and the same is true for cider in which apple PPO oxidises some of the procyanidins during fermentation (Goodenough & Lea, 1979). In the examples mentioned, retention of a suitable level of PPO during processing is very important for the quality of the final product.

In contrast to these examples of desirable effects of the activity of phenoloxidases in foods, we must now recall some of the adverse effects. From the point of view of the consumer, two aspects are significant. The first is simply the visual aspect. Fruits, e.g. apples or bananas, which brown rapidly on cutting or bruising, or vegetables such as globe artichokes or potatoes, which may brown prior to cooking or after minor injury, are simply unsightly and hence unappealing. The same applies to mushrooms incorrectly treated during cooking. Some of these aspects were reviewed by Mathew and Parpia (1971). These browning reactions are not necessarily detrimental to the final product and need not change its quality, nutritional value, shelf-life or flavour. However, as already mentioned the products of phenoloxidase activity—quinones—are highly reactive compounds. Consequently, they undergo a variety of reactions with other constituents of fruits and vegetables during cutting, drying, processing, preservation, etc. Such reactions include those of quinones with thiol groups, with amino acids, with phenolics, coumarins, sugars and flavour components of various kinds. The interaction of the products of phenolic oxidation, mediated by PPO, with such compounds can readily have an adverse effect on the products by, for example, producing off-flavours. For reviews, see Mathew and Parpia (1971) and Morton and McLeod (1986). The interaction of quinones with proteins can reduce their digestibility and palatability and reduce their nutritional value (Matheis & Whitaker, 1984; Pierpoint, 1985). From these few examples it is very clear that the presence of phenoloxidases in fruits and vegetables can profoundly affect the quality of a food product, not only because discoloured products have an adverse influence on customer appeal but also because the actual nutritional value of the food may be changed. Since the phenolics are so widespread in food products and often act as toxicants (Singleton, 1981), the enzyme which can modify their structure is of obvious importance in foods and food products. Attempts have been made to use PPO in a controlled fashion in order to obtain stable natural colours (Motoda, 1982*a,b*; Cantarelli, 1986) or to eliminate turbidity (Devreux & Jerumanis, 1983) in food products. Some browning phenomena which are observed during cooking are quite unrelated to PPO activity, but are due to other reactions, e.g. the Maillard reaction (Waller & Feather, 1983; Danehy, 1986).

PREVENTION OF BROWNING

Because enzymic browning phenomena in food products are very often undesirable, considerable effort has been devoted to finding means of preventing or reducing such enzymic browning.

The strategies which have been devised to achieve this can be summed up under the following general headings:

(1) The use of inhibitors of PPOs.
(2) The addition of reducing agents, which immediately reduce the oxidised phenolics, thereby preventing the occurrence of secondary reactions.
(3) The use of processing techniques by which the enzyme, which is primarily particulate, is rapidly separated from the normally soluble phenolic compounds.
(4) Reduction of the level of substrates in the tissues by selection and breeding for lines or cultivars in which the level of phenolics is low, or absent.
(5) Reduction of the level of PPOs by selection or breeding for cultivars with reduced level of enzyme activity or in which the enzyme is absent.

These five possible strategies will be briefly discussed.

The use of inhibitors

A large number of inhibitors of PPOs, competitive or noncompetitive or reacting with copper, are known (Mayer & Harel, 1979, 1981; Mayer, 1987) and new ones are constantly being added. However, the vast majority of such inhibitors is not entirely specific for PPOs. Many are unacceptable as food additives because of possible toxicity or other undesirable properties. The most commonly used inhibitors are sulphur dioxide or sodium chloride. The use of sulphur dioxide as a food additive is increasingly restricted by the laws limiting its use all over the world (e.g. Langdon, 1987). The use of sodium chloride is based on the inhibitory action of the halogen ion, which is pH dependent (Rouet-Mayer & Philipon, 1986). The addition of sodium chloride, although fundamentally innocuous, is often undesirable because the concentrations needed are quite high and because the taste of the products is inevitably altered. Another form of inhibition is heat inactivation of PPO and other enzymes by a blanching process. Obviously, this is inapplicable to fresh fruits and vegetables and therefore does not provide a general solution to the prevention of browning. Irradiation of fruits and vegetables does not solve the browning problem, indeed it may aggravate it. The changes in permeability of cell compartments resulting from irradiation can lead to greater contact between enzyme and substrate, which in the intact tissue are normally separated.

Use of reducing agents

The addition of a reducing agent to a fruit or vegetable product to prevent, or rather to reverse, oxidation and therefore reduce browning is quite common, the most frequent reducing agent used being ascorbic acid. This is regarded as permissible, but may not be desirable since the amounts of reducing agent required are large, in direct proportion to the enzyme activity present. Whether the massive accumulation of the oxidation products of ascorbic acid are in fact as innocuous, as is assumed, has not apparently been considered or studied. Sulphur dioxide in the form of bisulphite acts as a reducing agent in addition to acting as an inhibitor. Its action is pH dependent (Sayavedra-Soto & Montgomery, 1986; Taylor *et al.*, 1986). As already stated, its use is being severely restricted. Another effective reducing agent is cysteine, but the same considerations which apply to ascorbic acid also apply.

Separation between enzyme and substrate

In a number of situations, particularly juice production, e.g. from apples, it is possible to rapidly separate the PPO, which appears primarily in the subcellular particles of the pulp, from its substrates which are present in the juice itself. Such a rapid separation by filtration or centrifugation is essentially very effective where it can be used, and can often considerably reduce the extent of browning. It has been shown to be important in vinification (Sapis *et al.*, 1983; Yokotsuka *et al.*, 1988). It is also possible to actually remove the phenolic substrates by adsorption onto some more-or-less inert compound such as polyvinylpyrrolidone. Obviously, this necessitates the eventual removal of the adsorbent, but the procedure is basically simple, provided the adsorbent is sufficiently specific. The prevention of access to one of the other substrates of the browning reaction, oxygen, is also feasible by simply reducing the oxygen tension in the tissue or plant product. Such a prevention of access of oxygen is used in certain steps of the vinification process. It is effective, relatively inexpensive and very often quite simple. It must be remembered that the prevention of access of oxygen to intact fruits or vegetables can of course lead to undesirable results due to the onset of fermentative reactions in the tissue. However, the K_m of PPO for oxygen is much higher, by two to three orders of magnitude, than that of normal oxidative respiratory pathways.

Selection for low levels of enzyme or substrate

Selection for cultivars which have either low levels of enzyme or substrate is quite feasible and has been achieved in a number of cases. The techniques of modern plant breeding, together with those of molecular biology, and of propagation by cell or tissue cultures should make this a relatively easy task. It should be possible to obtain plant tissues deficient in, or with much reduced levels of, either enzyme or substrate or both, since browning is generally correlated with both, e.g. in olives (Sciancalepore, 1985) or tea (Lelyveld & Rooster, 1986), to cite recent examples.

However, we would like to suggest that the long-term effects of such selection may be totally self-defeating. It is now generally accepted that at least one of the biological functions of the PPO–phenolics interaction is in plant defence mechanisms. Such defence is either against various predators, insects or against microbial, fungal or even viral attack. The precise mechanism of such defence is still very much open to debate, but the basic idea is accepted by most plant biologists. Related to this idea is the concept that competition between plants and between plants and other organisms is closely linked to the great variety of secondary plant metabolites and the enzymes concerned in their metabolism. Consequently, it may be suggested that any reduction in the level of a secondary metabolite such as a phenolic compound or in the enzymes which can oxidise it, the PPOs, will have a direct or indirect effect on the ability of the plant to compete successfully, to deter predators or to resist disease. Such a reduction in competitiveness may, eventually, lead to reduced yield or in extreme cases total loss of product, due to the loss of the ability to withstand some previously unsuspected or less-prevalent disease. Because the phenolic–PPO reaction is apparently fairly nonspecific and its products have high biological activity they might afford some degree of nonspecific protection even against previously unknown disease factors. This aspect has not been investigated experimentally (see also discussion above on the determination of the function of PPOs). It seems, therefore, that this suggestion merits investigation.

The problem of prevention of browning in any given fruit or vegetable or their products clearly has not been solved. Its solution will require the simultaneous use of more than one of the strategies discussed above. Total prevention of browning may not only be impossible, but also undesirable. Instead, efforts should be devoted to a better appreciation of the fundamental function(s) of PPO in plant tissues.

ADDENDUM

Since the manuscript was submitted a number of papers have appeared which are significant to this topic. We would like to mention a few of these papers as adding to the subject reviewed.

Two recent papers review certain aspects of PPO and their significance in foods (Flurkey & Ingebrigtsen, 1989; Sapers & Hicks, 1989). An addition to the methods of purification of the enzyme comes from Ferrer-Sanchez *et al.*

(1989), who showed that by the use of Triton X-114 the chloroplastic enzyme can be readily purified, without contamination by phenolics.

Steffens *et al.* (1990) and Kowalski *et al,* 1990 have demonstrated the occurrence of two distinct PPOs, occurring respectively in the mesophyll cells and trichome hairs of *Solanum berthaultii.* Those two PPOs differ in molecular weight and apparently the trichomal one which can be easily solubilised is involved in the resistance of the plant to small insects. This trichomal enzyme has a role in insect entrappment and resistance to the Colorado beetle (Neal *et al.,* 1989). This PPO is present as two allelic isozymes.

Recent work by Flurkey (1989), suggests that four distinct PPO polypeptides are present in *Vicia,* which probably were derived from a common parent form, but underwent post-translational modification. This is the first such report for any PPO.

REFERENCES

Adnin, T. B. T. A., Augustin, M. A. & Ghazali, H. M. (1986). *Pertanika,* **9,** 219–24.
Allan, A. C. & Walker, J. R. L. (1988). *Phytochem.,* **27,** 3075–6.
Amorim, H. V. & Amorim, V. L. (1976). *ACS Symposium Series,* **47,** 27–56.
Anosike, E. O. & Ayaebene, A. O. (1981). *Phytochem.,* **20,** 2625–8.
Anosike, E. O. & Ojimelukwe, P. C. (1982). *J. Exper. Bot.,* **33,** 487–94.
Augustin, M. A., Ghazali, H. M. & Hashim, H. (1985). *J. Sci. Food Agric.,* **36,** 1259–65.
Badiani, M., Felici, M., Luna, M. & Artemi, F. (1983). *Anal. Biochem.,* **133,** 275–6.
Bajaj, K. L. & Kaur, G. (1986). *Trop. Sci.,* **26,** 121–8.
Baldini, V. L. S., Draetta, I. S. & Ghazali, E. H. (1982). *Coletanea do Instituto de Technologia de Alimentos,* **12,** 283–91.
Ben-Arie, R. & Or, E. (1986). *J. Amer. Soc. Hort. Sci.,* **111,** 395–9.
Bligny, R. & Douce, R. (1983). *Biochem. J.,* **209,** 489–96.
Blumberg, W. E. & Peisach, J. (1987). *Life Chem. Rep.* **5,** 5–36.
Bonner, W. D. (1957). *Ann. Rev. Plant Physiol.,* **8,** 427–52.
Bower, J. P. & Lelyveld, L. J. van (1985). *J. Hort. Sci.,* **60,** 545–7.
Brooks, D. W. & Dawson, C. R. (1966). In *The Biochemistry of Copper,* ed. J. Peisach, P. Aisen & W. E. Blumberg. Academic Press, New York, USA, pp. 343–57.
Butt, V. S. (1985). *Ann. Proc. Phytochem. Soc. Europe,* **25,** 349–66.
Cabanes, J., Garcia-Canovas, F., Tudela, J., Lozano, J. A. & Garcia-Carmona, F. (1987). *Phytochem.* **26,** 917–19.
Cantarelli, C. (1986). *Vini d'Italia,* **28,** 87–98.
Carlberg, I. (1987). Acta Universitatis Upsaliensis, *Comprehensive Summaries of Uppsala Dissertation from the Faculty of Science,* Uppsala, Sweden, 72.
Carlberg, I., Soederhall, K. & Erikson, T. (1985). *FEBS Lett.,* **187,** 295–8.
Coggon, P., Moss, G. A. & Sanderson, G. W. (1973). *Phytochem.,* **12,** 1947–55.
Danehy, J. P. (1986). *Adv. Food Res.,* **30,** 77–138.
Delauney, A. J., Tabaeizadeh, Z. & Verma, D. P. S. (1988). *Proc. Nat. Acad. Sci. USA,* **85,** 4300–4.
Devreux, A. F. O. & Jerumanis, J. (1983). US Patent US-4411914.
Drawert, F., El-Refai, A., Vellios, A., Radola, B. J. & Matzner, F. (1985). *Chem. Mikrobiol. Technol. Lebensm.,* **9,** 48–56.

Farver, O. & Pecht, I. (1984). In *Copper Proteins and Copper Enzymes* (Vol. 1), ed. R. Lontie. CRC Press, Boca Raton, FL, USA, pp. 183–214.
Ferreira, V. L. P., Graner, M., Bovi, M. L. A., Draetta, I. D. S., Paschoalino, J. E. & Shirose, I. (1982). *Coletanea do Instituto de Tecnologia de Alimentos,* **12,** 255–72.
Ferrer-Sanchez, A., Bru, R. & Garcia-Carmona, F. (1989). *Plant Physiol.,* **91,** 1481–7.
Flurkey, W. H. (1985). *Plant Physiol.,* **79,** 564–7.
Flurkey, W. H. (1986). *Plant Physiol.,* **81,** 614–18.
Flurkey, W. H. (1989). *Plant Physiol.,* **91,** 481–3.
Flurkey, W. H. & Ingebrigtsen, J. (1989). *ACS Symposium Series,* **40,** 44–54.
Flurkey, W. H. & Jen, J. J. (1980). *J. Food Biochem.,* **4,** 29–41.
Fujita, S. & Tono, T. (1982). *J. Jap. Soc. Food Sci. Technol.,* **29,** 214–20.
Germann, U. A., Muller, G., Hunziker, P. E. & Lerch, K. (1988). *J. Biol. Chem.,* **263,** 885–96.
Golan-Goldhirsh, A., Kahn, V. & Whitaker, J. R. (1986). *Is. J. Bot.,* **35,** 203–14.
Goodenough, P. W. & Lea, A. G. H. (1979). Report, Long Ashton Research Station, Long Ashton, UK, pp. 207–14.
Goodenough, F. W., Kessell, S. D., Lea, A. G. H. & Loeffler, T. (1983). *Phytochem.,* **22,** 359–63.
Goto, M., Minamide, T., Fujii, M. & Iwata, T. (1984). *J. Jap. Soc. Hort. Sci.,* **53,** 210–18.
Hanna, P. M., McMillin, D. R., Pasenkiewicz-Gierula, M., Antholione, W. E. & Reinhammar, B. (1988). *Biochem. J.,* **253,** 561–8.
Harkin, J. M. & Obst, J. R. (1973). *Experientia,* **29,** 381–5.
Hasegawa, S. & Maier, V. P. (1980). *J. Agric. Food Chem.,* **28,** 891–7.
Henry, E. W., O'Connor, M. N., Shnak, S. M. & Demorrow, J. M. (1981). *Photochem. Photobiophys.,* **3,** 69–76.
Holden, M. (1961). In *The Chemistry of Cocoa,* The British Food Manufacturing Industries Research Association, Scientific and Technical Survey No. 38, Leatherhead, Surrey, UK, pp. 31–40.
Huystee, R. B. van, & Chibbar, R. N. (1987). In *Isozymes-Current Topics in Biological and Medical Research* (Vol. 13), Alan R. Liss Inc., New York, pp. 155–79.
Ikediobi, C. O. & Oti, E. (1983). *J. Sci. Food Agric.,* **34,** 1123–9.
Ingebrigtsen, J. & Flurkey, W. H. (1988). *Phytochem.,* **27,** 1593–99.
Interesse, F. S., Ruggiero, P., D'Avella, G. & Lamparelli, F. (1980). *J. Sci. Food Agric.,* **31,** 459–66.
Interesse, F. S., Ruggiero, P., Lamparelli, F. & D'Avella, G. (1981). *Z. Lebensm. Unters. Forsch.,* **172,** 100–3.
Interesse, F. S., Ruggiero, P., Lamparelli, F. & D'Avella, G. (1982). *J. Agric. Food Chem.,* **30,** 1198–203.
Ito, S., Kato, T., Shinpo, K. & Fujita, K. (1984). *Biochem. J.,* **222,** 407–11.
Jain, J. C. & Takeo, T. (1984). *J. Food Biochem.,* **8,** 243–79.
Jarrah, A. Z. & Benjamin, N. D. (1982). *Date Palm J.,* **1,** 5–18.
Joel, D., Marbach, I. & Mayer, A. M. (1978). *Phytochem.,* **17,** 796–7.
Kahl, G., Gorz, A. & Schafer, W. (1988). *2nd Int. Cong. Plant Mol. Biol. Abstr.,* No. 95.
Kahn, V. (1985). *J. Food Sci.,* **50,** 111–15.
Kahn, V. & Andrawis, A. W. (1985*a*). *Phytochem.,* **24,** 905–8.
Kahn, V. & Andrawis, A. W. (1985*b*). *Phytochem.,* **24,** 909–13.
Kahn, V. & Andrawis, A. W. (1985*c*). *Phytochem.,* **24,** 915–20.
Kahn, V. & Pomerantz, S. H. (1980). *Phytochem.,* **19,** 379–85.
Katwa, L. C., Ramakrishna, M. & Raghavendra Rao, M. R. (1982). *J. Food Biochem.,* **6,** 217–28.
Kean, E. A. (1964). *Biochim. Biophys. Acta,* **92,** 602–4.
Keegstra, K. & Bauerle, C. (1988). *BioEssays,* **9,** 15–19.

Keilin, D. & Mann, T. (1940). *Nature,* **145,** 304.
Kidron, M., Harel, E. & Mayer, A. M. (1978). *Amer. J. Enol. Vitic.,* **29,** 30–5.
Kohorn, B. D. & Tobin, E. M. (1989). *Plant Cell,* **1,** 159–66.
Kowalski, S. P., Bamberg, J. B., Tingey, W. M. & Steffens, J. C. (1990). *J. Heredity,* **81:** 475–8.
Lamkin, W. M., Miller, B. S., Nelson, S. W. & Traylor, D. D. (1981). *Cereal Chem.,* **58,** 27–31.
Langdon, T. T. (1987). *Food Technol.,* **41,** 66–7.
Lanker, T., King, T. G., Arnold, S. & Flurkey, W. H. (1987). *Physiol. Plant.,* **69,** 323–9.
Lanker, T., Flurkey, W. H. & Hughes, J. P. (1988). *Phytochem.,* **27,** 3731–4.
Lax, A. R., Vaughn, K. C. & Templeton, G. E. (1984). *J. Heredity,* **75,** 285–7.
Lee, C. Y., Smith, N. L. & Pennesi, A. P. (1983). *J. Sci. Food Agric.,* **34,** 987–91.
Lelyveld, L. J. van & Rooster, K. de (1986). *J. Hort. Sci.,* **61,** 545–7.
Lerch, K. (1982). *J. Biol. Chem.,* **257,** 6414–19.
Lerch, K. (1987). *Life Chem. Rep.,* **5,** 221–34.
Lerch, K., Longoni, C. & Toreli, E. (1982). *J. Biol. Chem.,* **257,** 6408–13.
Lerner, H. R., Harel, E., Lehman, E. & Mayer, A. M. (1971). *Phytochem.,* **10,** 2637–40.
Levine, W. G. (1966). In *The Biochemistry of Copper,* ed. J. Peisach, P. Aisen & W. E. Blumberg. Academic Press, New York, USA, pp. 371–87.
Malmström, B. G. (1970). *Pure Appl. Chem.,* **24,** 393–406.
Malmström, B. G., Andreasson, L. E. & Reinhammar, B. (1975). In *The Enzymes* (Vol. XII), ed. P. D. Boyer. Academic Press, Orlando, FL, USA, pp. 557–79.
Mangalakumari, C. K., Sreedharan, V. P. & Mathew, A. G. (1983). *J. Food Sci.,* **48,** 604–6.
Marbach, I., Harel, E. & Mayer, A. M. (1983). *Phytochem.,* **22,** 1535–8.
Marbach, I., Harel, E. & Mayer, A. M. (1984). *Phytochem.,* **23,** 2713–17.
Marsh, D. R. & Galliard, T. (1986). *J. Cereal Sci.,* **4,** 241–8.
Mason, H. S. (1955). *Adv. Enzymol.,* **16,** 105–84.
Matheis, G. & Whitaker, J. R. (1984). *J. Food Biochem.,* **8,** 137–62.
Mathew, A. G. & Parpia, H. A. B. (1971). *Adv. Food Res.,* **19,** 75–145.
Mayer, A. M. (1987). *Phytochem.,* **26,** 11–20.
Mayer, A. M. (1989). *Phytochem.,* **28,** 311–17.
Mayer, A. M. & Harel, E. (1968). *Phytochem.,* **7,** 1253–6.
Mayer, A. M. & Harel, E. (1979). *Phytochem.,* **18,** 193–215.
Mayer, A. M. & Harel, E. (1981). In *Recent Advances in the Biochemistry of Fruits and Vegetables,* ed. J. Friend & M. J. C. Rhodes. Academic Press, London, UK, pp. 159–80.
Mayer, A. M., Harel, E. & Shain, Y. (1964). *Phytochem.,* **3,** 447–51.
Mayer, A. M., Harel, E. & Ben Shaul, R. (1966). *Phytochem.,* **5,** 783–9.
Meilgaard, I. D. & Peppard, T. L. (1986). In *Food Flavours* (Part B), 2nd impression, ed. I. D. Morton & A. J. McLeod. Elsevier, Amsterdam, The Netherlands, pp. 99–170.
Moore, B. M., Kang, B. & Flurkey, W. H. (1988). *Phytochem.,* **27,** 3735–7.
Morpurgo, L. (1987). *Life Chem. Rep.,* **5,** 277–88.
Morton, I. D. & McLeod, A. J. (eds) (1986). *Food Flavours* (Parts A and B), Elsevier, Amsterdam, The Netherlands.
Motoda, S. (1982*a*). *J. Jap. Soc. Food Sci. Technol.,* **29,** 11–15.
Motoda, S. (1982*b*). *J. Jap. Soc. Food Sci. Technol.,* **29,** 25–30.
Mowlah, G. & Itoo, S. (1982). *J. Jap. Soc. Food Sci. Technol.,* **29,** 413–17.
Mowlah, G., Takano, K., Kamoi, I. & Obara, T. (1982). *Lebensm.-Wiss. Technol.,* **15,** 207–10.
Nakamura, K., Amano, Y. & Kagami, M. (1983). *Amer. J. Enol. Vitic.,* **34,** 122–7.
Neal, J. J., Steffens, J. C. & Tingey, W. M. (1989). *Entol. Exp. Appl.,* **51,** 133–40.

Okamoto, A., Imagawa, H., Arai, Y. & Ozawa, T. (1988). *Agric. Biol. Chem.*, **52,** 2215–22.
Parish, R. W. (1972). *Eur. J. Biochem.*, **31,** 446–55.
Park, E. Y. & Luh, B. S. (1985). *J. Food Sci.*, **50,** 678–84.
Pierpoint, W. S. (1985). In *Ann. Proc. Phytochem. Soc. Europe* (Vol. 25), ed. C. F. van Sumere & P. J. Lea. Oxford Univ. Press, Oxford, UK, pp. 427–51.
Pilando, L. S., Wrolstad, R. E. & Heatherbell, D. A. (1985). *J. Food Sci.*, **50,** 1121–5.
Pomeranz, S. H. (1966). *J. Biol. Chem.*, **214,** 161–8.
Poux, C. (1966). *Ann. Technol. Agric.*, **15,** 149–58.
Pras, N., Wichers, H. J., Bruins, A. P. & Malingre, T. M. (1988). *Plant Cell Organ Cult.*, **13,** 15–26.
Quesnel, V. C. & Jugmohunsingh, K. (1970). *J. Food Sci.*, **21,** 537–45.
Reeves, S. G., Owuor, P. O. & Othieno, C. O. (1987). *Trop. Sci.*, **27,** 121–33.
Reinhammer, B. (1984). In *Copper Proteins and Copper Enzymes* (Vol. 3), ed. R. Lontie. CRC Press, Boca Raton, FL, USA, pp. 1–35.
Reinhammar, B. & Malmstrőm, B. G. (1981). In *Copper Proteins, Metal Ions In Biology* (Vol. 3), ed. T. G. Spiro. Wiley, New York, USA, pp. 107–49.
Rhee, J. K. & Iwata, M. (1982). *J. Jap. Soc. Hort. Sci.*, **51,** 362–8.
Ribereau-Gayon, J., Ribereau-Gayon, P. & Seguin, G. (1980). In *The Biology of Botrytis.* ed. J. R. Coley-Smith, K. Verhoeff & W. R. James. Academic Press, London, UK, pp. 251–74.
Richard, J. E. (1985). *J. Sci. Food Agric.*, **36,** 167–76.
Robb, D. A. (1984). In *Copper Proteins and Copper Enzymes* (Vol. 2.), ed. R. Lontie. CRC Press, Boca Raton, FL, USA, pp. 207–40.
Roberts, E. A. H. (1962). In *The Chemistry of the Flavanoids,* ed. T. A. Geisman. Pergamon Press, Oxford, UK, pp. 468–552.
Roberts, J. B. (1961). In *The Chemistry of Cocoa,* The British Food Manufacturing Industries Research Association, Scientific and Technical Survey No 38, Leatherhead, Surrey, UK, pp. 41–46.
Robertson, A. (1983*a*). *Phytochem.*, **22,** 883–7.
Robertson, A. (1983*b*). *Phytochem.*, **22,** 889–96.
Robertson, A. (1983*c*). *Phytochem.*, **22,** 897–903.
Rouet-Mayer, M.-A. & Philipon, J. (1986). *Phytochem.*, **25,** 2717–19.
Ruegg, C., Ammer, D. & Lerch, K. (1982). *J. Biol. Chem.*, **257,** 6420–6.
Sanchez de Medina, L., Plata, M. C., Martinez-Cayuela, M., Faus, M. J. & Gil, A. (1987). *Rev. Agric. Tec. Alimentos,* **26,** 529–38.
Sanderson, G. W. (1964*a*). *Biochim. Biophys. Acta,* **92,** 622–4.
Sanderson, G. W. (1964*b*). *J. Sci. Food Agric.*, **9,** 634–9.
Santana, M. C., Shlutzky, B. & Alonso, A. (1982). *Agrotecnia de Cuba,* **14,** 167–74.
Sapers, G. M. & Hicks, K. B. (1989). *ACS Symposium Series* **405,** 29–43.
Sapis, J. C., Macheix, J. J. & Cordonnier, R. E. (1983). *Amer. J. Enol. Vitic.*, **34,** 157–62.
Sayavedra-Soto, L. A. & Montgomery, M. W. (1986). *J. Food Sci.*, **51,** 1531–6.
Schmidt, G. W. & Mishkind, M. (1986). *Ann. Rev. Biochem.*, **55,** 879–912.
Sciancalepore, V. (1985). *J. Food Sci.*, **50,** 1194–5.
Shomer, I., Ben-Shalom, N., Harel, E. & Mayer, A. M. (1979). *Ann. Bot.*, **44,** 261–3.
Sidhu, A. S., Kaur, G. & Bajaj, K. L. (1982). *Vegetable Sci.*, **9,** 112–18.
Singleton, V. L. (1981). *Adv. Food Res.*, **27,** 149–242.
Smith, C. J. S., Watson, C. F., Ray, J., Bird, C. R., Morris, P. C., Schuch, W. & Grierson, D. (1988). *Nature,* **334,** 724–6.
Smith, D. M. & Montgomery, M. W. (1985). *Phytochem.* **24,** 901–4.
Soederhall, I. & Soederhall, K. (1989). *Phytochem.*, **28,** 1805–8.
Soederhall, K., Carlberg, I. & Erikssen, T. (1985). *Plant Physiol.*, **78,** 730–3.
Solomon, E. I. (1981). In *Copper Proteins*: *Metal Ions in Biology* (Vol. 3), ed. T. G. Spiro. Wiley, New York, USA, pp. 41–108.

Solomon, E. I. (1988). *ACS Symposium Series,* **372,** 116–50.
Solomon, E. I., Allendorf, M. D., Lung-Shan Kau, Pate, J. E., Spira-Solomon, D., Wilcox, D. E. & Porras, A. G. (1987). *Life Chem. Rep.,* **5,** 37–89.
Spruce, J., Mayer, A. M. & Osborne, D. J. (1987). *Phytochem.,* **26,** 2901–3.
Steffens, J. C., Kowalski, S. P. & Yu, H. (1990). In *Molecular Biology of the Potato,* ed. M. Vayda & W. Park. Wiley, New York, USA, pp. 103–12.
Sugumaran, M. (1986). *Biochem.,* **25,** 4489–92.
Suhaila, M. & Leong, S. H. (1987). *Pertanika,* **10,** 143–9.
Swain, T. (1962). In *The Chemistry of the Flavanoids,* ed. T. A. Geisman. Pergamon Press, Oxford, UK, pp. 513–52.
Taira, S., Sugiura, A. & Tomana, T. (1987). *J. Jap. Soc. Food Sci. Technol.,* **34,** 612–15.
Taylor, S. L., Highley, N. A. & Bush, R. K. (1986). *Adv. Food Chem.,* **30,** 1–76.
Tolbert, N. E. (1973). *Plant Physiol.,* **51,** 234–44.
Udayasekhara Rao, P. & Deosthale, Y. G. (1987). *J. Food Sci.,* **52,** 1549–51.
Vaughn, K. C. & Duke, S. O. (1981). *Protoplasma,* **108,** 319–27.
Vaughn, K. C. & Duke, S. O. (1984). *Physiol. Plant.,* **60,** 106–12.
Vaughn, K. C., Lax, R. & Duke, S. O. (1988). *Physiol. Plant.,* **72,** 659–65.
Vijay, S. (1986). *Ind. Food Packer,* **40,** 7–10.
Villeneuve, F., Cros, E. & Macheix, J. J. (1985). *Café-Cacao–Thé,* **29,** 113–20.
Walker, J. R. & McCallion, R. F. (1980). *Phytochem.,* **19,** 373–7.
Waller, G. R. & Feather, M. S. (eds) (1983). *ACS Symposium Series* **213.**
Walter, W. M. & Schadel, W. E. (1982). *J. Food Sci.,* **47,** 813–17.
Watson, R. A. & Flurkey, W. H. (1986). *J. Sci. Food Agric.,* **37,** 791–6.
Wichers, H. J., Peetsma, G. J., Malingre, T. M. & Huizing, H. J. (1984). *Planta,* **162,** 334–41.
Wisseman, K. W. & Lee, C. Y. (1980). *Amer. J. Enol. Vit.,* **31,** 206–11.
Wisseman, K. W. & Montgomery, M. W. (1985). *Plant Physiol.,* **78,** 256–62.
Yamaguchi, M., Henderson, H. M. & Hwang, P. M. J. (1969). *Anal. Biochem.,* **32,** 178–82.
Yokotsuka, K., Makino, S. & Singleton, V. L. (1988). *Amer. J. Enol. Vitic.,* **39,** 293–302.
Zawistowski, J., Blank, G. & Murray, E. D. (1987*a*). *Can. Inst. Food Sci. Technol. J.,* **20,** 162–5.
Zawistowski, J., Weselake, R. J., Blank, G. & Murray, E. D. (1987*b*). *Phytochem.,* **26,** 2905–7.

10

Peroxidases and their Significance in Fruits and Vegetables

DAVID S. ROBINSON
Procter Department of Food Science, University of Leeds, Leeds LS2 9JT, UK

INTRODUCTION

Products of the action of peroxidases in both horseradish and milk were first observed approximately 180 years ago: it was observed that extracts of guaiacum turned blue in the presence of, what we now know to be, peroxidase present in both horseradish and milk. Fundamental investigations began in 1920 with the work of Onslow (1920) who observed that the activity of oxidising enzymes varied with the ripeness of fruit. These early studies benefited from the observation that the enzyme could be detected by the use of guaiacum and hydrogen peroxide. It was also found by Willimot and Wokes (1926) that peroxidase was present in both the flavedo and albedo of citrus fruits. During the 1930s, Keilin and co-workers observed the importance of the haem component for peroxidase activity (Keilin & Mann, 1937). The colour changes of peroxidase in the presence of hydroperoxide, observed by these researchers, led to detailed studies by Chance (1951) who showed that the peroxidase–hydrogen peroxide complexes, compound 1 and compound 2, were important intermediates in peroxidase catalysed reactions (Chance, 1952).

The enzyme from horseradish has been the most extensively studied peroxidase because of its abundance in horseradish and, consequently, its availability. The elucidated mode of action of horseradish peroxidase together with knowledge of its molecular structure have served to illustrate, by way of example, the probable structure and role of many other peroxidases universally present in fruits and vegetables. Nevertheless, even today the precise roles and physiological functions of peroxidases in living plants and post-harvest products are still not fully understood.

PEROXIDASE ACTIVITY AND FOOD QUALITY

It has been suggested that peroxidase activity could be a possible parameter in the ripening and senescence of Golden Delicious apples (Gorin & Heidema,

1976). After harvesting, the peroxidase activity of apricots increased four- to five-fold whereas for comparison, the level of polyphenoloxidase activity increased by only 10% (Vamos-Vigyazo *et al.*, 1985). However, the exact physiological functions of the ubiquitous peroxidases remain obscure, although their roles in peroxidatic and oxidatic reactions, lignification and degradation of indoleacetic acid are well established (Fils *et al.*, 1985).

Physiological stress, wounding, fungal and viral infections all bring about changes in peroxidase isoenzymes. During these complex physiological changes the precise chemical reactions catalysed by peroxidases have not been fully elucidated; peroxidase can promote a large number of reactions and therefore has a degree of versatility not surpassed by any other similar enzyme. The picture is incomplete for a number of reasons. First, peroxidase does not exist as a single enzyme in fruits and vegetables; like many other plant enzymes, peroxidase activity is found in the form of a number of discrete isoenzymes, isoperoxidases, which are readily detected by isoelectric focusing. Secondly, there are many naturally occurring phenolic compounds in all plant tissues, each of which can be oxidised by peroxidase in the presence of small amounts of hydroperoxide. Thirdly, due to the diversity of compounds which are susceptible to oxidations catalysed by peroxidases, the range of products formed is very extensive. Consequently, food scientists have found it difficult to correlate perceived post-harvest changes, e.g. loss of flavour and texture, with specific chemical reactions which are generally thought to be catalysed by peroxidases. Furthermore, peroxidases can exist in different oxidation states and catalyse at least three different types of oxidation through a number of separate but often interlinked chemical reactions.

Peroxidase (EC 1.11.1.7) is a member of a large group of enzymes called the oxido-reductases and is considered to have an empirical relationship to off-flavours and off-colours in raw and unblanched vegetables. In post-harvest plant foods, peroxidases are believed to be responsible for deterioration in flavour, colour, texture and nutritional qualities of raw and processed foods (Burnette, 1977; Haard, 1977). A relationship between peroxidase activity and off-flavour development in peas is well established (Wagenknecht & Lee, 1958; Pinsent, 1962). Active enzyme can spoil fruits and vegetables at temperatures as low as -18°C and at low moisture levels, the development of off-flavours is often associated with the oxidising effect of peroxidases on indigenous lipids and the phenolic constituents of foods (Sessa & Anderson, 1981). For orange juice, it has been suggested that the significant negative correlation between peroxidase activities and the flavour scores of high- and low-yield orange juice indicates that it might be possible to use peroxidase activity as an index of potential adverse flavour for high-yield juice (Bruemmer *et al.*, 1976). An increased activity of peroxidase and content of pulp in the juice due to processing conditions was accompanied by lower quality of the juice. For string beans, a qualitative peroxidase test was found to be the best index of adequate blanching (Bedford & Joslyn, 1939). For scaled asparagus, Joslyn and Bedford (1940), using gum guaiac as substrate, found that flavour retention was closely related to the inactivation of peroxidase activity. Scalding of asparagus in water for 4 min at 92°C, or 3 min at 100°C, was sufficient for retention of flavour. Wagenknecht and Lee (1958) found that off-flavours developed when horseradish peroxidase was added in varying concentrations to pea slurries. Zoueil and Esselen (1959) also demonstrated that the addition of horseradish peroxidase to sterilised green-bean puree caused an increase in the concentration of acetaldehyde in stored samples. Similarly, earlier work by Arighi *et al.* (1936) and Joslyn (1949) showed the accumulation of volatile aldehydes in the tissues of raw and unblanched peas and Joslyn and David (1952) claimed that peas containing more than 10 ppm of volatile aldehydes were of objectionable flavour. Cherry and Ory (1973), working with various peanut cultivars, suggested that conditions in the mature peanut may be conducive to the synthesis of isoperoxidases during storage which may vary with seeds of cultivars grown in different geographical regions. Gardner *et al.* (1969) showed that peroxidase activity in maize varied with respect to anatomical part, and Davis (1942) reported that the greatest peroxidase activity was in the inner seed coat in citrus fruits. Peroxidase is now also suspected to have a role in the degradation of chlorophyll and senescence in plants and the ripening of fruits.

DISTRIBUTION OF PEROXIDASES IN CELLS

Much of the difficulty in understanding the effects of peroxidase activity in plant foods is due to the presence of multiple enzymes (isoperoxidases): early studies on peroxidase activity in plants demonstrated the existence of a complex system of isoenzymes (Shannon, 1968). Based on their isoelectric points, isoperoxidases have been roughly classified into acidic, neutral and basic groups: the most basic (cationic) isoperoxidases generally have isoelectric points in the pH range 8·0–10·0, while the acidic (anionic) isoperoxidases have isoelectric points in the pH range 3·5–6·0 (McLellan & Robinson, 1984; 1987*a,b*; Moulding *et al.*, 1987, 1988). The differences within a single group, e.g. between anionic A1 and A2 isoperoxidases, are not thought to be very significant. Several isoperoxidases can be allocated to a discrete number of genes. For a given group of isoperoxidases, for example the acidic isoenzymes, the minor differences in isoelectric point may represent small mutations and post-translation modification. Different electrophoretic patterns found for different anatomical parts of plants and more mature tissues may also reflect genetic variation and minor post-synthesis modification. Peroxidase activity is also expected to be controlled by genetic regulation of enzyme synthesis (Wijsman &Hendriks, 1986).

Clear solutions containing peroxidase activity are generally obtained by relatively low-speed centrifugation (e.g. 1500 g) of tissue homogenates in buffer solution. Such a solution is normally designated the soluble or cytoplasmic peroxidase fraction, although the solution may still contain suspended intracellular membranes, probably with bound peroxidases. The

1500 g pellet consists mainly of cell walls, possibly some large vacuoles, a few nuclei and any partially broken cells. After extensive washing of the pellet, the residual ionically-bound peroxidases can be extracted from the insoluble pellet using 1 M salt solutions (McLellan & Robinson, 1981; Moulding *et al.*, 1987; Ros Barcelo *et al.*, 1988). This simple fractionation procedure has proved valuable for the rapid separaton of mainly the cationic peroxidases which appear to be ionically bound to crude cell wall material. It seems that these ionically bound isoperoxidases are mainly of the cationic type and rarely contain a substantial quantity of anionic isoenzymes (Thomas *et al.*, 1981; McLellan & Robinson, 1983; Moulding *et al.*, 1987). However, a full complement of isoperoxidases, including both the anionic and cationic types, is present in the soluble fraction which would be expected to contain suspended intracellular endoplasmic reticulum and Golgi membranes. However, it is possible that the cationic peroxidases are not necessarily located in the cell walls in native plants since binding might arise during homogenisation and in that sense their location would be an artifact of cell breakage, as recently suggested by Schloss *et al.* (1987) and referred to by Goldberg *et al.* (1989). The localisation of bound peroxidases has not been sufficiently investigated, although Naveh *et al.* (1981) developed a chemiluminescent method for the quantitative analysis of bound peroxidases. Furthermore, Ros Barcelo *et al.* (1987) reported that cell wall isoperoxidases from etiolated lupin hypocotyls have acidic pI values. The mixture of extracted acidic peroxidases in the soluble fraction, possibly released during homogenisation, can be purified by the use of electrophoretic and chromatographic techniques.

The location, transport and biosynthesis of peroxidases has been examined by Gaspar *et al.* (1986). The enzymes seem to be located in all parts and organelles of the cell, although they may vary in number and activity with physiological changes. Fractionation of intracellular membranes by iso-pynic sucrose density gradient centrifugation (Thomas *et al.*, 1981) has shown that lupin anionic isoperoxidases are associated principally with membranes of the endoplasmic reticulum and to a lesser extent with the Golgi apparatus and plasmalemma-derived membranes. The locations are in accord with the biosynthesis of peroxidases on polysomes attached to the rough endoplasmic reticulum and with the pathway

$$\text{endoplasmic reticulum} \rightarrow \text{Golgi} \rightarrow \text{plasmalemma}$$

inferred from cytochemical probes for the intracellular transport of acidic lupin iso-enzymes to the cell wall. Cytochemical studies (Thomas *et al.*, 1981) have shown that tomato fruit peroxidase is localised on the inner membrane surface of the tonoplast and the cell walls, while other researchers have concluded that part of the tomato cellular peroxidase is bound to the cell wall (see Thomas *et al.*, 1981). Peroxidase was not found in the cytoplasm of tomato fruit cells but only on the cell walls and the inner-membrane surface on the tonoplasts (Thomas *et al.*, 1981). For tomatoes, it has been assumed that soluble peroxidases obtained on tissue extraction were derived from the tonoplast owing to the weak attachment of the enzymes to the membranes (Thomas *et al.*, 1981). The subcellular location of membrane-bound acidic isoperoxidases in etiolated lupin hypocotyls has been reported (Ros Barcelo *et al.*, 1988).

Following biosynthesis and transport of peroxidases, it has been claimed (Gaspar *et al.*, 1986) that the final locations of peroxidases are likely to be cell walls, intracellular free spaces and vacuoles. The transport of peroxidases towards their actual site of action may also be considered as a control mechanism of cellular peroxidase activity. It has been suggested that the binding of cathodic isoperoxidases is dependent on Ca^{2+}-dependent receptors (Penel, 1986). Other post-translational controls of activity are likely to involve several cations, including calcium, and the incorporation of sugar residues into at least some of the isoperoxidases. For instance, it has been established (Bakardjieva *et al.*, 1987) that calcium intensifies the release of enzyme protein from culture cells and increases peroxidase activity (Penel, 1986). Indeed, there are grounds to believe that Ca^{2+} are a specific and necessary factor for the secretory function of plant cells and that they play a significant role in the control of peroxidase activity. Fry (1986*b*) has claimed that it is primarily the molecular forms of peroxidases connected with some of the functions of cell walls like lignification which are released in the culture medium. The calcium-induced increase in peroxidase activity has been attributed to, especially, an increase in the activity of the cationic isoperoxidases and is thought to be under the control of intracellular calcium, initiated by stimuli-induced release of potassium (Gaspar *et al.*, 1985). The release of peroxidase activity by such a mechanism can occur within minutes of any physical damage or stress and, hence, can affect the quality of the product. However, the biosynthesis or the mechanism of formation of cationic isoperoxidases is not understood as these isoenzymes, unlike the acidic isoperoxidases, do not rapidly incorporate labelled protein precurors. Gaspar *et al.* (1986) suggested that a rapid increase of cationic peroxidase activity in many systems results from activation of enzymes already present in tissue rather than from de-novo protein synthesis. It has been pointed out by Gaspar *et al.* (1986) that because anionic isoperoxidases readily incorporate labelled amino acids, indicating a higher turnover, genetically regulated protein synthesis is involved in their control and therefore it seems, based on Gaspar *et al.* (1986), that the secretion of anionic isoperoxidases may be more dependent on the nature of the physiological process than is the case for cationic isoperoxidases. It has also been suggested (Gaspar *et al.*, 1986) that the enhanced activity of cationic isoperoxidases in some cases may have been due to a de-masking effect brought about by a rapidly decreased level of phenols or ascorbic acid. It has also been claimed (Castillo, 1986) that ozone treatment results in the simultaneous release of at least one natural substrate, i.e. ascorbic acid, and that basic peroxidases have a much higher affinity towards the substrate than the acidic isoenzymes. Given the complex-bound and facile states in which isoperoxidases exist in cells, it is to be expected that such enzymes will have a very variable catalytic action towards a wide range of compounds. Further, it is not surprising that food storage and the different methods of processing used in practice affect the action of these enzymes. Only a fuller understanding of both the specificity of the various iso-peroxidases in a wide range of fruit and vegetables and the chemistry of the degradation intermediates, which often occur as free radicals, will facilitate the production of quality-controlled products.

THERMOSTABILITY OF PEROXIDASE ACTIVITY IN A RANGE OF FRUITS AND VEGETABLES

Thermal inactivation of peroxidases has been studied for many years, because the action of the enzymes not only causes unfavourable colour and off-flavour changes during food preservation but also because the level of residual enzymic activity indicates the effectiveness of blanching treatments. Although thermostable enzymes may represent only a small proportion of the total enzymic activity in fresh foods, for blanched vegetables and fruits their effects on the quality may be quite substantial. The high correlations between off-flavour development in frozen vegetables during storage and the residual peroxidase activity (Nebesky *et al.*, 1951; Guyer & Holmquist, 1954; Labbee & Esselen, 1954) has made peroxidase activity useful as an index for the degree of blanching.

Effective blanching of some vegetables is particularly difficult, not only because of the presence of thermostable enzymes, but also because of either their large size, shape or density. For instance, for Brussels sprouts, due to their compact structure, inactivation of enzymes throughout the sprout may be particularly difficult to achieve (McLellan & Robinson, 1987*a*). Similarly, corn-on-the-cob is one of the most difficult vegetables to blanch effectively because of its large size and the structural characteristics of its different parts (Garrote *et al.*, 1987). Long blanch times are necessary to inactivate the enzymes at all points within the cob. Lee and Hammes (1979) determined significant correlations between residual peroxidase activity in the outer cob and kernels blanched at 100°C and the off-flavour development detected by a panel after 9 months storage at −18°C. Assuming a first-order reaction for the thermal inactivation of peroxidases, it was calculated, using linear regression equations, that to obviate off-flavour development it was necessary to reduce the residual peroxidase activity to less than 20% in the outer cob and 1% in the kernels (Garrote *et al.*, 1987). However, it was calculated that a temperature of 100°C for 30 min was necessary to attain stability during storage of all sizes of corn-on-the-cob.

The rates of inactivation and reactivation of peroxidase activity have been reported (Kermasha *et al.*, 1988) to be dependent on many factors, such as temperature and time of blanching, pH value and sodium chloride concentration. Bruemmer *et al.* (1976) suggested that the loss of peroxidase activity in pasteurised orange juice is pH-dependent. The changes in peroxidase activity of *Phaseolus vulgaris* cv. 'haricot seed' at selected stages of development, of blanching, storage and drying, have been reported (Kermasha *et al.*, 1988). These authors also found an increase in haricot seed peroxidase activity after frozen storage which could have been due to freeze damage. The results showed that the peroxidase activity increased rapidly during the first few hours after thawing and reached a maximum after 36 h. For air-drying, the peroxidase activity decreased to 25% of that in fresh green haricot seed, whereas storage of air-dried seeds for 3 years at 4°C had little effect on peroxidase activity. However, freeze-drying reduced the peroxidase activity by a dramatic 96% in haricot seed (Kermasha *et al.*, 1988). Cherry *et al.* (1972) reported that peanuts showing growth of *Aspergillus parasiticus* have an increased level of peroxidase activity. The thermostability of peroxidase activity has been investigated in other vegetables, e.g. turnips (Servant *et al.*, 1986) and cabbages (McLellan & Robinson, 1981, 1983, 1987*b*), and in fruits, e.g. apples (Moulding *et al.*, 1987, 1988), mangoes, pears (Moulding *et al.*, 1989), grapes (Robinson *et al.*, 1990) and citrus products (McLellan & Robinson, 1984). The flavour of mango is very delicate and is easily affected by thermal processing. Mango is marketed in the form of juice, pulp or slices canned in syrup, but the thermal processes followed at present for these canned products are mainly empirical (Nanjundaswamy *et al.*, 1973).

Heat inactivation studies have been carried out over a number of years on peroxidase activity extracted from a range of vegetables and fruits. For such investigations, a simple centrifugal separation of the isoenzymes into soluble and bound types has, for a number of commodities, enabled a rapid but partial fractionation of the isoperoxidases into anionic and cationic types. The cell-wall-bound peroxidase activity isolated in this way from Spring cabbage and Brussels sprouts has been shown to be due mainly to non-covalently bound cationic isoenzymes, whereas the soluble peroxidase activity consists of both the cationic and anionic isoenzymes (McLellan & Robinson, 1981). Generally, for *Brassica* spp. the soluble peroxidase activity has been shown to be more heat stable than the peroxidase activity associated with the cationic isoenzymes extracted from the ionically bound fractions (McLellan & Robinson, 1981, 1987*a*,*b*). Likewise, for apples (Moulding *et al.*, 1988), pears (Moulding *et al.*, 1989) and grapes (Robinson *et al.*, 1990), the extracted ionically-bound isoperoxidases, containing mainly cationic isoenzymes, seem to be more heat labile. Nevertheless, for oranges the bound peroxidase activity was more thermostable than that found in the soluble fraction (McLellan & Robinson, 1984). Thus, the extract of each of a range of fruits and vegetables certainly includes a mixture of isoperoxidases with a wide range of isoelectric points and varying thermostability. Attempts have been made by Chang *et al.* (1988) using electrophoretic methods, to identify, the more thermostable horseradish isoperoxidases. However, although it was claimed for the mixture of horseradish isoperoxidases that the most heat stable isoenzymes had pI values of between 7·5 and 9·0, their proposal was based only on a non-quantitative visual assessment of enzymic activity as revealed by staining after separation of the isoperoxidases by electrophoresis.

Mechanisms for heat inactivation

Most studies on the thermal inactivation of peroxidase have been made on the assumption that the mechanism of inactivation of the enzyme follows first-order reaction kinetics. However, the plots of residual activity against heating time generally show non-linearity which indicates a deviation from first-order kinetics. For example, the plots in Fig. 1 show considerable deviation from linearity, especially beyond the point where more than 50% of peroxidase activity has been destroyed at 76°C. Several reasons and mechanisms have been suggested for the non-linear loss of peroxidase activity with heating time, but so far none of the theories has been validated unambiguously. The most obvious, and therefore the most favoured, explanation for the deviation from linearity is one due to the claimed presence of a mixture of thermostable and heat-labile isoperoxidases (Yamamoto *et al.*, 1962; Lu & Whitaker, 1974; Ling & Lund, 1978; Naveh *et al.*, 1982; Yoon & Park, 1982; Lee & Pennesi, 1984). While it seems very likely that some of the deviation is probably accounted for by the presence of isoperoxidases with different susceptibilities to heat, this has not yet been clearly established as the main cause of the deviation from linearity. Furthermore, it has not been shown that purified individual thermostable isoperoxidases, as opposed to the isolated heat labile isoenzymes, exhibit linear first-order thermal-inactivation kinetics. To date, it has proved difficult to completely purify and isolate sufficient quantities of the individual thermostable isoperoxidases. For corn ears, Naveh *et al.* (1982) formulated a kinetic model to take account of a mixture of several isoperoxidases, with each assumed to be inactivated in a first-order manner but with different rate constants. An equation was given for calculating the total residual activity in the form of an integrated Arrhenius equation. It has been claimed (Naveh *et al.*, 1982) that there is a good fit between the model

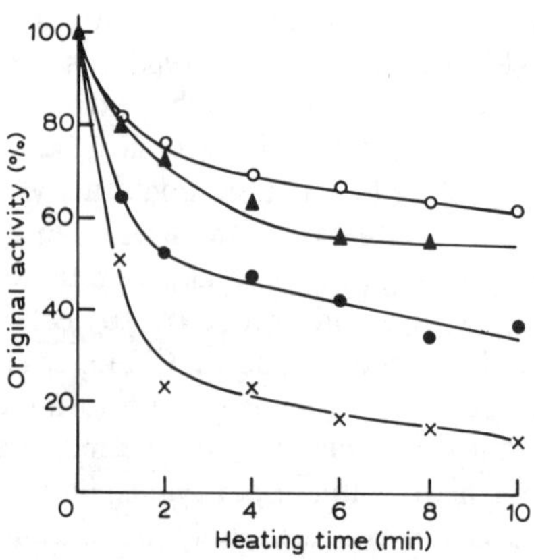

FIG. 1. Heat inactivation of orange juice soluble peroxidase. Heating at (○) 60°C, (▲) 65°C, (●) 70°C, and (×) 75°C. (From McLellan & Robinson (1984).)

described by the equation and corrected kinetic data which indicates that the model is sufficiently accurate for practical blanching calculations. The results suggest that the inactivation of peroxidase activity can be practically described by assuming two enzyme fractions with vastly different thermal stabilities. However, without biochemical proof of the existence of different isoenzymes with different heat stabilities, then the observed phenomena could also be due to the presence of isoperoxidases in different aggregation states which possess different thermostabilities. Indeed, other tenable explanations have been proposed for the deviation from linearity and include the formation of aggregates of the thermostable isoperoxidases which may possess an increased heat stability (Winter, 1971), or alternatively, a series-type enzyme inactivation kinetics induced by heat (Henley & Sadana, 1985). For other thermostable proteins it has been suggested that aggregation may be a stabilising mechanism (Merkler *et al.*, 1988). Likewise, aggregation could also equally well account for the claim that the thermostable fraction represented 7% of the enzymic activity with the heat labile fraction representing 93% of the initial enzymic activity, as calculated by Naveh *et al.* (1982). Although small quantities of a thermostable isoperoxidase have been isolated from cauliflower (Lee & Pennesi, 1984), the heat inactivation plots for the isolated isoperoxidases were still non-linear.

As referred to above, considerable effort is being made to isolate and purify isoperoxidases from a wide range of sources. Of the three isoenzymes isolated from cauliflower using hydrophobic chromatography, the major isoenzyme, C, which accounted for 94% of the total peroxidase activity, has been shown to be very heat labile (Lee *et al.*, 1984): 98% of its activity was lost in 15 min at 50°C. The more heat stable cauliflower isoenzyme, A, had a $t_{1/2}$ value of 30 min at 50°C compared to the more heat labile isoenzyme, C, for which $t_{1/2}$ at 50°C was 3 min. Although the logarithmic plots of enzyme activity against heating time were still non-linear, estimation of the inactivation energies revealed that for the more thermostable isoenzyme an inactivation energy of 89·46 kJ/mol was required for denaturation, whereas for the heat labile isoenzyme the inactivation energy was 194·88 kJ/mol. Although the thermostable isoenzyme, A, accounted for only a small proportion of the total enzymic activity, because of its greater heat resistance it could appreciably affect the quality of blanched cauliflower during storage.

McLellan and Robinson (1987*a*) reported the isolation of four isoperoxidases from Brussels sprouts by gel filtration and ion exchange chromatography and compared their heat stabilities. The homogeneity of the isoperoxidase preparations was confirmed by isoelectric focusing. The nomenclature used reflects the positions of the isoenzymes (anionic, A, and cationic, C) on polyacrylamide gels after isoelectric focusing (Fig. 2). For purified Brussels sprout isoperoxidases A1, A2 and C1, heat treatment for 10 min at 70°C resulted in a 70–90% loss of measurable peroxidase activity. Isoperoxidase C4 was considerably more heat labile, some 90% inactivation being achieved following heat treatment at 60°C for 7·5 min and it was inactivated in a more

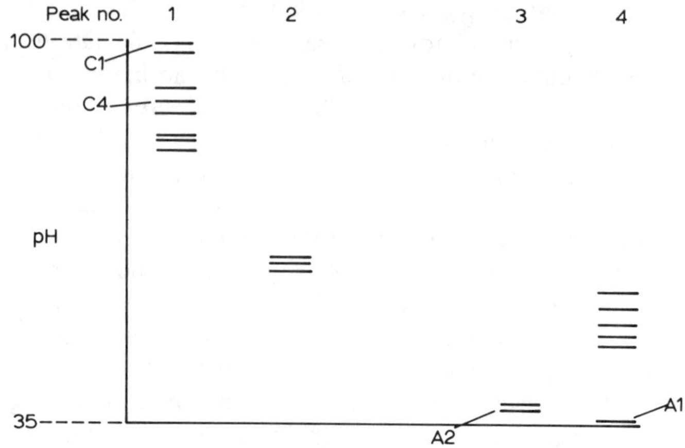

FIG. 2. Isoelectric focusing of peroxidase fractions obtained following ion-exchange chromatography on DEAE—Sephacel. Peak numbers refer to DEAE Sephacel elution profile. (From McLellan & Robinson (1987*a*).)

linear manner (Fig. 3) than the other isolated isoenzymes. The non-linear heat inactivation plots for the purified isoperoxidases A1, A2 and C1 (McLellan & Robinson, 1987*a*) may have arisen from an unknown property of only the more thermostable isoenzymes.

Similar non-linear plots have been found for isoenzymes purified from Spring cabbage. McLellan and Robinson (1987*b*) reported that purified Spring cabbage isoperoxidase A3.7, which was quite heat stable (only a 50% inactivation resulted from heating at 67°C for 3 min), still gave a non-linear heat inactivation plot. The degree of heat stability was comparable to that observed previously (McLellan & Robinson, 1981) when only 50% of the peroxidase activity present in crude soluble extracts of cabbage was destroyed after heating at 70°C for 3 min. Isoenzyme C9.9, which represented a large proportion of the total peroxidase activity, was, however, considerably more heat labile, and was 50% inactivated after heating for 5 min at 53°C. However, for both of these apparently pure isoperoxidases the heat inactivation plots showed non-linearity. Hence, although vegetable extracts do contain peroxidase isoenzymes with different heat stabilities, this does not fully explain the non-linear nature of heat inactivation plots for peroxidase. Thus, the non-linearity of the plots for the purified isoenzymes could be due to factors other than any undetected traces of other more stable isoperoxidases. It is generally observed that the higher the thermal stability of isoperoxidases, the more non-linear is the heat inactivation plot, an observation that is less likely to be accounted for by the presence of other isoperoxidases with even higher thermostabilities (McLellan & Robinson, 1987*a,b*). For purified apple isoperoxidases, the anionic isoperoxidases A1 and A2 were more thermostable than the cationic types and in particular isoperoxidase A2 retained at least 70% of its activity after heating at 70°C for 2 min (Moulding *et al.*, 1987).

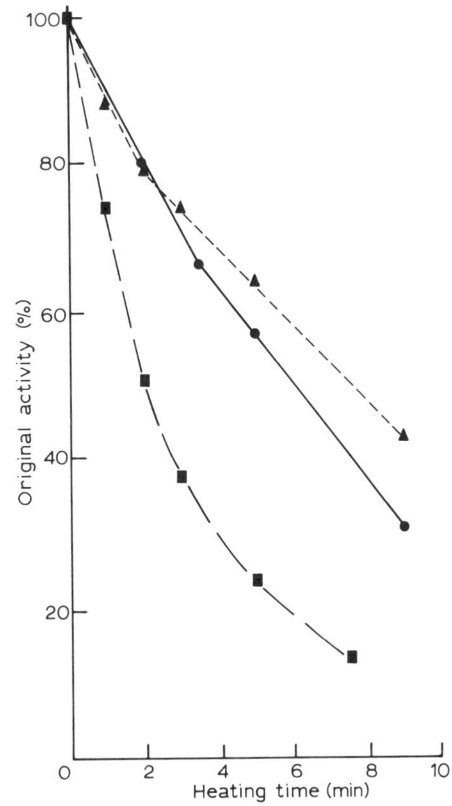

FIG. 3. Heat inactivation of Brussels sprout peroxidase iso-enzyme C4. (▲) 50°C, (○) 55°C, and (■) 60°C. (From McLellan & Robinson (1987*a*).)

Insight into the mechanism of heat inactivation can, in principle, be obtained by studying the effects of a number of experimental variables like pH, ionic strength, and the concentrations of different polyols and anions such as SCN^- on the kinetics of inactivation. Lu and Whitaker (1974) showed that the rate of inactivation of horseradish peroxidase decreased some eight-fold as the pH was increased from 4 to 7 and then two-fold as the pH was increased to 10. As added hematin reduced the rate of inactivation of peroxidase activity, it was claimed by Lu and Whitaker (1974) that the marked effect of pH on the rate of heat inactivation of peroxidase was associated with the release of hemin and that the apoprotein, devoid of haem, was much less heat stable than haem-protein complex. When studying the effect of ionic strength on the rate of inactivation, Lu and Whitaker (1974) observed that in the presence of sodium chloride there was a considerable decrease in the thermostability of horseradish peroxidase, with an eight-fold increase in the initial rate of inactivation of peroxidase in the presence of 0·6 M NaCl. Such a decrease in thermostability might be due to NaCl-induced changes in the conformation of the peroxidase and possibly the dissociation of thermostable aggregated molecules.

Tamura and Morita (1975), while studying the spectral changes on heating of concentrated solutions of peroxidases, observed decreases in absorbance at 404 nm (the maximum of the Soret band of peroxidase) and at 505 nm (Fig. 4). Cooling for 20 h at room temperature after treatment led to a recovery in the absorbance at 404 nm. Inactivation of the enzyme was marked at both pH 5 and 9, whereas at pH 7, heat inactivation at 60°C was slight. Tamura and Morita (1975) showed that the degree of regeneration was considerably lower at pH 5 and 9 than at pH 7 after prolonged heat treatments. The work of Tamura and Morita (1975) and the earlier investigations of Lu and Whitaker (1974) all support the idea that heat inactivation of peroxidases involves the dissociation of the heam moiety from the apoprotein, as well as the separate denaturation of the peroxidase protein. Indeed, Tamura and Morita (1975) demonstrated that peroxidase activity can be restored by the addition of prohemin to previously heat-treated apoperoxidase (Fig. 5). The increase in enzymic activity also coincided with an increase in absorbance in the Soret region at 404 nm. Therefore, as suggested by Lu and Whitaker (1974) and Tamura and Morita (1975), it seems likely that the inactivation of peroxidases during blanching involves the release of the haem moiety, possibly its degradation chemically, and the separate denaturation of the liberated apoprotein.

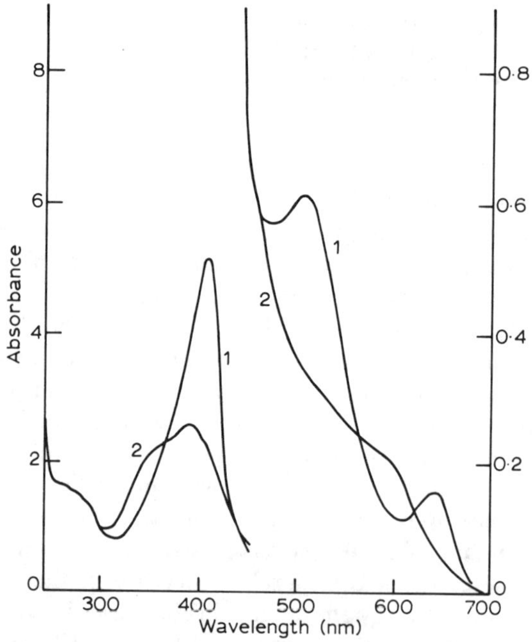

FIG. 4. Absorption spectra of heat-treated peroxidase at pH 7. 1: untreated, and 2: after heating for 30 min at 70°C. The enzyme concentration was 9·2 μM. (From Tamura & Morita (1975), reproduced by permission of the Japanese Biochemical Society.)

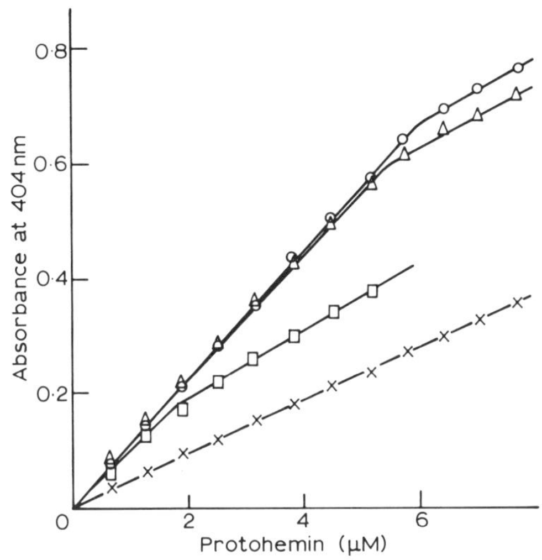

FIG. 5. Combination of protohemin with heat-treated apoperoxidase. Apoperoxidase (6·0 μM) dissolved in 2 ml of 0·1 M phosphate buffer, pH 7, was titrated spectrophotometrically by successive additions of 5 μl aliquots of 250 μM protohemin dissolved in 0·1 M NaOH in the dark. The apoenzyme was pre-treated by heating in buffers of various pH values for 30 min at 60°C. Apoenzyme pretreated at pH (□) 5, (○) 7, and (△) 9. (×) protohemin only. Untreated apoenzyme gave the same curve as that for pH-7 treatment. (From Tamura & Morita (1975), reproduced by permission of the Japanese Biochemical Society.)

In recent years there has been increased interest in the structural features, domains of proteins and small molecules, and additives generally that may be involved in determining thermostability; the wider aspects of the subject have been reviewed by O'Fagain *et al.* (1988). Since, generally, polyols and sugars have been found to protect the native conformation of proteins against thermal denaturation, the effects of polyhydric alcohols and sugars have recently been extended to studies on the denaturation of peroxidases. Chang *et al.* (1988) showed by the use of differential scanning calorimetry, that in the presence of sucrose the denaturation temperature of the protein increased. However, measurements of residual enzymic activity showed that in the presence of 10% sucrose the thermal stability of the enzymic activity was reduced. For a range of sugars tested, fructose was the most effective in reducing the thermostability of the enzymatic activity and it was suggested that this was due to the interaction of fructose, a reducing sugar, with the protein's amino acids. While attempting to explain these apparently contradictory results, Chang *et al.* (1988) suggested that the destabilising effect of low concentrations of sucrose might be due to an influence on the conformation of the enzyme's active site, which might then result in a change in thermostability, while the overall native structure of the enzyme remains unaltered below its denaturation temperature.

Regeneration of peroxidase activity

Regeneration of enzymic activity after heat denaturation is unusual for enzymes generally, although it is a well recognised property of peroxidases. The ability of peroxidases to regenerate after heat denaturation varies not only between different plant species but also between the isoenzymes that occur within a single variety. Restoration of peroxidase activity is generally observed over a period of a few hours after heat treatment of either test enzyme solutions or whole vegetables.

At the molecular level, the mechanism of regeneration of peroxidases is not understood. However, in general terms it is considered that regeneration must require a reversal of denaturation with concomitant changes in the conformational structure of the protein moiety. As it has been shown that added haematin increases the amount of restored enzymic activity (Lu & Whitaker, 1974), it has been generally accepted that the haem moiety can recombine with the apoprotein to form active enzyme. Lu and Whitaker (1974) observed maximum restoration of enzymic activity at pH 9. Tamura and Morita (1975) observed that, after heat treatment, cooling for 20 h at room temperature resulted in a recovery in the absorbance at 404 nm which indicates that the native molecule with bound haem can be reassembled under such conditions.

Lu and Whitaker (1974) showed that the percentage of original activity restored depended on the extent of inactivation and the process did not follow either first- or second-order kinetics but was complex. Both the rates of inactivation and reactivation of peroxidase were found to be pH dependent. Other factors and unknown components may also be required in some cases for the restoration of peroxidase activity, as McLellan and Robinson (1987*a*) found that purified Brussels sprout anionic isoperoxidases, unlike the crude preparations from Brussels sprouts and cabbage, did not regenerate following heat inactivation. On the other hand, for a purified apple isoperoxidase, A2, up to 80% of the original activity was restored when the heat-treated isoenzyme was held at 30°C (Moulding *et al.*, 1988). Cationic isoperoxidases in crude extracts from *Brassica* spp. and apples (McLellan & Robinson, 1981, 1987*a*; Moulding *et al.*, 1988) have been shown to be incapable of regeneration which implies that the structure of the protein may primarily determine this property. Naveh *et al.* (1982) showed an interesting linear relationship between residual activity immediately after heat treatment and subsequent regenerated peroxidase activity (Fig. 6). The large slope of approximately 3·0 for the graph implies that more drastic heat processing would not only increase inactivation but also markedly decrease the extent of regeneration. Also, the proportionality between increasing thermostability and the enzyme's ability to regenerate supports the general observation that the more thermostable enzymes tend to exhibit a greater ability to regenerate.

Commercial methods other than direct conductive heating, such as microwave and gamma-irradiation, can justifiably be expected to affect enzymic activity. The subject has been reviewed by Thomas (1987). Microwave heating

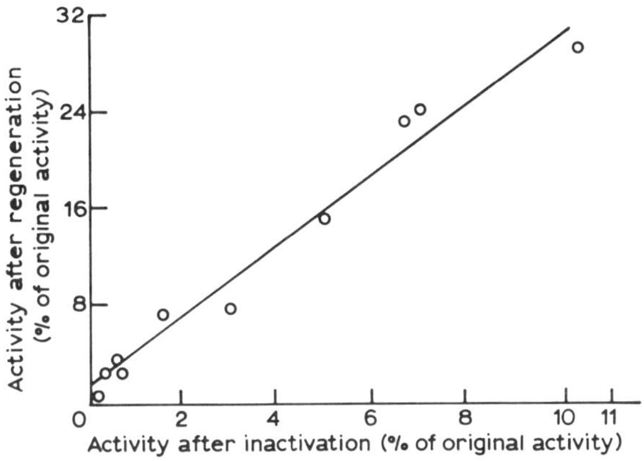

FIG. 6. Effect of extent of thermal inactivation on the reaction of peroxidase. (From Naveh *et al.* (1982), reproduced by permission of the American Chemical Society.)

causes rapid changes in temperature, not dependent on conduction, that might be particularly useful for heterogeneous foods. On the other hand, microwave radiation is not sufficiently intrinsically energetic to release electrons or generate radicals to cause inactivation of enzymes. Indeed, Lopez and Baganis (1971) showed that radio-frequency energy at 60 MHz had no significant effect on a number of enzymes studied, including peroxidases, aside from that caused by the heat effects of the radiation. Nevertheless, the potential for rapid heating and control that microwave processing offers may not yet have been sufficiently exploited for the blanching of plant foods. By way of example, Copson (1954) showed that microwave irradiation can be used to heat inactivate pectin methylesterase in orange juice without the adverse effects normally observed, provided that the juice was cooled quickly.

Irradiation of pure solutions of horseradish peroxidase using ^{60}Co gamma-rays caused a loss of enzymic activity which was highest in the presence of oxygen-saturated water and Metodiewa *et al.* (1987) suggested that hydroxy and peroxy radicals were responsible for the inactivation. It is reasonable to suppose that such inactivation is the result of secondary attack on peroxidase by free radicals produced from water during the irradiation process. The species involved are likely to include the hydroxy, peroxy and superoxide radicals. On the other hand, for irradiated citrus fruits the accumulation of phenolic compounds was thought to be due to enhanced activity of phenylalanine ammonia lyase which is a key enzyme for the biosynthesis of phenolics (Riov *et al.*, 1972). Increases of peroxidase activity one or two weeks after irradiation of citrus fruits have also been reported (Monselise *et al.*, 1969). For climacteric fruits, the major objective of radiation is to extend the post-harvest storage life by slowing down the rates of ripening and sensescence and hence in this case the advantage resides in the inhibition of the biosynthesis of a larger number of enzymes responsible for the ripening processes.

MOLECULAR STRUCTURE OF PEROXIDASE

Peroxidases in plants are haem-containing enzymes where the prosthetic group is protophorphyrin IX. Peroxidase spectra show absorption peaks at 640, 583, 548 and 498 nm. Purity is often assessed by the ratio of absorption (the RZ value) at 403 and 280 nm. However, purity can not be judged entirely on this criterion due to the dependence of absorption at 280 nm on the content of aromatic acids which would be expected to vary between haem proteins. At pH 2·4 and low concentrations of acetone, total detachment of the haem occurs with a decrease in the solubility of the apoprotein. Thus, acetone powder preparations of peroxidases are of doubtful value for investigating enzymatic activity.

Isoperoxidases do not differ significantly in size, with molecular weights in the range 40 000–50 000. Since the isoelectric points of the isoenzymes for a given species of plant vary considerably over a wide pH range, it is reasonable to expect that the isoenzymes differ substantially in amino acid content. For horseradish isoperoxidase C, which comprises the bulk of commercial preparations, there are 308 amino acid residues and two atoms of calcium per molecule. The structure of the protein includes four disulphide bonds in a single polypeptide chain that carries eight neutral carbohydrate side chains for a total molecular weight close to 42 000. Calcium contributes to maintaining the structural conformation of the protein. The primary structures of horseradish peroxidase isoenzyme C, turnip peroxidase TP7 and the amino acid sequence around the haem group for some turnip isoperoxidases have been claimed (Welinder, 1985). Certain amino acids, i.e. Arg-38, Phe-41, His-42, Tyr-185 and Leu-237, are believed to form part of the structure of the haem-binding pocket of horseradish peroxidase (Thanabal *et al.*, 1987).

Peroxidase catalysed oxidations

The haem-containing peroxidases can promote a large variety of reactions and therefore have a degree of versatility not exhibited by any other enzyme. Peroxidases can catalyse peroxidatic reactions, oxidatic reactions and hydroxylations. Peroxidases utilise both an oxidising substrate and a reducing substrate. The oxidising substrate is usually a peroxide or a peroxy acid, ROOH. Typically, peroxidases can use hydrogen peroxide, methyl-hydrogen peroxide and ethyl-hydrogen peroxide (Kermasha & Metche, 1988) as oxidising agents in the peroxidatic reaction. The distinctive absorption spectra due to the haem prosthetic group exhibits transient changes (Fig. 7) on mixing with hydrogen peroxide. The Soret band and visible spectra change markedly on the binding of peroxide and the consequent formation of compound I, which has a ferryl-type structure carrying an oxygen atom acquired from the peroxide. Compound I absorbs at 583 and 548 nm. Peroxidases cause the heterolytic cleaveage of the O—O bond. The iron–oxo compound (Fig. 7) is in the formal oxidation state of +5, the hydrogen peroxide having effectively removed two electrons from the iron atom of the enzyme.

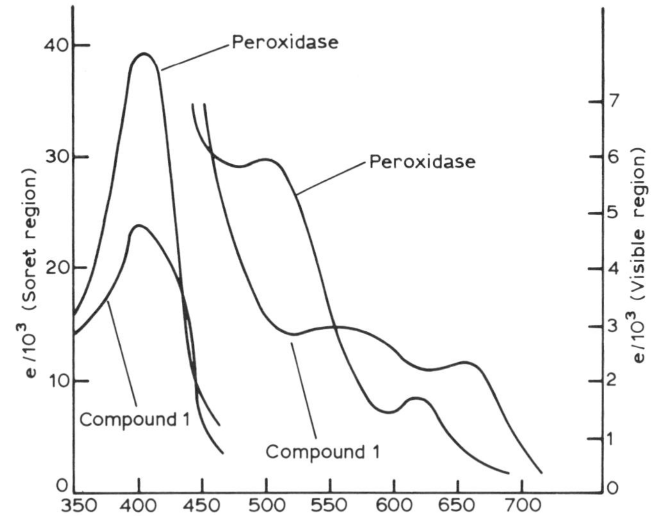

FIG. 7. The Soret and visible spectra of peroxidase and compound I. (From Robinson (1987).)

Peroxidatic reaction

The second stage of the peroxidatic reaction involves the reduction of compound I by a reducing donor substrate. Some typical donor substrates are listed in Table 1.

Generally, peroxidases are defined as enzymes that catalyse the following overall peroxidatic reaction:

$$ROOH + AH_2 \rightarrow H_2O + ROH + A^{\cdot}$$

TABLE 1
Action of Peroxidase–Hydrogen-Peroxide

Substrate	*Product*
Pyrogallol	Purpurogallin
Guaiacol	Tetraguaiacoquinone
Hydroquinone	Quinhydorne
Benzidine	*p*-Quinone di-imide
o-Phenylene diamine	Phenazine
Leucomalachite green	Malachite green
Catechol	*o*-Quinone
p-Cresol	Milky precipitate
o-Cresol	Green solution
m-Cresol	Flesh-coloured solution
Tyrosine	Yellow solution
Adrenaline	Reddish solution
Bilirubin	Biliverdin

(From Burnette (1977), reproduced by permission of the Institute of Food Technologists, Chicago, IL, USA.)

The first step of the initial oxidation of the donor substrate (AH_2) produces a free radical ($A^{\cdot}$). Compound I is reduced by the stepwise addition of two single electrons to form native peroxidase:

$$\underset{\text{compound I}}{\text{Per–Fe(V)=O}} + {}^{-}e + H^{+} \rightarrow \underset{\text{compound II}}{\text{Per–Fe(IV)=OH}}$$

$$\underset{\text{compound II}}{\text{Per–Fe(IV)—OH}} + {}^{-}e + H^{+} \rightarrow \underset{\text{native enzyme}}{\text{Per–Fe(III)}} \rightarrow H_2O$$

Compound II absorbs at 498 and 640 nm. Reduction of compound II results in the reformation of the native peroxidase in oxidation state +3. Both reductions require a single electron and a proton. Hydrogen abstraction readily occurs from phenolic substrates because of the resonance stabilisation of the resulting phenoxy radical. A broad range of reducing substrates usually convert compound I back to the native enzyme via the intermediate formation of compound II. The overall reaction involves a cycle of oxidation and reformation of the enzyme. The separate cationic free radicals ($A^{\cdot}$) derived from the donor substrate then react non-enzymically with each other to form dimers or higher polymers. At high concentrations of donor substrates, the liberated free radicals interact with each other to form polymers of the A-A type, as in the case of the oxidation of guaiacol or *o*-dianisidine used as test substrates. Because of the high reactivity of the free radicals, a large number of products may be formed for a single donor substrate. The peroxidases are usually assayed with artificial substrates from which they produce coloured products.

From the natural substrates, including the phenolic compounds, present in fruits and vegetables, a wide range of products is formed. The natural substrates in orange juice alone include ascorbic, caffeic, genistic and coumaric acids. Peroxidatic oxidation of ascorbic acid to form dehydroascorbic acid, which is relatively unstable, results in loss of nutritional value. However, because the concentration of hydrogen peroxide is generally low in citrus fruits and other plant foods, even though the concentration of ascorbate is relatively high, the rates of oxidation by this route are likely to be low. Consequently, the effects of peroxidase action are difficult to identify, become apparent only during prolonged storage of fresh and processed foods and may be confused with polyphenol oxidase-catalysed oxidations. Mono- and dihydroxyphenols and complex compounds like chlorogenic acid all seem to be potential substrates for peroxidases. Other potential natural phenolic substrates include the flavonoids from which it is claimed a large number of products are formed during peroxidase-catalysed reactions (Schreier & Miller, 1985). Other reducing agents for compound I include halides, the bisulphite ion and secondary aldehydes (Dunford, 1986). Specific ascorbic acid peroxidases (Gerbling *et al.*, 1984; Castillo *et al.*, 1986; Goldberg *et al.*, 1989) have been described which also seem to be active on the common test substrates for peroxidase. Although the ascorbate peroxidase enzyme may assist in detoxification by rapid removal of any hydrogen peroxidase, the enzyme's role is not known.

Oxidatic reaction

Oxidatic reactions catalysed by peroxidases accompanied by the uptake of oxygen occur with dihydroxyfumarate (DHF), ascorbic acid, thiols and hydroquinones. Investigations have been undertaken mainly with DHF (Halliwell & de Rycker, 1978). Based on the powerful inhibition of the reaction by superoxide dismutase (Halliwell, 1977), as opposed to hydroxyl radical scavengers, an oxidative scheme involving the superoxide radical and the continual oxidation of DHF has been widely accepted.

$$\text{peroxidase} + H_2O_2 \rightarrow \text{compound I}$$

$$\text{compound I} + \text{DHF} \rightarrow \text{DHF}^{\cdot} + \text{compound II}$$

$$\text{compound II} + \text{DHF} \rightarrow \text{DHF}^{\cdot} + \text{native peroxidase}$$

$$\text{DHF}^{\cdot} + O_2 \rightarrow \text{diketosuccinate} + O_2^{\overline{\cdot}}$$

$$\text{DHF} + O_2^{\overline{\cdot}} + 2H^+ \rightarrow \text{DHF}^{\cdot} + H_2O_2$$

One mole of O_2 is taken up per mol of DHF oxidised. The initial hydrogen peroxide is believed to be generated by the non-catalytic auto-oxidation of DHF:

$$\text{DHF} + O_2 \rightarrow \text{DHF}^{\cdot} + O_2^{\overline{\cdot}}$$

$$\text{DHF} + O_2^{\overline{\cdot}} + 2H^+ \rightarrow \text{DHF}^{\cdot} + H_2O_2$$

and oxidase reactions occur when radicals such as $\text{DHF}^{\cdot}$ can reduce oxygen directly. Peroxidase also reacts with $O_2^{\overline{\cdot}}$ to form an oxygenated ferroperoxidase (compound III):

$$\text{PER–Fe(III)} + O_2^{\overline{\cdot}} \rightarrow \underset{\text{compound III}}{\text{PER–Fe(II)—}O_2}$$

which can break-down to form the native peroxidase and $O_2^{\overline{\cdot}}$ and subsequently hydrogen peroxide.

Although high concentrations of DHF are not present in plants, a similar scheme is believed to operate for the oxidation by compounds I and II of at least NADH, pyridoxal compounds, thiols, indoleacetic acid (Srivastava & Huystee, 1977) and possibly ascorbic acid. Both $\text{NAD}^{\cdot}$ and the semi-dehydroascorbic radical can reduce oxygen directly.

Mason *et al.* (1957) first reported that horseradish peroxidase in the presence of reducing equivalents of dihydroxyfumarate and oxygen could catalyse the hydroxylation of a wide range of aromatic compounds. These include tyrosine, phenylalanine and other phenolic compounds. *p*-Coumaric acid is hydroxylated to caffeic acid in the presence of DHF and O_2 by peroxidase. Through the use of scavengers for both the superoxide and hydroxyl radicals, Dordick *et al.* (1986) showed that both radicals are involved and suggested that compound III, the oxyperoxidase (Fe(II)—O_2), is a key intermediate. Compound III is said to be formed from $O_2^{\overline{\cdot}}$, which itself is generated from the reduction of molecular oxygen by dihydroxyfumarate. The hydroxyl radical ($OH^{\cdot}$) is believed to be the hydroxylating species.

Oxidation of indole-3-acetic acid

It is generally accepted that the degradation of indole-3-acetic acid (IAA) by peroxidases controls the in-vivo concentration of this phytohormone in plants, although other enzymes may also catalyse the degradation of IAA (Reinecke & Bandurski, 1988). Diverse mechanisms have been proposed for the peroxidase-catalysed oxidation of IAA. The complete sequence of reactions, including the mechanism of oxygen activation, has been difficult to elucidate due to the reported involvement of small amounts of hydrogen peroxide and phenolic compounds and because peroxidases in the absence of hydrogen peroxide catalyse oxidatically the degradation of IAA. The assay for IAA is normally performed in the presence of manganese chloride and 2,4-dichlorophenol, which act as co-factors. Small amounts of added hydrogen peroxide have been claimed to abolish a lag period.

A reaction scheme proposed by Grambow (1986) accounts for the formation of indole-3-methanol(IM) and 3-hydroxymethyloxindole. After abstraction of an electron from IAA by peroxidase, the radical $IAA^{\cdot}$ is believed to be formed, which then decarboxylates to form an indole-3-methyl radical ($I^{\cdot}$).

$$\text{IAA} + \text{PER–Fe(III)} \rightarrow \text{IAA}^{\cdot} + \text{PER–Fe(II)}$$

$$\text{IAA}^{\cdot} \rightarrow \text{I}^{\cdot} + \text{CO}_2$$

The indole-3-methyl radical is then peroxidated by oxygenated peroxidase (compound III).

$$\text{I}^{\cdot} + \underset{\text{compound III}}{\text{PER–Fe(II)—O}_2} \rightarrow \text{IOO}^{\cdot} + \text{PER–Fe(III)}$$

The indole-3-methyl peroxide is reduced peroxidatically to IM.

$$\text{IOO}^{\cdot} + \text{PER–Fe(III)} \rightarrow \text{IM} + \underset{\text{compound I}}{\text{PER–Fe(IV)=O}}$$

Both IM and 3-methyleneoxindole (Fig. 8) are now thought to be the main products (Grambow, 1986). The formation of 3-hydroxymethyloxindole may occur via a non-enzymatic rearrangement of the hydroperoxide. The reaction scheme includes the proposal that phenolic compounds, if present in sufficient quantity, regenerate native peroxidase by the reduction of compounds I and II peroxidatically. The results indicate that hydroxymethyloxindole can not be formed from IM and that the routes are controlled by the availability of phenolic compounds to regenerate native peroxidase. Activation of oxygen to form compound III may also produce a source of superoxide radical at the expense of molecular oxygen which may then also be available for oxygenase reactions (Smith *et al.*, 1982).

Although the mechanism for the peroxidase-catalysed degradation of IAA was established for horseradish peroxidase, there are many reports on the

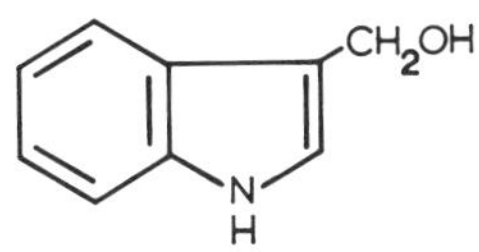

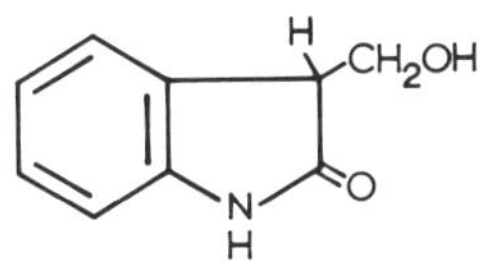

FIG. 8. Structures of indole-3-methanol and 3-methyleneoxindole.

oxidation of IAA by peroxidases from other sources. Since numerous peroxidase isoenzymes typically exist, often with different properties and substrate specificities, it is difficult to assign precise physiological roles to various isoperoxidases. Thomas *et al.* (1981) showed that IAA oxidase activity increases prior to the respiration climacteric. Since IAA is considered to be a naturally occurring retardant to fruit ripening, it is conceivable that IAA must be inactivated before fruit ripening can occur. Thomas *et al.* (1981) also reported an increase in IAA oxidase activity of tomato peroxidase during the ripening of tomatoes and that it appears that a capacity for IAA oxidation is developed during tomato fruit ripening. It has been suggested that cell wall-bound peroxidase may function in oxidising IAA as the phytohormone enters the cell. The IAA oxidase:peroxidase ratio in the bound fractions was found on average to be about 10-times greater than in the soluble fraction, thus indicating that the bound peroxidase has a greater capacity for oxidising IAA than the peroxidase found in the soluble fraction. Also, the isoenzyme patterns of the ionically bound fractions showed the appearance of two cationic peroxidases during fruit development, indicating that totally different isoenzyme forms are found in the cell wall. Brooks (1986) and Gaspar *et al.* (1982) indicated that basic isoperoxidases frequently have a higher capacity for IAA degradation than acidic isoperoxidases, although this is not consistently the case. Thus, the occurrence of different isoenzymes in the ionically and covalently bound fractions may explain the apparent greater capacity of the peroxidases in both fractions to oxidise IAA.

Oxidation of chlorophyll

Although two other enzymes, namely chlorophyllase and chlorophyll oxidase, are known to catalyse the degradation of chlorophyll, there is increasing circumstantial evidence that peroxidase plays a role in leaf senescence and the associated yellowing of fruits (Matile *et al.*, 1987; Abeles *et al.*, 1988). For instance, it has been claimed (Brennan & Frenkel, 1977; Laurière, 1983) that peroxidase activity and levels of hydrogen peroxide increase during the senescence and ripening of fruits, and Abeles *et al.* (1988) reported studies that implicate the biosynthesis of a 33 kDa peroxidase during ethene-induced senescence in cucumber cotyledons. Nevertheless, although Matile *et al.* (1987) possibly detected some non-green, pink-pigmented degradation products, the biochemistry of cholorophyll catabolism is virtually unknown.

There is also a claim (Gaspar *et al.*, 1986) that peroxidase has a role as an oxidase in the conversion of 1-aminocyclopropane-1-carboxylic acid to ethene and supporting evidence is quoted in Paul (1987). However, the evidence is not conclusive (Boyer & de Jaegher, 1986).

BIOCHEMICAL ROLE OF PEROXIDASES IN CELL WALLS

There have been numerous reports in the literature (e.g. Chibbar & van Huystee, 1984) with respect to the general involvement of peroxidases in lignin biosynthesis. Lignin is an abundant heterogeneous polymer which forms a three-dimensional network, mainly in plant cell walls, and thus contributes to the mechanical strength of the cells. Peroxidase is believed to play an important role in lignin biosynthesis by first catalysing the oxidation of coniferyl alcohol (Fig. 9) to form phenoxy radicals which then polymerise to form lignin, a strong, resistant polymer with a high proportion of ether bonds and methoxy groups. The formation of extensin and the cross-linking of diferulate (see Fig. 11) to hemicelluloses or pectin also requires peroxidase. Such strengthening of cell walls during the growth and maturation of plants largely accounts for the increased hardness and toughness in green beans, broad beans and other seed crops. Seed coat impermeability to water is important in the longevity of minute weed seeds in soil. During the maturation of seeds, peroxidase is instrumental in the conversion of soluble phenolics to insoluble lignin polymers. Egley *et al.* (1983) showed that peroxidase is instrumental in the development of water-impermeable seed coats in prickly sida seeds. Highest peroxidase activity of coat extracts was well correlated with

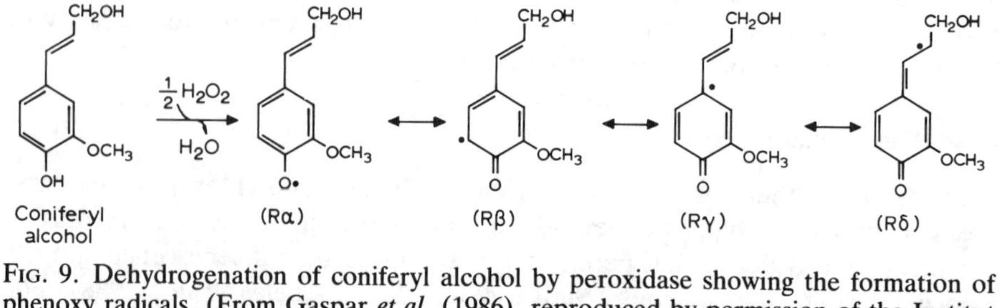

FIG. 9. Dehydrogenation of coniferyl alcohol by peroxidase showing the formation of phenoxy radicals. (From Gaspar *et al.* (1986), reproduced by permission of the Institut National de la Recherche Agronomique, France.)

the development stages for maximum conversion of soluble phenolics to soluble lignin polymers. Although seed extracts oxidised dihydroxyphenylalanine, this activity was eliminated by catalase, indicating that the oxidation of phenolics in the coat is catalysed by peroxidase. Peroxidase is thought to be one of the principal enzymes functioning in the deposition of lignin and vascular tissue (Ranadive & Haard, 1972). Studies on lignin formation in vascular regions of coleus plant have indicated that peroxidase is localised in the secondary and primary cell wall regions where lignin deposition occurs. It was also shown that the presence of calcium caused release of wall-bound peroxidases and a consequent decrease of lignin deposition. Kieffer pears, which contain a large number of sclerides, have been shown to contain more peroxidase than Bartlett fruit. Histochemical analysis with *o*-dianisidine and hydrogen peroxide showed that peroxidase activity was concentrated in the parenchyma cells surrounding grit cells and on the cell walls throughout the pulp. The lower calcium levels in fruits possessing excessive scleroid development were consistent with the thesis that calcium inhibits peroxidase localisation and accordingly controls peroxidase-catalysed lignin deposition in pear fruit. Many tropical fruits are susceptible to chill injury and it has been suggested that peroxidase may be involved in the commercially undesirable associated biochemical changes. Zauberman *et al.* (1988) showed that peroxidase and catalase activities in mango peel increased during the development of chill injury to much higher levels than in non-chilled fruits. Peroxidase activity was also enhanced in avocado fruit, another fruit known to undergo chill injury, that was stored at 5·5°C and under restricted ventilation. However, these empirical data do not provide an explanation for the mechanism of chill injury.

Peroxidase is now believed to play at least three different roles in cell wall construction (Gaspar *et al.*, 1986):

(1) diphenyl ether cross-links, e.g. iso-dityrosine in extensin,
(2) diferulate cross-links in hemicellulose and pectins, and
(3) lignin biosynthesis.

It has been shown that insolubilisation of extensin monomers is paralleled by the formation of iso-dityrosine (Fig. 10) acting as an interpeptide crosslink (Fry, 1982; Cooper & Varner, 1984). Also, in-vitro studies with horseradish peroxidase have shown that at low pH values iso-dityrosine is formed from tyrosine (Fry, 1986*a*). The second major class of wall polymer phenolic groups comprises those derived from *p*-coumaric acid and includes ferulic acid. It has been suggested that polysaccharide esters of ferulic acid may be polymerised by a peroxidase-catalysed reaction (Fry, 1986*b*) to form diferulate derivatives (Fig. 11).

Tracer experiments have shown that lignin is synthesised from phenylalanine or tyrosine via cinnamic acids and the corresponding hydroxycinnamyl alcohols. These compounds are directly dehydrogenated by peroxidase with hydrogen peroxide to produce phenoxy free radicals (Fig. 9). It has been

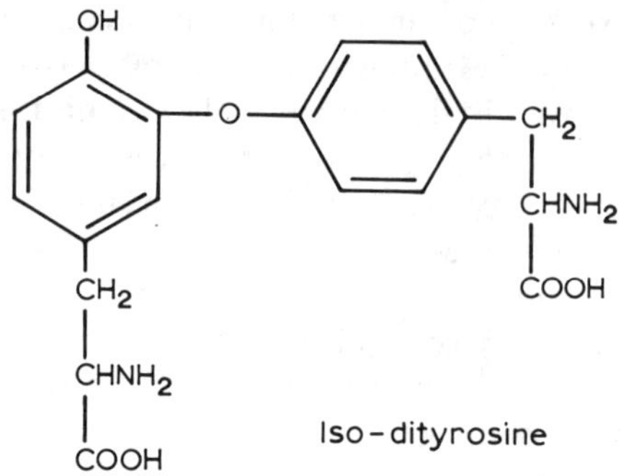

Iso-dityrosine

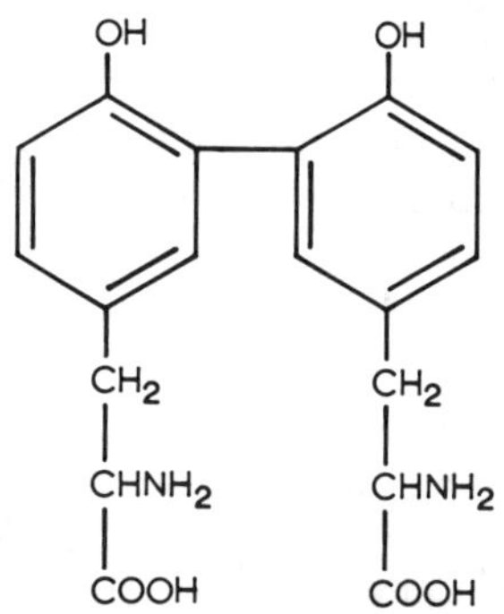

Dityrosine

FIG. 10. Structures of iso-dityrosine and dityrosine.

suggested that in living tissues the hydrogen peroxide might be formed in a cell wall at the expense of NAD (Gaspar *et al.*, 1986); this complex reaction involves the superoxide radical which is formed by the reduction of oxygen by NAD$^{\cdot}$ as intermediate. The formation of NAD$^{\cdot}$ is thought to be initiated through the oxidation of NADH by phenoxy radicals formed in the peroxidase-catalysed reaction in the presence of manganese. It has been claimed that cell walls may contain a bound maleic dehydrogenase which can form sufficient NADH to allow the subsequent synthesis of hydroperoxide (Gaspar *et al.*, 1986). This scheme can explain how hydrogen peroxide can be synthesised directly at the proper site without a requirement for transportation of toxic hydroperoxide from the cytoplasm to the cell wall. It is claimed that lignification is always accompanied by an increased activity of most anionic iso-peroxidases and a view has emerged during the last few years that lignin biosynthesis is associated with anionic peroxidases and the oxidation of indoleacetic acid by the cationic peroxidase isoenzymes.

Surprisingly, it has also been claimed that lignin degradation by white-rot fungi is due to the action of peroxidases (Evans, 1987). It is claimed that lignin peroxidase not only polymerises lignin model compounds but also degrades these compounds and therefore it has been proposed that there must be some

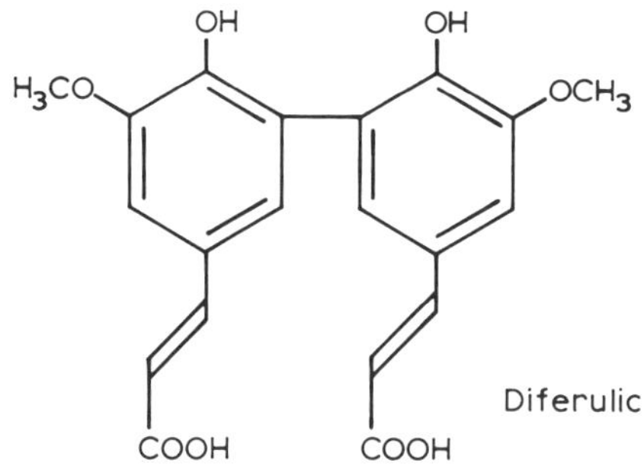

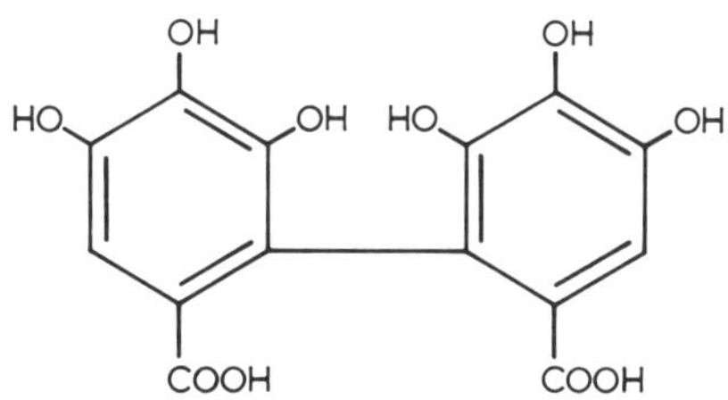

FIG. 11. Structures of diferulic acid and 'ellagic acid'.

regulating mechanism which determines whether the enzymes act in a polymerising or depolymerising direction if peroxidases are capable of catalysing reactions involved in both these functions. Studies by Mustranta (1987) have shown that wood-rotting fungi of the white-rot variety synthesise extracellular haem proteins that act as non-specific hydrogen peroxide-dependent peroxidases that are believed to have some role in lignin degradation.

REFERENCES

Abeles, F. B., Dunn, L. J., Morgens, P., Callahan, A., Dinterman, R. E. & Schmidt, J. (1988). *Plant Physiol.*, **87,** 609–15.

Arighi, A. L., Joslyn, M. A. & Marsh, G. L. (1936). *Ind. Engng Chem.*, **28,** 595–8.

Bakardjieva, N. T., Izvorska, N. D. & Hristova, N. (1987). *Comptes rendus l'Acadamie bulgare des Sciences,* **40,** 85–8.

Bedford, C. L. & Joslyn, M. A. (1939). *Ind. Engng Chem.*, **31,** 751–8.

Boyer, N. & de Jaegher, G. (1986). In *Molecular and Physiological Aspects of Plant Peroxidases,* ed. H. Greppin, C. Penel & Th. Gaspar. University of Geneva, Switzerland, pp. 47–60.

Brennan, T. & Frenkel, C. (1977). *Plant Physiol.*, **59,** 411–16.

Brooks, J. L. (1986). *Plant Physiol.*, **80,** 130–33.

Bruemmer, J. H., Roe, B. & Bowen, E. R. (1976). *J. Food Sci.*, **41,** 186–9.

Burnette, F. S. (1977). *J. Food Sci.*, **42,** 1–6.
Castillo, F. J. (1986). In *Molecular and Physiological Aspects of Plant Peroxidases*, ed. H. Greppin, C. Penel & Th. Gaspar. University of Geneva, Switzerland, pp. 419–26.
Castillo, F. J., Cotton, G., Kevers, C., Greppin, H., Gaspar, Th. & Driessche, Th. V. (1986). *Differentiation*, **33,** 17–23.
Chance, B. (1951). *Adv Enzymol.*, **12,** 153–90.
Chance, B. (1952). *Arch Biochem Biophys.*, **37,** 235–7.
Chang, B. S., Park, K. H. & Lund, D. B. (1988). *J. Food Sci.*, **53,** 920–3.
Cherry, J. P. & Ory, R. L. (1973). *Phytochem.*, **12,** 1581–3.
Cherry, J. P., Ory, R. L. & Mayne, R. Y. (1972). *J. Amer. Peanut Res. Educ. Assoc.*, **4,** 32–40.
Chibbar, R. N. & Huystee, R. B. van (1984). *Plant Physiol.*, **75,** 956–8.
Cooper, J. B. & Varner, J. E. (1984). *Plant Physiol.*, **76,** 414–17.
Copson, D. A. (1954). *Food Technol.*, 397–9.
Davis, W. B. (1942). *Amer. J. Bot.*, **29,** 252–4.
Dordick, J. S., Klibanov, A. M. & Marletta, M. A. (1986). *Biochem.*, **25,** 2946–51.
Dunford, H. B. (1986). In *Molecular and Physiological Aspects of Plant Peroxidases*, ed. H. Greppin, C. Penel & Th. Gaspar. University of Geneva, Switzerland, pp. 15–23.
Egley, G. H., Paul Jr, R. N., Vaughn, K. C. & Duke, S. O. (1983). *Planta*, **157,** 224–32.
Evans, C. S. (1987). *Process Biochem.*, **22,** 102–5.
Fils, B., Sauvage, F. X. & Nicolas, J. (1985). *Sciences des Aliments*, **5,** 217–32.
Fry, S. C. (1982). *Biochem. J.*, **204,** 449–55.
Fry, S. C. (1986*a*). In *Molecular and Physiological Aspects of Plant Peroxidases*, ed. H. Greppin, C. Penel & Th. Gaspar. University of Geneva, Switzerland, pp. 169–79.
Fry, S. C. (1986*b*). *Ann. Rev. Plant Physiol.*, **37,** 165–86.
Gardner, H. W., Inglett, G. E. & Anderson, R. A. (1969). *Cereal Chem.*, **46,** 626–34.
Garrote, R. L., Luna, J. A., Silva, E. R. & Bertone, R. A. (1987). *J. Food Sci.*, **52,** 232–5.
Gaspar, Th., Penel, C., Thorpe, T. & Greppin, H. (1982). University Geneva–Centre Bot. Publ. Geneva, Switzerland.
Gaspar, Th., Penel, C., Castillo, F. J. & Greppin, H. (1985). *Physiol. Plant*, **64,** 418–23.
Gaspar, Th., Penel, C. & Greppin, H. (1986). *Bull. Liaison Group Polyphenols*, **13,** 159–76.
Gerbling, K. P., Kelly, G. J., Fisher, K. H. & Latzlko, E. (1984). *J. Plant Physiol.*, **115,** 59–67.
Goldberg, R., Kevers, C. & Gaspar, T. (1989). *Biochem. Physiol. Pflanzen*, **184,** 155–61.
Gorin, N. & Heidema, F. T. (1976). *J. Agric. Food Chem.*, **24,** 200–1.
Grambow, H. J. (1986). In *Molecular and Physiological Aspects of Plant Peroxidases.* ed. H. Greppin, C. Penel & Th. Gaspar. University of Geneva, Switzerland, pp. 31–41.
Guyer, R. B. & Holmquist, J. W. (1954). *Food Technol.*, (Chicago), **8,** 547–50.
Haard, N. F. (1987). In *Enzymes in Food and Beverage Processing*, ed. R. L. Ory & A. J. St. Angelo. ACS Symposium Series 47, American Chemical Society, Washington DC, USA, pp. 143–71.
Halliwell, B. (1977). *Biochem. J.*, **163,** 441–8.
Halliwell, B. & Rycker, J. de (1978). *Photochem. Photobiol.*, **28,** 757–63.
Henley, J. P. & Sadana, A. (1985). *Enzym. Microbiol. Technol.*, **7,** 50–60.
Joslyn, M. A. (1949). *Adv. Enzymol.*, **9,** 613–52.

Joslyn, M. A. & Bedford, C. L. (1940). *Ind. Engng Chem.*, **32,** 702–6.
Joslyn, M. A. & David, J. H. (1952). *Quick Frozen Foods,* **15,** 51.
Keilin, D. & Mann, T. (1937). *Proc. Royal Soc., London,* **122B,** 119–33.
Kermasha, S. & Metche, M. (1988). *J. Food Sci.*, **53,** 247–52.
Kermasha, S., Alli, I. & Metche, M. (1988). *J. Food Sci.*, **53,** 1753–5.
Labbee, M. D. & Esselen, W. B. (1954). *Food Technol.* (Chicago), **8,** 50–4.
Laurière, C. (1983). *Physiol. Veg.*, **21,** 1159–77.
Lee, C. Y. & Hammes, J. K. (1979). *J. Food Sci.*, **44,** 785–7.
Lee, C. Y. & Pennesi, A. P. (1984). *J. Food Sci.*, **49,** 1616–17.
Lee, C. Y., Pennesi, A. P. & Dickson, M. H. (1984). *J. Agric. Food Chem.*, **32,** 18–21.
Ling, A. C. & Lund, D. B. (1978). *J. Food Sci.*, **43,** 1307–10.
Lopez, A. & Baganis, N. A. (1971). *J. Food Sci.*, **36,** 911–14.
Lu, A. T. & Whitaker, J. R. (1974). *J. Food Sci.*, **39,** 1173–8.
McLellan, K. M. & Robinson, D. S. (1981). *Food Chem.*, **7,** 257–66.
McLellan, K. M. & Robinson, D. S. (1983). *Phytochem.*, **22,** 645–7.
McLellan, K. M. & Robinson, D. S. (1984). *Food Chem.*, **13,** 139–47.
McLellan, K. M. & Robinson, D. S. (1987*a*). *Food Chem.*, **23,** 305–19.
McLellan, K. M. & Robinson, D. S. (1987*b*). *Food Chem.*, **26,** 97–107.
Mason, H. S., Onopryenko, I., & Buhler, D. (1957). *Biochim. Biophys. Acta,* **24,** 225–6.
Matile, P., Ginsburg, S., Schellenberg, M. & Thomas, H. (1987). *J. Plant Physiol.*, **129,** 219–28.
Merkler, D. J., Srikumar, K., Marchese-Ragona, S. P. & Wedler, F. C. (1988). *Biochim. Biophys. Acta,* **953,** 101–14.
Metodiewa, D., Gebicka, L. & Bachman, S. (1987). *J. Radioanal. Nucl. Chem. Lett.*, **119,** 87–93.
Monselise, S. P., Riov, J. & Kahan, R. S. (1969). In *Enzymological Aspects of Food Irradiation,* International Atomic Energy Agency, Vienna, Austria pp. 71–81.
Moulding, P. H., Grant, H. F., McLellan, K. M. & Robinson, D. S. (1987). *Int. J. Food Sci. Technol.*, **22,** 391–7.
Moulding, P. H., Singleton, D. E., McLellan, K. M. & Robinson, D. S. (1988). *Int. J. Food Sci. Technol.*, **23,** 343–51.
Moulding, P. H., Goodfellow, J., McLellan, K. M. & Robinson, D. S. (1989). *Inter J. Food Sci. Technol.*, **24,** 269–75.
Mustranta, A. (1987). *Appl. Microbiol. Biotechnol.*, **2**7, 21–6.
Nanjundaswamy, A. M., Saroja, S. & Ranganna, S. (1973). *Indian Food Packer,* **27,** 5–13.
Naveh, D., Mizrahi, S. & Kopelman, I. J. (1981). *J. Agric. Food Chem.*, **29,** 36–8.
Naveh, D., Mizrahi, S. & Kopelman, I. J. (1982). *J. Agric. Food Chem.*, **30,** 967–70.
Nebesky, E. A., Esselen, W. B. & Fellers, C. P. (1951). *Food Technol.* (Chicago), **5,** 110–13.
O'Fagain, C., Sheehan, H., O'Kennedy, R. & Kilty, C. (1988). *Process Biochem.*, 166–71.
Onslow, M. W. (1920). *Biochem. J.*, **14,** 535-9.
Paul, K. G. (1987). *J. Oral Pathol.*, **16,** 409–11.
Penel, C. (1986). In *Molecular and Physiological aspects of plant peroxidases.* ed. H. Greppin, C. Penel & Th. Gaspar. University of Geneva, Switzerland, pp. 155–64.
Pinsent, B. R. W. (1962). *J. Food Sci.*, **27,** 120–6.
Ranadive, A. S. & Haard, N. F. (1972). *J. Food Sci.*, **37,** 381–3.
Reinecke, D. M. & Bandurski, R. S. (1988). *Plant Physiol.*, **86,** 868–72.
Riov, J., Monselise, S. P., Goren, R. & Kahan, R. S. (1972). *Radiat. Res. Rev.*, **3,** 417–27.
Robinson, D. S. (1987). *Food—Biochemistry and Nutritional Value,* Longman Group UK Ltd, with John Wiley & Sons, Inc., New York.

Robinson, D. S., Bretherick, M. & Donnelly, J. K. (1990). *Int. J. Food Sci. Technol.*, **24**, 613–18.
Ros Barcelo, A., Munoz, R. & Sabater, F. (1987). *Physiologia Plantarum*, **71**, 448–54.
Ros Barcelo, A., Munoz, R. & Sabater, F. (1988). *Plant Physiol. Biochem.*, **26**, 575–83.
Schloss, P., Walter, C. & Mader, M. (1987). *Planta*, **170**, 225–9.
Schreier, B. & Miller, E. (1985). *Food Chem.*, **18**, 301–17.
Servant, I., Gilot, B. & Gibert, H. J. (1986). *J. Food Engng*, **5**, 287–300.
Sessa, D. J. & Anderson, R. L. (1981). *J. Agric. Food Chem.*, **29**, 960–5.
Shannon, L. M. (1968). *Ann. Rev. Plant Physiol.*, **19**, 187–210.
Smith, A. M., Morrison, W. L. & Milham, P. J. (1982). *Biochem.*, **21**, 4414–19.
Srivastava, Om. P. & Huystee, R. B. van (1977). *Phytochem.*, **16**, 1527–30.
Tamura, Y. & Morita, Y. (1975). *J. Biochem.*, **78**, 561–71.
Thanabal, V., De Ropp, J. S. & La Mar, G. N. (1987). *J. Amer. Chem. Soc.*, **109**, 7516–25.
Thomas, P. (1987). *CRC Crit. Rev. Food Sci. Nutr.*, **24**, 53–89.
Thomas, R. L., Jen, J. J. & Morr, C. V. (1981). *J. Food Sci.*, **47**, 158–61.
Vamos-Vigyazo, L., Gajzago, I., Nadudvari-Markus, V. & Kerek, M. M. (1985). *Qual. Plant Foods Hum. Nutr.*, **35**, 139–43.
Wagenknecht, A. C. & Lee, F. A. (1958). *Food Res.*, **23**, 25–31.
Welinder, K. G. (1985). *Eur. J. Biochem.*, **151**, 497–504.
Wijsman, H. J. W. & Hendriks, T. (1986). In *Molecular and Physiological Aspects of Plant Peroxidases*, ed. H. Greppin, C. Penel & Th. Gaspar. University of Geneva, Switzerland, pp. 109–23.
Willimott, S. G. & Wokes, F. (1926). *Biochem. J.*, **20**, 1008–101.
Winter, E. (1971). *Z. Lebensm.-unters. Forsch.*, **145**, 3–6.
Yamamoto, H. Y., Steinberg, M. P. & Nelson, A. I. (1962). *J. Food Sci.*, **27**, 113–19.
Yoon, J. R. & Park, K. H. (1982). *Korean J. Food Sci. Technol.*, **14**, 125–9.
Zauberman, G., Fuchs, Y., Rot, I. & Wexler, A. (1988). *Hort. Sci.*, **23**, 732–3.
Zoueil, M. E. & Esselen, W. B. (1959). *Food Res.*, **24**, 119–33.

11

Enzymes in Winemaking

J. C. VILLETTAZ

Ecole d'Ingénieurs du Valais, Food Technology Department, Rte du Rawyl 47, 1950 *Sion, Switzerland*

&

D. DUBOURDIEU

Université de Bordeaux, Institut d'Oenologie, Bordeaux, France

INTRODUCTION

It has been claimed that oenology is a science whereas winemaking must remain an art. ('L'oenologie est une science, la vinification doit rester un art'). This excellent statement is probably the shortest that tries to differentiate oenology from winemaking. It is a fact that the quality of the grapes varies from year to year, from region to region and even from one harvesting day to another. Each batch of grapes can be regarded as a particular case.

While the winemaker has no, or very little, influence on climatic conditions, he must, however, be in a position to master all the steps from the harvest of the grapes to the bottling of the wine. The production of high-quality wine can not be achieved by following standard recipes but rather by adapting the 'oenological laws' to each particular case.

Enzymes fulfil key functions in winemaking. Without enzymes, grape juice would never become wine. Not all enzymes occurring naturally in the grapes or produced by the microorganisms involved in fermentations are beneficial to wine quality; some of them are very detrimental. In order to get the best out of his grapes, the winemaker must be aware of the importance of certain enzymatic reactions. His role is to create the optimal operating conditions for getting the desirable enzymes working while at the same time inhibiting those detrimental to wine quality.

A cost- and quality-conscious winemaker must also be aware of the potentialities offered by the so-called 'commercial enzyme preparations'. These preparations have been developed in order to reinforce and complement the natural enzymes of the grapes. Judiciously used, they have a positive impact on juice and wine quality as well as on the economy of winemaking.

The present chapter is divided into two main parts. The first part deals with the enzyme activities occurring naturally in the grapes and microorganisms involved during fermentation. The second part is devoted to the use of commercial enzyme preparations. The future outlook for enzymes in winemaking is also briefly described.

ENDOGENOUS ENZYMES

Endogenous enzymes involved in biochemical transformations in winemaking have several origins: the grape berries, in certain cases fungi (e.g. *Botrytis cinerea*) and finally the microorganisms responsible for alcoholic and malolactic fermentations. This review completes the one made by Villettaz (1984).

Enzymatic activities of healthy grapes

Due to their technological relevance, the following four categories of enzymes have primarily interested oenologists:

(1) enzymes involved in the formation of C-6 volatile compounds
(2) pectinolytic enzymes
(3) polyphenoloxidases
(4) proteolytic enzymes

Enzymes involved in the formation of C-6 volatile compounds
Hexanal, *cis*-hexene-3-al and *trans*-hexene-2-al are volatile compounds responsible for the grassy taste of juices and wines. These compounds are formed in the presence of oxygen during the pre-fermentation phase, after the mechanical treatment of the grapes (crushing, de-stemming, pressing). Essentially, aldehydes and very low amounts of corresponding alcohols have been found in grape juice. During alcohol fermentation, hexanal and hexenals are almost entirely transformed into hexanol and hexenols by the yeasts.

The threshold limit (5 mg/litre) for the organoleptic perception of C-6 alcohols in a 12% ethanol solution is often higher than the levels usually encountered in wines (1–6 mg/litre). Nevertheless, it is supposed that these substances can reinforce the unpleasant grass-like aroma of wines, and are, therefore, regarded as detrimental to wine quality. The winemaker must consequently take all the necessary precautions to avoid the presence of these substances in the wine.

Works performed by Rapp *et al.* (1976), Cayrel *et al.* (1983) and Crouzet *et al.* (1985) have shown that the formation of C-6 compounds in juice requires the sequential action of four enzymes, namely: acyl hydrolases, lipoxygenase, a hydroperoxide-cleaving enzyme and an alcohol dehydrogenase (ADH) (Fig. 1).

The acyl hydrolases acts on the membrane lipids and liberate C-18 unsaturated fatty acids (linoleic and linolenic acids). These compounds serve as substrates for the lipoxygenase of the grapes. The lipoxygenase activity, starting from linoleic acid, leads to the formation of 80% of C-13 hydroperoxide and 20% of the C-9 isomer (see Table 1). This enzymatic activity

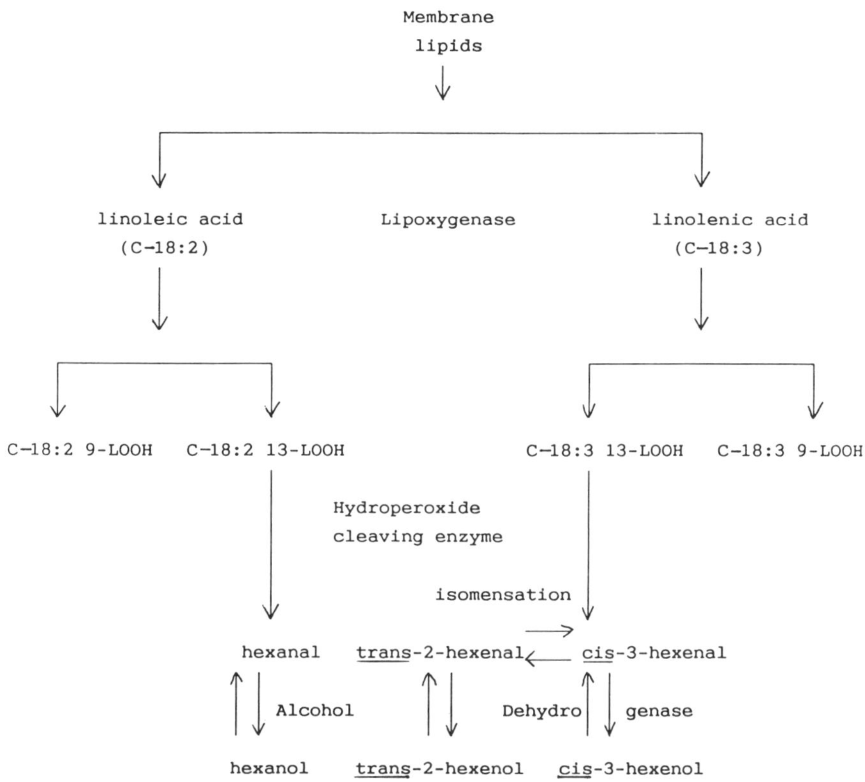

FIG. 1. Pathways for the formation of volatile compounds responsible for the green, grassy taste of the grapes (Crouzet *et al.*, 1985).

TABLE 1
Distribution (%) of Hydroperoxides of Linoleic Acid (Cayrel *et al.*, 1983)

Samples	*13-hydro peroxide cis-trans*	*13-hydro peroxide trans-trans*	*9-hydro peroxide cis-trans*	*9-hydro peroxide trans-trans*
Autoxidation of linoleic acid	30·9	20·6	30·6	17·6
Action of grape lipoxygenase on linoleic acid	73	6·6	15·8	4·1

TABLE 2
Comparison of the Grassy-Taste Potential and the Lipoxygenase Activity of Different Grape Varieties at Two Different Stages of Grape Maturity

Variety	*Grassy-taste potential (mg/kg)*	*Units lipoxygenase activity*
Carignan		
ripe	9·86	0·4
half-ripe	6·55	0·6
Cabernet Sauvignon		
ripe	5·23	0·39
half-ripe	16·25	0·56
Syrah		
ripe	9·49	0·31
half-ripe	11·13	0·4
Ugni blanc		
ripe	13·9	0·12
half-ripe	5·83	0·26

(Valentine, G., 1983, Unpublished Results)

decreases during ripening of the fruit (Table 2). However, it should be mentioned that the grassy taste potential is not only linked to the lipoxygenase activity, but also depends upon the grape variety. According to Valentin (1983), this potential increases between the beginning and the end of ripening for Carignan and Ugni blanc and decreases during this period in the case of Cabernet Sauvignon (Table 2).

The hydroperoxide-cleaving enzyme of grapes has been investigated recently by Valentin and Crouzet (1989); it has a molecular weight of 200 kDa. The hydroperoxide at position 13 (13-LOOH) is the only substrate used by this enzyme; the isomer at position 9 (9-LOOH) is not used as substrate. This result explains the absence of C-9 volatile compounds (nonenal and nonadienal isomers) in grape juices as well as the presence of C-6 aldehydes and alcohols. The latter originate from the action of the hydroperoxide-cleaving enzyme on the hydroxyperoxide 13-LOOH.

The enzymatic activity of the hydroperoxide-cleaving enzyme is twice as high in the skin as in the berry flesh.

The ADH of the grape has also been isolated and characterised (Molina *et al.*, 1986). This enzyme has the apparent structure of a dimer. It is formed by two identical subunits having a molecular weight of 45 kDa. Inhibition studies indicate that the grape ADH is a metallo-enzyme with SH groups essential for enzyme activity.

From the technological point of view, the pathways of formation of C-6 alcohols and aldehydes in grape juices and wines have been investigated by

TABLE 3
Influence of Mechanical Treatments on the Production of C-6 Alcohols and Aldehydes ($\mu g/kg$) (Cordonnier & Bayonove, 1981)

Treatment of the grapes	*Hexanal*	*Trans-2 hexenal*	*Hexanol*	*Cis-3-hexenol*	*Trans-2 hexenol*	*Alde-hydes*
Crushed and exposed to air for 10 min	918	5 700	78	61	89	28
First juice fraction from uncrushed grapes	91	62	72	37	47	1

Cordonnier and Bayonove (1981). Only traces of these compounds are found in the fruits. They are formed later on, under aerobic conditions, during the crushing of the grapes (Table 3). The stems, and especially the leaves, are known sources for C-6 compounds (Table 4).

Inhibition of the formation of these compounds by sulphur dioxide or by the absence of oxygen is effective only under experimental laboratory conditions.

The mechanical treatments applied to the grape mash are the determining factors responsible for the formation of the C-6 compounds. Crushing of the grapes leads to a greater formation of these compounds than the pressing of uncrushed grapes (Table 5). Crushing of grapes also influences the level of C-6 compounds in the different press fractions (Fig. 2).

In Sauvignon and Semillon dry white wines (Dubourdieu *et al.*, 1980; Lafon-Lafourcade *et al.*, 1980), the level of C-6 alcohols is influenced mainly by factors such as the ripening of the grapes and the crushing as well as by clarification of the juice (Tables 5 and 6).

If juice clarification is properly done (turbidity level below 200 Nephelometric Turbidity Units (NTU)), the skin contact during the pre-fermentation phase, even at 25°C, does not increase significantly the level of the C-6 alcohols of the wine (Dubourdieu *et al.*, 1986; Ollivier, 1987).

TABLE 4
Formation of C-6 Compounds ($\mu g/kg$), Contribution of the Berries, Stems and Leaves of Ugni Blanc (Cordonnier & Bayonove, 1981)

Crushing under aerobic conditions and inhibition with TCA after 10 min	*Hexanal*	*Trans-2-hexenal*	*Hexanol*	*Cis-2-hexenol*	*Trans-2-hexenol*
Berries	918	5 700	78	61	6 846
Berries + stems	1 014	6 798	159	115	8 245
Berries + leaves	1 218	14 289	133	164	15 990

TABLE 5
Influence of the Degree of Maturation and Crushing of the Grapes on the Concentrations of C-6 Alcohols (Hexanol + Hexenols) in Semillon Dry White Wines

Treatment	*C-6 alcohols (mg/litre)*	
	Maturation A	*Maturation B (10 days after A)*
Uncrushed	1·36	1·02
Crushed	2·12	1·47

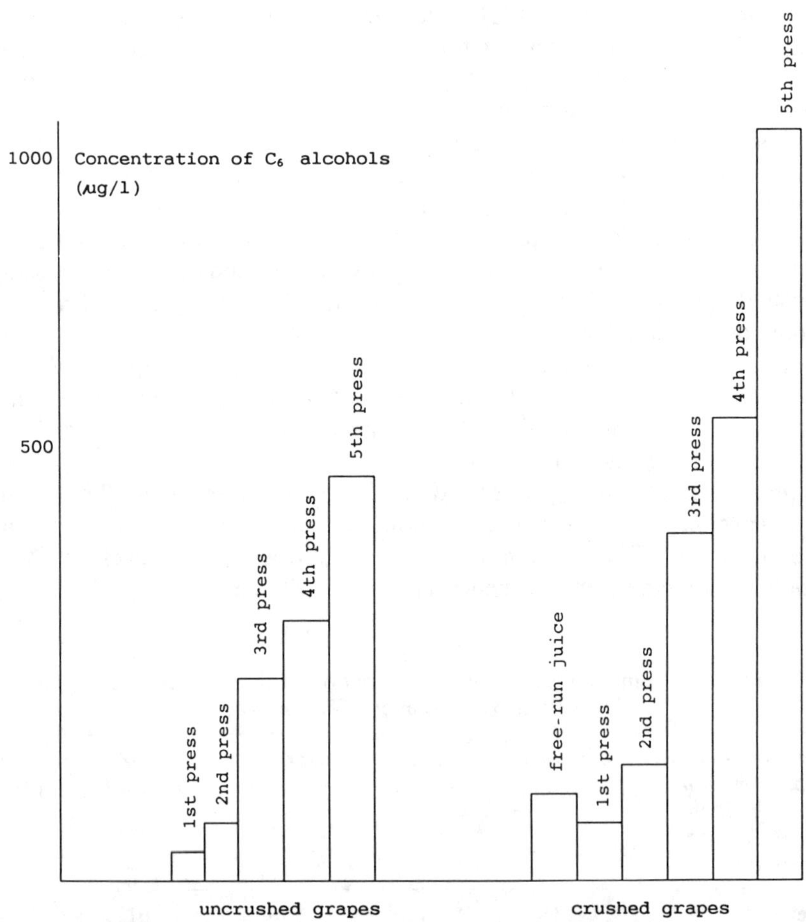

FIG. 2. Formation of C-6 alcohols (hexanol, 2-hexenol, 3-hexenol) during the pressing of crushed and uncrushed grapes (Cordonnier & Bayonove, 1981).

TABLE 6
Influence of Juice Clarification on the Concentrations of C-6 Alcohols in Wines

Treatment	*C-6 alcohols (mg/litre)*
Wine obtained from non-clarified juice	2·0
Wine obtained from clarified juice (natural settling of the cloud)	1·0
Wine obtained from filtered cloud sediments	0·9
Wine obtained from unfiltered cloud sediments	2·6

From the enzymatic point of view, the role of juice clarification certainly would be worth further detailed investigations. We only know so far that elimination of the cloud particles leads to an important reduction in the concentrations of the C-6 alcohols in wines. The C-6 aldehydes that are formed during juice extraction may be partially adsorbed on the cloud particles and therefore be removed by juice clarification. Also one may ask if the cloud particles, rich in C-18 fatty acids, contain lipoxygenase and hydroperoxide-cleaving activities that could generate C-6 volatile compounds.

Pectolytic enzymes

Very little work has been devoted so far to the study of pectolytic enzyme activities of healthy grapes (Flanzy & Bouzigues, 1959; Marteau *et al.*, 1963; Tarantola & Usseglio-Tomasset, 1963; Usseglio-Tomasset and Tarantola, 1963).

The presence of polygalacturonase (PG) and pectinesterase (PE) activities has been detected in mature grapes. Several authors have pointed out the relationship between the PE activity of the grape and the ability of the juice to clarify itself. Grassin (1987) studied the evolution of the PE and PG activities of Semillon and Merlot during the ripening (Fig. 3). After crushing of the grapes, crude enzyme extracts have been obtained by centrifugation followed by precipitation with 80% saturated solution of ammonium sulphate. It has been found that the PE and PG activities are very low in the green berries but these activities increase strongly at the start of ripening and decrease towards the end of maturation. Gel filtration on Sephacryl S-200 has indicated that the molecular weight of these enzymes is about 50–60 kDa.

It seems that the climatic conditions during ripening of the grapes influence the hydrolysis of pectic substances and therefore, later on, juice yield and juice clarification.

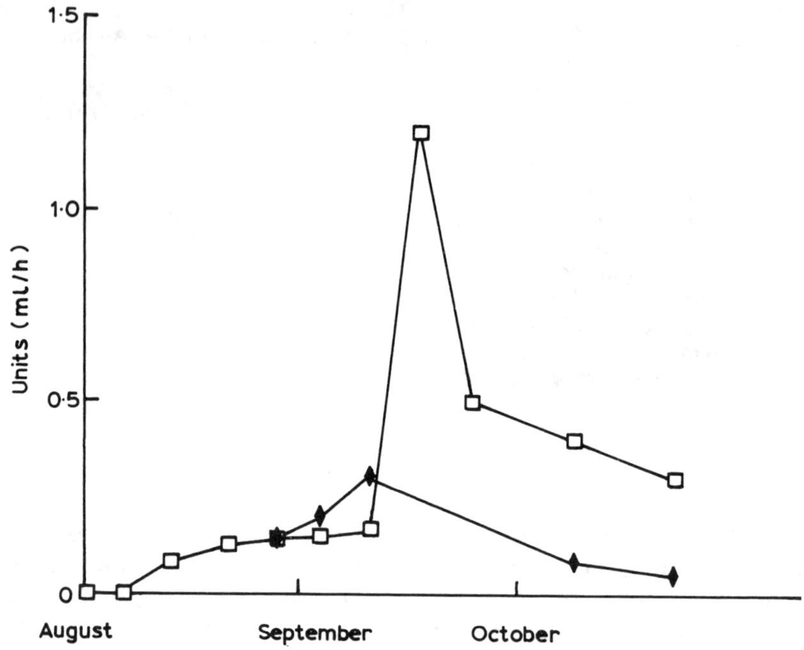

FIG. 3. Evolution of the (◆) PGs and (□) PEs activities of healthy grapes during their maturation.

Oxidative enzymes of the grapes

The oxidation of phenolic compounds in the juice by polyphenoloxidases (PPO) of the grapes starts at the crushing and pressing stage. This phenomenon was observed a long time ago. Surprisingly, it was only in the 1970s (Dubernet & Ribéreau-Gayon, 1973; 1974; Dubernet, 1974; Dubernet *et al.*, 1977) that it has been established that two enzymes having different properties are involved in this oxidation. These enzymes are called tyrosinase and laccase. The tyrosinase is present in healthy grapes while the laccase is of microbial origin and is found in juices only when the grapes are attacked by the fungus *Botrytis cinerea*. This enzyme will be described later.

The tyrosinase is bound to insoluble cellular compounds, such as chloroplasts. Its activity (variable with the type of grape) is fairly high in the green berries but decreases during maturation of the grapes (Kidron, 1976). A portion of the enzyme activity becomes soluble during juice extraction while the rest remains attached to the particles. The mechanical treatments of the grapes (crushing, pressing) as well as clarification of the juice (removal of cloud particles) have, therefore, an impact on the tyrosinase activity of the juices and subsequently on their oxidation potential.

The rate of oxygen consumption of grape juices shows marked variations (0·5–4·6 mg/litre min), the average level being about 2 mg/litre min. In this average situation, 8 mg dissolved oxygen/litre of juice (saturation at 25°C) are consumed within 4 min. Enzyme activity disappears within a short period; it has been observed that after three successive saturations with O_2, enzyme activity has almost disappeared.

Oxygen uptake also varies with juice temperature. At 30°C, the reaction is three times faster than at 10°C. Under winemaking conditions (pH, SO_2, alcohol), tyrosinase activity progressively disappears. The enzyme is also destroyed by oxidation products formed by itself. Wines made from healthy grapes generally are, at the end of the alcoholic fermentation, stable towards catalytic tyrosinase oxidation.

Recent work conducted by Gunata *et al.* (1987) has indicated that cafeoyl tartric acid, as well as *p*-coumaroyltartaric acid, are, in white grape juice, preferred substrates for tyrosinase. These substrates are oxidised in *o*-quinones. These oxidation compounds can then react further with glutathione, ascorbic acid or other polyphenols. They have also been found in polymerised forms (Cheynier *et al.*, 1986, 1989) (Fig. 4).

It seems that beside glutathione, other grape juice compounds act as a 'quinone trap'. Sulphur dioxide (4–8 g/h litre) is a strong inhibitor of tyrosinase activity. This product is generally used in white wine technology in the pre-fermentation phase in order to limit oxidation phenomena. Many years ago, Müller-Späth (1977) proposed that not only should the use of SO_2 be discontinued but the juice should be hyperoxygenated in order to reduce the level of oxidable polyphenols in white wines (Fig. 5). This technology has a certain amount of interest; however, the influence of hyperoxygenation on the volatile compounds responsible for the varietal characteristics of white wines needs to be investigated further.

Proteases of the grape

It is well known that grape proteins are responsible for the colloidal instability of white wines. The haze development that occurs in the bottle is

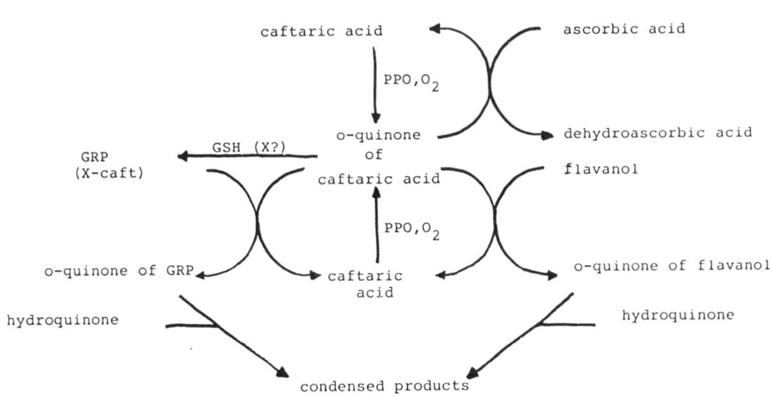

FIG. 4. Oxidation mechanisms in grape juices (Cheynier *et al.*, 1989).

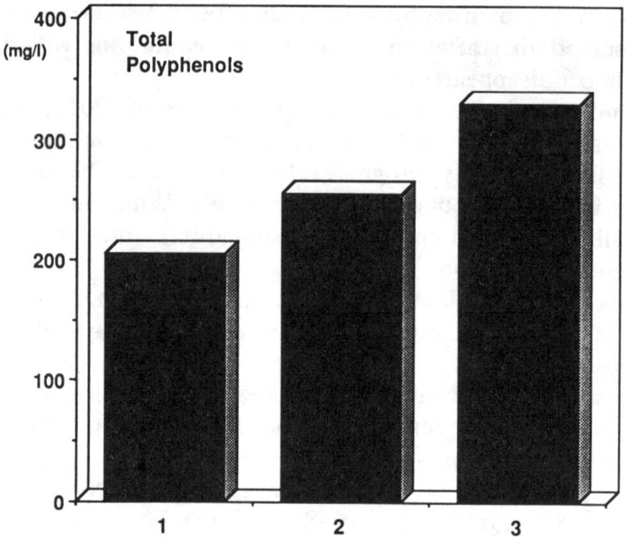

FIG. 5. Influence of SO_2 addition on the polyphenol level in a wine (Müller-Späth, 1989). (1) SO_2 addition before alcohol fermentation, (2) No SO_2 addition before alcohol fermentation, and (3) No SO_2 addition before alcohol fermentation; aeration (oxidation) of the juice.

generally due to variations of storage temperature or to the precipitation of protein by tannins extracted from the cork (Ribéreau-Gayon *et al.*, 1977).

The molecular weight of the thermolabile proteins varies with the grape variety, but it lies generally between 12 and 30 kDa (Hsu & Heatherbell, 1987; Dubourdieu *et al.*, 1988).

The grape berries possess proteolytic activity (Cordonnier & Dugal, 1968; Canonica & Ferrari, 1987) but this particular activity is rapidly inhibited by the ethanol formed during alcoholic fermentation and even by clarification of the juice (Feuillat *et al.*, 1980).

In spite of the recent work (see proteolytic enzymes of the yeasts) done in connection with the enzymatic hydrolysis of grape proteins, bentonite fining (Saywell, 1934; Ribéreau-Gayon, 1935; Blade & Boulton, 1988) remains today the only efficient method for stabilising white wines against protein precipitation.

Enzymatic activities of grapes infected by *Botrytis cinerea*

Botrytis cinerea is a parasite of the vine and several other plants. This fungus is responsible for one of the most feared diseases in viniculture, i.e. grey rot. Study of the extracellular enzymes of *B. cinerea* in infected plant tissues can help to understand the infection process of the fungus. Such a study is of considerable interest from the phytopathological point of view.

The action of *Botrytis* enzymes on the compounds of grapes leads to important modifications in, and consequences for, the fermentation, as well as on the composition of the wine. *B. cinerea* enzymes are, therefore, also of interest to the oenologist.

The main work performed so far on the extracellular enzymes of *B. cinerea* deals with the hydrolysis of cell walls and the oxidation of phenolic compounds. Different pectinases (Drawert & Kreff, 1978), a cellulase (Werhoeff & Warren, 1972), a laccase (Dubernet & Ribereau-Gayon, 1975; Dubenet et al., 1977), phospholipases (Shepard & Pitt, 1976), an esterase (Dubourdieu, 1982), an *N*-acetylglucosaminidase (Grassin *et al.*, 1987), and a proteinase (Heale & Movahedi, 1989) have been identified so far.

Grassin (1987) showed that the PEs and PG are produced mainly during the growth phase of the fungus, whereas the laccase and the *N*-acetylglucosaminidase are also produced during the stationary phase.

The level of PG activities in grape juices obtained from botrytised grapes is about 200-times higher than that present in juices obtained from healthy grapes. Due to this very high enzyme activity, the pectin level in such juices is usually extremely low. The use of exogenous pectinolytic enzymes is not necessary in these cases.

Grassin (1987) has also shown that immature berries contain a specific inhibitor for the endoPG of *B. cinerea.* The level of this inhibitor decreases during maturation of the grape. This could be the explanation of why green berries are resistant to *Botrytis* infection, while mature grapes are sensitive.

The laccase is an oxidation enzyme that oxidises the phenolic compounds in grape juices and wines. This enzyme is much more dangerous than the tyrosinase: it is responsible for the loss of colour as well as the browning of red wines and its activity on tannins and anthocyanin is about 30-times higher than that of tyrosinase. Moreover, the laccase shows high resistance towards SO_2. It is almost uninhibited by the quinones formed during the oxidation. This enzyme is, therefore, much more stable under winemaking conditions than tyrosinase.

Dubourdieu *et al.* (1984) and Grassin and Dubourdieu (1989) have proposed a rapid and selective colorimetric method for determination of the laccase activity in grape juices and young wines. This method has been integrated into a test kit which is now commercially available (Manfredini, 1989). The browning of red wines occurs when the laccase level is higher than 3 units/ml. The laccase level in a batch of grapes depends not only on the percentage of infected grapes but also on the state of *B. cinerea* development (Table 7).

Enzymes of the yeasts and lactic acid bacteria

In winemaking, the alcoholic fermentation is essentially performed by a single species of yeast, namely *Saccharomyces cerevisiae.* This species includes many types and a great diversity of strains.

The recent work performed on the enzymes of wine yeasts deals with the autolysis of the cells and with the transformation of some aroma precursors of the grape.

TABLE 7
State of *Botrytis cinerea* Development and Laccase Activity in Grape Juices

Variety	*State of* Botrytis cinerea *development*	*Laccase activity (U.LAC/ml)*
Cabernet Sauvignon	Sound berries	0·0
	Fully rotted berries	0·8
	With conidia	18·5
	Shrivelled berries	42·6
Sauvignon	Sound berries	0·0
	Fully rotted berries	20·3
	Shrivelled berries	56·0
Semillon	Sound berries	0·0
	Fully rotted berries	33·2
	Shrivelled berries	71·0

Yeast autolysis

The autolysis of yeast cells is defined as the enzymatic hydrolysis of intracellular biopolymers after the death of the cell. The diffusion of low-molecular-weight compounds into the extracellular medium is the result of this autolysis. This process takes place after the alcoholic fermentation when the sedimented yeasts are left in contact with the wine as is the case for certain high quality white wines (matured on the lees) as well as for the traditional champagne method. Feuillat and Charpentier (1982), as well as Silva *et al.* (1987), reported that such practices improve the organoleptic qualities of the wine.

During autolysis, amino acids and peptides resulting from proteolysis, glycoproteins formed on the degradation of cell walls, as well as some volatiles compounds, diffuse into the wine.

The work of Lurton (1988, 1989) is an original approach to the enzymatic phenomena that occur during proteolysis. It has been shown that under particular conditions (pH, temperature, ethanol level of the wines) the main enzymes involved are a protease A, an acid endopeptidase that produces peptides and a carboxypeptidase Y which hydrolyses proteins and peptides to amino acids. At pH 3·0, protease A plays a determinant role. Polypeptides (1–10 kDa) and oligopeptides (1 kDa) represent 75% of the total diffused nitrogen.

Llaubères (1988) and Llaubères *et al.* (1987) have studied the liberation of the mannoproteins of yeast cell walls during the storage of white wines over the lees. This liberation (a few hundred mg/litre) is due to the endo-β-1,3 glucanase activity of the yeast cell wall.

The amount of glycoprotein that diffuses into the wine depends essentially on the yeast strain, temperature, duration of the storage over the lees, and the contact of the wine with the biomass (resuspending of the yeast sediment).

The ageing of white wines over the lees leads to wines with a higher content of yeast colloids. During the ageing process some of the extracted macromolecules are hydrolysed. Therefore, the ageing conditions over the lees have an influence on the filterability of the wine.

Transformations of aroma precursors

Dubourdieu *et al.* (1988) and Darriet *et al.* (1988) have shown that, in fact, yeast cells or the periplasmic extract of *Saccharomyces cerevisiae* are able to hydrolyse, by means of a β-glucosidase, part of the bound terpenic substances of grape juice to volatile monoterpenes and sugars. The enzymatic action of the yeast enzymes on other aroma precursors of the grapes is presently being investigated.

The decarboxylation of the hydroxycianamic acids (*p*-coumaric and ferulic) of the grape juice by yeast decarboxylase (HCD) leads to the formation of volatile phenols (vinyl-4-phenol and vinyl-4-guiacol) during alcoholic fermentation. The organoleptical influence of these substances in wines is presently being investigated by Boidron *et al.* (1988).

The localisation of the HCD activity in the yeast cell wall, as well as its properties, have been described by Chatonnet *et al.* (1989) and Dubourdieu *et al.* (1989). The influence of the yeast strain on the HCD activity as well as its inhibition by some phenolic compounds of the grape explain the differences observed concerning the content of volatile phenols of wines.

Enzymes of the bacteria involved in the malolactic fermentation

During the so-called second fermentation of wine, the malolactic enzyme of lactic acid bacteria ensures the transformation of L-malic acid to L-lactic acid and CO_2. This enzyme activity is present in species of different origins. The malolactic enzymes of *Lactobacillus plantarum* and *L. mesenteroides* were studied by Schütz and Radler (1973), and were later purified by Lonvaud (1975) for *L. plantarum,* Lonvaud *et al.* (1982) for *L. mesenteroides,* Strasser de Saad (1983) for *L. murinus,* and Spettoli *et al.* (1984) for *L. oenos.* More recently, Renault (1987) has shown the existence of such an enzyme in *Lactococcus lactis*; this enzyme has been purified by Lonvaud-Funnel (unpublished).

The malolactic enzyme is often described as a tetrameric protein with a molecular weight of 230 kDa; its function requires the presence of Mn^{2+}, NAD^+ as well as L-malate in a sequential order. The pH optimum is about 5·9. Carboxylic acids, such as citric, tartaric and succinic acids, act as competitive inhibitors while lactic acid acts as a non-competitive inhibitor (Lonvaud *et al.*, 1982).

THE USE OF COMMERCIAL ENZYMES IN WINEMAKING

The natural enzyme activities of grapes are often too low to have a technological relevance. Reinforcement of the natural enzyme activity is therefore one of the reasons for adding commercial enzymes. Commercial enzymes are also used to fill a gap in cases where certain types of enzymatic activities are not present naturally.

The two types of enzyme preparations used in winemaking today are either pectinolytic enzymes or β-glucanases. Pectinolytic enzymes have been developed and used in the fruit juice industry for many decades and were introduced in winemaking in the 1970s. The use of β-glucanases is, however, much more recent; they were developed and tested in the early 1980s (Dubourdieu *et al.*, 1981*b*; Villettaz *et al.*, 1982).

Regardless of the type of commercial enzyme used, they are either applied to improve the extraction of juice or other valuable compounds from the grapes or to improve the clarification and filtration of juices and wines. The use of enzymes generally results in improvements in technology, quality, and economy.

Enzyme production

The enzyme preparations used in winemaking are produced from selected microorganisms, generally *Aspergillus* or *Trichoderma* spp. The production technology varies from one company to another. Two basic production methods are used: surface or submersed fermentation. The production technology of microbial enzymes has been reviewed by Aunstrup *et al.* (1979).

After production, under sterile conditions, the desired enzymes are recovered, purified several-fold, standardised and finally packed and stored. Enzyme preparations undergo stringent quality control tests.

In comparison to the β-glucanases, where the technologically relevant enzyme activities are well defined and characterised, the optimisation and quality control operations for pectinolytic enzyme preparations are extremely complex.

Pectinolytic enzyme preparations usually contain two groups of pectic enzymes, namely the de-esterifying enzymes (PEs) and the chain-splitting enzymes (depolymerases). The pectin esterases de-esterify pectin, producing methanol and pectic acid (Fig. 6a). The depolymerases split the glycosidic bonds of their preferred substrate, either by hydrolysis (e.g. endo- and exo-polygalacturonase) (Fig. 6b) or by β-elimination (e.g. pectin lyase) (Fig. 6c).

All these different activities have different specificity towards substates (high- or low-esterified-pectin, low- or high-molecular-weight pectin). Synergistic effects are also existing between the different types of pectic enzymes (e.g. between lyase and esterase). Pectic enzymes have been reviewed by Rombouts and Pilnik (1980), and in Chapter 7 of this volume.

Besides the pectic enzyme activities, commercial preparations contain several other enzymatic activities such as hemicellulases (arabanases, galactanases, etc.) or cellulases. These side activities are very important from the technological point of view. Synergistic reactions exist between the pectic enzymes and the so-called side activities. The ratio between the pectic enzymes

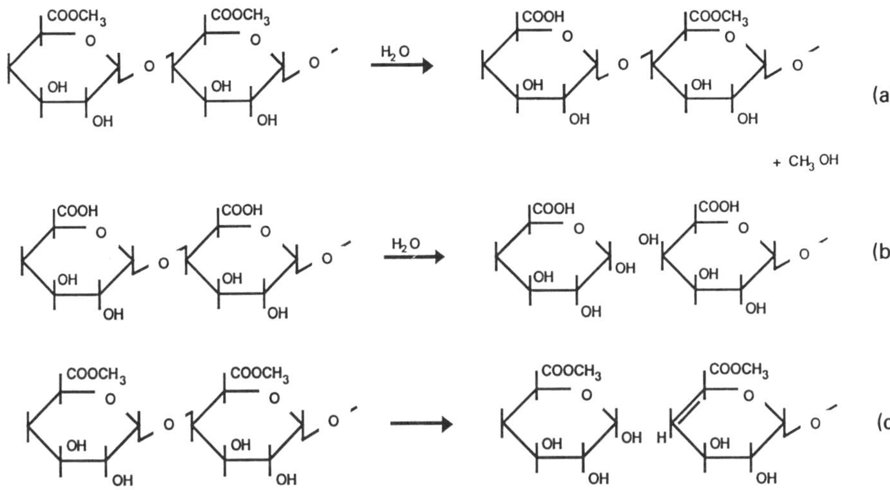

FIG. 6. Action pattern of different pectolic enzymes. (a) PE, (b) PG, and (c) pectin lylase.

and side activities varies from one commercial product to another. The enzyme production parameters (temperature, substrate, humidity, etc.) also influence this ratio and thus the technological characteristics of the enzyme preparations. This makes quality control a delicate, complicated and labour-intensive operation and the comparison between two commercial pectinolytic enzyme preparations very difficult.

Commercial preparations are available as liquid or powdered (granulated) form. Compared to the liquid ones, granulated products are much less sensitive to storage temperature; moreover, they do not necessitate the addition of preservatives. Liquid enzyme preparations should be stored in a cool place (4–10°C).

Prior to use, the enzyme preparations (usually highly concentrated products) should be dissolved, diluted and dispersed properly into the mash, juice or wine. It is strongly recommended to follow the instructions given by the enzyme producers.

The main producer of commercial enzymes for the wine industry is Novo Nordisk Ferment, Dittingen, Switzerland. Other major producers include Gist Brocades (Delft, Netherlands), Röhm (Darmstadt, Germany) and Miles (Elkhart, Indiana, USA).

This chapter describes the main applications for the two types of enzyme preparations used today in winemaking, namely the pectinolytic enzymes and the β-glucanases.

Pectolytic enzymes

The classical white- and red-wine technology is indicated in Figs 7 and 8. The stages where enzymes are usually added are marked with the sign E. The main applications are now described in detail.

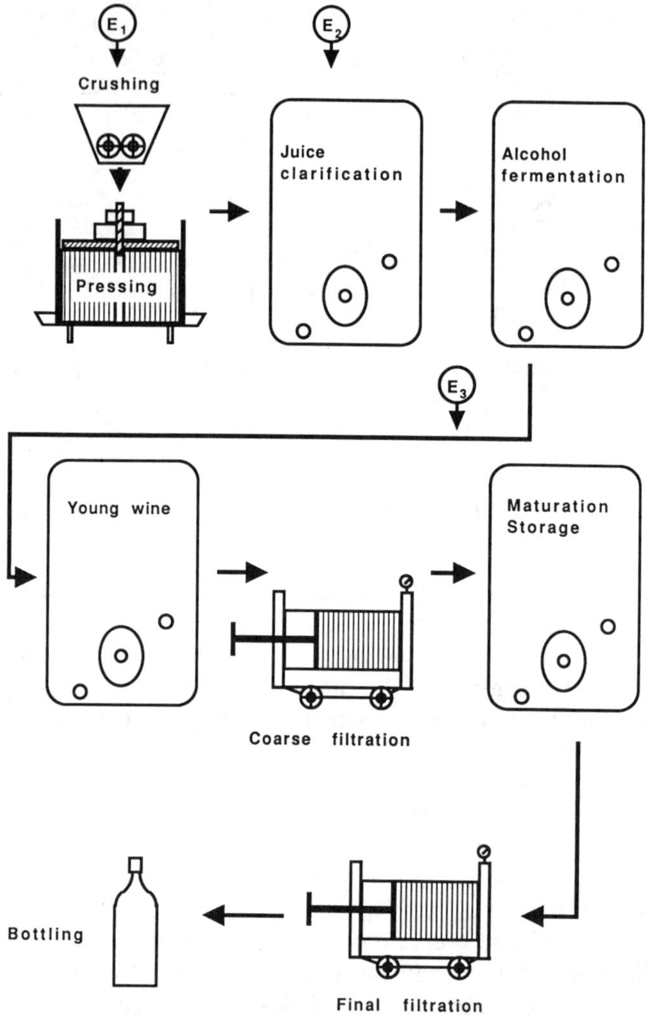

FIG. 7. Flow-sheet of white-wine production.

White wine technology (Fig. 7)
Juice extraction. Certain types of grapes, such as Muscat or Sylvaner, contain a high level of pectin which often presents problems at the pressing stage. Juice extraction takes a very long time and in spite of the high pressure applied, the juice yield obtained is generally low. In order to reduce the pressing time (increase press capacity) and obtain a satisfactory juice yield, the pectic substances of the grapes must be hydrolysed.

Technology: The enzyme preparation is first dissolved or diluted in a portion of juice (e.g. a 1% solution) and then added continuously in the crusher (Fig. 7, E1). In this way the enzyme preparation is well distributed into the mash.

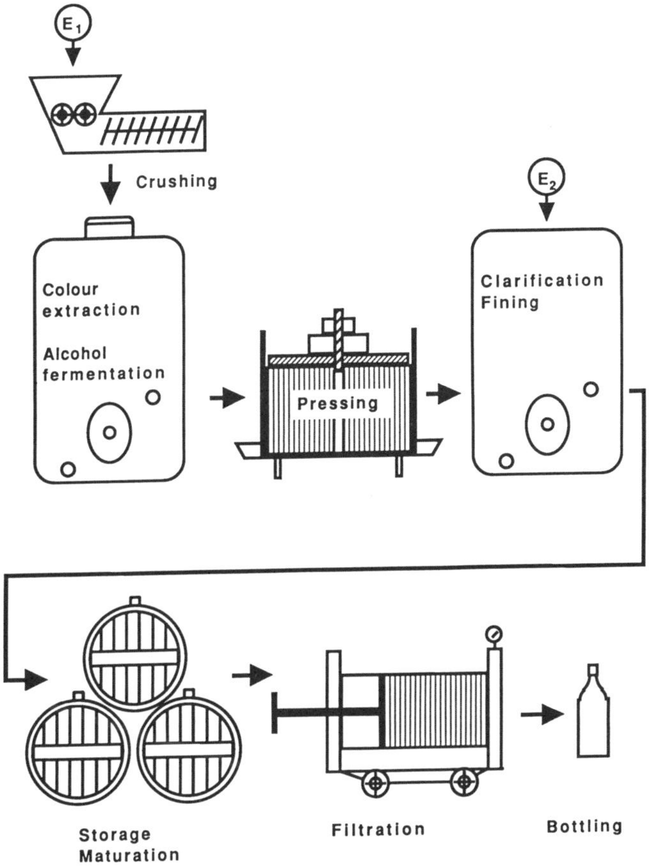

FIG. 8. Flow-sheet of red-wine production.

The contact time between enzyme and substrate is usually short (0·5–2·0 h). In cases where the grapes are extremely rich in pectin (e.g. Muscat grapes), or the mash temperature is low (10°C or lower), an adjustment of the enzyme dosage is indicated.

Results: The results obtained after enzyme treatment depend primarily on the type of grapes crushed: the higher the pectin level, the better the results will be. The amount of free-run juice (first quality fraction) and the press capacity will increase markedly.

To illustrate these results, we have chosen two examples: one involving a Sylvaner (low pectin level) and the other a Hungarian grape variety (high pectin level). The results obtained are indicated in Tables 8 and 9. As can be seen from both tables, the results obtained with enzyme treatment of the grape mash are substantial. The high quality juice fraction was increased by 33–40% and the total yield by 8–12%.

TABLE 8
Influence of Enzyme Treatment[a] on Juice Yield from a Sylvaner Mash

	Enzyme treated mash	*Control*
Trial conditions		
Pressure (10^5 Pa)		
start	0·2	0·2
end	1·8	1·8
Contact time (min)	55	45
Temperature (°C)	8	11
Batch size (kg)	13 000	13 000
Results		
Free-run juice (litres)	5 600	4 000
Juice yield (total) (litres	10 900	10 100
Pressing time (min)	130	120

[a] Vinozym, Novo Nordisk.

Juice/wine clarification and filtration. The production of high quality white wines requires good clarification of the juice prior to the start of the alcoholic fermentation. High molecular colloidal substances such as pectin, prevent the rapid sedimentation of cloud particles and thus considerably slow down clarification of the juice. The clarification process can be speeded up by reducing the viscosity of juice and therefore by hydrolysing pectic substances. *Endo*-PG activity is able to reduce the viscosity of a pectin solution by more than 50% by hydrolysing just 2–3% of the glycosidic bounds of the pectin chains.

Technology: In this case, enzymes are *added to the juice,* first after the press, in the so-called clarification tank, during the filling operation (Fig. 7, E2). An enzyme-treated mash usually does not require any further enzyme addition to clarify the juice. The enzymatic treatment facilitates excellent clarification of the juice within 4–8 h. Without enzyme treatment, juice clarification time

TABLE 9
Influence of Mash Treatment with Pectinolytic Enzymes[a] on Juice Yield in the Different Press Fractions (Batch Size 8000 kg Mash)

	Effective yield (%)		*Percentage % distribution*		
	Free-run juice	*Total juice yield*	*Free-run juice*	*1st–3rd press-fractions*	*4th + 5th-press-fractions*
Control mash	47·4	73·7	64·3	24·7	11·0
Enzyme treated mash	63·3	82·4	76·9	19·6	3·5

[a] Vinozym, Novo Nordisk.

would be at least twice this (8–16 h) and frequently without reaching the desired degree of clarification. Lower juice temperatures (e.g. 8–10°C) require adjustment of the enzyme dosage.

In order to obtain the best results, it is important to avoid contact of enzymes with certain fining agents that inactivate them (e.g. bentonite). If bentonite treatment is required, it should be performed on the clear juice after completion of the enzymatic treatment.

Results: Beside better and faster clarification, the enzyme treatment results in a more compact cloud sediment. An enzymatically-clarified juice leads to a clear wine that is easier to filtrate later on. The results summarised in Table 10 show the effects of juice treatment on the filterability of the wine. In this case, filterability was increased by a factor of two! This is a great advantage from the technological point of view (increased filtration capacity), as well as from the aspect of production cost (lower filtration cost, less wine losses, etc.)

Red wine technology (Fig. 8)
Improvement of clarification and filtration of press wines are two classical applications for pectolytic enzymes in red wine technology. In addition enzymes have also been used for colour extraction.

Clarification and filtration of press wines. During maceration (alcoholic fermentation on the skin), a portion of grape colloids diffuse into the young wine. Pressing of the mash (press wine) contributes to a greater extraction of the colloids. It is well known to every winemaker that press wines are much more difficult to clarify and filter than free-run wines.

The major part of the extracted high molecular colloids are composed of pectic substances and neutral polysaccharides such as arabans, galactans,

TABLE 10
Effect of an Enzymatic Juice Treatment[a] on the Filterability of Sylvaner Wine in Comparison to a Control (Kieselguhrfiltration, 10 m^2 surface)

Filtration time (h)	*Filtration behaviour*			
	Enzyme treated juice		*Control*	
	Pressure (10^5 Pa)	*Volume filtered (liters)*	*Pressure (10^5 Pa)*	*Volume filtered (liters)*
0·5	3	3 500	3	3 500
1·0	3	7 000	6[b]	6 500
1·5	3	10 500		
2·0	3	13 500		
2·5	6[b]	15 500		

[a] Vinozym, Novo Nordisk.
[b] Filtration run stopped, maximal pressure reached.

arabinogalactans (Villettaz, 1979; Villettaz *et al.*, 1981). In order to be able to clarify and filter press wines, it is necessary to hydrolyse these colloidal substances to small fragments.

Technology: The enzymes are introduced after the press, directly into the clarification tank (Fig. 8, E2). If the wines contain a very high polyphenol level, it is necessary to perform a gelatin treatment before the enzyme is added to the wine. In these cases, enzyme treatment is usually done 8–12 h after the gelatin treatment. The need to perform a gelatin treatment should always be based on enzyme clarification tests in the laboratory.

Pectin degradation can be followed by a simple test: pectic substances are insoluble in alcohol, a property exploited in their detection.

Test procedure: One part of wine to be tested (e.g. 5 ml) is mixed in a test tube with two parts of 96% ethanol. If pectin is present, it will flocculate within 15 min. The volume of the sediment and the size of the 'flocs' provide an estimate of the amount of pectin present, as well as on the degree of degradation. If no or very slight pectin depolymerisation has occurred, a dense precipitate will appear immediately and then rise to the top to form a plug. If pectin is partially degraded, finer 'flocs' are formed, which are not able to retain the air bubbles formed and do not float.

Results: The results obtained vary from wine to wine. Clarification usually takes less than 12 h. As far as the improvement of filterability is concerned, the results to be expected are of the same order or superior to those obtained with Sylvaner wine (see Table 10).

It is often observed that enzymatically-clarified press wines are even easier to filter than the corresponding untreated free-run wines. This is often the 'motivation' for extending the enzyme treatment to the free-run wines.

Colour extraction. The red pigments, which are located in the skin of the grape berries, are extracted during maceration. In certain cases the level of extraction is very low (wine with little colour) or takes a very long time (reduction of the maceration tank capacity). This problem is often encountered in the northern parts of Europe.

Thermovinification (heat treatment of the mash) has been proposed to solve these problems. The grape mash is heated in an exchanger at about 80°C and held at this temperature for 2 min. This permits a colour extraction within a very short time compared with the traditional method where maceration is done at 25–30°C for 1–3 weeks.

Thermovinification is also advantageously used on botrytised grapes to inactivate laccase activity (polyphenol oxidase). Enzymatic colour extraction has been proposed as an alternative technology to thermovinification. It has even been tried to combine the enzyme treatment with thermovinification to obtain bctter colour extraction.

Technology: In connection with thermovinification, enzyme is added either before and/or after heat treatment. In most of the cases, the enzymes are added after the heat treatment to the 'cooled mash'. The duration of the treatment depends on the temperature of the mash (1–2 h at 50°C, 2–4 h at 20°C).

After this contact time, the mash is pressed. Thermovinified wines usually require an additional enzyme treatment in order to ensure good clarification and filtration. Without such an enzyme treatment, thermovinified wines would be extremely difficult to clarify and filter (the heat treatment inactivates all natural enzymes).

In the classical technology (maceration at room temperature for 1–3 weeks) enzymes have also been used with good results. In this case, the enzyme treatment is done during maceration (Fig. 8, E1). Such a treatment enables a better colour yield to be obtained in a shorter maceration time.

Results: Thermovinification facilitates good colour yield within a very short period (a few minutes). However, it seems that the colour stability obtained with this technology is not always acceptable so that after maturation some red wines are still 'light' in colour.

As far as the classical technology is concerned, good results have been obtained with an enzyme treatment. Pectinolytic enzyme preparations with a broad spectrum of activities (hemicellulases) appear to give better results than the 'standard' pectinases.

In general, the results depend upon the variety of grape. Unfortunately, the results obtained so far are not always consistent and reproducible. Too little is known about the mechanisms of colour extraction and colour stability to ensure good results. Further basic research is necessary to master the 'colour problems'.

β-Glucanases

β-Glucanases are used to facilitate clarification and filtration. They have been developed especially to solve particular problems encountered in wines made from botrytised grapes. This type of enzyme is used both for white and red wines.

Botrytis *infection: clarification and filtration problems*

It is well known to every winemaker that wines made from *Botrytis cinerea* infected grapes always present clarification and filtration problems. One can differentiate between two types of *Botrytis* infection: the one highly desired (noble mould) for late-harvest wine production such as the world famous Sauternes, Tokaj, etc. and the absolutely undesired grey mould. Both types of infection lead, however, to clarification and filtration problems. These problems are due to the presence of a high molecular β-glucan (β-1,3 main chain, β-1,6 short side chains) synthesised by the fungus directly into the grape berries. This glucan has, erroneously, been called 'dextran' in the past; its precise structure has been characterised only recently (Dubourdieu *et al.*, 1981*a*).

During the crushing operation, the glucan passes into the juice and finally into the wine. It is not hydrolysed by the natural enzymes of the grapes or microorganisms involved during alcoholic fermentation. Due to the large molecular weight and the linear structure of this type of β-glucan, the presence of very small amounts (5–10 mg/litre) is enough to stabilise the cloud, prevent the natural clarification and create serious filtration problems. Several techniques have been used to solve these problems but all were unsuccessful. Recently, an enzymatic preparation has been especially developed to solve this old problem (Dubourdieu *et al.*, 1981*a*; Villettaz *et al.*, 1982).

Mechanism of the enzymatic glucan degradation

It has been found that endo-β-1,3 activity is not able to hydrolyse the *Botrytis* glucan. Only the exo-β-1,3 activity is successful in this respect. This type of hydrolysis mechanism makes glucan degradation a time-consuming reaction. The enzymatic degradation products of *Botrytis* glucan are glucose and gentiobiose. Gentiobiose is, in a second step, further degraded to monomers by a β-glucosidase. All these activities are present in the newly developed enzyme preparation.

Technology: The β-glucanase activities are adsorbed and partially inhibited by the yeast during alcoholic fermentation. The enzymatic treatment must therefore be performed between the first racking (once the alcoholic fermentation is over) and the first filtration (Fig. 7, E3; Fig. 8, E2). Complete glucan degradation usually takes 7–10 days. During this period, the temperature of the wine should not drop below 10°C. Contact with bentonite and certain fining agents such as blue fining must be avoided absolutely. If such a fining treatment is necessary, it should only be performed after complete glucan degradation. At this stage, these fining treatments can be performed under much better conditions since the 'protector colloid' has been enzymatically hydrolysed.

Polyphenols also inhibit the β-glucanases. It is therefore desirable in certain cases (red wines) to perform a gelatin treatment prior to enzyme addition.

The presence of *Botrytis* glucan can be detected by the following specific and very simple test: 10 ml of the wine to be tested are poured into a test tube and 5 ml of 96% alcohol (industrial or denatured alcohol) are added to the wine. The solution is then shaken lightly to ensure thorough mixing. If more than 15 mg glucan/litre are present in the wine, a filamentous precipitate will become visible within minutes.

In order to detect lower quantities (3 mg/litre and above), the following test has been developed: 5 ml of 96% alcohol are added to 5 ml of wine in a conical centrifuge/tube. The contents of the tube are mixed and left to stand at room temperature for 30 min. The mixture is then centrifuged at 3000 rpm (700–800 *g*) for 20 min. The supernatant is carefully decanted; the sediment is dissolved in 1·0 ml water and 0·5 ml 96% alcohol added. The formation of a filamentous precipitate indicates the presence of small amounts of *Botrytis* glucan.

This test is somewhat more laborious than the first one but is, nevertheless, strongly recommended since filtration problems start with the presence of 3–5 mg glucan/litre.

TABLE 11
Influence of Glucan Degradation[a] on the Natural Clarification of a Bordeaux Sweet White Wine (Glucan Level, 22 mg/litre) (Villettaz *et al.*, 1982)

	Turbidity units (NTU)	
	Enzyme treated wine	*Control wine*
After alcoholic fermentation (December 1980)	1 000	1 000
At enzyme treatment (August 1981)	800	800
10 days after enzyme treatment	400	800
3 weeks after enzyme treatment	200	800

[a] Glucanex, Novo Nordisk.

Results: Table 11 indicates the effect of enzymatic glucan degradation on wine clarification. The turbidity index was reduced by 50% within 10 days, and by 75% within 3 weeks. The turbidity of the control wine remained unchanged during the entire trial period. Both wines (control and enzyme treated) were filtered over Kieselguhr (4 m^2 filtration surface) under the same conditions.

In a single filtration run, 7200 litres of the enzyme-treated wine were filtered. However, filtration of the control wine had to be stopped after only 800 litres. In this case, enzyme treatment allowed filtration of nine times more wine per filtration run! The next example illustrates well the problems created by the presence of glucan during the different filtration steps.

A dry white wine containing 15–20 mg glucan/litre was enzymatically treated while a corresponding batch of the same wine was kept untreated as control. Before filtration, both wines were first fined with bentonite (80 g/hl) and then centrifuged. The first filtration was carried out over Kieselguhr. After filtration, the wines were stored separately for a few weeks and then filtered a second time. The second filtration was performed with filter pads (sterile brand). Table 12 indicates the filtration behaviour of both wines over Kieselguhr. The pad filtration behaviour of the pre-filtered wines is shown in Table 13.

As can be seen from these results, the technological and economical incidence of glucan degradation are very important. This trial also indicates clearly that mechanical treatments such as finings or centrifugation do not remove the glucan and therefore do not solve the filtration problems. This two-step filtration trial also demonstrates clearly that filtration does not remove all the glucan and that filtration problems are therefore encountered all the way through from the first until the last filtration step.

TABLE 12
Incidence of Glucanase Treatment[a] on the Filterability of a White Wine Containing 15–20 mg Glucan/Litre (Villettaz *et al.*, 1987)

Filtration time (h)	*Filtration behaviour*[b]			
	Enzyme treated wine		*Control wine*	
	Pressure (10^5 Pa)	*Volume filtered (litres)*	*Pressure (10^5 Pa)*	*Volume filtered (litres)*
0·5	0·7	5 500	2·8	3 500
1·0	0·8	10 500	4·8	7 000
1·5	0·8	17 500	5·2[c]	7 500
2·0	1·0	21 000		
2·5	1·1	25 500		
3·0	1·2	30 000		
3·5	1·3	34 500		
4·0	1·4	39 000[d]		

[a] Glucanex, Novo Nordisk.
[b] Filtration over Kieselguhr (10 m^2 filtration surface).
[c] Filtration run stopped, maximal pressure reached.
[d] Tank empty.

TABLE 13
Filtration of a Kieselguhr-Prefiltered Wine (Table 5) Over Sterile Filter Pads[a] (Villettaz *et al.*, 1987)

Filtration time (h)	*Filtration behaviour*			
	Enzyme treated wine[b]		*Control wine*[c]	
	Pressure (10^5 Pa)	*Volume filtered (litres)*	*Pressure (10^5 Pa)*	*Volume filtered (litres)*
0·5	0·20	400	0·8	400
1·0	0·20	800	1·0[a]	450
2·0	0·25	1 600		
3·0	0·30	2 400		
4·0	0·30	3 200		
5·0	0·40	4 000		
6·0	0·45	4 800		
7·0	0·50	5 600		
8·0	0·60	6 400		
9·0	0·80	7 200		
10·0	1·00[d]	8 000		

[a] 22 pads 400 × 40 cm.
[b] Glucan level before pad filtration: 0 mg/litre.
[c] Glucan level before pad filtration: 7–mg/litre.
[d] Filtration run stopped, maximal pressure reached.

FUTURE DEVELOPMENTS

One can assume that future developments in the application of enzymes in winemaking will be mainly in connection with colour extraction, colour stability, protein stabilisation of white wines, and aroma precursors.

It has been observed that pectinolytic enzyme preparations with a broad spectrum of side activities are superior to standard pectinases for colour extraction purposes. Trials have also indicated that a short enzymatic maceration time leads to wines with poor colour stability. The stabilising effect of certain types of polyphenols remain to be explored further, as well as the influence of certain side activities (hemicellulases, cellulases) present in pectinolytic enzyme preparations.

Protein stabilisation in white wines remains, in spite of the considerable amount of work done so far, an unsolved problem. All attempts made with fungal proteases have so far been less successful than bentonite treatment for achieving protein stability in white wines (Modra *et al.*, 1989). The replacement of bentonite by an enzyme treatment certainly remains a most desirable goal in winemaking.

Another interesting subject is the hydrolysis of bound terpenes to reinforce the aromatic profile of certain wines. It has been found that a large portion of the terpenic substances are bound to sugars (non-volatile) and therefore do not contribute to the aroma profile of the wines. Through the action of a β-glucosidase, it should be possible to release these terpenic substances.

It is possible by means of a glucose-oxidase–catalase treatment of grape juice to produce wines with low alcohol content or better balanced wines (better ratio between alcohol and acidity) (Villettaz, 1986). The sugar contained in grape juice is composed of *c.* 50% glucose and 50% fructose. By means of a treatment of the grape juice with a glucose-oxidase–catalase system, the glucose is converted to gluconic acid. Since the gluconic acid is not metabolised by the yeast during the course of the alcoholic fermentation, the resultant wine has a reduced alcohol content. The surplus gluconic acid formed has then to be eliminated or neutralised.

The same technology can also be used for the production of more balanced wines. In this case, the treated wine containing a high level of gluconic acid is used as an 'acid reserve' for blending purposes with acid-deficient wines. This process is especially suited for wines produced in hot climatic countries. In these regions the grape juices often have a high sugar level and simultaneously low acidity values. These technologies are not optimised yet and remain to be developed further.

REFERENCES

Aunstrup, K., Andersen, O., Falch, E. & Kjaer Nielsen, T. (1979). *Microbial Technology* (Vol. 1), ed. H. J. Peppler & D. Perlman. Academic Press, New York, USA, pp. 281–309.

Blade, H. & Boulton, R. (1988). *Amer. J. Enol. Vitic.*, **39**, 193–9.
Boidron, J. N., Chatonnet, P. & Pons, M. (1988). *Connaissance Vigne Vin*, **22** (4), 275–94.
Canonica, B. & Ferrari, G. (1987). Mémoire fin d'étude ENITA, Dijon, France.
Cayrel, A., Crouzet, J., Chan, H. W. S. & Price, K. R. (1983). *Amer. J. Enol. Vitic.*, **34**, 64–77.
Chattonet, P., Dubourdieu, D. & Boidron, J. N. (1989). *Connaissance Vigne Vin*, **1**, 59–62.
Cheynier, V., Rigaud, J. & Moutounet, M. (1989). In "*IV Symposium International d'Oenologie de Bordeaux*", Dunod, Paris. (To be published).
Cheynier, V., Trousdale, E., Singleton, V. L., Saigues, M. & Wyloe, R. (1986). *J. Agric. Food. Chem.*, **34**, 217–21.
Cordonnier, R. & Dugal, A. (1968). *Ann. Technol. Agric.*, **17**, 189–206.
Cordonnier, R. & Bayonove, C. (1981). *Connaissance Vigne Vin*, **15**, 169–286.
Crouzet, J., Nicolas, M., Molina, J. & Valentin, G. (1985). In *Progress in Flavour Research*, ed J. Adda. Elsevier Science Publishers, B.V., Amsterdam, The Netherlands, pp. 401–8.
Darriet, Ph., Boidron, J. N. & Dubourdieu, D. (1988) *Conn. Vigne. Vin.* **22,** No. 3, 189–95.
Drawert, F. & Krefft, M. (1978). *Chemie, Microbiol. Techn. Lebensmittel*, **5**, 189–92.
Dubernet, M. (1974). Thèse 3eme cycle, Université Bordeaux II, Bordeaux, France.
Dubernet, M. & Ribéreau-Gayon, P. (1973). *C.R. Acad. Sci.*, **277D**, 245–470.
Dubernet, M. & Ribéreau-Gayon, P. (1974). *Phytochem.*, **13**, I–85. I–087.
Dubernet, M. & Ribéreau-Gayon, P. (1975). *C.R. Acad. Sci.*, **280D**, I–313.
Dubernet, M., Ribéreau-Gayon, P., Lerner, H. R., Harel, R. & Mayer, A. M. (1977). *Phytochem.*, **16**, 191–3.
Dubourdieu, D. (1982). Thèse Doctorat, Université de Bordeaux II, Bordeaux, France.
Dubourdieu, D., Hadjinicolaou, D. & Betrand, A. (1980). *Connaissance Vigne Vin*, **14**, 247–61.
Dubourdieu, D., Fournet, B. & Ribéreau-Gayon, P. (1981*a*). *Cabrohydr. Res.*, **93**, 294–9.
Dubourdieu, D., Villettaz, J. C., Desplanques, C. & Ribéreau-Gayon, P. (1981*b*). *Connaissance Vigne Vin*, **15**, 161–77.
Dubourdieu, D., Grassin, C., Deruche, C. & Ribéreau-Gayon, P. (1984). *Connaissance Vigne Vin*, **18**, 237–52.
Dubourdieu, D., Ollivier, Ch. & Boidron, J. N. (1986). *Connaissance Vigne Vin*, **20**, 53–76.
Dubourdieu, D., Serrano, M., Vannier, A. C. & Ribéreau-Gayon, P. (1988). *Connaissance Vigne Vin*, **22**, 261–73.
Dubourdieu, D., Darriet, Ph. & Chatonnet, P. (1989). In *Actualités Oenologiques. IV Symposium International d'Oenologie de Bordeaux*. Dunod (Paris). pp. 151–9.
Feuillat, M. & Charpentier, C. (1982). *Amer. J. Enol. Vitic.*, **33**, 5–13.
Feuillat, M., Brillant, G. & Rochard, J. (1980). *Connaissance Vigne Vin*, **14**, 37–52.
Flanzy, M. & Bouziques, L. (1959). *Ann. Technol. Agric.*, **8**, 59–67.
Grassin, C. (1987). Thèse Doctorat, Université de Bordeaux II, Bordeaux, France.
Grassin, C. & Dubourdieu, D. (1989). *J. Sci. Food Agric.*, **48**, 369–76.
Grassin, C., Dubourdieu, D., Koh, K. H. & Ribéreau-Gayon, P. (1987). *CR Acad. Sci.*, **IIID**, 381–4.
Gunata, Y. Z., Sapis, S. C. & Moutounet, M. (1987). *Phytochem.*, **26**, 1573–5.
Heale, J. & Movahedi, S. (1989). In *Actualités Oenologiques 1989. IV Symposium International d'Oenologie de Bordeaux*. Dunod (Paris). pp. 127–32.
Hsu, J. C. & Heatherbell, D. A. (1987). *Amer. J. Enol. Vitic.*, **38**, 6–11.
Kidron, M. (1976). PhD Thesis, Hebrew University of Jerusalem, Jerusalem, Israel.

Lafon-Lafourcade, S., Dubourdieu, D., Hadjinicolaou, D. & Ribéreau-Gayon, P. (1980). *Connaissance Vigne Vin,* **14,** 127–38.
Llaubères, R. M. (1988). Thèse Doctorat, Université de Bordeaux II, Bordeaux, France.
Llaubères, R. M., Dubourdieu, D. & Villettaz, J. C. (1987). *J. Sci. Food Agric.,* **41,** 277–86.
Lonvaud, M. (1975). Thèse de 3ème cycle, Université de Bordeaux II, Bordeaux, France.
Lonvaud, M., Funel, A. & Strasser de Saad, A. M. (1982). *Appl. Environ. Microbiol.,* **43,** 357–61.
Lurton, L. (1988). Thèse Doctorat, Université de Dijon, Dijon, France.
Lurton, L. (1989). In *Actualités Oenologiques 1989. IV Symposium International d'Oenologie de Bordeaux.* Dunod (Paris). pp. 212–6.
Manfredini, M. (1989). *L'Enotechnico,* **9,** 93–6.
Marteau, G., Scheur, J. & Olivieri, C. (1963). *Ann. Technol. Agric.,* **12**(3), 155–76.
Modra, E. J., Wallace, W., Williams, P. J. & Lee, T. H. (1989). In *Actualités Oenologiques 1989. IV Symposium International d'Oenologie de Bordeaux.* Dunod (Paris). pp. 217–21.
Molina, J., Salles, C., Nicolas, M. & Crouzet, J. (1986). *Am. J. Enol. Vitic.* **37,** 69–173.
Müller-Späth, H. (1977). *Die Weinwirtschaft,* **6,** 15–20.
Müller-Späth, H. (1989). In "*Actualités Oenologiques.*" *VI Symposium d'Oenologie de Bordeaux.* Dunod (Paris). pp. 139–45.
Ollivier, C. (1987). Université de Bordeaux II, Bordeaux, France.
Rapp, A., Hastrich, H. & Engel, L. (1976). *Vitis* **15,** 183–92.
Renault, P. (1987). Thèse Docteur Ingénieur, INA, Paris-Grignon, Paris, France.
Ribéreau-Gayon, J. (1935). *Rev. Vitic.,* **82,** 367–8.
Ribéreau-Gayon, J., Peynaud, E., Ribéreau-Gayon, P. & Sudraud, P. (1977). *Sciences et Techniques du Vin* (No. 4), Dunod, Paris, France.
Rombouts, F. M. & Pilnik, W. (1980). *Microbial Enzymes and Bio Conversions, Economic Microbiology* (Vol. 5), Academic Press, New York, USA, pp. 227–82.
Saywell, L. G. (1934). *Ind. and Eng. Chem.,* **26,** 981–6.
Schütz, M. & Radler, F. (1973). *Arch. Mikrobiol.,* **91,** 183–202.
Shepard, V. & Pitt, D. (1976) *Phytochem,* **15,** 1465–70.
Silva, A., Fumi, M., Montesissa, G., Colombi, M. G. & Colagrande, O. (1987). *Connaissance Vigne Vin.,* **3,** 141–62.
Spettoli, P., Nutti, M. P. & Zamorani, A. (1984). *App. Environ Microbiol.,* **48,** 900–3.
Strasser de Saad, A. M. (1983). Thèse Université de Tucuman, Argentina.
Usseglio-Tomasset, L. & Tarantola, C. (1963). *Vitis,* **3,** 190–206.
Valentin, G. & Crouzet, J. (1989). In *Actualités Oenologiques 1989. VI Symposium International d'Oenologie de Bordeaux.* Dunod (Paris). pp. 133–8.
Verhoeff, R. & Warren, J. M. (1972). *Neth. J. Plant. Pathol.,* **78,** 179–85.
Villettaz, J. C. (1979). Thesis No. 6503, ETH, Zürich, Switzerland.
Villettaz, J. C. (1984). *Bulletin OIV,* **57,** 19–29.
Villettaz, J. C. (1986). *Proc. of the Sixth Australian Wine Industry Technical Conference,* Adelaide, Australia, pp. 125–8.
Villettaz, J. C., Amado, R. & Neukom, H. (1981). *Carbohydr. Polym.,* **1,** 101–5.
Villettaz, J. C., Dubourdieu, D. & Lefebvre, A. (1982). In *Use of Enzymes in Food Technology,* ed. P. Dupuy. Tec + Doc, Lavoisier, Paris, France, pp. 457–62.
Villettaz, J. C., Weber, J., Wurtz, R. & Cayrel, C. (1987). *Revue des Oenologues,* **45,** 25–8.

12

Enzymes in Fermented Vegetable and Legume Products

G. Campbell-Platt & P. E. Cook
Department of Food Science and Technology, University of Reading, PO Box 226, Whiteknights, Reading, Berkshire, RG6 2AP, UK

INTRODUCTION

Fermentations probably represent the oldest biotechnological processes employed by man, and the roots of this technology are lost in antiquity. Vegetable and legume fermentations have been documented as having been used for at least several thousand years and may have existed before crops became widely domesticated. Fermentation can be considered a mild or minimal form of food preservation (Campbell-Platt, 1988). It enables an extension of the storage life of plant foods, development of new textural and flavour properties and a reduction in levels of anti-nutritional factors and toxins. Enzymes play a key role in these processes. The majority are produced by the fermentative microorganisms but in some cases endogenous plant enzymes may also influence the properties of these foods.

FERMENTED VEGETABLES

Fermented vegetables and vegetable mixtures are eaten in many parts of the world as a major part of the diet (e.g. kimchi in Korea) or as snacks, condiments and cocktails. Many types of plants and plant parts are utilised for fermentation including bamboo shoots, cabbage, carrots, celery, cucumbers, olives, radish and turnips. Table 1 gives examples of some of these products, the geographic regions where they are produced and the microorganisms associated with the fermentations. More detailed information is provided by Beuchat (1978), Pederson (1979), Steinkraus (1983), Vaughn (1985) and Campbell-Platt (1987).

Although a wide range of vegetables are fermented, almost all undergo a lactic fermentation involving a succession of lactic acid bacteria and sometimes yeasts (Pederson, 1979; Vaughn, 1985). An apparent exception is the legume

TABLE 1
Fermented Vegetable Products, their Distribution, Substrates used[a] and Microorganisms Involved in the Fermentations[b]

Product and Geography	*Substrate (s)*	*Microorganisms Involved*
Sauerkraut (Europe, North America)	Cabbage (*Brassica oleracea*)	*Leuconostoc mesenteroides*, *Lactobacillus* spp., *Pediococcus cereviseae*
Egyptian pickled vegetables (Middle East)	Various vegetables	*L. mesenteroides*, *Lactobacillus* spp., Yeasts
Gundruk (Nepal)	Mustard (*B. juncea*) or Cauliflower (*B. oleracea*)	*Lactobacillus cellobiosus*, *Lactobacillus plantarum*, *Pediococcus pentosaceus*
Hum choy (China)	Cabbage (*B. chinensis*)	Lactic acid bacteria
Kimchi (Korea)	Korean cabbage (*B. oleracea*) Korean radish (*Raphanus* sp.) Cucumber (*Cucumber sativus*)	*Lactobacillus* spp., *L. mesenteroides*, *P. cereviseae*, Yeasts and moulds
Kushuk (Iraq)	Parboiled wheat (*Triticum* sp.) and Turnips (*B. campestris*)	*L. plantarum*, *Streptococcus lactis*, *Saccharomyces rouxii*
Malaysian pickles (Malaysia)	Cucumbers and Gerkins (*C. sativus*) Onion (*Allium ascalonicum*) Ginger (*Zingiber officinale*) Leek (*A. ampeloprasum*) Chilli (*Capsicum annuum*) Bamboo (*Bambusa glaucesens*) Mustard (*Brassica sp.*)	Lactic acid bacteria
Olives (Europe, North and South America)	Olive (*Olea europaea*)	*L. plantarum*, Other lactic acid. Bacteria, Yeasts
Pak-sian-dong (China)	*Gynandropsis pentaphylla* Bambooshoot (*Bambusa arundinacea*) Red onion (*A. ascalonicum*)	*Lactobacillus* spp.

TABLE 1 – *contd.*

Product and Geography	*Substrate (s)*	*Microorganisms Involved*
Cucumber pickles (World-wide)	Cucumber (*C. sativus*) Dill (*Anethum graveolens*) in some pickles	*P. cereviseae*, *Lactobacillus* spp. *L. mesenteroides*
Shoidon (India)	Bamboo (*Melocanna humilis*)	Unknown
Tsukemono (Japan)	Various (*Brassica juncea*, *B. pekinensis*, *B. rapa*, *B. rapa* var *sunkina* or *Rumex japonicus*)	Lactic acid bacteria

[a] Only the major ingredients are listed.
[b] Compiled from Beuchat (1978, 1983*a*), Steinkraus (1983), Hesseltine & Wang (1986), Campbell-Platt (1987).

Cassia obtusifolia in which the leaves are thought to be fermented by a *Rhizopus* mould and bacteria, predominantly of the genus *Bacillus* (Dirar, 1984). Detail of the biochemistry of carbohydrate metabolism in lactic acid bacteria is provided by Kandler (1983), and Pederson (1979) and Vaughn (1985) discuss these bacteria in relation to vegetable fermentations.

The post-harvest physiology of vegetables has been the subject of considerable study with attempts being made to prolong storage life and arrest undesirable metabolic effects following harvesting (Weichmann, 1987). Depending on the plant part and its maturity, the process of harvesting may represent a significant metabolic perturbation. Most unprocessed vegetables are slightly acidic (pH 5·5–6·8) and subject to fungal and bacterial spoilage; therefore, storage conditions are particularly important. Miller *et al.* (1987) found that post-harvest handling of cucumbers prior to pickling was critical since mechanical stresses, storage time and temperature altered production of endogenous pectin methylesterase and polygalacturonase which could result in softening in these products.

Cabbage and related *Brassica* crops are important vegetables in many fermentations (Table 1) and the sulphur-containing components play a significant role in their distinctive flavour and aroma. The enzyme myrosinase, a thioglucosidase, hydrolyses glucosinolates and thioglycosides and leads to the production of a range of compounds including isothiocyanates, thiocyanates, nitriles, glucose and sulphate, depending on the hydrolytic conditions, enzyme origin and presence of modifying components (Fenwick *et al.*, 1983; Ludford & Isenberg, 1987). Myrosinase is located in special cells, called idioblasts, which are separated from the substrate glucosinolates which are widespread in *Brassica* parenchymal tissues (Larsen, 1981). Physical disruption of tissues (cutting, crushing) results in enzymic action on glucosinolates with the production of volatile isothiocyanates and nitriles (Uda & Maeda, 1986).

These volatiles are important in flavour generation in these vegetables and their fermented products. Problems have occurred in industry where freezing and blanching treatments are used to inactivate polyphenol oxidases, peroxidases and lipoxygenase which might cause undesirable colour, texture and flavour changes in vegetables (Baardseth & Slinde, 1987; McLellan & Robinson, 1987). These treatments result in losses of flavour components although it may be possible to increase flavour by adding myrosinase to treated brassicas (Ludford & Isenberg, 1987). The flavour characteristics of fermented vegetables are important factors in acceptability, yet the role of endogenous plant enzymes and microbial enzymes has received little attention. Karki *et al.* (1983*a*) studied flavour components in fermented gundruk, a Nepalese unsalted mustard pickle, and found that all cyanides and isothiocyanates except allyl cyanide, were increased following fermentation and were the dominant flavour constituents. Daxenbichler *et al.* (1980) also found that glucosinolates were hydrolysed during fermentation of cabbage for sauerkraut production. The pH optimum for myrosinase is between 4 and 7, so the enzyme is likely to be active in the early stages of lactic fermentation depending on the amount of physical disruption to the plant tissues.

Ascorbic acid appears to play an important role in the activity of myrosinase by modifying the active site of the enzyme (Ohtsuru & Hata, 1979). Its effect is not thought to be due to the oxidation–reduction potential of ascorbic acid (Bjorkman, 1976). Levels of ascorbic acid either remain constant or increase during sauerkraut fermentation (Pederson, 1979) and they increase in fermented bamboo shoots (Pravabati & Singh, 1986). The low pH of fermented vegetables coupled with anaerobiosis may favour the stability of ascorbic acid. Kim *et al.* (1985*a*) found that ascorbic acid oxidase in cucumbers had an optimum pH of 6 and an optimum temperature of 40°C. In orange juice at pH 3·4, ascorbic acid destruction ceased and the enzyme appeared to be deactivated at pH 3·9. Ascorbic acid and other organic acids may play an important role in flavour generation in fermented vegetables. West *et al.* (1977) found that levels of allyl thiocyanate in coleslaw were more than six times those in shredded cabbage, suggesting that components in the dressing ingredients enhanced enzymic activity in the cabbage.

Although the lipid content of vegetables is generally low, modification by plant lipases and lipoxygenases and microbial lipases could be important in flavour generation. Vorbeck *et al.* (1983) reported an increase in non-esterified fatty acids during sauerkraut fermentation and similar findings for the fermentation of cucumbers were mentioned. The increase was ascribed to lipase-like activity from the bacteria involved in the fermentation, since pH was thought to limit the activity of plant enzymes. Karki *et al.* (1983*b*) also reported increases in free fatty acids in gundruk fermentation and suggested that they act as precursors of ester-like flavouring compounds in the fermentation.

Pectinolytic enzymes have received considerable attention because they can result in significant economic losses in fermented vegetable products, particularly cucumbers and olives. Less attention has been given to cellulolytic enzymes although they are present in plant materials (Heredia-Moreno & Fernández-Bolaños, 1984) and produced by microorganisms (Enari, 1983). Pectinolytic activity can result from endogenous plant enzymes, microbial enzymes secreted during growth and development of the vegetables, or growth of pectinolytic microorganisms during fermentation (Vaughn, 1985). Formation of pectinolytic enzymes in the plant tissues is usually restricted to a short period during ripening and these enzymes are found in many vegetables used for fermentation (Pederson, 1979; Fernández Díez, 1983, 1984; Vaughn, 1985; McFeeters, 1988). Preheating treatments at 55°C in the presence of 0·05 M calcium chloride was found to reduce polygalacturonase and pectinase activity and extend the shelf life of kimchi (Yook *et al.*, 1985) and calcium chloride treatments have been used to reduce softening of cucumber pickles (Buescher *et al.*, 1981; Tang & McFeeters, 1983). The time of calcium chloride addition, brining and storage time have been found to influence the characteristics of pectic substances and the degree of tissue softening (Hudson & Buescher, 1985).

Etchells *et al.* (1958) demonstrated that moulds could grow and secrete softening enzymes into cucumber flowers in the field and following brining, these enzymes could result in spoilage. This problem was later controlled by replacing the initial cover brine to remove flowers and enzymes, although this results in some losses of nutrients and buffering capacity (Vaughn, 1985). The concentration of salt used in brining is important because it is a major selective factor in the elimination of undesirable bacteria from the fermentation as well as in influencing pectinolytic activity. If the lactic acid bacteria are retarded or arrested, the pH of brined vegetables may remain high and pectinolytic bacteria of *Bacillus* spp., (Nagel & Vaughn, 1961; Davé & Vaughn, 1971; Karbassi & Vaughn, 1980), *Aeromonas, Escherichia, Erwinia* and *Klebsiella* (King & Vaughn, 1961; Hsu & Vaughn, 1969; Vaughn *et al.*, 1969) may become important in spoilage. All of these bacteria have been shown to degrade pickles tested *in vitro* using pure cultures of the bacteria and sterilised cucumbers in 5% sodium chloride at pH 5·5 or higher (Vaughn, 1985). Polygalacturonate lyases are produced by most of these bacteria, although *Erwinia* spp. produce in addition polygalacturonase, oligogalacturonate lyase and polymethylgalacturonate lyase (Fogarty & Kelly, 1983).

Extracellular polygalacturonase activity has also been reported for lactic acid bacteria in plant substrates containing pectins as carbon sources (Sakellaris *et al.*, 1988). Stamer and Stoyla (1968) and Juven *et al.* (1985) found that *Lactobacillus brevis* and *Leuconostoc mesenteroides* were able to grow with polygalacturonic acid as the sole carbon source and Sakellaris *et al.* (1988) reported that *Lactobacillus plantarum* could produce a polygalacturonase with an optimum pH of 4·5 and an optimum temperature of 30°C. The latter authors found that *L. plantarum* was able to grow only if glucose was added to the polygalacturonic acid and that its metabolism was altered, resulting in lower lactic acid production. It is not known if lactic acid bacteria are involved in pectin degradation in fermented vegetables, although *L. plantarum* has been reported as capable of growing inside fermenting cucumbers in addition to growing in the brine (Potts, 1985).

Yeasts and moulds are also capable of pectinolytic activity (Fogarty & Kelly, 1983) and result in spoilage of pickles in some cases, although the role of fungi in softening is not entirely clear (Pederson, 1979). Softening of pickles can be delayed by pasteurisation to inactivate pectinases and cellulases or by storage at low temperatures. Fermentations such as Korean kimchi, and Japanese suki pickles are carried out at lower temperatures than for pickled cucumber or sauerkraut production. Softening can also be reduced by blanching treatments (Gates & Costilow, 1981; Itabashi, 1985; Yook *et al.*, 1985). In some cases, plant phenolics such as those naturally present in grape leaves (Bell & Etchells, 1958) have been proposed as additives for controlling softening, but these are not approved for commercial situations (Vaughn, 1985). Natural secondary inhibitors of pectinolytic enzymes have been detected in olive fruits (Castillo Gómez *et al.*, 1979).

Minor ingredients of fermented vegetables include garlic, dill, various peppers and spices. Little information is available on how these influence fermentation or enzymic activities. In crushed garlic, alliin is hydrolysed by alliinase to allicin, a diallyl thiosulphinate, together with pyruvate and ammonia. A similar enzyme has been reported in *Bacillus subtilis* (Marakami, 1960). Cold fermentation of garlic using the endogenous enzyme is exploited in the production of garlic powders for health foods (Mohr, 1987). Garlic levels in kimchi appear to influence populations of lactic acid bacteria. Cho *et al.* (1988) found that increased garlic concentrations in the range of 0–6% resulted in increased growth of lactic acid bacteria. At concentrations of garlic greater than 2%, the concentration of aerobic bacteria did not increase as fast. The enzyme, alliinase, in garlic has an optimum pH in the range 5–8, whereas the enzyme from onion (used in some fermentations) has an optimum pH of about 8, and, unlike that from garlic, it is unstable below pH 4 (Whitaker, 1976).

Fermented juices and cocktails

Fermented vegetable juices and cocktails are receiving increasing attention for processing of waste and the development of novel fermented products. Zetelaki-Horvath and Andersson (1986) treated a mixture of beetroot, carrot, red pepper, horseradish and onion (47:20:20:8:5) with polygalacturonase. The mixture was fermented for 22 h at 30°C with *Lactobacillus plantarum*, *Lactobacillus helviticus* or one of three starters from kefir, yogurt or cheese production. The vegetable mixture fermented with *L. plantarum* was found to give the best sensory score. Lactobacilli have also been used in the fermentation of okra (*Hisbiscus esculentus*) juices (Kotzekidou & Roukas, 1987). Pectic enzymes are used commercially in maceration processes for fruit and vegetable purées and baby foods (Montenecourt & Eveleigh, 1985) and methods have been proposed for producing dietetic and diabetic fermented vegetable and vegetable fruit juice cocktails using endopolygalacturonase enzymes.

Although pectinolytic and cellulolytic enzymes are regarded as detrimental to fermented products such as cucumbers and olives, these enzymes could potentially result in greater exploitation of vegetable substrates. Reduction in cellular compartmentalisation, increase in surface area and release of nutrients could lead to greater microbial productivity during fermentation. Microorganisms and pectinolytic enzymes have already been exploited in the utilisation of beet juice for the production of useful pigments (Adams *et al.*, 1976; Lee *et al.*, 1982).

Vegetable and protein mixtures

Although vegetables are the major components in fermented products such as sauerkraut, cucumbers and olives, in some regions of the world, other ingredients may be added including cowpeas, soybeans, fish, oysters, chestnuts and mushrooms (Beuchat, 1978; Steinkraus, 1983, Campbell-Platt, 1987). Little is known about the microbiology or enzymology of these mixed products although they appear to involve a lactic fermentation. Lee *et al.* (1984) found that addition of protein supplements to vegetables for kimchi fermentation at 7°C resulted in a higher total acidity than in controls. The protein provided good buffering capacity and levels of ascorbic acid were higher after 3 weeks than in the unfermented product.

Fekete and Buechi (1972) described a novel process for mixed fermentation of milk with vegetable, cereal or fruit ingredients. The ingredients were added to the milk and heated to 80–100°C for 30 min followed by cooling to 45–48°C prior to fermentation. After fermentation using a lactic acid bacterial culture, the sour milk product was ripened at 4°C for 12–16 h. After pasteurisation at 110°C for 10 min and rapid cooling, the product was stable for 3–6 months.

FERMENTED LEGUMES

Legumes are one of the most important classes of foods and in South and East Asia, the Indian subcontinent and parts of Africa, fermented legume foods form a significant part of the human diet. The range of fermented legume foods is shown in Table 2 and detailed information on methods of preparation, microbiology and biochemical factors are provided by Steinkraus (1983), Wood (1985), Reddy *et al.* (1986*a*), Hesseltine and Wang (1986) and Campbell-Platt (1987). Legumes can be fermented alone, in combinations of legumes, legume-cereal mixtures, or products derived from legumes (e.g. soy-milk, tofu). Most fermentations utilise the seeds of legume although in Africa the leaves of *Cassia obtusifolia* are fermented to produce the condiment, kawal (Dirar, 1984). Although the microbiology and biochemistry of products such as soy-sauce have been extensively studied, enzymes have been characterised in detail in only a few cases.

TABLE 2
Fermented Products using Legume and Legume-Cereal Mixtures Together with the Substrates[a] and Microorganisms Involved[b]

Product and Geography	*Substrate (s)*	*Microorganisms*
Chee-fan (China)	Soy-bean (*Glycine max*)	*Mucor* sp., and *Aspergillus glaucus*
Dawadawa and Iru (West and Central Africa)	African locust bean (*Parkia biglobosa*) seeds	*Bacillus* spp., *Staphylococcus* sp.
Dhokla (India)	Bengal gram (*Cicer arietinum*) and wheat (*Triticum* sp.) or rice (*Oryza sativa*)	*Leuconostoc mesenteroides*, *Lactobacillus fermenti* *Streptococcus faecalis*
Dosai (India)	Black gram (*Vigna mungo*) and rice (*O. sativa*)	*L. mesenteroides*, *Lactobacillus* spp., *S. faecalis*, *Bacillus* spp., Yeasts
Idli (India)	Black gram (*V. mungo*) and rice (*O. sativa*)	*L. mesenteroides*, *Lactobacillus* spp., *Pediococcus cereviseae*, *Streptococcus* spp., Yeasts
Kawal (Sudan)	*Cassia obtusifolia* leaves	Lactic acid bacteria, *Bacillus* spp., *Rhizopus* sp.
Meitauza (China)	Soy-bean (*G. max*) okara (residue from tofu production)	*Actinomucor elegans*, *Mucor* spp.
Miso and related products (Southeast Asia, North America)	Soy-bean (*G. max*) seeds and various cereals	*A. oryzae*, *Aspergillus* spp., *Rhizopus* spp., *Saccharomyces rouxii*, *S. sojae*, *Lactobacillus delbruckii*, *Pediococcus acidilactici* *Pediococcus halophilus*

TABLE 2—*contd.*

Product and Geography	*Substrate (s)*	*Microorganisms*
Natto (Japan)	Soy-bean (*G. max*) seeds	*Bacillus* spp.
Kenima (Nepal)		
Thua nao (Thailand)		
Oncom (Java)	Groundnut (*Arachis hypogaea*), press-cake	*Neurospora intermedia*, *Rhizopus oligosporus*
Papadam (India, Pakistan)	Black gram (*V. mungo*) flour	*Candida krusei*, *Saccharomyces* spp.
Shoyu (soy-sauce) and related products (Southeast and East Asia, North America, Europe)	Soy-bean (*G. max*) seeds and various cereals	*A. oryzae*, *Aspergillus sojae*, *S. rouxii*, *P. halophilus*, *Pediococcus* spp. *Streptcoccus* spp., *Torulopsis* sp., *Lactobacillus* spp.
Soy-milk (East Asia, North America)	Soy-bean (*G. max*)	Lactic acid bacteria
Sufu (China, Taiwan)	Soy-bean (*G. max*) curd (tofu)	*A. elegans*, *Mucor* spp.
Tempe (Southwest Asia, North America, Europe)	Soy-bean (*Glycine max*) and others[c] seeds	*R. oligosporus*, *Rhizopus* spp.
Ugba (West Africa)	Oil bean (*Pentaclethra macrophylla*) seeds	*Bacillus* spp., Other bacteria
Waries (India)	Black gram (*V. mungo*) or Bengal gram (*C. arietinum*) flour	*Candida* spp., *Saccharomyces* sp., Bacteria

[a] Only the major components are listed.
[b] Compiled from Beuchat (1978, 1983*a*), Steinkraus (1983), Hesseltine & Wang (1986), Reddy *et al.* (1986*a*), Campbell-Platt (1987).
[c] Velvet bean (*Mucuna puriens*), Jack bean (*Canavalia ensiformis*), Winged bean (*Psophocarpus tetragonolubus*), Pigeon pea (*Cajanus cajan*), Broad bean (*Vicia faba*), Wild tamarind (*Leucaena leucocephala* and *L. glauca*), Chick-pea (*Cicer arietinum*), Sweet narrow leaved lupin (*Lupinus angustifolius*), Yellow pea (*Pisum sativum*), Cowpeas (*Vigna sinensis*), Bonavist bean or lablab (*Dolichos lablab*).

In most legume seeds, oxidative enzymes such as lipoxygenases may be active as well as lipases, acyl hydrolases, transferases and phospholipase (Galliard & Chan, 1980). These enzymes can lead to rancidity if not inactivated prior to fermentation. In soy-beans, there are two types of lipoxygenase with pH optima of 6·5 and 9·0, respectively. The purified lipoxygenases are rapidly denatured at 70°C although they are more resistant in food environments. Residual activity can lead to oxidation of fatty acids resulting in off-flavours, particularly in some soy-bean products. Treatments such as soaking, grinding and roasting will destroy most of the endogenous enzyme activity prior to fermentation. Changes in residual enzyme activity during fermentation do not appear to have been studied.

Microbial proteolytic enzymes associated with legume fermentations have received more detailed study than carbohydrases and lipases and in a number of cases these enzymes have been extracted and characterised. Yamamoto *et al.* (1972) and Impoolsub *et al.* (1981) examined the kinetics of proteolytic enzymes during the production of soy-sauce koji and found that maximum enzymic activity occurred after 48 h fermentation. The major proteases produced by the koji moulds *Aspergillus oryzae, A. sojae* and *A. tamarii* are alkaline types, although three types of acid protease, two neutral proteases and one semi-alkaline protease have also been isolated and characterised (Misaki *et al.*, 1970; Nakadai, 1977; Nasuno & Nakadai, 1977). Intracellular membrane-bound acid proteases have been identified as associated with the rough and smooth microsomes of *A. oryzae* (Tsujita & Endo, 1980).

Because the pH is nearly neutral in koji production and the early stages of mashing, it is thought that the acid proteases (optimum pH 4·0–4·5) play a minimal role in protein hydrolysis (Matsubara & Feder, 1971). The proteases isolated from koji are two neutral proteases with pH optima of 5·5–7·0 and alkaline proteases with optima at pH 8·0–10·5. The two neutral proteases are metalloproteases. One enzyme has an optimum pH of 7, being stable in the pH range 5·5–12·0 but is rapidly deactivated above 50°C (Kalisz, 1988). The other neutral protease shows optimum activity at pH 5·5–6·0. The enzyme is relatively heat stable, with only 30% loss of activity after 10 min at 90°C (Nakadai *et al.*, 1973*a*). The alkaline serine proteases comprise about 80% of the proteolytic activity produced by koji moulds (Kalisz, 1988) and show optimal activity at pH 7·0–8·5. These proteases show high stability between pH 4·5–9·0 and have similar properties to the alkaline proteases of *Bacillus* spp. (Godfrey & Reichelt, 1983). Tagami and Sugawara (1977), in a study of 109 koji moulds, found that the proportion of neutral to total protease activity varied from 10 to 20%, suggesting that alkaline proteases probably comprise the major fraction of proteolytic enzymes during koji fermentation.

Ebine (1968) compared the proteolytic activity of *Aspergillus* and *Rhizopus* moulds. The *Rhizopus* moulds differed from those of *Aspergillus* since proteolytic activity was maximal at pH 3 and almost undetectable at neutral pH. Wang and Hesseltine (1970) reported five active proteolytic fractions in culture filtrates of the tempe mould, *Rhizopus oligosporus*. All fractions responded similarly to inhibitors and showed maximal activity at pH 2·5–3·0, with stability in the range pH 2·2–6·0. All were rapidly denatured above pH 7. Since the pH increases from an initial acid pH during tempe fermentation, maximum proteolytic activity probably occurs during the early stages, although neutral proteases have been reported in *Rhizopus* spp. (Godfrey & Reichelt, 1983). Aspartyl acid proteases from *Rhizomucor miehei* are used extensively in the dairy industry as an alternative to bovine chymosin for the coagulation of milk proteins in cheese manufacture (Löffler, 1986; Kalisz, 1988). Beuchat and Basha (1976) studied protease production by *Neurospora intermedia* which is used in the production of oncom from peanut press-cake. Maximal proteolytic activity occurred at pH 6·0–7·5 and *N. intermedia* appeared to produce more protein hydrolysis of peanuts than *R. oligosporus* (Beuchat *et al.*, 1975).

Bacillus spp. are involved in a number of legume fermentations and proteolytic enzymes from these bacteria are of major significance commercially in protein, brewing, meat, photographic, leather and dairy industries (Kalisz, 1988). In fermentation of locust beans (*Parkia biglobosa*) for dawadawa production, *Bacillus* proteases represent the major enzymic activity, resulting in a 500% increase in free amino acids (Odunfa, 1986). Aderibigbe and Odunfa (1988) obtained three proteases from a strain of *Bacillus subtilis* isolated from fermented locust bean. A neutral protease and a serine protease were identified, the latter having a high degree of hydrophobicity. Because the beans are fermented in moist solid-state, the authors suggested that the hydrophobic property of the serine protease was due to a close association between *B. subtilis* cells and the beans. *B. subtilis* neutral metalloproteases and alkaline proteases are reported to be sensitive to inhibitors in cereals and legumes (Godfrey & Reichelt, 1983). Other *Bacillus* fermented foods include kenima, natto and probably ugba, and similar proteolytic enzymes are likely to be active in these fermentations.

Three types of aminopeptidase and four types of carboxypeptidase have been isolated from koji moulds and these enzymes appear to be more suitable for the liberation of legume amino acids than those of *Bacillus* or *Rhizopus* species (Nakadai & Nasuno, 1976, 1977; Iguchi & Nasuno, 1978). The aminopeptidases of *Aspergillus oryzae* show optimal activity at pH 7–9 and 40–60°C and are inactivated above 70°C (Maestracci *et al.*, 1988).

Leucine aminopeptidase is thought to be a key enzyme in the formation of glutamic acid during soy-sauce production (Iguchi & Nasuno, 1978). Glutamic acid is an important flavour enhancer in soy-sauce and can also be formed by the action of glutaminase on L-glutamine and α-methyl-DL-glutamine. Attempts have been made to isolate glutaminase hyper-productive mutants of koji moulds for use in soy-sauce and related products (Yamamoto & Hirooka, 1974).

Microbial lipolytic enzymes are active in many food fermentations and may be responsible for both desirable and undesirable changes in organoleptic properties. Natural inhibitors of microbial lipases have been reported in legume seeds, but most are either lost or inactivated by treatments prior to fermentation. Wagenknecht *et al.* (1961) examined lipolytic activity during tempe fermentation using *Rhizopus oryzae.* They found that after 69 h, one-third of the neutral fats had been hydrolysed to form predominantly linoleic and oleic acids and lesser amounts of palmitic, stearic and linolenic acids. Significantly, no linolenic or linoleic acids were detected in the uncooked and unfermented beans. Murata *et al.* (1967) reported an increase in the concentrations of oleic and palmitic acids, but decreases in stearic and linoleic acids following fermentation using *Rhizopus oligosporus.* In fermentation of peanuts using *Rhizopus delemar* and *R. oligosporus,* levels of oleic and linoleic acids were increased and palmitic acid levels reduced (Beuchat & Worthington, 1974).

Microbial lipases differ in their degree of positional specificity. Those from the fungi *Candida rugosa, Geotrichum candidum, Penicillium cyclopium* and the bacteria *Chromobacterium viscosum* and *Staphylococcus aureus* are non-specific which result in complete hydrolysis of triglycerides to free fatty acids and glycerol (Kilara, 1985). The lipases of the fungi *Aspergillus niger, Mucor javanicus, Rhizomucor miehei, Rhizopus arrhizus, Rhizopus delemar* and the bacterium *Pseudomonas fragi* are more specific in only attacking the 1, 3 positions. Most lipases appear to be non-preferential in their specificity towards fatty acids (Kilara, 1985; Sonnet, 1988).

Multiform lipases have been reported from *Mucor lipolyticus, Rhizopus arrhizus, Rhizopus delemar* and *Rhizopus microsporus* (Iwai & Tsujisaka, 1984). The production mechanisms of these lipases appear to be related to either interaction with phospholipids (*Rhizopus delemar*), differences in association/dissociation of subunits (*Rhizopus microsporus*) or post-secretory proteolysis by coexisting proteases (*Rhizopus arrhizus*) (Iwai & Tsujisaka, 1984; Boel *et al.*, 1988). Most fungal lipases show optimum activity at pH 5–8 and 35–50°C (Iwai & Tsujisaka, 1984; Kilara, 1985). Because fermented foods contain mixtures of enzymes, interaction between proteolytic and lipolytic enzymes is probable. Iwai (1963) found that in semi-solid culture on wheat bran, *Rhizopus delemar* produced low amounts of lipase whereas proteases and amylases were produced in large quantities. In liquid culture, no protease was detected whereas lipase was detected in substantial quantities. Interactions between the purified enzymes indicated that lipase was rapidly inactivated by the presence of protease and similar results have been observed with other fungi (Tsujisaka *et al.*, 1972). Many proteases are secreted as inactive zymogens to avoid damage to cell constituents but no known protective systems from lipases exist for microbial intracellular or membrane lipids (Boel *et al.*, 1988).

Lipases appear to be particularly sensitive to iron which can result in a reversible inhibition at low concentrations, and an irreversible inactivation at higher levels (Iwai & Tsujisaka, 1984). Iwai *et al.* (1970) found that 90% of *Aspergillus niger* lipase activity was lost in the presence of 8 ppm of iron ions in a reaction mixture at pH 5·6 and 30°C. Although soy-beans and other legumes are a good source of iron, chelation by phytates may limit bioavailability.

Phytates are reduced by physical processing and by fermentation. Phytases are produced by a number of microorganisms in fermented foods (Reddy *et al.*, 1982) and an increased availability of iron through their action may have an important influence on lipolytic activity in these systems.

Amylolytic activity has been studied in a number of legume-based fermented foods and is particularly important in those involving legume–cereal mixtures. α-Amylases are produced by both fungi (e.g. *Aspergillus oryzae*) and bacteria (e.g. *Bacillus* spp.) which are important in legume fermentations. These enzymes are only known to hydrolyse α-1,4 glucosidic bonds in amylose, amylopectin and glycogen. The *Bacillus* species produce both saccharifying and liquefying amylases which are optimally active at pH 4·8–6·5, and stable between pH 5·5–8·0 (Fogarty, 1983). Calcium is required for activity and appears to protect *Bacillus* α-amylases from proteases, low pH and heat (Fogarty, 1983).

Amyloglucosidase is found mostly in fungi, including the important food fermentative genera *Aspergillus, Neurospora* and *Rhizopus*. The enzyme hydrolyses the α-1,4-glucosidic bonds and at a slower rate, the α-1,6- and α-1,3 bonds in amylose, amylopectin and glycogen. The enzymes are usually optimally active at pH 4·5–5·0 and 40–60°C (Fogarty, 1983). Both α-amylase and amyloglucosidase may be produced in some fermentations and synergistic hydrolysis of starch in these systems seems likely.

α-Glucosidases hydrolyse the terminal non-reducing α-1,4 linked glucose residues to release D-glucose and are produced by bacteria, yeasts and moulds (Fogarty, 1983). These enzymes are important in reducing levels of flatulent oligosaccharides during some legume and legume–cereal fermentations (Reddy *et al.*, 1986*b*). Pectinolytic and cellulolytic enzymes have been little studied in legume fermentations. Charles and Gavin (1983) demonstrated that *Rhizopus oligosporus* produced pectinolytic enzymes during tempe fermentation and *Rhizopus arrhizus, Neurospora sitophila, Aspergillus sojae,* and *Bacillus* spp. have been reported to produce endopolygalacturonases, endopolymethylgalacturonate lyase, and endopolygalacturonate lyase, respectively (Fogarty & Kelly, 1983). There is little information on cellulolytic enzyme systems in fermented legumes although *Aspergillus oryzae* is reported to produce β-glucosidase (Mega & Matsushima, 1979) and some *Bacillus* species are capable of degrading cellulose (Enari, 1983).

FERMENTED SOY-MILK AND TOFU

Soy-bean products such as soy-milk, tofu and okara (tofu production waste) can be used to prepare fermented products and fermented legume milks have been produced from groundnuts (Swaminathan & Parpia, 1967; Beuchat & Nail, 1978; Bucker *et al.*, 1979) and winged bean (Smoot *et al.*, 1979). The fermentation of soy products differs markedly, the milk undergoing a lactic fermentation (Rao *et al.*, 1986), while tofu and okara undergo a mould fermentation, usually with *Actinomucor elegans* or *Mucor* spp. (Su, 1986). Before fermentation of soy-milk or tofu, pre-soaked soy-beans are usually hot-ground at above 80°C, since the presence of lipoxygenase activity results in the development of off-flavours in the products (Wang, 1987). Both proteolytic and lipolytic activities have been studied during soy-milk fermentation. Angeles and Marth (1971*a*) found that *Lactobacillus delbrueckii* and *Streptococcus thermophilus* were slightly proteolytic in soy-milk and *L. casei* and *L. delbrueckii* caused lipolysis after 14 days incubation (Angeles & Marth, 1971*b*). Law and Kolstad (1983) have reviewed the information available on proteolytic enzymes in dairy strains of lactic acid bacteria. Although oligosaccharide levels are reduced primarily by soaking soy-beans prior to fermentation, some *Lactobacillus* spp. produce α-galactosidase (Mital *et al.*, 1973) and are capable of utilising residual levels of some oligosaccharides present in soy-milk (Mital & Steinkraus, 1975).

In mould-fermentation of tofu for the production of sufu, the moulds produce protease, α-amylase, α-galactosidase and lipase activities (Liu & Chen, 1965; Wang, 1967; Chou *et al.*, 1988). Chou *et al.* (1988) found that growth and enzyme production by *Actinomucor taiwanensis* on tofu was influenced by temperature, humidity and incubation time with growth being favoured by 30°C and 97% relative humidity, and protease and lipase activities being maximal after 60 h fermentation at 25°C and 97% relative humidity. After fermenting tofu, the product (pehtze) is normally brined in 12–20% sodium chloride which is reported to release a mycelium-bound protease (Wang, 1967). Wang (1967) found that the protease of *Mucor hiemalis* had an optimum pH of 3·0–3·5 although Liu and Chen (1965) reported the optimum pH for *Actinomucor* sufu to be about 7 and Chou *et al.* (1988) detected neutral protease activity in *A. taiwanensis*.

ENZYMES AND WATER ACTIVITY

Water activity (a_w), or equilibrium relative humidity, is an important physiological parameter in food microbiology since it defines the lower limits which permit microbial growth. It also interacts with factors such as pH, temperature and food preservatives to cause the inhibition of microorganisms (Beuchat, 1983*b*; Sperber, 1983). Enzymes are known which are capable of activity at levels of a_w well below those known to support microbial growth (Acker & Kaiser, 1959; Andersson, 1980) provided that a monomolecular layer of water is present (James & Augenstein, 1966; Hahn-Hägerdal, 1986). In non-aqueous systems, water may only be required to maintain the molecular conformation in a form which is suitable for enzymic activity (Zaks & Klibanov, 1984).

Prior and Kenyon (1980) found that in *Pseudomonas fluorescens*, the Entner–Doudoroff pathway was functional at a_w below 0·8 although the minimum a_w for growth of the organism was 0·957. Kim *et al.* (1985*b*) observed that when *Trichoderma reesei* was grown on wheat the optimum substrate–water content for cellulolytic activity and mycelial growth were different. Grajek and Gervais (1987) studied the influence of a_w on the production of enzymes by *Trichoderma viride* in fermenting sugar-beet pulp and in enzyme extracts. In both cases, the use of sodium chloride to reduce the a_w resulted in a rapid reduction in polygalacturonase activity. Water activities below 0·98 resulted in a rapid reduction in the activities of polygalacturonase, D-xylanase and β-glucosidase. Sodium chloride appeared to affect the catalytic activity in a different way to the non-electrolyte a_w depressors, sorbitol and glycerol. Measures (1975) found that phosphohexose isomerase, isocitrate dehydrogenase and aldolase of three bacterial species were more susceptible to inhibition by sodium chloride than whole bacterial cells.

In most food fermentations, there is free water available for microbial growth and enzymic activity, although few authors have examined a_w during fermentation. Mucoraceous moulds require a relative humidity of at least 90% and grow best at humidities near to 100% (Wang *et al.*, 1974). Decreased relative humidity was found to reduce both growth and enzyme activity of *Actinomucor taiwanensis* grown on tofu for sufu production (Chou *et al.*, 1988). Sodium chloride is added to many vegetable fermentations at levels of 2·25–9% and would be expected to have some effects on a_w, depending on the water content of the vegetables being fermented. Miso contains 6·1–13% sodium chloride (Ebine, 1986); soy-sauce, 8·9–18·5% (Yokotsuka, 1986); and inyu, a fermented black-bean sauce, 25–30% (Pierson, *et al.*, 1986). The addition of salt after dawadawa fermentation appears to prevent further fermentation and proteolytic activity by *Bacillus subtilis* (Odunfa, 1986). Impoolsub *et al.* (1981) found that high concentrations of sodium chloride (18–20%) were inhibitory to alkaline and neutral proteases of *Aspergillus flavus* var *columnaris*. Activities were only 10–15% of those without addition of sodium chloride and Nakadai *et al.* (1973*b*,*c*) obtained similar results with *Aspergillus oryzae*. In mould-fermented tofu (sufu), a brining stage with 12–20% sodium chloride is reported to release fungal proteases from mycelium into the curd (Wang, 1967).

In solid-substrate fermentations, enzymes may be exposed to high concentrations of substrates and products of hydrolysis (sugars, amino acids, fatty acids) which could influence the a_w in these systems. The presence of high concentrations of enzymes, products and substrates in industrial starch hydrolysis using amyloglucosidase can result in reversion, with the equilibrium being shifted towards synthesis of higher molecular weight saccharides from D-glucose (Fogarty, 1983). Other examples of reversion include plastein formation from low-molecular-weight peptides (Ward, 1983) and glycerides from fatty acids (Iwai *et al.*, 1964). In the case of glyceride synthesis, the hydrolytic and synthetic reactions of fungal lipases were controlled by the quantity of water present in the reaction mixture (Iwai *et al.*, 1964) and a similar effect of water appears to influence reversion of amyloglucosidase (Hehre *et al.*, 1969). Addition of a_w depressors such as sodium chloride, coupled with high substrate and product concentrations, may create conditions under which reversion could occur, although this possibility does not seem to have been examined in the context of fermented foods.

In some *Bacillus* fermentations (kenima, natto), there is extensive production of a highly viscous polymer of fructose and glutamic acid. Production of the polymer is reported to be stimulated by amino acids and sucrose in the soy-beans and by additions of monosodium glutamate (Ohata, 1986). *Leuconostoc mesenteroides* sometimes produces the polymer, dextran, during sauerkraut fermentation (Pederson, 1979). Mattiasson and Hahn-Hägerdal (1982) suggested that microbial cells immobilised in gel matrices are exposed to an environment with reduced a_w due to altered diffusion properties of cell metabolites. It is conceivable that *Bacillus* cells embedded in a micro-environment of polymeric material may experience reduced a_w and that extracellular proteases and carbohydrases will show altered properties under these conditions.

TOXINS AND ANTI-NUTRITIONAL FACTORS

Legumes and vegetables contain a number of toxins and anti-nutritional factors which in many cases are lost as a result of preparation or fermentation. Glucosinolates are precursors of compounds with goitrogenic action, and there is concern about problems of toxicity from the consumption of large quantities of *Brassica* crops (Heaney & Fenwick, 1985). Daxenbichler *et al.* (1980) found that during sauerkraut fermentation the glucosinolates were hydrolysed in the first two weeks and tests for goitrin and isothiocyanates were negative. Karki *et al.* (1983*a*) found that isothiocyanate levels decreased during a 6 day gundruk fermentation using mustard (*Brassica juncea*).

The phenolic glycoside, oleuropein, and its aglycone are primarily responsible for the bitterness of olive fruits which is removed by treatment with sodium hydroxide prior to fermentation (Vaughn, 1985). It is not known if residual glycosides are degraded during fermentation but oleuropein is known to be an inhibitor of lactic fermentation of green olives (Juven *et al.*, 1968; Juven & Henis, 1970). Juven *et al.* (1972) found that degradation products of oleuropein are surface active and result in losses of potassium, inorganic phosphate and glutamate from cells of *Lactobacillus plantarum*.

Legumes contain a number of toxic and inhibitory factors including phytohaemagglutinins (lectins), trypsin–chymotrypsin inhibitors, amylase inhibitors, saponins, goitrogenic substances, phenolic compounds, flatulents and oestrogens (Sgarbieri & Whitaker, 1982; Gupta, 1987). Phytohaemagglutinins may comprise 2–10% of the protein of legume seeds and trypsin–chymotrypsin inhibitors 0·2–2% (Sgarbieri & Whitaker, 1982). Although many anti-nutritional and toxic factors are lost or destroyed during soaking and cooking treatments, there can be considerable variability, particularly in heat stability, so that some may be present during legume fermentations. Cooking is not used prior to idli fermentation although levels of haemagglutinins, phytic acid, trypsin inhibitors and oligosaccharides are lower following fermentation (Reddy *et al.*, 1986*b*). Rao (1978) isolated a strain of *Leuconostoc mesenteroides* which was found to hydrolyse haemagglutinins from soy-beans, red kidney beans, navy beans and black beans. A mixture of enzymes including protease, 4-aspartyl glucosylamine amidohydrolase, β-*N*-acetyl glucosaminidase and α-D-mannosidase are secreted by this strain and were thought to be responsible for the haemagglutinin hydrolysis.

Cooking soy-beans, followed by fermentation with *Rhizopus oligosporus* to produce tempe, may reduce the levels of trypsin and chymotrypsin inhibitors (Hesseltine *et al.*, 1963) although Zamora and Veum (1979) found higher levels in tempe than in unfermented soy-beans. Padhye and Salunkhe (1978) found that chymotrypsin inhibitory activity in idli was eliminated completely by fermentation and steaming, but anti-trypsin activity was not. More studies are needed on the mechanisms involved in the reduction of these anti-nutritional factors, particularly on the role of pre- and post-fermentation processing and the activities of microbial enzymes during fermentation.

Flatulent oligosaccharides (raffinose, stachyose and verbascose) appear to be present at reduced levels in many fermented legume products. Soaking of uncooked legumes results in losses of flatulent oligosaccharides. Wang (1987) reported that soy-beans soaked for 18 h at 25°C lost 47·5% of the raffinose and 32·4% of the stachyose. Boiling also reduces oligosaccharides, ultrafiltration and immobilised fungal galactosidase have been used commercially to reduce oligosaccharide levels in soy-milk products. Pectinases and β-glucosidases are ineffective in eliminating oligosaccharides during protein isolation from unfermented soybeans (Godfrey & Reichelt, 1983).

Phytic acid forms complexes with multivalent cations and proteins in plant seeds and insolubility at physiological pH appears to render some complexed minerals biologically unavailable (Reddy *et al.*, 1982). Levels of phytic acid range from 0·14–2·05% in cereals and legumes, and account for 18–88% of total phosphorus (Reddy *et al.*, 1982). Processes such as soaking, cooking, milling and fermentation can result in an appreciable reduction in the levels of phytic acid. Table 3 shows levels of phytic acid in fermented and unfermented legumes and legume cereal mixtures. In all cases, fermentation results in a reduction in levels of phytic acid. In the case of peanut press-cake fermentation to produce oncom, the degree of phytic acid reduction appears to depend on the particular mould used for the fermentation. Phytase (*meso*-inositol hexaphosphate phosphohydrolase) is found in animals, plants and microorganisms (Cosgrove, 1966). The optimum temperature and pH together with kinetic properties of microbial and legume phytases are shown in Table 4. Most phytases show optimal activity at a temperature of 45°C and in the pH range 4–6, although there are marked differences in substrate specificity. Although phytases may be active during legume fermentations their nutritional significance is difficult to assess because inositol phosphates formed during hydrolysis may influence bioavailability of some elements (Fardiaz & Markakis, 1981).

TABLE 3
Levels of Phytic Acid (mg/g) in Legume and Legume Cereal Mixtures Before and After Fermentation

Food	*Unfermented*	*Fermented*	*Reduction (%)*	*Reference*
Dawadawa	150	75	50	Odunfa (1986)
Dhokla	60	6·5	89	Desai & Salunkhe (1986)
Idli	174	113	35	Rajalakshmi & Vanaja (1967)
Soyidli	66	42·5	36	Ramakrishnan *et al.* (1976)
Khaman	211	135	36	Rajalakshmi & Vanaja (1967)
Oncom[a]	13·6	7	49	Fardiaz & Markakis (1977)
Oncom[b]	13·6	0·5	96	Fardiaz & Markakis (1977)
Tempe	12·3	9·6	22	Sudarmadji & Markakis (1977)

[a] Oncom using *Rhizopus oligosporus*.
[b] Oncom using *Neurospora intermedia*.

TABLE 4
Temperature and pH Optima and Kinetic Properties of Fungal, Bacterial and Legume Phytases

Source	*pH Optimum*	*Optimum Temperature (°C)*	K_m (mM)	V_{max} *(μmol P_i/min/ml)*	*Reference*
Aspergillus oryzae	4·5	45–50	0·47	ND	Wang *et al.* (1980)
Rhizopus oligosporus[a]	4·5	55	0·15	0·076	Sutardi & Buckle (1988)
Rhizopus oligosporus[b]	4·5	55	0·17	0·034	Sutardi & Buckle (1988)
Saccharomyces cereviseae	4·6	45	0·21	0·005	Nayini & Markakis (1984)
Enterobacter aerogenes	4–5	45–50	0·135	ND[c]	Greaves *et al.* (1967)
Soy-bean (*Glycine max*)	4·8	60	2·4	0·22	Sutardi & Buckle (1986)
Navy bean (*Phaseolus vulgaris*)	5·3	50	0·018	0·20	Lolas & Markakis (1977)

[a] Extracellular.
[b] Intracellular.
[c] ND—no data.

Many legumes and cereals can become contaminated with mycotoxins as a result of the growth of field and storage fungi. Aflatoxins are fairly heat stable although they are sensitive to UV radiation and can be destroyed by aqueous solutions of acids, bases and oxidising agents (Marth & Doyle, 1979). Groundnuts appear to be particularly susceptible to mycotoxin contamination and Veen and Steinkraus (1970) detected levels of aflatoxin B_1 of between 0·23 and 5 mg/kg in raw peanut press-cake. Veen *et al.* (1968) analysed five samples of fermented peanut press-cake (oncom) and found levels of aflatoxin between 0·09 and 0·8 mg/kg. They found that during fermentation *Neurospora intermedia* and *Rhizopus oligosporus* reduced the aflatoxin content of peanut press-cake by 50 and 70%, respectively. Additionally, they demonstrated that *Neurospora intermedia* reduced aflatoxin content by 50% in Sabouraud dextrose agar containing the toxin. The ability to degrade aflatoxin has been reported for a number of *Rhizopus* species (Jarvis, 1971; Cole *et al.*, 1972) and some bacteria (Ciegler *et al.*, 1966). Although the mechanism of detoxification is not understood, enzymatic (Ashworth *et al.*, 1965; Doyle & Marth, 1979; Marth & Doyle, 1979; Yousef & Marth, 1981) and non-enzymatic degradation (Ciegler *et al.*, 1966) have been postulated. Doyle and Marth (1979) concluded that a peroxidase in *Aspergillus parasiticus* was responsible for the majority of aflatoxin degradation observed in their studies with this species. Tests using commercially available lactoperoxidase demonstrated that this enzyme could catalyse the reactions necessary to bring about aflatoxin degradation.

Further studies are needed on the enzymic degradation of aflatoxins and on whether a similar mechanism of breakdown occurs in non-mycotoxigenic food fermentative fungi.

RESEARCH NEEDS

Although a considerable amount of work has been done on the microbiology and chemistry of fermented foods, relatively little appears to be known about the behaviour of enzymes in many of these systems. Most work has concentrated on isolating the key organisms and studying the properties of their enzymes *in vitro*. It may be difficult to relate the properties of purified enzymes to these food systems where plant and microbial constituents may influence activity. Tonita and Kim (1965) found that the heat resistance of Taka-amylase A was greater in the presence of starch and its hydrolysis products, and enzymes such as lipoxygenase and pectin methylesterase appear to be more stable when associated with plant extracts or tissue components (Aylward & Haisman, 1969). What is needed is experimentation, where purified and well characterised enzymes are added to fermentation systems at appropriate times under defined environmental and microbiological conditions. It is only by performing such perturbation type experiments that we are likely to gain an understanding of the properties and role of particular enzymes in these food systems.

The majority of information on enzymes in vegetable fermentations concerns the spoilage problems due to pectinolytic activity. Little information is available on the role of enzymes in flavour generation in these products. Chemical analysis of fatty acids, amino acids, glucosinolates and isothiocyanates and related compounds have been provided but the enzymes associated with the production or hydrolysis of these products do not appear to have been studied during fermentation.

More research is needed on fermented vegetable and protein mixtures and on the enzymic processes occurring in these products. Some fermented legumes products have been studied in detail (soy-sauce, miso) but many others have received little attention. The area of enzymic detoxification of legume components and mycotoxins is in need of more study, particularly as this may have important implications for nutrition and food safety.

REFERENCES

Acker, L. & Kaiser, H. (1959). *Z. Lebensm. Unters. Forsch.*, **110,** 349–56.

Adams, J. P., Elbe, J. H., von & Amundsom, C. H. (1976). *J. Food Sci.*, **41,** 78–81.

Aderibigbe, E. Y. & Odunfa, S. A. (1988). *J. Appl. Bact.*, **65,** 361–9.

Andersson, R. E. (1980). *Appl. Environ. Microbiol.*, **39,** 36–40.

Angeles, A. G. & Marth, E. H. (1971*a*). *J. Milk Food Technol.*, **34,** 124–8.

Angeles, A. G. & Marth, E. H. (1971*b*). *J. Milk Food Technol.*, **34,** 69–73.

Ashworth, Jr, L. J., Shroeder, H. W. & Langley, B. C. (1965). *Science*, **148,** 1228–9.

Aylward, F. & Haisman, D. R. (1969). *Adv. Food Res.*, **17,** 1–76.

Baardseth, P. & Slinde, E. (1987). *Norw. J. Agric. Sci.*, **1,** 111–17.

Bell, T. A. & Etchells, J. L. (1958). *Bot. Gaz.*, **119,** 192–96.

Beuchat, L. R. (1978). *Food and Beverage Mycology.* AVI Publishing Co., Westport, CT, USA.

Beuchat, L. R. (1983*a*). In *Biotechnology* (Vol. 5), ed. H. J. Rehm & G. Reed, Verlag Chemie, Weinheim, Germany, pp. 477–528.

Beuchat, L. R. (1983*b*). *J. Food Prot.*, **46,** 135–41.

Beuchat, L. R. & Basha, S. M. M. (1976). *Eur. J. Appl. Microbiol.*, **2,** 195–203.

Beuchat, L. R. & Nail, B. J. (1978). *J. Food Sci.*, **43,** 1109–12.

Beuchat, L. R. & Worthington, R. E. (1974). *J. Agric. Food Chem.*, **22,** 509–12.

Beuchat, L. R., Young, C. T. & Cherry, J. P. (1975). *Can. Inst. Food Sci. Technol. J.*, **8,** 40–5.

Bjorkman, R. (1976). In *The Biology and Chemistry of the Cruciferae*, ed. A. J. McLeod & B. M. G. Jones. Academic Press, New York, USA, pp. 191–205.

Boel, E., Huge-Jensen, B., Christensen, M., Thim, L. & Fiil, N. P. (1988). *Lipids*, **23,** 701–6.

Bucker Jr, E. R., Mitchell Jr, J. H. & Johnson, M. B. (1979). *J. Food Sci.*, **44,** 1934–8.

Buescher, R. W., Hudson, J. M. & Adams, J. R. (1981). *Lebensm.-Wiss. u. -Technol.*, **14,** 65–9.

Campbell-Platt, G. (1987). *Fermented Foods of the World: A Dictionary and Guide.* Butterworths, London, UK.

Campbell-Platt, G. (1988). *Food Sci. Technol. Today*, **2,** 95–103.

Castillo Gómez, J., Minguez Mosquera, M. I., Cabrera Martin, J. & Fernández Diez, M. J. (1979). *Grasas y Aceites (Seville)*, **30,** 11.

Charles, M. & Gavin, J. R. (1983). In *Handbook of Indigenous Fermented Foods*, ed. K. H. Steinkraus. Marcel Dekker, New York, USA, pp. 74–87.

Cho, N. C., Jhon, D. Y., Shin, M. S., Hong, Y. H. & Lim, H. S. (1988). *Korean J. Food Sci. Tech.*, **20**, 231–5.
Chou, C.-C., Ho, F.-M. & Tsai, C.-S. (1988). *Appl. Environ. Microbiol.*, **54**, 688–92.
Ciegler, A., Lillehoj, E. B., Peterson, R. E. & Hall, H. H. (1966). *Appl. Microbiol.*, **14**, 934–9.
Cole, R. J., Kirksey, J. W. & Blankenship, B. R. (1972). *J. Agric. Food Chem.*, **20**, 1100–2.
Cosgrove, D. J. (1966). *Rev. Pure Appl. Chem.*, **16**, 209–25.
Davé, B. A. & Vaughan, R. H. (1971). *J. Bacteriol.*, **108**, 166–74.
Daxenbichler, M. E., Van Etten, C. H. & Williams, P. H. (1980). *J. Agric. Food Chem.*, **28**, 809–11.
Desai, B. B. & Salunkhe, D. K. (1986). In *Legume-based Fermented Foods*, ed. N. R. Reddy, M. D. Pierson & D. K. Salunkhe. CRC Press, Boca Raton, FL, USA, pp. 161–71.
Dirar, H. A. (1984). *Economic Bot.*, **38**, 342–9.
Doyle, M. P. & Marth, E. H. (1979). *Eur. J. Appl. Microbiol.*, **7**, 211–17.
Ebine, H. (1968). *Season Sci.*, **15**, 10–18.
Ebine, H. (1986). In *Legume-based Fermented Foods*, ed. N. R. Reddy, M. D., Pierson & D. K. Salunkhe. CRC Press, Boca Raton, FL, USA, pp. 47–68.
Enari, T.-M. (1983). In *Microbial Enzymes and Biotechnology*, ed. W. M. Fogarty. Applied Science Publishers, London, UK, pp. 183–223.
Etchells, J. L., Bell, T. A., Monroe, R. J., Masley, P. M. & Demain, A. L. (1958). *Appl. Microbiol.*, **6**, 427–40.
Fardiaz, D. & Markakis, P. (1981). *J. Food Sci.*, **46**, 523–5.
Fekete, K. & Buechi, W. (1972). Swiss Patent 521092 (*Chem Abst.*, **77**, 60350p).
Fenwick, G. R., Heaney, R. K. & Mullin, W. J. (1983). *CRC Crit. Rev. Food Sci. Nutr.*, **18**, 123–201.
Fernández Diez, M. J. (1983). In *Biotechnology* (Vol. 5), ed. H. J. Rehm & G. Reed. Verlag Chemie, Weinheim, Germany, pp. 379–97.
Fernández Diez, M. J. (1984). In *Proc. of IUFOST International Symposium on Chemical Changes during Food Processing* (Vol. 1), ed. S. B. Poveda, Instituto de Agroquimica y technologia de Alimentos, Valencia, Spain, pp. 301–18.
Fogarty, W. M. (1983). In *Microbial Enzymes and Biotechnology*, ed. W. M. Fogarty. Applied Science Publishers, London, UK, pp. 1–92.
Fogarty, W. M. & Kelly, C. T. (1983). In *Microbial Enzymes and Biotechnology*, ed. W. M. Fogarty. Applied Science Publishers, London, UK, pp. 131–82.
Galliard, T. & Chan, H. W.-S. (1980). In *The Biochemistry of Plants* (Vol. 4), ed. P. K. Stumpf & E. E. Conn. Academic Press, London, UK, pp. 131–61.
Gates, K. & Costilow, R. N. (1981). *J. Food Prot.*, **46**, 274–7
Godfrey, T. & Reichelt, J. (1983). *Industrial Enzymology*. Macmillan, London, UK.
Grajek, W. & Gervais, P. (1987). *Enzyme Microb. Technol.*, **9**, 658–62.
Greaves, M. P., Anderson, G. & Webley, D. M. (1967). *Biochem. Biophys. Acta*, **132**, 412–18.
Gupta, Y. (1987). *Plant Foods Human Nutr.*, **37**, 201–28.
Hahn-Hägerdal, B. (1986). *Enzyme. Microb. Technol.*, **8**, 322–7.
Heaney, R. K. & Fenwick, G. R. (1985). *World Crops: Production, Utilisation, Description*, **11**, 40–9.
Hehre, E. J., Okada, G. & Genghof, D. S. (1969). *Arch. Biochem. Biophys.*, **135**, 75–89.
Heredia-Moreno, A. & Fernández-Bolaños, J. (1984). In *Proc. of International Symposium on Chemical Changes during Food Processing* (Vol. 1), ed. S. B. Poveda. Instituto de Agroquimica y Technologia de Alimentos, Valencia, Spain, pp. 319–29.

Hesseltine, C. W. & Wang, H. L. (1986). *Indigenous Fermented Foods of Non-Western Origin* (Mycologia Memoir No. 11), J. Cramer, Berlin, Germany.
Hesseltine, C. W., Decamargo, R. & Rachis, J. J. (1963). *Nature,* **200,** 1226–7.
Hsu, E. J. & Vaughn, R. H. (1969). *J. Bacteriol.,* **98,** 172–81.
Hudson, J. M. & Buescher, R. W. (1985). *J. Food Biochem.,* **9,** 211–29.
Iguchi, N. & Nasuno, S. (1978). *J. Ferment. Technol.,* **56,** 645–55.
Impoolsub, A., Bhumiratana, A. & Flegel, T. W. (1981). *Appl. Environ. Microbiol.,* **42,** 619–28.
Itabashi, M. (1985). *J. Jap. Soc. Food Sci. Technol.,* **32,** 124–6.
Iwai, M. (1963). *Bull. Osaka. Municipal Tech. Res. Inst.,* **39,** 63–70.
Iwai, M. & Tsujisaka, Y. (1984). In *Lipases,* ed. B. Borgström & H. L. Brockman. Elsevier, Oxford, UK, pp. 443–69.
Iwai, M., Tsujisaka, Y. & Fukomoto, J. (1964). *J. Gen. Appl. Microbiol.,* **10,** 13–22.
Iwai, M., Tsujisaka, Y. & Fukomoto, J. (1970). *J. Gen. Appl. Microbiol.,* **16,** 81–90.
James, L. K. & Augenstein, L. G. (1966). *Adv. Enzymol.,* **28,** 1–40.
Jarvis, B. (1971). *J. Appl. Bact.,* **34,** 199–213.
Juven, B. & Henis, Y. (1970). *J. Appl. Bact.,* **33,** 721–32.
Juven, B., Samish, Z. & Henis, Y. (1968). *Israel J. Agric. Res.,* **18,** 137–8.
Juven, B., Henis, Y. & Jacoby, B. (1972). *J. Appl. Bact.,* **35,** 559–67.
Juven, B., Lindner, P. & Weisslowicz, H. (1985). *J. Appl. Bact.,* **58,** 533–8.
Kalisz, H. M. (1988). *Adv. Biochem. Engng/Biotechnol.,* **36,** 1–65.
Kandler, O. (1983). *Antonie van Leeuwenhoek,* **49,** 209–24.
Karbassi, A. & Vaughn, R. H. (1980). *Can. J. Microbiol.,* **26,** 377–84.
Karki, T., Itoh, H. & Kozaki, M. (1983*a*). *Lebensm.-Wiss. u.-Technol.,* **16,** 203–8.
Karki, T., Itoh, H., Kluchi, K., Ebine, H. & Kozaki, M. (1983*b*). *Lebensm.-Wiss. u.-Technol.,* **16,** 167–71.
Kilara, A. (1985). *Process Biochem.,* **20,** 35–45.
Kim, J. W., Park, E. S. & Yoon, S. (1985*a*). *Korean J. Nutr.,* **18,** 312–17.
Kim, J. H., Hosobuchi, M., Kishimoto, M., Seki, T., Yoshida, T., Taguchi, H. & Ryu, D. D. Y. (1985*b*). *Biotechnol. Bioengn,* **27,** 1445–50.
King Jr, A. D. & Vaughn, R. H. (1961). *J. Food Sci.,* **26,** 635–43.
Kotzekidou, P. & Roukas, T. (1987). *J. Food Sci.,* **52,** 487–8, 490.
Larsen, P. O. (1981). In *Biochemistry of Plants* (Vol. 7), ed. E. F. Conn. Academic Press, London, UK, pp. 501–25.
Law, B. A. & Kolstad, J. (1983). *Antonie van Leeuwenhoek* **49,** 225–45.
Lee, Y. N., Wiley, R. C., Sheu, M. J. & Schlimme, D. V. (1982). *J. Food Sci.,* **47,** 465–71.
Lee, H. S., Ko, Y. T. & Lim, S. J. (1984). *Korean J. Nutr.,* **17,** 101–7.
Liu, P. W. & Chen, S. H. (1965). *J. Chinese Agric. Chem. Soc.,* **3,** 110–16.
Löffler, A. (1986). *Food Technol.,* **40,** 63–70.
Lolas, G. M. & Markakis, P. (1977). *J. Food Sci.,* **42,** 1094–106.
Ludford, P. M. & Isenberg, F. M. R. (1987). In *Postharvest Physiology of Vegetables,* ed. J. Weichmann. Marcel Dekker, New York, USA, pp. 497–522.
Maestracci, M., Bui, K., Thiéry, A., Arnaud, A. & Galzy, P. (1988). *Adv. Biochem. Engng/Biotechnol.,* **36,** 67–115.
Marakami, F. (1960). *Vitamin,* **20,** 126–31.
Marth, E. H. & Doyle, M. P. (1979). *Food Technol.,* **33,** 81–7.
Mattiasson, B. & Hahn-Hägerdal, B. (1982). *Eur. J. Appl. Microbiol. Biotechnol.,* **16,** 52–5.
Matsubara, H. & Feder, J. (1971). In *The Enzymes* (Vol. 3), ed. D. D. Boyer. Academic Press, New York, USA, pp. 721–95.
McFeeters, R. F. (1988). In *Chemistry and Function of Pectins,* ed. M. L. Fishman & J. J. Jen. ACS Symposium Series 310, American Chemical Society, Washington, DC, USA, pp. 217–29.

McLellan, K. M. & Robinson, D. S. (1987). *Food Chem.*, **26**, 97–107.
Measures, J. C. (1975). *Nature*, **257**, 398–400.
Mega, T. & Matsushima, Y. (1979). *J. Biochem.*, **85**, 335.
Miller, A. R., Dalmasso, J. P. & Kretchman, D. W. (1987). *J. Amer. Soc. Hort. Sci.*, **112**, 666–71.
Misaki, T., Yamada, M., Okazaki, T. & Sawada, J. (1970). *Agric. Biol. Chem.*, **34**, 1383–92.
Mital, B. K. & Steinkraus, K. H. (1975). *J. Food Sci.*, **40**, 114–18.
Mital, B. K., Shallenberger, R. S. & Steinkraus, K. H. (1973). *Appl. Microbiol.*, **26**, 783–8.
Mohr, E. (1987). *Gordian*, **87**(10), 195–6.
Montenecourt, B. S. & Eveleigh, D. E. (1985). In *Gene Manipulations in Fungi*, ed. J. W. Bennett & L. L. Lasure. Academic Press, New York, USA, pp. 491–512.
Murata, K., Ikehata, H. & Myamoto, T. (1967). *J. Food Sci.*, **32**, 580–6.
Nagel, C. W. & Vaughn, R. H. (1961). *Arch. Biochem. Biophys.*, **93**, 344–52.
Nakadai, T. (1977). *J. Jap. Soy Sauce Res. Inst.*, **3**, 99–104.
Nakadai, T. & Nasuno, S. (1976). *J. Ferment. Technol.*, **54**, 872–84.
Nakadai, T. & Nasuno, S. (1977). *J. Ferment. Technol.*, **55**, 273–6.
Nakadai, T., Nasuno, S. & Iguchi, N. (1973*a*). *Agric. Biol. Chem.* **37**, 2703–8.
Nakadai, T., Nasuno, S. & Iguchi, N. (1973*b*). *Agric. Biol. Chem.* **37**, 2685–94.
Nakadai, T., Nasuno, S. & Iguchi, N. (1973*c*). *Agric. Biol. Chem.*, **37**, 2695–701.
Nasuno, S. & Nakadai, T. (1977). *J. Jap. Soy Sauce Res. Inst.*, **3**, 309–13.
Nayini, N. R. & Markakis, P. (1984). *Lebensm.-Wiss. u.-Technol.*, **17**, 24–6.
Odunfa, S. A. (1986). In *Legume-based Fermented Foods*, ed. N. R. Reddy, M. D. Pierson & D. K. Salunkhe. CRC Press, Boca Raton, FL, USA, pp. 137–89.
Ohata, T. (1986). In *Legume-based Fermented Foods*, ed. N. R. Reddy, M. D. Pierson & D. K. Salunkhe, CRC Press, Boca Raton, FL, USA, pp. 85–117.
Ohtsura, M. & Hata, T. (1979). *Biochim. Biophys. Acta.*, **567**, 384–91.
Padhye, V. W. & Salunkhe, D. K. (1978). *J. Food Biochem.*, **2**, 327–47.
Pederson, C. S. (1979). *Microbiology of Food Fermentations* (2nd edn). AVI Publishing Co., Westport, CT, USA.
Pierson, M. D., Reddy, N. R. & Odunfa, S. A. (1986). In *Legume-based Fermented Foods*, ed. N. R. Reddy, M. D. Pierson & D. K. Salunkhe. CRC Press, Boca Raton, FL, USA, pp. 219–31.
Potts, E. A. (1985). *Dissertation Abst. Int.*, **B46**, (2) 367.
Pravabati, D. S. & Singh, H. T. (1986). *J. Food Sci. Technol. India*, **23**, 338–9.
Prior, B. A. & Kenyon, C. P. (1980). *J. Appl. Bact.*, **48**, 211–22.
Rajalakshmi, R. & Vanaja, K. (1967). *Brit. J. Nutr.*, **21**, 467–73.
Ramakrishnan, C. V., Parekh, L. J., Akolkar, P. N., Rao, G. S. & Bhandari, S. (1976). *J. Plant Foods*, **2**, 15.
Rao, G. S. (1978). PhD dissertation, Maharaja Sayajirao University of Baroda, India (cited by Reddy *et al.*, 1986*b*).
Rao, D. R., Pulusani, S. R. & Chawan, C. B. (1986). In *Legume-based Fermented Foods*, ed. N. R. Reddy, M. D. Pierson & D. K. Salunkhe. CRC Press, Boca Raton, FL, USA, pp. 119–34.
Reddy, N. R., Sathe, S. K. & Salunkhe, D. K. (1982). *Adv. Food Res.* **28**, 1–92.
Reddy, N. R., Pierson, M. D. & Salunkhe, D. K. (1986*a*). *Legume-based Fermented Foods*, CRC Press, Boca Raton, FL, USA.
Reddy, N. R., Pierson, M. D. & Salunkhe, D. K. (1986*b*). In *Legume-based Fermented Foods*, ed. N. R. Reddy, M. D. Pierson & D. K. Salunkhe. CRC Press, Boca Raton, FL, USA, pp. 145–60.
Sakellaris, G., Nikolaropoulos, S. & Evangelopoulos, A. E. (1988). *J. Appl. Bact.*, **65**, 397–404.
Sgarbieri, V. C. & Whitaker, J. R. (1982). *Adv. Food Res.*, **28**, 93–166.

Smoot, L. A., Garcia, V. V. & Pierson, M. D. (1979). *J. Food Prot.*, **42,** 839.
Sonnet, P. E. (1988). *JAOCS,* **65,** 900–4.
Sperber, W. H. (1983). *J. Food Prot.*, **46,** 142–50.
Stamer, J. R. & Stoyla, B. D. (1968). *Appl. Microbiol.*, **16,** 536–7.
Steinkraus, K. H. (1983). *Handbook of Indigenous Fermented Foods,* Marcel Dekker, New York, USA.
Su, Y.-C. (1986). In *Legume-based Fermented Foods,* ed. N. R. Reddy, M. D. Pierson & D. K. Salunkhe, CRC Press, Boca Raton, FL, USA, pp. 69–83.
Sudarmadji, S. & Markakis, P. (1977). *J. Sci. Food Agric.*, **28,** 381–3.
Sutardi & Buckle, K. A. (1986). *J. Food Biochem.*, **10,** 197–216.
Sutardi & Buckle, K. A. (1988). *Int. J. Food Micro.*, **6,** 67–79.
Swaminathan, M. & Parpia, H. A. B. (1967). *World Rev. Nutr. Diet.* **8,** 184.
Tagami, H. & Sugawara, S. (1977). *J. Jap. Soy Sauce Res. Inst.* **3,** 111–115.
Tang, H. C. L. & McFeeters, R. F. (1983). *J. Food Sci.*, **48,** 66–70.
Tonita, G. & Kim, S. S. (1965). *Nature,* **205,** 46–8.
Tsujisaka, Y., Iwai, M. & Tominaga, Y. (1972). In *Fermentation Technology Today,* ed. G. Terui, Society of Fermentation Technology, Osaka, Japan, pp. 315–20.
Tsujita, Y. & Endo, A. (1980). *Biochem.* (*Tokyo*), **88,** 113.
Uda, Y. & Maeda, Y. (1986). *Agric. Biol. Chem.*, **50,** 205–8.
Vaughn, R. H. (1985). In *Microbiology of Fermented Foods* (Vol. 1), ed. B. J. B. Wood. Elsevier Applied Science, London, UK, pp. 49–109.
Vaughn, R. H., King Jr, A. D., Nagel, C. W., Ng, H., Levin, R. E., Macmillan, J. D. & York II, G. K. (1969). *J. Food Sci.*, **34,** 224–7.
Veen, A. G. van, & Steinkraus, K. H. (1970). *J. Agric. Food Chem.*, **18,** 576–8.
Veen, A. G., van, Graham, D. C. W. & Steinkraus, K. H. (1968). *Cereal Sci. Today,* **13,** 96–9.
Vorbeck, M. L., Albury, M. N., Mattick, L. R., Lee, F. A. & Pederson, C. S. (1983). *J. Food Sci.*, **28,** 495–502.
Wagenknecht, A. C., Mattick, L. R., Lewin, L. M., Hand, D. B. & Steinkraus, K. H. (1961). *J. Food Sci.*, **26,** 373–6.
Wang, H. L. (1967). *J. Bacteriol.*, **93,** 1794–99.
Wang, H. L. (1987). *Soyfoods,* **1,** 9–17.
Wang, H. L. & Hesseltine, C. W. (1970). *Arch. Biochem. Biophys.*, **140,** 459–63.
Wang, H. L., Vespa, J. B. & Hesseltine, C. W. (1974). *Appl. Microbiol.*, **27,** 906–11.
Wang, H. L., Swain, E. W. & Hesseltine, C. W. (1980). *J. Food Sci.*, **45,** 1262–6.
Ward, O. P. (1983). In *Microbial Enzymes and Biotechnology,* ed. W. M. Fogarty. Applied Science Publishers, London, UK, pp. 251–317.
Weichmann, J. (1987). *Postharvest Physiology of Vegetables,* Marcel Dekker, New York, USA.
West, L. G., Badenhop, A. F. & McLaughlin, J. L. (1977). *J. Agric. Food Chem.*, **25,** 1234–8.
Whitaker, J. R. (1976). *Adv. Food Res.*, **22,** 73–133.
Wood, B. J. B. (1985). *Microbiology of Fermented Foods* (2 Vols), Elsevier Applied Science, London, UK.
Yamamoto, Y. & Hirooka, H. (1974). *J. Ferment. Technol.*, **52,** 564–9.
Yamamoto, Y., Yanagida, F. & Suminoe, K. (1972). *J. Food Sci. Technol.* (*Nikon Shokuhin Kogyo Gakkai-shi*), **20,** 109–12.
Yokotsuka, T. (1986). *Adv. Food Res.*, **30,** 195–329.
Yook, C., Chang, K., Park, K. H. & Ahn, S. Y. (1985). *Korean J. Food Sci. Tech.*, **17,** 447–53.
Yousef, A. E. & Marth, E. H. (1981). *J. Food Prot.*, **44,** 736–41.
Zaks, A. & Klibanov, A. M. (1984). *Science,* **224,** 1249–51.
Zamora, R. G. & Veum, J. L. (1979). *J. Nutr.*, **109,** 1333–9.
Zetelaki-Horvath, K. & Andersson, R. (1986). *Acta Alimentaria,* **15,** 265–79.

13

Enzymes Important in Flavor Development in the Alliums

John R. Whitaker & Mendel Mazelis

Department of Food Science and Technology, University of California, Davis, California 95616, USA

INTRODUCTION

Sulfur-containing compounds are the major flavor constituents of some plants of the Amaryllidaceae, Cruciferae and Leguminosae families, as well as of some microorganisms. Well-known examples are onions (*Allium cepa*), garlic (*A. sativum*), leek (*A. porrum*) and chives (*A. schoenoprasum*), lily (*Tulbaghia violacea*), cabbage (*Brassica oleracea capitata*), cauliflower (*B. oleracea botrytis*) and rutabaga (*B. napobrassica*), acacia (*Acacia farnesiana*), *Albizzia lophanta* seeds, shiitake mushrooms (*Lentinus endodis*), and microorganisms such as *Penicillium Corynibiferum* and *Pseudomonas* spp. These plants and microorganisms all have in common a number of sulfur-containing compounds that are derivatives of cysteine. Some, such as the *Brassica,* also contain glucosinolates.

SULFUR COMPOUNDS OF INTACT ONION

Sulfur-containing compounds of the intact (whole) onion are shown in Table 1. Note that all 26 compounds contain one or more commonly occurring amino acids, with cysteine a component of all except methionine (no. 2), methionine sulfoxide (no. 3), and γ-L-glutamyl-L-methionine (no. 17). Compounds 4–21, except for no. 16 (cycloalliin), and no. 17 (γ-glutamyl-L-methionine), are alkyl derivatives of cysteine. The cysteine may be as the sulfoxide (compounds 12–15), converted to the *N*-γ-L-glutamyl derivative (compounds 17–21), a part of the tripeptide glutathione (compounds 22–26) or oxidized to the disulfide (compounds 23 and 25). The alkyl portion of the alkylcysteine derivatives may be as the carboxyalkyl derivative (compounds 8–11) or be decarboxylated (compounds 4–7, 12–15, 18–20).

TABLE 1
Sulfur Compounds of Intact Onion

Compound	*Concentration (μg/g onion wet weight)*
1. L-Cystine[a,b]	3·6[c]
2. L-Methionine[a]	4·2[c]
3. L-Methionine sulfoxide[a]	—
4. *S*-Methyl-L-cysteine[a,b,d]	15·8[c]
5. *S*-Propyl-L-cysteine[b]	—
6. *trans*-*S*-(1-Propenyl)-L-cysteine[a,b]	7·8[c]
7. *trans*-*S*-(2-Propenyl)-L-cysteine (*S*-allylcysteine)[a,e]	—
8. *S*-(Carboxymethyl)-L-cysteine[b]	—
9. *S*-(2-Carboxyethyl)-L-cysteine[b]	—
10. *S*-(2-Carboxypropyl)-L-cysteine[a,b]	—
11. *S*-(Carboxy-isopropyl)-L-cysteine[b]	—
12. (+)-*S*-(Methyl-L-cysteine sulfoxide[a,b,e,f,g]	200[f]
13. (+)-*S*-Propyl-L-cysteine sulfoxide[a,b,e,f,g]	50[f]
14. *trans*-(+)-*S*(1-Propenyl)-L-cysteine sulfoxide[a,b,f,g]	40[f]
15. (+)-*S*-Allyl-L-cysteine sulfoxide[b]	—
16. Cycloalliin[a,b,f]	2500[f]
17. γ-L-Glutamyl-L-methionine[a,f]	50[f]
18. γ-L-Glutamyl-*S*-methyl-L cysteine[a,f]	50[f]
19. γ-L-Glutamyl-*S*-methyl-L-cysteine sulfoxide[h]	—
20. γ-L-Glutamyl-*trans*-(+)-(1-propenyl)-L-cysteine sulfoxide[a,f]	1300[f]
21. γ-Glutamyl-*S*-(2-carboxypropyl)-L-cysteinylglycine[j]	—
22. Glutathione[h]	—
23. Glutathione cysteine disulfide[h]	—
24. *S*-(2-Carboxypropyl)glutathione[a,b,f]	330[f]
25. Glutathione-γ-glutamylcysteine disulfide[h]	—
26. *S*-Sulfoglutathione[h]	—

[a] Matikkala & Virtanen (1967); separated by two-dimensional electrophoresis and chromatography.
[b] Granroth (1968).
[c] Neutral and basic extract (Matikkala & Virtanen, 1967).
[d] *S*-Methylcysteine content of four varieties of onions ranged from 9·3 to 19·3 μg/g of onions wet weight (Kuon & Bernhard, 1963).
[e] Carson & Wong (1961*a*).
[f] Ettala & Virtanen (1962); amino acid analyzer data.
[b] Belo (1972).
[h] Virtanen (1969).

ORIGIN OF THE SULFUR COMPOUND

Are the sulfur-containing compounds of Table 1 interrelated? How are the cysteine derivatives formed in the onion? Investigations on the pathways of biosynthesis of the sulfur compounds in *Allium* are limited almost exclusively to determining the compounds formed when radiolabelled intermediates are fed to a plant. Essentially no research has been done with tissue homogenates, cell-free extracts or isolated enzyme systems. Based on the time-course of radiolabelling of intermediates and postulations by other workers, Granroth (1970) proposed the pathway shown in Fig. 1. The biosynthesis of cysteine from sulfide and serine is experimentally well-documented in higher plants (Thompson, 1967). The reactions involved are shown in eqns (1)–(3).

$$SO_4^{2-} \xrightarrow[H_2O]{+2H} SO_3^{2-} \xrightarrow[3H_2O]{+6H} S^{2-} \quad (1)$$

$$\text{acetyl-CoA} + \text{serine} \rightarrow O\text{-acetylserine} + \text{CoA} \quad (2)$$

$$O\text{-acetylserine} + HS^- \rightarrow \text{L-cysteine} + \text{acetate} \quad (3)$$

A sulfite reductase (hydrogen sulfide:(acceptor) oxidoreductase; EC 1.8.99.1) was purified from *Allium odorum* by Tamura (1965). This enzyme reduces SO_3^{2-} to S^{2-} (eqn (1)). Reduction of SO_4^{2-} to SO_3^{2-} is via the ATP + adenosine-3′-phosphate-5′-phosphosulfate pathway. The enzyme, serine acetyltransferase (acetyl-CoA: L-serine *O*-acetyltransferase; EC 2.3.1.30), that catalyzes the reaction shown in eqn (2) has been reported in extracts of a

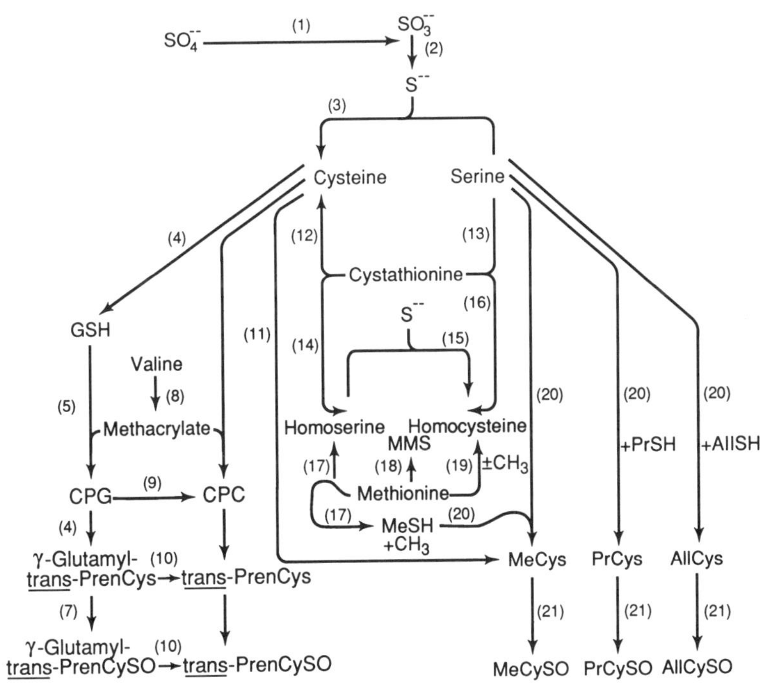

FIG. 1. Proposed pathway for biosynthesis of some sulfur compounds in *Allium* species. GSH, glutathione; CPG, *S*-(2-carboxypropyl)glutathione; CPC, *S*-(2-carboxypropyl)-L-cysteine; MMS, *S*-methylmethionine sulfonium ion. (Adapted from Granroth (1970), and reproduced by permission of the Finnish Academy of Sciences.)

number of higher plants (Smith & Thompson, 1969). Cysteine synthase [*O*-acetyl-L-serine acetate-lyase (adding hydrogen sulfide); EC 4.2.99.8; also called *O*-acetylserine sulfhydrylase] that catalyzes eqn (3) has been found in onion (Granroth, 1974) as well as other higher plants.

The alkylcysteines are presumably formed via several pathways (Fig. 1). Methyl-L-cysteine is most likely formed from methionine and cysteine as shown by feeding the plant (methyl-^{14}C) methionine (eqn (4); reaction 20 in Fig. 1).

$$\text{(methyl-}^{14}\text{C)methionine} + \text{L-cysteine} \rightarrow (^{14}\text{C})S\text{-methyl-L-cysteine} + \text{homocysteine} \quad (4)$$

There is evidence that some *S*-methylcysteine is formed via thiomethyl transfer from the reaction of serine with methionine (reaction 17 in Fig. 1) and from a transulfuration via cystathionine (reaction 13 in Fig. 1).

S-Propyl-L-cysteine and *S*-alkyl-L-cysteine are probably synthesized via the reactions in eqns (5) and (6) (reaction 20 in Fig. 1).

$$\text{L-serine} + \text{propylmercaptan} \rightarrow S\text{-propyl-L-cysteine} \quad (5)$$

$$\text{L-serine} + \text{allylmercaptan} \rightarrow S\text{-allyl-L-cysteine} \quad (6)$$

Evidence for these pathways comes from feeding ^{14}C-serine and various mercaptans (methyl, ethyl, propyl, allyl, benzyl). The corresponding (^{14}C)*S*-alkylcysteine is formed rapidly. Feeding ^{14}C-cysteine did not result in labelled alkylcysteines (Granroth, 1970, 1974). The biosynthetic pathways of the alkylmercaptans are not known.

In contrast to the pathways above, the propenyl residue of *trans*-(+)-*S*-(1-propenyl)-L-cysteine comes from valine, as shown in Fig. 1 (reactions 5 and 8). More details of the reactions are shown in Fig. 2. Feeding experiments with onions have shown that valine is deaminated to α-ketoisovaleric acid which in turn is decarboxylated to isobutyryl-CoA and isomerized to methacrylyl-CoA. Methylacrylic acid condenses with cysteine (reaction 5 in Fig. 2) to give 2-carboxypropyl-L-cysteine. Elimination of the 2-carboxy group gives *trans*-(+)-*S*-(1-propenyl)-L-cysteine. The individual enzymes involved in reactions 1–6 in Fig. 2 have not been studied.

Feeding the various *S*-alkyl- and *S*-alkenylcysteines to onions or garlic leads rapidly to formation of the corresponding sulfoxides (reaction 20 in Fig. 1). Nothing is known about the oxido-reductase(s) involved. The role of *S*-(2-carboxypropyl)glutathione in the biosynthesis of *trans*-(+)-*S*(1-propenyl)-L-cysteine sulfoxide is not clear (reaction 9 in Fig. 1). The origin and role of the other carboxyalkylcysteine derivatives found in *Allium* species (Table 1; compounds 8–11) are not known.

The γ-glutamyl peptides (compounds 17–21 in Table 1) are undoubtedly formed via transpeptidation reactions involving glutamine, the various *S*-alkylcysteines and *S*-alkylcysteine sulfoxides and glutamyltransferase. Very small amounts of these compounds are present in the growing tissue but large amounts of the *S*-alkylcysteine sulfoxides are present as glutamyl derivatives in

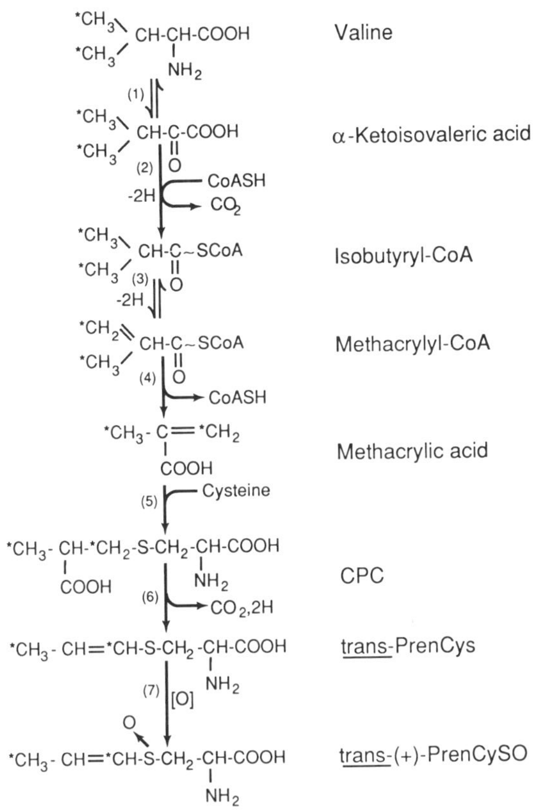

FIG. 2. Proposed pathway for biosynthesis of *trans*-(+)-(1-propenyl)-L-cysteine sulfoxide (*trans*(+)-PrenCySO) from valine and cysteine. *, ^{14}C label; CPC, *S*-(2-carboxypropyl)-L-cysteine; *trans*-PrenCys, *trans*-*S*-(1-propenyl)-L-cysteine. (Adapted from Granroth (1970), and reproduced by permission of the Finnish Academy of Sciences.)

the dormant bulb (see concentrations of compounds 14 and 20 in Table 1). The low levels of *trans*-(+)-*S*-(1-propenyl)-L-cysteine sulfoxide, relative to (+)-*S*-propyl-L-cysteine sulfoxide may be a result of conversion of the former compound to cycloalliin (no. 16) during the isolation procedure.

FLAVOR AND AROMA COMPOUNDS OF THE ALLIUM

None of the 26 compounds in Table 1 from the intact onion has the flavor and aroma of the *Allium* species. Immediately upon crushing or cutting there is a rapid increase in flavor and aroma and the chemical composition of the *Allium* species changes markedly, as shown for crushed onions (Table 2).

TABLE 2
Components of Crushed Onion (Allium Cepa)

Compound	*Compound*	*Compound*
1. Methyl alcohol[a]	17. Dipropyl sulfide[h]	29. Methyl propyl trisulfide[b,i]
2. Ethyl alcohol[b]	18. Dimethyl disulfide[b,c,g,i,j]	30. Dipropyl trisulfide[b,i]
3. *n*-Propyl alcohol[a,b]	19. Methyl propyl disulfide[b,c,g,i,j]	31. Methyl 1-propenyl trisulfide[i,k]
4. Isopropyl alcohol[b]	20. Methyl *cis*-1-propenyl disulfide[c,i]	32. Methyl 1-propenyl trisulfide[i,e]
5. Ethanal[a–c]	21. Methyl *trans*-1-propenyl disulfide[c,i]	33. Propyl trisulfide[i]
6. Propanal[a–d]	22. Allyl methyl disulfide[c,g,i,j]	34. 1-Propenyl propyl trisulfide[i,k]
7. *n*-Butanal[b]	23. Dipropyl disulfide[a-d,g,i,j]	35. 1-Propenyl propyl trisulfide[i,l]
8. 2-Methylbutanal[c]	24. *cis*-1-Propenyl propyl disulfide[c,i]	36. Methyl methanethiosulfinate[h,m]
9. 2-Methylpentanal[c]	25. *trans*-1-Propenyl propyl disulfide[c,i]	37. Propyl propanethiosulfinate[h,m]
10. 2-Methyl-2-pentenal[c,d]	26. Allyl propyl disulfide[c,g,i,j]	38. 1-Propenyl-1-propenethiosulfinate[h]
11. Acetone[b]	27. Diallyl disulfide[c,g,j]	39. Methyl methanethiosulfonate[n]
12. Methyl ethyl ketone	28. Dimethyl trisulfide[b,i]	40. Propyl methanethiosulfonate[n]
13. Sulfur dioxide[a]		41. Propyl propanethiosulfonate[n]
14. Hydrogen sulfide[a,b]		42. 2,4-Dimethylthiophene[c]
15. Propanethiol[a,b,d,e]		43. 3,4-Dimethylthiophene[c,i]
16. Diallyl sulfide[f,g]		

[a] Niegisch & Stahl (1956); common yellow onion, minced while frozen, distillation at room temperature under vacuum.
[b] Carson & Wong (1961b); 'Sunspice' onions, vacuum and steam distillation (at 25°C) of freshly chopped onions.
[c] From head space of Dutch or Egyptian cut onions (Boelens *et al.*, 1971).
[d] Kawakishi *et al.* (1971); 'Sensyu Yellow' onions, from head space of blended onions.
[e] From aspiration of chopped onions (Challenger & Greenwood, 1949).
[f] From head space of some freshly cut cultivars (Bernhard *et al.*, 1964).
[g] 'Southport White Globe', from head space analysis (Saghir *et al.*, 1966).
[h] Belo (1972). Formed at low temperatures in onion extracts but immediately rearranges to zwiebelanes (Bayer et al., 1989; Block and Bayer, 1990).
[i] From onion oil produced commercially and representing mixtures of varieties, maturity, and sizes (Brodnitz *et al.*, 1969).
[j] From head space of 53 lots of fresh and dried onions (Bernhard, 1968).
[k] Probably *cis* (Brodnitz *et al.*, 1969).
[l] Probably *trans* (Brodnitz *et al.*, 1969).
[m] Following reaction with thiamine (Fujiwara *et al.*, 1955; Lukes, 1971).
[n] Dutch or Egyptian onions extracted with dichloromethane (Boelens *et al.*, 1971).

Some 43 compounds that do not exist in the intact onion are formed rapidly. These include alcohols, aldehydes and ketones and a series of sulfides, disulfides, trisulfides, thiosulfinates, thiosulfonates and thiophenes. There must surely be some relationship among these compounds, as well as those of Table 1, since they are all methyl, ethyl, propyl and propenyl derivatives.

Formation of the flavor and aroma of crushed or cut *Allium* species has been known for hundreds of years, of course. The first serious attempts to study the compounds formed were probably around 1840. Wertheim (1844, 1845) and later Semmler (1892) isolated some volatile compounds from steam-distilled garlic oil; these included diallyldisulfide and smaller amounts of diallyltrisulfide and diallylpolysulfide. Rundquist (1909) was the first to attempt to isolate the precursor compound(s).

In a classic set of experiments, Stoll and Seebeck (1951) investigated the conversion of (+)-*S*-allyl-L-cysteine sulfoxide ((+)-*S*-(2-propenyl)-L-cysteine sulfoxide; alliin) to diallylthiosulfinate (allicin) by an enzyme found in garlic. They proposed that the following reaction was involved:

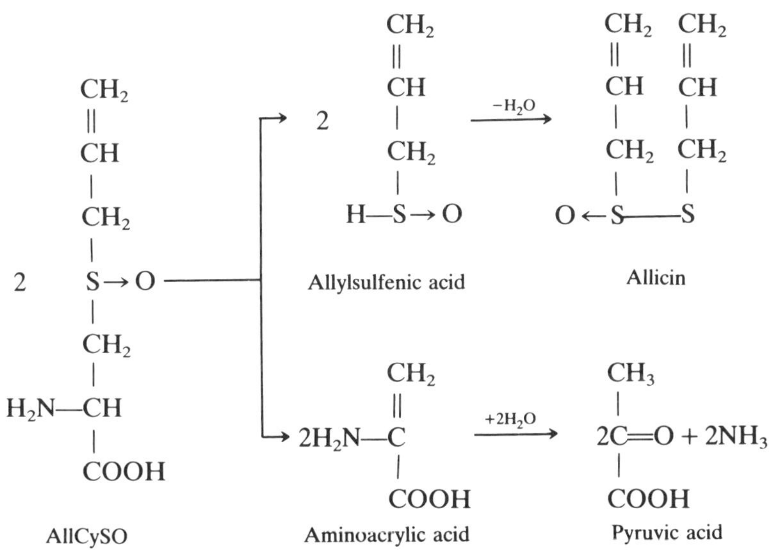

(7)

The enzyme, named alliinase by Stoll and Seebeck, has been given several names. These include alkylcysteine sulfoxide lyase, cysteine sulfoxide lyase, C-S-lyase and alliin lyase. The Commission on Enzymes assigned the systematic name, alliin alkyl-sulfenate-lyase (EC 4.4.1.4), with a trivial name of alliin lyase or alliinase. We shall call the enzyme alliin lyase.

Alliin lyase catalyzes the first step in the reaction leading to the products allylsulfenic acid and aminoacrylic acid (eqn (7)). Allyl 2-propenethiosulfinate (allicin), pyruvate and ammonia are formed by non-enzymatic reactions. Both the alkanesulfenic acid and the allyl 2-propenethiosulfinate are quite reactive and go on to form other compounds (Table 2; see Block, 1985; Block *et al.*,

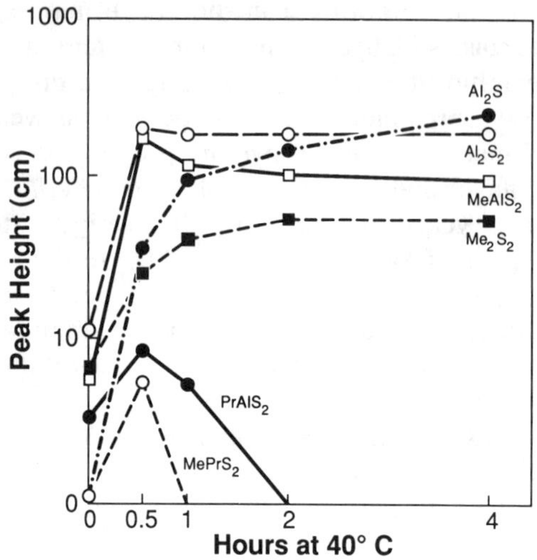

FIG. 3. Effect of incubation time on head space composition of chopped California Late garlic. Note the semilogarithmic plot. Al, allyl; Me, methyl; Pr, propyl. (From Saghir *et al.* (1964), and reproduced by permission of the American Society of Horticulturists.)

1984, 1986, 1988 for compounds found in garlic and their reactivities, and Kuo *et al.*, 1990 for compounds in scallions and Welsh onions). Figure 3 shows the rate of formation of diallylsulfide and several dialkyl disulfides from chopped garlic. Figure 4 shows how aldehydes, sulfinates, sulfonates, sulfides, disulfides and trisulfides can arise from the initial formation of the alkanesulfenic acid produced by alliin lyase. Propanethial S-oxide was first demonstrated by Brodnitz & Pascale (1971) and later unequivocally demonstrated to have the conformation (*Z*)-propanethial *S*-oxide by Block *et al.* (1979, 1980). More recent work on products derived from 1-propenesulfenic acid (and (*Z*)-propanethial *S*-oxide) include that of Bayer *et al.* (1989*a*, *b*), Block & Bayer (1990) and Morimitsu & Kawakishi (1990). Mixed alkyl and alkenyl derivatives are readily formed via catalytic amounts of sulfhydryl compounds in the tissue as shown by eqns (8) and (9) (cysteine used as example). The CH_3SOH will then catalyze further sulfhydryl-disulfide interchange reactions.

$$CH_3—\overset{\overset{\displaystyle O}{\|}}{S}—S—CH_3 + HS—Cys \rightarrow CH_3—\overset{\overset{\displaystyle O}{\|}}{S}—S—Cys + CH_3—SH \qquad (8)$$

As far as is known, the reactions following alkanesulfenic acid formation do not require enzymes. Dipropyl thiosulfonate (threshold 1·5 ppb (where billion is 10^9)) and methyl propyl thiosulfonate (1·7 ppb) give fresh onion aroma, while methyl(1-propenyl)disulfide (6·3 ppb), dipropyl disulfide (3·2 ppb) and 1-propenyl propyl disulfide (2·2 ppb) give cooked onion aroma and 2,4-dimethylthiophene (3 ppm) gives fried onion aroma (Boelens *et al.*, 1971).

Aldehydes and alcohols, formed on crushing or cutting *Allium* species, are also important flavor constituents. As shown in Fig. 5, aldehydes are formed

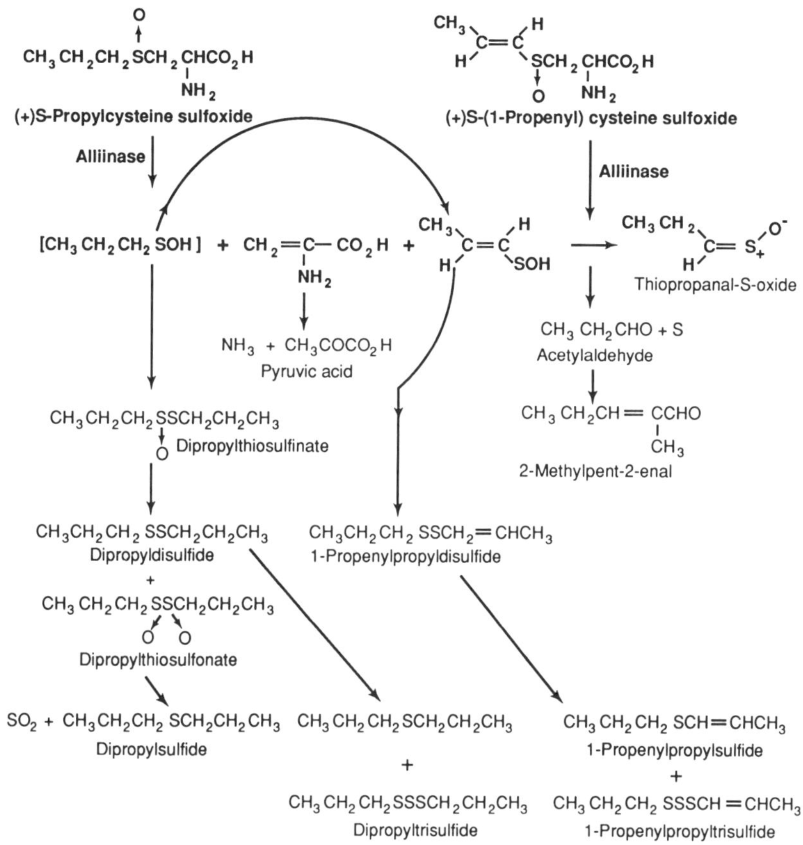

FIG. 4. Enzymatic (bold type) and non-enzymatic production of flavor compounds from the precursors (+)*S*-propyl-L-cysteine sulfoxide and (+)*S*-(1-propenyl)-L-cysteine sulfoxide in *Allium* species. (From Fenwick & Hanley (1985), and reproduced by permission of CRC.)

from both the alkane (alkene) sulfenic acid and pyruvate (requiring pyruvate decarboxylase). The initial aldehydes undergo aldol condensation reactions to give a number of aldehydes. Alcohol dehydrogenase, with the cofactor NADH, can reduce the aldehydes to the corresponding alcohols (see Table 2).

ALLIIN LYASES

The principal substrate for alliin lyases is different from one *Allium* species to another. Freeman and Whenham (1975), in a thorough survey of 27 *Allium* species and cultivars, distinguished three classes of plants based on the *S*-alkyl(alkenyl) substituent of the principal alliin lyase substrate; these are: class 1, *S*-1-propenyl-; class 2, *S*-2-propenyl-; and class 3, *S*-ethyl-. Onions, leek, chives, shallots and rakkyo belong to class 1; class 2 includes garlic, Chinese chives, wild onion and ainunegi (caucus, *A. victorialis*; a Japanese

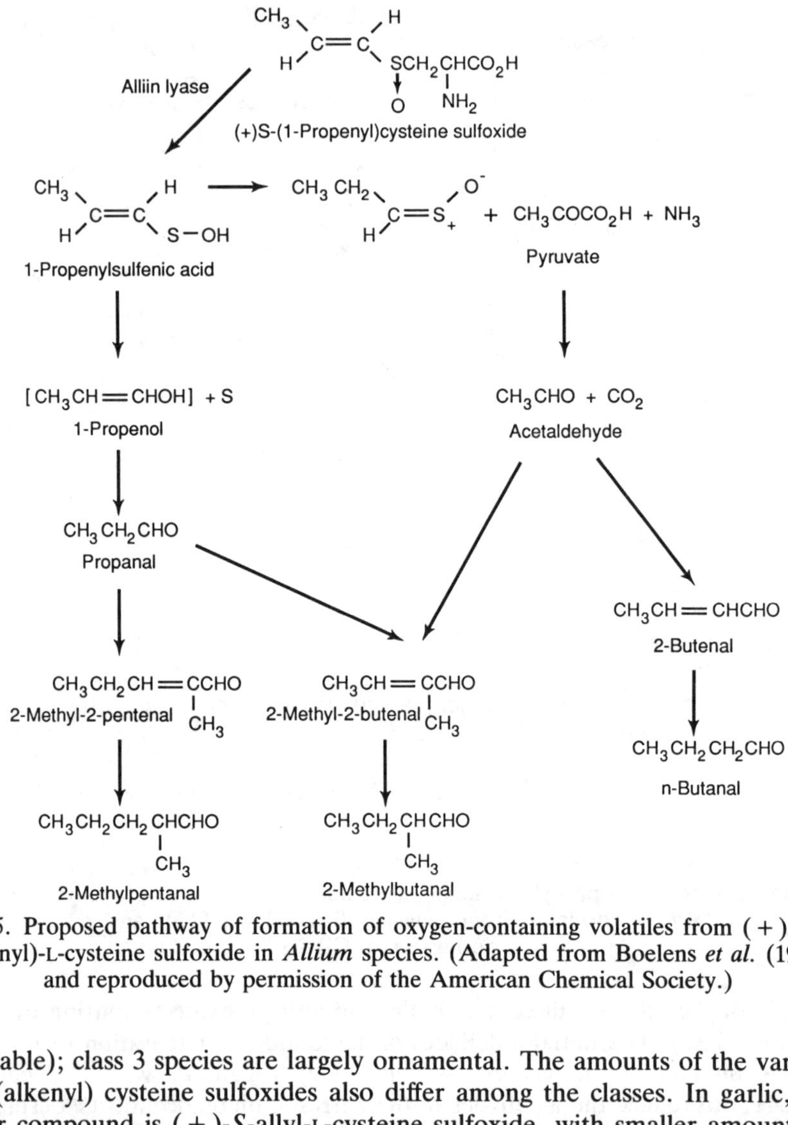

FIG. 5. Proposed pathway of formation of oxygen-containing volatiles from (+)*S*-(1-propenyl)-L-cysteine sulfoxide in *Allium* species. (Adapted from Boelens *et al.* (1971), and reproduced by permission of the American Chemical Society.)

vegetable); class 3 species are largely ornamental. The amounts of the various alkyl(alkenyl) cysteine sulfoxides also differ among the classes. In garlic, the major compound is (+)-*S*-allyl-L-cysteine sulfoxide, with smaller amounts of (+)-*S*-methyl-L-cysteine sulfoxide and *trans*-(+)-*S*-(1-propenyl)-L-cysteine sulfoxide (E. Block, pers comm.). In onion, the major compound is *trans*-(+)-*S*-(1-propenyl)-L-cysteine sulfoxide, with smaller amounts of (+)-*S*-methyl-L-cysteine sulfoxide and (+)-*S*-propyl-L-cysteine sulfoxide. Z-Propanethial S-oxide is the lachrymator in onions (Brodnitz and Pascale, 1971; Block *et al.*, 1979, 1980). *Trans*-1-propenesulfenic acid, if formed, has never been isolated.

While Stoll and Seebeck (1951) studied some of the properties of alliin lyase in crude extracts, it was not until the 1960s that partially purified enzymes were

TABLE 3
Some Properties of Onion, Garlic and Leek Alliin Lyases

Property	*Onion* (A. cepa)	*Garlic* (A. sativum)	*Leek*[f] (A. porrum)
Molecular weight	200 000[a]; 150 000[b] 220 000[c]	85 000[a]; 130 000[d] 108 000[e]	386 000; 580 000
Subunit weight (No.)	50 000(4)[a] 50 000(3)[b] 52 000(4)[c]	42 000(2)[a] 65 000(3)[c] 54 000(2)[e]	48 000 (8 or 12)
Specific activity (Units/mg)	289;[b] 126[a]	267[a]; 490[e]	56
pH optimum	7·6;[b] 7·0[c]	6·5[a,e]	8·0
pI	8·0[a]	6·2[c]; 7·0[a]; 4·9[e]	—
Carbohydrate (%)	5·8[b]; 4·6[a]	5·5[a]	7
PALP/subunit	1[a,b]	1[a,c]	1

[a] Nock & Mazelis (1987).
[b] Tobkin & Mazelis (1979); pH optimum dependent on buffer.
[c] Fujita *et al.* (1990).
[d] Kazaryan & Goryachenkova (1979).
[e] Jansen *et al.* (1989a).
[f] Won & Mazelis (1989).

obtained and not until the 1980s were pure enzymes available (Nock & Mazelis, 1986, 1987, 1989; Won & Mazelis, 1989; Jansen *et al.*, 1989*a*; Fujita *et al.*, 1990). Table 3 lists some of the properties of onion, garlic and leek alliin lyases. In common, they all are glycoproteins and are pyridoxal phosphate-requiring enzymes. However, they differ in other properties. The onion enzyme contains four subunits of 50 000 Da each (200 000 mol. wt), the garlic enzyme is probably a dimer of two subunits of 42 000 Da each (85 000 mol. wt), while the leek enzyme has 8 to 12 subunits of 48 000 Da each that form two very large molecular species of 386 000 and 580 000 Da. The pH optima of the onion and leek enzymes are near 8 (Welsh onion alliin lyase pH optimum was reported to be 7 (Fujita *et al.*, 1990)) while the pH optimum for the garlic enzyme is at 6·5. The pI of the garlic enzyme is lower than that of the onion enzyme.

The amino acid compositions of the garlic and onion alliin lyases are similar as purified by Nock and Mazelis (1987) (Table 4). The major differences are in the histidine, 1/2 cystine and tryptophan contents. The amino acid composition of onion alliin lyase reported by Tobkin and Mazelis (1979) is somewhat different from that reported by Nock & Mazelis (1987). The specific activity of the enzyme preparation of Tobkin and Mazelis was 2·3-times that of Nock and Mazelis.

The *N*-terminal amino acid sequences of the three enzymes are different. The sequence of the leek enzyme is Ala–Asp–Thr–Ile–Val. . . (Won & Mazelis, 1989), the garlic enzyme is Lys–Met–Thr–Trp . . . (Nock & Mazelis,

TABLE 4
Amino Acid Composition (%)[a] of Garlic and Onion Alliin Lyase[b]

Amino acid	*Garlic*	*Onion*[c]
Aspartic acid	11·6	11·1(9·0)[c]
Threonine	6·8	5·5(6·0)
Serine	7·3	7·7(6·3)
Glutamic acid	11·0	9·5(11·5)
Proline	4·3	4·2(1·2)
Glycine	7·0	8·2(7·1)
Alanine	6·9	6·0(7·1)
Valine	6·4	6·9(6·5)
1/2 Cystine	1·6	0·4(0·7)
Methionine	0·6	0·6(2·2)
Isoleucine	3·4	4·8(4·7)
Leucine	6·8	6·9(7·1)
Tyrosine	5·5	5·2(4·7)
Phenylalanine	4·3	4·4(4·2)
Lysine	7·2	6·0(7·1)
Histidine	2·0	7·0(1·7)
Arginine	4·9	4·4(3·3)
Tryptophan	2·2	1·2(5·4)

[a] Percentages based on the mols amino acid per mol protein values.
[b] Nock & Mazelis (1987).
[c] Values in parentheses from Tobkin & Mazelis (1979).

1989) while the *N*-terminal of onion enzyme is blocked (Nock & Mazelis, 1989).

The alliin lyases from onion, garlic and leek had 5·5–7% carbohydrate (Table 5). The glyco components contained arabinose, xylose, mannose, galactose and glucose. The major monosaccharide unit was mannose in all three enzymes; however, the percentage varied from 43 to 78%. The enzymes differed significantly from each other in monosaccharide composition. The three enzymes eluted quite differently from a Concanavalin A–Sepharose

TABLE 5
Carbohydrate Composition of Garlic, Onion and Leek Alliin Lyases

	Percentage of total released				
Plant	*Arabinose*	*Xylose*	*Mannose*	*Galactose*	*Glucose*
Garlic[a]	9	5	43	8	34
Onion[a]	1	1	78	10	10
Leek[b]	3	5	49	27	17

[a] Nock & Mazelis (1987).
[b] Won & Mazelis (1989).

4B column. Garlic alliin lyase was eluted at 10 mM α-methyl-D-mannoside maltoside while the onion and leek enzymes were eluted at 100 mM and 300 mM α-methyl-D-mannoside respectively (Won & Mazelis, 1989).

The onion and garlic alliin lyases are immunochemically distinct (Nock & Mazelis, 1989). There was no cross-reaction of the garlic enzyme rabbit polyclonal antibodies with the onion enzyme, and an inconsistent amount of cross-reaction of the garlic enzyme with onion enzyme rabbit polyclonal antibodies.

The onion, garlic and leek alliin lyases have specificity for the *S*-alkyl- and *S*-alkenylcysteine sulfoxides. The *S*-alkyl- and *S*-alkenylcysteines are not substrates; in fact they are competitive inhibitors (see below). The three enzymes used the *S*-methyl-, *S*-ethyl-, *S*-propyl-, *S*-2-propenyl- and *S*-butylcysteine (not determined for leek enzyme) sulfoxides as substrates (Table 6). In terms of K_m and K_m/V_{max}, *S*-propylcysteine sulfoxide was the best substrate for the onion enzyme, while *S*-2-propenylcysteine sulfoxide was the best substrate for the garlic and leek enzymes.

The *S*-alkyl- and *S*-alkenylcysteines are inhibitors of onion and garlic alliin lyases (Schwimmer *et al.*, 1964; Mazelis & Crews, 1968). The K_i values for *S*-methyl-L-cysteine, *S*-ethyl-L-cysteine and *S*-propyl-L-cysteine were 33, 18 and 5 mM, respectively, for garlic alliin lyase (Mazelis & Crews, 1968) while the K_m values for the *S*-methyl-, *S*-ethyl- and *S*-propyl-L-cysteine sulfoxides were 15, 6 and 3 mM, respectively (see Table 6). The K_i and K_m values changed in the same direction for the two sets of compounds; the K_i and K_m values are not very different. By contrast, the *S*-alkyl(alkenyl)cysteines are substrates for alliin lyases from *Acacia farnesiana* (Mazelis & Creveling, 1975), *Albizzia lophanta* seeds (Schwimmer & Kjaer, 1960), shiitake mushroom (*Lentinus edodis*) (Iwami *et al.*, 1975) and *Pseudomonas* spp. (Nishizuka, 1971). Rutabaga (*Brassica napobrassica*) (Mazelis *et al.*, 1967), cauliflower (*B. oleracea botrytis*) (Lim *et al*, 1989*a*) and broccoli (*B. oleracea botrytis*) (Lim *et al.*, 1989*b*) use free cystine, as well as possibly cysteine, as substrates. Cystine is a much preferred substrate.

The onion, garlic and leek alliin lyases are pyridoxal phosphate-requiring enzymes. Evidence includes the characteristic spectra of pyridoxal phosphate enzymes (Fig. 6) with an absorbance peak at 430 nm, the requirement for pyridoxal phosphate for full activity of the garlic and leek enzymes and the nature of the products formed. Pyridoxal phosphate binding to leek alliin lyase appears to be much less tight (Won & Mazelis, 1989) than that of the onion and garlic enzymes (Nock & Mazelis, 1987) since the cofactor had to be added to the buffer at all steps in purification. The onion alliin lyase appears to bind pyridoxal phosphate even tighter than the garlic enzyme, since L-cysteine could not form a thiazolidine ring with the aldehyde group to give an absorbance peak at 330 nm, as could the garlic enzyme (Fig. 6) (Nock & Mazelis, 1986). Jansen *et al.* (1989*b*) concluded, based on yellow color and rotenone inhibition, that their purified garlic enzyme contained an unknown flavin coenzyme. The rotenone inhibition could not be confirmed (Ho & Mazelis, unpublished data).

TABLE 6

Substrate Specificities of Onion, Garlic and Leek Alliin Lyases

Substrates (cysteine sulfoxides)	*Onion*[a]			*Garlic*[b]			*Leek*[c]		
	K_m *(mM)*	V_{max} *(units/mg)*[d]	V_{max}/K_m *(units/mg . mM)*	K_m *(mM)*	V_{max} *(units/mg)*	V_{max}/K_m *(units/mg . mM)*	K_m *(mM)*	V_{max} *(units/mg)*	V_{max}/K_m *(units/mg . mM)*
S-Methyl-	9·1	128	14	15	76	5·1	55	13	0·24
S-Ethyl-	12(5·8)[e]	20	1·7	6(5·7[f]; 4·9[g])	86(17)[g]	14(3·5)[g]	13	33	2·5
S-Propyl-	6·2(4)[h]	132	21	3(1·1)[g]	30(8·3)[g]	10(7·5)[g]	—	13	—
S-Allyl-	7·1	147	21	6(3·3[f]; 2·2[g]; 0·5[i])	123(20)[g]	20(9·1)[g]	10	60	6·0
S-Butyl-	5·7	62	11	5	50	10	—	—	—

[a] Brosin (1969); 5 mM phosphate buffer, pH 7·5, in 20% glycerol at 0°C. V_{max} values calculated for saturating substrate concentrations, based on K_m.
[b] Mazelis & Crews (1968); 23°C, pH 6·5.
[c] Won & Mazelis (1989); 25°C, pH 8·0.
[d] Units = 1 mole of pyruvate formed per min per ml reaction.
[e] Fujita *et al.* (1990).
[f] Nock & Mazelis (1986); 23°C, pH 6·5.
[g] Jansen *et al.* (1989*a*).
[h] Schwimmer & Mazelis (1963); 37°C, pH 8·4.
[i] Kazaryan & Goryachenkova (1979); 23°C, pH 6·5.

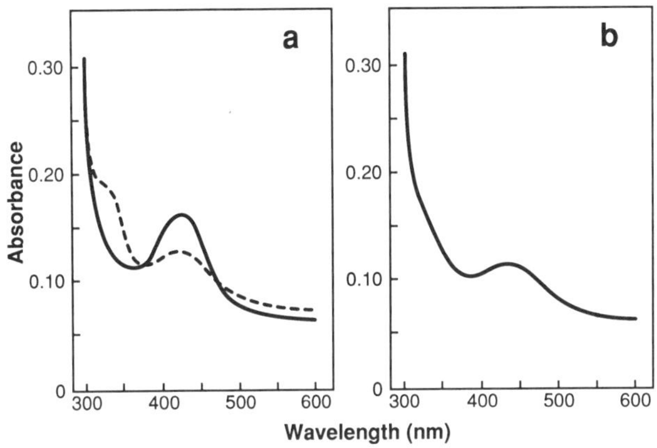

FIG. 6. Absorbance spectrum of (*a*) garlic and (*b*) onion allium lyase in the (- - -) presence and (——) absence of cysteine. The solutions contained 1 mg/ml of protein and 3 mM L-cysteine (when added). The spectrum of onion alliin lyase was the same in the presence and absence of cysteine. (Nock & Mazelis (1987), and reproduced by permission of the American Society of Plant Physiologists.)

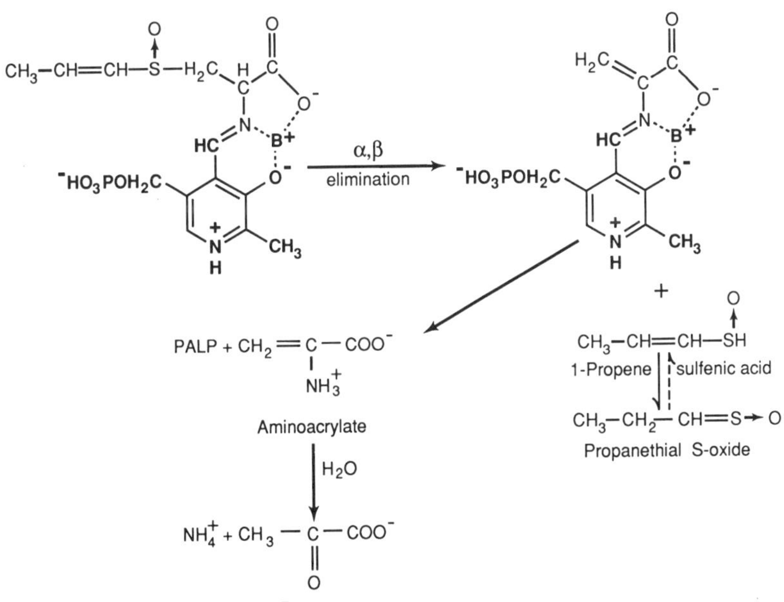

FIG. 7. Proposed mechanism of the conversion of (+)*S*-(1-propenyl)-L-cysteine sulfoxide to products by alliin lyase. The enzyme is represented by the cofactor, pyridoxal phosphate and B^+, an acidic group on the enzyme (in bold type).

A proposed mechanism for alliin lyases based on the requirement for pyridoxal phosphate is shown in Fig. 7. The enzyme, represented by pyridoxal phosphate (bold), forms a Schiff's base with the α-amino group of the cysteine portion of the substrate. Also important is a protonated amino acid side (B^+) of the enzyme which coordinates with the O^- group of the pyridoxal phosphate and the O^- of the carboxyl group and the amino N of the cysteine residue. The enzyme then catalyzes the β-elimination of the 1-propenyl sulfoxide portion of the substrate to give the initial products of the reaction, aminoacrylate and 1-propenelsulfenic acid (Schwimmer, 1964; Schwimmer & Friedman, 1972).

γ-GLUTAMYL PEPTIDASE AND γ-GLUTAMYL TRANSPEPTIDASE

Intact, mature onions contain substantial amounts of γ-glutamyl-*S*-alkyl(alkenyl)cysteines and γ-glutamyl-*S*-alkyl(alkenyl)cysteine sulfoxides while growing onions do not. This implies the γ-glutamyl derivatives are turned over rapidly in the growing plant and that appropriate enzymes for removal of the γ-glutamyl group exist. The enzymes might be industrially important in increasing the flavor potential of commercially dehydrated onions (Schwimmer, 1971).

The reactions catalyzed by γ-L-glutamyl peptidase (a hydrolase) and γ-L-glutamyl transpeptidase (γ-glutamyltransferase; (γ-glutamyl)-peptide:amino acid γ-glutamyltransferase; EC 2.3.2.2) are shown in eqns (10) and (11), where propenylCySO is *trans*-(+)-(1-propenyl)-L-cysteine sulfoxide.

$$\gamma\text{-L-glutamylpropenylCySO} \xrightarrow[\gamma\text{-glutamyl peptidase}]{+H_2O} \text{L-glutamic acid} + \text{propenylCySO} \quad (9)$$

$$\gamma\text{-L-glutamylpropenylCySO} \xrightarrow[\gamma\text{-glutamyl transpepidase}]{+\text{amino acid}} \gamma\text{-L glutamyl-NHCHRCOOX} + \text{propenylCySO} \quad (10)$$

Both enzymes are widely distributed in nature. γ-Glutamyl transpeptidase is present in large amounts in mammalian kidney (Orlowski & Meister, 1970) and the enzyme has been extracted from sprouting onion bulbs (Austin & Schwimmer, 1971) and other higher plants (Thompson *et al.*, 1964). Matikkala and Virtanen (1965*a,b*) reported the presence of a γ-glutamyl peptidase in sprouting onion bulbs and in germinating chive seeds. γ-Glutamyl transpeptidase preparations from onion (Schwimmer & Austin, 1971) and from kidney (Schwimmer, 1971) have been shown to enhance the flavor quality of dehydrated onions and garlic. A patent on their use has been granted (Schwimmer, 1973).

SUMMARY

Members of the *Allium* family include a large number of sulfur-containing compounds; the majority of these compounds are alkyl- and alkenylcysteines and derivatives. The alkyl- and alkenylcysteine sulfoxides are flavor precursors. In onion, the major compound is *trans*-(+)-*S*-(1-propenyl)-L-cysteine sulfoxide, while in garlic it is (+)-*S*-allyl-L-cysteine sulfoxide. Not much is known about the enzymes responsible for biosynthesis of these compounds. When onion tissue is damaged the enzyme, alliin lyase, acts on the alkyl- and alkenylcysteine sulfoxides, to produce primarily (*Z*)-propanethial *S*-oxide, pyruvate and ammonia. The sulfenic acids rapidly react to form thiosulfinates (fresh onion and garlic flavors), thiosulfonates and sulfides, disulfides, trisulfides and possibly other compounds. Recently, the alliin lyases of onion, garlic and leek have been purified to homogeneity. All are pyridoxal phosphate-containing glycoproteins which act by a β-elimination mechanism. Jansen *et al.* (1989*b*) reported that garlic alliin lyase is a flavin-containing enzyme but no role for the coenzyme was suggested. Onion and leek alliin lyases have pH optima near 8, while the garlic enzyme has maximum activity at pH 6·5. The enzymes differ in molecular weights and subunit composition and in *N*-terminal amino acid sequence. The amino acid compositions of the onion and garlic alliin lyases are quite similar except for histidine, 1/2 cystine and tryptophan contents, yet they are completely different immunochemically.

ACKNOWLEDGEMENT

The help of Eric Block on chemical nomenclature and recently published papers is gratefully acknowledged.

REFERENCES

Austin, S. J. & Schwimmer, S. (1971). *Enzymologia,* **40,** 273–85.

Bayer, T., Wagner, H., Block, E., Grisoni, S., Zhao, S. H. & Neszmelyi, A. (1989*a*). *J. Amer. Chem. Soc.* **111,** 3085–6.

Bayer, T., Breu, W., Seligmann, O., Wray, V. & Wagner, H. (1989*b*). *Phytochem.,* **28,** 2373–7.

Belo Jr, P. S. (1972). PhD Thesis, Michigan State University, East Lansing, Michigan, USA.

Bernhard, R. A. (1968). *J. Food Sci.,* **33,** 298–304.

Bernhard, R. A., Saghir, A. R., Jacobsen, J. V. & Mann, L. K. (1964). *Arch. Biochem. Biophys.,* **107,** 137–40.

Block, E. (1985). *Sci. Amer.* **252,** 114–19.

Block, E. & Bayer, T. (1990). *J. Amer. Chem. Soc.* **112,** 4584–5.

Block, E., Penn, R. E. & Revelle, L. K. (1979). *J. Amer. Chem. Soc.* **101,** 2200–1.

Block, E., Revelle, L. K. & Bazzi, A. A. (1980). *Tetrahedron Lett.* **21,** 1277–80.

Block, E., Ahmad, S., Jain, M. K., Crecely, R. W., Apitz-Castro, R. & Cruz, M. R. (1984). *J. Amer. Chem. Soc.* **106,** 8295–6.

Block, E., Ahmad, S., Catalfano, J., Jain, M. K. & Apitz-Castro, R. (1986). *J. Amer. Chem. Soc.* **108,** 7045–55.
Block, E., Iyer, R., Saha, C. & Belman, S. (1988). *J. Amer. Chem. Soc.* **110,** 7813–27.
Boelens, M., de Valois, P. J., Wobben, H. J. & van der Gen, A. (1971). *J. Agric. Food Chem.*, **19,** 984–91.
Brodnitz, M. H., Pollock, C. L. & Vallon, P. P. (1969). *J. Agric. Food Chem.*, **17,** 760–3.
Brosin, J. W. (1969). MS Thesis, University of California, Davis, California, USA.
Challenger, F. & Greenwood, D. (1949). *Biochem. J.*, **44,** 87–91.
Carson, J. F. & Wong, F. F. (1961*a*). *J. Org. Chem.*, **26,** 4997–5000.
Carson, J. F. & Wong, F. F. (1961*b*). *J. Agric. Food Chem.*, **9,** 140–3.
Ettala, T. & Virtanen, A. I. (1962). *Acta Chem. Scand.*, **16,** 2061–3.
Fenwick, G. R. & Hanley, A. B. (1985). *CRC Crit. Rev. Food Sci. Nutr.*, **22,** 273–377.
Freeman, G. G. & Whenham, R. J. (1975). *J. Sci. Food Agric.*, **26,** 1529–43.
Fujita, M., Endo, M. & Sano, M. (1990). *Agric. Biol. Chem.* **54,** 1077–9.
Fujiwara, M., Yoshimura, M. & Tsuno, S. (1955). *J. Biochem.*, (Tokyo), **42,** 591–601.
Granroth, B. (1968). *Acta Chem. Scand.* **22,** 3333–5.
Granroth, B. (1970). *Ann. Acad. Sci. Fenn.* (*Ser. A*2) **154,** 1–71.
Granroth, B. (1974). *Acta Chem. Scand.* **B28,** 813–14.
Iwami, K., Yasumoto, K. & Mittsuda, H. (1975). *Agric. Biol. Chem.*, (Japan) **39,** 1947–55.
Jansen, H., Müller, B. & Knobloch, K. (1989*a*). *Planta Medica* **55,** 434–9.
Jansen, H., Müller, B. & Knobloch, K. (1989*b*). *Planta Medica* **55,** 440–5.
Kawakishi, S., Namiki, K., Nishimura, H. & Namiki, M. (1971). *J. Agric. Food Chem.*, **19,** 166–9.
Kazaryan, R. A. & Goryachenkova, E. V. (1978). *Biochemistry,* (Russian), **43**(10), 1502–8.
Kuo, M.-C., Chien, M. & Ilo, C.-T. (1990). *J. Agric. Food Chem.* **38,** 1378–81.
Kuon, J. & Bernhard, R. A. (1963). *J. Food Sci.*, **28,** 291–304.
Lim, M. H., Chen, A. O., Pangborn, R. M. & Whitaker, J. R. (1989*a*). *Proc. 7th World Congr. Food Sci. & Technol,* 1987. International Union of Food Science and Technology, Singapore.
Lim, M. H., Velasco, P. J., Pangborn, R. M. & Whitaker, J. R. (1989*b*). In *Quality Improvements of Fruits and Vegetables,* ed. J. J. Jen. ACS Symposium Series. American Chemical Society, Washington, DC, USA, **405,** 72–83.
Lukes, T. M. (1971). *J. Food Sci.*, **36,** 662–4.
Matikkala, E. J. & Virtanen, A. I. (1965*a*). *Acta Chem. Scand.*, **19,** 1258–61.
Matikkala, E. J. & Virtanen, A. I. (1965*b*). *Acta Chem. Scand.*, **19,** 1261–2.
Matikkala, E. J. & Virtanen, A. I. (1967). *Acta Chem. Scand.*, **21,** 2891–3.
Mazelis, M. & Creveling, R. K. (1975). *Biochem. J.*, **147,** 485–91.
Mazelis, M. & Crews, L. (1968). *Biochem. J.*, **108,** 725–30.
Mazelis, M., Beimer, N., & Creveling, R. K. (1967). *Arch. Biochem. Biophys.* **120,** 371–8.
Morimitsu, Y. & Kawakishi, S. (1990). *Phytochem.* **29,** 3435–3439.
Niegisch, W. D. & Stahl, W. H. (1956). *Food Res.*, **21,** 657–65.
Nishizuka, Y. (1971). *Methods Enzymol.*, **17,** 470–4.
Nock, L. P. & Mazelis, M. (1986). *Arch. Biochem. Biophys.*, **249,** 27–33.
Nock, L. P. & Mazelis, M. (1987). *Plant Physiol.*, **85,** 1079–83.
Nock, L. P. & Mazelis, M. (1989). *Phytochem.*, **28,** 729–31.
Orlowski, M. & Meister, A. (1970). *Proc. Nat. Acad. Sci. USA,* **67,** 1248–55.
Rundquist, C. (1909). *Pharm. Notisblad* **18,** 323–33.
Saghir, A. R., Mann, L. K., Bernhard, R. A. & Jacobsen, J. V. (1964). *Proc. Amer. Soc. Hort. Sci.*, **84,** 386–98.

Saghir, A. R. B., Mann, L. K., Ownbey, M. & Berg, R. Y. (1966). *Amer. J. Bot.*, **53**, 477–84.
Schwimmer, S. (1964). *Biochim. Biophys. Acta*, **81**, 377–85.
Schwimmer, S. (1971). *J. Agric. Food Chem.*, **19**, 980–3.
Schwimmer, S. (1973). US Patent 3,725,085.
Schwimmer, S. & Austin, S. J. (1971). *J. Food Sci.*, **36**, 1081–85.
Schwimmer, S. & Friedman, M. (1972). *Flavor Ind.*, (March) 137–45.
Schwimmer, S. & Kjaer, A. (1960). *Biochim. Biophys. Acta*, **42**, 316–24.
Schwimmer, S. & Mazelis, M. (1963). *Arch. Biochem. Biophys.*, **100**, 66–73.
Schwimmer, S., Ryan, C. A. & Wong, F. F. (1964). *J. Biol. Chem.*, **239**, 777–82.
Semmler, F. W. (1892). *Arch. Pharm.*, **230**, 434–43.
Smith, I. K. & Thompson, J. F. (1969). *Biochem. Biophys. Res. Commun.*, **35**, 939–45.
Stoll, A. & Seebeck, E. (1951). *Advan. Enzymol.*, **11**, 377–400.
Tamura, G. (1965). *J. Biochem.*, (Tokyo), **57**, 207–14.
Thompson, J. F. (1967). *Annu. Rev. Plant Physiol.*, **18**, 59–84.
Thompson, J. F., Turner, D. H. & Gering, R. K. (1964). *Phytochem.*, **3**, 33–46.
Tobkin, H. E. & Mazelis, M. (1979). *Arch. Biochem. Biophys.*, **193**, 150–7.
Virtanen, A. I. (1969). *Qual. Plant. Mater. Veg.*, **18**, 8–28.
Wertheim, T. (1844). *Ann.* **51**, 289–315.
Wertheim, T. (1845). *Ann.* **55**, 297–364.
Won, T. & Mazelis, M. (1989). *Physiol. Plant.*, **77**, 87–92.

14

Significance of Enzymes in Individual Vegetables

J. B. Adams
Campden Food and Drink Research Association, Chipping Campden, Gloucestershire GL55 6LD, UK

INTRODUCTION

Vegetables are an important source of nutrients, vitamins and fibre in the human diet. However, the consumption of vegetables depends greatly on their sensory quality. This is determined by the colour and general appearance of the vegetable, its flavour acceptability as judged by aroma and taste, and its textural properties, particularly the firmness to touch of the raw material, breakdown on cooking and mouth-feel when eating. Each quality parameter can be influenced by endogenous enzyme activity either because the first-formed products have a direct effect on quality or because they react further, under enzyme control or via non-enzymatic catalysis, to give compounds which then affect quality. This chapter will emphasise the enzymatic aspects of quality changes and will relate this to commercial processing practice wherever possible.

The importance of vitamins in vegetables is recognised by the inclusion of a section on enzyme-initiated changes. Finally, changes in toxic substances, which can influence the nature and degree of processing required for some vegetables, are covered by reference to the enzyme-catalysed degradation of two important groups of naturally-occurring constituents.

COLOUR CHANGES

Perception of food colour is a function of how the food is presented and viewed in relation to the conditions and quality of illumination. The degree of light scattering, absorption and reflection are particularly important. Pigments have a primary role to play in the perceived colour but the physical state of the pigment and the presence and state of non-pigmented substances can have highly significant secondary effects. For instance, the colour of green vegetables changes from a dull to a bright green on blanching. This is not, for the most part, caused by a change in composition of the chlorophyll pigments but is due to changes in their physical state within the chloroplasts, to expulsion of air and to increases in translucency of intercellular components as well as that of surrounding tissue.

As pigments have a primary influence on colour, it is perhaps not surprising that the effect of enzymes on vegetable colour has been concerned almost entirely with discolorations caused either by the formation of pigmented compounds or by the degradation of the naturally-occurring pigments.

Phenol oxidation

When vegetables are damaged during handling or packaging, cell disruption leads to the mixing of vacuolar phenols with cytoplasmic polyphenol oxidase and atmospheric oxygen. This causes the formation of quinones which can react non-enzymatically with other constituents, particularly amino compounds, to form coloured pigments (melanins). These generally arise at the injury site and vary in hue from yellow to red-brown through to dark brown. Browning is most obvious in vegetables which are white or cream-coloured in the fresh state, such as mushrooms and potatoes. A related phenomenon is the physiological browning which occurs during the growth or storage of some vegetables under adverse conditions. Internal browning in Brussels sprouts, tip-burn and russet-spotting in lettuce, internal brown spot in potatoes and 'cloud' in tomatoes are examples of this type of browning which are due to a combination of genetic and environmental factors that eventually lead to cellular disruption and tissue necrosis. The oxidation of phenols, catalysed by polyphenol oxidase, may be involved in the final browning reactions (Ke & Saltveit, 1989).

Mushrooms

In the edible, cultivated mushroom, *Agaricus bisporus,* the browning reaction is most intense at the cap epidermis as a result of the higher concentration of phenol substrates and polyphenol oxidase (Burton, 1986, Moore *et al.*, 1988). The rest of the mushroom cap can still turn brown but to a lesser extent. In addition to the active forms of the enzyme, which can initiate browning immediately, mushrooms contain a considerable amount of latent activity (Yamaguchi *et al.*, 1970). A large increase occurs in the latter on storage and ageing as the small 'buttons' develop into large 'flats'.

Control of browning in mushrooms can be achieved by taking special care in picking and subsequent handling and by using techniques which are designed to restrict the polyphenol oxidase activity or to destroy the enzyme completely. Cool storage at 5°C is the technology most extensively used for maintaining fresh mushroom quality primarily because it suppresses maturation. Low temperature storage also reduces the activity of both forms of polyphenol oxidase and lowers the rate of phenol oxidation and brown discoloration (Murr & Morris, 1975).

Modifications to the composition and pressure of the atmosphere above mushrooms can influence the discoloration reactions. Murr and Morris (1974) found that storage in nitrogen (0% oxygen) reduced both the polyphenol oxidase activity and mushroom surface discoloration. At 2% oxygen, or higher, discoloration was similar to the control. High levels of carbon dioxide (25 and 50% at 20% oxygen) also appeared to increase the brown discoloration although polyphenol oxidase activity was partly inhibited. Developments in modified atmosphere packaging, comprisng a highly permeable microporous 'window' in a relatively impermeable over-wrapping film, have also been shown to reduce the rate at which mushrooms brown as a result of the lowered level of oxygen in the headspace of the pack (Burton *et al.*, 1987). According to Robitaille and Badenhop (1981), mushrooms may be stored under hyperbaric conditions without any injurious effect so long as depressurisation is gradual. The inclusion of carbon monoxide was found to reduce browning, presumably because it inhibited polyphenol oxidase by combining with the copper at the active site.

Mushrooms for freezing are often blanched for a few minutes in near-boiling water to destroy enzymes, and a simple test for polyphenol oxidase using a catechol solution may be employed to test the adequacy of the treatment. Unfortunately, blanching causes shrinkage and solids loss and alternative techniques have therefore been investigated. Some success has been claimed for vacuum packaging (Steinbuch, 1979; Prestamo & Fuster, 1982) and dips in metabisulphite have been shown to be partially effective (Prestamo & Fuster, 1982; Gormley, 1986). The uptake of sulphite requires careful monitoring as a result of its possible toxicity and hypersensitivity reactions (Taylor *et al.*, 1986). Some countries have set legal limits on the amount which is acceptable in frozen mushrooms (50 ppm as SO_2 in the UK, for example). Using purified mushroom polyphenol oxidase, Golan-Goldhirsch and Whitaker (1984) demonstrated that dithiothreitol and glutathione are more effective inhibitors than sodium bisulphite or ascorbic acid. Short chain peptides were suggested by Mayer and Harel (1979) as suitable inhibitors, and Kahn (1985) showed that diglycine and triglycine were indeed effective in lowering the final level of coloured melanins formed by the action of mushroom polyphenol oxidase on DL-dihydroxyphenylalanine. Free amino acids were also found to be inhibitory. The effectiveness of amino acids and peptides in the prevention of browning during the processing of mushrooms has yet to be established.

During the dehydration of mushrooms, the activity of polyphenol oxidase can be severely restricted by performing the initial phase of drying at a relatively high temperature (75°C). Thermal inactivation of the enzyme would be expected to be significant at this temperature and the rapid loss of water and consequent reduction in water activity may also play a role in the reduction of enzyme activity (Vámos-Vigyázó, 1981).

Gamma-irradiation has generally been shown to retain the original mushroom skin colour reduce the rate of browning. However, the effect does not appear to involve loss of polyphenol oxidase activity (Thomas, 1988).

Other edible species of mushrooms, such as the Japanese forest mushroom, Shiitake (*Lentinus edodes*), and the Oyster mushroom (*Pleurotus ostreatus*), are also known to contain polyphenol oxidase. Oddson and Jelen (1981) have reported that polyphenol oxidase activity and the concentration of its substrates are lower in the *Pleurotus* mushroom, *P. florida,* than in *A. bisporus.* As a result, the rate of tissue browning in the *Pleurotus* species was much reduced. The properties of *Pleurotus* mushrooms have been reviewed by Bano and Rajarathnam (1988) who point out that most species have a brown-pink tint which would aid in masking minor changes in browning.

Potatoes

Two main types of discolorations occur in raw potatoes: browning and blackspot. Browning occurs as a result of severe bruising, when the damage extends through the outer skin ('shatter' bruising), and on peeling, slicing and dicing potatoes prior to heat processing. Blackspot occurs sub-epidermally in the vascular region of the tissue as a blue-grey zone which can take several hours or days to appear after relatively minor damage. Low temperatures during lifting, handling and storage increase blackspot and it is therefore advisable to raise storage temperatures before handling (Salunkhhe *et al.,* 1989) and to process the potatoes as rapidly as possible. Potatoes which have lost water either during growing, harvesting or storage and have become soft, and material grown under low levels of potassium are generally prone to blackspot. Also, certain cultivars appear to be more susceptible than others (Hughes, 1980). Whilst there is good evidence that the brown discoloration occurs as a result of polyphenol oxidase activity on tyrosine (Matheis, 1987), the involvement of the enzyme in blackspot formation is less clear. Schaller and Amberger (1974) were unable to demonstrate any relationship between blackspot and polyphenol oxidase activity using factorial and multiple regression analysis. In addition, potato cultivars with high levels of ascorbic acid tend to have low incidences of brown discoloration because of the inhibitory effect of ascorbic acid on phenol oxidation, whereas ascorbic acid and susceptibiltiy to blackspot do not appear to be related (Thornton & Workman, 1987). According to Stark *et al.* (1985), blackspot formation is related to the concentration of phenolic substrates rather than to differences in the quantity or properties of the polyphenol oxidase. This would imply that other enzymes, such as peroxidase, may be involved in the oxidation of the phenols (Weaver and Hautala, 1970).

A rapid method for assessing the susceptibility of potato cultivars to blackspot has been developed by Pavek *et al.* (1985). Tuber samples were abrasively peeled for 30 s and the amount of enzymatic discoloration evaluated after 24 h. The results showed that immature potatoes, dug whilst the plants were still green, were more resistant to blackspot than were mature potatoes. Other methods, such as shaking a standard number of potatoes for a short time, have also proved to be useful for assessing blackspot shortly before stores are emptied (Meijers, 1987). However, Kunkel *et al.* (1986), using a bruising method to induce blackspot, have observed that susceptibility varies greatly both within a tuber and among tubers. Thornton and Beaver (1986) found that immersing potatoes in a catechol solution for 5 min and then holding in air can be used to detect 'shatter' bruising, whilst blackspot susceptibility can be detected more rapidly by holding potatoes at elevated temperatures after the initial bruising.

The extent of brown discoloration observed on bruising or cutting potatoes depends mainly on the concentration and type of phenolic compounds, the concentration and the substrate specificity of the polyphenol oxidase, the concentration of naturally occurring inhibitors of browning (especially ascorbic acid), oxygen concentration, pH and temperature (Matheis, 1987). The concentration of phenolic compounds appears to depend on location and climate whereas the concentration of polyphenol oxidase and ascorbic acid depends mainly on the cultivar. Polyphenol oxidase activity and concentrations of phenolic compounds are generally highest in the outer parts of potato tubers although detailed distribution patterns are cultivar-dependent. On storage, both polyphenol oxidase activity and phenols increase in the outer parts of potatoes whilst the enzyme activity decreases in the inner parts. However, changes in the rate of browning during storage cannot be correlated with changes in a single variable, whether it be polyphenol oxidase activity or the concentration of phenols or ascorbic acid. Application of potassium fertilisers is known to decrease the rate of potato browning. Concomitantly, decreases in the concentrations of polyphenol oxidase and phenolic compounds are found along with increases in ascorbic acid concentration. The reverse is found for nitrogen fertilisation. The potential for a given cultivar to discolour can therefore be minimised by applying fertilisers containing optimum ratios of potassium to nitrogen. The control of browning by the application of fertilisers is limited, however, and it is clearly desirable to breed potato cultivars with low concentrations of phenols, polyphenol oxidase and high concentrations of ascorbic acid, irrespective of growing conditions. In the USA, nematode-resistant cultivars have been bred which are lower in phenolic compounds and show a reduced tendency to discolour (Gosselin & Mondy, 1986), whilst in the UK, clones of wild potato species have been found to exhibit low enzymatic browning potentials (Gubb *et al.*, 1989). However, until cultivars are generally available with low-browning tendencies, and which retain the other desirable quality attributes of potatoes, practical measures for controlling polyphenol oxidase activity will continue to be required to reduce browning in raw potatoes prepared for subsequent processing.

Ascorbic acid is probably the major, naturally occurring inhibitor of enzymatic browning in potatoes as a result of its ability to reduce the initial oxidation products, the *o*-quinones, back to *o*-diphenols, and possibly also because it can inhibit polyphenol oxidase directly. The concentration of endogenous ascorbic acid is insufficient to prevent browning in prepared raw potatoes and further amounts need to be added as a dip or soak, possibly in conjunction with citric acid as a sequestering agent and acidulant (Langdon, 1987). Sulphites may also be used to inhibit browning and are more effective in the presence of citric acid. Francis and Amla (1961) found that potato strips dipped in a 2500 ppm sulphur dioxide solution for 30 s absorbed about 50 ppm and that this was generally sufficient to delay browning for approximately 2 weeks at 4°C. Potato strips treated with 5000 ppm sulphur dioxide were significantly softer than control samples after this time. The minimum concentration of sulphur dioxide required to prevent browning will, of course, vary with the source and cultivar according to the factors outlined above. Polyphosphates, such as hexametaphosphate or acid pyrophosphate, are effective browning inhibitors although it has been suggested that ethylenediaminetetra-acetic acid (EDTA) is more cost-effective (Anon., 1984). However, EDTA is not accepted as an additive in all countries. Many other compounds have been found to inhibit potato browning but are not used in practice because of their relatively high costs or adverse effects on quality and safety of the potato product. The most certain way of preventing browning is to inactivate the polyphenol oxidase by heating the potato for a sufficient time at a high temperature, as occurs in blanching or sterilisation processes. This is sometimes performed after treatment with, or in the presence of, an inhibitor such as citric acid, ascorbic acid or a sulphite and is particularly effective for sliced, diced or chipped material. The main disadvantage of heating is that the potato softens, and may slough, causing the final product to appear overcooked. Softening may be reduced by incorporating calcium salts in the inhibitor mixture and/or by using relatively long-time–low-temperature heat treatments to activate the pectinmethylesterase, the action of which can enhance the resistance of the tisssue to cell separation (*vide infra*).

Gamma-irradiation, which is used to prevent sprouting, apparently increases the browning in potatoes as a result of cell damage, solubilisation of cell-bound enzyme, increased substrate accessibility and lack of quinone reduction by ascorbic acid (Vámos-Vigyázó, 1981). Increased browning is dependent on the radiation dose, the harvest time and on the length of post-harvest storage prior to treatment. The effect of gamma-irradiation on polyphenol oxidase activity is uncertain due to factors associated with the raw material and the radiation dose (Thomas, 1984).

Other vegetables

Brown discolorations that occur in other vegetables are generally caused by the oxidation of phenolic compounds catalysed by polyphenol oxidase and are influenced by similar factors to those discussed for mushrooms and potatoes. Control can be achieved by inhibiting or inactivating the enzyme.

Parsnips. Parsnips are white when harvested but turn light brown soon after, even without the physical damage which occurs on harvesting, cleaning and packaging. Browning is greatest in the tissue near the surface where polyphenol oxidase activity is highest (Chubey & Dorrell, 1972). The total phenolic content is lowest on the root surface and highest in the vascular cylinder region, whilst the enzyme is almost absent in the vascular tissue. Chlorogenic acid is uniformly distributed throughout the roots and is probably not the principal browning substrate. On storage of peeled roots, surface browning increases along with total phenolic compounds, chlorogenic acid and polyphenol oxidase activity (Dorrell & Chubey, 1972).

Carrots. Enzymatic browning in carrots is restricted to the surface where the concentration of phenolic compounds is highest. Chlorogenic acid is the main browning substrate with tyrosine of secondary importance (Chubey & Nylund, 1969). Polyphenol oxidase is less active than in parsnips. In addition, it appears to be more evenly distributed throughout the root although the activity does depend on cultivar (Schaller *et al.*, 1985).

Green beans. Mechanical harvesting can lead to browning of the broken ends of green beans, a severe problem if the raw material is held for extended periods prior to canning or freezing (Sistrunk *et al.*, 1989). Some cultivars are more prone to browning than others, suggesting that the problem may be alleviated by breeding and selection. Holding beans in elevated carbon dioxide atmospheres inhibits the increase in phenols which is related to browning, whilst having no affect on polyphenol oxidase activity. Sulphur dioxide reduces browning when applied in the field immediately after harvesting green beans.

Yam. In yam tubers, brown discoloration is most intense at the stem end where the concentration of phenol compounds is highest (Onayemi, 1986). The middle of the yam is most suited to the preparation of light-coloured products because of its relatively low content of phenols. The main browning substrate in yam appears to be (+)-catechin (Ozo & Caygill, 1986). The species *Dioscorea rotundata* shows less tendency to brown because of its low concentration of the flavonoid substrate whilst *D. dumetorum* has low browning potential apparently as a result of its low polyphenol oxidase activity.

Artichoke. Artichoke heads harvested in winter months can be stored in good condition at 20 or 4°C, but the shelf life diminishes markedly for spring-harvested material due to internal blackening. Lattanzio *et al.* (1989) have suggested that this may be associated with increased phenol biosynthesis but have questioned the subsequent involvement of polyphenol oxidase. Soaking heads in 1% solutions of citric or ascorbic acid delayed the discoloration reaction.

Beetroot. The occurrence of a 'black ring' discoloration in canned beetroot slices has been associated with polyphenol oxidase activity in cells partly disrupted by steam peeling (Im *et al.*, 1990). Delays after cooling and slicing allowed the formation of quinones which could then participate in secondary reactions during heat sterilisation to yield dark-coloured pigments. Recommendations to minimise the discoloration are to use short steam peeling treatments, to eliminate delays before retorting or, if a delay is anticipated, to hold the peeled and sliced beetroot in water or dilute brine.

Chlorophyll changes

The colour of green vegetables is due mainly to the pigments chlorophyll a (blue-green) and chlorophyll b (yellow-green). Acidification of the chlorophylls causes removal of magnesium and the formation of pheophytins a and b, both of which are olive-brown. Hydrolytic removal of the phytol group converts the chlorophylls to chlorophyllides, which remain green, and the pheophytins to pheophorbides which are olive-brown. The chlorophylls and pheophytins are lipophilic due to the presence of the phytol group, while the chlorophyllides and pheophorbides, without phytol, are hydrophilic. Enzymatic degradation of the chlorophylls can be particularly important in colour changes in green vegetables whilst chlorophyll synthesis can cause an unacceptable greening in other vegetables, especially potatoes.

Greening of potatoes

On exposure to light, the outer parenchyma layers of potatoes become green as a result of chlorophyll formation. Many of the enzymes involved in the biosynthesis of chlorophyll are known and the influence of light at least partly understood. Studies on chlorophyll formation in cold-stored potatoes have indicated that the generally-accepted biosynthetic pathway is operative (Ramaswamy & Nair, 1974). Chlorophyll a appears to be synthesised first and chlorophyll b at a later stage. The glycoalkaloids, α-solanine and α-chaconine, are also formed at the same time in an independent process. In abnormally high amounts (>20 mg/100 g fresh weight), the latter compounds give rise to a bitter taste and are thought to be toxic (Morgan & Coxon, 1987).

Numerous factors influence the rate of greening (Jadhav & Salunkhe, 1975). Potato cultivar and maturity have a significant effect with immature and small potatoes apparently showing a greater tendency towards greening than mature and large ones. In general, intense light, high temperatures during light exposure, and low storage temperatures following exposure to light, enhance the rate of greening. The longer the duration of light exposure, the greater is the accumulation of chlorophyll. Once formed in potato tubers, chlorophyll appears to be stable during subsequent storage in the dark (Harkett, 1975).

The simplest way to prevent chlorophyll formation is to store potatoes in the dark. Alternatives to dark storage, such as light-absorbing films, various dips and coatings, modified atmospheres and ionising radiation have been evaluated (Schwimmer, 1981) but only partial success has been achieved. Kadkade (1983) has shown that illuminating potatoes with far-red light at 750 nm did not stimulate chlorophyll formation. In addition, no increase in glycoalkaloid formation was observed and sprouting was reduced. In the long-term, the development of cultivars which are resistant to greening may be the only satisfactory answer to the problem.

Chlorophyllide formation
The enzyme, chlorophyllase, catalyses the hydrolysis of chlorophyll to chlorophyllide and phytol with retention of green colour. Activity has been found in tomatoes and many green vegetables, and its effect on colour studied most intensively in peas and spinach. Buckle and Edwards (1970) found evidence for phytol-free chlorophyll derivatives in frozen peas although greater interest has been shown in the apparent stability of the enzyme during heat processing. It has been suggested that chlorophyllide is more stable than chlorophyll and proposals in the form of patents have been made to utilise spinach chlorophyllase activity by low temperature blanching prior to canning (Thomas, 1928; Lesley & Shumate, 1937). However, the experimental evidence for the greater stability of chlorophyllides is not conclusive (Clydesdale & Francis, 1968).

At present, there is insufficient evidence to claim that chlorophyllase activity may be able to play a role in the stabilisation of green colour in heat-processed vegetables. In the early work, the identification and quantification of chlorophyllides by solvent extraction and spectrophotometry was prone to error. More recently, it has been found that high performance liquid chromatography (HPLC) can be used to resolve chlorophyllides from chlorophyll (Shioi & Beale, 1987) and methods have been developed which allow the continuous assay of chlorophyllase (Lambers & Terpstra, 1985). Thus, it should now be feasible to correlate the activity of the enzymes with the presence of the chlorophyllide products in appropriate vegetables or model systems.

Chlorophyll degradation
Raw green vegetables undergo changes during freezing and frozen storage which lead to unattractive yellow or olive-brown hues. Early work on peas, such as that of Dietrich *et al.* (1957), attributed these colour changes to the conversion of chlorophyll to pheophytin brought about by increased acidity due to lipase activity, or by the natural acidity of the vegetable. Later, the colour change in peas was considered to be due to the action of lipoxygenase on free fatty acids formed by lipase action, and that this was, in some way, responsible for the removal of magnesium from chlorophyll (Wagenknecht & Lee, 1958). The activity of lipase in frozen peas was confirmed by Bengtsson and Bosund (1966) who found that linoleic and linolenic acids were liberated preferentially. More recently, a phospholipase has been detected in peas which releases fatty acids from phospholipids (Henderson *et al.*, 1984). Buckle and Edwards (1970) showed that pheophytin was indeed formed during the storage of frozen peas and also that there was a loss of total chlorophyll. In addition, they demonstrated that the rate of chlorophyll degradation was reduced by lowering the temperature from −9·4 to −23·3°C and by storage under nitrogen instead of oxygen. Lipoxygenase was active in the early stages of storage but almost completely destroyed after 20 weeks at −9·4°C, perhaps as a result of reaction with the lipid oxidation products. Chlorophyll bleaching is known to occur in model systems in coupled oxidations with linoleic acid catalysed by soy-bean

lipoxygenase (Orthoefer & Dugan, 1973; Hildebrand & Hymowitz, 1982), and it is feasible that this is the cause of the destruction of the pigment in frozen peas. Similar involvement of lipid oxidation in chlorophyll degradation has been suggested in frozen green beans (Walker, 1964). The current weight of evidence strongly suggests that chlorophyll degradation in frozen, unblanched green vegetables is linked to enzymatically catalysed lipid oxidation. However, further studies are needed in order to suggest a convincing mechanism of chlorophyll degradation. In particular, identification of the colourless chlorophyll derivatives is required and measurements are needed to relate lipase and lipoxygenase activities to the formation of the chlorophyll degradation products.

In living plants, a possible correlation between chlorophyll concentration and lipoxygenase activity has been inferred by Klein *et al.* (1985) who reported that during the germination of mustard and pea seedlings in the dark and in the light, chlorophyll synthesis was inversely related to lipoxygenase activity. On senescence, the disappearance of chlorophyll may be associated with the lipoxygenase-catalysed co-oxidation of the pigment and some evidence has been presented for this in studies on the yellowing of spinach leaves (Yamauchi *et al.*, 1986).

Changes in carotenoids

The importance of changes in carotenoids is most evident in vegetables which contain relatively high concentrations in the edible portions such as tomatoes and carrots. During the ripening of tomatoes, the chlorophylls are rapidly degraded and in the early stages, the yellow pigments, β-carotene and the xanthophylls, are formed. Subsequently, the red pigment, lycopene, rapidly accumulates and makes a major contribution to the colour of ripe tomatoes (Davies & Hobson, 1981). In carrots, carotene levels increase slowly during chill storage, apparently following the biosynthetic pathways established for tomatoes (Lee, 1986). This increases the level of provitamin A but the effect on the perceived colour of stored carrots is unknown.

The possibility of enzyme-catalysed degradation of carotenoids in vegetables has been appreciated for some time, particularly as a result of studies on model systems with soy-bean lipoxygenase and polyunsaturated fatty acids, such as linoleic acid. Co-oxidation of the pigments to colourless products occurs under these conditions, catalysed mainly by the Type 2 lipoxygenases. The latter enzymes occur in many vegetables, and it has been suggested that the efficiency of blanching in stabilising carotene in frozen vegetables is due to inactivation of lipoxygenase (Schwimmer, 1981). The apparent stability of carrot colour may result from carrots being among the small group of vegetables which have very low lipoxygenase activities (Pinsky *et al.*, 1971; Baardseth & Slinde, 1980; Schaller & Vámos-Vigyázó, 1986). The presence of phenols in vegetables may reduce carotenoid degradation by inhibiting lipoxygenase activity (Ozmianski & Lee, 1990). A comparison of carotenoid degradation in raw or frozen, unblanched vegetables with that which occurs in model systems is required in order to verify the involvement of lipoxygenase. The importance of such work would lie in the effect on the nutritional quality of vegetables as sources of essential fatty acids as much as on the colour quality.

Betalain degradation

The water-soluble betalain pigments of the red beet comprise the red betacyanins and the yellow betaxanthins localised within the vacuole of intact cells. Because of increasing restrictions on the use of artificial food colours, the red-violet betacyanin, betanin, has come into prominence as an alternative red food dye. However, application of betanin has been restricted by its generally low stability. Factors that affect the thermal stability of betacyanin pigments during the processing of red beets have been studied by Elbe *et al.* (1981) but information regarding the enzymatic degradation is limited. Wasserman and Guilfoy (1983) have demonstrated that hydrogen peroxide is formed by isolated beet cell walls and have suggested that a peroxidase is involved in the decolorisation of betanin. The enzyme and the products of betanin degradation have yet to be identified and characterised, so there remains considerable doubt regarding the true nature of the decolorisation reaction. Nevertheless, the activity of the enzyme must be considered as an important factor in the extraction of raw beets and during storage and utilisation of beet juices in food products. The enzyme has been shown to be present in beet juice extracted by a diffusion process (Lashley & Wiley, 1979).

FLAVOUR CHANGES

The flavour of foods is a human response to a complex mixture of stimuli perceived by the taste and odour senses. As with other foods, the flavour of vegetables can be attributed to both volatile and non-volatile compounds. When fresh (unprocessed) vegetables are consumed raw, the volatile compounds make a relatively high contribution to flavour as a result of enzyme activity during mastication. When consumed after normal cooking, the non-volatiles may have a greater influence on flavour as the volatile-related enzymes will have been heat-denatured and any pre-formed volatile compounds will have been partly lost through evaporation. Overcooking can lead to the non-enzymatic formation of new compounds with flavours which are not typical, either of the raw or of the normally cooked vegetable.

During harvesting, handling and preparation prior to commercial preservation, vegetables are subject to a significant degree of damage. This causes disruption of the cells comprising the vegetable tissue and the mixing of enzymes and substrates which are physically separated in undamaged material. As a result, off-flavours can be generated which may be carried through into the processed product. Clearly, reducing both the degree of damage and the time prior to heat inactivation or chemical inhibition of the enzymes in vegetables should cause a diminution in off-flavour. Chill storage can also be helpful by reducing enzyme activity. However, long-term storage at chill temperatures can, in some cases, trigger a sequence of enzyme-catalysed reactions which increases quality deterioration. The sweetening of some vegetables which occurs at low temperatures is an example of the latter problem.

The ensuing discussion will be concerned with the enzymatic degradation of certain key constituents, occurring naturally, which have a vital influence on the development of flavour (and off-flavour) in vegetables.

Lipid degradation

Physical disruption of vegetable tissue initiates autolytic changes catalysed by endogenous enzymes amongst which are those causing hydrolytic and oxidative degradation of lipids. The enzymatic hydrolysis of lipids leads to free fatty acids which invariably cause unacceptable flavours, whereas the oxidation of polyunsaturated fatty acids may produce volatile carbonyl compounds with either desirable or undesirable flavour properties, depending on whether the vegetable is consumed in the raw state or has been processed.

Cucumbers

The characteristic flavour of raw cucumber is due mainly to C-9 unsaturated carbonyls released on disruption by cutting or chewing. These compounds are not present in the intact tissue and are not formed if the enzymes are inactivated before disruption occurs. The major flavour component is *trans*-2,*cis*-6-nonadienal which is formed from linolenic acid as a result of the consecutive action of four enzymes: lipid acyl hydrolase, lipoxygenase, hydroperoxide lyase and *cis*-3,*trans*-2-enal isomerase (Phillips *et al.*, 1979). Other saturated and unsaturated aldehydes may also contribute to the cucumber flavour profile, some of which are derived from linoleic acid by a similar pathway. In addition, C-12–C-18 aldehydes are present in the volatile fraction of cucumber homogenates (Kemp, 1975; Galliard & Matthew, 1976) which arise as a result of the activity of an α-oxidation enzyme system. The significance of the latter compounds is unknown, however, despite attempts to evaluate the contribution made by tetradecanal to cucumber flavour (Baardseth & Slinde, 1987).

Attempts to reduce chilling injury in cucumbers by dipping in 6% calcium chloride for 30 min have been shown to lower the levels of *trans*-2,*cis*-6-nonadienal and *trans*-2-nonenal, while hexanal, *trans*-2-hexenal and 2-pentylfuran were increased (Geduspan & Peng, 1986). 6-Nonenal and tetradecanal showed no clear trend.

Tomatoes

The characteristic flavour of raw tomatoes is determined predominantly by volatile substances, some of which are derived from lipid degradation.

Although some 400 volatile components have been identified in tomatoes, their relative importance for tomato flavour is not clearly understood (Petró-Turza, 1986/87). Buttery *et al.* (1987) have attempted to evaluate the odour contributions made by the major volatiles by determining the ratio of their concentrations in fresh tomatoes to the odour threshold values. This work has suggested that *cis*-3-hexenal, *β*-ionone, 1-penten-3-one, hexanal, *cis*-3-hexenol, *trans*-2-hexenal, 2- and 3-methylbutanol, 2-isobutylthiazole and 6-methyl-5-hepten-2-one should be major contributors. Indeed, an aqueous solution of the eight compounds with the highest odour contributions at concentrations found in tomatoes was judged to be very similar to fresh sliced tomato by a sensory panel. The formation of *cis*-3-hexenal and hexanal is known to be caused by the consecutive action of acyl hydrolase (and possibly phospholipase D), lipoxygenase and hydroperoxide lyase on tomato lipids (Galliard *et al.*, 1977). An important distinction between volatile formation in cucumber and tomato is that in the former, both the 9- and 13-hydroperoxide isomers are cleaved to yield the C-9 and C-6 aldehydes, respectively. In tomatoes, lipoxygenase activity leads to the formation of the 9-hydroperoxides mainly but only the 13-hydroperoxides are degraded, giving rise to the C-6 aldehydes alone. The fate of the 9-hydroperoxides is unknown. *Trans*-2-hexenal can be formed from *cis*-3-hexenal as a result of isomerase activity while *cis*-3-hexenol arises by virtue of alcohol oxidoreductase activity. All the C-6 aldehydes and alcohols contribute a 'green' note to the flavour of fresh tomatoes. During the production of tomato paste, *cis*-3-hexenal and hexanal concentrations are greatly diminished (Buttery *et al.*, 1990), thereby causing a loss of fresh tomato aroma.

The reactions which lead to the formation of the other volatile compounds of putative importance to tomato flavour are not understood although Stevens (1970) has suggested that *β*-ionone may arise in a coupled oxidation of *β*-carotene with polyunsaturated fatty acids catalysed by lipoxygenase while 6-methyl-5-hepten-2-one may be formed in a similar way from lycopene. Buttery *et al.* (1988) have offered some preliminary evidence consistent with this view, although a high lycopene strain of tomatoes did not show a significantly higher concentration of 6-methyl-5-hepten-2-one.

Many factors, such as cultivar, climate, conditions of cultivation, natural or artificial ripening, stage of ripeness and storage conditions are known to have an effect on fresh tomato flavour but relatively little is known concerning the effect on the flavour volatiles. A problem, to date, has been that some of the volatiles are degraded by enzymes before or during their isolation, making quantitative analysis impossible. Buttery *et al.* (1987) have found that calcium chloride can be used to deactivate such enzyme systems and have investigated the effect of chill storage of fresh, ripe tomatoes on the volatile composition. Their results have indicated that storage at 2°C lowered the concentration of *cis*-3-hexenal, which is generally considered to be the main contributor to the 'green' flavour note. The concentrations of other volatiles were also lowered, suggesting that a reduction in flavour volatiles may be a major cause of the lack of flavour in tomatoes stored under chill conditions prior to sale. The reason for the lower levels of volatiles in the chilled tomatoes was not investigated in this study although it was proposed to be the result of normal tomato metabolism.

Heat processing, as in canning and in juice purée or ketchup preparation, will reduce the fresh tomato flavour not only as a result of evaporation of the most volatile components (hexanal, hexenal and hexenols) but also because of the thermal inactivation of the enzymes involved in their generation (Sieso & Crouzet, 1977). Non-enzymatic degradation of polyunsaturated fatty acids can yield some of these volatiles but apparently not in sufficient quantities or in the right proportions to simulate the original tomato flavour.

Mushrooms

A major contribution to the flavour of mushrooms comes from the volatile compounds 1-octen-3-ol and 1-octen-3-one (Pyysalo & Suihko, 1976). These are derived from the 10-hydroperoxide isomer of linoleic acid, presumably formed as a result of lipoxygenase activity (Wurzenburger & Grosch, 1984). The formation and degradation of the 10-hydroperoxide differentiates the mushroom from other vegetables in which the 9- and 13-hydroperoxides are invariably formed and degraded. The $R(-)$-enantiomer of 1-octen-3-ol has the typical mushroom aroma whilst $S(+)$1-octen-3-ol has a mouldy, grassy flavour note (Mosandl *et al.*, 1986). Recently, Fischer and Grosch (1987) have presented data showing that whilst 1-octen-3-ol is present in much larger amounts in *Psalliota bispora* mushrooms than 1-octen-3-one, the ketone has a higher aroma value and is thus a more important flavour volatile. The conversion of 1-octen-3-one into 1-octen-3-ol and 3-octanone in homogenates of *Agaricus bisporus* mushrooms has been studied by Chen and Wu (1984) who found that the formation of the octenol was independent of pH in the range 5·0–9·0, whilst the octanone formation was favoured at pH 8·0–9·0. Reductase enzymes were considered to be responsible for the conversion.

Only limited data are available on the effects of processing on the flavour volatiles in mushrooms. Dehydration generally causes massive losses of 1-octen-3-ol in *A. bisporus* mushrooms (Maga, 1981*a*) whilst cooking leads to an increase in both the alcohol and the ketone, especially the latter (Maga, 1981*b*). There may thus exist some potential for the improvement of mushroom flavour by means of suitable heat treatments. However, high concentrations of the alcohol and ketone can cause the aroma to be too intense and it is then objectionable. Blanching apparently causes a loss of desirable flavour in the oyster mushroom (*Pleurotus ostreatus*), whereas omitting the blanch results in a strong off-flavour after 3 months at −30°C (Gormley & O'Riordan, 1976).

Soy-beans

Volatile components. A major problem concerning soy-bean flavour, which limits the number of applications of soy materials, is the undesirable 'green' and 'beany' notes caused largely by aldehydes such as hexanal, *cis*-3-hexenal and *trans*-2-hexenal. These compounds arise as a result of the activity of acyl hydrolase, lipoxygenase, hydroperoxide lyase, hydroperoxide isomerase and alcohol dehydrogenase, the final composition of volatile carbonyls depending on the relative activities of the enzymes. Three main lipoxygenase isoenzymes are involved in hydroperoxide formation. The linoleic and linolenic acid 13-hydroperoxides produced by the activity of the lipoxygenase-2 isoenzyme are specifically cleaved by a hydroperoxide lyase to yield hexanal and *cis*-3-hexenal respectively (Matoba *et al.*, 1985*a,b*; Olías *et al.*, 1990). The lipoxygenase-3 isoenzyme appears to lower the hexanal concentration, perhaps by interfering with the hydroperoxide degradation reactions (Hildebrand *et al.*, 1990). Alcohol dehydrogenase may be responsible for the reduction of some of the aldehydes to alcohols (Matoba *et al.*, 1989; Tihanyl *et al.*, 1989) leading to a less undesirable flavour because of the higher alcohol threshold values.

As part of a comprehensive review on soy flavour, MacLeod and Ames (1988) have evaluated the sensory contribution which various volatile compounds make to aroma. They consider that many of the compounds which have been associated with odour defects in soy products are due to lipid oxidation reactions. As well as the 'green' and 'beany' notes, such defects include 'hydrogenated', 'straw-like', 'paint-like', 'oily/fatty', 'fishy', 'cucumber', 'mushroom' and 'potato'. The influence of enzymes in the development of the various odours is not clear, however, and autoxidation reactions are probably involved in some cases.

The problem of improving soy flavour has been approached from three general directions: inhibitions and/or inactivation of lipoxygenase, removal of undesirable flavour components and flavour masking (MacLeod & Ames, 1988). Inhibition of lipoxygenase activity by the naturally occurring soy-bean antioxidants is limited and this has led to studies on the effect of natural and synthetic antioxidants and inhibitors on the activity of the enzyme. It may be possible to inhibit lipoxygenase by soaking dry beans in solutions containing antioxidants. At present, however, most of the methods developed commercially for controlling lipoxygenase activity depend on heat inactivation of the enzyme. In general, the thermal procedures use whole soy-beans with dry or steam heating as the sole method of lipoxygenase inactivation. Enzyme activity is greatly enhanced if beans are damaged or crushed. The heat treatment should be as mild as possible in order to minimise insolubilisation of the soy proteins and this can be achieved by combinations of heating with aqueous alcohol or under controlled pH conditions. Breeding lipoxygenase-deficient soy-bean mutants has been shown to reduce off-flavour levels while maintaining protein, amino acid, carbohydrate and lipid contents similar to those of normal soy-beans. The development of cultivars with a sub-normal content of polyunsaturated fatty acids has also shown promise in this area.

The removal of undesirable flavour components can be performed using heat, solvents, pH adjustment, enzymes or other physical or chemical methods. Enzymatic methods, using proteolytic enzymes to break the bonds between the soy proteins and the volatile components, have had some success but may cause problems of bitterness, due to the formation of bitter peptides, and loss of protein functionality. More effective methods involve the use of aldehyde dehydrogenase to oxidise the aldehydes to the corresponding acids which have higher flavour threshold values.

The off-flavour in soy-beans can be masked by the use of suitably flavoured additives or by using precursors which will give an appropriate flavour on heat processing. However, a major difficulty with any attempt to mask the off-flavour is that it remains in the soy-bean product and can eventually penetrate the masking flavours.

Nonvolatile components of soy-beans. A bitter taste has been found in whole soy-beans which increases with bean maturation and with lipoxygenase activity. Various non-volatile oxygenated derivatives of fatty acids have been implicated in its development, particularly certain trihydroxy unsaturated acids. Baur *et al.* (1977) showed that the main components responsible for the formation of bitterness from linoleic acid were 9,12,13-trihydroxy-octadec-10-enoic acid and 9,10,13-trihydroxyoctadec-11-enoic acid. Moll *et al.* (1979) presented evidence for the additional presence in stored soy flour of a trihydroxyoctadecadienoic acid obtained by oxidation of linolenic acid. Oxidation of soy lecithin also gives rise to bitterness. This has been attributed to a combination of chemical composition of the oxygenated fatty acid residues and the surface-active properties of the oxidised lecithin molecules under aqueous conditions (Sessa *et al.*, 1977). The trihydroxy acids, either free or esterified in triglycerides or phospholipids, are evidently formed from the 9- and 13-hydroperoxides arising from lipoxygenase activity. The subsequent action of a peroxidase or a hydroperoxide isomerase leads to the formation of epoxy-hydroxy unsaturated fatty acids which then hydrolyse to the corresponding trihydroxy acids.

Fatty acid dimers formed from the hydroperoxides have also been implicated in the development of bitterness in soy-bean oil (Evans *et al.*, 1960).

Green peas

Immature green peas (*Pisum sativum*) develop characteristic off-flavours if held for long periods before blanching and freezing. This occurs as a result of the bruising which takes place during the mechanical separation of the peas from the pods and vine. Off-flavour development also occurs on freezing unblanched peas, probably due to further damage to cells caused by ice crystallisation and recrystallisation. The latter off-flavour is described as 'hay-like' and may be perceived as a different off-flavour to that caused by the vining operation. Many lipid-derived volatile components have been detected in frozen, unblanched peas, including alkanols, alkanals, alkenals, alkadienals and alkadienones but no single compound has been found to be responsible for the flavour deterioration (Pattee *et al.*, 1982). It is probable that the volatile compounds are formed by the actions of phospholipid acyl hydrolase and lipoxygenase (Henderson *et al.*, 1984). Hydroperoxide lyase, hydroperoxide isomerase and alcohol dehydrogenase may also be involved.

An association between off-flavour in peas and lipoxygenase activity has been demonstrated by adding back the purified enzyme to blanched material (Lee & Wagenknecht, 1958; Williams *et al.*, 1986). A study on the distribution of lipoxygenase in peas has shown that the cotyledons contain significantly higher specific activities than the skins whilst the linoleic and linolenic acid substrates are evenly distributed (Eriksson, 1967). Off-flavour formation in peas can be prevented by blanching sufficiently to inactivate lipoxygenase completely. The enzyme has a similar stability to catalase during pea blanching whilst peroxidase is more stable and reactivates during storage at −10°C (Rhee & Watts, 1966).

Green beans

Significant contributions to the acceptable flavour of green beans (*Phaseolus vulgaris*) are made by the unsaturated volatile compounds, 1-octen-3-ol, *cis*-3-hexenol and *trans*-2-hexenal, and perhaps also by the saturated compounds, hexanol and hexanal (de Lumen *et al.*, 1978). On freezing raw green beans, bitter and rancid off-flavours develop during storage although the volatile and non-volatile components which cause this to occur have not yet been identified. Blanching in near-boiling water for a sufficient time to inactivate the lipoxygenase completely has been shown to prevent off-flavour formation (Adams, 1983). Appreciable peroxidase activities may remain after blanching without flavour deterioration (Böttcher, 1975; Baardseth, 1978; Adams, 1983). Green bean lipoxygenase has been partly characterised (Adams & Ongley, 1989) and hydroperoxide isomerase activity has been detected in the bean seeds (Kermasha & Metche, 1987). Alcohol dehydrogenase is also known to be active (de Lumen *et al.*, 1978). Further work is clearly needed to identify the off-flavour compounds in frozen green beans and to relate their formation to the activity of enzymes known to catalyse lipid degradation.

Potatoes

Potatoes held under nitrogen can develop a cucumber-like odour (Samotus & Schwimmer, 1963). This is probably caused by the formation of the volatile *trans*-2, *cis*-6-nonadienal from linolenic acid which is present as a significant proportion of the total potato lipid (Galliard, 1973). 2,6-Nonadienal has been shown to be the most abundant carbonyl formed through purified potato lipoxygenase-mediated oxidation of linolenic acid (Grosch *et al.*, 1976). Knowledge of the complete range of enzymes involved in the production of the cucumber off-flavour in potatoes is not available at present. However, a lipolytic acyl hydrolase has been purified and partly characterised (Galliard, 1971) which may be the cause of the enhanced fatty acid formation during the production of potato granules (Lilja & Lingnert, 1989). A lipoxygenase which specifically yields the 9-hydroperoxide from linolenic acid has also been identified (Galliard & Phillips, 1971; Berkeley & Galliard, 1976). The presence of hydroperoxide lyase or isomerase does not appear to have been studied.

It has been suggested that *trans*-2, *cis*-6-nonadienal is formed enzymatically during the normal cooking of potatoes in boiling water. The cucumber off-flavour is not detected because the nonadienal undergoes double bond hydration and retro-aldol condensation to *cis*-4-heptenal which contributes to the boiled potato flavour (Josephson & Lindsay, 1987).

Carrots

The flavour of raw carrots is generally quite stable, possibly due to the low level of lipoxygenase activity. However, unblanched carrots can develop a soapy off-flavour during frozen storage; this may be caused by the release of free fatty acids from lipids. Baardseth and Slinde (1983) have proposed that palmitic acid may make a significant contribution as a result of the activity of palmitoyl-CoA hydrolase. In addition, carrot root is known to contain active lipolytic acyl hydrolase enzymes which can liberate free fatty acids on tissue disruption (Dalgarno & Birt, 1963) and these enzymes may also contribute to the off-flavour in frozen, unblanched material. A mild heat treatment is usually sufficient to stabilise carrot flavour during frozen storage. In the case of small, whole carrots it has been found that steam-peeling alone can be sufficient to stabilise flavour, as well as colour and texture (Adams, 1982). Residual peroxidase activity had no perceptible effect on carrot quality.

Glucosinolate degradation

Of the organosulphur compounds which occur in non-allium vegetables, the glucosinolates have been the most extensively studied. They have the structure shown in Fig. 1 (I) which can be degraded to various volatile and non-volatile compounds with great flavour potential. Degradation is initiated by the hydrolytic removal of glucose, catalysed by the enzyme thioglucosidase (common name, myrosinase). The aglucone formed (Fig. 1, II) is unstable and undergoes a Lossen-type rearrangement in which the R-group transfers from the carbon to the nitrogen atom of the same molecule. Further reactions then occur which depend on the chemical nature of the side chain and factors which modify the action of the thioglucosidase. Glucosinolate degradation in food and food plants has been comprehensively reviewed by Fenwick *et al.* (1983*a*).

Brassica vegetables contain significant, though variable, quantities of glucosinolates (see Table 1) which can break down on mixing with the enzyme during cell disruption. The resulting degradation products make a major contribution to the raw vegetable flavour which is most clearly understood in the case of Brussels sprouts, although there are many gaps in our knowledge even here. In Brussels sprouts there are four main glucosinolates (Heaney & Fenwick, 1980*a*). These are sinigrin (R = 2-propenyl), gluconapin (R = 3-butenyl), progoitrin (R = 2-hydroxy-3-butenyl) and glucobrassicin (R = 3-indolylmethyl). The main products found as a result of enzyme-catalysed degradation are

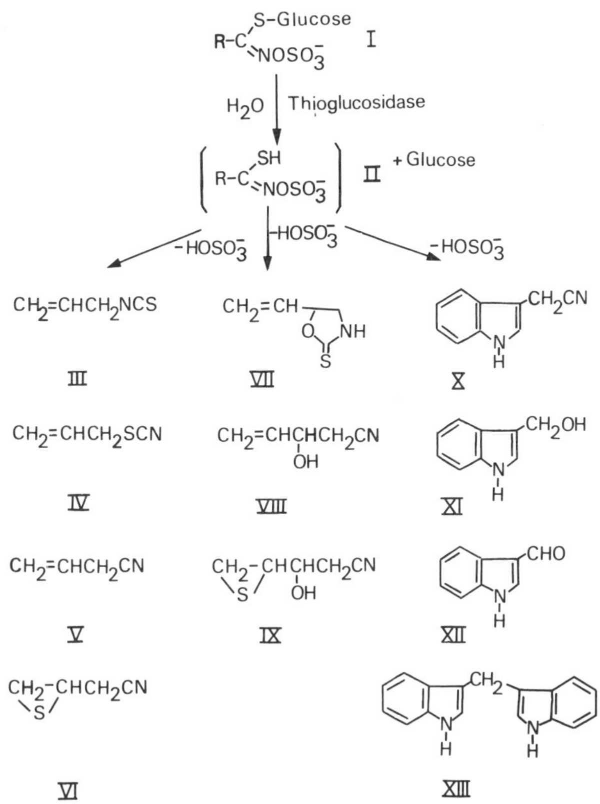

FIG. 1. Degradation of glucosinolates in Brussels sprouts initiated by thioglucosidase.

shown in Fig. 1 except for gluconapin which gives similar products to sinigrin. The significance of each of these glucosinolates, and their degradation products, for the flavour of Brussels sprouts is now considered.

Sinigrin

Sinigrin is a bitter glucosinolate and makes an important contribution to the bitterness of Brussels sprouts (Fenwick *et al.*, 1983*b*). On thioglucosidase-catalysed hydrolysis, the volatile 2-propenyl isothiocyanate (Fig. 1, III) is generally the major degradation product. It imparts a pungent, mustard-like flavour when raw sprouts are consumed. In practice, the relative flavour contributions from sinigrin and the isothiocyanate will depend on the thioglucosidase activity. Ascorbic acid, at low levels, is known to activate mustard thioglucosidase and to inhibit it at high levels and similar behaviour has been demonstrated for the Brussels sprouts enzyme (Springett & Adams, 1989). At the ascorbic acid concentration in sprouts, the thioglucosidase appears to be optimally active although the observed rate of isothiocyanate formation will clearly depend on the sinigrin and active enzyme concentrations. Whilst the

TABLE 1[a]
Contents of the Major Glucosinolates for Fresh *Brassica* Vegetables

Brassica	Glucosinolate (μg/g Fresh Weight)							
	1[b]	*2*[c]	*3*[d]	*4*[e]	*5*[f]	*6*[g]	*7*[h]	*Total*
White	35–590	—	—	46–270	0–144	—	51–510	
cabbage	74–414	0–28		23–890		8–83	45–631	420–1 560
Savoy	0–160	—	—	70–420	1–40	—	300–526	
cabbage	125–646	16–53	—	334–1 290	—	24–126	340–971	1 210–2 960
Red cabbage	6–102	19–64	—	22–143	150–390	—	155–330	
Brussels sprouts	11–1 560	30–500	—	—	—	40–990	220–1 110	600–3 900
Cauli-flower	10–627	0–28	0–295	6–419	—	0–101	73–789	610–1 140

[a] From Heaney & Fenwick (1987); with permission.
[b] 2-Propenyl glucosinolate.
[c] 3-Butenyl glucosinolate.
[d] 3-Methylthiopropyl glucosinolate.
[e] 3-Methylsulphinylpropyl glucosinolate.
[f] 4-Methylsulphinylbutyl glucosinolate.
[g] 2-Hydroxy-3-butenyl glucosinolate.
[h] 3-Indolylmethyl glucosinolate.

relative glucosinolate concentrations in Brussels sprouts have been shown to be cultivar dependent (Heaney & Fenwick, 1980*b*), the absolute levels of individual glucosinolates can vary considerably, depending on factors such as soil type, fertiliser, moisture levels and plant density.

In marked contrast to the isothiocyanate, the 2-propenyl thiocyanate (Fig. 1, IV) is not pungent and possesses a garlic-like odour. It has yet to be detected in Brussels sprouts, either as a sensory note or as a volatile component identified by gas chromatography–mass spectrometry techniques.

The 2-propenyl nitrile (Fig. 1, V) has been detected in Brussels sprouts, although it may be formed by a non-enzymatic or thermal process from sinigrin (MacLeod & MacLeod, 1970). Its significance for the flavour of sprouts is unknown.

Evidence for the presence of 1-cyano-2,3-epithio-propane (Fig. 1, VI) in Brussels sprouts has been presented but the flavour contribution made by this compound, *per se,* is not known at present (Springett & Adams, 1988). Its formation is under the control of an epithiospecifier protein which has no glucosinolate-degrading activity in itself but which, in the presence of thioglucosidase and ferrous ion, can cause the intramolecular transfer of sulphur to the double bond of the propenyl side chain. A possible mechanism for this reaction is that the nitrogen of the unstable aglucone (Fig. 1, II) forms a co-ordination bond with ferrous ion in the epithiospecifier protein, thus preventing the Lossen rearrangement.

Progoitrin

Progoitrin is not in itself a bitter glucosinolate (Fenwick *et al.*, 1983*b*) but after thioglucosidase-catalysed hydrolysis, Lossen rearrangement and cyclisation leads to the formation of goitrin (5-vinyl oxazolidine-2-thione; Fig. 1, VII) which is intensely bitter (Fenwick & Griffiths, 1981). A problem of particular concern to processors of frozen Brussels sprouts in the UK has been the development of excessive bitterness as a result of goitrin formation. Both optical isomers of goitrin have been found to be of similar bitterness by taste panels, although some tasters are 'bitter-blind' to the laevo-rotatory isomer, a genetically linked characteristic akin to that shown by phenylthiocarbamide (Shepard, 1961).

Evaluation of the role played by the nitrile (Fig. 1, VIII) and the epithionitrile (Fig. 1, IX) in Brussels sprouts flavour has yet to be made.

Glucobrassicin

Little is known about the enzyme-catalysed degradation of glucobrassicin in Brussels sprouts or about the contribution made by the glucosinolate and its breakdown products to sprouts' flavour. The isothiocyanate is unstable (Hanley & Parsley, 1990), yielding indole-3-acetonitrile (Fig. 1, X) and sulphur under acidic conditions and indole-3-carbinol (Fig. 1, XI) and thiocyanate ion at neutral pH values. The indole-3-carbinol appears to be unstable under autolytic conditions, giving rise to indole-3-aldehyde (Fig. 1, XII), and 3,3-diindolylmethane (Fig. 1, XIII) (Bradfield & Bjeldanes, 1987*a*). In the presence of ascorbic acid, it yields an adduct, ascorbigen. The glucobrassicin degradation products are only slightly volatile but may have distinct flavour characteristics. Indole itself has an intense faecal odour but in highly dilute solutions the odour is pleasant, leading to the use of indole in perfumery.

Frozen Brussels sprouts are generally considered to lack flavour, partly as a result of the inactivation of the thioglucosidase during the blanch process. Omitting the blanch entirely is not a feasible alternative, however, as a characteristic 'paint' off-flavour develops during the first few days of frozen storage. A similar off-flavour is also known to occur in other *Brassica* vegetables, particularly cauliflower florets, which have been frozen without blanching. The cause of this off-flavour is not known although contributions from lipid oxidation products, and possibly from the degradation products of sulphur-containing amino acids, are expected. By blanching Brussels sprouts to leave some residual thioglucosidase, a more acceptable flavour may be achieved if the enzyme is able to catalyse the breakdown of glucosinolates during the remainder of the freezing process. However, the thioglucosidase activity appears to be highest in the outer regions of the sprout (Springett & Adams, 1989), implying that a much reduced heat treatment could be required to leave sufficient active enzyme. This would risk leaving other residual enzyme activities which could have a detrimental effect on product quality. A greater knowledge is required of the factors which influence the desirable and undesirable flavours in Brussels sprouts and other *Brassica* vegetables, so that optimal processes may be suggested.

Degradation of sulphur-containing amino acids

Pyridoxal phosphate-dependent carbon–sulphur (C–S) lyases are found in *Brassica* and *Allium* vegetables where they catalyse the initial breakdown of *S*-alkyl-L-cysteine sulphoxides which are present as major non-protein amino acids (Fig. 2). *Brassica* vegetables also contain a C–S lyase which can cleave cystine and this enzyme has been isolated and purified from broccoli (*Brassica oleracea* var *botrytis*) by Hamamoto and Mazelis (1986). The degradation of *S*-alkyl-cysteine sulphoxide and cystine catalysed by the brassica C–S lyase is shown in Fig. 2. The alkyl thiosulphinates or their unstable sulphenic acid precursors can apparently react further, especially on heating, to give rise to alkyl sulphides which have potent flavour properties (Shankaranarayana *et al.*, 1982). The further reactions of the cysteine persulphide do not appear to have been studied, at present, and in general there is a lack of information regarding the factors that influence the secondary breakdown of the sulphur-containing amino acids in relation to brassica flavour.

In the Shiitake mushroom (*Lentinus edodes*), a C–S lyase is responsible for the cleavage of desglutamyl-lentinic acid (Iwami *et al.*, 1975; Iwami & Yasumoto, 1980). A range of sulphur compounds are subsequently formed, of which three cyclic compounds (lenthionine, 1,2,4-trithiolane and 1,2,4,5-tetrathiane) appear to contribute most to the characteristic mushroom aroma (Chen & Ho, 1986*a*,*b*).

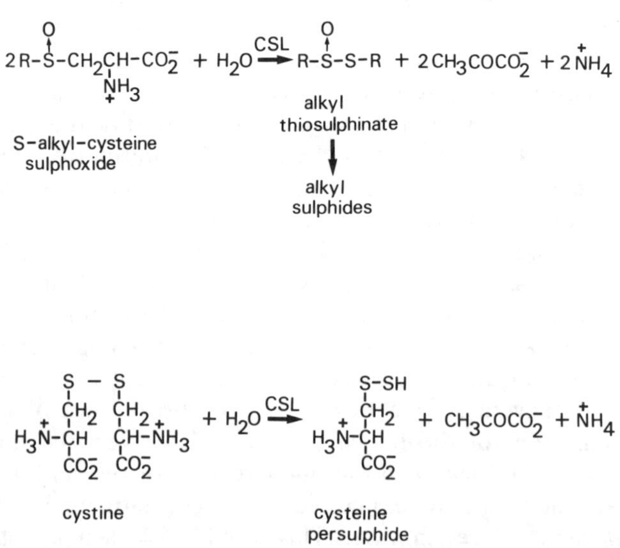

FIG. 2. Degradation of sulphur-containing amino acids in *Brassica* vegetables initiated by carbon–sulphur lyase (CSL).

Starch–sugar interconversion

During the maturation of actively metabolising vegetables, a net conversion of sucrose to starch occurs. The extent of this conversion has an important influence on the acceptability of some vegetables, particularly green peas (Lee, 1989) and sweet corn (Wiley *et al.*, 1989), when consumed in an immature state. The sucrose level has a direct effect on the sweet flavour, whilst the starch content affects textural attributes such as mealiness. Mature pea seeds with a wrinkled appearance have a higher sucrose content and are sweeter than round peas. The elevated sucrose level probably arises because one isoform of starch–branching enzyme is not produced, leading to higher water content and greater cell size in the developing peas. On maturation, the seeds lose a larger proportion of their volume than occurs in the round peas and the wrinkled appearance is due to the testa not shrinking with the cotyledons (Smith, 1988).

Conversion of starch to sucrose and reducing sugars during low-temperature storage of certain vegetables can also have a serious effect on quality. The most intensively studied example of this is the low temperature sweetening of potatoes which occurs on storage at less than 10°C (Fig. 3). The sweetening makes the tubers unpalatable and the high levels of reducing sugars can lead to excessive browning by non-enzymatic reaction with amino acids during processing of the potatoes into chips or crisps. Considerable variation in sweetening occurs in the response of different potato cultivars to low temperatures and, even within a single cultivar, the degree of sweetening can vary with the conditions of growth. Translucent zones may also develop in centrally located pith-parenchyma cells during storage which are characterised by reduced starch and increased sucrose and glucose levels (Sowokinos *et al.*, 1985). Besides low temperature, handling and other stresses can lead to elevated levels of free sugar in stored potatoes (Sowokinos *et al.*, 1987).

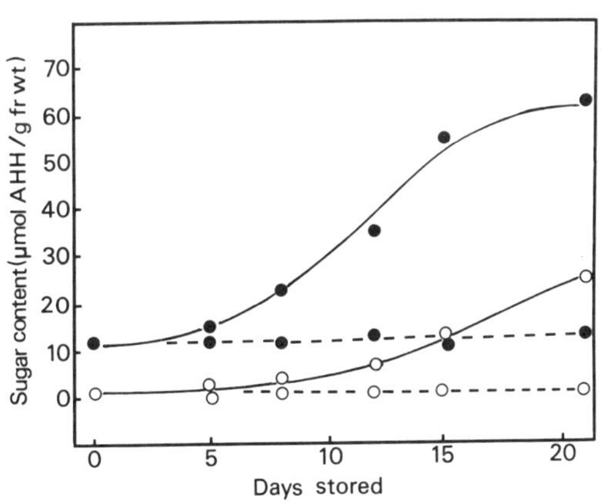

FIG. 3. Sugar content of potato tubers (cv Record) during storage (anhydrohexose (AHH)/g fresh weight), (—) 2°C, (– – –) 10°C, (●) sucrose, (○) reducing sugar. (From Pollack & ap Rees (1975), reproduced by permission of Pergamon Press plc.)

The overall pathway of starch breakdown during the sweetening of potatoes is understood in general outline but there is considerable doubt regarding the factors which regulate it. Morrell and ap Rees (1986) have proposed that starch breakdown is phosphorolytic rather than hydrolytic. The conversion of the hexose phosphates to sucrose phosphate is then catalysed by the enzyme sucrose phosphate synthase. Subsequently, sucrose phosphate phosphatase catalyses the hydrolysis of sucrose phosphate to sucrose and inorganic phosphate. The normal breakdown of the hexose phosphates by glycolysis appears to be restricted at low temperatures, possibly due to the cold-lability of key regulatory enzymes of glycolysis, such as phosphofructokinase (ap Rees *et al.*, 1981; Hammond *et al.*, 1990) or fructose-6-phosphate phosphotransferase (Es & Hartmans, 1987). Hydrolysis of sucrose to glucose and fructose during low-temperature sweetening may possibly be mediated by invertase. During potato development, sucrose, translocated from the leaves, is partitioned on arrival in the tubers and may be converted to starch because the invertase is apparently inhibited. Exposure to low temperatures evidently leads to further amounts of invertase being synthesised and the simultaneous disappearance of the inhibitor, so that sucrose is converted to the reducing sugars (Pressey & Shaw, 1966; Pressey, 1972). Sucrose synthase may also make a significant contribution to sucrose breakdown (Morrell & ap Rees, 1986).

Other factors which may be involved in starch breakdown include the resistance of the amyloplast membrane to enzymatic degradation, and the starch granule composition as determined by the relative amounts of amylose and amylopectin (Barichello *et al.*, 1990.)

Potatoes are stored in the cold to prevent sprouting, weight loss and general deterioration before being processed. In order to reduce the sweetening and non-enzymatic browning effects, it is normal to condition cold-stored material for several weeks at room temperature, depending on cultivar and cultural history. This causes synthesis of starch from sucrose and removal of the reducing sugars which control the extent and intensity of browning. Considerable research has been, and continues to be, carried out to develop potato cultivars which can be processed directly from cold storage, without conditioning. As result of this effort, selections have now been found which produce acceptably coloured chips after storage at 4°C (Coffin *et al.*, 1987; Brown *et al.*, 1990; Ehlenfeldt *et al.*, 1990).

Starch-to-sugar conversion in cold-stored potatoes can be slowed down by employing controlled atmospheres containing less than 3% O_2, although such treatments may increase sprouting (Burton, 1982). Storage of potatoes in 2·5% O_2 at low temperature has been shown to lower the amount of reducing sugars formed and to improve chip colour of some cultivars relative to storage in air (Sherman & Ewing, 1983; Schwobe & Parkin, 1990).

Inhibition of potato sprouting by alternative means to cold-storage has been employed. Chemical inhibitors appear to be satisfactory for long-term storage of potatoes at 7–10°C so long as good storage practices are employed (Thomas,

1984). Unfortunately, most sprouting inhibitors suppress the development of wound periderm in potato tubers and may lead to an increased incidence of rot in storage. Moreover, the use of chemical inhibitors give rise to residues in or on the potatoes and there is increasing concern about the safety of such residues.

Ionising radiation of the correct dose is sufficient to inhibit sprouting of potatoes, although the development of wound periderm is suppressed in this case also and sugar content may increase during storage depending on the potato cultivar, the time of irradiation and the storage conditions. The rise in sugar content may only be temporary, however, with a return to normal levels during further storage (Thomas, 1984).

Blanching potato chips in hot water can be used to lower the reducing sugar content by leaching. This limits the extent of non-enzymatic browning at the chip surface where conditions are particularly favourable for its occurrence.

Nucleotide formation

Nucleotide flavour enhancers have been found in vegetables such as potatoes, peas, asparagus, tomatoes and mushrooms. 5′-Guanosine monophosphate may be of particular significance, at least partly by virtue of its effect in lowering the odour threshold of certain volatile components (Schinneller *et al.*, 1972). Ribonucleases which specifically catalyse the hydrolysis of the 3′-ribose-phosphate bonds in ribonucleic acid, are probably the major enzymes leading to 5′-nucleotide formation. The action of these enzymes appears to be potentiated during the early stages of cooking of many vegetables. More recently, work with potatoes has suggested that processing at 80°C enhances the enzymatic liberation of 5′-guanosine monophosphate (Nguyen *et al.*, 1987*a,b*).

TEXTURE CHANGES

Texture is a most important quality determinant in vegetables and its optimisation can only be achieved through application of a sound knowledge of enzyme-induced changes. The textural characteristics which consumers appear to value most are crispness and firmness. Vegetables often become objectionable by becoming limp, mealy, rubbery or excessively soft (or hard), even when their colour and flavour are not seriously affected. The factors which contribute most to texture are turgor (the force exerted on the cell membrane by intracellular fluid), cell wall strength and intercellular adhesion. Turgor is usually lost on processing as a result of damage to the cell membranes caused by heat (as in blanching or sterilisation processes) or by ice-crystal formation if the vegetable is frozen without a prior heating step.

An example of the latter situation is the freezing of tomatoes where the cell walls are not sufficiently rigid to withstand the stress caused by the membrane rupture and the tissue collapses. On the other hand, commodities with well-developed, fibrous cell walls, such as green beans and carrots, withstand heating or freezing processes quite well. The membranes lyse, yet major textural changes do not occur.

The factors which affect the strength of the cell wall and the adhesion of cells to one another are only partly understood and research is continuing in this area. However, it is generally considered that pectins play a highly significant role and in certain circumstances lignins may be important.

Pectin degradation

The pectins are polysaccharides composed mainly of methyl esters of galacturonic acid, α-(1→4) linked, with occasional in-chain interruptions by rhamnose, and with side chains containing arabinose, galactose and other sugars. They exist in the cell wall and between cells in all plant tissues, acting as an important part of a matrix which also contains hemicelluloses, proteins, lignins, lower molecular-weight solutes and water. Cellulose fibrils are embedded in this matrix in the region of the cell wall. The strength of the intercellular bonds depends greatly on the activity of three groups of pectin-degrading enzymes. These are: pectinesterase, pectate hydrolase (other name, polygalacturonase) and the lyase enzymes. Pectinesterase catalyses the partial or complete hydrolysis of pectin to pectate and methanol. Divalent metal ions, particularly calcium ions, can then cross-link blocks of carboxylate groups in the pectate molecules, thereby enhancing the resistance of the tissue to cell separation. This cross-linking effect of calcium has been generally accepted although it may not represent the true picture under all circumstances (McFeeters & Fleming, 1989). Exopectate hydrolase catalyses the hydrolysis of pectates and pectins to galacturonic acid, whilst the endo-enzyme catalyses random hydrolysis within the pectin chain. The latter enzyme, in particular, causes a reduction in strength of the pectin glue and consequent softening of the tissue. The lyases also cause cleavage of the pectin chain and lead to tissue softening. However, they are not strictly relevant to the present discussion, being largely confined to plant pathogenic organisms whose presence results in deterioration of all aspects of food quality.

Tomatoes

During the ripening of tomatoes, softening of the tissue occurs which is associated with an increase in solubility of the pectic substances. Pectate hydrolase probably plays an important role in this solubilisation as its activity increases markedly during ripening (Davies & Hobson, 1981) and significant increases in galacturonic acid also accompany the ripening process (Gross, 1986). Further evidence for the involvement of the enzyme has been presented by Crookes and Grierson (1983) who found that the application of purified tomato pectate hydrolase iso-enzymes to mature green tomato tissue duplicated the ultrastructural changes which occur on normal ripening. Two iso-enzyme forms have been identified, one of which has a relatively high molecular weight as a result of interaction between one of the iso-enzymes and a 'converter' protein. The relative contributions made by the two forms to pectin solubilisation and the role of the 'converter' protein are matters of current investigation. However, evidence has been presented that the low molecular weight form is the only endogenous pectate hydrolase in tomatoes (Pressey, 1988). This is in agreement with the finding of a single pectate hydrolase gene reported by Bird *et al.* (1988). The 'converter' protein may act by anchoring the enzyme to the cell wall (Knegt *et al.*, 1988). The precise mechanism of solubilisation of the pectic substances remains obscure, particularly regarding the relationship of the pectate hydrolase activity to the pectinesterase activity and to Ca^{2+} binding. Pectinesterase de-esterifies the pectin, thereby giving a preferred substrate for pectate hydrolase. However, Ca^{2} binding to de-esterified pectin reduces the hydrolase activity (Buescher & Hobson, 1982). Thus, the observed pectate hydrolase activity may depend on the extent to which certain, perhaps critical, regions of the pectin chain have been de-esterified and have subsequently cross-linked with Ca^{2+}. The fact that Ca^{2+} also appears to be a requirement for pectate hydrolase activity (Pressey, 1987) suggests that Ca^{2+} can regulate the amount of pectate hydrolysis which occurs.

A considerable amount of research has been devoted to producing tomato strains which remain firm during ripening but which develop the good colour and flavour characteristics of normal tomatoes. Three mutants, 'never ripe', 'ripening inhibitor' and 'non-ripening' have been known for some time which ripen extremely slowly and consequently have very long shelf lives. The gene for pectate hydrolase is expressed only at very low levels in these mutants and does not respond to exogenous ethylene (Della Penna *et al.*, 1987; Knapp *et al.* 1988). However, there is considerable commercial potential in the use of F_1 material resulting from hybridising normal and non-ripening lines, especially when the normal line has high compositional quality and good flavour and colour (Richardson & Hobson, 1987). The main disadvantage of this approach is the lack of pigmentation of the ripe tomatoes although the introduction of genes which cause increased carotenoid synthesis may be an answer (Davies & Hobson, 1981). An alternative, and possibly more promising, approach has been reported whereby the synthesis of the pectate hydrolase during tomato growth was down regulated by antisense RNA (Sheehy *et al.*, 1988; Smith *et al.*, 1988; 1990). Tomatoes containing the antisense RNA developed similar amounts of lycopene pigment to untransformed control material but substantially reduced levels of pectate hydrolase were synthesised at all stages of ripening (Fig. 4).

In the preparation of comminuted tomato products such as juices, soups,

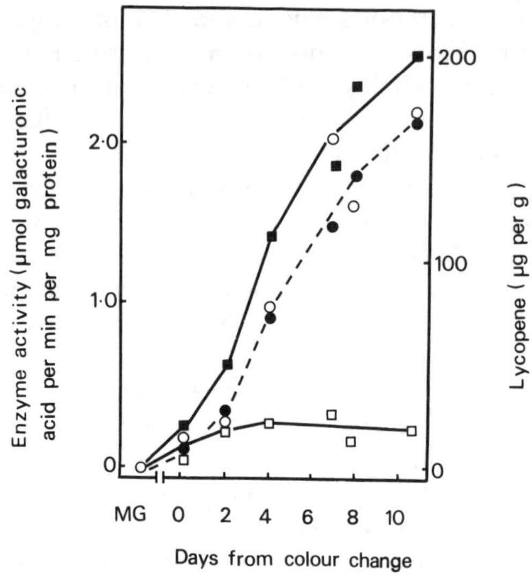

FIG. 4. Pectate hydrolase and lycopene synthesis in genetically transformed tomatoes. MG samples from mature green tomatoes; pectate hydrolase activity: (■) control, (□) transformed; lycopene content: (●) control, (○) transformed. (From C. J. S. Smith *et al.* (1988), reproduced by permission of Macmillan Magazines Ltd.)

ketchups and pastes, two alternative methods are employed during the blending stage. The 'cold-break' procedure involves scalding the tomatoes to loosen the skins prior to their removal and thereafter the temperature never exceeds 60°C. At this temperature, the pectin-degrading enzymes are active, leading to large amounts of cellulose fibrils and cell walls being liberated into the serum phase. When separated from the cell walls, these fibrils become entangled in a manner which can contribute to an increase in the consistency of the final product. Thus, in this case, the contribution of the cellulose may override that of the pectins which are partly broken down by the pectolytic enzymes. The main advantages of the 'cold-break' procedure are the improved retentions of colour and ascorbic acid. In the 'hot-break' procedure, which has been more widely adopted, tomatoes are heated to temperatures in excess of 80°C as quickly as possible. This inactivates the pectolytic enzymes and thereby stabilises the pectins, which can then have a major influence on the consistency of the tomato product.

Chou and Kokini (1987) compared tomato pastes prepared by the 'cold-break' and 'hot-break' procedures and found that they led to pectins with completely different properties. As a result of pectolytic enzyme activity, the 'cold-break' pectin had a much lower degree of esterification and lower molecular weight than the 'hot-break' pectin and gave a greatly reduced intrinsic viscosity. In addition, the rheological data suggested that the 'hot-break' pectin had a random-coil conformation whilst the 'cold-break' pectin was more rigid. Evidence from Xu *et al.* (1986) has indicated that the higher the break temperature the greater is the apparent viscosity of tomato juices and pastes. Part of this effect was attributed to the higher degree of disruption of cell structure at the high temperatures. At the highest break temperature of 107°C, one cultivar was found, by scanning electron microscopy, to contain no remaining cell structure.

A variant of the 'hot-break' procedure is the 'acid-hot-break' (Schwimmer, 1981). This apparently causes solubilisation of protopectin, the water-insoluble parent substance, from the tomato cell walls which leads to liberation of further quantities of cellulose fibrils. The method does not appear to be widely used compared with the 'hot-break' procedure, however.

Potatoes

Texture is often considered the most important sensory property of cooked potatoes. Cultivars can be classified either as 'mealy' or 'waxy'. On cooking, 'mealy' cultivars give a dry-appearing tissue that crumbles readily and has a granular feel on the tongue. They are preferred by consumers for baking and mashing and by industry for chipping and frying. 'Waxy' cultivars are used for potato salads and creamed potatoes as they have a moist appearance on cooking and disintegrate less. 'Mealiness' can be influenced by a number of factors which affect starch granule size and composition, especially cultivar, specific gravity and storage temperature. It can be viewed as the subjective perception of the flow characteristics of cooked potato tissue and appears to be controlled primarily by the starch or solids content of the tuber rather than the pectic substances of the cell wall and middle lamella.

'Sloughing' or 'breakdown' of cooked potatoes occurs independently of 'mealiness' and is probably due to an excessive hydration of the pectic substances. This increases the thickness of the cell wall and reduces the viscosity of the cell wall matrix to such an extent that spontaneous separation of cells takes place (Warren & Woodman, 1974). When potatoes are heated to 60–70°C, firming of the tissue occurs which can lead to a more desirable texture in canned and frozen products, and which prevents sloughing in dehydrated mashed potato (Hadziyev & Steele, 1979). Bartolome and Hoff (1972) have suggested that firming is due to pectinesterase activity which is at its optimum at these temperatures. Firming occurs by metal bridge formation between the liberated carboxyl groups and Ca^{2+} and Mg^{2+} which have diffused from the cell interior. On the other hand, Moledina *et al.* (1981) have proposed that the loss of semi-permeability of cell membranes on heating leads to the liberation of Ca^{2+} (from starch-phosphate groups) which can then diffuse into the cell wall and middle lamella, and cross-link with the pectin. The latter hypothesis implies that pectinesterase activity is not essential for firming so long as sufficient blocks of free carboxylate groups are present in the correct orientation on the pectin chains for cross-linking to occur. Further work is clearly needed to distinguish between these two hypotheses.

Green beans

Textural problems associated with green beans (*Phaseolus vulgaris*) include stringiness or toughness related to overmaturity, loss of crispness due to overblanching and sloughing of the skins in cooked and frozen products (Schwimmer, 1981; Sistrunk *et al.*, 1989). Stringiness and toughness may be caused by lignification reactions, whilst changes in pectic substances have been implicated in the softening and sloughing phenomena. As with potatoes, heating green beans at 60–70°C causes a firming of the tissue. Blanching in this temperature range has been employed extensively as a means of firming canned green beans, and has been proposed as a method of improving the texture of the frozen product (Steinbuch, 1976). In the latter case, the low-temperature blanch is followed by a short blanch at a higher temperature (90–95°C) to inactivate residual enzymes which may affect quality during frozen storage (two-stage blanching). Unfortunately, low-temperature blanches have been shown to cause colour loss in green beans (Adams *et al.*, 1983). This is not a significant problem with canned green beans as the colour is lost in any case as a result of the non-enzymatic degradation of chlorophyll during the canning process. Such colour loss can be masked by the addition of artificial green pigments. However, with frozen beans the loss of colour may be a serious drawback to the commercial exploitation of the two-stage blanch technique.

As with potatoes, pectinesterase activity has been implicated in the firming of green bean tissue but, again, unequivocal evidence that firming only occurs as a consequence of such activity is lacking. Redistribution of Ca^{2+} within the tissue on membrane breakdown may be equally important. Added Ca^{2+} has been shown to be useful in correcting textural defects in green beans (Buren *et al.*, 1960).

Lignification reactions

Lignin is a heterogeneous, high molecular weight aromatic polymer consisting of randomly-linked *p*-hydroxy-cinnamyl alcohols. It is a product of the oxidation and polymerisation of polyphenolic compounds catalysed by cell wall-bound peroxidase. Lignin is deposited mainly in the secondary cell wall, but may also be present in the primary cell wall and middle lamella, covalently bound to hemicellulose components. Lignification commonly occurs in higher plants, leading to increased structural integrity and reduced water permeation across cell walls. Edible plant tissue typically contains primary cell walls only, with lignin in low concentrations except in the specialised vascular and structural tissue cells. On cutting vegetables, de-novo synthesis of lignin begins. Tissues with lignified cell walls have diminished swelling power and increased compressive strength which results in an undesirably tough and fibrous texture.

Asparagus
Lignification of fibrovascular tissue occurs within hours of harvesting asparagus spears at ambient temperatures (Haard *et al.*, 1974). This leads to a progressive toughening of the spears which limits their storage life. Using a shear-press, Powers and Drake (1980) found that toughening increased whether asparagus was held in the shade or exposed to the sun, although significantly higher values were found in spears held in the sun or cut 'all green'. Hsiao *et al.* (1981) have demonstrated that storage at 4°C greatly reduces the rate of lignification. Holding asparagus spears in solutions of *N*-(phosphonomethyl) glycine (an inhibitor of aromatic amino acid synthesis) has also been shown to reduce lignification and toughening (Saltveit, 1988). The effect increased with storage time (at 2·5°C) and additive concentration but decreased with distance from the cut end. Depending on the time of harvest and length of storage, lg/litre *N*-(phosphonomethyl)glycine increased the usable portion of the spear from 40 to 60%. The use of this substance as a practical means of reducing the rate of toughening of asparagus spears would probably be unacceptable despite its low mammalian toxicity. However, the results of this work may suggest other means of control.

Non-enzymatic reactions also contribute to asparagus toughening as the blanched vegetable, containing neither peroxidase nor polyphenol oxidase, can give significant increases in shear force values on storage (Smith & Stanley, 1989).

Legumes
Hardness is considered a fundamental textural quality of cooked beans and other legumes. Hardness defects have been classified as 'hardshell' when the legume seeds do not absorb sufficient water on soaking and fail to soften during cooking, and 'hard-to-cook' when enough water is absorbed but softening still does not occur (Stanley & Aguilera, 1985). The textural defects have been studied extensively and methods for their control have been reviewed (Aguilera & Stanley, 1985).

In the case of 'hard-to-cook' beans, extended storage at high temperatures and high relative humidity is known to exacerbate the problem. However, the chemical reactions responsible are not completely understood. The most widely accepted explanation relates to the insolubilisation of the pectic substances due to the enzyme, phytase. Phytase activity reduces the chelation potential of naturally-occurring phytate which allows Ca^{2+} to cross-link with the pectic substances in the cell wall and middle lamella. It is proposed that this restricts cell separation on cooking and thereby causes hardening. Alternatively, or perhaps additionally, lignification reactions may have a significant role to play in determining hardness. Using black beans (*Phaseolus vulgaris*), Hincks and Stanley (1987) have presented microstructural and histological evidence for lignin accumulation in the corners of the primary cell walls, the secondary cell wall and the middle lamella in samples which had been stored under high-temperature–high-humidity conditions. The lignin was proposed to act either by physically restraining cell separation or by restricting hydration due to its hydrophobic nature. However, the involvement of peroxidase is not apparent as higher temperature heat treatments than necessary to inactivate the enzyme are required to destroy the hardening mechanism (Rivera *et al.*, 1989).

Other vegetables

Yam. Yam tubers begin to harden a few days after harvesting and two or three weeks later hardening is so pronounced that the tubers are inedible, even after extended cooking. Ultrastructural studies have shown that the parenchyma cell walls rapidly become completely lignified, starting from the corners of the cells around intercellular spaces and proceeding along the walls (Sealy *et al.*, 1985).

Sweet potato. In the interior of sweet potato, a textural disorder known as 'hardcore' can be induced by chilling at 2°C and then holding at higher temperatures. It appears as hard tissue after cooking which is not softened by extending the cooking time. Occurrence is strongly cultivar dependent (Buescher *et al.*, 1976). Phenylpropanoid metabolism is known to be stimulated during cold treatments of sweet potato (Rhodes *et al.*, 1981), suggesting that lignin may form rapidly when the temperature is raised. However, the contribution made by lignin to 'hardcore' is uncertain, at present, and cross-linking of demethylated pectins may be equally significant (Buescher & Balmoori, 1982).

VITAMIN CHANGES

Despite the fact that vegetables can be significant sources of water-soluble and fat-soluble vitamins in the diet, information on enzyme-catalysed changes is sparse. The main reason for this probably lies in the relatively slow development of accurate assays for many of the vitamins and, indeed, the identification of which chemical forms are biologically active and to what extent. This situation should improve considerably in the near future as a result of the application of modern chromatographic methods, such as HPLC, which can resolve the different forms. These methods will allow the study of the chemical changes which occur in vitamins during vegetable handling and storage and would provide the starting point for the detection of enzymes which cause such changes.

Vitamin C

Much information has been amassed on the chemistry of vitamin C (L-ascorbic acid) in relation to foods generally where non-enzymatic oxidation reactions (autoxidations) can evidently explain the majority of losses (Liao & Seib, 1988). Dehydroascorbic acid is the first stable oxidation product and is a key intermediate in both the enzymatic and non-enzymatic degradation. It has a vitamin C activity equal to that of L-ascorbic acid but this can be lost by hydrolysis to 2,3-diketogulonic acid under neutral or alkaline conditions. Wills *et al.* (1984) used HPLC to demonstrate the presence of dehydroascorbic acid in a range of raw vegetables at the point of sale and it would seem feasible that enzyme-catalysed oxidations were involved. Identifying the enzyme may prove to be difficult, however, as at least three enzymes are present in vegetables which could potentially lead to the formation of dehydroascorbic acid. These are ascorbate oxidase, peroxidase and polyphenol oxidase. Ascorbate oxidase utilises aerial oxygen to oxidise the vitamin directly whilst peroxidase requires hydrogen peroxide generated by other enzyme systems. Both enzymes have been associated with cell walls. Polyphenol oxidase activity causes the oxidation of phenols to quinones which can then be reduced back to the phenols by ascorbic acid; dehydroascorbic acid is formed as a result.

Other enzymes which can cause loss of vitamin C activity without forming dehydroascorbic acid are ascorbate-2,3-dioxygenase and thioglucosidase. The former incorporates oxygen into ascorbic acid leading to the production of L-threonate and oxalate. The enzyme does not appear to have been reported in foods, however. Thioglucosidase activity on indole glucosinolate in *Brassica* vegetables gives unstable indole compounds which can react with ascorbic acid to give ascorbigen. This may cause a reduction in vitamin C as ascorbigen has been reported as having only 15–20% of the biological activity of L-ascorbic acid (Matano & Kato, 1967).

Evidence has been presented that both increases and decreases in ascorbic acid occur in potatoes depending on the nature and extent of mechanical injury (Mondy & Leja, 1986; Mondy *et al.*, 1987). Slicing or cutting into strips caused an increase within the first two days of storage, whilst bruising led to rapid loss of the vitamin and was associated with blackspot development. Increases in ascorbic acid were also found in bruised potatoes after the initial loss whilst unbruised control tubers showed a regular decrease. Unfortunately, chromatographic identification of ascorbic acid was not performed in this work which therefore leaves some doubt as to the true nature of the potato constituents causing the observed changes.

Vitamin B_1

In the case of the Shiitake mushroom (*Lentinus edodes*), γ-glutamyltransferase and C–S lyase have been shown to be involved in the formation of a factor which reacts with vitamin B_1 (thiamin) and thereby interferes with the thiochrome assay (Iwami *et al.*, 1982). The factor is formed from lentinic acid, the precursor of lenthionine which was previously mentioned as an important Shiitake aroma compound.

The use of sulphiting agents to reduce enzymatic browning in mushrooms may have a deleterious effect on vitamin B_1 levels due to chemical cleavage by the sulphite anion to biologically inactive compounds (Wedzicha, 1984).

TOXIN CHANGES

Many vegetables contain natural constituents which are potentially toxic to humans in the sense that they can produce an adverse physiological response, especially when eaten raw, or unprocessed, or in excessively large amounts (see Table 2). Heat processes, such as those used in blanching, cooking and canning, are useful in reducing the levels of many toxins as a result of the heat denaturation of enzymes and proteins and the leaching effect of the aqueous environment. In many cases, the enzyme-catalysed reactions involved in toxin formation and degradation have not been identified and characterised. The following discussion has therefore been restricted to the glucosinolates and cyanogenic glucosides where enzyme effects have been extensively studied.

Glucosinolates

The importance of the glucosinolates and their degradation products for the flavour of *Brassica* vegetables has been discussed previously. Some of the degradation products are also known to be toxic. In particular, the nitriles (Fig. 1, VIII and IX) are known to cause damage to the kidneys and livers of rats (Tookey *et al.*, 1980). Also, thiocyanate ion and goitrin (Fig. 1, VII) can cause an anti-thyroid effect in animals and in man. Thiocyanate ion exerts its effect by competing with iodine for thyroidal transport, thereby decreasing iodide uptake and formation of the hormone, thyroxine. The effect is evident only in iodine-deficient diets. In contrast, goitrin is considered to interfere with thyroxine synthesis, its effect not being alleviated by dietary iodide (Fenwick *et al.*, 1983*a*). The isothiocyanates (Fig. 1, III) have also been found to exert an

TABLE 2
Toxic Constituents of Vegetables

Toxin (or Precursor)	*Vegetable*	*Reported Effect*
Coumarins	Many: stress metabolites formed on cutting, bruising or microbial attack	Liver damage
Cyanogenic glycosides	Legumes, cassava, bamboo shoots	Respiratory inhibition
S-Methyl-cysteine sulphoxide	*Brassica* vegetables	Anaemia in ruminant animals
Glycoalkaloids	Potatoes	Gastrointestinal disturbances, neurological disorders
Lathyrogens	Grass pea (Chickling vetch)	Male lower limb paralysis
Lectins	Legumes	Growth inhibition
Proteinase inhibitors	Legumes, potatoes, sweet potatoes	Growth inhibition
Vicine, convicine	Broad bean	Fatigue, jaundice

anti-thyroid effect although this has been disputed recently (Lewerenz *et al.*, 1988).

Under certain circumstances, nitrite can react with goitrin or the indole glucosinolate degradation products to yield nitroso-compounds which exhibit direct-acting mutagenicity (Heaney & Fenwick, 1987). Nitrate is known to be present in leafy vegetables and could be reduced to nitrite *in vivo*. However, studies on the extent of formation and metabolism of the nitroso-compounds are necessary before a reliable estimation of risk can be made.

The hydrolysis products of certain aromatic and indolyl glucosinolates are known to inhibit chemical carcinogenesis in rodents, putatively as a result of an increase in the activity of xenobiotic-metabolising enzymes, such as cytochrome P-448 dependent monooxygenase (Wattenberg, 1979; Wattenberg, 1983). Studies on the contribution of indolylic compounds to the anticarcinogenic effect have indicated that the glucobrassicin-derived indole compounds (Fig. 1, X–XIII) are not directly responsible for the increase in hepatic monooxygenase activity in mice fed on dried vegetable material (Bradfield & Bjeldanes, 1987*b*). 1-methoxyindole-3-aldehyde (from the hydrolysis of the co-occurring neoglucobrassicin) was demonstrated to have a more potent effect, although other unidentified compounds could also be important.

Very few *Brassica* vegetables are eaten in the raw state and heat processsing (cooking, or blanching prior to freezing) probably inactivates most of the endogenous thioglucosidase enzyme. Thus, the greater part of the glucosinolates ingested from this source will be intact, and the toxic or beneficial effects of the hydrolysis products may be of no real significance. Sones *et al.* (1984) have estimated a mean daily intake of approximately 50 mg total glucosinolates from the consumption of raw cruciferous vegetables in the UK. Large regional variations were observed, however, as well as differences among income groups and dependency on the time of the year. Based upon such variations and the large variation in glucosinolates in the vegetables, it was estimated that a minority of the population probably consumes in excess of 300 mg total glucosinolate per day. Little is known about the extent and nature of the in-vivo metabolism of intact glucosinolates in humans and animals so that the importance of dietary intake is difficult to assess.

Cyanogenic glycosides

The cyanogenic glycosides are β-glycosides of 2-hydroxynitriles which release hydrogen cyanide by enzymatic action, when the cell structure of cyanophoric plant tissues is disrupted. In contrast to the glucosinolates, the cyanogenic glycosides are widely distributed in plants. Their physiological role is not clearly understood although it is generally considered that they play a significant part in repelling or inhibiting herbivores, pathogens or competitors with subsequent selection favouring the better-protected species (Conn, 1980). Two groups of enzymes are involved in the production of hydrogen cyanide: a β-glycosidase which catalyses hydrolysis of the glycosides to D-glucose and a

2-hydroxynitrile, and a lyase which cleaves the hydroxynitrile to yield an aldehyde or ketone and hydrogen cyanide. It is generally believed that substrate and enzyme are spatially separated in the intact cell, thereby preventing their interaction until the cell structure is destroyed. The enzymes have not been well characterised, however, and in this regard, the situation is similar to glucosinolate hydrolysis catalysed by thioglucosidase.

Hydrogen cyanide is lethal to man at levels between 0·5 and 3·5 mg/kg body weight as a result of its strong affinity for cytochrome oxidase, a key respiratory enzyme (Montgomery, 1980). Smaller amounts of cyanide can be converted to thiocyanate through the action of thiosulphate thiotransferase, present mainly in the liver. As previously noted, thiocyanate has an anti-thyroid effect when consumed in an iodine-deficient diet, and long-term ingestion of cyanophoric foods, such as cassava, can lead to goitre. The ultimate fate of intact cyanogenic glycosides in man is largely unknown, although, in certain circumstances, the microbial flora of the intestine may play a part in their decomposition.

Cassava

Cassava is a staple food crop in tropical regions with more than 300 million people depending on it as an energy source. It has many advantages, being tolerant of poor soils, high-yielding and relatively free from insect pests and diseases (Balagopalan *et al.*, 1988). However, cassava has two major disadvantages. One is the highly perishable nature of the tubers which begin to deteriorate, both physiologically and microbially, a day or two after harvest. Physiological deterioration leads to vascular discoloration (streaking) which usually commences at the site of tuber damage inflicted during harvesting and handling operations. Microbial deterioration also occurs at damaged sites and in tubers which already show physiological deterioration. The second drawback is the existence of the cyanogenic glucosides, linamarin (2-(β-D-glucopyranosyl-oxy)-isobutyronitrile) and lotaustralin (2-(β-D-glucopyranosyl-oxy)-methylbutyronitrile) in the cassava tissue. Linamarin accounts for 93% of the total cyanogenic glucoside and lotaustralin 7%. Hydrolysis of the glucosides is catalysed by the β-D-glucohydrolase, linamarase, followed by a lyase-catalysed cleavage of the hydroxynitriles which yields hydrogen cyanide (Fig. 5). There appears to be considerable variation in the content of hydrogen cyanide in cassava due to many factors such as growing conditions, soil, moisture, temperature and age of the plant (Okezie & Kosikowski, 1982) although some allowance should be made for the fact that hydrogen cyanide determinations may have been subject to interferences especially with the commonly-used alkaline picrate method (Arihantana & Buckle, 1986). Cassava has often been classified as 'sweet' or 'bitter' according to the amount of cyanogenic glucosides present in it, although a definite correlation has not been found (Balagopalan *et al.*, 1988). 'Bitter' cultivars are generally used as raw material for starch production except in Africa where they are strongly

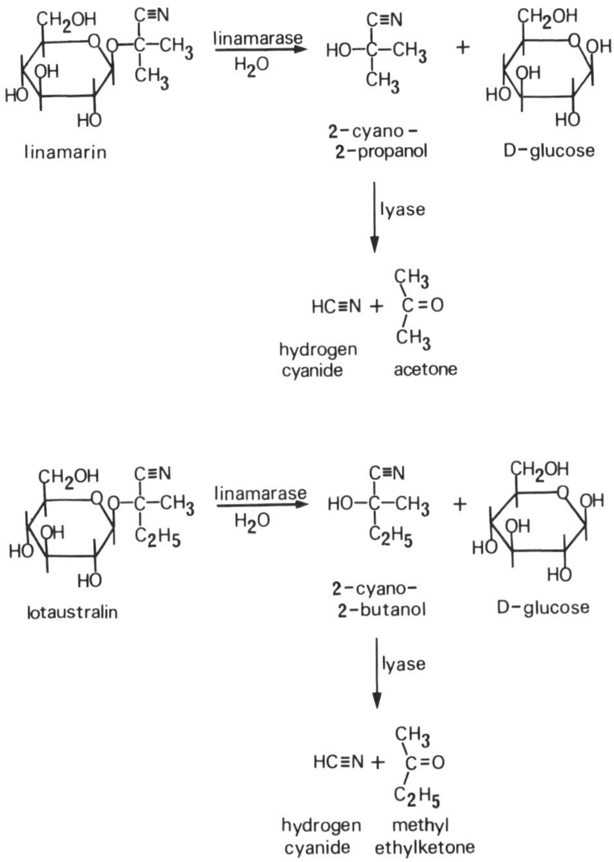

FIG. 5. Enzymatic degradation of linamarin and lotaustralin in cassava.

preferred for human consumption. Goitre and ataxic neuropathy have been reported as consequences of cassava consumption.

Although cassava is eaten as a fresh vegetable in some parts of the tropics, a large proportion of the total crop is processed to reduce the cyanogenic glucoside and hydrogen cyanide content. The cyanides can be reduced to safe levels by the traditional methods of processing and preparing cassava for consumption. These methods include peeling, grating, fermentation, frying, roasting, steaming, boiling or drying. Nambisan and Sundaresan (1985) found that the most effective method for removing the toxic glucosides was to crush the fresh tuber and sun-dry. Less than 5% of the original level was retained, perhaps as a result of linamarase activity in the crushed material. Most of the hydrogen cyanide and ketone produced would be lost on sun-drying. In contrast, baked, fried and steamed tubers retained greater than 80% of the initial glucoside level. Extensive soaking of cassava roots in water (retting) followed by sun-drying has also been shown to be effective in removing cyanide (Ayernor, 1985). Dried cassava products are intrinsically more stable than the fresh roots, the main forms being meal, flour, chips and starch. Cassava chips and starch are mainly industrial products that are little used for direct human consumption, but they account for most of the cassava that enters international trade. Wastes from meal manufacture contain high cyanide concentrations which are unsuitable for direct discharge into rivers. Biological treatment, including prolonged aeration, has been shown to be successful in lowering cyanide levels (Branco, 1979). Current efforts to modernise and increase the output of cassava products through mechanised processing could pose toxic hazards to process workers. This is especially true in a factory environment where hydrogen cyanide levels could build up unless steps are taken to provide adequate ventilation.

An alternative to reducing cyanide levels by processing is to decrease the quantity of cyanogenic glucosides synthesised and stored in the cassava tubers using plant-breeding techniques (Makame *et al.*, 1987) or by genetic engineering as proposed by Bunders *et al.* (1989). A possible disadvantage of this approach is that the new cassava cultivars may be less resistant to attack by insects such as locusts.

Lima beans

Many legumes contain cyanogenic glucosides. In particular, the lima bean (*Phaseolus lunatus*), also known as the butter bean, contains a relatively high level of linamarin. The lima bean, which is widely grown in tropical and subtropical countries, generally contains linamarin equivalent to less than 20 mg hydrogen cyanide/100 g fresh weight and is therefore considered safe for human consumption. However, some dark-coloured cultivars have at least ten times this amount and their consumption has been associated with fatalities (Montgomery, 1980). The fact that these occurred even when the beans had been boiled for extended periods suggests that in-situ enzymatic degradation of the cyanogenic glucosides was not involved. Other possibilities are either that chemical degradation occurred, or that the fatal amounts of hydrogen cyanide were produced from the intact glucosides by β-glucosidases of the intestinal flora. Prolonged soaking and traditional cooking have been shown to reduce cyanide levels in lima beans significantly (Ologhobo *et al.*, 1984; Akpapunam, 1985). Nevertheless, the absolute safety of cooked lima beans remains open to question.

CONCLUDING COMMENTS

The relationships between endogenous enzymes and changes which can occur in the major attributes of vegetable quality have been reviewed in this chapter. While a considerable amount of information is available, especially on the commercially important vegetables, many gaps in our knowlege still persist. This lack of knowlege implies that the processes used by industry to control enzyme activity are often determined and applied with less than complete understanding of the underlying enzymology. Vegetables are increasingly appreciated for their health-giving properties and, as a result, the range of types of raw and processed vegetables has increased greatly in recent years. It is vital, therefore, to acquire the basic knowledge on quality-related enzymes. Otherwise, vegetables will be harvested, prepared and processed in ways which destroy a relatively high proportion of their nutritional value and sensory quality and, perhaps, encourage the formation of toxins. Ultimately, this could lead to reduced consumption which would surely be a backward step in the efforts for a healthier population.

REFERENCES

Adams, J. B. (1982). *Campden Food and Drink Research Association Technical Memorandum* No. 309, CFDRA, Gloucestershire, UK.

Adams, J. B. (1983). *Campden Food and Drink Research Association Technical Memorandum* No. 322, CFDRA, Gloucestershire, UK.

Adams, J. B. & Ongley, M. H. (1989). *Food Chem.*, **31**, 57–71.

Adams, J. B., Hill, D. & Gaze, R. R. (1983). *Campden Food and Drink Research Association Technical Memorandum* No. 339, CFDRA, Gloucestershire, UK.

Aguilera, J. M. & Stanley, D. W. (1985). *J. Food Proc. Pres.*, **9**, 145–69.

Akpapunam, M. A. (1985). *J. Food Sci.*, **50**, 1191–2.

Anon. (1984). *Food Engng*, **56**(7), 57.

ap Rees, T., Dixon, W. L., Pollock, C. J. & Franks, F. (1981). In *Recent Advances in the Biochemistry of Fruit and Vegetables*, ed. J. Friend & M. J. C. Rhodes. Academic Press, London, UK, pp. 41–61.

Arihantana, M. B. & Buckle, K. A. (1986). *J. Food Tech.*, **21**, 189–97.

Ayernor, G. S. (1985). *J. Food Tech.*, **20**, 89–96.

Baardseth, P. (1978). *Food Chem.*, **3**, 271–82.

Baardseth, P. & Slinde, E. (1980). *Food Chem.*, **5**, 169–74.

Baardseth, P. & Slinde, E. (1983). *J. Sci. Food Agric.*, **34**, 1257–62.

Baardseth, P. & Slinde, E. (1987). In *Flavour Science and Technology*, ed. M. Martens, G. A. Dalen & H. Russwurm Jr., John Wiley and Sons, Chichester, UK, pp. 337–40.

Balagopalan, C., Padmaja, G., Nanda, S. K. & Moorthy, S. N. (1988). *Cassava in Food, Feed, and Industry*. CRC Press, Inc., Boca Raton, FL, USA.

Bano, Z. & Rajarathnam, S. (1988). *CRC Crit. Rev. Food Sci. Nutr.*, **27**, 87–158.

Barichello, V., Yada, R. Y., Coffin, R. H. & Stanley, D. W. (1990). *J. Food Sci.*, **55**, 1054–9.

Bartolome, L. G. & Hoff, J. E. (1972). *J. Agric. Food Chem.*, **20**, 266–70.

Baur, C., Grosch, W., Wieser, H. & Jugel, H. (1977). *Z. Lebensm. Unters. Forsch.*, **164**, 171–6.

Bengtsson, B. L. & Bosund, I. (1966). *J. Food Sci.*, **31**, 474–81.

Berkley, H. D. & Galliard, T. (1976). *Phytochem.*, **15**, 1481–4.

Bird, C. R., Smith, C. J. S., Ray, J. A., Moureau, P., Bevan, M. W., Bird, A. S., Hughes, S., Morris, P. C., Grierson, D. & Schuch, W. (1988). *Plant Mol. Biol.*, **11**, 651–62.

Böttcher, H. (1975). *Nahrung*, **19**, 173–9.

Bradfield, C. A. & Bjeldanes, L. F. (1987*a*). *J. Agric. Food Chem.*, **35**, 46–9.

Bradfield, C. A. & Bjeldanes, L. F. (1987*b*). *J. Agric. Food Chem.*, **35**, 896–900.

Branco, S. M. (1979). *Prog. Water Technol.*, **11,** 51–4, 617–18.
Brown, J., Mackay, G. R., Bain, H., Griffith, D. W. & Allison, M. J. (1990). *Potato Res.*, **33,** 219–27.
Buckle, K. A. & Edwards, R. A. (1970). *J. Sci. Food Agric.*, **21,** 307–12.
Buescher, R. W. & Balmoori, M. R. (1982). *J. Food Biochem.*, **6,** 1–11.
Buescher, R. W. & Hobson, G. E. (1982). *J. Food Biochem.*, **6,** 147–60.
Buescher, R. W., Sistrunk, W. A. & Kasaian, A. E. (1976). *J. Amer. Soc. Hort. Sci.*, **101,** 516–19.
Bunders, J., Broerse, J. & Stolp, A. (1989). *Trends Biotechnol.*, **7**(1), S16–24.
Buren, J. P. van, Moyer, J. C., Wilson, W. B. & Hand, D. B. (1960). *Food Technol.*, **14,** 233–6.
Burton, K. S. (1986). *The Mushroom Journal,* **158,** 68–70.
Burton, K. S., Frost, C. E. & Nichols, R. (1987). *Biotechnol Lett.*, **9,** 529–34.
Burton, W. G. (1982). *Post-Harvest Physiology of Food Crops,* Longman Inc., New York, USA.
Buttery, R. G., Teranishi, R. & Ling, L. C. (1987). *J. Agric. Food Chem.*, **35,** 540–4.
Buttery, R. G., Teranishi, R., Ling, L. C., Flath, R. A. & Stern, D. J. (1988). *J. Agric. Food Chem.*, **36,** 1247–50.
Buttery, R. G., Teranishi, R., Ling, L. C. & Turnbaugh, J. G. (1990). *J. Agric. Food Chem.*, **38,** 336–40.
Chen, C. C. & Ho, C. T. (1986*a*). *J. Chromatogr.*, **356,** 455–9.
Chen, C. C. & Ho, C. T. (1986*b*). *J. Agric. Food Chem.*, **34,** 830–3.
Chen, C. C. & Wu, C. M. (1984). *J. Agric. Food Chem.*, **32,** 1342–44.
Chou, T. D. & Kokini, J. L. (1987). *J. Food Sci.*, **52,** 1658–64.
Chubey, B. B. & Dorrell, D. G. (1972). *J. Amer. Soc. Hort. Sci.*, **97,** 107–9.
Chubey, B. B. & Nylund, R. E. (1969). *Can. J. Plant Sci.*, **49,** 421–6.
Clydesdale, F. M. & Francis, F. J. (1968). *Food Technol.*, **22,** 793–6.
Coffin, R. H., Yada, R. Y., Parkin, K. L., Grodzinski, B. & Stanley, D. W. (1987). *J. Food Sci.*, **52,** 639–45.
Conn, E. E. (1980). In *Secondary plant products,* ed. E. A. Bell & B. V. Charlwood. Springer-Verlag, Berlin, Germany, pp. 461–92.
Crookes, P. R. & Grierson, D. (1983). *Plant Physiol.*, **72,** 1088–93.
Dalgarno, L. & Birt, L. M. (1963). *Biochem. J.*, **87,** 586–96.
Davies, J. N. & Hobson, G. E. (1981). *CRC Crit. Rev. Food Sci. Nutr.*, **15,** 205–80.
Della Penna, D., Kates, D. S. & Bennett, A. B. (1987). *Plant Physiol.*, **85,** 502–7.
de Lumen, B. O., Stone, E. J., Kazeniac, S. J. & Forsythe, R. H. (1978). *J. Food Sci.*, **43,** 698–702, 708.
Dietrich, W. C., Lindquist, F. E. & Boggs, M. M. (1957). *Food Technol.*, **11,** 485–7.
Dorrell, D. G. & Chubey, B. B. (1972). *J. Amer. Soc. Hort. Sci.*, **97,** 110–11.
Ehlenfeldt, M. K., Lopez-Portilla, D. F., Boe, A. A. & Johansen, R. H. (1990). *Amer. Potato J.*, **67,** 83–91.
Elbe, J. H. von, Schwartz, S. J. & Hildebrand, B. E. (1981). *J. Food Sci.*, **46,** 1713–15.
Eriksson, C. E. (1967). *J. Food Sci.*, **32,** 438–41.
Es, A. van & Hartmans, K. J. (1987). In *Storage of Potatoes: Post-Harvest Behaviour, Store Design, Storage Practice, Handling,* ed. A. Rastovski, A. van Es *et al.*, Pudoc, Wageningen, The Netherlands, pp. 79–113.
Evans, C. D., Frankel, E. N., Cooney, P. M. & Moser, H. A. (1960). *J. Amer. Oil Chem. Soc.*, **37,** 452–6.
Fenwick, G. R. & Griffiths, N. M. (1981). *Z. Lebensm. Unters. Forsch.*, **172,** 90–2.
Fenwick, G. R., Heaney, R. K. & Mullin, W. J. (1983*a*). *CRC Crit. Rev. Food Sci. Nutr.*, **18,** 123–201.
Fenwick, G. R., Griffiths, N. M. & Heaney, R. K. (1983*b*). *J. Sci. Food Agric.*, **34,** 73–8.
Fischer, K. H. & Grosch, W. (1987). *Lebensm.-Wiss. Technol.*, **20,** 233–6.

Francis, F. J. & Amla, B. L. (1961). *Amer. Potato J.*, **38,** 89–94.
Galliard, T. (1971). *Biochem. J.*, **121,** 379–90.
Galliard, T. (1973). *J. Sci. Food Agric.*, **24,** 617–22.
Galliard, T. & Matthew, J. A. (1976). *Biochim. Biophys. Acta,* **424,** 26–35.
Galliard, T. & Phillips, D. R. (1971). *Biochem. J.*, **124,** 431–8.
Galliard, T., Matthew, J. A., Wright, A. J. & Fishwick, M. J. (1977). *J. Sci. Food Agric.*, **28,** 863–8.
Geduspan, H. S. & Peng, A. C. (1986). *J. Food Sci.*, **51,** 852–3.
Golan-Goldhirsch, A. & Whitaker, J. R. (1984). *J. Agric. Food Chem.*, **32,** 1003–9.
Gormley, T. R. (1986). *Farm Food Res.*, **17,** 71–3.
Gormley, T. R. & O'Riordan, F. (1976). *Lebensm.-Wiss. Technol.*, **9,** 75–8.
Gosselin, B. & Mondy, N. I. (1986). *J. Food Sci.*, **51,** 251–2.
Grosch, W., Laskawy, G. & Weber, F. (1976). *J. Agric. Food Chem.*, **24,** 456–9.
Gross, K. C. (1986). *Phytochem.*, **25,** 373–6.
Gubb, I. R., Hughes, J. C., Jackson, M. T. & Callow, J. A. (1989). *Ann. Appl. Biol.*, **114,** 579–86.
Haard, N. F., Sharma, S. C., Wolfe, R. & Frenkel, C. (1974). *J. Food Sci.*, **39,** 452–6.
Hadziyev, D. & Steele, L. (1979). *Adv. Food Res.*, **25,** 55–136.
Hamamoto, A. & Mazelis, M. (1986). *Plant Physiol,* **80,** 702–6.
Hammond, J. B. W., Burrell, M. M. & Kruger, N. J. (1990). *Planta,* **180,** 613–16.
Hanley, A. B. & Parsley, K. R. (1990). *Phytochem.*, **29,** 769–71.
Harkett, P. J. (1975). *Abstr. Conf. Papers 6th triennl. Conf. Eur. Assn. Potato Res.*, Wageningen, pp. 179–180.
Heaney, R. K. & Fenwick, G. R. (1980*a*). *J. Sci. Food Agric.*, **31,** 785–93.
Heaney, R. K. & Fenwick, G. R. (1980*b*). *J. Sci. Food Agric.*, **31,** 794–801.
Heaney, R. K. & Fenwick, G. R. (1987). In *Natural Toxicants in Food: Progress and Prospects,* ed. D. H. Watson. Ellis Horwood Ltd, Chichester, UK, pp. 76–109.
Henderson, H. M., Kanhai, J. & Eskin, N. A. M. (1984). *Food Chem.*, **13,** 129–37.
Hildebrand, D. F. & Hymowitz, T. (1982). *J. Agric. Food Chem.*, **30,** 705–8.
Hildebrand, D. F., Hamilton-Kemp, T. R., Loughrin, J. H., Ali, K. and Anderson, R. A. (1990). *J. Agric. Food Chem.*, **38,** 1934–6.
Hincks, M. J. & Stanley, D. W. (1987). *J. Food Biochem.*, **11,** 41–58.
Hsiao, P. H., Su, J. C. & Yung, H. Y. (1981). *J. Chinese Agric. Chem. Soc.*, **19**(1/2), 12–28.
Hughes, J. C. (1980). *Ann. Appl. Biol.*, **96,** 344–5.
Im, J.-S., Parkin, K. L. & Elbe, J. H. von (1990). *J. Food Sci.*, **55,** 1042–5, 1059.
Iwami, K. & Yasumoto, K. (1980). *Agric. Biol. Chem.*, **44,** 3003–4.
Iwami, K., Yasumoto, K. & Mitsuda, H. (1975). *Agric. Biol. Chem.*, **39,** 1947–55.
Iwami, K., Yasumoto, K., Mizusawa, H. & Mitsuda, H. (1982). *J. Agric. Chem. Soc. Japan,* **56,** 905–10.
Jadhav, S. J. & Salunkhe, D. K. (1975). *Adv. Food Res.*, **21,** 307–54.
Josephson, D. B. & Lindsay, R. C. (1987). *J. Food Sci.*, **52,** 328–31.
Kadkade, P. G. (1983). In *Abstracts of the Sixth Symposium of the International Society for Tropical Root Crops,* International Potato Center, Lima, Peru, p. 74.
Kahn, V. (1985). *J. Food Sci.*, **50,** 111–15.
Ke, D. & Saltveit, Jr M. E. (1989). *J. Amer. Soc. Hort. Sci.*, **114,** 789–94.
Kemp, T. R. (1975). *J. Amer. Oil Chem. Soc.*, **52,** 300–2.
Kermasha, S. & Metche, M. (1987). *J. Sci. Food Agric.*, **40,** 1–10.
Klein, B. P., King, D. & Grossman, S. (1985). *Adv. Free Rad. Biol. Med.*, **1,** 309–43.
Knapp, J., Moureau, P., Schuch, W. & Grierson, D. (1988). *Plant Mol. Biol.*, **12,** 105–16.
Knegt, E., Vermeer, E. & Bruinsma, J. (1988). *Physiol. Plant.*, **72,** 108–14.
Kunkel, R., Gardner, W. H. & Holstad, N. M. (1986). *Amer. Potato J.*, **63,** 13–23.
Lambers, J. W. J. & Terpstra, W. (1985). *Biochim. Biophys. Acta,* **831,** 225–35.

Langdon, T. T. (1987). *Food Technol.*, **41**(5), 54, 66, 67.
Lashley, D. & Wiley, R. C. (1979). *J. Food Sci.*, **44**, 1568–9.
Lattanzio, V., Linsalata, V., Palmieri, S. & Sumere, C. F. van (1989). *Food Chem.*, **33**, 93–106.
Lee, C. Y. (1986). *Food Chem.*, **20**, 285–93.
Lee, C. Y. (1989). In *Quality and Preservation of Vegetables*, ed. N. A. M. Eskin. CRC Press, Inc., Boca Raton, FL, USA, pp. 159–83.
Lee, F. A. & Wagenknecht, A. C. (1958). *Food Res.*, **23**, 584–90.
Lesley, B. E. & Shumate, J. W. (1937). *US Patent* 2,097,198.
Lewerenz, J.-J., Plass, R., Bleyl, D. W. R. & Macholz, R. (1988). *Nahrung*, **32**, 723–8.
Lilja, M. & Lingnert, H. (1989). *Food Chem.*, **31**, 267–77.
Liao, M. L. & Seib, P. A. (1988). *Food Chem.*, **30**, 289–312.
MacLeod, G. & Ames, J. (1988). *CRC Crit. Rev. Food Sci. Nutr.*, **27**, 219–400.
MacLeod, A. J. & MacLeod, G. (1970). *J. Food Sci.*, **35**, 734–8.
Maga, J. A. (1981*a*). *J. Agric. Food Chem.*, **29**, 1–4.
Maga, J. A. (1981*b*). *J. Food Proc. Pres.*, **5**, 95–101.
Makame, M., Akoroda, M. O. & Hahn, S. K. (1987). *J. Agric. Sci., UK*, **109**, 605–8.
Matano, K. & Kato, N. (1967). *Acta Chem. Scand.*, **21**, 2886–7.
Matheis, G. (1987). *Chem. Mikrobiol. Technol. Lebensm.*, **11**, 33–41.
Matoba, T., Hidaka, H., Kitamura, K., Kaizuma, N. & Kito, M. (1985*a*). *J. Agric. Food Chem.*, **33**, 856–8.
Matoba, T., Hidaka, H., Narita, H., Kitamura, K., Kaizuma, N. & Kito, M. (1985*b*). *J. Agric. Food Chem.*, **33**, 852–5.
Matoba, T., Sakurai, A., Taninoki, N., Saitoh, T., Kariya, F., Kuwahata, M., Yukawa, N., Fujino, S. & Hasegawa, K. (1989). *J. Food Sci.*, **54**, 1607–10.
Mayer, A. M. & Harel, E. (1979). *Phytochem.*, **18**, 193–215.
McFeeters, R. F. & Fleming, H. P. (1989). *J. Agric. Food Chem.*, **37**, 1053–9.
Meijers, C. P. (1987). In *Storage of Potatoes: Post Harvest Behaviour, Store Design, Storage Practice, Handling*, ed. A. Rastovski, A. van Es *et al.*, Pudoc, Wageningen, The Netherlands, pp. 148–74.
Moledina, K. H., Haydar, M., Ooraikul, B. & Hadziyev, D. (1981). *J. Sci. Food Agric.*, **32**, 1091–102.
Moll, C., Biermann, O. & Grosch, W. (1979). *J. Agric. Food Chem.*, **27**, 239–43.
Mondy, N. I. & Leja, M. (1986). *J. Food Sci.*, **51**, 355–7.
Mondy, N. I., Leja, M. & Gosselin, B. (1987). *J. Food Sci.*, **52**, 631–3.
Montgomery, R. D. (1980). In *Toxic Constituents of Plant Foodstuffs* (2nd edn) ed. I. E. Liener. Academic Press, New York, USA, pp. 143–60.
Moore, B. M., Byoung Kang & Flurkey, W. H. (1988). *Phytochem.*, **27**, 3735–7.
Morgan, M. R. A. & Coxon, D. T. (1987). In *Natural Toxicants in Foods: Progress and Prospects*, ed. D. H. Watson. Ellis Horwood Ltd, Chichester, UK, pp. 221–30.
Morrell, S. & ap Rees, T. (1986). *Phytochem.*, **25**, 1073–6.
Mosandl, A., Heusinger, G. & Gessner, M. (1986). *J. Agric. Food Chem.*, **34**, 119–22.
Murr, D. P. & Morris, L. L. (1974). *J. Amer. Soc. Hort. Sci.*, **99**, 155–8.
Murr, D. P. & Morris, L. L. (1975). *J. Amer. Soc. Hort. Sci.*, **100**, 16–19.
Nambisan, B. & Sundaresan, S. (1985). *J. Sci. Food Agric.*, **36**, 1197–203.
Nguyen, T. T., Palcic, M. M. & Hadziyev, D. (1987*a*). *J. Chromatog.*, **388**, 189–99.
Nguyen, T. T., Palcic, M. M. & Hadziyev, D. (1987*b*). *J. Chromatog.*, **391**, 257–65.
Oddson, L. & Jelen, P. (1981). *Can. Inst. Food Sci. Technol. J.*, **14**, 36–41.
Okezie, B. O. & Kosikowski, F. V. (1982). *CRC Crit. Rev. Food Sci. Nutr.*, **17**, 259–75.
Olías, J. M., Rios, J. J., Valle, M., Zamora, R., Sanz, L. C. & Axelrod, B. (1990). *J. Agric. Food Chem.*, **38**, 624–30.
Ologhobo, A. D., Fetuga, B. L. & Tewe, O. O. (1984). *Food Chem.*, **13**, 117–28.
Onayemi, O. (1986). *J. Food Sci.*, **51**, 161–4.

Orthoefer, F. T. & Dugan, Jr L. R. (1973). *J. Sci. Food Agric.*, **24**, 357–65.
Ozmianski, J. & Lee, C. Y. (1990). *J. Agric. Food Chem.*, **38**, 688–90.
Ozo, O. N. & Caygill, J. C. (1986). *J. Sci. Food Agric.*, **37**, 283–8.
Pattee, H. E., Salunkhe, D. K., Sathe, S. K. & Reddy, N. R. (1982). *CRC Crit. Rev. Food Sci. Nutr.*, **17**, 97–139.
Pavek, J., Corsini, D. & Nissley, F. (1985). *Amer. Potato J.*, **62**, 511–17.
Petró-Turza, M. (1986/87). *Food Rev. Int.*, **2**, 309–51.
Phillips, D. R., Matthew, J. A., Reynolds, J. & Fenwick, G. R. (1979). *Phytochem.*, **18**, 401–4.
Pinsky, A., Grossman, S. & Trop, M. (1971). *J. Food Sci.*, **36**, 571–2.
Pollock, C. J. & ap Rees, T. (1975). *Phytochem.*, **14**, 613–17.
Powers, J. R. & Drake, S. R. (1980). *J. Food Sci.*, **45**, 509–10.
Pressey, R. (1972). *J. Food Sci.*, **37**, 521–3.
Pressey, R. (1987). *Phytochem.*, **26**, 1867–70.
Pressey, R. (1988). *Planta*, **174**, 39–43.
Pressey, R. & Shaw, R. (1966). *Plant Physiol.*, **41**, 1657–61.
Prestamo, G. & Fuster, C. (1982). *Refrigeration Sci. Technol.*, **4**, 290–6.
Pyssalo, H. & Suihko, M. (1976). *Lebensm.-Wiss. Technol.*, **9**, 371–3.
Ramaswamy, N. K. & Nair, P. M. (1974). *Plant Sci. Lett.*, **2**, 249–56.
Rhee, K. S & Watts, B. M. (1966). *J. Food Sci.*, **31**, 675–9.
Rhodes, M. J. C., Wooltorton, L. S. C. & Hill, A. C. (1981). In *Recent Advances in the Biochemistry of Fruits and Vegetables*, ed. J. Friend & M. J. C. Rhodes. Academic Press, London, UK, pp. 193–220.
Richardson, C. & Hobson, G. E. (1987). *J. Sci. Food Agric.*, **40**, 245–52.
Rivera, J. A., Hohlberg, A. I., Aguilera, J. M., Plhak, L. C. & Stanley, D. W. (1989). *Can. Inst. Food Sci. Technol.*, **22**, 270–5.
Robitaille, H. A. & Badenhop, A. F. (1981). *J. Food Sci.*, **46**, 249–53.
Saltveit, Jr M. E. (1988). *J. Amer. Soc. Hort. Sci.*, **113**, 569–72.
Salunkhe, D. K., Desai, B. B. & Chevan, J. K. (1989). In *Quality and Preservation of Vegetables*, ed. N. A. M. Eskin. CRC Press, Inc., Boca Raton, FL, USA, pp. 1–52.
Samotus, B. & Schwimmer, S. (1963). *J. Food Sci.*, **28**, 1–5.
Schaller, K. & Amberger, A. (1974). *Qual. Plant. Plant Foods Hum. Nutr.*, **24**, 191–8.
Schaller, A. & Vámos-Vigyázó, L. (1986). *Lebensm. Biotechnol.*, **3**, 143–4.
Schaller, A., Csüdör-Pavisa, A. & Vámos-Vigyázó, L. (1985). *Lebensm. Biotechnol.*, **2**, 169–72.
Schinneller, D. J., Dougherty, R. H. & Biggs, R. H. (1972). *J. Food Sci.*, **37**, 935–7.
Schwimmer, S. (1981). *Source Book of Food Enzymology*. The AVI Publishing Company, Inc., Westport, CT, USA.
Schwobe, M. A. & Parkin, K. L. (1990). *J. Food Proc. Pres.* **14**, 241–52.
Sealy, L., Renaudin, S., Gallant, D. J., Bouchet, B. & Brillouet, J. M. (1985). *Food Microstructure*, **4**, 173–81.
Sessa, D. J., Gardner, H. W., Kleiman, R. & Weisleder, D. (1977). *Lipids*, **12**, 613–19.
Shankaranarayana, M. L., Raghavan, B., Abraham, K. O. & Natarajan, C. P. (1982). In *Food Flavours. Part A. Introduction*, ed. I. D. Morton & A. J. Macleod. Elsevier Scientific Publishing Company, Amsterdam, The Netherlands, pp. 169–281.
Sheehy, R. E., Kramer, M. & Hiatt, W. R. (1988). *Proc. Nat. Acad. Sci., USA*, **85**, 8805–9.
Shepard, II T. H. (1961). *J. Clin. Invest.*, **40**, 1751–7.
Sherman, M. & Ewing, E. E. (1983). *J. Amer. Soc. Hort Sci.*, **108**, 129–33.
Shioi, Y. & Beale, S. I. (1987). *Anal. Biochem.*, **162**, 493–9.
Sieso, V. & Crouzet, J. (1977). *Food Chem.*, **2**, 241–52.

Sistrunk, W. A., Gonzalez, A. R. & Moore, K. J. (1989). In *Quality and Preservation of Vegetables,* ed. N. A. M. Eskin, CRC Press, Inc., Boca Raton, FL, USA, pp. 185–215.
Smith, A. M. (1988). *Planta,* **175,** 270–9.
Smith, C. J. S., Watson, C. F., Ray, J., Bird, C. R., Morris, P. C., Schuch, W. & Grierson, D. (1988). *Nature,* **334,** 724–6.
Smith, C. J. S., Watson, C. F., Morris, P. C., Bird, C. R., Seymour, G. B., Gray, J. E., Arnold, C., Tucker, G. A., Schuch, W., Harding, S. & Grierson, D. (1990). *Plant Mol. Biol.,* **14,** 369–79.
Smith, J. L. & Stanley, D. W. (1989). *J. Food Biochem.,* **13,** 271–87.
Sones, K., Heaney, R. K. & Fenwick, G. R. (1984). *J. Sci. Food Agric.,* **35,** 712–20.
Sowokinos, J. R., Lulai, E. C. & Knoper, J. A. (1985). *Plant Physiol.,* **78,** 489–94.
Sowokinos, J. R., Orr, P. H., Knoper, J. A. & Varns, J. L. (1987). *Amer. Potato J.,* **64,** 213–26.
Springett, M. B. & Adams, J. B. (1988). *J. Sci. Food Agric.,* **46,** 211–19.
Springett, M. B. & Adams, J. B. (1989). *Food Chem.,* **33,** 173–86.
Stanley, D. W. & Aguilera, J. M. (1985). *J. Food Biochem.,* **9,** 277–323.
Stark, J. C., Corsini, D. L., Hurley, P. J. & Dwelle, R. B. (1985). *Amer. Potato J.,* **62,** 657–66.
Steinbuch, E. (1976). *J. Food Technol.,* **11,** 313–16.
Steinbuch, E. (1979). *J. Food Technol.,* **14,** 321–3.
Stevens, M. A. (1970). *J. Amer. Soc. Hort. Sci.,* **95,** 461–4.
Taylor, S. L., Higley, N. A. & Bush, R. K. (1986). *Adv. Food Res.,* **30,** 1–76.
Thomas, W. E. (1928). *US Patent* 1,685,511.
Thomas, P. (1984). *CRC Crit. Rev. Food Sci. Nutr.,* **19,** 327–79.
Thomas, P. (1988). *CRC Crit. Rev. Food Sci. Nutr.,* **26,** 313–58.
Thornton, R. E. & Beaver, G. (1986). *Chipper/Snacker,* **43**(7), 48.
Thornton, M. K. & Workman, M. (1987). *Hort. Sci.,* **22,** 455–456.
Tihanyl, K., Talbot, B., Brzezinski, R. & Thirion, J. P. (1989). *Phytochem.,* **28,** 1335–8.
Tookey, H. L., van Etten, C. H. & Daxenbichler, M. E. (1980). In *Toxic Constituents of Plant Foodstuffs* (2nd edn), ed. I. E. Liener. Academic Press, New York, USA, pp. 103–42.
Vámos-Vigyázó, L. (1981). *CRC Crit. Rev. Food Sci. Nutr.,* **15,** 49–127.
Wagenknecht, A. C. & Lee, F. A. (1958). *Food Res.,* **23,** 25–31.
Walker, G. C. (1964). *J. Food Sci.,* **29,** 383–8.
Warren, D. S. & Woodman, J. S. (1974). *J. Sci. Food Agric.,* **25,** 129–38.
Wasserman, B. P. & Guilfoy, M. P. (1983). *Phytochem.,* **22,** 2653–6.
Wattenberg, L. W. (1979). In *Environmental Carcinogenesis, Occurrence, Risk Evaluation and Mechanisms,* ed. P. Emmelot & E. Kriek, Elsevier, Amsterdam, The Netherlands, pp. 241–63.
Wattenberg, L. W. (1983). *Cancer Res.,* **43,** 2448s–53s.
Weaver, M. L. & Hautala, E. (1970). *Amer. Potato J.,* **47,** 457–68.
Wedzicha, B. L. (1984). *Chemistry of Sulphur Dioxide in Foods,* Elsevier Applied Science Publishers, London, UK.
Wiley, R. C., Schales, F. D. & Corey, K. A. (1989). In *Quality and Preservation of Vegetables,* ed. N. A. M. Eskin. CRC Press, Inc., Boca Raton, FL, USA, pp. 121–57.
Williams, D. C., Lim, M. H., Chen, A. O., Pangborn, R. M. & Whitaker, J. R. (1986). *Food Technol.,* **40,** 130–40.
Wills, R. B. H., Wimalasiri, P. & Greenfield, H. (1984). *J. Agric. Food Chem.,* **32,** 836–8.
Wurzenberger, M. & Grosch, W. (1984). *Biochim. Biophys. Acta,* **794,** 25–30.

Xu, S-Y., Shoemaker, C. F. & Luh, B. S. (1986). *J. Food Sci.*, **51,** 399–402, 407.
Yamaguchi, M., Hwang, P. M. & Campbell, J. D. (1970). *Can. J. Biochem.*, **48,** 198–202.
Yamauchi, N., Iida, S., Minamide, T. and Iwata, T. (1986). *J. Japan. Soc. Hort. Sci.*, **55,** 355–62.

15

Enzymes in Fruits

MICHAEL KNEE
Department of Horticulture, The Ohio State University, 2001 Fyffe Court, Columbus, Ohio 43210, USA

ROBERT E. PAULL
Department of Plant Molecular Physiology, University of Hawaii at Manoa, 3190 Maile Way, Honolulu, Hawaii, 96822, USA

RUTH BEN ARIE
Agricultural Research Organization, The Volcani Center, Institute for Technology and Storage of Agricultural Products, PO Box 6, Bet Dagan 50250, Israel

JOHN S. HAWKER
Department of Botany, The University of Adelaide, GPO Box 498, Adelaide, South Australia 5001, Australia

INTRODUCTION

The main purpose of this chapter is to review knowledge of the occurrence and role of enzymes in fruit. Particular attention is given to enzymes involved in ripening. These affect the qualities which make fruits more or less attractive to consumers. Enzymes with a role in fruit processing are mentioned less often. Apart from oxidative enzymes, endogenous activities are rarely troublesome in processing. Exogenous enzymes, in particular pectinase, are frequently used in juice and purée manufacture. Accounts of their use are often highly empirical and there is little to say in terms of enzymology (Baumann, 1981). Future product development may rely more on the use of enzymes to modify specific components of fruit products to improve quality. The use of an immobilized enzyme to remove limonene from citrus products may be a forerunner of much biotechnology (Hasegawa & Delton, 1983). Only a few sources of commercial enzymes are found among fruits; these are the proteases, papain from papaya and bromelain from pineapple. Actinidin from kiwi fruit may be similarly exploited in the future.

Without doubt, research on tomato has been at the forefront of fruit physiology and biochemistry in recent years. The tomato is particularly suitable, both for traditional breeding and for genetic manipulation. The rapid advance from experimental to practical manipulation is evidenced by the book '*Tomato Biotechnology*' (Nevins & Jones, 1987). The apple and the grape account for a substantial proportion of the chapter. This is not simply a result of the prejudice of the authors and does not mean that these fruits are necessarily more interesting than others. The more substantial literature relating to these fruits probably results from their importance in the horticultural economy of western nations. The question arises, why all of this work was done. No doubt many investigators hoped that knowledge of some aspect of fruit metabolism could lead to the ability to control it for practical benefit. Subsequently, the manipulation of enzyme activities in plants has become a reality. Many of the fruits in this chapter are borne on woody perennial plants, which are among the least amenable species for genetic manipulation. Nevertheless, they are high-value commodities for which consumers have sophisticated quality requirements. Equally sophisticated post-harvest technology has been developed, especially for the pome fruits. In the long run, transgenic fruit trees may be commercially developed to satisfy particular quality requirements. This may prove more economically and socially acceptable than the development of further chemicals to modify fruit quality. If this manipulation is to be a reality, more knowledge of the enzymology of fruit development will be required.

The publication of '*The Biochemistry of Fruits and Their Products*' (Hulme, 1970, 1971*a*) was a landmark in fruit physiology, and the two volumes are still a valuable reference source. This chapter deals mainly with the more recent literature, but mentions that published before 1971 when necessary for clarity, or when reinterpretation is called for.

The well-known difficulties in extracting enzymes from plant tissues in general have been reviewed, with their potential solutions (Loomis, 1969, 1974; Rhodes, 1977). Fruits present some extreme examples of these difficulties and often call for special measures to overcome them. Thus, the next section deals with the contributory factors in the problem of enzymic extraction from fruits.

EXTRACTION OF ENZYMES FROM FRUITS

Low protein content

One of the fundamental difficulties is the low protein content of fruits. Typically, their total protein is 1 mg/g fresh weight, or less (Hansen, 1971). Soluble protein is a fraction of this and, in most extraction procedures, the concentration will be further diluted by addition of buffer. Thus, proteins are particularly vulnerable to surface denaturation in the extract, particularly if extra surface is generated by frothing during homogenization. One remedy is to concentrate proteins at an early stage of preparation, for example by precipitation with ammonium sulphate. Alternatively, an excess of enzymically inactive protein, such as bovine serum albumin or casein, may be added. These displace fruit proteins from the liquid surface and, in addition, provide an alternative reactant with phenols in the mixture (see below). Other poly-electrolytes, such as soluble polyvinyl pyrrolidone (PVP), may protect against surface denaturation.

Acids

With a few exceptions, the cell vacuolar saps of fruits are pH 3–3·5 (Ulrich, 1971). The concentrations of organic acids are 50–100 mM or higher. Since they are di- and tri-basic, these acids show multiple p*K* values and buffering capacity over a wide range of pH values. In the intact cell, the acid is localized mainly in the vacuole, separated from most of the enzymes in the cytoplasm or cell wall which are maintained at higher pH values. Immediately on disruption of the tissue, these proteins are exposed to the acid which often accelerates loss of enzymic activity. One remedy which is frequently adopted is to titrate with base until a desired pH is attained. This leads to the opposite problem that high pH values occur locally before complete mixing of the added base. If proteins are reasonably stable under acid conditions, the pH value can be raised slowly by dialysis or gel filtration with an appropriate buffer. If, as is often the case, minimum exposure to acid is desired, the extraction medium must be buffered. Unless molar concentrations of buffer salts can be used, there will always be a drop in pH value when fruit tissue is homogenized in buffer. One can compensate for this by using buffers higher than the desired final pH value. For example, two volumes of 0·2 M buffer (tricine or phosphate) at pH 9·0 with unit weight of apple tissue attains a final pH of 7·5. The precise formula has to be determined by trial and error.

Phenols

The total phenol content of fruits is typically 1–5-times the protein content on a mass basis (Buren, 1971). The excess of phenol is higher on a molar basis, and although phenols vary in their reactivity towards proteins, the presence of members of this class of compounds accounts for many of the problems of fruit enzymology. Phenolic hyydroxyl groups form hydrogen bonds with the carbonyl function of peptide bonds in proteins. This may occur particularly in hydrophobic regions of the protein which might include the active site of an enzyme (Ozawa, *et al.*, 1987). Simple monophenols, like the coumaric acids, do not form stable complexes; this is a property of polycyclic and polyhydroxy phenols. In their oxidized form, as quinones, phenols are more reactive and may form covalent linkages with proteins. If no precautions are taken against the action of phenols on proteins, no enzyme activity may be recovered; sometimes, as with immature apple fruit, no soluble protein may be recovered (Knee, unpublished data). The problem usually diminishes with fruit development. Goldstein and Swain (1963) concluded that the condensation of phenolic compounds increased with ripening of several fruits. However, Ozawa, *et al.* (1987), finding that there was little change in phenol composition, suggested that a change in another fruit component affected the reactivity of the phenols. This change could be solubilization of pectin which can form weak complexes with phenols. Whatever its underlying cause, the diminishing reactivity of phenolics can lead to a misleading impression of increasing enzyme activity.

The avoidance of phenol inactivation involves a choice of three strategies. The first is to include a complexing agent to compete with protein for the reactive molecules. The usual agents are insoluble polymers such as PVP, and charged (Dowex AG1-X8) or uncharged (Amberlite XAD) resins. These are most efficient when hydrated for at least 1 h before use. If they are included during homogenization, fines or even soluble fragments may be generated. This problem has to be weighed against the desirability of protection from the moment of tissue disruption. Some authors recommend addition to the homogenate. The amount of absorbent needed in relation to weight of tissue has to be determined by experiment. Homogenization becomes difficult when the dry weight of material exceeds 10% of the total volume of homogenate. Smith and Montgomery (1985) found that AG1-X8 was the most effective protectant in the extraction of phenolase from pears, but suggest that a mixture of this resin with PVP is to be preferred. Soluble PVP can also be used and was found to be a more effective protectant than the insoluble form and other polymers in the extraction of malic enzyme from apples (Rhodes, 1977). Polysaccharide-based gel filtration media will also remove phenolic compounds from protein extracts.

Other strategies used in the extraction of enzymes aim to prevent oxidation of phenols to more reactive quinones. This may be achieved by inhibiting phenolase, usually by chelation of copper from its active site. Cyanide, diethyldithiocarbamate (DIECA), or EDTA can be used, but are only applicable if the enzyme under study is not inhibited by these materials.

The final strategy is to provide sufficient reducing agent to reverse quinone formation for the duration of the extraction procedure. Metabisulfite has been used as a reductant but organic compounds are more commonly employed. Apart from ascorbate, these are thiols varying in complexity from mercapto-ethanol through cysteine to dithioerythritol. The effectiveness of these agents has rarely been compared, but Driessche *et al.* (1984) found that thiourea was preferable to DIECA or metabisulfite for the extraction of enzymes from phenol-rich plant tissues.

Pectin

The frequent occurrence of high concentrations of soluble pectin in fruits is more of a nuisance than a danger in enzyme extraction. The polymer can cause difficulties in handling extracts, particularly if concentration is attempted before pectin is removed. Procedures such as centrifugation, ultrafiltration and column chromatography are hindered by the viscosity of pectin solutions.

However, specific measures to overcome these problems are rarely mentioned. Occasionally, calcium salts have been used in extraction media to maintain the pectin in an insoluble form (Frenkel *et al.*, 1969). Pectinesterase is usually present in fruit extracts and, if cations are present as activators, it will cause flocculation of pectin. The pectin can then be removed by centrifugation (Bartley, 1978). Depending upon its molecular weight and charge, pectin will often precipitate at 30% saturation with ammonium sulphate. Since most enzymes do not precipitate at this concentration, this can provide a basis for eliminating pectin prior to enzyme purification.

Other procedures

Some of the problems of enzyme extraction can be avoided by adopting particular tissue preparation procedures. One little used procedure is to infiltrate protective agents into tissues before homogenization. This is accomplished by evacuation of a container filled with tissue submerged in medium. Neutralizing and phenol-protective agents permeate intercellular spaces as the vacuum is released and are immediately present when cells are later disrupted. Another way of minimizing the contact of enzymes with acids and phenols is to start by freezing tissue in liquid nitrogen. After grinding, the tissue is added, while still frozen, to the extraction medium, containing suitable protectants. Finally, acids and phenols may be removed by homogenization and extraction with aqueous acetone. Usually, this is cooled to −20°C or lower and may include further protective agents such as buffers (Bartley, 1974) or thiols (Loomis, 1969).

Evaluating methods

Typical procedures for enzyme extraction from fruits will include some means of neutralizing acidity, a phenol-complexing agent and a reducing compound. Innumerable variations of procedure are possible and evaluating them can seem unrewarding, since researchers usually have other objectives in view. Often, a limited range of options will be tested and the main criterion will be achieving maximal activity from a given tissue sample. The simple expedient of mixing two types of tissue prior to extraction should be more widely adopted, especially if a difference in activity between tissues (at different stages of development or from different genotypes) is to be demonstrated. Improved extraction procedures will be needed if the measured activity for the mixture is less than the mean of activities for the separate tissues.

Some alterations or damage to protein may leave enzyme activity unimpaired, or even enhanced. However, the possibility of these more subtle modifications is rarely considered when evaluating extraction procedures. Loss of allosteric properties or the generation of forms with altered electrophoretic or chromatographic mobilities becomes increasingly important when regulation of enzyme activity is studied.

The tissues of many fruits soften markedly during development. This can affect the efficiency of cell breakage and recovery of protein. For this reason, specific activity can be a better basis for expression of results than fresh weight. However, this presupposes that a reliable estimate of protein content is available. Phenolic compounds interfere in most protein assays and traditional methods such as UV absorbance and the assay of Lowry *et al.* (1951) are useless for crude enzyme preparations from fruits. Most investigators now use a dye binding assay (Bradford, 1976), although Agustin and Beltran (1982) report that this assay gave higher protein estimates than a Kjeldahl digestion unless protein was precipitated with trichloroacetic acid after extraction in the presence of EDTA, mercaptoethanol and PVP.

Organelle preparations

The problems of extraction of soluble enzymes from fruits apply equally to organelles. Mitochondria have been the most frequent subject of study for over 30 years. Great efforts were expended in obtaining functional organelles and almost the only separation method employed has been differential centrifugation. Fruit preparations are less homogeneous than those from other plant sources. Yields of organelles from fruit tissue are low and the need to process a large bulk of tissue leads to the use of high shear techniques for tissue disruption. Consequently, the organelles are often functionally inferior to those from more favorable sources.

ACTINIDIACEAE

Kiwi fruit (*Actinidia deliciosa*)

Various enzymes

Production of kiwi fruit is expanding in many countries and the fruit is attracting increasing attention from researchers. The existence of a thiol-proteinase ('actinidin', EC 3.4.22.14) in the ripe fruit has long been known (Arcus, 1959) and the amino acid sequence of this protein shows extensive homology with papain (from papaya) (Carne & Moore, 1978). The protease is a potential problem in the isolation of other enzymes. Soda *et al.* (1986) found that inhibition of the protease by leupeptin was required in order to detect polygalacturonase (PG) activity in the ripe fruit. No PG was detected in the unripe fruit and although Soda *et al.* (1986) concluded that it is an endo-PG (EC 3.2.1.15), the loss of viscosity of polygalacturonate caused by the enzyme was slow, suggesting that exo-PG (EC 3.2.1.67) was also present.

Kiwi fruit is exceptional as a source of membrane vesicles which convert 1-aminocyclopropane-1-carboxylic acid to ethylene with properties consistent with in-vivo activity: high affinity for substrate, stereochemical discrimination between isomers of a substrate analogue and response to inhibitors (Mitchell *et al.*, 1988).

ANACARDIACEAE

Mango (*Mangifera indica*)

General

During growth, mango fruits accumulate starch and to a lesser extent soluble solids (Kasantikul, 1983; Quintana *et al.*, 1984). Changes during growth and ripening have been reviewed (Hulme, 1971*b*; Krishnamurthy & Subramanyam, 1973; Subramanyam *et al.*, 1975; Lakshminarayana, 1980). Young mango fruit are astringent, acidic, and rich in vitamin C, whereas ripe mangos are sweet, rich in provitamin A, moderate in vitamin C and highly aromatic (Lakshminarayana, 1980).

Carbohydrate metabolism

The principal free sugars present in mango fruit are glucose, fructose and sucrose (Medlicott & Thompson, 1985). The concentration of sucrose increases during ripening from about 1% to between 11 and 17%, depending upon the cultivar (Palaniswamy *et al.*, 1974), so that it becomes the predominant sugar in the ripe fruit (Vazquez-Salinas & Lakshminarayana, 1985). An invertase (EC 3.2.1.26) which increases in the peel of chilled fruit (Chhatpar *et al.*, 1971), was partially purified and shown to have a double temperature optimum; one at 0°C, the other at 37°C (Chhatpar & Modi, 1974). Invertase was also found in the fruit pulp by Sen *et al.* (1985).

Amylase activity (EC 3.2.1.2) increases as the fruit grows and then declines as it approaches maturity (Sen *et al.*, 1985). The amylase activity peaked at the same time as starch content (Tandon & Kalra, 1983). However, an increase in amylase activity does occur in mature fruit detached and allowed to ripen (Fuchs *et al.*, 1980). The inconsistency between these findings could result from the presence of a proteinaceous amylase inhibitor (Mattoo & Modi, 1970*a*; Fuchs *et al.*, 1980).

Glucose-6-phosphate dehydrogenase (EC 1.1.1.49), 6-phosphogluconic dehydrogenase (EC 1.1.1.44) and malic enzyme (malate NADP-oxidoreductase; EC 1.1.1.40) increase during ripening (Modi & Reddy, 1967; Dubery *et al.*, 1984).

Organic acid metabolism

Malic enzyme activity and the capacity of mitochondria from mango fruit to oxidize succinate first increased three-fold to four-fold and then decreased as ripening progressed (Kane *et al.*, 1978). A cytosolic malate dehydrogenase (EC 1.1.1.37) from mango has a pI of 6·86, a pH optimum of 6·8 and an activation energy of 49·98 kJ/mol. The enzyme has low cysteine and tryptophan contents and shows positive cooperativity between binding of malate and the allosteric regulators, citrate, isocitrate and α-ketoglutarate; thus, these precursors of malate activate the enzyme, whereas the product, oxaloacetate, is inhibitory (Dubery & Schabort, 1981; Dubery *et al.*, 1984). The average molecular weight is 258 kDa (Dubery & Schabort, 1981).

Citrate lyase (EC 4.1.3.6) activity increases considerably during ripening and is stimulated by glucose, fructose, and fatty acids (Mattoo & Modi, 1970*b*). This enzyme may provide acetyl CoA and oxaloacetate for synthetic processes. The pH optimum is near 7·0. Acid phosphatase (EC 3.1.3.2) and glutamine synthetase (EC 6.1.3.2) have also been detected in mango fruit (Mattoo & Modi, 1970*a*).

Cell wall metabolism

β-1,4 Glucanase (EC 3.2.1.4) and PG (EC 3.2.1.15) both increase 50-fold from very low levels in green fruit as the fruit softens during ripening. Pectinesterase (EC 3.1.1.11), however, declines (Roe & Bruemmer, 1981).

Phenol metabolism

Laccase (EC 1.10.3.2) is characteristic of the mango fruit resin (Joel *et al.*, 1978). Polyphenoloxidase (EC 1.10.3.1; PPO) has been detected in the mesocarp and exocarp (Brekke *et al.*, 1975; Katwa *et al.*, 1982). The activity is higher in the exocarp and the activity in both tissues declines to trace levels in ripe fruit. At the same stage of fruit development cultivars have widely differing activities (van Lelyveld, 1975). The purified polyphenoloxidase (126-fold) exists as a single form with pI 4·1 and molecular weight 137 kDa, and contains a single copper atom per molecule. It is particularly active on catechol, and thiol reagents are effective inhibitors. Activity is optimal at pH 5·5 and 46°C. Fifty percent of activity is lost at 65°C in 25·5 min (Katwa *et al.*, 1982).

Other enzymes

Peroxidase (EC 1.11.1.7) activity increases as the fruit matures (Lelyveld, 1975). The activity is cell wall bound (Brekke *et al.*, 1975), and at least seven isozymes have been detected with quantitative variations in activity between cultivars (Lelyveld, 1978). Ascorbic acid oxidase and catalase have been reported (Brekke *et al.*, 1975; Lelyveld, 1975). A heat labile proteinaceous inhibitor of peroxidase has also been found in mango (Matoo & Modi, 1970*b*) though whether it functions *in vivo* has not been shown. Aspartate glutamate aminotransferase (EC 2.6.1.1) activity increased, then decreased following the pattern of rise and fall in respiration in the ripening fruit, whereas the activity of glutamic decarboxylase (EC 4.1.1.15) continued to increase during ripening (Krishnamurthy *et al.*, 1971).

ANNONACEAE

***Annona* fruit (*Annona spp*.)**

Various enzymes

Ripening of this climacteric fruit is characterized by a dramatic degradation of starch by an α-amylase (EC 3.2.1.1) (Paull *et al.*, 1983; Paull, 1990), and increases in sugars and organic acids. Flesh browning, probably associated with polyphenoloxidase activity, has been observed in most *Annona* fruit. PPO (EC 1.10.3.1) has been isolated and purified 70-fold from cherimoya epicarp (Martinez-Cayuela *et al.*, 1988). This enzyme has a molecular weight of 160 kDa and is present as multiple forms. Activity was stabilized by cysteine and mercaptoethanol, but ascorbic acid was ineffective.

Multiple forms of various enzymes have been used in the characterization of cherimoya and atemoya varieties (Ellstrand & Lee, 1987).

AVERRHOACEAE

Carambola (*Averrhoa carambola*)

Cell wall metabolism

An exo-PG (EC 3.2.1.67) with optimum activity at pH 5·2 has been partially purified from carambola (Ghazali & Leong, 1987).

BROMELIACEAE

Pineapple (*Ananas comosus*)

Various enzymes

Changes in phenols, especially post-harvest and during storage, have been associated with chilling injury of pineapple (Paull & Rohrbach, 1985; Bartholomew & Paull, 1986). PPO (EC 1.10.3.1) is low at harvest but increases dramatically following storage at 8°C in sensitive fruit (Lelyveld & Bruyn, 1977; Teisson & Combres, 1979). Peroxidase (EC 1.11.1.7) did not appear to be related to chilling-injury symptom development. No ascorbic acid oxidase has been detected (Lelyveld & Bruyn, 1977). The PPO is reported to have a pH optimum near 5·0 and a temperature optimum near 45°C. The PPO is stable to heat when extracted but loses over 50% of its activity in 20 min at 60°C (Teisson, 1979). Thiol and bisulfite inhibit activity. Catalase (EC 1.11.1.6) and indole acetic acid oxidase have also been detected but changes in activity during ripening have not been reported.

Fruit proteinase declines during ripening (Lodh *et al.*, 1973). The major cysteine proteinase is bromelain (EC 3.4.22.4), a glycoprotein. Two minor proteinases have also been isolated (Ota *et al.*, 1972). The fruit bromelain has a molecular weight of 31 kDa, an isoelectric point of 4·6 and optimum for activity at pH 8. The bromelain in the stem is a 28 kDa protein with pI of 9·6 and optimum activity at pH 5–6. The sequence around the active cysteine of both enzymes is the same but the fruit enzyme has an extra alanine residue at its amino terminus (Yamada *et al.*, 1976). Fruit bromelain shows a 20% cross reactivity with antibodies to stem bromelain (Sasaki *et al.*, 1973).

CARICACEAE

Papaya (*Carica papaya*)

Carbohydrate metabolism

Early in papaya fruit development glucose is the major sugar. At about 110 days after anthesis the seeds and pulp begin to change color, and sucrose concentration increases from about 18% of total sugars to about 80% at 135 days (Akamine & Goo, 1971; Chan *et al.*, 1979). Earlier reports on the sugar composition of ripe papaya show less sucrose and higher levels of glucose and fructose (Jones & Kubota, 1940; Selvaraj *et al.*, 1982) and can now be explained by the action of invertase during tissue homogenization (Chan & Kwok, 1976). This enzyme activity increases 15-fold during fruit ripening (Pal & Selvaraj, 1987).

Invertase (EC 3.2.1.26) has been partially purified and characterized (Chan & Kwok, 1976; Lopez *et al.*, 1988). The optimum pH is approximately 4·5 for this β-fructofuranosidase, with an optimum temperature of 40°C (Chan & Kwok, 1976). The apparent molecular weight, as determined by gel filtration, is 275 kDa (Chan & Kwok, 1976) and 52 kDa by gel electrophoresis (Lopez *et al.*, 1988). The higher value is possibly due to aggregation at low salt concentrations. The enzyme is inhibited by fructose but not by glucose. Most (99%) of the invertase activity is lost in 2 min at 60°C.

Little or no starch has been detected in papaya mesocarp tissue (Chan *et al.*, 1979; Selvaraj *et al.*, 1982). The small increase (1–2%) in total soluble sugars during the ripening of detached fruit support this conclusion (Chen, 1964). Recently, amylase activity has been reported to increase transiently during ripening of papaya fruit (Pal & Selvaraj, 1987). It is possible that there is some starch and amylase in the cells near the exocarp. Further work is necessary to clarify the presence of starch and the role of amylase in papaya fruit.

Organic acid metabolism

The total titratable acidity of ripe papaya puree is 1·54 meq/100 g, with a pH in the range 5·0–5·5. Malic, citric and ascorbic acids account for about 80% of the total, with α-ketoglutaric and volatile acids as minor components (Chan *et al.*, 1971). The slight increase in titrable acidity during ripening (de Arriola *et al.*, 1975; 1980) could derive in part from a rise in free galacturonic acid. Non-volatile acidity, especially citric acid, decreases during fruit maturation with little change during the ripening phase (Selvaraj *et al.*, 1982). Ascorbic acid increases three-fold during fruit ripening. The low climacteric rise in papaya (Paull & Chen, 1983) and low contribution of acidity to fruit flavor could account for the lack of studies of enzymes involved in organic acid metabolism. Malic dehydrogenase (EC 1.1.1.37) declines during fruit maturation and ripening as does NADH reductase (EC 1.6.99.3) (Pal & Selvaraj, 1987).

Cell wall metabolism

Papaya fruit soften to an edible stage in 6–12 days from harvest at the color break stage. Water soluble pectin declines rapidly after the 75% skin yellow stage (Chen, 1964), but acid-soluble pectin remains constant. This is partly confirmed by more recent findings which also show an initial increase in water soluble pectin during ripening (de Arriola *et al.*, 1975). However, peak activities of xylanase (EC 3.2.1.8) and PG occur at an earlier stage, when the fruit is about 40–60% yellow (Paull & Chen, 1983). On the other hand, pectin esterase and β-1,4 glucanase (EC 3.2.1.4) both continue to increase as the fruits ripen, and only decline when the overripe fruit begins to break down.

The pectin esterase (EC 3.1.1.11), which is present throughout ripening (Paull & Chen, 1983), has a pH optimum at 7·5–8·0, is activated by 0·2 M sodium chloride, and is inhibited about 10% by 10% sucrose (Chang *et al.*, 1965; Lourenco & Catutani, 1984). The enzyme is heat stable, retaining *c.* 80% of its original activity after 60 min at 50°C. Kinetic data suggest that polygalacturonic acid is a competitive inhibitor (Lourenco & Catutani, 1984). The molecular weight is reported as 53 kDa.

Two PGs have been isolated: an endo-PG (EC 3.2.1.15) with a molecular weight *c.* 164 kDa, and an exo-PG (EC 3.2.1.67), with a molecular weight of *c.* 34 kDa (Chan & Tam, 1982). The pH optimum for both partially purified enzymes is 4·6, with an optimum temperature of 45°C. Up to 80% of total activity is lost in 10 min at 85°C (Chan *et al.*, 1981; Chan & Tam, 1982). Activity is highest in the endocarp.

Ethylene synthesis

The highest level of ethylene forming enzyme (EFE) activity was found in the exocarp of 75% ripe fruit (Chan *et al.*, 1989). In mature green fruit and non-senescent tissue of ripe fruit, the majority was in the placenta and dorsal bundle. As the fruit ripens from the inside out, there is a progressive decline in EFE activity in the placental tissue and a rise and decline of EFE in dorsal bundles. Chilling (Chan *et al.*, 1985) and heat treatments used for insect disinfestation (Chan, 1986; Paull & Chen, 1990) both lead to a temporary decline in EFE activity. The level of EFE remains constant until the late stages of ripening (Paull & Chen, 1990). The level of 1-aminocyclopropane-1-carboxylic acid (ACC) is initially low in fruit mesocarp tissue during ripening, increasing three-fold when the peak of ethylene synthesis occurs (Paull & Chen, 1990). The results suggest that at the peak of ethylene synthesis the EFE activity may be limiting.

Isothiocyanate metabolism

The fruit latex contains *c.* 10% of the dry weight as thioglucoside, e.g. benzylglucosinate (Tang, 1971). This compound is hydrolyzed by the enzyme thioglucosidase (EC 3.2.3.1), which is in the parenchyma rather than in the latex (Tang, 1973), to benzyl isothiocyanate (BITC). The concentration of BITC decreases in the fruit pulp but increases in the seeds as the fruit matures and ripens. Free BITC is detected by fruit head space analysis, implying that it is a normal flavor metabolite (Tang, 1971). Injury to the fruit induces the enzymatic hydrolysis of benzylglucosinate. It has been suggested that BITC inhibits ethylene synthesis in papaya tissue discs (Patil & Tang, 1974). However, there have been no subsequent reports of BITC inhibiting ethylene synthesis in other fruit tissue. Other possible roles for BITC include the inhibition of papain (Tang, 1974), resistance to disease (Patil *et al.*, 1973), and inhibition of fruit-fly egg lay and development (Seo & Tang, 1982).

Proteinases

Latices from the leaves, petioles, stems and fruit of papaya contain a number of basic thiol proteinases, including papain (EC 3.4.22.2), chymopapains A and B (EC 3.4.22.6), and papaya peptidase A (EC 3.4.23.6). Much of the earlier confusion about the number of proteinases present in papaya latex resulted from the use of dried instead of fresh latex as a source of enzyme. Amino acid sequences of the catalytic sites have been determined for all except peptidase A.

Papain has been the most intensively studied because of its importance in brewing, food and pharmaceutical industries. The enzyme is very stable and easily available. Its polypeptide chain contains 212 amino-acid residues, and it has a molecular weight of 23 kDa. Three-dimensional structures have been determined by X-ray diffraction (Kamphuis *et al.*, 1984) and complete sequences of papain (Mitchel *et al.*, 1970) and chymopapain A (Robinson, 1975) have been published. Essential to the active site is a cysteine at position 25 in the sequence of both enzymes, and six other cysteines help to maintain the conformation of the enzyme through disulfide bridges (Baker & Drenth, 1987). Serine and tryptophan lie on either side of the cysteine in papain and glutamine and tryptophan in chymopapain. Both papain and chymopapain include a sequence of asparagine, histidine and alanine at the active site (Brocklehurst *et al.*, 1987). The chymopapains are glycoproteins. The latex proteinases could protect the fruit against insect or fungal attack. The digestive juice of a predatory insect may catalyze the conversion of propapain to the active enzyme (Lowe, 1976). During fruit ripening the proteinase activity in mesocarp tissue declines (Paull & Chen, 1983) as the latifers begin to breakdown. Interestingly, the latex contains a low molecular weight polypeptide papain inhibitor which binds most effectively above pH 5·5 (Baines *et al.*, 1982).

Other enzymes

Ripe papaya contains an acid phosphatase (EC 3.1.3.2) (Carreno & Chan, 1982). The partially purified enzyme has a pH optimum at 6·0, a temperature optimum of 37°C, K_m of 1·0 mM and an apparent molecular weight of 120 kDa. The enzyme lost 70% of its activity when heated at 60°C for 8 min. Activity increases on ripening and has been held responsible for the generation of off-flavours during processing of fruit purée (Pal & Selvaraj, 1987).

Catalase (EC 1.11.1.6) activity increases during fruit maturation and ripening (Pal & Selvaraj, 1987). The enzyme has been purified 32-fold from ripe fruit (Chan *et al.*, 1978), and has a pH optimum at 6·1, and a molecular weight of 160 kDa. It is inactivated by acidification to pH 3·5 and by heat, losing 90% of activity after 3 min at 60°C. Peroxidase (EC 1.11.1.7) increases slightly during ripening (Pal & Selvaraj, 1987) and there are at least four peroxidase isozymes in quarter-ripe fruit (Sawato, 1969).

Light and light plus exogenous nitrate induce nitrate reductase (EC 1.6.6.2) activity in the exocarp of green and mature green fruit (Menary & Jones, 1972). As the fruit ripened, the activities greatly declined in the endocarp, mesocarp and exocarp. No activity was detected in the seed. The decline in activity is probably associated with the decline of xylem transport of nitrate to the fruit as it reaches the mature green stage.

The activity of phenylalanine ammonia lyase (EC 4.3.1.5) remains constant during ripening of papaya fruit but increases two-fold 24 h after irradiation (100 krad) (Tan & Lam, 1985). PPO (EC 1.10.3.1) has been reported to increase during the late ripening stage (Pal & Selvaraj, 1987) and could account for a decline in phenol content (Tan & Lam 1985). The concentrations of phenolic compounds, especially chlorogenic acid, show a relationship to fungal resistance in the fruit (Tan *et al.*, 1982).

CUCURBITACEAE

Cucumber (*Cucumis sativus*)

Cell wall metabolism

Pressey and Avants (1975) reported that the PG of fresh market cucumbers is a terminal cleavage enzyme (exo-PG, EC 3.2.1.67), with a molecular weight of 55 kDa. However, McFeeters *et al.* (1980) found an endo-PG (EC 3.2.1.15) with a molecular weight of 35 kDa in ripe fruits of a pickling variety. The activity of this enzyme increased after a transient rise in ethylene production by the detached fruits and was stimulated by ethylene treatment of immature fruit.

Miller *et al.* (1989) observed a rise in the activity of a number of cell wall degrading enzymes when cucumber fruits attained about half their final size. The specific activities of PG (endo- or exo-), β-1,4 glucanase (EC 3.2.1.4 or 21), xylanase (EC 3.2.1.8 or 37) and pectinesterase increased in the mesocarp but remained more constant in carpel tissue. Mesocarp tissue was softest around the time of maximum enzyme activity, whereas the carpel softened continuously through development. In this work, changes in enzyme activity were not correlated with any change in ethylene synthesis. Miller *et al.* (1987) found that 'mechanical stress' induced increases in PG and xylanase activities.

Pectinesterase activity can survive blanching of cucumbers and can therefore cause de-esterification of pectin during pickling (McFeeters *et al.*, 1985). This could have beneficial effects on the texture of the final product, but the complexities of the pickling process are discussed in Chapter 12.

Phenol metabolism
Cucumber fruits contain low concentrations of phenolic compounds but phenoloxidase (EC 1.10.3.1) activity is present (Miller & Kelley, 1989). Peroxidase (EC 1.11.1.7) activity increases during normal fruit development, but is greatly enhanced by mechanical stress; stress results in enhancement of the activities of existing iso-enzymes and the appearance of additional forms (Miller & Kelley, 1989). Peroxidase activity survives the pickling process (Buescher *et al.*, 1987).

Ethylene synthesis
The increase in ethylene synthesis reported by McFeeters *et al.* (1980) has not been observed by other authors (Miller *et al.*, 1989). Chilled cucumbers produce ethylene rapidly on return to warmer temperatures. The ethylene precursor, ACC, accumulates at low temperatures and enhanced activity of ACC synthase (EC 4.4.1.14) has been observed in chilled cucumbers (Wang & Adams, 1982). However, this metabolism is an indicator rather than a cause of injury. As in other plant tissues (Wright, 1974), chilling injury of cucumbers appears to result from failure of cells to retain water (Fukushima *et al.*, 1977).

Lipid metabolism
Some acclimation occurs in cucumbers held at low temperature. This is accompanied by enrichment of cellular lipid with less saturated fatty acids (Wang & Baker, 1979), and may be an example of the enhancement of desaturase activity by low temperature (Harris & James, 1969).

Disruption of cucumber tissue leads to hydrolysis of lipids by enzymes including lipolytic acyl hydrolase (EC 3.1.1.3). An oxidation system then converts fatty acids to long-chain aldehydes. The unsaturated acids, linoleic and linolenic, are specifically oxidized by lipoxygenase (EC 1.13.11.12); the fatty acid hydroperoxide lyases in cucumber form both six and nine carbon aldehydes from lipoxygenase products (Galliard *et al.*, 1976). The mixture of aldehydes formed is largely responsible for the flavor of fresh cucumber.

Melon (*Cucumis melo*)

Carbohydrate metabolism
The predominant sugar synthesized in melon leaves is stachyose (Fig. 1). It has been shown to be transported via the stem to the fruit, but fruit tissue contains little or no stachyose (Hughes & Yamaguchi, 1983). Presumably, the fruit contains an enzyme system for metabolizing this oligosaccharide. Gross and Pharr (1982) showed that an extract from the fruit peduncle of half-grown fruits was capable of converting galactose-1-phosphate and fructose to sucrose *in vitro* in the presence of fructose, evidently via UDP-galactose-

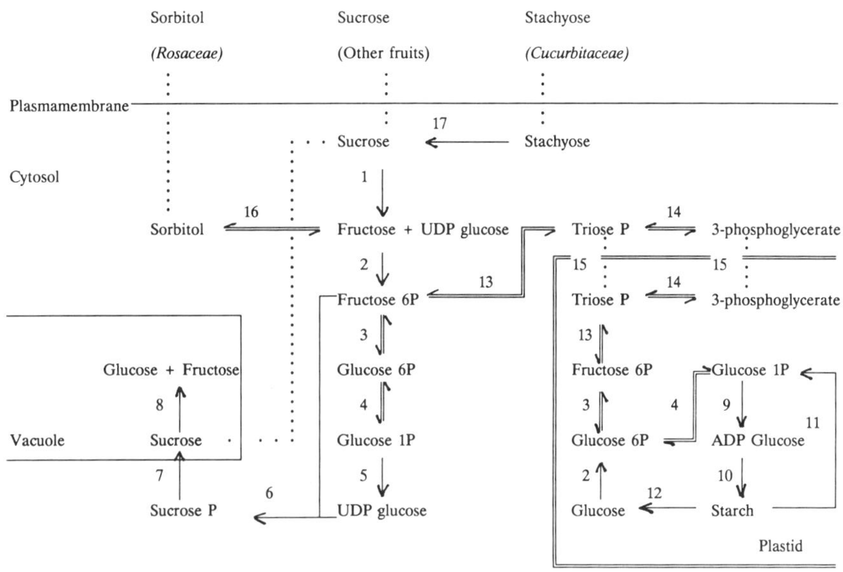

FIG. 1. General scheme of sugar metabolism in fleshy fruits.

1.	Sucrose synthase	EC 2.4.1.13*
2.	Hexokinase	EC 2.7.1.1*
3.	Phosphohexose isomerase	EC 5.3.1.9*
4.	Phosphoglucomutase	EC 2.7.5.1
5.	UDP glucose pyrophosphorylase	EC 2.7.7.9*
6.	Sucrose phosphate synthase	EC 2.4.1.14*
7.	Sucrose phosphatase	EC 3.1.3.24*
8.	Invertase	EC 3.2.1.26*
9.	ADP glucose pyrophosphorylase	EC 2.7.7.27*
10.	ADP glucose-starch glucosyl transferase	EC 2.4.1.21*
11.	Phosphorylase	EC 2.4.1.1*
12.	α-Amylase and	EC 3.2.1.1*
	β-Amylase	EC 3.2.1.2*
13.	See Fig. 2	
14.	Glyceraldehyde-phosphate dehydrogenase	EC 1.2.1.12
	Phosphoglycerate kinase	EC 2.7.2.3
15.	Phosphate translocator	
16.	Sorbitol dehydrogenase	EC 1.1.1.14*
17.	α-Galactosidase	EC 3.2.1.22*

* Activities detected in fruits.

pyrophosphorylase (EC 2.7.7.10), UDP-galactose-4-epimerase (EC 5.1.3.2) and sucrose synthase (EC 1.4.1.13). This would also imply that β-galactosidase (EC 3.2.1.23) and galactokinase (EC 2.7.1.6) are present to complete the pathway of stachyose conversion to sucrose.

Sucrose is the major free sugar found in the ripe melon fruit; its level continuously rises in the mesocarp during fruit development, while glucose and fructose remain either at a constant level or show a tendency to decrease in concentration (Lingle & Dunlap, 1987; McCollum *et al.*, 1988). The same authors have shown that the increase in sucrose concentration is closely correlated to a decline in acid invertase (EC 3.2.1.26) activity to very low levels in the mature fruit. However, acid invertase activity bore no correlation to reducing sugar content during fruit development. Sucrose synthase activity increased throughout fruit development but the direction of its activity *in vivo* was not established as synthetic or degradative. Lingle and Dunlap (1987) did record an increase in sucrose phosphate synthase (EC 2.4.1.14) activity, which was correlated with the accumulation of sucrose in the ripening fruit, as well as with the different levels of this sugar found in the green and orange sections of the mesocarp tissue. The enzyme catalyzing the final step to sucrose, sucrose phosphate phosphatase (EC 3.1.3.24), has not been detected in melon fruits.

The sucrose content of melons after harvest decreases, but there is little change in the levels of the two free reducing sugars, glucose and fructose (Lingle *et al.*, 1987). The decrease in sucrose was correlated with an increase in sucrose synthase activity in some, but not all, storage regimes. The activity of sucrose phosphate synthase in stored fruit generally declined. Very low acid invertase activity was detected, but considerable invertase activity at neutral pH was measured at a fairly constant level throughout storage. In all probability, in the harvested fruit, the combined activities of all three enzymes are involved in the turnover of sucrose, the final level of which is determined by the equilibrium established between their relative levels of activity and therefore the activity of no individual enzyme can be highly correlated to the changes occurring in sucrose levels in the harvested fruits.

Cell wall metabolism

Though muskmelon fruit soften noticeably with ripening, no cell wall changes have been observed which could account for this textural change (Lester & Dunlap, 1985). Pectin esterase and β-1,4-glucanase activities were measured in the mature fruit, but tended to decline with softening. No PG activity was detected by the viscometric method of assay.

EBENACEAE

Persimmon (*Diospyros kaki*)

Phenol metabolism

The unique characteristic of all persimmon cultivars, whether astringent or non-astringent, is their high condensed tannin content. In the non-astringent cultivars, the soluble tannin is immobilized before the fruit becomes soft and ripe. In the astringent cultivars the tannin remains soluble until the fruit is overripe, unless it is treated to induce the tannin to polymerize or coagulate, and become insoluble. The natural or induced change in tannin solubility does not appear to be a direct result of enzymic activity. It has been suggested, with supporting in-vitro evidence, that the soluble tannin polymerizes by reaction with acetaldehyde to form acetals (Matsuo & Ito, 1982). Acetaldehyde itself derives from enzymic reactions, either in the developing seeds of the maturing fruit (Sugiura & Tomana, 1983; Taira *et al.*, 1986), or under severely reduced O_2 levels (Pesis & Ben-Arie, 1984). An important enzyme involved in this reaction is alcohol dehydrogenase (EC 1.1.1.1; ADH). High ADH activity was detected in the mesocarp of persimmons (Nakamura, 1973*a*), but there was no correlation between the level of enzyme activity and the acetaldehyde or ethanol content of the fruits (Nakamura, 1973*b*).

Tannins are polyphenolic compounds and may serve as substrates for PPO (EC 1.10.3.1). This enzyme has been implicated in the browning which sometimes occurs in persimmon pulp tissue, following the removal of astringency (Sugiura *et al.*, 1975). Ethanol treatment of astringent persimmons, while still on the tree, induced total browning of the flesh with a concomitant rise in PPO activity (Sugiura *et al.*, 1985).

A different hypothesis has been proposed to explain the loss of astringency in ripening fruits. Ozawa *et al.* (1987) suggest that in some cases the tannins are immobilized by complexation with water-soluble pectin fragments, resulting from enzymic degradation of cell walls in the softening fruit. The mature persimmon becomes extremely soft when fully ripe and non-astringent, but very little is known about cell-wall degradation during softening. Pectin esterase (EC 3.1.1.11; PE) activity increased in astringent persimmon fruit following treatment with ethephon (Awad, 1985). A climacteric rise in respiration occurred at the same time but no increase in PE, respiration or softening occurred in untreated fruit. However, Awad (1985) did not use any protective agents against tannins during the enzyme extraction procedure and this could account for the different activities measured in the non-astringent ethephon-treated fruit, and the astringent control. A comparison of enzyme activity during the softening of astringent and non-astringent 'Triumph' persimmons (Ben-Arie—unpublished data) showed no change in PE activity as the fruit softened; in this work the enzyme was extracted in the presence of PVP or polyethylene glycol. However, endo-PG (EC 3.2.1.15) activity increased and there was a correlation between fruit firmness and PG activity ($R^2 > 0{\cdot}6$). After storage of the fruit, correlations were also found between β-1,4 glucanase (EC 3.2.1.4) activity and fruit softening ($R^2 > 0{\cdot}5$), but at harvest this was generally not the case (Roizman, Ben-Arie and Blumenfield, unpublished data).

LAURACEAE

Avocado (*Persea americana*)

General

The avocado is unusual among fleshy fruits in that it accumulates lipid and cell division continues through fruit development. The distinctive physiology and biochemistry of avocadoes have been reviewed (Biale & Young, 1971; Biale, 1975; Ahmed & Barmore, 1980). Some varieties of this climacteric fruit remain attached but do not ripen on the tree for six months or more after maturity. However, West Indian cultivars abscise and begin to ripen when they reach maturity (Spalding, 1976). Multiple forms of various enzymes in avocado are proving useful as markers in genetic studies (Torres *et al.*, 1978; Torres, 1980).

Carbohydrate metabolism

The plastids of immature fruit mesocarp contain starch grains, whose size and number diminish as the fruit ripens (Pesis, *et al.*, 1978). Amylase (EC 3.2.1.2) activity is high in immature fruit, and decreases with growth, but increases again on ripening. Its action in degrading starch probably contributes to the climacteric rise in respiration and associated metabolic activity, since lipid is not utilized for these purposes. Three hexokinases (EC 2.7.1.1) of similar molecular weight (85 kDa) are involved in glucose and fructose utilization in glycolysis (Copeland & Tanner, 1988). The glucose-phosphorylating activity is associated with mitochondria, but two highly specific fructokinases (EC 2.7.1.4) are soluble enzymes with different kinetic properties. ATP is the preferred phosphoryl donor for all three enzymes and the increase in this during the climacteric was recently confirmed using intact fruit and [^{31}P] NMR (Bennett *et al.*, 1987). However, contrary to earlier suggestions (Solomos & Laties, 1974), Bennett *et al.* (1987) found that the activity of phosphofructokinase (EC 2.7.1.11) increased only a little during the climacteric. Increased glycolytic flux seemed to depend on activation of pyrophosphate: fructose-6-phosphate phosphotransferase (EC 2.7.1.90) caused by a 90% rise in the concentration of fructose-2,6-bis-phosphate.

Respiration

Extensive studies of the tricarboxylic acid cycle and mitochondrial activity in the 1960s (Biale & Young, 1971) were followed by investigation of the role of the cyanide-insensitive pathway in the climacteric (Solomos & Laties 1974). Theologis and Laties (1978) concluded that the fruit's capacity for cyanide-insensitive respiration was not engaged, even at the climacteric peak. Nevertheless, Moreau and Romani (1982) found that the climacteric was associated with a rise in the malate oxidizing capacity of purified mitochondrial preparations. This capacity was measured in state 4 mitochondria and involved malic enzyme (EC 1.1.1.39) and the alternate oxidase. Under other conditions malate was oxidized through malate dehydrogenase (EC 1.1.1.37) and the cytochrome system. The relevance of tissue slice and mitochondrial studies to the intact fruit is an open question, and the pathway operative during the climacteric is still undefined.

The activity of succinoxidase (EC 3.9.9.1) is highest in pre-climacteric fruit and declines as the fruit ripens (Kosiyachinda & Young, 1977). This continuous change in mitochondrial membrane activity is not associated with any change in membrane density during ripening (Dallman *et al.*, 1988) though ultrastructural changes in avocado mitochondria during ripening have been recorded (Platt-Aloia & Thomson, 1981). An increase in buoyant density is found for the Golgi and plasma membrane during avocado fruit ripening. Part of these changes may be associated with a decrease in the membrane glycolipid:protein ratios (Dallman *et al.*, 1988).

Ethylene synthesis

Ethylene synthesis in avocado tissue is insensitive to the inhibitor of ACC synthase (EC 4.4.1.14), aminoethoxyvinylglycine (AVG; Baker *et al.*, 1978). However, the operation of the ACC pathway was later confirmed and, although AVG is taken up by the tissue, it is thought not to reach the site of ACC synthase activity (Baker *et al.*, 1982). The mesocarp is the sole site of ethylene synthesis during the climacteric but the seed coat of young fruitlets produces ethylene at a high rate (Adato & Gazit, 1977) and recently it has been shown that this tissue and the cotyledons contain high ACC synthase activity (Sitrit *et al.*, 1987). There appears to be higher activity of EFE in mature pericarp tissue (Sitrit *et al.*, 1987). The seed coat has very low EFE activity and much of the ACC it produces is probably conjugated to form the malonyl derivative.

A cell free extract from avocado forms ketomethylthiobutyric acid and hydroxythiobutyric acid from methylthioribose in the presence of ATP, or methylthioribose phosphate alone (Kushad *et al.*, 1983). Methylthioribose kinase (EC 2.7.1.100) has been detected in avocado tissue (Kushad *et al.*, 1982) and these reactions are involved in the 'salvage pathway' for synthesis of methionine established by Miyazaki and Yang (1987) in avocado mesocarp tissue. This pathway recycles the byproducts of the action of ACC synthase, so that ethylene synthesis can continue.

Phenol metabolism

More than 20 phenolic compounds have been found in the mesocarp but most are unidentified (Golan *et al.*, 1977). Only the first enzyme of biosynthesis, phenylalanine ammonia lyase (EC 4.3.1.5), has been reported in avocado fruit (Lelyveld & Bower, 1984). Most studies of phenol metabolism relate to meso- and endocarp discoloration which are major post-harvest problems of avocado. Chilling is a frequent cause of gray or dark-brown discoloration of the mesocarp (Lelyveld & Bower, 1984). Pre-harvest water stress (Bower & Lelyveld, 1985; Cutting *et al.*, 1988), harvest delay, and postharvest waxing (Lelyveld & Bower, 1984), all aggravate the problem, whereas calcium dips (Chaplin & Scott, 1980; Rensburg & Engelbrecht, 1986) and low oxygen storage (Spalding & Reeder, 1975) reduce symptom development.

Variation in the discoloration of avocado cultivars depends on their PPO (EC 1.10.3.1) activity (Kahn, 1975), and their content of phenol substrate (Golan *et al.*, 1977; Sharon-Raber & Kahn, 1983). PPO is found as a soluble enzyme and associated with microbodies but not in plastids or mitochondria (Sharon & Kahn, 1979). During ripening, activity shifts from the soluble to the particulate fraction; soluble activity can be enhanced by the addition of sodium dodecyl sulphate (Kahn, 1977*a*). Five iso-enzymes were separated according to their molecular weights (14–400 kDa) and one of these (mol. wt 28 kDa) was further separated into four to six components by electrophoresis (Dizik & Knapp, 1970). Different iso-enzyme patterns were encountered in varieties with different browning potentials (Kahn, 1976*a*). However, partially purified enzyme preparations from these varieties showed similar optima at pH 6, substrate specificity, heat stability, sensitivity to inhibitors and distribution of activity between soluble, latent and particulate fractions (Kahn, 1977*a,b*). However, differences in the properties of individual iso-enzymes may have been obscured in the mixture.

Non-substrate phenols can inhibit (*p*-cresol, *p*-coumaric and protocatechuic acids) or promote (resorcinol and phloroglucinol) avocado PPO activity (Kahn 1976*b*). Thus, two varieties with the same *o*-diphenol content may show different browning potential, depending on the other phenols present as well as their PPO activity (Golan *et al.*, 1977).

PPO, *in vitro,* can hydroxylate monophenols; this 'tyrosinase' activity is inhibited by the quinones produced by the oxidation of *o*-diphenols, unless a reducing agent such as ascorbate is present to inhibit this reaction. Tyrosinase activity starts after a lag which can be abolished by adding hydrogen peroxide and Kahn (1983) suggests that peroxide and catalase levels may affect the browning potential of intact tissue. Lelyveld and Bower (1984) observed correlations of peroxidase as well as PPO activity with the browning of avocadoes in restricted ventilation. It seems that browning of cut surfaces and of intact fruit during storage may be influenced by different factors.

Catalase (EC 1.11.1.6) and peroxidase (EC 1.11.1.7) activities are not related to the browning potential of cut tissue (Sharon-Raber & Kahn, 1983; Zauberman *et al.*, 1985). *Trans*-cinnamate-4-monoxygenase (EC 1.14.13.11) has been isolated from avocado mesocarp microsomes (Karasaki *et al.*, 1983).

Lipid metabolism

Avocado fat contents can vary from 1–2% to >30% (Biale & Young, 1971; Ahmed & Barmore, 1980). The commercial significance of the oil content has led to its use as a maturity index (Lewis, 1978; Lewis *et al.*, 1978). Fatty acid composition varies with cultivar, stage of ripening, anatomical region of the fruit and growing area (Ahmed & Barmore, 1980; Ratovohery *et al.*, 1988). The major fatty acid is always oleic followed by palmitic and linolenic acids (Tango *et al.*, 1972; Ahmed & Barmore, 1980).

Fat-synthesizing enzymes are active in the mesocarp tissue (Simoni *et al.*, 1967; Caughey & Kekwick, 1982). Acetyl-CoA carboxylase (EC 6.4.1.21) appeared to be located in the plastid lamellae (Mohan & Kekwick, 1980). Acyl carrier protein (ACP) has been isolated and purified and found to be similar to that from spinach in its reaction with synthetases. The molecular weight is 11 kDa which is slightly larger than the spinach protein (10 kDa). Fatty acyl-thiokinase (EC 6.2.1.3) was isolated from avocado and catalyzed the synthesis of long chain *S*-acyl coenzyme A (Galliard & Stumpf, 1968).

Avocado plastids have yielded two fatty acid synthetase systems, a soluble system forming stearate from malonyl CoA and a particulate system forming palmitate and oleate from acetate or acetyl CoA. The preparations required fortification by *Escherichia coli* ACP (Weaire & Kekwick, 1975*a,b*). Harwood and Stumpf (1972) obtained a fatty acid synthetase system from whole avocado mesocarp which synthesized palmitate from malonyl-CoA. NADH and NADPH specific 3-oxoacyl-ACP reductases (EC 1.1.1.100), enoyl-ACP reductase (EC 1.3.1.9) and ACP-malonyl transferase (EC 2.3.1.39) have been separated from the plastid soluble synthase system and found to resemble the procaryotic type II enzymes of *E. coli* and *Euglena*. The molecular weight of malonyltransferase is about 40 kDa, that of enoyl ACP reductase is about 62 kDa and NADH and NADPH specific ACP reductase have molecular weights of 168 and 46 kDa, respectively. The pH optimum for the enzymes is between 6·8 and 7·5, except for malonyltransferase which has an optimum near pH 8·0. The NADH-specific ACP reductase is more tightly bound to the membrane than the other enzymes (Caughey & Kekwick, 1982).

Knowledge of lipid synthesis has developed slowly since the demonstration of microsomal glycerol-3-phosphate acyl transferase (EC 2.3.1.15), phosphatidate phosphatase (EC 3.1.3.4) and diacyl glycerol acyltransferase (EC 2.3.1.20) by Barron and Stumpf (1962). Enzymes subsequently detected have been shown to be located in oil bodies (Mancha & Garcia, 1982). These included stearyl-ACP desaturase (EC 1.14.99.6) and an acyl thioesterase of low specificity; acyl-ACPs were better precursors for triacyl glycerol synthesis than acyl CoAs (Shine *et al.*, 1976). In addition, Mancha and Garcia (1982) reported an acyl-ACP hydrolase (EC 3.1.2.14) with a pH optimum of 9·5, and a molecular weight of 70–80 kDa, which hydrolyzed oleyl ACP faster than other acyl ACPs. Also present are two acyl-CoA hydrolases (EC 3.1.2.2) with optimum activity at pH 8·4–8·6, molecular weights of 70–80 kDa and 200 kDa and a broad substrate specificity.

Fruit ripening results in increased levels of lipid and free fatty acids (Tango *et al.*, 1972). There has, however, been no report of a fat degrading system during ripening (Biale & Young, 1971). A lipoxygenase (EC 1.13.11.12) has been purified from peel tissue and has optimum activity at 36°C and pH 6·8 and a molecular weight of 74 kDa (Marcus *et al.*, 1988). The pI is 5·9, close to that for other lipoxygenases, with many similarities in amino acid composition. The enzyme was investigated for its role in the degradation during ripening of an antifungal diene which resembled linoleic acid. Lipoxygenase increased 80% in specific activity in ripening fruit and the enzyme probably occurs in the mesocarp also.

Cell-wall metabolism

Ultrastructural studies suggest degradation of the middle lamella and cell-wall matrix (i.e. pectin degradation) as well as disorganization and loss of fibrillar material (i.e. cellulose) (Pesis *et al.*, 1978; Platt-Aloia *et al.*, 1980). β-1,4 Glucanase, PG and PE have been reported in ripening avocado fruit (Awad & Young, 1979). PG is not detectable in the preclimacteric fruit, but increases continuously during the climacteric and the post-climacteric phases (Reymond & Phaff, 1965; Barash & Khazzam, 1970; Zauberman & Schiffman-Nadel, 1972) and apparently before the increase in ethylene synthesis (Fuchs *et al.*, 1986). β-1,4 Glucanase is present early in ripening; its increase in activity coincides with that in respiration, but precedes the increase in PG activity (Pesis *et al.*, 1978; Awad & Young, 1979). The early presence of the cell wall hydrolases has revived the idea that their (oligosaccharide) products might be a signal for other aspects of ripening (Fuchs *et al.*, 1986), but their own regulation requires some explanation.

The level of PE (EC 3.1.1.11) declines from a maximum value at the time of picking to a low level early in the climacteric (Zaubermann & Schiffman-Nadel, 1972; Gertman & Fuchs, 1974; Awad & Young, 1980). PE was maximally extracted with 0·4 M sodium chloride and the salt slightly stimulated the assay (Awad & Young, 1980). It has been reported that the enzyme is reversibly inactivated by ethylene at 6°C (Gertman & Fuchs, 1974) and these authors questioned the usual assumption that its activity is required before PG can degrade pectin. However, changes in pectin composition during ripening support sequential activity of the enzymes (Dolendo *et al.*, 1966; Rouse & Barmore, 1974).

PG activity spreads from the blossom end towards the stem end during ripening (Reymond & Phaff, 1965). The enzyme causes random hydrolysis of polygalacturonic acid with a pH optimum of 5·5, and is partially inhibited by ammonium, potassium, and phosphate ions. There is some evidence for a natural inhibitor of avocado PG in mesocarp tissue, which may be a free fatty acid. The enzyme loses 60% of its activity after 10 min exposure to 50°C (Barash & Khazzam, 1970). Since this work, little attention has been given to PG in avocadoes; the focus has been on the glucanase, whose role is more securely established here than in any other fruit.

The avocado β-1,4 glucanase (EC 3.2.1.4) has a pI of 4·7 (Awad & Lewis, 1980), and although it is often referred to as 'cellulase' it does not degrade cellulose (Hatfield & Nevins, 1986). The enzyme does hydrolyse carboxymethylcellulose and mixed (1,3) (1,4)-β-D-glucans from *Avena* endosperm cell walls but it will only hydrolyse where there is an adjacent 1,4 linkage and a minimal chain length of four residues. It releases mainly galactose- and arabinose-containing fragments from fruit cell wall preparations. More xylose and glucose were released from walls isolated from ripe than from unripe fruit and the substrate *in vivo* may be a xyloglucan.

The increase in the glucanase during ripening results from an increase in the corresponding mRNA at the onset of ripening (Christoffersen *et al.*, 1984). The mature protein is a glycoprotein with a molecular weight of 54 kDa (Bennett & Christoffersen, 1986). The precursor of this protein has two high-mannose oligosaccharide side chains which are possibly modified during transport via the endoplasmic reticulum and Golgi to the cell wall. Multiple forms of the glucanase were detected immunologically after two-dimensional PAGE (Tucker *et al.*, 1985) and a Southern blot analysis of digested genomic DNA suggested a multigene family (Tucker *et al.*, 1987).

Other enzymes
Arginine decarboxylase (EC 4.1.1.19), a key enzyme in polyamine metabolism, has been isolated from mesocarp tissue (Winer *et al.*, 1984). The pH optimum is between 8 and 9 at 60°C in the presence of 1·2 mM manganese chloride, 2 mM dithiothreitol and 0·06 mM pyridoxal phosphate. The heat stability of the avocado enzyme is exceptional; spermine caused less than 5% inhibition of the enzyme from fully developed mesocarp and 65% inhibition of enzyme from 4-week-old fruitlet seed coat. Most of the activity is present in the seed coat and declines as the coat shrivels in the maturing fruit. This is accompanied by a decline of putrescine, spermidine and spermine in the mesocarp. Since these polyamines inhibit ethylene synthesis in discs of both pre- and post-climacteric fruit, Winer and Apelbaum (1986) attributed to them a role in the regulation of fruit ripening.

MUSACEAE

Banana (*Musa acuminata*)

Carbohydrate metabolism
Starch accounts for about 25% of the fresh weight of pre-climacteric banana fruits, whereas sugars account for less than 1%. When the fruit ripens, the starch is converted to sucrose, glucose and fructose in about five days, and a number of authors have examined the enzymological basis for this transformation. Electrophoretic separations revealed the presence of three phosphorylase (EC 2.4.1.1) iso-enzymes and two iso-enzymes each of α-amylase (EC 3.2.1.1) and β-amylase (EC 3.2.1.2) at all stages of fruit ripening, but conventional assays showed little or no amylase activity in preclimacteric fruit (Young *et al.*, 1975).

Singh and Sanwal (1975) also found three phosphorylases and two of the iso-enzymes were more strongly inhibited by an extract of phenolics from immature fruit than by an extract from mature fruit. This could account for the increase in activity of phosphorylase during ripening, observed by Surendranathan and Nair (1973), who used no protective agents in their extraction medium. Using cysteine and PVP, Areas and Lajolo (1981) recorded almost constant phosphorylase activity during ripening; the activity of this enzyme was sufficient for the observed rate of starch degradation. However, in a different variety of banana, Chitarra and Lajolo (1981) observed a decline in phosphorylase activity and concluded that α-amylase was responsible for starch degradation. Young *et al.* (1975) postulated that phenolic inhibitors prevented starch degradation in the pre-climacteric banana, but it is likely that the enzyme and the inhibitors exist in separate cellular compartments in the intact fruit. Singh and Sanwal (1973) reported allosteric control of banana phosphorylase by tyrosine and ATP, but, more generally, enzymes of starch degradation from higher plants do not appear to be subject to fine control (Preiss, 1982). Coarse control, by synthesis of hydrolytic or phosphorolytic enzymes, does not seem to operate in the banana and regulation of starch concentrations by the rate of synthesis (with constant degradation) seems unlikely. Conversion of starch to sugars is common to many ripening fruits; it is perhaps most dramatic in the banana, and this fruit could be the best experimental subject for elucidation of the underlying controls.

Although Young *et al.* (1975) found that the climacteric increase in respiration preceded a decline in starch concentrations, these two events seemed to be more closely associated in other papers (Surendranathan & Nair, 1973; Areas & Lajolo, 1981; Chitarra & Lajolo, 1981). Salminen and Young (1975) associated the climacteric rise with an increase in the activity of (ATP dependent) phosphofructokinase (EC 2.7.1.11), resulting from a loss of control by fructose-6-phosphate. This led to an increase in fructose-1,6-bisphosphate which other authors have observed. Recently, Beaudry *et al.* (1987) have argued that the activity of this enzyme is inadequate to support the five-fold increase in respiration that occurs, and invoke the alternate pyrophosphate-fructose-6-phosphate phosphotransferase (EC 2.7.1.90) as the regulator of glycolytic flux (Fig. 2). A transient increase in the activity of this enzyme at the onset of ethylene production was associated with an increase in a possible activator, fructose-2,6-biphosphate. The decline in the concentration of this compound later in ripening would favor the action of fructose bisphosphatase (EC 3.1.3.11). Involvement of this enzyme in the pathway from starch to sugar would be required if the amyloplast exported triose-phosphate, rather than hexose(-phosphate). Aldolase (EC 4.1.2.13) activity has been observed to remain constant during ripening (Lal *et al.*, 1974).

Terra *et al.* (1983) assumed that sucrose would be formed directly from glucose-1-phosphate and demonstrated activities of relevant enzymes (UDP glucose pyrophosphorylase, EC 2.7.7.9, and sucrose synthase, EC 2.4.1.13) which were constant throughout fruit ripening. Sucrose synthesis (Fig. 1) is more likely to involve sucrose-phosphate synthase (EC 2.4.1.14) which was found in the fruit by Shukla *et al.* (1973), who also detected ADP-glucose pyrophosphorylase (EC 2.7.7.27), required for starch synthesis. Both of these groups detected invertase (EC 3.2.1.26) and this activity increased with ripening (Terra *et al.*, 1983). Purified banana invertase is a glycoprotein, containing 25% carbohydrate, with mannose and glucosamine predominant, and a molecular weight of 220 kDa (Sum *et al.*, 1980).

Organic acid metabolism

As in other fruits, oxygen uptake by mitochondria from bananas remains coupled to phosphorylation during ripening; succinic dehydrogenase (EC 1.3.99.1) and mitochondrial protein did not change during ripening (Haard &

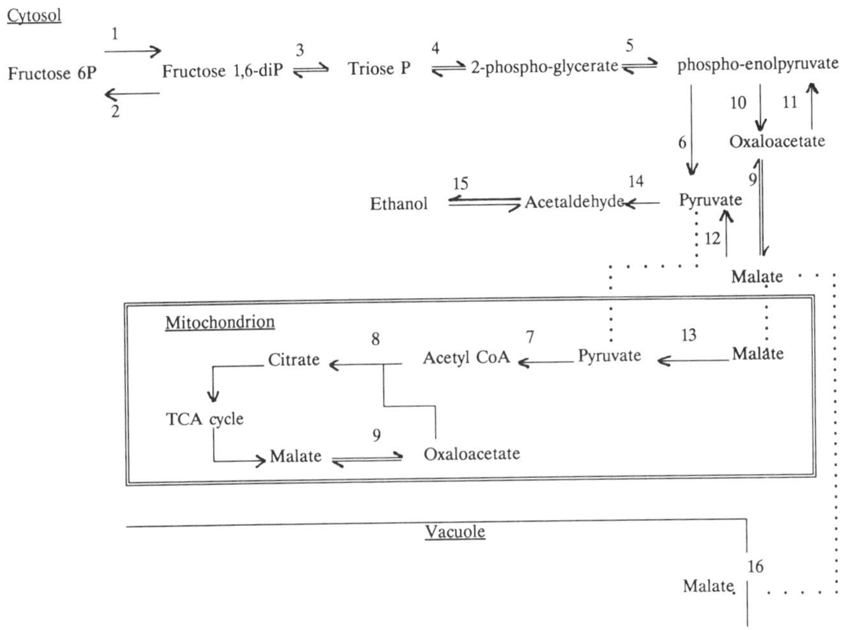

FIG. 2. General scheme of organic acid metabolism in fleshy fruits.

1.	6-Phosphofructokinase or	EC 2.7.1.11*
	Pyrophosphate-fructose-6-phosphate 1-phosphotransferase	EC 2.7.1.90*
2.	Fructose-bisphosphatase	EC 3.1.3.11*
3.	Aldolase	EC 4.1.2.13*
	Triose phosphate isomerase	EC 5.3.1.1
4.	Glyceraldehyde-phosphate dehydrogenase	EC 1.2.1.12
	Phosphoglycerate kinase	EC 2.7.2.3
	Phosphoglycerate mutase	EC 5.4.2.1
5.	Enolase	EC 4.2.1.11
6.	Pyruvate kinase	EC 2.7.1.40
7.	Pyruvate dehydrogenase	EC 1.2.4.1 EC. 2.3.1.12 EC. 1.6.4.3
8.	Citrate synthase	EC 4.1.3.7
9.	Malate dehydrogenase	EC 1.1.1.37*
10.	Phosphoenolpyruvate carboxylase	EC 4.1.1.31*
11.	Phosphoenolpyruvate carboxykinase	EC 4.1.1.39*
12.	NADP malic enzyme	EC 1.1.1.40*
13.	NAD malic enzyme	EC 1.1.1.39
14.	Pyruvate decarboxylase	EC 4.1.1.1*
15.	Alcohol dehydrogenase	EC 1.1.1.1*
16.	Malate translocator (hypothetical)	

* Activities detected in fruits.

Hultin, 1970). In agreement with these observations, Theologis and Laties (1978) found no evidence that the (non-phosphorylating) alternate oxidase participated in oxygen uptake in ripening bananas. However, Skakoun and Daussant (1975), in a pioneering immunological study, observed increases in malate dehydrogenase (EC 1.1.1.37) iso-enzymes, both in activity and antigenic reaction in ripening bananas. Surendranathan and Nair (1976) obtained radiochemical evidence for the operation of the glyoxylate pathway in bananas; this was enhanced after irradiation and key enzymes in the pathway (malate synthase, EC 4.1.3.2 and isocitrate lyase, EC 4.1.3.1) were detected in irradiated fruit. Lal *et al.* (1974) found increases in transaminase (EC 2.6.1.2) activity with glutamate as donor and pyruvate or oxaloacetate as acceptor, during ripening.

Cell wall metabolism

Hultin and Levine (1965) reported an increase in PE activity during ripening, but Brady (1976), who took precautions against phenols, found essentially constant activity. The multiple forms of the enzyme encountered by Hultin *et al.* (1966) and Markovic *et al.* (1975) may also have resulted from interactions with phenols. Brady (1976) obtained two iso-enzymes with pI values at 8·9 and 9·4, but similar molecular weight, salt activation and inhibition by polyols. Markovic *et al.* (1975) reported the presence in banana of enzyme activities which released reducing sugar from polygalacturonate and cleaved digalacturonic acid. They described these as endo- and exo-PG (EC 3.2.1.15 and EC 3.2.1.67), respectively, but give little other information about properties or changes in activity with ripening.

Phenol metabolism

There have been many studies on PPO (EC 1.10.3.1) activity in bananas; the enzyme exists in multiple forms and crude preparations have the capacity to hydroxylate monophenols as well as to oxidize polyphenols (Deacon & Marsh, 1971; Montgomery & Sgarbieri, 1975). A partially purified enzyme, which still contained four iso-enzymes, did not oxidize tyrosine, but had a high affinity for dopamine, the predominant polyphenol in the fruit (Galeazzi & Sgarbieri, 1981). In contrast, Thomas and Janave (1986) associated *p*-cresol (monophenol) oxidizing activity with several iso-enzymes. Phenoloxidase activity changes little during fruit ripening (Montgomery & Sgarbieri, 1975) and variation in the browning of damaged tissue is correlated with ascorbate and dopamine levels rather than with oxidase activity (Weaver & Charley, 1974).

Peroxidase activity increases during the ripening of bananas; the increase is mainly in cell-wall bound enzyme (Haard, 1973).

Other enzymes

One of the most pronounced increases in enzyme activity recorded in ripening bananas is that in phosphatase (EC 3.1.3.2) (de Leo & Sacher, 1970); the role of this largely membrane-bound enzyme has never been explained. Skakoun and Daussant (1975) recorded an increase of ADH (EC 1.1.1.1); apparently ethanol does accumulate in ripening bananas (Hyodo *et al.*, 1983). Looney and Patterson (1967) found an increase in chlorophyllase (EC 3.1.1.14) on ripening.

MYRTACEAE

Guava (*Psidium guajava*)

Various enzymes

As guavas ripen they lose starch and accumulate glucose and fructose, but little sucrose (Augustin & Osman, 1988). This is consistent with an increase in invertase (EC 3.2.1.26) and amylase (EC 3.2.1.1) activities during ripening (Mowlah & Itoo, 1982).

Pectin content and PE (EC 3.1.1.11) activity increase during growth of the fruit, and both decline during ripening (Shastri & Shastri, 1975).

Guava has a reputation for high ascorbic acid content (350 mg/g) but the presence of ascorbic acid oxidase (EC 1.10.3.3) (Jimenez, 1947) leads to a risk of loss of the vitamin during processing of fruit pulp (Jimenez, 1947; Rodriguez *et al.*, 1971).

As fruit ripening progresses, astringency disappears and there is a decrease in polyphenolics, concomitant with an increase in polyphenoloxidase (EC 1.10.3.1) activity (Mowlah & Itoo, 1982). The highest levels of activity are found around the inner flesh and surrounding the vascular bundles (Jimenez, 1947). Peroxidase also has a similar tissue distribution.

ROSACEAE

Apple (*Malus pumila*)

Carbohydrate metabolism

Sorbitol is the main translocated carbohydrate in apple, as in other members of the Rosaceae (Fig. 1), but the fruit accumulates fructose, sucrose and glucose (in order of importance) as soluble reserves, as well as starch (Beruter, 1985). Immature apple fruits contain an acid invertase (EC 3.2.1.26), which is mainly wall-bound (Yamaki & Ishikawa, 1986), and they should be able to utilize sucrose. Later, sorbitol metabolizing enzymes predominate. In leaves, sorbitol-6-phosphate dehydrogenase (EC 1.1.1.140; S6PDH) is involved in the conversion of photosynthate (fructose) to translocate (Yamaki & Ishikawa, 1986). Some photosynthesis occurs in immature apple fruit (Jones, 1981); they contain S6PDH but this activity declines with growth (Beruter, 1985; Yamaki & Ishikawa, 1986). The sorbitol dehydrogenase (EC 1.1.1.14; SDH) activity, which predominates later in development, converts sorbitol to fructose without the expenditure of ATP required prior to action of S6PDH. Beruter (1985) reported that NAD-dependent SDH appears in Golden Delicious apples in July, when enzyme activity increases ten-fold in one week before declining up to late September. Yamaki and Ishikawa (1986) found NAD-SDH activity in Jonagold apples from May onwards. Activity was higher in June and September than in August, whether quoted on a fresh weight or on a per fruit basis. They also report an NADP-dependent SDH, which increased throughout development to become the predominant activity. This was correlated with fructose accumulation, whereas in the work of Beruter (1985), the NAD-SDH seemed most likely to be responsible. Beruter (1985) could not detect sorbitol oxidase as a soluble enzyme; Yamaki and Ishikawa (1986) found it mainly as a cell-wall-bound enzyme, and always as a minor activity in relation to the dehydrogenases.

Organic acid metabolism

Apple fruits contain phosphoenol pyruvate (PEP) carboxylase. With this activity the fruit could function as a C4 plant; the carboxylase could trap CO_2 in a 4 carbon acid pool from which malic enzyme could release CO_2 for photosynthetic fixation (Blanke *et al.*, 1987). However, photosynthesis in the fruit is a small fraction of its carbon economy (Jones, 1981) and the demand on a system for trapping and regenerating carbon dioxide should be low. The main role of PEP carboxylase is likely to be the accumulation of organic acids, especially malate in the vacuole.

Slices of apple tissue decarboxylate malate to pyruvate with an NADP malic enzyme (EC 1.1.1.40; ME) (Dilley, 1962), pyruvate to acetaldehyde with carboxylase (EC 4.1.1.1), and reduce the acetaldehyde to ethanol with ADH (EC 1.1.1.1) (Hulme & Rhodes, 1971) (Fig. 2). Since apple ADH accepts NADPH as a cofactor (Bartley & Hindley, 1980), NADP can be regenerated so that ME action can continue. If slices are cut from apples at different times during their passage through the respiration climacteric, a correlation is observed between stimulation of CO_2 production by malate applied to the slices and whole fruit respiration (Knee, 1971). However, whole apples do not accumulate acetaldehyde or ethanol and it is unlikely that the pathway stimulated in slices occurs in the intact fruit.

The NADP-ME is a soluble enzyme with a requirement for manganese and other properties similar to the enzyme from other plant sources (Dilley, 1966). Hulme *et al.* (1963) reported that mitochondria from apple peel lacked ME activity. However, they used NADP for their assays and an NAD-dependent ME (EC 1.1.1.39) is more typical of plant mitochondria, accounting for their capacity to utilize malate. Indeed, Hulme's group showed that apple mitochondria release CO_2 from malate as rapidly as from succinate (Hulme *et al.*, 1963). The higher respiratory quotient (RQ) of mitochondria with malate (Hulme *et al.*, 1963) than with succinate is consistent with participation of ME rather than malate dehydrogenase (EC 1.1.1.37) acting alone.

Malate is a major substrate of respiration in ripening apples (Fidler, 1951). The soluble ME is likely to be involved in its metabolism but the pyruvate produced would be metabolized by pyruvic decarboxylase only when its concentration exceeded the capacity of pyruvic dehydrogenase (EC 1.2.4.1). This would happen when slices are incubated with excess malate. In the intact tissue, malic enzyme may fit into a metabolic cycle which is enhanced during ripening. NADPH produced by ME activity could be utilized in synthetic processes. In animal and microbial systems, malic enzymes participate in shuttle mechanisms for acids and reduced nucleotides between cytoplasm and mitochondria. Since plant mitochondria have external NAD(P)H and malate oxidizing activities, the need for shuttle mechanisms is not obvious (Lance, 1981).

Great advances in the techniques of preparation of mitochondria were made in the early 1960s (Hulme *et al.*, 1964). The molar ratio of phosphorylation to oxygen uptake was between two and three and these appeared to remain constant during fruit ripening at 12°C (Jones *et al.*, 1965). Respiratory control ratio (RCR) is a more exacting criterion of functional integrity and this shows some decline during storage of apples at 0°C (Lin *et al.*, 1985).

Various authors have tried to relate the effects of high CO_2 concentration, used in storage of apples, to effects on mitochondrial enzymes. Apples in high CO_2 atmospheres accumulate succinate (Hulme, 1956), and CO_2 inhibits succinate dehydrogenase (EC 1.3.99.1) from other plant sources (Bendall *et al.*, 1960). Shipway and Bramlage (1973) argue for a more general effect of CO_2 on mitochondrial enzymes. Much of their data is consistent with an immediate effect on succinic dehydrogenase; CO_2 had less effect on oxidation of substrates remote from succinate in the TCA cycle (citrate, α-keto glutarate and NADH). The observed inhibition of pyruvate oxidation can be explained by the lack of the required oxaloacetate when the TCA cycle was blocked at succinate. Stimulation of malate oxidation by CO_2 raises again the possibility of distinct pathways of metabolism for this substrate. Interestingly, a lower RCR was observed with malate than other substrates except NADH (and citrate, but citrate oxidation was slow). This is consistent with the reported ability of plant mitochondria to oxidize these substrates using an 'alternate pathway' to bypass phosphorylation sites (Lance, 1981). Perhaps malic enzyme activity was favored by high CO_2 conditions; there is evidence for this in other plant tissues (Wager, 1974), although high CO_2 usually inhibits the decline of malate in apples (Fidler & North, 1967).

Cell wall metabolism

The only cell wall degrading enzymes which have been convincingly demonstrated in apple fruits are pectinesterase (EC 3.1.1.11), exo-PG (EC 3.2.1.67) and β-galactosidase (EC 3.2.1.23). Many other fruits contain endo-PG (EC 3.2.1.15), which is thought to cause them to soften as they ripen

(Bartley & Knee, 1982). The absence of this activity from apples is also indicated by the properties of the soluble pectin which accumulates on ripening. Its molecular weight is at least as high as that of pre-existing wall-bound and soluble pectins (Knee, 1978). The reported occurrence of alcohol-soluble galacturonic acid in apples (Liang *et al.*, 1982) appears to be an error arising from use of a non-specific colorimetric procedure and is not supported by earlier chromatographic studies (Ash & Reynolds, 1954). Observations with the electron microscope that cell walls in unripe apple tissue treated with endo-PG (EC 3.2.1.15) and cellulase (EC 3.2.1.4) resembled walls from ripe apples (Ben Arie *et al.*, 1979) are not conclusive evidence that the same enzymes were active during ripening. Perhaps most intriguing is the evidence from nuclear spin dynamics for the increased mobility of pectic carboxyl groups in fruit cell walls on ripening (Irwin *et al.*, 1984). In principle, this means only that some restraint on movement was lost, although the authors eliminated a number of possible explanations and favored PG as the agent of the change.

The exo-PG can be extracted at high ionic strength and resembles the enzyme from other fruits in its pH optimum (4·5–5), activation by calcium and molecular weight (58 kDa) (Bartley, 1978). Miyairi *et al.* (1975) purified PE from apple to homogeneity, and found it to resemble enzymes from other fruit sources in terms of pH optimum (7·3) and cation activation (0·2 M Na^+ or 0·05 M Ca^{2+}). The β-galactosidase from apple is unusual in its ability to release galactose from polymers and from plant cell walls (Bartley, 1974). Most of the activity remains in the wall fraction after extraction in 5 mM phosphate buffer and the wall-bound activity can be detected by incubation of apple discs with 4-nitrophenyl-β-D-galactoside (Bartley, 1977). Activity is present in unripe (pre-climacteric) fruit and increases by about 50% on ripening (Bartley, 1977). Its presence accounts for the loss of galactose from fruit cell walls, which can precede other aspects of ripening (Knee, 1973). The significance of this change in cell wall structure is not clear.

The persistence of high molecular weight pectin and the absence of autolytic activity from apples leads to problems in processing ripe fruit (Pollard & Timberlake, 1971). Addition of microbial PG is necessary for good yields of juice. If a clarified juice is made, this may develop a haze unless pectin has been adequately degraded or an agent such as honey is added to aid clarification. Apparently, protein in the honey coprecipitates with pectin (McLellan *et al.*, 1985).

Phenol metabolism

Phenylalanine ammonia lyase (EC 4.3.1.5; PAL) has been studied in apples mainly in relation to anthocyanin synthesis in the peel (Faragher & Chalmers, 1977). However, PAL is the first enzyme in the pathway specifically committed to the synthesis of phenolic compounds in general and anthocyanin is only a small fraction of the total phenols in the apple. Thus, it is not surprising that a green apple variety can contain as much PAL as a red variety, and in a red variety the rise in PAL occurs after the development of color (Blankenship & Unrath, 1988). The primary control factor for induction of PAL in apples appears to be light (Siegelman & Hendricks, 1958). Ethylene induces a further increase in enzyme activity in tissue held in the light and various forms of stress probably enhance PAL activity via an effect on ethylene production (Faragher & Chalmers, 1977). When PAL activity changes there is often an opposite change in the activity of a PAL inactivating system (Tan, 1980).

In addition to a complex mixture of phenolic compounds, apples contain PPO (EC 1.10.3.1) activity, so that when tissue is cut or disrupted, browning occurs. The enzyme is reported to be present as a single protein with a molecular weight of 26 kDa with optimum activity at pH 4·5 (Goodenough *et al.*, 1983) or two iso-enzymes with optima at pH 6·2 (Satjawatcharaphong *et al.*, 1983). However, Smith and Montgomery (1985) present evidence that binding of phenols during extraction may raise the pH optimum of PPO from pears. The high K_m for catechol (0·22 M) reported by Satjawatcharaphong *et al.* (1983), together with the high pH optimum, are both suggestive of denaturation. Goodenough *et al.* (1983) reported a K_m for 4-methylcatechol of 3·2 mM. In view of the complexity of substrates and the possible modification of the enzyme, it is not surprising that it has not proved consistently possible to relate browning to total phenol or enzyme content (Coseting & Lee, 1987).

Moulding *et al.* (1987) described properties of peroxidase (EC 1.14.18.1) in relation to discoloration of apple products. As is common with plant tissues, they found multiple forms of the enzyme with pH optima around 5·2 and 5·9. They drew attention to the heat stability of peroxidase and were able to detect activity in a commercial apple juice. According to Goodenough *et al.* (1983), the pH optimum of PPO is close to the natural pH of apple juice and this enzyme also showed high heat stability.

Microsomal activities

Most of the lipoxygenase (EC 1.13.11.12) is present in the microsomal fraction (Kim & Grosch, 1979). Although apple is a poor source in fresh weight terms, lipoxygenase has high specific activity. This enzyme is higher in the core and peel of the fruit than in the cortex and it has been associated with membrane disruption leading to core browning disorders (Feys *et al.*, 1980). Apple lipoxygenase forms mainly 13-hydroperoxides from linoleic or linolenic acids (Kim & Grosch, 1979). These could be cleaved to hexanal and 2-*trans*-hexanal, respectively, by hydroperoxide lyase. These compounds have been detected in apple juice and are thought to contribute a ‘green’ or unripe flavor to the product (Flath *et al.*, 1967).

Also in the microsomal fraction is an ATPase (EC 3.6.1.3) (Lurie & Ben Arie, 1983). This was found to be associated with a plasma membrane-like fraction on a sucrose density gradient. It was inhibited by Mg^{2+} and other divalent cations and its substrate, unusually, appeared to be free ATP. Its activity increased two-fold on ripening, but its role remains unexplained.

When the microsomal fraction from apples is precpitated by acetone, proteins can be extracted from the pellet with buffer. These include a calcium-activated protein kinase and calmodulin which can be separated by hydrophobic interaction chromatography. The isolated kinase does not require calmodulin and its regulatory role in the tissue remains to be explored (Battey & Venis, 1988).

Ethylene synthesis

Earlier suggestions linking peroxidase or lipoxygenase with ethylene synthesis in apples (Galliard *et al.*, 1968) are now discounted. A crucial link in the final elucidation of the pathway of ethylene synthesis was the discovery of the intermediate ACC in apple tissue (Adams & Yang, 1979). However, the enzymology of its formation has been studied mainly in other tissues. The enzyme which converts *S*-adenosyl methionine to ACC, ACC synthase (EC 4.4.1.14), has been found in apples (Bufler & Bangerth, 1983), as has the enzyme responsible for synthesis of the derivative, malonyl ACC (Mansour *et al.*, 1986). Malonylation continues during ripening but may have a greater role in regulating ethylene synthesis during growth of the fruit (Knee, 1985).

Aliphatic metabolism

As mentioned above, apple ADH can utilize NADH or NADPH for reduction of aldehydes but only NAD in the direction of oxidation of alcohol (Bartley & Hindley, 1980). Activity diminishes with increasing carbon number in the aldehyde series so that the rate with hexanal is only 5% of that with acetaldehyde. However, this activity is adequate for rates of alcohol production required for synthesis of volatile esters of ripening fruit.

The only other known enzyme involved in ester metabolism in apple is carboxylic ester hydrolase (EC 3.1.1.1). Bartley and Stevens (1981) obtained evidence for multiple forms with a range of molecular weights and specificities for substrates from acetate to hexanoate. These were distinct from lipolytic acylhydrolase (EC 3.1.1.3) which could be detected after a different preparation procedure. Goodenough and Entwhistle (1982) purified a single, high molecular weight esterase, the activity of which appeared to increase during fruit development.

Pear (*Pyrus* spp)

General

The close botanical relationship of the pome fruits suggests that the enzymology of pears and apples should be similar. This section will highlight the distinctive features of pears rather than recapitulating information already given for apples.

Carbohydrate metabolism

Both apples and pears accumulate starch during early fruit development, but convert their reserves to soluble sugars in the later stages of growth (Fig. 1). Amylases have been more extensively studied in pears than in apples. Two β-amylases (EC 3.2.1.2) and an α-amylase (EC 3.2.1.1) were separated by gel permeation of detergent extracts of Passe Crassane pears (Pech *et al.*, 1973). β-Amylase activity is present throughout fruit development, whereas α-amylase activity increases on ripening after most of the starch has disappeared (Latche *et al.*, 1975). It is possible that phosphorylase (EC 2.4.1.1), which has been detected in apples (Clements, 1971), is more important in starch mobilization than the amylases (Preiss, 1982) (Fig. 1).

Cell wall metabolism

The effects of storage in high carbon dioxide atmospheres on fruit metabolism have been further elucidated in research on pears. The activities of ATP-dependent phosphofructokinase (EC 2.7.1.11) and the pyrophosphate-fructose-6-phosphate phosphotransferase (EC 2.7.1.90) decline under 10% CO_2 (Kerbel *et al.*, 1988). This could account for diminished respiration, and it would be interesting to know whether there is a change in the quantity of enzyme protein or a modulation of activity.

European pears (*Pyrus malus*) soften more extensively than apples. In addition to an exo-PG (EC 3.2.1.67) which may be similar to the apple enzyme, they contain an endo-PG (EC 3.2.1.15) (Pressey & Avants, 1976). Gel permeation chromatography data suggest a higher molecular weight (160 kDa) for this enzyme than for other endo-PGs (Bartley & Knee, 1982). The enzyme can release uronic acid from fruit cell walls (Pressey & Avants, 1976) and the rise in activity on ripening accounts for the simultaneous decline in degree of polymerization of pectin in the fruit (Bartley *et al.*, 1982). Ahmed and Labavitch (1980) report the presence of α- and β-galactosidases (EC 3.2.1.22 and 3.2.1.23), α- and β-glucosidases (EC 3.2.1.20 and 3.2.1.21), β-mannosidase (EC 3.2.1.24), α-arabinosidase (EC 3.2.1.55) and β-xylosidase (3.2.1.37) in pears, but could not detect endo-β-1,4 glucanase (EC 3.2.1.4). Yamaki and Matsuda (1977) detected endoglucanases with pH optima at 5·5 and 7·0 in Oriental pears (*Pyrus serotina*). The acidic form predominated in immature fruit, whereas the neutral enzyme increased on ripening, as did a β-glucosidase. Endo-PG and PE activities were also present in immature fruit and their activities declined to low levels before increasing again on ripening. β-Galactosidase activity also increased with ripening, but this enzyme was present throughout fruit development. The observation of Yamaki and Matsuda (1977) of endoglycosidase activities in immature fruits is exceptional, although most other authors have not examined fruit so early in development.

Pears, unlike apples, contain clusters of lignified ‘stone cells’, which are resistant to wall-degrading enzymes and survive as ‘gritty’ areas in the ripe fruit. Wall-bound peroxidase (EC 1.11.1.17) is probably responsible for lignification. In the so called ‘Yuzuhada disease’, an unusually high proportion of total peroxidase is wall-bound and the fruit flesh remains firm on ripening (Radanive & Haard, 1972).

Lipid metabolism

Pears, like apples, contain membrane-bound lipoxygenase (EC 1.13.11.12), which forms mainly 13-hydroperoxides from linoleic and linolenic acids (Kim & Grosch, 1978). Surprisingly, a hydroperoxide lyase isolated from pears by the same authors (Kim & Grosch, 1981), was specific for linoleate 9-hydroperoxides and formed nonenal and oxononanoic acid.

Ethylene synthesis

Pears are often slow to ripen if held continuously at temperatures around 20°C. Exposure to temperatures around 0°C for about a month stimulates ethylene synthesis and the pears ripen rapidly on return to 20°C (Leblond, 1975). The period of chilling required varies according to the variety of pear. The increase in ethylene at the low temperature is associated with an increase in wound-inducible ACC synthase (EC 4.4.1.14) (Knee, 1987), but activity of this enzyme in the whole fruit remains low until fruit has ripened for at least 2 days at a higher temperature (Knee, 1987).

Stone fruits (cherry, apricot, plum, peach) (*Prunus* spp)

Carbohydrate metabolism

As in other Rosaceae, sorbitol is translocated in *Prunus* species, but sucrose, glucose and fructose are accumulated in the fruit. Peach fruits contain an apoplastic invertase (EC 3.2.1.26), the activity of which varies little during development (Ugalde *et al.*, 1988), and a cytoplasmic sucrose synthase (EC 3.4.1.13) (Moriguchi & Yamaki, 1988).

Organic acid metabolism

The stone fruits include the non-climacteric cherry as well as climacteric types (apricot, peach, plum). The fact that NADP-ME (EC 1.1.1.40) activity increases at the time of color development in cherries raises questions about the linkage of this enzyme with the climacteric phenomenon (Hartmann, 1975).

Cell wall metabolism

Cell wall degradation has been more extensively studied in peaches than in the other stone fruits. Endo-PG activity develops and the soluble pectin is increasingly retarded on gel permeation chromatography in ripening 'Elberta' peaches (Pressey *et al.*, 1971). 'Clingstone' varieties of peach remain firm on ripening and are used mainly for canning; they contain exo-PG (EC 3.2.1.67) activity, but little endo-PG (EC 3.2.1.15) and their soluble pectin remains low. The softer 'free stone' varieties, which are eaten fresh, contain both endo- and exo-PG and develop more soluble pectin than clingstone peaches (Pressey & Avants, 1978). Peaches, nectarines and apricots remain firm during storage at

0–1°C. After about 4 weeks storage, if the fruits are transferred to 20–25°C to ripen, they develop a 'woolly' texture. This is a major constraint in storage of this type of fruit. The ability to develop endo-PG activity on ripening is lost during refrigerated storage (Buescher & Furmanski, 1978; Ben Arie & Sonego, 1980). The constitutive PE (EC 3.1.1.11) causes a decline in esterification of pectin at any storage temperature (Ben Arie & Lavee, 1971). Endo-PG would normally degrade the de-esterified pectin to form water-soluble products. Low PG activity leads to the formation of chelator-soluble rather than water-soluble pectin (Ben Arie & Lavee, 1971; Buescher & Furmanski, 1978).

Peaches also contain β-1,4-glucanase (EC 3.2.1.4) activity (Hinton & Pressey, 1974). This activity increases while the fruit is still growing on the tree, and changes little on ripening. It may be involved in some early modification of wall structure, which precedes softening, but its natural substrate is not known.

Fruit cell wall-degrading enzymes are readily inactivated by heat and do not cause problems in processing. The stability of some fungal PGs increases above 80°C (Archer & Fielding, 1975). Consequently, activity may survive canning of infected fruit and this accounts for softening of canned apricots (Strand *et al.*, 1981).

Other enzymes

The loss of anthocyanin that occurs on disruption of cherry fruit tissue is probably caused by cooxidation of the pigment with simpler phenols by PPO (EC 1.10.3.1) (Pifferi & Cultrera, 1974). Oxidase activity declines during peach development and a distinct laccase (EC 1.10.3.2) increases as the fruit ripens (Harel *et al.*, 1970).

A fraction of the activity of β-glucosidase (EC 3.2.1.21) in the kernels of stone fruits can survive canning and cause a slow hydrolysis of amygdalin, also present in the seed. Cherries are often preserved intact with their kernels, but the small amount of cyanide released on hydrolysis of amygdalin is not a hazard to consumers (Stoewsand & Anderson, 1973).

Strawberry (*Fragaria ananassa*)

Cell wall metabolism

An early report of the presence of endo-PG (EC 3.2.1.15) in strawberries (Gizis, 1964) has not been substantiated. Woodward (1972) observed a decline in the viscosity of pectin on ripening, but Huber (1984) found no change in the molecular weight measured by gel permeation. PG can be produced by the surface microflora and latent infective organisms of strawberry fruits. Trace quantities of the enzyme have caused breakdown of fruits stored in sulphite preservative (Harris & Dennis, 1981), and could well account for degradation of pectin in senescent fruits.

Barnes and Patchett (1976) found no PG activity, but reported that β-1,4-glucanase (EC 3.2.1.4) increased on ripening. More recently, Given (pers. comm.) found that glucanase activity was present at all stages of fruit development. There are substantial changes in the solubility of structural polysaccharide and texture during ripening of strawberries (Knee *et al.*, 1977). However, the underlying enzymology is not understood. Suggestions that softening could result from withdrawal of calcium from the cell wall have yet to be corroborated (Neal, 1965).

Phenol metabolism

Hyodo (1971) observed a correlation between anthocyanin concentration and PAL (EC 4.3.1.5) activity during strawberry fruit development. In contrast to many other plant tissues, the levels of both components were unaffected by light. Given *et al.* (1988*a*) showed that strawberry PAL is a tetramer of a single 72 kDa polypeptide. In-vivo labeling and immunoprecipitation suggest that PAL is formed *de novo* on ripening, and inhibitor studies indicate that this increase is required for anthocyanin synthesis. Uridine diphosphate glucose-flavonoid *0*-glucosyltransferase (EC 2.4.1.91) was also observed to increase (Given *et al.*, 1988*b*). Anthocyanin synthesis may involve the coordinated induction of a number of enzymes in the pathway of biosynthesis. How induction is regulated in this tissue is not understood.

Other enzymes

The microsomal lipoxygenase (EC 1.13.11.12) of strawberries forms 9- and 13-hydroperoxides in the ratio of 2:8 (Kim & Grosch, 1978). ATPase (EC 3.6.1.3) is also mainly microsomal but occurs in other cellular fractions; it increases in activity in all fractions during ripening and, like the apple enzyme, is inhibited by divalent metal ions (Ben Arie & Faust, 1980).

SAPINDACEAE

Litchi (*Litchi chinensis*)

Various enzymes

Acid phosphatase increases while alkaline phosphatase declines in the fleshy aril as the fruit pericarp develops its red color (Prasad & Jha, 1979); the significance of these changes is unclear. A major criterion of quality in litchi is the bright red pericarp, but this can rapidly turn brown after harvest. PPO activity increases as browning develops in the pericarp. The PPO has been partially purified (Li & Yan, 1963; Tan & Li, 1984). It is specific to polyphenols, and loses activity rapidly below pH 6·8, although its pH optimum has not been determined. Optimum activity occurred at 35°C and 50% of its activity was lost at 45°C in 25 min. The enzyme is strongly inhibited by sodium bisulfite and sodium diethyldithiocarbamate (Li & Yan, 1963).

SOLANACEAE

Tomato (*Lycopersicon esculentum*)

Carbohydrate metabolism

The fruit contains more acid invertase (EC 3.2.1.26) activity than other parts of the tomato plant, and this has been held to account for its preferential accumulation of carbohydrate (Manning & Maw, 1975). Enzyme activity rises during ripening. Other authors observed a decline in activity during senescence and there was immunological evidence for the accumulation of inactive enzyme in overripe fruit (Nakagawa *et al.*, 1980).

As in other plant tissues, ATP-dependent phosphofructokinase (EC 2.7.1.11) in immature tomatoes is an oligomeric enzyme (mol. wt, 180 kDa), and its activity is subject to allosteric modulation by a number of metabolites (Isaac & Rhodes, 1982). During ripening, a low-molecular-weight (60 kDa) form increases and this is not regulated by fructose-6-phosphate and other metabolites (Isaac & Rhodes, 1987). Dissociation of the oligomeric enzyme occurs *in vitro* and is promoted by a small increase in pH (from pH 7·5 to 8·0) and the presence of phosphate. The implication is that increased glycolytic flux is mediated by deregulation of fructose-1,6-biphosphate synthesis, and this results from a change in the properties of an existing enzyme, rather than de-novo protein synthesis (Isaacs & Rhodes, 1987).

Organic acid metabolism

Early work on tomatoes showed that during ripening mitochondria remain coupled as in other fruits, but there was some decline of mitochondrial activity (Hobson, 1970). This is difficult to reconcile with the occurrence of the climacteric rise in respiration. More recently, Jeffery *et al.* (1984) reported that the activities of malate dehydrogenase (EC 1.1.1.37) and citrate synthase (EC 4.1.3.7) in the 20 000*g* supernatant of tomato homogenates decline during storage or ripening of the fruit. The occurrence of soluble citrate synthase is unusual and data are not available for either activity in the mitochondria. There is a transient increase in the activity of NADP-linked ME (EC 1.1.1.40) at the onset of ripening. This enzyme was purified to apparent homogeneity (Goodenough *et al.*, 1985) but its role in metabolism remains to be explained for tomatoes as much as for other fruits.

Plastid enzymes

In many ripening fruits, plastids change from the green photosynthetic state to a yellow, orange or red chromoplast. Although this is a conspicuous feature of ripening, and we often judge ripeness by color, little is known of the enzymology of the process. Losses of both the large and small subunits of ribulose bisphosphate carboxylase (Rubisco, EC 4.1.1.39) have been demonstrated in chloroplast preparations from ripe tomato fruit (Bathgate *et al.*, 1985). Other, undefined proteins declined and some new proteins appeared in the course of ripening. A changing pattern of plastid protein synthesis was also indicated when tissue at different stages was incubated with [^{35}S] methionine (Bathgate *et al.*, 1985). Fewer mRNA transcripts for the large subunit of Rubisco were isolated from plastic preparations from ripening tomato fruit than from the unripe fruit by Piechulla *et al.* (1985). However, transcripts for the 32 kDa 'herbicide binding protein' were found throughout ripening. There was some evidence in this work that nuclear and plastid transcription of plastid proteins are not coordinately regulated during ripening. Further studies indicated a more rapid decline of the small subunit than of the large subunit of Rubisco (Livne & Gepstein, 1988). In this study, superoxide dismutase (EC 1.15.1.1) proved to be one of the more persistent plastid proteins. Fluctuations in the activity of this enzyme during development and under heat stress have been correlated with the fruit's ability to withstand photochemical damage leading to 'sunscald' (Rabinovitch & Sklan, 1980).

According to Ngernprasirtsiri *et al.* (1988), progressive methylation of the plastid genome closely parallels the transcriptional changes observed by Piechulla *et al.* (1985). The latter authors reported that plastid transcripts of subunits of the ATPase (EC 3.6.1.34) complex seemed to be processed differently in fruit than in leaves. These studies give some idea of the complexity of regulation of protein synthesis that awaits further research.

Cell wall metabolism

The endo-PG (EC 3.2.1.15) of tomato is, without doubt, the most intensively studied fruit enzyme. No activity has been detected in the green, unripe fruit, but after the rise in ethylene synthesis, and as color develops PG activity increases in the fruit pericarp (Su *et al.*, 1984). The enzyme is present as a high-molecular weight form (PG I, 100 kDa) and two low-molecular-weight forms (PG IIa and b, 46 kDa) (Ali & Brady, 1982). All of these are formed from the same polypeptide and conversion of PGII to PG I occurs *in vitro* in the presence of a factor from unripe fruit (Tucker *et al.*, 1981*a*). PG I is the first form to be detected at the onset of fruit ripening (Tucker *et al.*, 1981*b*) but Pressey (1986) has argued that it is formed from PG II during enzyme extraction.

The protein has been sequenced, as has its mRNA, which carries a 5′ sequence, possibly coding for a signal peptide; perhaps this directs secretion of the enzyme across a membrane, and is cleaved in the process (Grierson *et al.*, 1986). The mature protein is also glycosylated (Moshrefi & Luh, 1983). Processing of the polypeptide has been followed *in vitro* (Della Penna & Bennett, 1988). Like the enzyme, PG mRNA is absent from mature green fruit, and accumulates on ripening (Grierson *et al.*, 1985).

Tomato PG II liberates uronide from cell walls isolated from immature fruits. Rapid degradation requires removal of calcium from the walls (Brady *et al.*, 1987). Cell walls examined in the electron microscope after enzyme treatment resemble those from naturally ripened fruit (Crookes & Grierson,

1983). In the intact fruit, the molecular weight of polyuronide falls by about 50% on ripening (Seymour & Harding, 1987).

The isolation of the gene for PG has led to attempts to manipulate enzyme activity. In one set of experiments, normal tomato plants were transformed with a DNA construct carrying an inverted sequence of part of the structural gene with a promoter sequence from cauliflower mosaic virus. In transformed plants, there was constitutive expression of 'antisense' RNA, whereas expression of PG mRNA was suppressed and PG activity itself reduced by up to 90% in the ripening fruit. Preliminary indications are that normal ripening, including fruit softening, still occurred (Smith *et al.*, 1988). Tomatoes contain more PG than many other fruits (Hobson, 1962), and it may be that only a fraction of this activity is required for normal softening.

In other experiments, the PG gene was introduced with an ethylene-inducible promoter into a non-ripening mutant, '*rin*'. The original mutant phenotype lacks PG and the fruit does not soften. Treatment of fruit from transformed plants with the ethylene analogue, propylene, induced PG activity. This PG appeared to be active *in vivo* because soluble uronides were formed in propylene-treated fruit. However, the fruit did not soften, change color or produce ethylene as wild type fruit would (Giovannoni *et al.*, 1989). Earlier, it was postulated that PG occupied a pivotal role in tomato ripening (Tigchelaar *et al.*, 1978). More recent work shows that, far from being the initiating agent for all aspects of fruit ripening, PG may not wholly account for fruit softening.

Tomatoes also contain exo-PG (EC 3.2.1.67) and PE (EC 3.1.1.11) activity. Hobson (1981) calculated that enough esterase was present to hydrolyse all of the pectin methyl ester groups in the fruit in four minutes. Although activity increases on ripening, it is present at all stages of fruit development and it must be regulated by some other factor than enzyme synthesis.

One of three β-galactosidases in tomato is capable of degrading polymers and may account for the loss of galactose from the cell wall during ripening (Pressey, 1983). Ripe tomato fruit contain endo-β-1,4-glucanase (EC 3.2.1.21) and β-glucosidase (EC 3.2.1.21) activities (Sobotka & Stelzig, 1974). Endoglucanase activity has also been observed in the early stages of fruit development. It declines as the fruit expands (Hobson, 1968) so that it is not detectable at the mature green stage and increases in both non-ripening and normal fruits about 40 days after anthesis, when the normal fruits are ripening (Poovaiah & Nukaya, 1979). In other plant tissues, different forms of glucanase are associated with expansion and separation of cells (Reid *et al.*, 1974). Perhaps the glucanases at different stages of tomato development will prove to be distinct, but the presence of the enzyme in the non-ripening tomato raises questions about its role and the regulation of its activity.

Attention to tomato PG has predominated over research on other enzymes which might have a role in fruit softening. Often, a potential role has been dismissed because the enzyme in question was found in non-ripening mutants; this showed that the enzyme was not sufficient for fruit softening, but did not show that it was not involved. Perhaps the same could now be said for PG and it is timely to apply to the other enzymes the molecular techniques which are unraveling the role of PG.

Ethylene synthesis

The key enzyme in ethylene synthesis, ACC synthase (EC 4.4.1.14), was first detected in tomato tissue (Boller *et al.*, 1979). Although a rise in activity accompanies ripening, higher activity is found a few hours after wounding fruit tissue. Thus, sliced pericarp has been used as a source of enzyme for purification. The enzyme appears to turn over rapidly *in vivo* and has proved labile *in vitro* (Acaster & Kende, 1983). This is partly explained by the observation that the substrate, *S*-adenosyl methionine, reacts with ACC synthase and inactivates it (Satoh & Yang, 1988). Nevertheless, the ACC synthase from tomato has been highly purified and monoclonal antibodies raised against it (Bleecker *et al.*, 1986). So far, these have served to follow induction of synthesis of the protein in wounded tissue (Bleecker *et al.*, 1988). In the near future, immunological and nucleic acid probes should permit further study and manipulation of the regulation of ethylene synthesis during normal fruit development.

Phenol metabolism

Phenoloxidase (EC 1.10.3.1) is generally regarded as a plastid-localised enzyme and the sharp decline in its activity on ripening (Rhodes & Wooltorton, 1977) may be a further example of the loss of plastid proteins (Bathgate *et al.*, 1985). Browning of cut surfaces is not a problem in handling tomatoes, and in enzyme extraction from tomatoes, often no precautions are taken against the reaction of phenolic compounds with proteins. The major phenolic compound present is chlorogenic acid, which appears to be innocuous for enzyme activity. Phenol metabolism in tomatoes is enhanced when they are stored below 15°C. Below 10°C, this leads to the development of abnormal flavors, which render the fruit unsaleable. Chilling leads to an increase in the activity of PAL (EC 4.3.1.5) and hydroxycinnamoyl CoA: quinate hydroxycinnamoyl transferase (Rhodes & Wooltorton, 1977). PAL is the first enzyme of the pathway committed to phenol biosynthesis in general, and the transferase is the final enzyme for chlorogenate synthesis. Other enzymes, such as 4-coumarate-CoA-ligase (EC 6.2.1.12) and cinnamic-acid 4-hydroxylase (EC 1.1.13.11) and aromatic alcohol dehydrogenase are more constant in their activities.

Lipid metabolism

Physical disruption of tomato tissue leads to rapid loss of polar lipids and triacylglycerol. Phospholipase D (EC 3.1.4.4) and acyl hydrolase (EC 3.1.1.3) appear to be active. Linoleate and linolenate are the major fatty acyl substituents of the lipid but these do not accumulate as free fatty acids. They are oxidized by lipoxygenase (EC 1.13.11.12) mainly to the 9-hydroperoxide, but the minor 13-hydroperoxide is further degraded to six-carbon aldehydes and 12-oxo-dodecenoate. The aldehydes, hexanal from linoleate and 3-hexenal from linolenate, contribute to the 'fresh' or 'green' flavour of tomatoes (Galliard *et al.*, 1977).

VITACEAE

Grape (*Vitis vinifera*)

Carbohydrate metabolism

In commercial grape varieties the main, if not only, translocated carbohydrate is sucrose, but its hydrolysis products, glucose and fructose, accumulate in the fruit in about equal concentrations as soluble reserves (Hawker, 1985). Only a small amount of sucrose occurs in the fruit, probably as transit carbohydrate, and starch is almost completely absent. Much work has been done on the distribution of sugars in the grape (Coombe *et al.*, 1987), but there has been less work on the enzymes involved. However, from studies of enzyme activities of developing grape berries, the pathway in Fig. 1 was proposed for transfer of sucrose to the vacuole (Hawker, 1985).

As with other fruits, extraction of enzymes from grapes requires the use of enzymically inactive protein (bovine serum albumin), high concentrations of buffer, antioxidants and complexing agents in order to counteract the effects of low enzyme concentrations, organic acids and denaturation of enzymes by phenols and tannins (Downton & Hawker, 1973). In the proposed pathway, sucrose arriving at the pericarp cells is converted to UDP glucose and fructose by sucrose synthase, and after a series of reactions, including sucrose phosphate as an intermediate, sucrose is hydrolysed by invertase in the vacuole. Evidence suggests that invertase is a soluble enzyme located in the vacuole, despite earlier reports that it was a cell wall enzyme. The berries of some grape species contain more sucrose than the commercial varieties and it would be interesting to determine whether they contain less invertase (Lott & Barrett, 1967). Intensive research has not demonstrated unequivocally the mechanism of sucrose transport across the tonoplast in any plant tissue. Both sucrose proton co-transport and a group translocation mechanism have been proposed (Hawker, 1985; Niemitz & Hawker, 1988). The preparation of vacuoles from the fragile pericarp cells of grape berries is difficult, if not impossible, and studies on the enzymology of sugar transport in this tissue have not progressed further (Brown & Coombe, 1984).

Grape berries contain enzymes of starch metabolism (Fig. 1), including ADP glucose pyrophosphorylase (EC 2.7.7.27), starch synthase (EC 2.4.1.21), phosphorylase (EC 2.4.1.1), and amylase (EC 3.2.1.1) (Downton & Hawker, 1973). Activities of the first two enzymes are much lower in fruit than in the leaves. Since they are involved in starch synthesis this is consistent with the virtual absence of starch from the berries. Similar activities of the remaining enzymes are found in fruit and leaves.

The conversion of sugars to acids is an important process in grape berries. Concentrations of glycolytic intermediates, acid components and adenosine nucleotides have been measured in developing fruit. Comparison of mass action ratios with equilibrium constants for the reactions catalyzed by phosphofructokinase (EC 2.7.1.11) and pyruvate kinase (EC 2.7.1.40) showed these to be non-equilibrium reactions. This indicates that these are regulatory steps in glycolysis in the grape, as in other organisms. In contrast, the reactions catalyzed by phosphoglucose isomerase (EC 5.3.1.9), fructose bisphosphatase (EC 3.1.3.11) and aldolase (EC 4.1.2.13) were in equilibrium (Ruffner & Hawker, 1977).

Organic acid metabolism

Low concentrations of citric, succinic, fumaric, formic, acetic, glycolic, lactic, aconitic, quinic, shikimic, and mandelic acids are found in grapes. The concentrations of the major acids, malic and tartaric, are affected by climate, particularly temperature, and commercially it is extremely important to grow grapes with the optimum amounts of the acids to allow the production of acceptable food and drink products. Despite their chemical resemblance, malic and tartaric acids are unrelated metabolically (Ruffner, 1982*a,b*). Both acids are synthesized in the berries, most likely from carbohydrate precursors. In the first phase of fruit growth, malate is produced mainly by the carboxylation of phosphoenol pyruvate derived from sucrose by glycolysis. The reactions are catalyzed by PEP carboxylase and malate dehydrogenase (Fig. 2).

PEP carboxylase activity is high in the first phase of growth as malic acid concentration increases but is low after the onset of ripening when malic acid decreases (Hawker, 1969). Only in the very early stages of grape development is there sufficient Calvin cycle activity to provide phosphoenol pyruvate via ribulose-1,5-bisphosphate carboxylase. Because of the strong inhibition of the metabolism of sucrose through glycolysis after the onset of ripening, carbohydrates are replaced by malate as the respiratory substrate. It is not known whether this is because active permeation of the acid through the tonoplast makes malate more available or because of increased activity of catabolizing enzymes. In contrast to malic acid, tartaric acid is hardly metabolized after its formation in young tissue.

Unlike malic acid, tartaric acid does not incorporate ^{14}C from radioactive CO_2 in the dark and even in the light, radioactivity is only detectable in tartaric acid after an extended time. This shows that the compound is a secondary product. Pretaric acid (1,2-dihydroxymethyl hydrogen L-(+)-tartrate) and ascorbic acid are possible precursors. Whether tartrate is further metabolized, and if so, by what pathway, is unclear, as is the whole question of its physiological role. Enzymes of tartrate synthesis and dissimilation in the grape berry are not known. On the other hand, malate is used not only as a respiratory substrate after the onset of ripening, but is also converted to sugar by gluconeogenesis (Ruffner *et al.*, 1983). Degradation occurs via NADP-dependent ME (EC 1.1.1.40) followed by respiration of the resulting pyruvate while conversion of malate to sugars occurs by formation of oxaloacetate by malate dehydrogenase (EC 1.1.1.37), subsequent conversion of oxaloacetate to PEP by PEP-carboxykinase (EC 4.1.1.39) and then reversal of glycolysis (Fig. 2).

In view of the importance of NAD-ME (EC 1.1.1.39) in most plant mitochondria and the presence of a malate transporter (Wiskich & Dry, 1985), it seems possible that in grape berries at least part of the malate could be degraded by NAD-ME in the mitochondria, but the enzyme has not, so far, been detected in grapes. Grapes do not have a climacteric rise in respiration, but they do show increased RQ values after the onset of ripening (Hawker, 1969). In contrast to apples, grapes do not show a consistent large increase in the activities of NADP-ME or pyruvate decarboxylase (EC 4.1.1.1) at or after the onset of ripening. However, NADP-ME is almost certainly the major enzyme involved in malate degradation in the grape (Fig. 2) and there is substantial activity at all stages of development of the berry. The decrease in acid components during ripening is caused mainly by a decline in malic acid because tartaric acid is relatively inert. Despite the in-vitro reversibility of the reaction of NADP-ME, alanine feeding experiments to grapes before and after the onset of ripening indicated that the reverse reaction is not a major contributor to grape malate synthesis (Ruffner *et al.*, 1984).

In order to understand the effects of temperature during growth on the acid content of grapes, ME, PEP carboxylase, PEP carboxykinase, glucose-6-P dehydrogenase (EC 1.1.1.49), phosphofructokinase and isocitrate dehydrogenase (EC 1.1.1.42) activities were measured (Ruffner *et al.*, 1976). PEP carboxylase was shown to be the major enzyme of malate synthesis, and NADP-ME and PEP carboxykinase to be the enzymes involved in the degradation of malate during ripening. Grapes grown at high temperature have lower acidity levels, at maturity, and these were attributed to a changed pattern of acid breakdown rather than to reduced malic acid accumulation. However, enzymes of malate consumption do not increase sufficiently with rising temperatures to explain low malic acid levels. Changes in permeation rates of malate across the tonoplast in grapes grown at different temperatures seem a more likely explanation.

Nitrogen metabolism

The levels of arginine in dormant canes and mature fruits have been used to evaluate the nitrogen reserves of grape vines because arginine plays a primary role in the storage of nitrogen in *Vitis vinifera.* All of the enzymes of the ornithine urea cycle (Krebs urea cycle, Krebs–Henseleit cycle) have been characterized from the grapevine and described in a series of three papers (see Roubelakis & Kliewer, 1978). This was the first report of all enzymes of the cycle occurring in one plant tissue and presumably they are involved in the storage and subsequent remobilization of nitrogen in the form of arginine and urea. The enzymes are ornithine carbamoyl transferase (EC 2.1.3.3), arginosuccinate synthetase (EC 6.3.4.5), arginosuccinate lyase (EC 4.3.2.1) and arginase (EC 3.5.3.7). It seems likely that grapes also contain urease (EC 3.5.1.5) to release ammonia from urea. Aspartase (EC 4.3.1.1) which catalyses the reversible deamination of L-aspartic acid to fumaric acid and ammonia is usually assayed by the well-known spectrophotometric method. However, because of the high phenolic content of their grape samples, Robin *et al.* (1987) used a radioactive method to assay aspartase. Their technique precluded the interference of fumarase, aspartate transaminase and malate dehydrogenase in the assays.

Cell wall metabolism

Grape berries soften dramatically at the onset of ripening, but apart from meager reports of β-1,4-glucanase and PE activity in one variety (Hawker & Walker, 1978), nothing is known of the enzymology of this important process.

Phenol metabolism

In spite of the importance of anthocyanins in red grapes, little or nothing is known of the enzymology of their synthesis. Colorless phenols in white grapes are also extremely important in determining the quality and color of both dried grapes and wine. Oxidation of phenols to quinones and subsequent polymerization produces yellow-brown tannins. A mutant of sultana, Bruce's Sport (Antcliff & Webster, 1962) contains only 10% of the normal *o*-diphenol oxidase (EC 1.10.3.1) activity (Radler, 1964). Both the mutant and normal grape contain similar amounts of phenolic compounds according to chemical analysis of extracts and staining of sections (Hawker *et al.*, 1972). The normal sultana browns upon drying, whereas Bruce's Sport remains a pale golden yellow.

Commercially, grape berries of cv sultanina (sultana, Thompson seedless) are sun-dried after dipping in an oil emulsion to produce pale golden-brown sultanas (Australia), sun-dried untreated to produce dark brown raisins in the USA or shade-dried to produce green naturals or Soyagi (Afghanistan). There is evidence that the browning of untreated grapes is caused partly by the action of *o*-diphenol oxidase (non-browning of Bruce's Sport) and partly by non-enzymic browning (Grncarivec & Hawker, 1971). When berries were partly covered with a commercial oil emulsion and dried in an oven at 40°C, treated parts became yellow, whereas untreated parts were bluish-violet. The browning reaction was prevented in isolated skins by sodium diethyldithiocarbamate, which chelates copper required for activity of the oxidase. These results support the hypothesis that sultanas become brown during drying under mild conditions when cell compartmentation breaks down allowing *o*-diphenol oxidase to act on the substrates released.

Aroma metabolism

Grape berries contain a diversity of volatile compounds which impart aroma and flavor to grapes used for eating and wine-making. Much of the potential flavor of grapes is, however, not expressed because the compounds are present as aroma-inactive glycosidic forms. Interest in methods of releasing the flavored aglycones from the non-flavored glycosidic compounds has resulted in studies on both endogenous and exogenous glycosidases (Aryan *et al.*, 1987). Two β-glucosidases (EC 3.2.1.21), a β-galactosidase (EC 3.2.1.23) and low activities of two α-glucosidases (EC 3.2.1.22) occur in Muscat of Alexandria grapes. The β-glucosidases (as well as those of Chardonnay, Gewurztraminer and Muscat Hamburg) have optimal activity of pH 5·0, are inhibited by glucose and are relatively tolerant to 10% alcohol, at least in the short term. Their activity increases with increasing maturity of the grapes. The enzymes are incapable of hydrolysing sugar conjugates of tertiary alcohols and thus glycosides of some of the most flavor-important monoterpenes, e.g. linalool, remain intact. In the intact fruit once tertiary alcohols are glycosylated they are irreversibly lost as flavor compounds. Only non-specific exogenous glucosidases or chemical hydrolysis will free the aglycone following crushing of the grapes. Aryan *et al.* (1987) concluded that the properties of the grape and other known glycosidases make enzymic release of aglycones impractical and non-enzymic methods for glycoside hydrolysis are the only practical alternative.

REFERENCES AND BIBLIOGRAPHY

Acaster, M. A. & Kende, H. (1983). *Plant Physiol.*, **72**, 139–45.
Adams, T. O. & Yang, S. F. (1979). *Proc. Nat. Acad. Sci., USA*, **76**, 170–4.
Adato, I. & Gazit, S. (1977). *J. Exper. Bot.*, **28**, 644–9.
Agosti, J. A. & Beltran, J. P. (1982). *Anal. Biochem.*, **127**, 368–71.
Ahmed, A. E. R. & Labavitch, J. M. (1980). *Plant Physiol.*, **65**, 1014–16.
Ahmed, E M. & Barmore, C. R. (1980). In *Tropical and Subtropical Fruits: Composition, Properties and Uses*, ed. S. Nagy & P. E. Shaw. AVI Publishing Inc., Westport, CT, USA, pp. 121–56.
Akamine, E. K. & Goo, T. (1971). *Hort. Sci.*, **6**, 567–8.
Ali, Z. M. & Brady, C. J. (1982). *Aust. J. Plant Physiol.*, **9**, 155–69.
Antcliff, A. J. & Webster, W. J. (1962). *Aust. J. Exper. Agric.*, **2**, 97–100.
Archer, S. A. & Fielding, A. M. (1975). *J. Food Sci.*, **40**, 423–4.
Arcus, A. C. (1959). *Biochim. Biophys. Acta*, **33**, 242–4.
Areas, J. A. & Lajolo, F. M. (1981). *J. Food Biochem.*, **5**, 19–37.
Aryan, A. P., Wilson, B., Strauss, C. R. & Williams, P. J. (1987). *Amer. J. Enol. Vitic.*, **38**, 182–8.
Ash, A. S. F. & Reynolds, T. H. (1954). *Aust. J. Biol. Sci.*, **7**, 435–43.
Augustin, M. A. & Osman, A. (1988). *Pertanika*, **11**, 45–50.
Awad, M. (1985). *J. Food Sci.*, **50**, 1643–5.
Awad, M. & Lewis, L. N. (1980). *J. Food Sci.*, **45**, 1625–8.
Awad, M. & Young, R. E. (1979). *Plant Physiol.*, **64**, 306–8.
Awad, M. & Young, R. E. (1980). *J. Amer. Soc. Hort. Sci.*, **105**, 638–41.
Baines, B. S., Kierstan, M. P. J. & Brocklehurst, K. (1982). *Biol. Soc. Trans.*, **10**, 171–2.
Baker, E. N. & Drenth, J. (1987). In *Biological Macromolecules and Assemblies.* ed. F. A. Jurnak & A. McPherson. John Wiley & Sons, New York, USA, pp. 314–68.

Baker, J. E., Lieberman, M. & Anderson, J. D. (1978). *Plant Physiol.*, **61,** 886–8.
Baker, J. E., Anderson, J. D., Adams, D. O., Apelbaum, A. & Lieberman, M. (1982). *Plant Physiol.*, **69,** 93–7.
Barash, I. & Khazzam, S. (1970). *Phytochem.*, **9,** 1189–97.
Barnes, M. F. & Patchett, B. J. (1976). *J. Food Sci.*, **41,** 1392–5.
Barron, E. J. & Stumpf, P. K. (1962). *Biochim. Biophys. Acta*, **60,** 329–37.
Bartholomew, D. P. & Paull, R. E. (1986). In *CRC Handbook of Fruit Set and Development*, ed. S. P. Monselise. CRC Press, Boca Raton, FL, USA, pp. 371–88.
Bartley, I. M. (1974). *Phytochem.*, **13,** 2107–11.
Bartley, I. M. (1977). *J. Exp. Bot.*, **28,** 943–8.
Bartley, I. M. (1978). *Phytochem.*, **17,** 213–16.
Bartley, I. M. & Hindley, S. J. (1980). *J. Exper. Bot.*, **31,** 449–59.
Bartley, I. M. & Knee, M. (1982). *Food Chem.*, **9,** 47–58.
Bartley, I. M. & Stevens, W. M. (1981). *J. Exper. Bot.*, **32,** 741–51.
Bartley, I. M., Knee, M. & Casimir, M. A. (1982). *J. Exper. Bot.*, **33,** 1248–55.
Bathgate, B., Purton, M. E., Grierson, D. & Goodenough, P. W. (1985). *Planta*, **165,** 197–204.
Battey, N. H. & Venis, M. A. (1988). *Anal. Biochem.*, **170,** 116–22.
Baumann, J. W. (1981). In *Enzymes and Food Processing*, ed. G. G. Birch, N. Blakebrough & K. J. Parker. Applied Science, London, UK, pp. 129–47.
Beaudry, R. M., Paz, N., Black, C. C. & Kays S. J. (1987). *Plant Physiol.*, **85,** 277–82.
Ben Arie, R. & Faust, M. (1980). *Phytochem.*, **19,** 1631–6.
Ben Arie, R. & Lavee, S. (1971). *Phytochem.*, **10,** 531–8.
Ben Arie, R. & Sonego, L. (1980). *Phytochem.*, **19,** 2553–5.
Ben Arie, R., Kislev, N. & Frenkel, C. (1979). *Plant Physiol.*, **64,** 197–202.
Bendall, D. S., Ranson, S. C. & Walker, D. A. (1960). *Biochem. J.*, **76,** 220–5.
Bennett, A. B. & Christoffersen, R. E. (1986). *Plant Physiol.*, **81,** 830–5.
Bennett, A. B., Smith, G. M. & Nichols, B. G. (1987). *Plant Physiol.*, **83,** 973–6.
Beruter, J. (1985). *J. Plant Physiol.*, **121,** 331–41.
Biale, J. B. (1975). In *Postharvest Biology and Handling of Fruits and Vegetables*, ed. N. F. Haard & D. K. Salunkhe. AVI Publishing Co., Westport, CT, USA, pp. 5–18.
Biale, J. B. & Young, R. E. (1971). In *The Biochemistry of Fruits and Their Products*, ed. A. C. Hulme. Academic Press, New York, pp. 1–63.
Blanke, M. M., Hucklesby, D. P., Nolton, B. A. & Lenz, F. (1987). *Phytochem.*, **26,** 2475–6.
Blankenship, S. M. & Unrath, C. R. (1988). *Phytochem.*, **27,** 1001–3.
Bleecker, A. B., Kenyon, W. H., Somerville, S. C. & Kende, H. (1986). *Proc. Natl Acad. Sci. USA*, **83,** 7755–9.
Bleecker, A. B., Robinson, G. & Kende, H. (1988). *Planta*, **173,** 385–90.
Boller, T., Herner, R. C. & Kende, H. (1979). *Planta*, **145,** 293–303.
Bower, J. P. & Lelyveld, L. J. van (1985). *J. Hort. Sci.*, **60,** 545–7.
Bradford, M. M. (1976). *Anal. Biochem.*, **72,** 248–54.
Brady, C. J. (1976). *Aust. J. Plant Physiol.*, **3,** 163–72.
Brady, C., McGlasson, B. & Speirs, J. (1987). In *Tomato Biotechnology*, ed. D. J. Nevins & R. A. Jones. Alan R. Liss, Inc., New York, pp. 279–88.
Brekke, J. E., Cavaletto, C. G., Stafford, A. E. & Chan, H. T. (1975). *Mango: Processed Products*. Agric. Res. Service, US Dept. of Agric., Washington DC, USA.
Brocklehurst, K., Willenbrock, F. & Salih, E. (1987). In *Hydrolytic Enzymes*, ed. A. Nueberger & K. Brocklehurst. Elsevier, Amsterdam, The Netherlands, pp. 39–152.
Brown, S. C. & Coombe, B. G. (1984). *Biochem. Physiol. Pflanzen.*, **179,** 157–71.
Buescher, R. W. & Furmanski, R. J. (1978). *J. Food Sci.*, **43,** 264–6.
Buescher, R. W., McGuire, C. & Skulman, B. (1987). *J. Food Sci.*, **52,** 228–9.

Bufler, G. & Bangerth, F. (1983). *Physiol. Plant.*, **58,** 486–92.
Buren, J. van (1971). In *The Biochemistry of Fruits and Their Products* (Vol. 13), ed. A. C. Hulme. New York, USA, pp. 269–304.
Carne, A. & Moore, C. H. (1978). *Biochem. J.*, **173,** 73–83.
Carreno, R. & Chan, H. T. (1982). *J. Food Sci.*, **47,** 1498–500.
Caughey, I. & Kekwick, R. G. O. (1982). *Eur. J. Biochem.*, **123,** 553–61.
Chan, H. T. (1986). *J. Food Sci.*, **51,** 581–3.
Chan, H. T. & Kwok, S. C. M. (1976). *J. Food Sci.*, **41,** 320–3.
Chan, H. T. & Tam, S. Y. T. (1982). *J. Food Sci.*, **47,** 1478–83.
Chan, H. T., Chang, T. S. K., Stafford, A. E. & Brekke, J. E. (1971). *J. Agric. Food Chem.*, **19,** 263–5.
Chan, H. T., Tam, S. Y. T. & Koide, R. T. (1978). *J. Food Sci.*, **43,** 989–91.
Chan, H. T., Hibbard, K. L., Goo, T. & Akamine, E. K. (1979). *Hort. Sci.*, **14,** 140–1.
Chan, H. T., Tam, S. Y. T. & Seo, S. T. (1981). *J. Food Sci.*, **46,** 190–1.
Chan, H. T., Sanxter, S. & Couey, H. M. (1985). *Hort. Sci.*, **20,** 1070–2.
Chan, H. T., Couey, H. M. & Sakai, W. S. (1991). *Proc. 7th World Congress of Food Science and Technology, Singapore,* Singapore Inst. Food Science and Technology (in press).
Chang, L. W. S., Morita, L. L. & Yamamoto, H. Y. (1965). *J. Food Sci.*, **30,** 218–22.
Chaplin, G. R. & Scott, K. J. (1980). *Hort. Sci.*, **15,** 514–15.
Chen, N. K. L. (1964). *Bot. Bull. Acad. Sinica,* **V,** 89–99.
Chitarra, A. B. & Lajolo, F. M. (1981). *J. Amer. Soc. Hort. Sci.*, **106,** 579–84.
Chhatpar, H. S. & Modi, V. V. (1974). *Phytochem.*, **13,** 693–6.
Chhatpar, H. S., Mattoo, A. K. & Modi, V. V. (1971). *Phytochem.*, **10,** 1007–9.
Christoffersen, R. E., Tucker, M. L. & Laties, G. C. (1984). *Plant Molecular Biol.*, **3,** 385–91.
Clements, R. L. (1971). In *The Biochemistry of Fruits and Their Products* (Vol. I), ed. A. C. Hulme. Academic Press, London, UK, pp. 159–77.
Coombe, B. G., Bovio, M. & Schneider, A. (1987). *J. Exper. Bot.*, **38,** 1789–98.
Copeland, L. & Tanner, G. J. (1988). *Physiol. Plant.*, **74,** 531–6.
Coseting, M. Y. & Lee, C. Y. (1987). *J. Food Sci.*, **52,** 985–9.
Crookes, P. R. & Grierson, D. (1983). *Plant Physiol.*, **72,** 1088–93.
Cutting, J. G. M., Bower, J. P. & Wolstenholme, B. N. (1988). *J. Hort. Sci.*, **63,** 509–15.
Dallman, T. F., Eaks, I. L., Thomson, W. W. & Nothnager, E. A. (1988). *J. Amer. Soc. Hort. Sci.*, **113,** 729–33.
Deacon, W. & Marsh, H. V. (1971). *Phytochem.*, **10,** 2915–24.
de Arriola, M. C., de Madrid, M. C. & Rolz, C. (1975). *Proc. Trop. Reg. Amer. Soc. Hort. Sci.*, **19,** 97–109.
de Arriola, M. C., Calzada, J. F., Menchu, J. F., Rolz, C. & Garcia, R. (1980). In *Tropical and subtropical fruits,* ed. S. Nagy & P. E. Shaw. AVI Publishing, Inc., Westport, CT, USA, pp. 316–40.
de Leo, P. & Sacher, J. A. (1970). *Plant Physiol.*, **46,** 208–11.
Della Penna, D. & Bennett, A. B. (1988). *Plant Physiol.*, **86,** 1057–63.
Dilley, D. R. (1962). *Nature,* **196,** 387.
Dilley, D. R. (1966). *Plant Physiol.*, **41,** 214–20.
Dizik, N. S. & Knapp, F. W. (1970). *J. Food Sci.*, **35,** 282–5.
Dolendo, A. L., Luh, B. S. & Pratt, H. K. (1966). *J. Food Sci.*, **31,** 332–6.
Downton, W. J. S. & Hawker, J. S. (1973). *Phytochem.*, **12,** 1557–63.
Driessche, E. van, Beeckmans, S., Dejuegere, R. & Kanarek, L. (1984). *Anal. Biochem.*, **141,** 184–8.
Dubery, I. A. & Schabort, J. C. (1981). *Biochim. Biophys. Acta,* **662,** 102–10.
Dubery, I. A., Schabort, J. C. & Cloete, F. (1984). *Int. J. Biochem.*, **16,** 417–22.
Ellstrand, N. C. & Lee, J. M. (1987). *Sci. Hort.*, **32,** 25–31.

Faragher, J. D. & Chalmers, T. J. (1977). *Aust. J. Plant Physiol.*, **4,** 133–41.
Feys, M., Naesens, W., Tobbach, P. & Maes, E. (1980). *Phytochem.*, **19,** 1009–11.
Fidler, J. L. (1951). *J. Exper. Bot.*, **2,** 41–64.
Fidler, J. L. & North, C. J. (1967). *J. Hort. Sci.*, **42,** 207–21.
Flath, R. A., Black, D. R., Guadagni, D. G., McFadden, W. M. & Schultz, T. M. (1967). *J. Agric. Food Chem.*, **15,** 29–35.
Frenkel, C., Klein, I. & Dilley, D. R. (1969). *Phytochem.*, **8,** 945–55.
Fuchs, Y., Pesis, E. & Zauberman, G. (1980). *Sci. Hort.*, **13,** 155–60.
Fuchs, Y., Zauberman, G. & Akerman, M. (1986). *Acta Hort.*, **179,** 801–4.
Fukushima, T., Yamazaki, M. & Odazima, T. (1977). *Sci. Hort.*, **6,** 323–31.
Galeazzi, M. A. M. & Sgarbieri, V. C. (1981). *J. Food Sci.*, **46,** 1404–6.
Galliard, T. & Stumpf, P. K. (1968). *Biochem. Prep.*, **12,** 66–9.
Galliard, T., Hulme, A. C., Rhodes, M. J. C. & Wooltorton, L. S. C. (1968). *FEBS Lett.*, **1,** 283–6.
Galliard, T., Matthew, J. A., Fishwick, M. J. & Wright, A. J. (1976). *Phytochem.*, **15,** 1647–50.
Galliard, T., Matthew, J. A., Wright, A. J. & Fishwick, M. J. (1977). *J. Sci. Food Agric.*, **28,** 863–8.
Gertman, E. & Fuchs, Y. (1974). *Plant and Cell Physiol.*, **15,** 501–5.
Ghazali, H. M. & Leong, C. K. (1987). *Food Chem.*, **24,** 147–57.
Giovannoni, J. J., DellaPenna, D., Bennett, A. B. & Fischer, R. L. (1989). *Plant Cell*, **1,** 53–63.
Given, N. K., Venis, M. A. & Grierson, D. (1988*a*). *J. Plant Physiol.*, **133,** 25–30.
Given, N. K., Venis, M. A. & Grierson, D. (1988*b*). *J. Plant Physiol.*, **133,** 31–7.
Gizis, E. J. (1964). PhD Thesis Oregon State University, Oregon, USA.
Golan, A., Kahn, V. & Sadovski, A. Y. (1977). *J. Agric. Food Chem.*, **25,** 1253–60.
Goldstein, J. L. & Swain, T. (1963). *Phytochem.*, **2,** 371–83.
Goodenough, P. W. & Entwhistle, T. G. (1982). *Eur. J. Biochem.*, **127,** 145–9.
Goodenough, P. W., Kessell, S., Lea, A. G. H. & Loeffler, T. (1983). *Phytochem.*, **22,** 359–63.
Goodenough, P. W., Prosser, I. M. & Young, R. E. (1985). *Phytochem.*, **24,** 1157–62.
Grierson, D., Slater, A., Speirs, J. & Tucker, G. A. (1985). *Planta*, **163,** 263–71.
Grierson, D., Tucker, G. A., Keen, J., Ray, J., Bird, C. R. & Schuch, W. (1986). *Nucleic Acids Res.*, **14,** 8595–603.
Grncarivec, M. & Hawker, J. S. (1971). *J. Sci. Food Agric.*, **22,** 270–2.
Gross, K. C. & Pharr, D. M. (1982). *Plant Physiol.*, **74,** 531–6.
Haard, N. F. (1973). *Phytochem.*, **12,** 555–60.
Haard, N. F. & Hultin, H. O. (1970). *J. Food Sci.*, **35,** 751–6.
Hansen, E. (1971). In *The Biochemistry of Fruits and Their Products* (Vol. 1), ed. A. C. Hulme. Academic Press, New York, USA, pp. 147–58.
Harel, E., Mayer, A. M. & Lerner, H. R. (1970). *J. Sci. Food Agric.*, **21,** 541–3.
Harris, J. E. & Dennis, C. (1981). *J. Sci. Food Agric.*, **31,** 1164–72.
Harris, P. & James, A. T. (1969). *Biochem. J.*, **112,** 325–30.
Hartmann, C. J. R. (1975). In *Facteurs et Regulation de la Maturation des Fruits*, ed. R. Ulrich. CNRS, Paris, France, pp. 235–40.
Harwood, J. L. & Stumpf, P. K. (1972). *Arch. Biochem. Biophys.*, **148,** 282–90.
Hasegawa, S. & Delton, U. A. (1983). *J. Agric. Food Chem.*, **31,** 178–80.
Hatfield, R. & Nevins, D. J. (1986). *Plant Cell Physiol.*, **27,** 541–52.
Hawker, J. S. (1969). *Phytochem.*, **8,** 19–23.
Hawker, J. S. (1985). In *Biochemistry of Storage Carbohydrates in Green Plants*, ed. P. M. Dey & R. A. Dixon. Academic Press, New York, USA, pp. 1–51.
Hawker, J. S. & Walker, R. R. (1978). *Amer. J. Enol. Vitic.*, **29,** 172–6.
Hawker, J. S., Buttrose, M. S., Soeffky, A. & Possingham, J. V. (1972). *Vitis*, **11,** 189–92.

Hinton, D. M. & Pressey, R. (1974). *J. Food Sci.*, **39**, 783–5.
Hobson, G. E. (1962). *Nature*, **195**, 804–5.
Hobson, G. E. (1968). *J. Food Sci.*, **33**, 588–91.
Hobson, G. E. (1970). *Phytochem.*, **9**, 2257–63.
Hobson, G. E. (1981). In *Recent Advances in the Biochemistry of Fruits and Vegetables*, ed. J. Friend & J. J. C. Rhodes. Academic Press, London, UK, pp. 123–32.
Huber, D. J. (1984). *J. Food Sci.*, **49**, 1310–15.
Hughes, D. L. & Yamaguchi, M. (1983). *Hort Sci.*, **18**, 739–40.
Hulme, A. C. (1956). *Nature*, **178**, 218–19.
Hulme, A. C., (ed.) (1971*a*). *The Biochemistry of Fruits and Their Products* (Vol. I & II), Academic Press, London, UK.
Hulme, A. C. (1971*b*). In *The Biochemistry of Fruits and Their Products* (Vol. 2), ed. A. C. Hulme. Academic Press, London, pp. 233–54.
Hulme, A. C. & Rhodes, M. J. C. (1971). In *The Biochemistry of Fruits and Their Products* (Vol. II), ed. A. C. Hulme. Academic Press, London, UK, pp. 333–73.
Hulme, A. C., Jones, J. D. & Wooltorton, L. S. C. (1963). *Proc. Roy. Soc. (B)*, **158**, 514–35.
Hulme, A. C., Jones, J. D. & Wooltorton, L. S. C. (1964). *Phytochem.*, **3**, 173–88.
Hultin, H. O. & Levine, A. S. (1965). *J. Food Sci.*, **30**, 917–21.
Hultin, H. O., Sun, B. & Bulger, J. (1966). *J. Food Sci.*, **31**, 320–7.
Hyodo, H. (1971). *Plant Cell Physiol.*, **12**, 989–91.
Hyodo, H., Ikeda, N., Nagatani, N. & Tanaka, K. (1983). *J. Jap. Soc. Hort. Sci.*, **52**, 196–9.
Irwin, P. L., Pfeffer, P. E., Gerasimowicz, W. V., Pressey, R. & Sams, C. E. (1984). *Phytochem.*, **23**, 2239–42.
Isaac, J. E. & Rhodes, M. J. C. (1982). *Phytochem.*, **21**, 1553–6.
Isaac, J. E. & Rhodes, M. J. C. (1987). *Phytochem.*, **26**, 649–53.
Jeffery, D., D., Smith, C., Goodenough, P., Prosser, I. & Grierson, D. (1984). *Plant Physiol.*, **74**, 32–8.
Jimenez, M. A. (1947). *Food Res.*, **12**, 300–10.
Joel, D. M., Marbach, I. & Mayer, A. M. (1978). *Phytochem.*, **17**, 796–7.
Jones, H. G. (1981). *J. Exper. Bot.*, **32**, 1203–10.
Jones, J. D., Hulme, A. C. & Wooltorton, L. S. C. (1965). *New Phytol.*, **64**, 158–67.
Jones, W. W. & Kubota, H. (1940). *Plant Physiol.*, **15**, 711–17.
Kahn, V. (1975). *J. Sci. Food Agric.*, **26**, 1319–24.
Kahn, V. (1976*a*). *Phytochem.*, **15**, 267–72.
Kahn, V. (1976*b*). *J. Food Sci.*, **41**, 1011–12.
Kahn, V. (1977*a*). *J. Sci. Food Agric.*, **28**, 233–9.
Kahn, V. (1977*b*). *J. Food Sci.*, **42**, 38–43.
Kahn, V. (1983). *Phytochem.*, **10**, 2155–9.
Kamphuis, I. G., Kalk, K. H., Swarte, M. B. A. & Drenth, J. (1984). *J. Mol. Biol.*, **179**, 233–56.
Kane, O., Marcellin, P. & Mazliak, P. (1978). *Plant Physiol.*, **61**, 634–8.
Karasaki, Y., Ikeuchi, K. & Higashi, K. (1983). *J. UOEH*, **5**, 329–35.
Kasantikul, D. (1983). Master's Thesis, Kestsart University, Bangkok, Thailand.
Katwa, L. C., Ramakrishna, M. & Raghavendra Rao, M. R. (1982). *J. Food Biochem.*, **6**, 217–28.
Kerbel, E. L., Kader, A. A. & Romani, R. J. (1988). *Plant Physiol.*, **86**, 1205–9.
Kim, I. S. & Grosch, W. (1978). *Z. Lebensmittel. U. Forsch.*, **167**, 324–6.
Kim, I. S. & Grosch, W. (1979). *J. Agric. Food Chem.*, **27**, 243–6.
Kim, I. S. & Grosch, W. (1981). *J. Agric. Food Chem.*, **29**, 1220–5.
Knee, M. (1971). *J. Sci. Food Agric.*, **22**, 368–71.
Knee, M. (1973). *Phytochem.*, **12**, 1543–9.
Knee, M. (1978). *Phytochem.*, **17**, 1257–60.

Knee, M. (1985). *J. Exp. Bot.*, **36,** 670–8.
Knee, M. (1987). *J. Exp. Bot.*, **38,** 1724–33.
Knee, M., Sargent, J. A. & Osborne, D. J. (1977). *J. Exp. Bot.*, **28,** 377–96.
Kosiyachinda, S. & Young, R. E. (1977). *Plant and Cell Physiol.*, **16,** 595–602.
Krishnamurthy, S. & Subramanyam, H. (1973). *Trop. Sci.*, **15,** 167–93.
Krishnamurthy, S., Patwardhan, M. V. & Subramanyan, H. (1971). *Phytochem.*, **10,** 2577–81.
Kushad, M. M., Richardson, D. G. & Ferro, A. J. (1982). *Biochem. Biophys. Res. Comm.*, **108,** 167–73.
Kushad, M. M., Richardson, D. G. & Ferro, A. J. (1983). *Plant Physiol.*, **73,** 257–61.
Lakshminarayana, S. (1980). In *Tropical and Subtropical Fruits: Composition Properties and Uses,* ed. S. Nagy & P. E. Shaw. AVI Publishing Inc., Westport, CT, USA, pp. 184–257.
Lal, R. K., Garg, M. & Krishnan, P. S. (1974). *Phytochem.*, **13,** 2365–70.
Lance, C. (1981). In *Recent Advances in the Biochemistry of Fruits and Vegetables,* ed. J. Friend & M. J. C. Rhodes. Academic Press, London, UK, pp. 63–87.
Latche, A., Pech, J. C., Diarra, A. & Fallot, J. (1975). In *Facteurs et Regulation de la Maturation des Fruits,* ed. R. Ulrich. CNRS, Paris, France, pp. 299–306.
Leblond, C. (1975). *Physiol. Veg.*, **13,** 651–65.
Lelyveld, L. J. van (1975). *Agroplantae,* **7,** 51–4.
Lelyveld, L. J. van (1978). *Z. Pflanzenphysiol.*, **89,** 449–52.
Lelyveld, L. J. van & Bruyon, J. A. de (1977). *Agrochemophysica,* **9,** 1–6.
Lelyveld, L. J. van & Bower, J. P. (1984). *J. Hort. Sci.*, **59,** 257–63.
Lester, G. E. & Dunlap, J. R. (1985). *Sci. Hort.*, **26,** 323–31.
Lewis, C. E. (1978). *J. Sci. Food Agric.*, **29,** 857–66.
Lewis, C. E., Morris, R. & O'Brien, K. (1978). *J. Sci. Food Agric.*, **29,** 943–9.
Li, M.-C. & Yan, J.-L. (1963). *Acta Bot. Sin.*, **11,** 329–37.
Liang, Y. F., Ao, L. D., Wang, M. X. & Yi, W. J. (1982). *Acta Bot. Sin.*, **24,** 143–6.
Lin, T. Y., Shewfelt, R. L. & Wilkinson, R. E. (1985). *J. Food Sci.*, **50,** 1227–9.
Lingle, S. E. & Dunlap, J. R. (1987). *Sci. Hort.*, **84,** 386–9.
Lingle, S. E., Lester, G. E. & Dunlap, J. R. (1987). *Hort Sci.*, **22,** 917–19.
Livne, A. & Gepstein, S. (1988). *Plant Physiol.*, **87,** 239–43.
Lodh, S. B., Divakar, N. G., Chadha, K. L., Melanta, K. R. & Selvaraj, Y. (1973). *Ind. J. Hort.*, **30,** 381–3.
Loomis, W. D. (1969). In *Methods in Enzymology* (Vol. XIII), ed. J. M. Lowenstein. Academic Press, New York, USA, pp. 555–63.
Loomis, W. D. (1974). In *Methods in Enzymology* (XXXI), ed. S. Fleischer & L. Packer. Academic Press, New York, USA, pp. 528–44.
Looney, N. E. & Patterson, M. E. (1967). *Nature,* **214,** 1245–6.
Lopez, M. E., Vattuone, M. A. & Sampietro, A. R. (1988). *Phytochem.*, **27,** 3077–81.
Lott, R. V. & Barrett, H. (1967). *Vitis,* **6,** 257–68.
Lourenco, E. J. & Catutani, A. T. (1984). *J. Sci. Food Agric.*, **35,** 1120–7.
Lowe, G. (1976). *Tetrahedron,* **32,** 291–302.
Lowry, O. M., Rosebrough, N. J., Farr, A. L. & Randall, R. J. (1951). *J. Biol. Chem.*, **193,** 265–75.
Lurie, L. & Ben Arie, R. (1983). *Phytochem.*, **22,** 49–52.
Mancha, M. & Garcia, J. M. (1982). In *Biochemistry and Metabolism of Plant Lipids,* ed. J. F. G. M. Wintermans & P. J. C. Kuiper. Elsevier North Holland Biomedical Press, Amsterdam, The Netherlands, pp. 243–6.
Manning, K. & Maw, G. A. (1975). *Phytochem.*, **14,** 1965–9.
Mansour, R., Latche, A., Vaillant, V., Pech, J. C. & Reid, M. S. (1986). *Physiol. Plant.*, **66,** 495–502.
Marcus, L., Prusky, D. & Jacoby, B. (1988). *J. Sci. Food. Agric.*, **29,** 943–9.

Markovic, O., Heinrichova, K. & Lenkey, B. (1975). *Coll. Czech. Chem. Comm.*, **40**, 769–74.
Martinez-Cayuela, M., de Medina, L. S., Faus, J. J. & Gil, A. (1988). *J. Food Sci.*, **53**, 1191–4.
Matsuo, T. & Ito, S. (1982). *Agric. Biol. Chem.*, **46**, 683–8.
Mattoo, A. K. & Modi, V. V. (1970*a*). *Enzymologia*, **39**, 237–47.
Mattoo, A. K. & Modi, V. V. (1970*b*). *Biochem. Biophys. Res. Comm.*, **39**, 895–904.
McCollum, T. G., Huber, D. J. & Cantiliffe, D. J. (1988). *J. Amer. Soc. Hort. Sci.*, **113**, 399–403.
McFeeters, R. F., Bell, T. A. & Fleming, M. P. (1980). *J. Food Biochem.*, **4**, 1–16.
McFeeters, R. F., Fleming, M. P. & Thompson, R. L. (1985). *J. Food Sci.*, **50**, 201–5.
McLellan, M. R., Kime, R. W. & Lind, L. R. (1985). *J. Food Sci.*, **50**, 206–8.
Medlicott, A. P. & Thompson, A. K. (1985). *J. Sci. Food Agric.*, **36**, 561–6.
Menary, R. C. & Jones, R. H. (1972). *Aust. J. Biol. Sci.*, **25**, 531–42.
Miller, A. R. & Kelley, T. J. (1989). *Hort. Sci.*, **24**, 650–2.
Miller, A. R., Dalmasso, J. P. & Kretchman, D. W. (1987). *J. Amer. Soc. Hort. Sci.*, **112**, 666–71.
Miller, A. R., Dalmasso, J. P. & Kretchman, D. W. (1989). *Can. J. Bot.*, **67**, 817–21.
Mitchel, R. E., Chaiken, I. M. & Smith, E. L. (1970). *J. Biol. Chem.*, **245**, 3485–92.
Mitchell, T., Porter, A. J. R. & John, P. (1988). *New Phytol.*, **109**, 313–19.
Miyairi, K., Okuno, T. & Sawai, K. (1975). *Bull. Fac. Agric. Hirosaki Univ.*, **24**, 22–30.
Miyazaki, J. H. & Yang, S. F. (1987). *Plant Physiol.*, **84**, 277–81.
Modi, V. V. & Reddy, V. V. R. (1967). *Ind. J. Exper. Biol.*, **5**, 233–5.
Mohan, S. B. & Kekwick, G. O. (1980). *Biochem. J.*, **187**, 667–6.
Montgomery, M. W. & Sgarbieri, V. C. (1975). *Phytochem.*, **14**, 1245–9.
Moreau, F. & Romani, R. (1982). *Plant Physiol.*, **70**, 1385–90.
Moriguchi, T. & Yamaki, S. (1988). *Plant Cell Physiol.*, **29**, 1361–6.
Moshrefi, M. & Luh, B. S. (1983). *Eur. J. Biochem.*, **135**, 511–14.
Moulding, P. H., Grant, M. F., McLellan, K. M. & Robinson, D. S. (1987). *Int. J. Food Sci. Technol.*, **22**, 391–7.
Mowlah, G. & Itoo, S. (1982). *J. Jap. Soc. Food Sci. Technol.*, **29**, 413–17.
Nakagawa, M., Iki, K., Mirata, M., Ishigami, S. & Ogura, N. (1980). *Phytochem.*, **19**, 195–7.
Nakamura, R. (1973*a*). *J. Jap. Soc. Food Sci. Tech.*, **20**, 524–8.
Nakamura, R. (1973*b*). *J. Jap. Soc. Food Sci. Tech.*, **20**, 520–36.
Neal, G. E. (1965). *J. Sci. Food Agric.*, **16**, 604–11.
Nevins, D. J. & Jones, R. A. (1987). *Tomato Biotechnology*. Alan R. Liss, Inc., New York, USA.
Ngernprasirtsiri, J., Kobayashi, M. & Azakawa, T. (1988). *Plant Physiol.*, **88**, 16–20.
Niemietz, C. & Hawker, J. S. (1988). *Aust. J. Plant Physiol.*, **15**, 359–66.
Ota, S., Horie, K., Hagino, F., Hashimoto, C. & Date, H. (1972). *J. Biochem.*, **71**, 817–30.
Ozawa, T., Lilley, T. H. & Haslam, E. (1987). *Phytochem.*, **26**, 2937–42.
Pal, D. K. & Selvaraj, Y. (1987). *J. Hort. Sci.*, **62**, 117–24.
Palaniswamy, K. P., Muthukrishan, C. & Shanmugavelu, K. G. (1974). *Ind. Food Packer*, **28**, 12–19.
Patil, S. S. & Tang, C. S. (1974). *Plant Physiol.*, **53**, 585–8.
Patil, S. S., Tang, C. S. & Hunter, J. E. (1973). *Plant Dis. Rep.*, **57**, 86–9.
Paull, R. E. (1990). *Acta Hort.*, **269**, 277–81.
Paull, R. E. & Chen, N. J. (1983). *Plant Physiol.*, **72**, 382–5.
Paull, R. E. & Chen, N. J. (1990). *J. Amer. Soc. Hort. Sci.*, **115**, 623–31.
Paull, R. E. & Rohrbach, K. G. (1985). *J. Amer. Soc. Hort. Sci.*, **110**, 100–5.

Paull, R. E., Deputy, J. & Chen, N. J. (1983). *J. Amer. Soc. Hort. Sci.*, **108**, 931–4.
Pech, J. C., Bonneau, G. & Fallot, J. (1973). *Phytochem.*, **12**, 209–305.
Pesis, E. & Ben-Arie, R. (1984). *J. Food Sci.*, **49**, 896–9.
Pesis, E., Fuchs, Y. & Zauberman, G. (1978). *Plant Physiol.*, **61**, 416–19.
Piechulla, B., Chonoles Imlay, K. R. & Gruissem, W. (1985). *Plant Mol. Biol.*, **5**, 373–84.
Pifferi, P. G. & Cultrera, R. (1974). *J. Food Sci.*, **39**, 786–91.
Platt-Aloia, K. A. & Thomson, W. W. (1981). *Ann. Bot.*, **48**, 451–6.
Platt-Aloia, K. A., Thomson, W. W. & Young, R. E. (1980). *Bot. Gaz.*, **141**, 366–73.
Pollard, A. & Timberlake, C. F. (1971). In *The Biochemistry of Fruits and Their Products* (Vol. II), ed. A. C. Hulme. Academic Press, London, UK, pp. 573–622.
Poovaiah, B. W. & Nukaya, A. (1979). *Plant Physiol.*, **64**, 534–7.
Prasad, U. S. & Jha, O. P. (1979). *Indian J. Plant Physiol.*, **22**, 84–6.
Preiss, J. (1982). *Ann. Rev. Plant Physiol.*, **33**, 431–54.
Pressey, R. (1983). *Plant Physiol.*, **71**, 132–5.
Pressey, R. (1986). *Hort. Sci.*, **21**, 1183–5.
Pressey, R. & Avants, J. K. (1975). *J. Food Sci.*, **40**, 937–9.
Pressey, R. & Avants, J. K. (1976). *Phytochem.*, **15**, 1349–51.
Pressey, R. & Avants, J. K. (1978). *J. Food Sci.*, **43**, 1415–17.
Pressey, R., Hinton, D. M. & Avants, J. K. (1971). *J. Food Sci.*, **36**, 1070–3.
Quintana, E. G., Nanthachai, P., Hiranpradit, H., Mendoza, D. B. & Ketsa, S. (1984). In *Mango*: *Fruit Development, Postharvest Physiology and Marketing in ASEAN*, ed. D. B. Mendoza & R. B. H. Wills. ASEAN Food Handling Bureau, Kuala Lumpur, Malaysia, pp. 21–7.
Rabinowitch, H. D. & Sklan, D. (1980). *Planta*, **148**, 162–7.
Radanive, A. S. & Haard, N. F. (1972). *J. Food Sci.*, **37**, 381–3.
Radler, F. (1964). *J. Sci. Food Agric.*, **15**, 864–69.
Ratovohery, J. V., Lozano, Y. F. & Gaydou, E. M. (1988). *J. Agric. Food Chem.*, **36**, 287–93.
Reid, P. D., Strong, M. G., Lew, F. & Lewis, L. N. (1974). *Plant Physiol.*, **53**, 732–7.
Rensburg, E. van & Englebrecht, A. H. P. (1986). *J. Food Sci.*, **51**, 1067–70.
Reymond, D. & Phaff, H. J. (1965). *J. Food Sci.*, **30**, 266–73.
Rhodes, M. J. C. (1977). In *Regulation of Enzyme Synthesis and Action in Higher Plants*, ed. H. Smith. Academic Press, London, UK, pp. 245–69.
Rhodes, M. J. C. & Wooltorton, L. S. C. (1977). *Phytochem.*, **16**, 655–9.
Robin, J.-P., Romieu, C., Sauvage, F.-X., Nicol, M.-Z. & Flanzy, C. (1987). *Plant Physiol. Biochem.*, **25**, 797–804.
Robinson, G. W. (1975). *Biochem.*, **14**, 3695–700.
Rodriguez, R., Agarwal, P. C. & Saha, N. K. (1971). *Indian Food Packer*, **25**, 5–12.
Roe, B. & Bruemmer, J. H. (1981). *J. Food Sci.*, **46**, 186–9.
Roubelakis, K. A. & Kliewer, W. M. (1978). *Plant Physiol.*, **62**, 344–47.
Rouse, A. H. & Barmore, C. R. (1974). *Hort. Sci.*, **9**, 36–7.
Ruffner, H. P. (1982*a*). *Vitis*, **21**, 247–59.
Ruffner, H. P. (1982*b*). *Vitis*, **21**, 346–58.
Ruffner, H. P. & Hawker, J. S. (1977). *Phytochem.*, **16**, 1171–5.
Ruffner, H. P., Hawker, J. S. & Hale, C. R. (1976). *Phytochem.*, **15**, 1877–80.
Ruffner, H. P., Brem, S. & Malipiero, U. (1983). *Acta Hort.*, **139**, 123–8.
Ruffner, H. P., Possner, D., Brem, S. & Rast, D. M. (1984). *Planta*, **160**, 444–8.
Salminen, S. O. & Young, R. E. (1975). *Plant Physiol.*, **55**, 45–50.
Sasaki, M., Kato, T. & Iida, S. (1973). *J. Biochem.*, **74**, 635–7.
Satjawatcharaphong, C., Rymal, K. S., Dosier, W. A. & Smith, R. C. (1983). *J. Food Sci.* **48**, 1879–80.
Satoh, S. & Yang, S. F. (1988). *Plant Physiol.*, **88**, 109–14.
Sawato, M. (1969). Master's Thesis, University of Hawaii, Hawaii, USA.

Selvaraj, Y., Pal, D. K., Subramanyam, M. D. & Iyer, C. P. A. (1982). *J. Hort. Sci.*, **57**, 135–43.
Sen, S., Chatterjee, B. K. & Roy, H. B. (1985). *Ind. Biol.*, **17**, 25–8.
Seo, S. T. & Tang, C. S. (1982). *J. Econ. Entomol.*, **75**, 1132–5.
Seymour, G. B. & Harding, S. E. (1987). *Biochem. J.*, **245**, 463–6.
Sharon, O. & Kahn, V. (1979). *Physiol. Plant.*, **45**, 227–34.
Sharon-Raber, O. & Kahn, V. (1983). *J. Food Sci.*, **48**, 1874–5.
Shastri, P. N. & Shastri, N. K. (1975). *J. Food Sci. Technol.*, **12**, 42–3.
Shine, W. E., Mancha, M. & Stumpf, P. K. (1976). *Arch. Biochem. Biophys.*, **172**, 110–16.
Shipway, M. R. & Bramlage, W. J. (1973). *Plant Physiol.*, **51**, 1095–8.
Shukla, R. N., Singh, S., Das, N., Baijal, M. & Sanwal, G. G. (1973). *Phytochem.*, **12**, 979–85.
Siegelman, H. W. & Hendricks, S. B. (1958). *Plant Physiol.*, **33**, 185–90.
Simoni, R. D., Criddle, R. S. & Stumpf, P. K. (1967). *J. Biol. Chem.*, **242**, 573–81.
Singh, S. & Sanwal, G. G. (1973). *Biochim. Biophys. Acta*, **309**, 280–8.
Singh, S. & Sanwal, G. G. (1975). *Phytochem.*, **14**, 113–18.
Sitrit, Y., Blumenfeld, A. & Riov, J. (1987). *Physiol. Plant.*, **69**, 504–10.
Skakoun, A. & Daussant, J. (1975). In *Facteurs et Regulation de la Maturation des Fruits*, ed. R. Ulrich. CNRS, Paris, France, pp. 281–9.
Smith, C. J. S., Watson, C. F., Ray, J., Bird, C. R., Morris, P. C., Schuch, W. & Grierson, D. (1988). *Nature*, **234**, 724–6.
Smith, D. M. & Montgomery, M. W. (1985). *Phytochem.*, **24**, 901–4.
Sobotka, F. E. & Stelzig, D. A. (1974). *Plant Physiol.*, **53**, 759–63.
Soda, I., Hasegawa, T., Suzuki, T. & Ogura, N. (1986). *Agric. Biol. Chem.*, **50**, 3191–2.
Solomos, T. & Laties, G. C. (1974). *Plant Physiol.*, **54**, 506–11.
Spalding, D. H. & Reeder, W. F. (1975). *Phytopath.*, **65**, 458–60.
Spalding, D. H. (1976). In *Proc. First International Tropical Fruit Short Course. The Avocado*. Eds J. W. Sauls, R. L. Phillips and L. K. Jackson, Florida Cooperative Extension Service, University of Florida, Gainesville, pp. 109–13.
Stoewsand, G. S. & Anderson, J. L. (1973). *J. Food Sci.*, **38**, 1256.
Strand, L. L., Ogawa, J. M., Bose, E. & Rumsey, J. W. (1981). *J. Food Sci.*, **46**, 498–500.
Su, L., McKeon, T., Grierson, D., Cantwell, M. & Yang, S. F. (1984). *Hort. Sci.*, **19**, 576–8.
Subramanyam, H., Krishnaumurthy, S. & Parpia, H. A. B. (1975). *Adv. Food Res.*, **21**, 223–305.
Sugiura, A. & Tomana, T. (1983). *Hort. Sci.*, **18**, 319–21.
Sugiura, A., Harada, H. & Tomana, T. (1975). *J. Jap. Soc. Hort. Sci.*, **44**, 265–9.
Sugiura, A., Taira, S., Ryugo, K. & Tomana, T. (1985). *Nippon Shokuhin Kogyo Gakkaishi*, **32**, 586–9.
Sum, W. F., Rogers, P. J., Jenkin, I. D. & Guthrie, R. D. (1980). *Phytochem.*, **19**, 399–401.
Surendranathan, K. K. & Nair, P. M. (1973). *Phytochem.*, **12**, 241–9.
Surendranathan, K. K. & Nair, P. M. (1976). *Phytochem.*, **15**, 371–3.
Taira, S., Sugiura, A. & Tomana, T. (1986). *J. Jap. Soc. Hort. Sci.*, **55**, 228–34.
Tan, S. C. (1980). *Aust. J. Plant Physiol.*, **7**, 159–67.
Tan, S. C. & Lam, P. F. (1985). *ASEAN Food J.*, **1**, 134–6.
Tan, S. C., Teo, S. W. & Adam, A. G. (1982). *Sains Malaysian*, **11**, 21–31.
Tan, X.-J. & Li, Y.-B. (1984). *Acta Phytophysiol. Sinica*, **10**, 339–46.
Tandon, D. D. & Kalra, S. K. (1983). *J. Hort. Sci.*, **58**, 449–53.
Tang, C. S. (1971). *Phytochem.*, **10**, 117–21.
Tang, C. S. (1973). *Phytochem.*, **12**, 769–73.

Tang, C. S. (1974). *J. Food Sci.*, **39,** 94–6.
Tango, J. S., da Costa, S. I., Antunes, A. J. & Figueiredo, I. B. (1972). *Fruits,* **27,** 143–6.
Teisson, C. (1979). *Fruits,* **34,** 245–61.
Teisson, C. & Combres, J. C. (1979). *Fruits,* **34,** 315–39.
Terra, N. N., Garcia, E. & Lajolo, F. M. (1983). *J. Food Sci.*, **48,** 1097–100.
Theologis, A. & Laties, G. C. (1978). *Plant Physiol.*, **62,** 249–55.
Thomas, P. & Janave, M. T. (1986). *J. Food Sci.*, **51,** 384–7.
Tigchelaar, E. C., McGlasson, W. C. & Buescher, R. W. (1978). *Hort. Sci.*, **13,** 508–12.
Torres, A. M. (1980). *J. Amer. Soc. Hort. Sci.*, **105,** 614–19.
Torres, A. M., Diedenhofen, U., Bergh, B. O. & Knight, R. J. (1978). *Amer. J. Bot.*, **65,** 134–9.
Tucker, G. A., Robertson, N. G. & Grierson, D. (1981*a*). *Eur. J. Biochem.*, **112,** 87–90.
Tucker, G. A., Robertson, N. G. & Grierson, D. (1981*b*). *Eur. J. Biochem.*, **112,** 119–24.
Tucker, M. L., Christoffersen, R. E., Woll, L. & Laties, G. G. (1985). In *Ethylene and Plant Development,* ed. J. A. Roberts & G. A. Tucker. Butterworths, London, UK, pp. 163–71.
Tucker, M. L., Durbin, M. L., Clegg, M. T. & Lewis, L. N. (1987). *Plant Mol. Biol.*, **9,** 197–203.
Ugalde, T. D., Chalmers, D. J. & Jerie, P. H. (1988). *Aust. J. Plant Physiol.*, **15,** 377–83.
Ulrich, R. (1971). In *The Biochemistry of Fruits and Their Products* (Vol. I), ed. A. C. Hulme. Academic Press, New York, USA, pp. 89–118.
Vasquez-Salinas, C. & Lakshiminarayana, S. (1985). *J. Food Sci.*, **50,** 1646–8.
Wager, H. G. (1974). *J. Exp. Bot.*, **25,** 338–51.
Wang, C. Y. & Adams, D. O. (1982). *Plant Physiol.*, **69,** 424–7.
Wang, C. Y. & Baker, J. G. (1979). *Plant and Cell Physiol.*, **20,** 243–51.
Weaire, P. J. & Kekwick, R. G. O. (1975*a*). *Biochem. J.*, **146,** 425–37.
Weaire, P. J. & Kekwick, R. G. O. (1975*b*). *Biochem. J.*, **146,** 439–45.
Weaver, C. & Charley, H. (1974). *J. Food Sci.*, **39,** 1200–2.
Winer, L. & Apelbaum, A. (1986). *J. Plant Physiol.*, **125,** 223–33.
Winer, L., Vinkler, C. & Apelbaum, A. (1984). *Plant Physiol.*, **76,** 233–7.
Wiskich, J. T. & Dry, I. B. (1985). In *Encyclopedia of Plant Physiology, New Series,* (Vol. 18), ed. R. Douce & D. A. Day. Springer Verlag, Berlin, Germany, pp. 281–313.
Woodward, J. R. (1972). *J. Sci. Food Agric.*, **23,** 465–73.
Wright, M. (1974). *Planta,* **120,** 63–9.
Yamada, F., Takahashi, N. & Murachi, T. (1976). *J. Biochem.*, **79,** 1223–34.
Yamaki, S. & Ishikawa, K. (1986). *J. Amer. Soc. Hort. Sci.*, **111,** 134–7.
Yamaki, S. & Matsuda, K. (1977). *Plant Cell Physiol.*, **18,** 81–93.
Young, R. E., Salminen, S. & Sornsrivichai, P. (1975). In *Facteurs et Regulation de la Maturation des Fruits,* ed. R. Ulrich. CNRS, Paris, France, pp. 271–9.
Zauberman, G. & Schiffman-Nadel, M. (1972). *J. Amer. Soc. Hort. Sci.*, **97,** 313–15.
Zauberman, G., Fuchs, Y. & Akerman, M. (1985). *Sci. Hort.*, **26,** 261–5.

16

Enzymes in Citrus Fruits

JESÚS CHAMARRO, JOSÉ MIGUEL ALONSO & JOSÉ LUIS GARCIA-MARTINEZ
Instituto de Agroquímica y Tecnología de Alimentos (CSIC), Jaime Roig 11, 46010 Valencia, Spain

INTRODUCTION

World production of citrus reached a record 64·5 million tonnes in 1988/89. An estimated 22 million tonnes, almost 35% of global output, were transformed into citrus products. Oranges represented 90% of this amount, frozen concentrated orange juice being the major product (FAO, 1990).

Most citrus fruits of commercial interest, i.e. sweet orange (*Citrus sinensis*), sour orange (*C. aurantium*), mandarin (*C. reticulata*), lemon (*C. limonia*), grapefruit (*C. grandis*) and lime (*C. aurantifolia*), belong to the genus *Citrus* from the family Rutaceae.

The citrus fruit is botanically an hesperidium and can be divided into flavedo, albedo and carpel segments. It also has a central axis and a button-like receptacle with an abscission layer at the stem end of the fruit (Figs 1 and 2). The flavedo is the outermost part of the peel or rind. It has a cutinized surface, partly covered with waxes, and contains numerous stomata. Essential oils accumulate in the oil glands and chlorophyll and/or carotenoid pigments in the chromatophores. Beneath the flavedo, the albedo is a white, spongy tissue layer of variable thickness, rich in cellulose, hemicellulose, lignin, pectins and phenolic compounds. The central axis of the fruit has a structure and composition similar to that of the albedo. The edible portion, sometimes called the pulp, is made up of the carpels or segments, which are separated by carpellary walls. Within the carpel are located the juice vesicles and seeds. The membranes of segments and juice vesicles have similar chemical constituents to the albedo. Most of the sugars, mainly sucrose, glucose and fructose, and nearly all the citric acid occurs in the juice, which also contains nitrogenous compounds, lipids, phenolics, vitamins and minerals.

Citrus fruits are non-climacteric and do not ripen after harvest. Respiration rate of ripe fruits is essentially stable, and endogenous levels of ethylene are very low. Eaks (1970) showed that the respiratory increase in citrus fruits, in

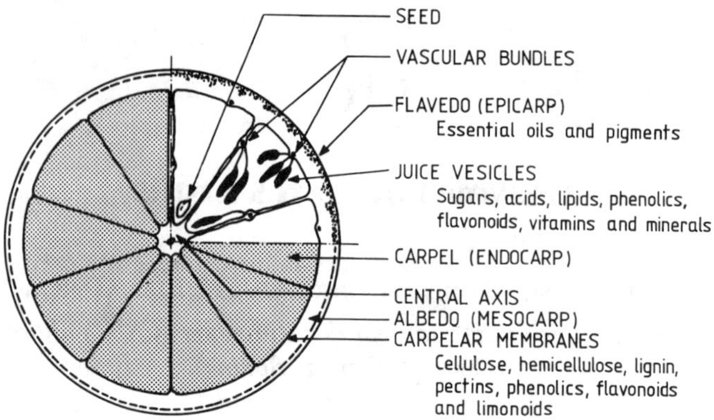

FIG. 1. Schematic representation of the equatorial section of a citrus fruit.

response to exogenous ethylene, was reversible, and that they responded to repeated stimulation.

In the last few years, after the classic review of Joslyn and Pilnik (1961), some general publications dealing with citrus enzymes have been published (Blundstone *et al.*, 1971; Ting & Attaway, 1971; Bruemmer, 1980; Sinclair, 1984). Most of the recent information on this topic is related to technological aspects of fresh and processed fruits, such as abscission, degreening, stress processes, cloud stability, and bitterness of juices. Consequently, this approach to citrus enzymes has been orientated to emphasize the implications of enzymes on quality, handling and processing, rather than to a description according to the principles of classic enzymology.

PIGMENT CHANGES IN CITRUS FRUITS

Colour changes in the peel during ripening and degreening

When citrus fruits mature, the green colour of the peel starts to disappear due to chlorophyll degradation. Simultaneously, carotenoid content increases in a

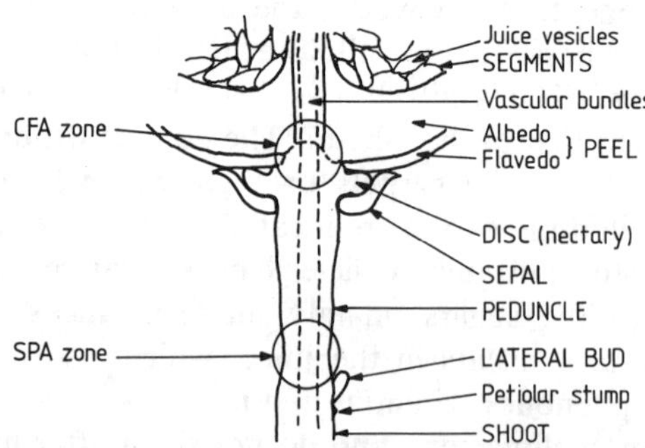

FIG. 2. Schematic illustration of the shoot–peduncle abscising (SPA) and calix–fruit abscising (CFA) zones of citrus. (Adapted from Goren & Huberman, 1976).

process that continues after total chlorophyll destruction, and they confer on the fruit its characteristic colour.

Colour change in citrus fruits is affected by climatic and growing conditions. Minimum temperatures in the range 7–13°C are required for normal colour development (Young & Erickson, 1961). The process is retarded by conditions that stimulate vigorous growth, such as irrigation and high nitrogen fertilization (Grierson *et al.*, 1986).

Under certain conditions, many cultivars mature internally while the colour of the peel is still green and unattractive to the consumer. This often occurs in subtropical climates where minimum temperatures are not adequate to attain a good colour. In this case, treatment with ethylene gas, usually 5–10 μl/litre at 20–25°C, or with ethylene-releasing compounds, such as ethephon (2-chloroethylphosphonic acid), stimulates the degradation of chlorophyll and, simultaneously, the biosynthesis of carotenoids (Purvis & Barmore, 1981). The specific mode of ethylene action in chlorophyll degradation is not well known.

Chlorophyll degradation

Chlorophyllase

The activity of chlorophyllase (chlorophyll chlorophyllidohydrolase, EC 8.1.1.14) in the rind of citrus fruits increases substantially following ethylene treatment, with a concomitant decrease in chlorophyll content. This suggests that the effect of ethylene on chlorophyll degradation is mediated through the enhancement of chlorophyllase activity (Barmore, 1975; Shimokawa *et al.*, 1978*b*). The activation of chlorophyllase by ethylene has a lag period of 6–12 h from commencement of treatment.

Although it is generally accepted that chlorophyllase is required for chlorophyll degradation, Purvis (1980) observed that a high level of chlorophyllase activity may not necessarily cause degreening. Purvis and Barmore (1981) found that ethylene in conjunction with chlorophyllase is necessary for chlorophyll degradation, and, on the other hand, the high chlorophyllase levels found during regreening of Valencia oranges suggest that the enzyme could also be involved in chlorophyll synthesis (Aljubiri *et al.*, 1979).

Chlorophyll degradation by chlorophyllase in various tissues has been associated with the appearance of dephytilated forms, such as chlorophyllide and pheophorbide, followed by further degradative reactions, particularly the removal of magnesium (Matile *et al.*, 1989). HPLC analysis of extracts from senescing citrus peel shows that chlorophyllide a and other polar dephytilated derivatives accumulate in fruits treated with ethylene, coinciding with the decrease in chlorophyll. This provides additional evidence about the in-vivo function of chlorophyllase in this system (Amir-Shapira *et al.*, 1987).

Chlorophyllase from *Citrus unshiu* has been purified to homogeneity by hydrophobic chromatography. The mol. wt of the enzyme was estimated as 110 kDa by gel filtration and 27 kDa by SDS–PAGE, indicating that the enzyme could be a tetramer. The optimum pH is 7·0 and the K_m for chlorophyll a is 2·65 μM (Shimokawa, 1982).

Chlorophyll peroxidase

Chlorophyll peroxidase (EC 1.11.1.7) extracted from the flavedo of oranges catalyses the destruction of chlorophyll by H_2O_2 in a reaction apparently coupled with peroxidation of certain phenols. Study of the characteristics of this reaction led Huff (1982) to suggest a possible role of peroxidases in chlorophyll degradation in citrus. It has been proposed that peroxidases are involved in the degradation of chlorophyll (Matile, 1980; Yamauchi & Minamide, 1985); however, recent studies (Abeles & Dunn, 1989; Abeles *et al.*, 1989) indicate that peroxidases do not participate in this reaction.

Other chlorophyll-degrading mechanisms

In addition to chlorophyllase, ethylene also affects other enzyme systems (Abeles, 1973), several of which could alter the structural integrity of chloroplasts, thus contributing indirectly to the overall chlorophyll degradation (Barmore, 1975).

Disruption of the lamellar membranes on the chloroplast is a primary process during degreening, either induced by ethylene (Shimokawa *et al.*, 1978*a*), or when it occurs naturally on the tree (El-Zeftawi & Garret, 1978). No appreciable change or loss of chlorophyll is observed until ultrastructural changes in the chloroplast are apparent (Purvis, 1980). Whether ethylene participates directly (i.e. interacting with the thylakoid membranes to release chlorophyll molecules and making them accessible to chlorophyllase), according to the in-vitro observations of Amir-Shapira *et al.* (1986), or whether it induces other enzymes whose activity precedes chlorophyllase activity in chlorophyll degradation, is not certain (Purvis & Barmore, 1981). Cycloheximide but not chloramphenicol inhibits the ethylene-enhanced chlorophyll degradation and represses the enhancement of chlorophyllase activity. The contrasting effect of these inhibitors on protein synthesis suggests that, for ethylene-enhanced chlorophyll degradation, the de-novo synthesis of cytoplasmic enzyme(s) is essential and chloroplast-dependent enzyme synthesis does not play a role (Shimokawa *et al.*, 1978*b*).

Regreening

Late-maturing cultivars, such as Valencia Late oranges, after attaining the characterisic colour of ripe fruit in winter, start to regreen in spring, coinciding with the increase in temperature and light intensity. Although the content of chlorophyll a is higher than that of chlorophyll b until colour break, chlorophyll a values drop below those of chlorophyll b once the fruit starts to colour. This has been attributed to the transformation of chlorophyll a to chlorophyll b with increasing light intensity (El-Zeftawi, 1977). It has also been observed that, during colour break, chlorophyll b is more resistant to degradation than chlorophyll a, and this has been attributed, in part, to a stronger binding capacity of chlorophyll b to proteins, which would lead to lower enzymatic degradation. This could explain the difficulty in degreening regreened oranges using ethylene (Jahn & Young, 1976). These results are in contrast to those of Shimokawa (1979), who reported that chlorophyll b is degraded preferentially in Satsuma mandarins, regardless of whether they are ripened on the tree or treated with ethylene. This latter observation has been confirmed recently by in-vivo spectroscopic measurement and has been attributed to the differences in the relative orientation between chlorophyll a and b in the chlorophyll a–b complex (Shimokawa, 1990).

Carotenoids

Application of ethylene to citrus fruits after harvest increases carotenoid biosynthesis. In some cultivars, cryptoxantin, β-citraurin and violaxantin, which enhance orange and red colours, are preferentially accumulated (Stewart & Wheaton, 1972). However, optimum degreening temperatures are considerably higher than those for carotenoid biosynthesis, although sometimes, in practice, this aspect is not considered, since ethylene treatment is primarily used to degrade the chlorophyll (Wheaton & Stewart, 1973). There is almost no information in the literature on the citrus enzymes that participate in these reactions.

Treatment of several cultivars of citrus fruits with 2-(4-chlorophenylthio)triethylamine hydrochloride (CPTA) caused a large increase in total carotene content. Lycopene (Ψ,Ψ-carotene), which does not normally accumulate in citrus, becomes a major pigment in CPTA-treated fruits. The effect is believed to be caused by depression of a gene that regulates the synthesis of specific enzymes and inhibition of the cyclases in the biosynthetic pathway of carotenoids. Several other derivatives of triethylamine have an effect similar to CPTA in citrus (Poling *et al.*, 1975). Synthesis of phytofluene and lycopene in the flavedo of lemons induced by treatment with the substituted tertiary amine 2-(4-methylphenoxy)triethylamine (MPTA), has been attributed to the regulation of the synthesis of isoprenoids through the control of the expression of genes that code for the enzymes in the carotene pathway (Benedict *et al.*, 1985).

ABSCISSION

Abscission of citrus fruits occurs normally at two different stages of growth: early in development during fruit-set (the so-called June drop), and at the end of fruit ripening. Both the June drop and mature fruit abscission may seriously limit the yield of many citrus cultivars. Control of the latter is also of interest in order (a) to improve fruit marketability, because the abscission scar is the entry zone for many fungal and bacterial infections; (b) to avoid premature fruit drop; and (c) to facilitate mechanical harvesting of fruits for processing. Citrus fruits abscise at the shoot–peduncle abscising (SPA) zone during the first 6–8 weeks after setting (Fig. 2). Afterwards, and particularly at maturity, the fruits abscise only at the calyx–fruit abscising (CFA) zone (Huberman *et al.*, 1988).

The activities of cellulase (EC 3.2.1.4) and polygalacturonase (EC 3.2.1.15) in the abscission zones increase before and during abscission, and are localized mainly in the abscission layers (Rasmussen, 1973; Greenberg *et al.*, 1975). During ripening of early- and late-maturing oranges, the increase in the contents of ethylene and abscisic acid in the fruit are correlated with the increase on cellulase activity in the separation zone (Rasmussen, 1975). The role of hydrolytic enzymes in the abscission process, however, is not clear, since the rise in their activity is not always associated with abscission. For instance, in mature fruits, ethylene enhances, and 2,4-D delays, the formation of cellulase and polygalacturonase at the SPA and CFA zones, at a time when no abscission occurs in the SPA zone (Goren & Huberman, 1976; Huberman & Goren, 1979). The effects of both growth regulators on almost all endo- and exo-cellular cellulase isoenzymes were similar (Huberman & Goren, 1979). This excludes the possibility, at least for cellulase, that only some enzymatic forms would be involved in the abscission process due to their subcellular location in the tissue.

SUGAR METABOLISM

Mature citrus fruits store large amounts of sugars, mainly glucose, fructose and sucrose, and their content is an important parameter of fruit quality (Ting, 1980). Subcellular distribution in protoplasts from Valencia orange juice vesicles shows that the vacuole contains 75% of the fructose and glucose and 100% of the sucrose (Echeverria & Valich, 1988).

Photosynthates are transported in citrus as sucrose (Kriedemann, 1969). It is believed that accumulation of sugars in storage tissues is regulated by different enzymes, of which invertase (EC 3.2.1.26), in particular, has been investigated. Acid and alkaline invertases are present in immature juice vesicles of Satsuma mandarin fruit, at a stage when their sugar content is low (Kato & Kubota, 1978). In maturing and mature juice vesicles, alkaline invertase is also detected, but acid invertase is absent. In immature juice vesicles, almost all of the acid-invertase activity is found in the soluble fraction, whereas alkaline invertase is firmly bound to the cell wall. The proportion of soluble alkaline invertase increases with fruit development, suggesting a change in the association between that enzyme and the cell wall (Kato & Kubota, 1978).

In citrus fruits, transport cells (vascular bundles plus segment epidermis) are physically separated from sink cells (multicellular juice vesicles which are free of phloematic cells) (Koch, 1984) (Fig. 1). This allows investigation, in a sink organ, of the role of sucrose-metabolizing enzymes in the transport, separately from their role in the storage and/or utilization, of sucrose. Soluble invertase is the most active enzyme in extracts from transport and sink tissues of very young fruits, but nearly disappears prior to the onset of sugar accumulation in the juice vesicles. The activity of sucrose synthase (EC 2.4.1.1) is much greater in phloem unloading and transport tissues than in adjacent sink tissues when juice vesicles grow most rapidly; the activity of sucrose synthase is also higher than any other sucrose-metabolizing activity in transport tissue, whereas sucrose-phosphate synthase (EC 2.4.1.14) and alkaline invertase are the most active enzymes in the rapidly growing juice vesicles. The physiological function of high sucrose-synthase activity in the transport tissues during rapid sucrose import remains to be determined (Lowell *et al.*, 1989).

Using protoplasts from Valencia orange juice vesicles, it has been shown that the activities of acid and neutral invertases, UDP-glucose pyrophosphorylase (EC 2.7.7.9), 6-phosphofructokinase (EC 2.7.1.11) and 6-phosphofructokinase pyrophosphate (EC 2.7.1.90) were present only in the cytoplasm (Echeverria & Valich, 1989). Only 5% of total phosphohexoisomerase (EC 5.3.1.8–9) and 29% of total phosphoglucomutase (EC 5.4.2.2) were found in the vacuoles. The different localization of sugars (vacuole) and enzymes involved in sugar metabolism (cytoplasm) suggests that the sugars located in the vacuoles are the major source of carbohydrate supplying energy to the mature juice-vesicle cells (Echeverria & Valich, 1989).

During storage of Valencia orange fruits, the activities of hexokinase (EC 2.7.1.1), 6-phosphofructokinase (glycolytic enzymes) and those of sucrose synthase and UDP-glucose pyrophosphorylase (enzymes involved in sucrose biosynthesis) increase in the juice. On the other hand, the activity of fructose-bisphosphatase (EC 3.1.3.11) (a regulatory enzyme of the gluconeogenesis pathway) remains essentially constant (Echeverria & Valich, 1989).

RESPIRATION

Citrus fruits do not show a rise in respiration during maturation, though low respiration activity results in internal metabolite changes. One of the most noticeable changes is the accumulation of ethanol and acetaldehyde, which begins during the development of the fruit and rapidly increases throughout maturity (Roe *et al.*, 1984). Ripening of citrus fruits is characterized by a decline in aerobic respiration (Hirai & Ueno, 1977) and an increase in anaerobic metabolism (Davis, 1970), associated with the accumulation of ethanol in the juice vesicles (Roe *et al.*, 1984).

A scheme of the different respiratory pathways in citrus is shown in Fig. 3. Pyruvate, the substrate for aerobic metabolism, after being oxidized by pyruvate dehydrogenase (EC 1.2.2.2) to acetyl-CoA (which enters the TCA cycle by condensation with oxalacetate to form citrate), is also a substrate for anaerobic metabolism in citrus through pyruvate decarboxylase (EC 4.1.1.1.) and alcohol dehydrogenase (EC 1.1.1.1.) (Roe & Bruemmer, 1974). Thus, pyruvate dehydrogenase and pyruvate decarboxylase compete for pyruvate at a metabolic branch point. In Hamlin oranges, at the time of rapid ethanol

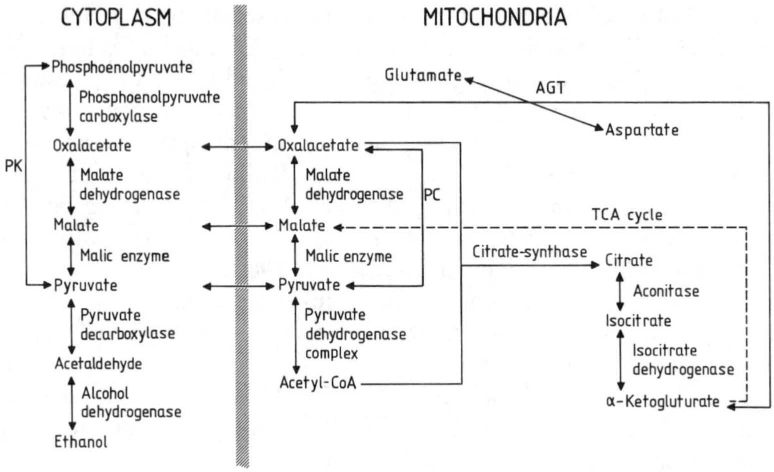

FIG. 3. Respiration and organic acid metabolism in citrus fruit. PC—pyruvate carboxylase, PK—pyruvate kinase, AGT—aspartate-glutamate transaminase. (Adapted from Roe *et al.*, 1984).

accumulation in the fruit, the activities of pyruvate decarboxylase and alcohol dehydrogenase increase markedly, and the level of pyruvate declines, suggesting that the increase in enzyme levels activates the conversion of pyruvate to ethanol (Roe *et al.*, 1984).

During the ripening of citrus fruits, there is an increase in the ratio of NADH–NAD in the juice vesicles (Bruemmer & Roe, 1985) associated to the accumulation of ethanol. As the fruit matures, the capacity of NADH oxidation by mitochondria isolated from juice vesicles does not change, and experiments using KCN and salicylhydroxamic acid (SHAM) (inhibitor of the KCN-insensitive respiration) indicate that ripening does not change the contribution of the alternative oxidase to total respiration. This suggests that the inability of the NADH-oxidase pathway to maintain the redox equilibrium stimulates the operation of the supplementary anaerobic pathway. It has been proposed that pyruvate dehydrogenase functions as a regulatory enzyme of pyruvate metabolism in the citrus juice vesicles by controlling the aerobic–anaerobic transition (Bruemmer & Roe, 1985).

ACID METABOLISM

The soluble solids–acid ratio is a major criterion of citrus juice quality. In acid fruits, such as lemon and lime, the concentration of acids in the fruit determines, to a large extent, the value of the crop. Citric and malic acids are the most abundant organic acids in citrus fruits (Bruemmer *et al.*, 1977; Sasson & Monselise, 1977; Roe *et al.*, 1984). In the juice vesicles of most citrus fruits, citric and malic acids are dominant, whereas, in the peel, malic and malonic acids predominate (Ting & Attaway, 1971; Sasson & Monselise, 1977). Both malic and citric acids accumulate as the fruit grows, though at different rates, and their ratio in the juice has been proposed as a parameter to determine the stage of maturity of the fruit during ripening and storage (Sasson & Monselise, 1977). Vacuoles in the juice cells contain most of the organic acids (Bruemmer *et al.*, 1977). In protoplasts from Valencia orange juice vesicles, approximately 70% of the malate and 90% of the citrate are vacuolar (Echeverria & Valich, 1988). The relatively high content of citric acid in the cytoplasm seems to be a peculiarity of citrus (Echeverria & Valich, 1988).

Oxalacetate (in equilibrium with phosphoenolpyruvate, pyruvate, malate, citrate and aspartate) occupies a central position in the metabolism of organic acids in citrus (Fig. 3), and its concentration appears to limit their metabolic rate (Bruemmer *et al.*, 1977; Roe *et al.*, 1984). Phosphoenolpyruvate carboxylase (EC 4.1.1.31) catalyses a major CO_2-fixing reaction in orange juice vesicles, and probably controls the production of oxalacetate (from phosphoenolpyruvate) in the cytoplasm for the synthesis of citrate and malate (Huffaker & Wallace, 1959; Young & Biale, 1968). The activity of phosphoenolpyruvate carboxylase increases markedly during the early stages of orange fruit growth (Roe *et al.*, 1984). L-Aspartate:2-oxoglutarate aminotransferase (EC 2.6.1.1), which has been partially purified from lemon juice vesicles (Parekh & Sakariah, 1971), can be another source of oxalacetate. The importance of pyruvate carboxylase (EC 6.4.1.1) for the synthesis of oxalacetate in citrus is not known.

Citric acid is apparently formed in citrus through the TCA cycle. The reaction catalysed by citrate synthase (EC 4.1.3.7) (condensation of oxalacetate and acetyl-CoA to give citric acid) is essentially irreversible. The low activity of the enzyme, compared to the rate at which its substrates may be produced in citrus fruits, makes the reaction a suitable point for metabolic control (Bruemmer *et al.*, 1977). The presence of citrate synthase has been demonstrated in orange (Srere & Senkin, 1966), and mitochondria from both sour lemon (Bogin & Wallace, 1966) and grapefruit fruits (Bruemmer *et al.*, 1977), and its activity increases notably during the ripening of Hamlin oranges (Roe *et al.*, 1984). In lemon fruits, accumulation of citrate is related to the increase in citrate-synthase activity and to a decrease in citrate-degrading enzymes (aconitase hydratase, EC 4.2.1.3; and isocitrate dehydrogenase (NAD^+), EC 1.1.1.41) (Ramakrishnan & Varma, 1959). Citrate synthase and isocitrate dehydrogenase levels seem also to control, at least partially, the accumulation of citric acid in orange fruits (Bruemmer *et al.*, 1977). During storage of Hamlin oranges, however, no change was detected in the activity of aconitase hydratase and isocitrate dehydrogenase ($NADP^+$) (EC 1.1.1.42) in the juice (Echeverria & Valich, 1989). It has been suggested that citrate accumulation in lemon fruit can be controlled by citramaleate, an inhibitor of aconitase hydratase (Bogin & Wallace, 1966). Isocitrate dehydrogenases (both NAD^+ and $NADP^+$ dependent) have been isolated from citrus juice vesicles (Bruemmer *et al.*, 1977).

Malate dehydrogenase (EC 1.1.1.37) catalyses the reduction of oxalacetate to malate in the cytoplasm and mitochondria. In orange juice vesicles, mitochondrial malate dehydrogenase is relatively constant, whereas the activity of cytoplasmic malate dehydrogenase increases during fruit maturation (Roe *et al.*, 1984). In mitochondria, malate dehydrogenase competes with citrate synthase for oxalacetate.

The levels of malate are also regulated by oxaloacetate-decarboxylating malate dehydrogenases (malic enzymes), which catalyse the oxidative decarboxylation of malate to pyruvate using NADH (EC 1.1.1.38) or NADPH (EC 1.1.1.40) as cofactors. The reaction is readily reversible and, therefore, may be a source of malate in citrus juice vesicles through dark CO_2 fixation (Bean & Todd, 1960). Both malic enzymes are present in citrus fruits (Hirai, 1978), and it has been shown that the activity of NADPH-dependent malic enzyme increases steadily during growth (Roe *et al.*, 1984) and during storage of orange fruits (Echeverria & Valich, 1989).

STABILITY OF JUICES AND CONCENTRATES

Fresh citrus juice contains finely-divided particles in suspension, constituting an unstable colloidal system, commonly known as 'cloud'. The cloud is composed mainly of pectins, proteins and lipids, and it contains most of the characteristic flavour and colour. A few hours after juice extraction, the cloud coalesces and citrus juice appears as a two-phase system composed of a flocculant sediment and a clear liquid, unacceptable to the consumer (Baker & Bruemmer, 1972).

The instability of the cloud has been attributed to pectinesterase (pectin pectylhydrolase, EC 3.1.1.11). This enzyme is capable of de-esterifying the pectin (Joslyn & Pilnik, 1961). Then, the low-methoxyl pectin formed reacts with calcium ions from the juice and coagulates, causing the cloud loss. To prevent this problem, the juice must be pasteurized to inactivate the enzyme at temperatures near 90°C, which sometimes cause off-flavours (Kew & Veldhuis, 1960), and juice concentrates must be stored at −20°C to keep degradative processes to a minimum (Olson, 1968).

Citrus juice concentrates gellify under certain circumstances. Gel formation is due to the reaction of demethylated pectin with divalent cations, by the same mechanism that causes cloud precipitation (see Blundstone *et al.*, 1971).

In addition to other factors, the structure of pectin molecules, including the kind and number of the side chain of neutral sugars, exerts a great influence on the extent of pectin hydrolysis by pectinesterase (Matsuura, 1987).

Polygalacturonase is present at very low levels, and therefore it is unlikely that it affects cloud stability (Riov, 1975).

Pectinesterase

Pectinesterase is associated with the cell-wall fraction, and is located mainly in the peel, and vesicle juice membranes. Upon extraction of citrus fruit, some of the enzyme is released into the juice.

Multiple forms of pectinesterase have been described in the literature; some of them differ remarkably in molecular weight, specific activity, optimum pH and temperature, affinity for the pectin substrate and heat stability. This has led to the idea that pectinesterase isoenzymes may play different roles in cloud loss (Versteeg *et al.*, 1980).

Versteeg *et al.* (1978) separated two isoenzymes from Navel oranges, accounting for 90% of the total activity. Both isoenzymes had a mol. wt of 36 kDa and isoelectric points of 10·05 (pectinesterase I) and over 11 (pectinesterase II). Pectinesterase I had an optimum pH of 7·6 and a K_m of 0·083 mg of pectin/ml, while pectinesterase II had an optimum pH of 8·0 and a K_m of 0·0046 mg pectin/ml. Pectinesterase II is much more active than pectinesterase I at low substrate concentrations and is more inhibited by polygalacturonate. In addition to these, Versteeg *et al.* (1980) purified a high-molecular-weight isoenzyme (54 kDa) with a pI of 10·0, which accounts for only 6% of the total activity. While pectinesterase I and II are quite temperature-sensitive, the high-molecular-weight isoenzyme requires 90°C for inactivation. This isoenzyme is very active at low pH and in juices chilled at 5°C, and it is thought that it is largely responsible for the gelation which may occur in concentrates produced by cut-back processes (addition of fresh juice to concentrates to incorporate flavours). These three isoenzymes have been found in very different citrus species and cultivars. Several other forms of pectinesterase have been detected by isoelectric focusing: for instance, 10 different isoenzymes in grapefruit (Rombouts *et al.*, 1982). These results are in contrast with those of Korner *et al.* (1980), who found that Navel-orange pectinesterase was constituted by two enzyme fractions, which apparently possess similar catalytic properties with an optimum pH around 7·5 and maximum activity at 60°C.

During extraction with alkaline buffers, a fraction of pectinesterase is not solubilized, but it is partially released after the degradation of the cell wall by pectinase and cellulase. Several characterization assays of this fraction suggest that there is an increase in solubilization of a heat-stable pectinesterase isoenzyme after cell-wall degradation (Wicker *et al.*, 1988).

Inactivation of pectinesterases

Treatment of citrus juices and concentrates is required to inactivate native enzymes, principally pectinesterase. Although the addition of oligogalacturonic acids has been recognized to inactivate pectinesterase (Termote *et al.*, 1977), the method most commonly used is pasteurization.

Thermal inactivation of pectinesterase depends on the characteristics of the product, such as the nature of the isoenzymes, total solids content, pH, viscosity. Variations in product specifications will necessitate different time–temperature requirements. Nath and Ranganna (1977) showed that the processing time for pectinesterase inactivation increased with pH, and a similar effect was observed with increasing total solids concentration up to 35° Brix. In contrast, a decrease in the rate of inactivation of pectinesterase occurred at 40 and 50° Brix, suggesting that the enzyme may be protected against inactivation at higher degrees Brix. The effect of pH on thermal inactivation was also different at higher total solid concentrations; for example, increasing the pH from 4·0 to 7·0 caused the rate of inactivation of pectinesterase to increase 4·5-fold in 20°-Brix orange juice (Marshall *et al.*, 1985). Heat inactivation in juice pulp is non-linear, suggesting the presence of fractions of pectinesterase with different heat stabilities (Wicker & Termelli, 1988).

Pectinesterase inactivation has been evaluated for different citrus juices after the studies initiated by Rouse and Atkins (1952). However, extrapolation of these results to other particular situations is difficult because the data were not expressed in the manner conventionally associated with thermal-process evaluations. Eagerman and Rouse (1976) tried to overcome this by obtaining an accurate evaluation of the effectiveness of pasteurization which would be helpful in improving quality and reducing fuel costs; charts of pasteurization conditions for three orange and one grapefruit cultivars commonly used in processing were also provided.

Depolymerizing enzymes

It was assumed for a long time that soluble pectin formed a colloidal matrix which supported the particles, therefore stabilizing the juice cloud. However, Baker and Bruemmer (1969) showed that soluble pectin was not necessary for cloud support, since a stable suspension of orange juice particulates could be made in water. On the contrary, insoluble pectates formed after pectin demethylation occlude the cloud particles and remove them from the juice by precipitation. Because pectin is the cause of cloud destabilization, controlled degradation of pectin by addition of commercial pectinases (Baker & Bruemmer, 1972) and other pectic depolymerizing enzymes (Krop, 1974; Krop & Pilnik, 1974*a*), which stabilize the cloud by hydrolysing pectins to soluble pectates, has been proposed as an alternative to heat denaturation. In addition to cloud stabilization, pectinases have also shown to be very effective in reducing the viscosity of juices and concentrates with little or no change in standard citrus juice quality (Braddock, 1981). The decrease in viscosity that can be achieved with pectinases makes it technically feasible to increase the concentration level of citrus juices from 65 to 72° Brix, with a sharp reduction in storage and transport-fuel requirements (Crandall & Beilock, 1985; Crandall *et al.*, 1986; Crandall *et al.*, 1990).

The possibility that other polymers could be implicated in cloud stability has also been investigated (Biggs & Pollard, 1970; Krop, 1974; Krop & Pilnik, 1974*b*; Krop *et al.*, 1974), but no clear conclusions were obtained.

Hesperidin

Some citrus species and cultivars are particularly rich in the glucoside, hesperidin, a flavanone derivative. This is the case of Shamouti oranges and some cultivars of mandarins.

The cloud of Shamouti orange juice (Mizrahi & Berk, 1970) and the syrup of canned mandarin segments contain hesperidin crystals which, upon precipitation, form an unattractive sludge. Hesperidin also crystallizes on the segments as white flecks. The addition of a heat-stable hesperidinase to the syrups before pasteurization (Ito *et al.*, 1969; Okado & Ono, 1969) maintains the clarity of syrups by converting the hesperidin to insoluble compounds that do not crystallize under these conditions.

BITTER PRINCIPLES IN CITRUS

Types of bitter principles

Citrus fruits contain bitter principles that belong to two chemically different groups: limonoids and flavonoids. The flavedo, the central bundle and the segment covers of the fruit contain both kinds of bitter principles, whereas the juice vesicles contain only flavonoids (Chandler & Nicol, 1975) (Fig. 1).

Limonoids are a group of chemically related triterpenes (Maier *et al.*, 1980), limonin (Fig. 4) and its precursor nomilin (nomilin is considerably less bitter than limonin) being the most important. Fruits differ not so much in the type of limonoids that they contain, but in their relative amounts.

Flavonoids are C-15 compounds arranged C-6–C-3–C-6; according to the degree of oxidation of the central group, they are classified as flavones, flavanones, flavonols, anthocyanins, etc. (Rouseff, 1980). In citrus, flavonoids usually occur as glycosides, and affect the quality of the fruit and juice due to the taste of flavanone 7-neohesperidosides (flavanone glycosides consisting of the disaccharide rhamnose-1,2-glucose attached to the position 7 of the favanone aglycone). The best representative is naringin (Fig. 5), a bitter principle of grapefruit (Fellers, 1989), which gives to grapefruit juice and segment preparations their characteristic taste. However, an excess of naringin content reduces consumer acceptability.

Metabolism of limonoids

Our understanding of limonoid metabolism in citrus has been advanced by comparative biochemical studies of limonoid metabolism in bacteria and citrus (Hasegawa & Maier, 1981, 1983). Figure 4 shows a general scheme of limonoid metabolism in citrus and bacteria. Substantial progress has been made recently on the biosynthesis of limonoids in citrus, leading to the conclusion that deacetylnomilinate, nomilin and ubacunone (Hasegawa & Herman, 1985, 1986; Herman & Hasegawa, 1985; Herman *et al.*, 1987) are initial precursors

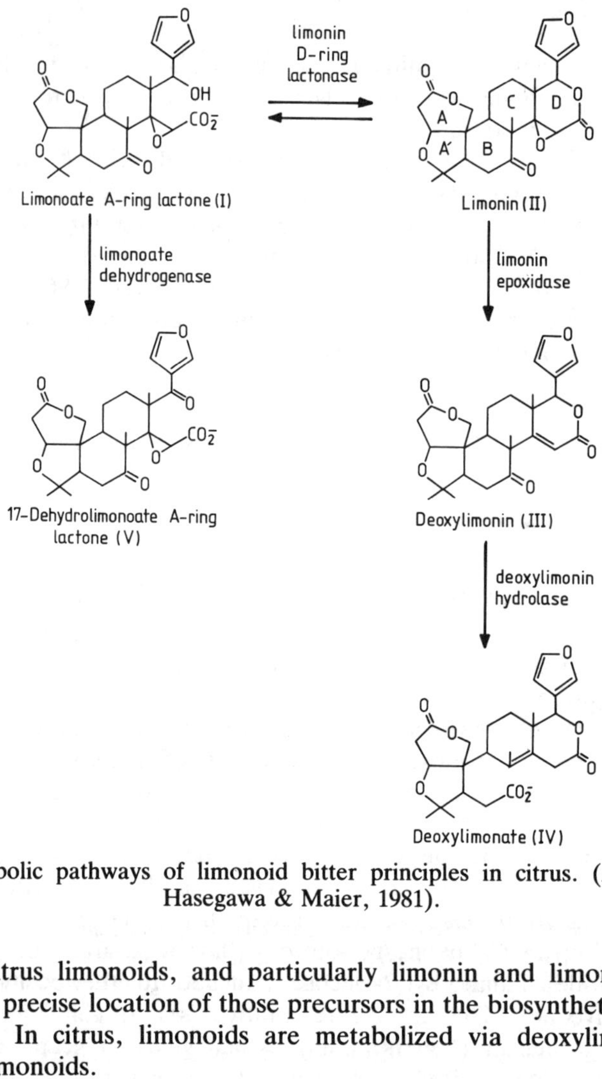

FIG. 4. Metabolic pathways of limonoid bitter principles in citrus. (Adapted from Hasegawa & Maier, 1981).

of all the citrus limonoids, and particularly limonin and limonoate A-ring lactone. The precise location of those precursors in the biosynthetic pathway is still unclear. In citrus, limonoids are metabolized via deoxylimonoids and 17-dehydrolimonoids.

Limonoate A-ring lactone (I) is thought to be synthesized in the leaves, translocated to the fruits, then to the seeds, where it is converted to limonin (II) and stored (Maier *et al.*, 1980). However, recent findings (Hasegawa *et al.*, 1986*a*) suggest that the stem may also play an important role in the biosynthesis of limonoid precursors. The content of limonoate A-ring lactone (I) in the fruit decreases progressively during ripening, and it is almost undetectable in many citrus cultivars at harvest. The enzyme, limonine-D-ring-lactonase (EC 3.1.1.36), catalyses the reaction I to II (Fig. 4) optimally at pH 6·0, and the reverse reaction (II to I) at pH 8·0. The lactonase isolated from grapefruit seeds is heat resistant (30% activity remains after 5 min at 100°C). Limonine-D-ring-lactonase is involved both in the biosynthesis and

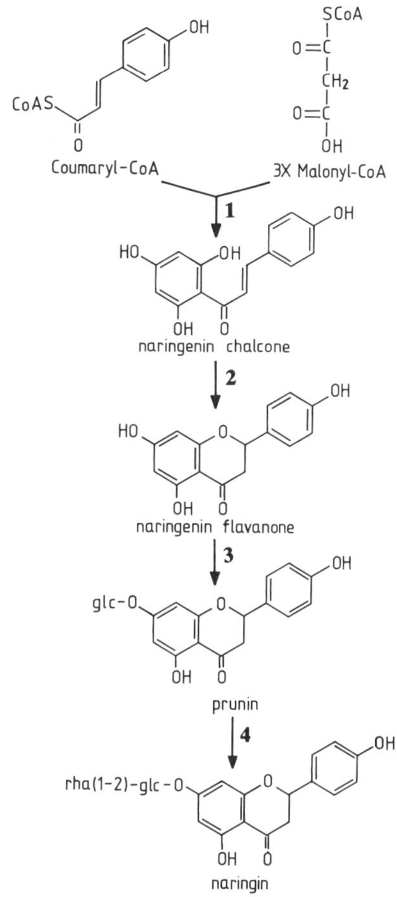

FIG. 5. Biosynthetic pathway of naringin in citrus: 1, chalcone synthase; 2, chalcone isomerase; 3, UDP-glucose: flavanone-7-*O*-glucosyl transferase; 4, UDP-rhamnose flavanone-glucoside rhamnosyl-transferase. (Adapted from Lewinsohn *et al.*, 1989).

biodegradation of limonoids. Limonin epoxidase and deoxylimonin hydrolase metabolize limonin through the deoxylimonin pathway.

The enzyme limonoate dehydrogenase converts limonoate A-ring lactone (I) to non-bitter 17-dehydrolimonoate A-ring lactone (V) (Fig. 4). It requires Zn^{2+} and protectors of sulphydryl groups for activity, and uses NAD as a cofactor.

Removal of limonoid bitterness

The intact citrus fruit contains very little limonin. However, the non-bitter precursor limonoate A-ring lactone (Fig. 4) (originally present in the insoluble parts of the fruit and solubilized during juice extraction and processing) is converted to limonin under the acidic conditions of the juice ('delayed bitterness'). This problem is increased in citrus cultivars, such as those of the Navel group, in which the levels of limonoate A-ring lactone are still relatively high at the time of fruit harvesting. In these cases, limonoid juice bitterness can be prevented by delaying harvesting until appropriate low levels of limonoate A-ring lactone in the fruit are attained. Other possibilities of avoiding juice bitterness are to degrade limonoate A-ring lactone in the juice before it is metabolized to limonin, and to remove limonin from the juice once it is has been produced.

The biosynthesis of nomilin, a bitter principle as well as a precursor of limonin, is inhibited by the application of auxins and abscisic acid (Hasegawa *et al.*, 1986*b*). Nomilin acetyl-lyase, which converts bitter nomilin to non-bitter ubacunone, has been purified from *Corynebacterium fascians* (Herman *et al.*, 1985). The enzyme is of potential interest for the degradation of nomilin in certain citrus juices.

Limonoate dehydrogenases of bacterial origin (*Pseudomonas* and *Arthrobacter globiformis*) have been assessed for the prevention of limonin bitterness in juices (Brewster *et al.*, 1976), but their use has several disadvantages: for example, (a) the pH optimum is 8·5–9·5, and therefore its activity in the natural environment of the juice is limited (it is necessary to make pH adjustment); and (b) the enzyme acts on the open-ring lactone, which exists only for a short time in the juice before it is converted to limonin. Treatment of oranges with ethylene accelerates limonoate A-ring lactone metabolism (Maier *et al.*, 1973).

Removal of juice bitterness due to limonin has been achieved by using immobilized *Acinetobacter* spp. in dialysis bags (Vaks & Lifshitz, 1981), and *Arthrobacter globiformis* (Hasegawa *et al.*, 1982; 1983) and *Corynebacterium fascians* (Hasegawa *et al.*, 1985) in acrylamide gels. The last species has the advantage of synthesizing constitutive limonoid-metabolizing enzymes rather than inducible ones.

Biosynthesis and metabolism of flavonoids—removal of flavonoid bitterness

Very little is known about the biosynthetic pathways to flavanone-glycosides in citrus. Recent data suggest that the flavonoid pathway in citrus is similar to that found in other species (Lewinsohn *et al.*, 1989; McIntosh & Mansell, 1990), i.e. chalcones are biosynthesized before their glucosylation (Fig. 5). Both young leaves (Fisher, 1968) and developing young fruits (Berhow & Vandercook, 1989) have been shown to be capable of synthesizing naringin from phenylalanine. The content of naringin in the grapefruit increases constantly during the growth of the fruit (Jourdan *et al.*, 1985).

Naringinase, an enzyme of microbial origin which is actually a mixture of α-L-rhamnosidase (EC 3.2.1.40) and β-glucosidase (EC 3.2.1.21), is very effective in converting pure naringin to naringenin (a non-bitter compound). However, the activity of naringinase when added to the juice is inhibited by glucose from the juice, and this inhibition can be counteracted by the simultaneous addition of emulsin to the juice (see Chandler & Nicol, 1975). Commercial preparations of the enzyme are usually heavily contaminated with pectinase (EC 3.2.1.15), which produces undesirable loss of cloud in the juice. This can be avoided by differential heat inactivation ('in can' debittering): the crude enzyme is added to the can before processing for 10 min at 80°C to inactivate pectinase; naringinase remains active under these conditions and gives a non-bitter juice in 2 h and non-bitter segments in 2 weeks.

Naringinase immobilized by occluding the protein in hollow fibres (Olson & Gray, 1981), and by chemical attachment of the enzyme to a preformed polymer (Ono *et al.*, 1978) or to derivatized glass (Jimeno *et al.*, 1987), has also been assessed for its ability to reduce naringin bitterness of grapefruit juices.

STRESS PROCESSES IN CITRUS FRUITS

Citrus fruits, after harvest, undergo a series of metabolic changes, leading to loss of quality, deterioration and finally death. Some of these changes are a consequence of the normal physiological processes of the fruit; others correspond to abnormal or traumatic situations produced by wounding, extreme temperatures, insect damage, pathological or physiological diseases, chemicals, etc. Stress processes are of most economical importance because they account for a major part of the post-harvest losses that range from 25 to 50% of the crop.

Stress processes are very different in nature and complexity. Most of the literature about citrus enzymes in relation to stress processes deals with the effects of wounding and pathogens.

Wounding

Ethylene and wounding

Citrus fruits are typically non-climacteric, and they produce small amounts of ethylene when they reach maturity. However, excised albedo discs of mature Satsuma mandarins evolve high amounts of ethylene. This has been attributed to wound-induced ethylene production, and evidence has been provided that methionine is the source of ethylene in the excised discs (Hyodo, 1977, 1978). The biosynthesis of ethylene is markedly inhibited by rhizobitoxine, an inhibitor of ACC synthase (*S*-adenosyl-L-methionine methyl thioadenosine-lyase, EC 4.4.1.14) and by cycloheximide (Hyodo & Nishino, 1981), suggesting that wound ethylene is produced by the mechanism described by Yang (1980).

Concentrations of cycloheximide lower than 10 μM affect neither the accumulation of ACC nor ACC-synthase activity, although they significantly inhibit ethylene production. This indicates that the ethylene-forming enzyme, which catalyses the conversion of ACC to ethylene, is more sensitive to cycloheximide than the synthesis of ACC (Riov & Yang, 1982*a*).

Wound-induced ethylene from citrus flavedo discs is inhibited by exogenous ethylene. This inhibition does not affect the conversion of ACC to ethylene, but decreases ACC formation through repression of the synthesis of ACC synthase, induced by wounding and inhibition of its activity (Riov & Yang, 1982*b*). Treatment of grapefruit flavedo with ethylene markedly enhanced the activity of extractable ACC malonyl transferase, indicating that auto-inhibition of ethylene production results not only from suppression of ACC synthesis, but also from promotion of ACC malonylation. Both processes reduce the availability of ACC for ethylene synthesis (Liu *et al.*, 1985).

Very little is known about the enzyme system that catalyses the last step of ethylene biosynthesis, i.e. the transformation of ACC to ethylene. Shimokawa (1983) purified an enzyme from *Citrus unshiu* fruits, with a mol. wt of 40 kDa, that catalyses in-vitro ethylene formation from ACC in the presence of pyridoxal phosphate, IAA, Mn^{2+} and 2,4-dichlorophenol. It exhibits IAA oxidase, but not guaiacol-peroxidase activity. It has a specific activity of 0·025 mol/min/mg protein, a K_m for ACC of 2·8 mM and its optimum pH is 5·7. The in-vivo participation of this enzyme in the last step of ethylene biosynthesis has not yet been demonstrated.

Phenylalanine ammonia-lyase and wounding

L-Phenylalanine ammonia-lyase (EC 4.3.1.5) catalyses the elimination of ammonia from L-phenylalanine to form *trans*-cinnamic acid. This reaction represents a branching point between the primary and secondary metabolic pathways of higher plants. It is considered that L-phenylalanine ammonia-lyase controls phenyl-propanoid metabolism, which produces substrates for several important biochemical pathways in plants, including the formation of phenolic compounds such as lignin, a structural component of plant cell walls, and phytoalexins (see Ecker & Davis, 1987).

L-Phenylalanine ammonia-lyase from developing orange fruit has been purified by hydrophobic chromatography. The protein has a pI of 3·85 and a mol. wt of 330 kDa (Dubery & Schabort, 1986). The enzyme exhibits negative homotropic effects between the substrate-binding centra, and binding of L-phenylalanine to the enzyme is characterized by two K_m values, $K_m{}^H = 13\ \mu M$ and $K_m{}^L = 52\ \mu M$. It is subjected to product inhibition by *trans*-cinnamate, but the effect of allosteric effectors and inhibitors seem to be of greater importance in the short-term regulation of phenyl-propanoid metabolism in *Citrus sinensis*. The activity of the enzyme is modulated by end products of diverging metabolic pathways (Dubery & Schabort, 1988).

Phenylalanine ammonia-lyase activity decreases exponentially with increasing size of orange fruit (Dubery & Schabort, 1986). It is low in the flavedo of mature grapefruit, but excision of flavedo discs increases it and, simultaneously, ethylene production. Exogenous ethylene also stimulates it in the flavedo of mature grapefruit; such activity rapidly decreases when fruit is removed from the ethylene atmosphere. Cycloheximide inhibits both phenylalanine ammonia-lyase activity and ethylene production, indicating that protein synthesis is needed to sustain both processes (Riov *et al.*, 1969).

Healing of minor injuries in Valencia oranges occurs rapidly at 30°C and 90–95% relative humidity. It is characterized by a marked increase in the activity of phenylalanine ammonia-lyase and deposition of lignin. Phenylalanine ammonia-lyase is confined mainly to injured flavedo, while un-injured peel and albedo tissues exhibit little activity. Inoculation with *Penicillium digitatum* immediately after flavedo injury disrupts the increase in activity of phenylalanine ammonia-lyase. The synthesis of both phenylalanine ammonia-lyase and lignin in mature fruit is strongly inhibited by cycloheximide, but not by actinomycin D (Ismail & Brown, 1979).

Pathogens

Ethylene and pathogens

Ethylene biosynthesis induced by pathogens shows some differences from that induced by wounding. The biosynthesis of ethylene by grapefruit inoculated with *Penicillium digitatum* appears to originate, during the first stages of the infection, mostly from methionine, and is largely of host origin. It is inhibited by aminoethoxyvinyl glycine, an inhibitor of ACC-synthase biosynthesis. However, as infection develops, the conversion of ACC to ethylene is inhibited, coinciding with a large increase in ethylene production in the infected tissue. Ethylene produced in the infected parts of the fruit is synthesized mostly by the fungus from glutamic acid (Achilea *et al.*, 1985).

The biochemical mechanisms by which pathogen infection induces ACC-synthase activity remains unknown. Baldwin and Biggs (1988) found that ACC and ethylene are induced, in Valencia oranges, by injecting into the peel the enzyme mixture, Pectolyase, which contains pectolytic enzymes obtained from *Aspergillus japonicus*. Cycloheximide partially inhibits pectolyase-induced ethylene production. Regulation of ethylene synthesis by cell-wall digestion products would be a possible mechanism by which plant cells could recognize pathogen invasion.

The role of ethylene in plant–pathogen interaction in citrus fruits, as well as in other vegetable tissues, is not well understood. Ethylene can stimulate the growth of pathogens or induce resistance. It has been shown in different host–pathogen systems that ethylene-induced resistance in citrus fruits does not develop rapidly enough to inhibit infection, except when ethylene treatment precedes pathogen inoculation (Brown & Barmore, 1977; El-Kazzaz *et al.*, 1983). Ethylene-induced pathogen resistance in citrus fruits has been associated to an increase in the synthesis of various compounds such as L-phenylalanine ammonia-lyase, lignin, phenolics and polyphenol oxidase (El-Kazzaz *et al.*, 1983).

Induction of L-phenylalanine ammonia-lyase activity in yellow orange and grapefruit peel by ethylene is inhibited when the fruit is inoculated with *Penicillium digitatum*. Other fungi, which do not produce ethylene, have a similar effect. These findings suggest the existence of a mechanism, not mediated by ethylene, by which fungi are able to prevent the defence response of host cells to invasion (Lisker *et al.*, 1983).

Tissue maceration by pectic enzymes during fungal attack
Pectolytic enzymes, mainly endopolygalacturonase and pectin transeliminase, which cleave randomly α-1,4-glucosidic bonds, are produced by various fungi and participate in tissue maceration by degrading the middle lamella of the host tissue during pathogenesis (Bateman & Millard, 1966). However, this is not a universal mechanism, and fungal attack and development can be produced using different strategies.

Infection of green mould, a soft rot caused by *Penicillium digitatum*, could not be associated with changes of endopolygalacturonase or pectin transeliminase in Valencia oranges (Barmore & Brown, 1979). The degradation of the middle lamella was attributed to the cooperative action of an exopolygalacturonase secreted by the pathogen, and the pectinesterase released by the damaged cells of the fruit. In this case, pectinesterase would demethylate the pectin to pectic acid, which is a prerequisite for exopolygalacturonase activity.

A different mechanism seems to be used by closely related species of *Penicillium digitatum* in grapefruit infection. An endopolygalacturonase (EC 3.2.1.15) has been isolated and purified from grapefruit (*Citrus paradisi*) peel infected with *Penicillium italicum* with a mol. wt of 36 kDa and an optimum pH of 5·0. The purified enzyme macerates citrus mesocarp *in vitro*. Dissolution of the middle lamella by the endopolygalacturonase is considered to be a major factor contributing to hyphal penetration (Barmore & Brown, 1981).

The decay of Valencia oranges produced by *Diplodia natalensis* has been associated with two endopolygalacturonase isoenzymes, with mol. wts of 64 and 50 kDa, apparently of fungal origin. Pectin degradation is aided by pectinesterase, which causes demethylation of pectin in the host cell walls (Barmore *et al.*, 1984). Endopolygalacturonase is inhibited at low pH by a proteinaceous inhibitor which binds reversibly to the enzyme and competes with the substrate. Activity of the enzyme would be high at the site of infection, because of a nearly neutral pH and high substrate concentration, which competes with the inhibitor, while it would be strongly inhibited in the exudate because of the low pH and depletion of substrate. This may be important in preventing the spread of this decay from an infected to a healthy fruit in contact with it (Barmore & Nguyen, 1985).

OTHER ENZYMES

L-*Ascorbate oxidase* (EC 1.10.3.3). The activity of this enzyme decreases in lemon as the fruit matures, and it is undetectable in fruits larger than 2·0 cm in diameter (Vines & Oberbacher, 1963). Therefore, the destruction of vitamin C observed in the juice of mature citrus fruits and stored concentrates may not be due to the activity of ascorbate oxidase.

Catalase (EC 1.11.1.6) *and peroxidase* (EC 1.11.1.7). Circumstantial evidence suggests that H_2O_2 is a regulator of the acid content of citrus juice, and that the levels of H_2O_2 depend on the activities of catalase and peroxidase (Clark & Wallace, 1963). It has been shown that Valencia orange juice contains numerous compounds which serve as substrates for peroxidase, and that the reaction rates are limited by the low concentration of H_2O_2 in the juice (Bruemmer *et al.*, 1976). Peroxidases from Valencia Late orange have been shown to be composed of at least four isoenzymes of pI ranging from 2·5 to 10·5 (Catalá & Chamarro, 1987). Thermal stability of peroxidase compared to most other enzymes has made it useful as an index of pasteurization effectiveness (Eskins *et al.*, 1971). Soluble and bound peroxidases have been obtained from juice and albedo of sour orange: the bound peroxidases were particularly heat-stable and regenerated when held at 30°C following heat inactivation (McLellan & Robinson, 1984). Peroxidase activity in the flavedo from Satsuma mandarin decreases during maturation and levels off at senescence; this rapid decrease in enzyme activity is inhibited by application of GA_3 and accelerated by ethylene (Kuraoka *et al.*, 1979). The anionic peroxidase from Washington Navel orange flavedo oxidizes 3-indoleacetic acid in the presence of H_2O_2, suggesting a role for H_2O_2 as a modulator of IAA degradation by peroxidase (Chamarro & Molina, 1989).

Esterases. Much of the flavour and aroma of citrus juices depends on the presence of esters, and therefore the activity of esterases is important in citrus juice processing (Sinclair, 1984). An acetyl esterase (EC 3.1.1.6) was isolated from grapefruit, orange and lemon tissues, the highest activity occurring in the flavedo of Navel orange (Jansen *et al.*, 1947). An enzyme of similar characteristics has also been found in Valencia orange juice, its activity decreasing rapidly on standing at 25°C (Bruemmer & Roe, 1975). This enzyme produces aroma changes in the juice which can be prevented by pasteurization.

Acid phosphatase (EC 3.1.3.2). This is widely distributed in citrus tissues, though its activity in the juice is much less than in other parts of the fruits (Axelrod, 1947). The activity of acid phosphatase decreases rapidly upon extraction and with storage of the juice (Bruemmer & Roe, 1975). It has been proposed that phosphatase activity may be used as an indicator of the effectiveness of juice pasteurization (Sinclair, 1984).

CONCLUDING REMARKS

Substantial progress has been made in various aspects of the enzymology of citrus fruits and citrus products in recent years. The improvement in techniques for enzyme purification has allowed the purification to homogeneity and characterization of enzymes such as pectinesterases, chlorophyllase and phenylalanine ammonia-lyase, and improved our knowledge about their location within the fruit. In general, however, greater effort is required in the characterization of enzymes, their location, within the fruit and at the cellular level, and their changes during fruit development. There are still considerable gaps in our knowledge about the physiological significance of different isoenzymes and their regulatory mechanisms. The biochemistry of processes such as degreening and host–pathogen interaction are also poorly understood.

The use of exogenous enzymes such as pectinases in cloud stabilization and in the improvement of the technology of citrus juice concentrates offers promising prospects. Immobilized enzymes and microorganisms for citrus juice debittering are interesting alternatives to less specific technologies, such as those based on the use of polymeric adsorbents and ion-exchange resins, that can affect other juice components.

Techniques of in-vitro cell and tissue culture have been developed (Tisserat *et al.*, 1989), and protoplast from citrus juice vesicle cells have been obtained (Echeverria, 1987), opening interesting perspectives as research tools for the study of the physiology and biochemistry of citrus fruits.

Techniques for the manipulation of isolated protoplasts are of great economic potential. Protoplast culture systems have been used to obtain somatic hybrids and cybrids (fused protoplasts which have the nuclear genome of one of the partners, and the mitochondrial genome of the other) of several citrus species. These protoplasts have been successfully regenerated into whole plants, indicating new ways for the improvement of citrus fruits by the incorporation of favourable breeding traits that would be problematic, laborious or impossible by conventional breeding (Vardi & Galun, 1988). Production of transformants by direct DNA transfer has been attempted as an alternative means of genetic manipulation of citrus protoplasts (Kobayashi & Uchimiya, 1989). Hopefully, the development of these techniques will allow, in the near future, access to an increasing diversity of genes previously limited to sexually compatible plant species, opening a wide range of possibilities for producing 'tailor made' fruits with the desired characteristics for the fresh market and for processing.

REFERENCES

Abeles, F. B. (1973). In *Ethylene in Plant Biology*. ed. F. B. Abeles. Academic Press Inc, New York, USA, pp. 200–6.
Abeles, F. B. & Dunn, L. J. (1989). *J. Plant. Growth Regul.*, **8,** 319–25.
Abeles, F. B., Hershberger, W. L. & Dunn, L. J. (1989). *Plant Physiol.*, **89,** 664–8.
Achilea, O., Fuchs, Y., Chalutz, E. & Rot, I. (1985). *Physiol. Plant Pathol.*, **27,** 55–63.
Aljubiri, H., Huff, A. & Hshieh, M. (1979). *Plant Physiol.*, **63,** S-73.
Amir-Shapira, D., Goldschmidt, E. E. & Altman, A. (1986). *Plant Sci.*, **43,** 201–6.
Amir-Shapira, D., Goldschmidt, E. E. & Altman, A. (1987). *Proc. Nat. Acad. Sci. USA*, **84,** 1901–5.
Axelrod, B. (1947). *J. Biol. Chem.*, **167,** 57–72.
Baker, R. A. & Bruemmer, J. H. (1969). *Proc. Fla State Hort. Soc.*, **82,** 215–20.
Baker, R. A. & Bruemmer, J. H. (1972). *J. Agric. Food Chem.*, **20,** 1169–73.

Baldwin, E. A. & Biggs, R. H. (1988). *Physiol. Plant.*, **73**, 58–64.
Barmore, C. R. (1975). *Hort. Sci.*, **10**, 595–6.
Barmore, C. R. & Brown, G. E. (1979). *Phytopathology*, **69**, 675–8.
Barmore, C. R. & Brown, G. E. (1981). *Phytopathology*, **71**, 328–31.
Barmore, C. R. & Nguyen, T. K. (1985). *Phytopathology*, **75**, 446–9.
Barmore, C. R., Snowden, S. E. & Brown, G. E. (1984). *Phytopathology*, **74**, 735–7.
Bateman, D. F. & Millard, R. L. (1966). *Annu. Rev. Phytopathology*, **4**, 119–46.
Bean, R. C. & Todd, G. W. (1960). *Plant Physiol.*, **35**, 425–9.
Benedict, C. R., Rosenfield, C. L., Mahan, J. R., Madhavan, J. R. & Yokoyama, H. (1985). *Plant Sci.*, **41**, 169–73.
Berhow, M. A. & Vandercook, C. E. (1989). *Phytochemistry*, **28**, 1627–30.
Biggs, R. H. & Pollard, J. E. (1970). *Proc. Fla State Hort. Soc.*, **83**, 314–18.
Blundstone, H. A. W., Woodman, J. S. & Adams, J. B. (1971). In *The Biochemistry of Fruits and their Products* (Vol. II), ed. A. C. Hulme. Academic Press, London, UK, pp. 543–72.
Bogin, E. & Wallace, A. (1966). *Biochem. Biophys. Acta*, **28**, 190–2.
Braddock, R. J. (1981). *Proc. Fla State Hort. Sci.*, **94**, 270–3.
Brewster, L. C., Hasegawa, S. & Maier, V. P. (1976). *J. Agric. Food Chem.*, **24**, 21–4.
Brown, G. E. & Barmore, C. R. (1977). *Phytopathology*, **67**, 120–3.
Bruemmer, J. H. (1980). In *Citrus Nutrition and Quality*, ed. S. Nagy & J. A. Attaway ACS Symposium Series 143. Chemical Society, Washington, DC, USA, pp. 155–6.
Bruemmer, J. H. & Roe, B. (1975). *Proc. Fla State Hort. Soc.*, **88**, 300–3.
Bruemmer, J. H. & Roe, B. (1985). *Phytochemistry*, **24**, 2105–6.
Bruemmer, J. H., Roe, B. & Bowen, E. K. (1976). *J. Food Sci.*, **41**, 186–9.
Bruemmer, J. H., Busling, B. & Roe, B. (1977). *Proc. Int. Soc. Citriculture*, **3**, 712–16.
Catalá, C. & Chamarro, J. (1987). *Rev. Agroquim. Tecnol. Aliment.*, **27**, 509–18.
Chamarro, J. & Molina, I. (1989). *J. Food Biochem.*, **13**, 361–75.
Chandler, B. V. & Nicol, K. J. (1975). *CSIRO Food Res. Q.*, **35**, 79–88.
Clark, R. B. & Wallance, A. (1963). *Proc. Amer. Soc. Hort. Sci.*, **83**, 322–31.
Crandall, P. G. & Beilock, R. P. (1985). *Proc. Fla State Hort. Soc.*, **98**, 200–6.
Crandall, P. G., Chen, C. S., Marcy, J. E. & Martin, F. G. (1986). *J. Food Sci.*, **51**, 1017–20.
Crandall, P. G., Davis, K. Z. & Baker, R. A. (1990). *Food Technol.* **44**, 126–9.
Davis, P. L. (1970). *Proc. Fla State Hort. Soc.*, **83**, 294–8.
Dubery, I. A. & Schabort, J. C. (1986). *Biochem. Int.*, **13**, 57–9.
Dubery, I. A. & Schabort, J. C. (1988). *Int. J. Biochem.*, **20**, 217–22.
Eagerman, B. A. & Rouse, A. H. (1976). *J. Food Sci.*, **41**, 1396–7.
Eaks, J. L. (1970). *Plant Physiol.*, **45**, 334–8.
Echeverria, E. (1987). *J. Amer. Soc. Hort. Sci.*, **112**, 393–96.
Echeverria, E. & Valich, J. (1988). *Phytochemistry*, **27**, 73–6.
Echeverria, E. & Valich, J. (1989). *J. Amer. Soc. Hort. Sci.*, **114**, 445–9.
Ecker, J. R. & Davis, R. W. (1987). *Proc. Nat. Acad. Sci. USA*, **84**, 5202–6.
El-Kazzaz, M. K., Chordas, A. & Kader, A. A. (1983). *J. Amer. Soc. Hort. Sci.*, **108**, 618–21.
El-Zeftawi, B. M. (1977). *J. Hort. Sci.*, **52**, 127–34.
El-Zeftawi, B. M. & Garret, R. G. (1978). *J. Hort. Sci.*, **53**, 215–23.
Eskins, N. A. M., Henderson, H. M. & Townsend, R. J. (1971). In *Biochemistry of Foods*, ed. N. A. M. Eskins, H. M. Henderson & R. J. Townsend. Academic Press, New York, USA, pp. 109–51.
FAO (1990). CCP: CI/ST/90, Rome, 55 pp.
Fellers, P. J. (1989). *J. Sci. Food Agric.*, **49**, 389–404.
Fisher, S. F. (1968). *Phytochemistry*, **7**, 769–71.
Goren, R. & Huberman, M. (1976). *Physiol. Plant.*, **37**, 123–30.
Greenberg, J., Goren, R. & Riov, J. (1975). *Physiol. Plant.*, **34**, 1–7.

Grierson, W., Cohen, E. & Kitagawa, H. (1986). In *Fresh Citrus Fruits*, ed. W. Grierson, E. Cohen & H. Kitagawa. AVI Publishing Co. Inc. Westport, CO, USA, pp. 253–74.

Hasegawa, S. & Herman, Z. (1985). *Phytochemistry*, **24**, 1973–74.

Hasegawa, S. & Herman, Z. (1986). *Phytochemistry*, **25**, 2523–4.

Hasegawa, S. & Maier, V. P. (1981). *Proc. Int. Soc. Citric*, Editorial Committee, 1981 Int. Citrus Congress Organizing Committee, Tokyo, Japan, pp. 918–41.

Hasegawa, S. & Maier, V. P. (1983). *Food Technol.*, **37**, 73–7.

Hasegawa, S., Patel, M. N. & Snyder, R. C. (1982). *J. Agric. Food Chem.*, **30**, 509–11.

Hasegawa, S., Pelton, V. A. & Bennett, R. D. (1983). *J. Agric. Food Chem.*, **31**, 1002–4.

Hasegawa, S., Vandercook, C., Choi, G., Herman, Z. & Ou, P. (1985). *J. Food Sci.*, **50**, 330–2.

Hasegawa, S., Herman, Z., Orme, E. & Ou, P. (1986*a*). *Phytochemistry*, **25**, 2783–5.

Hasegawa, S., Maier, V. P., German, Z. & Ou, P. (1986*b*). *Phytochemistry*, **25**, 1323–5.

Herman, Z. & Hasegawa, S. (1985). *Phytochemistry*, **24**, 2911–13.

Herman, Z., Hasegawa, S. & Ou, P. (1985). *J. Food Sci.*, **50**, 118–20.

Herman, Z., Bennett, R. D., Ou, P., Fong, C. H. & Hasegawa, S. (1987). *Phytochemistry*, **26**, 2247–50.

Hirai, M. (1978). *Phytochemistry*, **17**, 1507–10.

Hirai, M. & Ueno, I. (1977). *Plant Cell Physiol.*, **18**, 791–99.

Huberman, M. & Goren, R. (1979). *Physiol. Plant.*, **45**, 189–96.

Huberman, M., Zamski, E. & Goren, R. (1988). *Israel J. Bot.*, **37**, 107–21.

Huff, A. (1982). *Phytochemistry*, **21**, 261–5.

Huffaker, R. C. & Wallace, A. (1959). *J. Amer. Soc. Hort. Sci.*, **74**, 348–59.

Hyodo, H. (1977). *Plant Physiol.*, **59**, 111–13.

Hyodo, H. (1978). *Plant Cell Physiol.*, **19**, 545–51.

Hyodo, H. & Nishino, T. (1981). *Plant Physiol.*, **67**, 421–3.

Ismail, M. A. & Brown, G. E. (1979). *J. Amer. Soc. Hort. Sci.*, **104**, 126–9.

Ito, S., Izumi, Y. & Murata, N. (1969). *Canners' J.*, **48**, 225–8.

Jahn, O. L. & Young, R. (1976). *J. Amer. Soc. Hort. Sci.*, **101**, 416–18.

Jansen, E. F., Jang, R. & MacDonnell, R. R. (1947). *Arch. Biochem.*, **15**, 415–31.

Jimeno, A., Manjon, A., Canovas, M. & Iborra, J. L. (1987). *Process Biochem.*, **22**, 13–16.

Joslyn, M. A. & Pilnik, W. (1961). In *The Orange*, ed. W. B. Sinclair. University of California, Division of Agricultural Science, Riverside, California, pp. 373–435.

Jourdan, P. S., McIntosh, C. A. & Mansell, R. L. (1985). *Plant Physiol.*, **77**, 903–8.

Kato, T. & Kubota, S. (1978). *Physiol. Plant*, **42**, 67–72.

Kew, T. J. & Veldhuis, M. K. (1960). *Proc. Fla State Hort. Soc.*, **73**, 293–7.

Kobayashi, S. & Uchimiya, H. (1989). *Jpn. J. Genet.*, **64**, 91–7.

Koch, K. E. (1984). *Plant Cell Environ.*, **7**, 647–53.

Korner, B., Zimmermann, G. & Berk, Z. (1980). *J. Food Sci.*, **45**, 1203–6.

Kriedemann, E. (1969). *Aust. J. Agric. Res.*, **20**, 291–300.

Krop, J. J. P. (1974). Doctoral thesis. Centre of Agricultural Publishing and Documentation, Wageningen, The Netherlands.

Krop, J. J. P. & Pilnik, W. (1974*a*). *Lebensm.-Wiss. Technol.*, **7**, 121–4.

Krop, J. J. P. & Pilnik, W. (1974*b*). *Lebensm.-wiss. Technol.*, **7**, 125–6.

Krop, J. J. P., Pilnik, W. & Faddegon, J. M. (1974). *Lebensm.-Wiss. Technol.*, **7**, 50–52.

Kuraoka, T., Iwasaki, K. & Ishii, T. (1979). *J. Japan Soc. Hort. Sci.*, **47**, 437–42.

Lewinsohn, E., Britsch, L., Mazur, Y. & Gressel, J. (1989). *Plant Physiol.*, **91**, 1323–8.

Lisker, N., Cohen, L., Chalutz, E. & Fuchs, Y. (1983). *Physiol. Plant Pathol.*, **22**, 331–8.

Liu, Y., Hoffman, N. E. & Yang, S. F. (1985). *Planta,* **164,** 565–8.
Lowell, C. A., Tomlinson, P. T. & Koch, K. E. (1989). *Plant Physiol.,* **90,** 1394–402.
Maier, V. P., Brewster, L. C. & Hsu, A. C. (1973). *J. Agric. Food Chem.,* **21,** 490–5.
Maier, V. P., Hasegawa, S., Bennett, R. D. & Echols, L. C. (1980). In *Citrus Nutrition and Quality,* ed. S. Nagy & J. A. Attaway, ACS Symposium Series 143. American Chemical Society, Washington, DC, USA, pp. 63–82.
Marshall, M. R., Marcy, J. E. & Braddock, R. J. (1985). *J. Food Sci.,* **50,** 220–2.
Matile, P. (1980). *Z. Pflanzenphysiol.,* **99,** 475–8.
Matile, P., Duggelin, T., Shellenberg, M., Rentsch, P., Bartlink, R., Peisker, C. & Thomas, H. (1989). *Plant Physiol. Biochem.,* **27,** 595–604.
Matsuura, Y. (1987). *Agric. Biol. Chem.,* **51,** 1675–7.
McIntosh, C. A. & Mansell, R. L. (1990). *Phytochemistry,* **29,** 1533–8.
McLellan, K. M. & Robinson, D. S. (1984). *Food Chem.,* **13,** 139–47.
Mizrahi, S. & Berk, Z. (1970). *J. Sci. Food Agric.,* **21,** 250–3.
Nath, N. & Ranganna, J. (1977). *J. Food Technol.,* **12,** 411–19.
Okado, S. & Ono, M. (1969). US Patent 3, 484, 255.
Olson, R. L. (1968). In *Low Temperature Biology of Foodstuffs,* ed. J. Hawthorn and E. J. Roffe. Pergamon Press, London, p. 381.
Olson, A. C. & Gray, G. M. (1981). *J. Agric. Food Chem.,* **29,** 1298–301.
Ono, M., Tosa, T. & Chibata, I. (1978). *Agric. Biol. Chem.,* **42,** 1847–53.
Parekh, L. J. & Sakariah, K. K. (1971). *Enzymologia,* **41,** 47–52.
Poling, S. W., Hsu, W. S. & Yokoyama, H. (1975). *Phytochemistry,* **14,** 1933–8.
Purvis, A. C. (1980). *Plant Physiol.,* **66,** 624–7.
Purvis, A. C. & Barmore, C. R. (1981). *Plant Physiol.,* **68,** 854–6.
Ramakrishnan, C. V. & Varma, T. N. S. (1959). *Men. Ind. Bot. Soc.,* **2,** 79–87.
Rasmussen, G. K. (1973). *Plant Physiol.,* **51,** 626–8.
Rasmussen, G. K. (1975). *Plant Physiol.,* **56,** 765–7.
Riov, J. (1975). *J. Food Sci.,* **40,** 201–2.
Riov, J. & Yang, S. F. (1982*a*). *J. Plant Growth Regul.,* **1,** 95–104.
Riov, J. & Yang, S. F. (1982*b*). *Plant Physiol.,* **69,** 687–90.
Riov, J., Monselise, S. P. & Kahan, R. S. (1969). *Plant Physiol.,* **44,** 631–5.
Roe, B. & Bruemmer, J. H. (1974). *J. Agric. Food Chem.,* **22,** 285–8.
Roe, B., Davis, P. L. & Bruemmer, J. H. (1984). *Phytochemistry,* **23,** 713–17.
Rombouts, F. M., Versteeg, A., Karman, H. & Pilnik, W. (1982). In *Use of Enzymes in Food Technology: Symposium International,* ed. P. Dupuy, Technique et Documentation-Lavoisier, Paris, pp. 483–87.
Rouse, A. H. & Atkins, C. D. (1952). *Food Technol.,* **6,** 291–4.
Rouseff, R. L. (1980). In *Citrus Nutrition and Quality,* ed. S. Nagy & J. A. Attaway. ACS Symposium Series 143, American Chemical Society, Washington, DC, USA, pp. 83–108.
Sasson, A. & Monselise, S. P. (1977). *J. Amer. Soc. Hort. Sci.,* **102,** 331–6.
Shimokawa, K. (1979). *Scientia Horticulturae,* **11,** 253–5.
Shimokawa, K. (1982). *Phytochemistry,* **21,** 543–5.
Shimokawa, K. (1983). *Phytochemistry,* **22,** 1903–8.
Shimokawa, K. (1990). *Phytochemistry,* **29,** 1725–8.
Shimokawa, K. & Tominaga, N. (1972). *Agric. Hortic.,* **47,** 95–6.
Shimokawa, K., Sakanoshita, A. & Horiba, K. (1978*a*). *Plant Cell Physiol.,* **19,** 229–36.
Shimokawa, K., Shimada, S. & Yaeo, K. (1978*b*). *Scientia. Horticulturae,* **8,** 129–35.
Sinclair, W. B. (1984). In *The Biochemistry and Physiology of the Lemon and other Citrus Fruits,* ed. W. B. Sinclair. University of California, Oakland, CA, pp. 444–70.
Srere, E. A. & Snekin, J. (1966). *Nature,* **212,** 506–7.

Stewart, I. & Wheaton, T. A. (1972). *J. Agric. Food Chem.*, **20,** 448–9.
Termote, F., Rombouts, F. M. & Pilnik, W. J. (1977). *J. Food Biochem.*, **1,** 15–34.
Ting, S. V. & Attaway, J. A. (1971). In *The Biochemistry of Fruits and their Products* (Vol. II), ed. A. C. Hulme. Academic Press, London, UK, pp. 107–69.
Ting, S. V. (1980). In *Citrus Nutrition and Quality*, ed. S. Nagy and J. A. Attaway. ACS Symp. Ser. 143, Washington D.C., pp. 3–24.
Tisserat, B., Vandercook, C. E. & Berhow, M. (1989). *Food Technol.*, **43,** 95–100.
Vaks, B. & Lifshitz, A. (1981). *J. Agric. Food Chem.*, **29,** 1258–61.
Vardi, A. & Galun, E. (1988). *Scientia Horticulturae*, **37,** 217–30.
Versteeg, C., Rombouts, F. M. & Pilnik, W. (1978). *Lebensm.-Wiss. Technol.*, **11,** 267–74.
Versteeg, C., Rombouts, F. M., Spaansen, C. M. & Pilnik, W. (1980). *J. Food Sci.*, **45,** 969–71, 998.
Vines, H. M. & Oberbacher, M. F. (1963). *Plant Physiol.* **38,** 333–7.
Wheaton, T. A. & Stewart, I. (1973). *J. Amer. Soc. Hort. Sci.*, **98,** 337–40.
Wicker, L. & Temelli, F. (1988). *J. Food Sci.* **53,** 162–4.
Wicker, L., Varssallo, M. R. & Echevarria, E. J. (1988). *J. Food Sci.*, **53,** 1171–4.
Yamauchi, N. & Minamide, T. (1985). *J. Jpn Soc. Hort. Sci.*, **54,** 265–71.
Yang, S. F. (1980). *Hort. Sci.*, **15,** 238–43.
Young, R. E. & Biale, J. B. (1968). *Planta*, **81,** 253–63.
Young, L. B. & Erickson, L. C. (1961). *Proc. Amer. Soc. Hort. Sci.*, **78,** 197–200.

Index